Climate Change

Climate Change
Observed Impacts on Planet Earth

Second Edition

Edited by

Trevor M. Letcher

Emeritus Professor, School of Chemistry, University of KwaZulu-Natal,
Durban, South Africa

ELSEVIER

AMSTERDAM • BOSTON • HEIDELBERG • LONDON • NEW YORK • OXFORD
PARIS • SAN DIEGO • SAN FRANCISCO • SINGAPORE • SYDNEY • TOKYO

Elsevier
Radarweg 29, PO Box 211, 1000 AE Amsterdam, Netherlands
The Boulevard, Langford Lane, Kidlington, Oxford OX5 1GB, UK
225 Wyman Street, Waltham, MA 02451, USA

ISBN: 978-0-444-63524-2

British Library Cataloguing-in-Publication Data
A catalogue record for this book is available from the British Library

Library of Congress Cataloging-in-Publication Data
A catalog record for this book is available from the Library of Congress

For information on all Elsevier publications
visit our website at http://store.elsevier.com/

Working together
to grow libraries in
developing countries

www.elsevier.com • www.bookaid.org

Contents

PART 1   A GEOLOGICAL HISTORY OF CLIMATE CHANGE

PART 2   INDICATORS OF CLIMATE CHANGE

CHAPTER 11   Sea Life (Pelagic Ecosystems) 167

Martin Edwards

CHAPTER 12   Changes in Coral Reef Ecosystems................... 183

Martin J. Attrill, Nicola L. Foster

CHAPTER 13   Marine Biodiversity and Climate Change 195

Boris Worm, Heike K. Lotze

PART 3 MODELLING CLIMATE CHANGE

PART 4 POSSIBLE ROLES IN CAUSING CLIMATE CHANGE

PART 5 ENGINEERING, SOCIETAL AND FORESTRY ASPECTS OF CLIMATE CHANGE

List of Contributors

Babatunde J. Abiodun
Climate System Analysis Group, Department of Environmental and Geographical Sciences, University of Cape Town, Cape Town, South Africa

Akintayo Adedoyin
Department of Physics, University of Botswana, Gaborone, Botswana

André Aptroot
ABL Herbarium, Gerrit van der Veenstraat, Soest, The Netherlands

Martin J. Attrill
Marine Institute, Plymouth University, Plymouth, UK

Jonathan Bamber
University of Bristol, School of Geographical Sciences, Bristol Glaciology Centre, Bristol, UK

Jeremy R. Brammer
Department of Natural Resource Sciences, McGill University, Montreal, QC, Canada

Virginia Burkett
U.S. Geological Survey, Sunrise Valley Dr., Reston, VA, USA

Marcela E.S. Cáceres
Departamento de Biociências, Universidade Federal de Sergipe, Av. Vereador Olimpio Grande s/n, Itabaiana, Sergipe, Brazil

Shabtai Cohen
Department of Environmental Physics and Irrigation, Institute of Soil, Water and Environmental Sciences, Agricultural Research Organization, The Volcani Centre, Bet Dagan, Israel

E.D. De Wolf
Department of Plant Pathology, Kansas State University, Manhattan, KS, USA

Lev I. Dorman
Head of Cosmic Ray and Space Weather Centre with Emilio Segrè Observatory on Mt Hermon, affiliated to Tel Aviv University, Golan Research Institute, and Israel Space Agency, Qazrin, Israel; Chief Scientist of Cosmic Ray Department of IZMIRAN Russian Academy of Science, Moscow, Troitsk, Russia

Martin Edwards
Sir Alister Hardy Foundation for Ocean Science, Plymouth, UK; Marine Institute, University of Plymouth, Plymouth, UK

P.D. Esker
Escuela de Agronomía, Universidad de Costa Rica, San Pedro Montes de Oca, San José, Costa Rica

Wolfgang Fiedler
Department of Migration and Immuno-Ecology, Max Planck Institute for Ornithology, Radolfzell, Germany; Department of Biology, University of Konstanz, Konstanz, Germany

Helen S. Findlay
Plymouth Marine Laboratory, Plymouth, Devon, UK

Nicola L. Foster
Marine Institute, Plymouth University, Plymouth, UK

K.A. Garrett
Institute for Sustainable Food Systems and Plant Pathology Department, University of Florida, Gainesville, FL, USA; Department of Plant Pathology, Kansas State University, Manhattan, KS, USA

Roland Gehrels
Department of Environment, University of York, York, UK

Luis Gimeno
Departamento de Física Aplicada, Faculty of Sciences, University of Vigo, Ourense, Spain

L. Gomez-Montano
Department of Plant Pathology, Kansas State University, Manhattan, KS, USA

Jim Haywood
University of Exeter and Met Office Hadley Centre, Exeter, UK

Murray M. Humphries
Department of Natural Resource Sciences, McGill University, Montreal, QC, Canada

Torsten Kanzow
Fachbereich Klimawissenschaften, Alfred-Wegener-Institut für Polar- und Meeresforschung, Bremerhaven, Germany; Universität Bremen, Germany

Sally A. Keith
Center for Macroecology, Evolution and Climate, Natural History Museum of Denmark, Copenhagen, Denmark

Alica Košuthová
Institute of Botany, Department of Cryptogams, Slovak Academy of Science, Bratislava, Slovak Republik; Faculty of Science, Department of Botany and Zoology, Masaryk University, Brno, Czech Republik

Uta Krebs-Kanzow
Fachbereich Klimawissenschaften, Alfred-Wegener-Institut für Polar- und Meeresforschung, Bremerhaven, Germany

Rattan Lal
Carbon Management and Sequestration Center, The Ohio State University, Columbus, OH, USA

Anna Lawrence
Forest Research, Midlothian, UK

Heike K. Lotze
Department of Biology, Dalhousie University, Halifax, NS, Canada

Lucas J. Lourens
Department of Earth Sciences, Faculty of Geosciences, Utrecht University, Utrecht, The Netherlands

Audrey M. Maran
Department of Biological Sciences, Bowling Green State University, Bowling Green, OH, USA

Nova Mieszkowska
The Marine Biological Association of the UK, Plymouth, UK; School of Environmental Sciences, University of Liverpool, Liverpool, UK

Mike D. Morecroft
Natural England, Worcester, UK

Robert J. Nicholls
Faculty of Engineering and the Environment, University of Southampton, Southampton, UK; Tyndall Centre for Climate Change Research, University of Southampton, Southampton, UK

Bruce Nicoll
Forest Research, Midlothian, UK

M. Nita
Department of Plant Pathology, Physiology, and Weed Science, AHS Jr. AREC, Virginia Polytechnic Institute and State University, Winchester, VA, USA

Shannon L. Pelini
Department of Biological Sciences, Bowling Green State University, Bowling Green, OH, USA

Ben Powell
School of Mathematics, University of Bristol, Bristol, UK

Thomas Reichler
Department of Atmospheric Sciences, University of Utah, Salt Lake City, UT, USA

David Schroeder
Centre for Polar Observation and Modelling, Department of Meteorology, University of Reading, Reading, UK

A.H. Sparks
International Rice Research Institute (IRRI), Los Baños, Laguna, Philippines

Gerald Stanhill
Department of Environmental Physics and Irrigation, Institute of Soil, Water and Environmental Sciences, Agricultural Research Organization, The Volcani Centre, Bet Dagan, Israel

Norbert J. Stapper
Büro für Ökologische Studien, Verresbergerstraße, Monheim am Rhein, Germany

Georgiy Stenchikov
Division of Physical Sciences and Engineering, King Abdullah University of Science and Technology, Thuwal, Saudi Arabia

Peter Thorne
Nansen Environmental and Remote Sensing Centre, Bergen, Norway; Department of Geography, Maynooth University, Maynooth, Ireland

Ricardo M. Trigo
Faculdade de Ciências, Instituto Dom Luiz (IDL), Universidade de Lisboa, Lisbon, Portugal

Richard P. Tuckett
School of Chemistry, University of Birmingham, Edgbaston, Birmingham, UK

Carol Turley
Plymouth Marine Laboratory, Plymouth, Devon, UK

Daniel A. Vallero
Pratt School of Engineering, Duke University, Durham, NC, USA

Martin Visbeck
Fachbereich Ozeanzirkulation und Klimadynamik, GEOMAR Helmholtz-Zentrum für Ozeanforschung Kiel, Germany

Peter Wadhams
Department of Applied Mathematics and Theoretical Physics, University of Cambridge, Cambridge, UK

Mark Williams
Department of Geology, University of Leicester, Leicester, UK

Colin Woodroffe
School of Earth and Environmental Sciences, University of Wollongong, NSW, Australia

Boris Worm
Department of Biology, Dalhousie University, Halifax, NS, Canada

Jan Zalasiewicz
Department of Geology, University of Leicester, Leicester, UK

Preface

Since the first edition of *Climate Change: Observed Impacts on Planet Earth* was published in 2009, the evidence of a changing climate has become even more apparent. As a result, the time had come to put these new developments into a new book. This, together with the interest shown in the first edition, has culminated in a second edition of *Climate Change*. Many new chapters have been added, and chapters for the first edition have been updated to highlight new evidence that our climate is changing.

The evidence that our climate is warming is overwhelming. This evidence comes not only from land and sea surface temperature records but also from indicators such as the coverage of Arctic sea ice – all of which, and much more, is discussed in this book. Most scientists in the world now accept that anthropogenic activities and specifically the emissions of greenhouse gases are responsible for the major part of the observed warming. May 9, 2013, was an auspicious day for the warming of the planet, when it was reported by both the National Oceanic and Atmospheric Administration (NOAA) and the Scripps Institute of Oceanography that the daily mean concentration of CO_2 in the atmosphere at Mauna Loa laboratory exceeded 400 μmol mol^{-1} (400×10^{-6}) 400 ppm for the first time in millions of years.

This book, like the earlier edition, is not intended to compete with the Intergovernmental Panel on Climate Change (IPCC) reports but offers support through a different approach. Many of the authors were not involved in recent assessments of the IPCC, and here they present fresh evaluations of the evidence testifying to a problem that was described by Sir David King in the first edition as the most severe calamity our civilization has yet to face.

Unlike other books of similar title, this book has the advantage that the chapters have once again each been written by world-class scientists and engineers working in their respective fields. As a result, the new volume presents a balanced picture across the whole spectrum of climate change. With this line-up of authoritative and well-researched topics, any doubts about whether climate change is taking place or not will be immediately dispelled on reading this book.

The new edition is divided into five sections:

- A Geological History of Climate Change
- Indicators of Climate Change
- Modelling of Climate Change
- Possible Causes of Climate Change
- Engineering, Forestry and Societal Aspects of Climate Change.

This edition contains 33 chapters as opposed to the 25 in the first edition. The new chapters include:

- Global Surface Temperatures
- Arctic Sea Ice
- Antarctic Sea Ice Changes and Their Implications
- Land Ice: Indicator and Integrator of Climate Change
- Statistical Modelling of Climate Change
- A Modelling Perspective of Future Climate
- Aerosols and Carbon Black in the Atmosphere
- Climate Change and Agriculture
- Engineering Aspects of Global Change
- Climate Impacts and Adaptations in Forest Management: Science, Policy and Practice

The audience we hope to reach are: policy makers in local and central governments; students, teachers, researchers, professors, scientists, engineers and managers working in fields related to climate change and future energy options; editors and newspaper reporters responsible for informing the public; and the general public who need to be aware of the impending disasters that a warmer Earth will bring. A summary is provided at the beginning of each chapter for those interested in a brief synopsis, and copious references are provided for those wishing to study each chapter topic in greater detail.

The IPCC assessments have produced two basic conclusions: firstly, that the current climate changes are unequivocal, and secondly, that this is largely due to the emission of greenhouse gases resulting from human activity. This book re-enforces these two conclusions and the chapters on 'Indicators of Climate Change' and on the 'Possible Causes of Climate Change' are particularly relevant. Furthermore, the section on 'Modelling of Climate Change' further supports these conclusions through simulations of past climate changes and projections of future climate changes.

Projections of our global warming indicate that the temperature will exceed the 2°C global average regarded by many scientists as the upper limit in temperature within the next 50 years. If we do not take action to halt this rise in temperature, we must expect the serious consequences of extreme weather: droughts, floods, winds and storms. The book is a clarion call to humans to take immediate action to reduce the amount of CO_2 that we are pumping into the atmosphere, which arguably can best be accomplished by reducing our dependency on fossil fuels. We must strive to stop burning coal and oil in our power stations with the ultimate aim of keeping most of the fossil fuel in the ground and find new, renewable ways of producing electricity and propelling our vehicles.

The book has been supported by the International Union of Pure and Applied Chemistry (IUPAC) and the IUPAC logo appears on the cover of the book. The International System of Quantities (SI units) has been used throughout the book, and where necessary other units are given in parentheses. Furthermore, the authors have rigorously adhered to the IUPAC notation and spelling of physical quantities. For example the symbols for minute, hour, day and year, are min, h, d and a, respectively, are used. Also, the relationship between a physical quantity and its unit is given by: physical quantity = number times unit, and by example: temperature = 270 K or rearranged to give: 270 = temperature/K. This relationship makes the statement concerning units unambiguous.

In true IUPAC style, the chapters have been written by authors from many countries including Australia, Botswana, Brazil, Canada, Costa Rica, Denmark, Germany, Israel, the Netherlands, Norway, Philippines, Russia, Slovak Republic, South Africa, Spain, the Kingdom of Saudi Arabia, the United Kingdom and the United States of America.

The success of the book ultimately rests with the 54 authors and co-authors, and as editor I should like to thank all of them for their cooperation and their highly valued, willing and enthusiastic contributions. I would also like to thank: Dr Laura McConnell, President of IUPAC's Division of Chemistry and the Environment, who has assisted in editing the chapters; and Professor Ron Weir of the IUPACs Interdivisional Committee on Terminology, Nomenclature and Symbols (ICTNS) who has advised on the correct use of physical quantities, units and symbols. Finally, my thanks are due to Rowena Prasad and Candice Janco of Elsevier whose expertise steered this book to its publication.

Trevor M. Letcher
Laurel House
Fosse way
Stratton on the Fosse
BA3 4QN, United Kingdom
April 2015

A GEOLOGICAL HISTORY OF CLIMATE CHANGE

CLIMATE CHANGE THROUGH EARTH'S HISTORY

1

Jan Zalasiewicz, Mark Williams

Department of Geology, University of Leicester, Leicester, UK

CHAPTER OUTLINE

1. INTRODUCTION

Earth's climate is now changing in response to an array of anthropogenic perturbations, notably the release of greenhouse gases; an understanding of the rate, mode and scale of this change is now of literally vital importance to society. There is currently intense study of current and historical (i.e. measured) changes in both perceived climate drivers and the Earth system response. Such studies typically lead to climate models that, in linking proposed causes and effects, are aimed at allowing prediction of climate evolution over an annual to centennial scale.

However, the Earth system is complex and imperfectly understood, not least as regards resolving the effect of multiple feedbacks in the system and of assessing the scale and importance of leads, lags and thresholds ('tipping points') in climate change. There is thus a need to set modern climate studies within a realistic context by examining the preserved history of the Earth's climate in the rock succession. Such study cannot provide precise replicas of the unplanned global experiment that is now underway (for the sum of human actions represents a geological novelty). However, it is providing an increasingly detailed picture of the nature, scale, rate and causes of past climate change and of its wider effects [1], regarding, for instance, sea level and biota. Imperfect as it is, it provides an indispensable context for modern climate studies, not least as a provision of ground truth for computer models (see below) of former and present climate.

Aspects of climate that are recorded in strata include temperature and seasonality [2,3], humidity/aridity [4], and wind direction and intensity [5]. Classical palaeoenvironmental indicators such as glacial tills, reef limestones and desert dune sandstones have in recent years been joined by a plethora of other proxy indicators. These include many biological (fossilized pollen, insects, marine algae) and chemical proxies (e.g. Mg/Ca ratio in biogenic carbonates). Others are isotopic: oxygen isotopes provide information on temperature and ice volume; carbon isotopes reflect global biomass and inputs (of methane or carbon dioxide) into the ocean/atmosphere system; strontium and osmium are proxies for weathering, and the latter, with molybdenum also, for oceanic oxygenation levels. Other proxies include recalcitrant organic molecules: long-chain algal-derived alkenones as sea temperature indicators [6] and isorenieratane as a specific indicator of photic zone anoxia [7]. These and many other proxies are summarized in [8]. Levels of greenhouse gases such as carbon dioxide and methane going back to 800 ka can be measured in ice cores [9]. For older periods, indirect measurements are made, based on proxies such as leaf stomata densities [10], palaeosol chemistry [11], boron isotopes [12] and alkenones [13]; estimates of greenhouse gas concentrations have also been arrived at by modelling [14,15].

2. CLIMATE MODELS

Since the 1960s, computer models of climate have been developed that provide global and regional projections of future climate and reconstructions of deep time climate. Some of these models are used to simulate conditions during ancient icehouse climates, whilst others examine warm intervals of global climate, such as during the Mesozoic and Early Cenozoic greenhouse [16]. The most widely applied computer simulations of palaeoclimate are general circulation models (GCMs). The increasing complexity of these models has followed the exponential growth in computer power.

GCMs divide the Earth into a series of grid boxes. Within each of box variables important for the prediction of climate are calculated, based upon the laws of thermodynamics and Newton's laws of motion. At progressive time steps of the model, the reaction between the individual grid boxes is calculated. GCM simulations rely on establishing key boundary conditions. These conditions include solar intensity, atmospheric composition (e.g. level of greenhouse gases), surface albedo, ocean heat transport, geography, orography, vegetation cover and orbital parameters. In general, the boundary conditions are more difficult to establish for increasingly older time periods. Thus, orbital parameters may be established with high precision in a computer model for a short time interval of the Pliocene [17]. But for much older time periods, for example the glacial world of the late Ordovician, the rock record is much less complete, and it is difficult to constrain most of the boundary conditions with a reasonable degree of precision [18].

Geological data (e.g. sedimentology, palaeontology) are essential to 'ground truth' climate models, to establish whether they are providing a realistic reconstruction of the ancient world and also to provide data for calibrating boundary conditions for the models. Of major importance for GCM palaeoclimate reconstructions is accurate information about sea surface temperatures (SSTs), as these provide a strong indication of how ocean circulation is working. The most extensive deep time (pre-Quaternary) reconstruction of SSTs is that of the United States Geological Survey PRISM Group [19]. This global dataset has been used for calibrating a range of climate model scenarios for the 'mid Pliocene warm period' (a.k.a. 'Mid Piacenzian Warm Period') and also includes an extensive catalogue of terrestrial data [20]. Warm periods of the Pliocene are often cited as a useful comparison (though definitely not an analogue) for the path of late twenty-first century climate [16].

3. LONG-TERM CLIMATE TRENDS

Earth's known climate history, as decipherable through forensic examination of sedimentary strata, spans some 3.8 billion years (3.8×10^9 a), to the beginning of the Archaean (Fig. 1). The previous

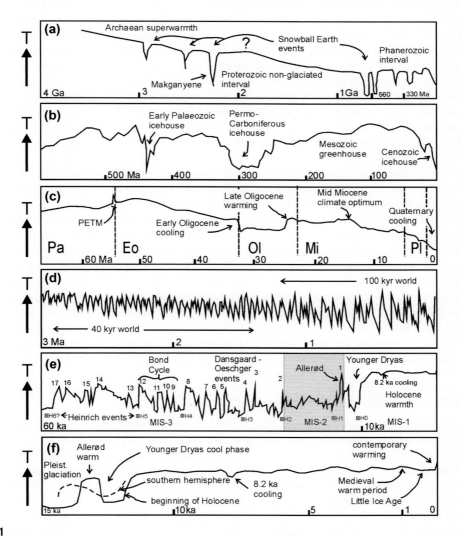

FIGURE 1

Global climate variation at six different timescales. *(Data adapted from sources including* [8,28,56,74,88,117].*)* On the left side of the figure, the figure 'T' denotes relative temperature. Note that the line denoting 'T' is derived from $\delta^{18}O$ from benthic foraminifera for the Cenozoic time slices (c–e), but for the intervals with polar ice, this line will record a combination of ice volume and temperature change.

history, now generally assigned to the Hadean Eon, is only fragmentarily recorded as occasional ancient mineral fragments contained within younger rocks, particularly of highly resistant zircon dated to nearly 4.4 billion years (4.4×10^9 a) ago [21] and thus stretching back to very nearly the beginning of the Earth at 4.56 billion years ago [22]. The chemistry of these very ancient fragments hints at the presence of a hydrosphere even at that early date, though one almost certainly disrupted by massive meteorite impacts [23]. Certainly, by the beginning of the Archaean, oceans had developed, and there was an atmosphere sufficiently reducing to allow the preservation of detrital minerals such as pyrite and uraninite in river deposits that would not survive in the presence of free oxygen [24].

From then until the present time, Earth's climate has remained within narrow temperature limits that have allowed the presence of abundant liquid water, water vapour and variable amounts of water ice, the last of these (when present) generally accumulating at high latitudes and/or high altitudes. This is despite widely accepted astrophysical models suggesting that the sun has increased its luminosity by some 20% since the early Archaean and contrasts sharply with the history of our planetary neighbours: Venus now having a surface temperature of *ca.*400°C with a dense anhydrous atmosphere dominated by carbon dioxide (representing approximately the amount of carbon that on Earth is bound up in rock form as carbonates and hydrocarbons); and Mars with an early history of running surface water (roughly during the Earth's Hadean Eon) and subsequently being essentially freeze-dried.

Hypotheses to explain the Earth's climate stability (which has allowed inter alia a continuous lineage of living organisms) have included such as the Gaia hypothesis [25], in which the totality of the Earth's biota operate to maintain optimum conditions for their existence (via feedback mechanisms that involve such factors as albedo and atmospheric composition). Currently, it is thought that terrestrial silicate weathering (a largely abiotic mechanism) is an important factor in Earthly homeostasis [26]. Thus, as temperatures rise through an increase in greenhouse gases, increased reaction rates of rainwater (i.e. dilute carbonic acid) with rock – allied to increased humidity from enhanced evaporation rates – will cause drawdown of carbon dioxide, thus lowering temperatures [27]. Similarly, as greenhouse gas levels and temperatures fall, diminished rates of weathering will allow carbon dioxide levels to rise and so warm the Earth's climate. The silicate weathering mechanism operates on timescales of hundreds of thousands to millions of years, with greenhouse gas levels having fallen throughout Earth history as the sun's luminosity has increased. At shorter timescales, this mechanism may be overridden by other factors, to allow the production of climate states that are hotter or colder than the long-term average.

4. EARLY CLIMATE HISTORY

At long timescales, Earth's (post-Hadean) climate history can be broadly divided into: *greenhouse* (or *hothouse*) states, when the Earth's climate was generally warm, with little or no polar ice; and *icehouse* states with substantial high/mid (and sometimes low latitude) ice masses over land and ocean. Ability to resolve the duration and timing of these states becomes increasingly better as the geological record becomes younger, with a gulf, in particular, between a Phanerozoic record (from 0.542 Ga) that is highly resolved because of an abundant fossil content and a Precambrian record in which dating and correlation are based upon sporadic radiometric dates and, increasingly, chemical and event stratigraphy. Similarly, the Cenozoic glaciation is much better resolved than previous Phanerozoic glaciations.

Some of the earliest indications of climate in the Archaean hint at a very warm world (Fig. 1(a)): silicon and oxygen isotopes on Archaean rocks have been interpreted to suggest temperatures of some 50°C–80°C [28,29], though these results have been challenged by other proxy studies that have suggested more temperate conditions at that time [30,31]. Whether extremely hot or merely warm overall, most of the Archaean seems to have equated to a greenhouse world. Various means of maintaining high temperatures in the face of a faint early sun have been suggested; they include high-carbon dioxide and high-methane atmospheres [32,33], an organic-rich haze [34] and simply an increase in density through higher nitrogen pressure [35].

Around the end of the Archaean and the beginning of the Proterozoic, there is evidence, in the form of widespread glaciations [36–38] of global cooling. This may be linked with the 'Great Oxygenation Event' that took place at approximately this time, *ca.*2.4 billion years ago [39], a plausible cooling mechanism being removal by oxidation of the powerful greenhouse gas methane from the atmosphere [40]. These early glaciations were succeeded by a globally stable temperate climate (termed the 'boring billion' by some). Toward the end of the Proterozoic came the extreme 'Snowball Earth' glaciations (Fig. 1(a)) that define the 'Cryogenian Period' (now widely used as a geological time period but not yet formally defined and ratified). Stratigraphic and palaeomagnetic evidence suggest widespread ice sheets in at least two pulses (Sturtian 740–660 million years ago and Marinoan 660–635 million years ago [41]) that reached into tropical palaeolatitudes, with ice present on all main continents. Budyko [42] suggested a theoretical basis for a snowball glaciation, showing that if ice extended to within 30° latitude of the equator, the ice albedo effect would produce a positive feedback mechanism allowing ice sheets to expand to the equator. It has been proposed, controversially, that ice encased the entire globe (the 'hard snowball' variant) [43], preventing exchange between land/oceans and atmosphere. This has been disputed, with opponents preferring 'slush ball', 'zipper-rift' or 'high tilt' Earth models [41] leaving significant areas of ocean ice-free.

Whichever version is nearer the truth, these appear to have been extreme excursions of the Earth system, with deglaciation being rapid, perhaps 'catastrophic', and marked by the deposition of unique 'cap carbonate' deposits – dolomites and limestones that, worldwide, immediately overlie the glacial deposits. Deglaciation mechanisms commonly involve crossing thresholds in greenhouse gas concentrations, perhaps coupled to changes in surface ice-albedo caused by the accumulation of continental and volcanic dust [44]. In the 'hard snowball' model deglaciation takes the form of volcanic carbon dioxide being prevented from dissolving in the ocean or reacting with rock (because of their carapace of ice), and hence building up to levels high enough to cause rapid ice melt, with acid rain then reacting rapidly with newly exposed bedrock to generate alkalinity that precipitated as carbonates. In the 'slush ball' model, deglaciation hypotheses include massive release of methane, with at least local isotopic evidence of methane release accompanied by ice melt [45]. Perhaps in support of a 'slush ball' or alternative glacial hypothesis, some GCMs do not replicate the conditions in which a 'hard' Snowball Earth could develop, even with very low levels of atmospheric carbon dioxide prescribed [46].

5. PHANEROZOIC GLACIATIONS

Glaciations during the Phanerozoic were less extreme, neither reaching the equator nor being associated with postglacial cap carbonates. This seemingly more equable climate might possibly be linked to the emergence of the metazoan biosphere through the late Proterozoic and earliest Phanerozoic, and its influence on global geochemical cycles. Three main glaciations took place (Fig. 1(b)): a late Ordovician/early Silurian 'Early Palaeozoic Icehouse' (*ca.*455 Ma–425 Ma, with an end-Ordovician glacial maximum [47,48]); a long-lived Permo-Carboniferous glaciation (*ca.*325 Ma–270 Ma [49]), with ice covering much of the palaeocontinent Gondwana (leaving widespread traces in South America and Africa, then over the South Pole); and the current glaciation, which began in the southern hemisphere through the Eocene-Oligocene Epoch boundary interval (*ca.*34 Ma) with ice growing on Antarctica [50] and developed into a full-scale bipolar glaciation toward the end of the Pliocene Period and the beginning of the Quaternary Period, at *ca.*2.6 Ma, with the significant expansion of northern hemisphere ice [51].

Each of these glaciations took place in different contexts, particularly as regards the carbon cycle. The Early Palaeozoic Icehouse took place in the absence of a well-developed terrestrial flora or of widespread well-developed (and hence carbon-rich) soils. Hence, the oceans and marine sediments were of prime importance in carbon storage, with the intermittent anoxia of those oceans perhaps playing a key role as thermostat, episodically enhancing carbon sequestration that led to cooling [47]. In the Carboniferous, the explosive growth and widespread burial of plants on deltaic/coastal plain sediments (subsequently becoming coal) has long been considered key in driving down atmospheric carbon dioxide and leading to glaciation [52], while continental rearrangement to alter patterns of ocean currents and hence global heat transport seems to have provided longer-term control [53].

6. THE MESOZOIC – EARLY CENOZOIC GREENHOUSE

The Palaeozoic switches between greenhouse and icehouse give invaluable (and increasingly well-resolved) information on the mode and rate of climate change. However, it is the temporal background to, and the development of, the current glaciation that offers the most resolved history and the best clue to causal and controlling mechanisms. This is partly because of a biota that is closer to the present one and hence more interpretable, but crucially because there is a widespread oceanic record (buried under the present ocean floors) to accompany that from land and continental seas; Palaeozoic ocean deposits, by contrast, have almost all been obliterated through subduction, with only rare fragments being preserved by obduction on to destructive continental margins.

Mesozoic and early Tertiary climate was generally in 'greenhouse' mode with little (but generally some) polar ice, widespread epicontinental seas and ocean circulation driven by salinity rather than temperature differences (and hence more sluggish than today's, with a tendency to anoxia). Within this broad pattern, there were warmer and colder intervals. Fossil evidence shows that high latitudes, in particular, were considerably warmer during this interval, with extensive near-polar forests.

This interval includes brief (0.1 Ma–0.2 Ma) climate 'spikes' in which sudden temperature rises were accompanied by biotic changes and marked changes in carbon isotopes. These changes suggest massive (thousands of gigatonnes) transfer of carbon from rock reservoirs to the atmosphere/ocean system with the consequence of ocean acidification as well as warming [54]. The best known of these [55,56] were in the Toarcian Age of the Jurassic Period (*ca.*183 Ma) and at the boundary of the Palaeocene and Eocene epochs (*ca.*55 Ma). The most likely mechanisms seem to be some initial warming (perhaps from volcanic carbon dioxide) that triggered large-scale dissociation of methane hydrates from the sea floor [57], although baking of coal basins by igneous intrusions [58] may also be implicated. By whichever mechanism, the relevance for contemporary global warming is clear as, while humankind has not yet released as much carbon (*ca.* 600 Gt), it has done so as more quickly [59,60]. Re-equilibration of climate following the spikes was likely achieved via silicate weathering [27,61] while burial of organic carbon may also have played a significant role [62].

7. DEVELOPMENT OF THE CENOZOIC ICEHOUSE

The development of the Cenozoic icehouse took place as a series of steps (Fig. 1(c)), with relatively rapid transitions between one climate state and the next, strongly suggesting the common operation of thresholds or 'tipping points' [63]. The early Oligocene inception is clearly seen as an isotopic and Mg/Ca signal, in benthic foraminifera [64], of ocean cooling and deacidification [65] linked to the growth of substantial ice on Antarctica. Two mechanisms have been invoked, that in reality were likely interrelated: the separation of South America from Antarctica to open the Drake Passage and hence to allow a continuous circum-Antarctic cold current [66]; and a steep drop in carbon dioxide levels from about two times to four times the present-day levels [50,67,68] with subsequent changes in ocean circulation [69].

Subsequent Cenozoic history includes mid-Miocene warming, possibly associated with release of carbon dioxide to the atmosphere via volcanism or meteorite impact [70] during which tundra conditions were developed at high southern latitude within 1500 km of the South Pole [71], and late mid-Miocene cooling (often termed the 'Monterey event'; see Ref. [72]), which may have been influenced by drawdown of carbon dioxide from the atmosphere or by changes to ocean heat transport that triggered ice sheet growth and cooling [73].

The subsequent Pliocene Epoch marks the final phase of late Cenozoic climate. The early and mid-Pliocene represent conditions that overall were somewhat warmer than present, with global ice volumes less and global surface temperatures perhaps 2–3°C warmer. The last phase of this warmer world was the 'mid-Pliocene warm period' (mid-Piacenzian warm period) some three million years ago [17]. Following this interval, global temperatures decreased, ice volumes increased, and the amplitude of glacial-interglacial oscillation also increased [74], heralding the intensification of Northern Hemisphere Glaciation (NHG). As the last interval of warmth, the 'mid-Pliocene warm period' has received significant attention as a possible comparison for the path of future global warming [17].

The intensification of NHG that is characteristic of the Quaternary Period [75] was marked by the growth of substantial ice in the northern polar region [76] and reorganization of the North Atlantic Ocean current system [77]. It is associated with ice-rafted debris appearing in North Atlantic Ocean floor deposits, together with the beginning of substantial loess accumulation in central Europe and China, the drying of Africa to create extensive savannah areas and other global phenomena. This event may partly reflect a further carbon dioxide threshold [78], with strontium isotope evidence of increased rock weathering, not least from uplift phases of the Himalayas [79]. However, there is strong evidence to suggest the importance of enhanced ice growth rather than simply temperature, with the development of the 'snow gun' hypothesis [80] in which the bringing of a warm moisture-laden ocean current against a cold North American continent led to increased snow precipitation and ice formation on that continent, and hence (via increased albedo and other feedbacks) to further cooling.

8. ASTRONOMICAL MODULATION OF CLIMATE

Over the last 40 years, an astronomical pacemaker for the Quaternary 'Ice Age' has been established beyond doubt, comprising variations in orbital eccentricity ('stretch'), axial tilt and precession ('wobble') with dominant periodicities of roughly a 100 000 a, 40 000 a and 20 000 a, respectively [63].

These produced small variations in the amount and seasonal distribution of sunlight reaching the Earth that, when amplified by various feedback mechanisms – notably via variations in atmospheric greenhouse gas concentrations – led to the well-established pattern of Quaternary glacial/interglacial and stadial/interstadial changes. This mechanism was famously championed in the early twentieth century by Milutin Milankovitch [81], fell out of favour because the timing of individual glaciations as deduced from the fragmentary terrestrial record did not seem to fit, and then was triumphantly vindicated by analysis using oxygen isotopes from fossil foraminifera, which reflected temporal variations in ambient temperature and ice volume of the more complete ocean record [82,83].

The exploitation of Milankovitch cycles has subsequently developed in various directions. It has become a stratigraphic tool for dating and correlation, not only in the Quaternary but in Tertiary and yet older strata [84], where a longer, 400 ka, orbital 'stretch' cycle is used as a more or less invariant 'pulse' that can be exploited stratigraphically and even quasi-formalized [85]. This in turn has led to the realization that climate in greenhouse as well as icehouse times was modulated by astronomical forcing, with variations in humidity/aridity and biological productivity producing patterns that, although more subtle than those produced by large ice volume changes, are nonetheless recognizable.

Also, the detailed expression of Milankovitch cycles has come under scrutiny. Astronomical calculation can precisely reveal insolation variations and hence predict the climate patterns that should result. The observed patterns from the stratal record depart from this in several ways. Firstly, they typically show a 'sawtooth' pattern rather than the predicted temporally symmetrical one; thus, individual glacial phases tend to develop slowly but finish abruptly. Secondly, the periodicity that is expected to be dominant is not always so, as will be seen below. Thirdly, and particularly in cold phases, there are marked, higher-frequency 'sub-Milankovitch' climate cycles that have been well described (also see below) but have not yet had adequate explanation.

9. MILANKOVITCH CYCLICITY IN QUATERNARY (PLEISTOCENE) CLIMATE HISTORY

The Quaternary displays a marked progression of overall climate state that may be regarded as an intensification of the glacial signature through time. The early Quaternary is dominated by the 40 ka axial tilt signal. About a million years ago, this gave way to dominance by the 100 ka orbital eccentricity cycle that has persisted to the present (Fig. 1(d)). This dominance has yet to be explained satisfactorily, for it would not be predicted from consideration of calculated insolation patterns over this interval, in which the eccentricity signal should be small. Suggested explanations have included the evolution of the ice sheet/substrate system to resonate (i.e. most easily grow and decay) to a 100 ka periodicity [86,87]; these explanations are tentative, for detailed models linking ice volume to insolation remain elusive [88]. The dominance by eccentricity has been accompanied by colder glacial maxima and warmer interglacial peaks, and it is this interval that has seen the greatest advances of ice and in general represents the 'ice ages' of vernacular usage.

The past million years includes a detailed record of atmospheric composition as well as temperature, in the form of the ice core data extracted from Greenland and Antarctica (with some ice core data of shorter duration from mountain glacier ice elsewhere) [89]. The longest current record is from Antarctica, extending to about 800 000 years ago [9,90], and planned drilling is aimed at extending the record to beyond a million years ago, and so into the '40 kiloyear world'. The Greenland record goes back to some 130 000 years ago [91], and so just into the last interglacial phase; but it is of high resolution because of a greater rate of snowfall, and is of great value in also allowing detailed comparison with the southern hemisphere, given the different climate behaviour of the hemispheres at short timescales (see below).

The combination of atmospheric composition records with climate proxy records (through hydrogen and oxygen isotope data, dust concentrations and so on) is extremely powerful (indeed, unique in the geological record); but, it is not precisely calibrated because ice data directly relates to deposition, while the gas data relates to the time of final closure of air bubbles in the ice, some distance down in the snowpack. The uncertainty that stems from this is small but important, because the correlation of carbon dioxide and methane levels with temperature is so close that questions of cause and effect have arisen. The consensus now is that astronomically driven insolation thresholds lead to small temperature rises, leading to carbon dioxide/methane increases that then strongly amplify the temperature rises [89].

The glacial-to-interglacial difference seen in the ice core records is about 100 ppm (from $ca.180$–280 ppm $p\mathrm{CO}_2$, respectively; where ppm refers to μmol mol^{-1}) representing several hundred gigatonnes of carbon that must be stored somewhere during glacial phases. Terrestrial storage via increased plant growth is unlikely, given the diminution of vegetated land during glacials, though storage in carbon-rich permafrost soils ('yedoma') has been mooted [92]. Ocean storage is generally considered more likely, and it is tempting to link this with the enhanced dust supply noted in the ice core records, that would fertilize open ocean waters and enhance carbon drawdown via increased plankton growth. However, ocean sediment records of barium (a proxy for plankton productivity) do not generally show increases during glacial phases. One means of combining low plankton productivity and increased trapping of carbon dioxide is to have a more stratified glacial ocean, limiting nutrient supply from below because of a stronger surface water 'lid' and also storing more dissolved carbon dioxide at depth [93]. There is evidence for such a model in the form of glacial-phase benthic foraminifer tests containing excess 'old' (i.e. radiocarbon-poor) carbon [94], with a plausible carbon dioxide sink/source being the deep water of the Southern Ocean [95].

10. QUATERNARY SUB-MILANKOVITCH CYCLICITY

Examination of high-resolution Quaternary records suggests significant climate variability that takes place on a sub-Milankovitch scale, a variability that is particularly marked in the cold phases that make up the bulk of the record (Fig. 1(e)). Thus, the cold phase that separates the present interglacial and the preceding (Eemian) one comprises not only five precession cycles but also 26 well-marked temperature oscillations, termed Dansgaard–Oeschger (D–O) cycles. These are most clearly expressed in the northern hemisphere, where they comprise rapid warming (of 8°C–16°C over Greenland) followed by slower cooling [96], to produce what are essentially a succession of interstadial and stadial units that average some 1470 a in duration [97]. The D-O cycles may be grouped into larger Bond cycles, terminated by intermittent (every several thousand years) Heinrich events [98]: iceberg 'armadas' released from the Laurentide and Scandinavian ice sheets marking episodes of partial collapse (Fig. 1(e)). The Heinrich events led to distinctive gravel-rich layers within sea floor sediments (brought in from melting icebergs), metre-scale rises in global sea level and rapid northern hemisphere cooling. They have one-to-one counterparts in the southern hemisphere but more muted ones (about 1°C–3°C in Antarctica) that are in partial anti-phase (Fig. 1(f)) (northern cold coinciding with southern warming) rather than in completely 'see-saw' fashion [99,100]. The causal mechanisms of the D-O cycles and related phenomena remain unclear, having been ascribed to changes in solar luminosity [101] and also to 'binge-purge' cycles of the great ice sheets [102].

The transition into the current Holocene interglacial was complex, thus, glacial conditions in the northern hemisphere were terminated at *ca.*14.5 ka, with rapid deglaciation ushering in the millennial-scale Allerød warm phase, itself terminated by rapid cooling into the Younger Dryas cold interval, also lasting about a thousand years. This finished abruptly at 11.8 ka, when temperatures in the northern hemisphere rose by *ca.*5°C in about a decade, ushering the warm and relatively stable conditions of the Holocene.

The reversal into the Younger Dryas has been ascribed to a major meltwater flood from the Laurentide ice sheet into the north Atlantic, putting a low-salinity 'lid' on the north Atlantic, hence stopping the formation of the cold, dense (high-salinity) North Atlantic Deep Water current and its ultimate return flow, the North Atlantic Drift ('Gulf Stream') [103–105]; eventual restart of this oceanic circulation pattern brought warmth once more back to the region. As with the D-O cycles, correlation with the southern hemisphere was complex, partly out of phase, and it is debated whether the climate changes were driven from the north or the south [106,107].

11. THE HOLOCENE

The Holocene is simply the latest of the many interglacial phases of the Quaternary; it is now longer than the preceding three interglacials by some 2000 years–3000 years [9], but only one-third of the length of the preceding one, OIS 11 [108], which lasted one-and-a-half rather than half a precession cycle; it is still unclear to which style of interglacial the Holocene 'naturally' belongs, on astronomical grounds (and thus what its 'natural' duration might be). Its duration to date has also been linked with the slow rise in atmospheric carbon dioxide levels from [260 to 280 ppm (parts per million)], ascribed (controversially) to preindustrial forest clearance by humans [109].

To date, though, other than a brief northern-hemisphere cooling event at 8.2 ka (also ascribed to a melt water pulse from the decaying ice sheets: [110]), the Holocene has seen remarkable stability of temperature and sea level, even when compared with other interglacials. Climate variation within it includes subdued, millennial-scale temperature oscillations of 1°C or so, examples being the 'Medieval warm period' and succeeding 'Little Ice Age'. Sea level variations of 1–2 m have been suggested [111] though more likely did not exceed a few decimeters [112]. As with the D-O cycles, the cause of the Holocene climate oscillations is obscure. Other shorter-period variations include the ENSO/El Nino events and the North Atlantic Oscillation (of a few years periodicity each); as with the millennial scale variations, these have far-reaching global impacts on such factors as regional rainfall patterns via a series of global teleconnections (e.g. [113]).

12. CLIMATE OF THE ANTHROPOCENE

About two centuries ago, human population rose above a billion (it is now over 7 billion). Widespread industrialization, powered by fossil fuels, also started then and continues to this day – indeed, is currently accelerating [114,115]. The sum total of physical, chemical and biological changes associated with this has led to the concept of the Anthropocene, a geological interval dominated by human activity [116]; if considered as a formal stratigraphic unit at an Epoch level [117–119], it follows that the Holocene has terminated.

Climate drivers of the Anthropocene are already well outside Holocene norms, for instance in the marked increase in greenhouse gas levels (now higher than in preindustrial times by the amount separating glacial and interglacial phases of the past), the change being considerably more rapid than either glacial-to-interglacial changes [120] or more arguably those associated with, say, the Toarcian and Paleocene-Eocene thermal maximum events [59,62,121]. In this respect, the Earth is in a non-analogue state. There is, too, the changing nature of carbon sinks associated with land-use changes and the diminishing albedo associated with rapidly waning Arctic sea ice. The approximately 1°C rise in global average temperature and the 30-cm rise in sea level may be confidently ascribed to human effect [122], but even with this, the Earth currently remains essentially within the limits of Quaternary interglacial climate. However, the current greenhouse warming, acting on this already warm phase, will, within decades to a few centuries, usher in a climate state that is unprecedented within the Quaternary. The long-term effect, if median predictions of the Intergovernmental Panel on Climate Change [123] come to pass, may resemble the brief 'super-interglacial' spike suggested by Broecker [124], the normal Quaternary Milankovitch cyclical climate changes subsequently resuming. Alternatively, modelled changes to the long-term carbon balance, together with threshold effects, suggest perturbation to at least several glacial cycles [125]. In whichever scenario (but particularly in the latter), the effects of the current warming will have striking and geologically long-lasting effects.

13. CONCLUSIONS

The history of Earth's climate system, as deduced from forensic examination of strata, has shown a general very long-term stability, which has probably been maintained by a complex interaction between the biosphere, atmosphere, hydrosphere, cryosphere and lithosphere. Superimposed on this overall stability have been a variety of climate perturbations on timescales ranging from multimillion year to sub-decadal, inferred to have been driven, amongst others, by variations in palaeogeography, greenhouse gas concentrations, astronomically forced insolation and interregional heat transport. Current anthropogenic changes to the Earth system, particularly as regards changes to the carbon cycle, are geologically significant. Their effects may likely include the onset of climate conditions of broadly pre-Quaternary style such as those of the 'mid-Pliocene warm period', with higher temperatures (particularly at high latitudes), substantially reduced polar ice cover, and modified precipitation and biotic patterns.

REFERENCES

[1] J. Zalasiewicz, M. Williams, The Goldilocks Planet: The Four Billion Year Story of Earth's Climate, Oxford University Press, Oxford, 2012, 336 pp.
[2] H.J. Dowsett, M.M. Robinson, D.K. Stoll, K.M. Foley, A.L.A. Johnson, M. Williams, C.R. Riesselman, Phil. Trans. R. Soc. A371 (2013). http://dx.doi.org/10.1098/rsta.2012.0524, 20120524.
[3] N. Clark, M. Williams, D. Hill, P. Quilty, J. Smellie, J. Zalasiewicz, M. Leng, M. Ellis, Naturwissenschaften 100 (2013) 699–722, http://dx.doi.org/10.1007/s00114-013-1075-9.
[4] H. Hasegawa, R. Tada, X. Jiang, S. Imsamut, P. Charusiri, N. Ichinnorov, Y. Khand, Clim. Past 8 (2012) 1323–2012.
[5] C.M. Rowe, D.B. Loope, R.J. Oglesby, R. Van der Voo, C.E. Broadwater, Science 318 (2007) 1284.

[6] K.T. Lawrence, T.D. Herbert, P.S. Dekens, A.C. Ravelo, in: M. Williams, A. Haywood, F.J. Gregory, D.N. Schmidt (Eds.), The Micropalaeontological Society, The Geological Society, London, 2007, pp. 539–562 (Special Publication).

[7] F. Kenig, J.D. Hudson, J.S. Sinninghe Damsté, B.N. Popp, Geology 32 (2004) 421–424.

[8] A.P.M. Vaughan, in: M. Williams, A. Haywood, F.J. Gregory, D.N. Schmidt (Eds.), The Micropalaeontological Society, The Geological Society, London, 2007, pp. 5–59 (Special Publication).

[9] E. Brook, Nature 453 (2008) 291–292.

[10] W.M. Kürschner, J. van der Burgh, H. Visscher, D.L. Dilcher, Mar. Micropaleontol. 27 (1/4) (1996) 299–331.

[11] D.O. Breecker, Z.D. Sharp, L.D. McFadden, Proc. Natl. Acad. Sci. U.S.A. 107 (2010) 576–580.

[12] P. Pearson, M. Palmer, Nature 406 (2000) 695–699.

[13] Y.G. Zhang, M. Pagani, Z. Liu, S.M. Bohaty, R. DeConto, Phil. Trans. R. Soc. A371 (2013), 20130096.

[14] R.A. Berner, Geochim. Cosmochim. Acta 70 (2006) 5653–5664.

[15] D.L. Royer, R.A. Berner, J. Park, Nature 446 (2007) 530–532.

[16] A. Haywood, A. Ridgewell, D. Lunt, D. Hill, M. Pound, H. Dowsett, A. Dolan, J. Francis, M. Williams, Phil. Trans. R. Soc. Lond. A369 (2011) 933–956.

[17] A.M. Haywood, A.M. Dolan, S.J. Pickering, H.J. Dowsett, E.L. McClymont, C.L. Prescott, U. Salzmann, D.J. Hill, S.J. Hunter, D.J. Lunt, J.O. Pope, P.J. Valdes, Phil. Trans. R. Soc. A371 (2013), 20120515.

[18] T.R.A. Vandenbroucke, H.A. Armstrong, M. Williams, F. Paris, J.A. Zalasiewicz, K. Sabbe, J. Nolvak, T.J. Challands, J. Verniers, T. Servais, Proc. Natl. Acad. Sci. 107 (2010) 14983–14986.

[19] H.J. Dowsett, M. Robinson, A. Haywood, U. Salzmann, D. Hill, L. Sohl, M. Chandler, M. Williams, K. Foley, D. Stoll, Stratigraphy 7 (2010) 123–139.

[20] U. Salzmann, A. Haywood, D. Lunt, P. Valdes, D. Hill, Global Ecol. Biogeogr. 17 (2008) 432–447.

[21] S.E. Wilde, J.W. Valley, W.H. Peck, C.M. Graham, Nature 409 (2001) 175–178.

[22] J. Baker, M. Bizzarro, N. Wittig, J. Connelly, H. Haack, Nature 436 (2005) 1127–1131.

[23] E.G. Nisbet, N.H. Sleep, Nature 409 (2001) 1083–1091.

[24] B. Rasmussen, R. Buick, Geology 27 (1999) 115–118.

[25] J.E. Lovelock, Nature 344 (1990) 100–102.

[26] M.J. Bickle, Terra Nova 8 (3) (1996) 270–276.

[27] A.S. Cohen, A.L. Coe, S.M. Harding, L. Schwark, Geology 32 (2004) 157–160.

[28] F. Robert, M. Chaussidon, Nature 443 (2006) 969–972.

[29] L.P. Knauth, Palaeogeogr. Palaeoclim. Palaeoceanogr. 219 (2005) 53–69.

[30] M.T. Hren, M.M. Tice, C.P. Chamberlain, Nature 462 (2009) 205–208.

[31] R.E. Blake, S.J. Chang, A. Lepland, Nature 464 (2010) 1029–1032.

[32] H. Ohmototo, Y. Watanabe, K. Kumazawa, Nature 429 (2004) 395–399.

[33] T.W. Lyons, Nature 429 (2004) 359–360.

[34] A. Newton, Nat. Geosci. 3 (2010) 458.

[35] C. Goldblatt, M.W. Claire, T.M. Lenton, A.J. Matthews, A.J. Watson, K.J. Zahnle, Nature Geosci. 2 (2009) 891–896.

[36] G.M. Young, V. von Brunn, D.J.C. Gold, W.E.L. Minter, J. Geol. 106 (1998) 523–538.

[37] S. Polteau, J.M. Moore, H. Tsikos, Precambrian Res. 148 (2006) 257–274.

[38] I.A. Hilburn, J.L. Kirschvink, E. Tajika, R. Tada, Y. Hamano, S. Yamamoto, Earth Planet. Sci. Lett. 409 (2005) 175–178.

[39] R.A. Kerr, Science 308 (2005) 1730–1732.

[40] R.E. Kopp, et al., PNAS 102 (2005) 11131–11136.

[41] I.J. Fairchild, M.J. Kennedy, J. Geol. Soc. Lond. 164 (2007) 895–921.

[42] M.I. Budyko, Tellus 21 (1969) 611–619.

[43] P.F. Hoffman, A.J. Kaufman, G.P. Halverson, D.A. Schrag, A Neoproterozoic snowball earth, Science 281 (1998) 1342–1346.

[44] D.S. Abbot, R.T. Pierrehumbert, J. Geophys. Res. 115 (2010) D03104.

[45] M. Kennedy, D. Mrofka, C. von der Borch, Nature 453 (2008) 642–645.

[46] L.E. Sohl, M.A. Chandler, in: M. Williams, A. Haywood, F.J. Gregory, D.N. Schmidt (Eds.), The Micropalaeontological Society, The Geological Society, London, 2007, pp. 61–80 (Special Publication).

[47] A.A. Page, J.A. Zalasiewicz, M. Williams, L.E. Popov, in: M. Williams, A. Haywood, F.J. Gregory, D.N. Schmidt (Eds.), The Micropalaeontological Society, The Geological Society, London, 2007, pp. 123–156 (Special Publication).

[48] S. Finnegan, K. Bergmann, J.M. Eiler, D.S. Jones, D.A. Fike, I. Eisenman, N.C. Hughes, A.K. Tripati, W.W. Fischer, Sci. Express (January 2011), Science 331 (2011) 903–906, http://dx.doi.org/10.1126/science.1200803.

[49] C.R. Fielding, T.D. Frank, J.L. Isbell, Special Paper of the Geological Society of America, Resolving the Late Paleozoic Ice Age in Time and Space, vol. 441, 2008.

[50] R.M. DeConto, D.M. Pollard, Nature 421 (2003) 245–249.

[51] G. Bartoli, M. Sarnthein, M. Weinelt, H. Erlenkeuser, D. Carbe-Schönberg, D.W. Lea, Earth Planet. Sci. Lett. 237 (2005) 33–44.

[52] T.D. Frank, L.P. Birgenheier, I.P. Montanez, C.R. Fielding, M.C. Rygel, Late Palaeozoic climate dynamics revealed by comparison of ice-proximal stratigraphic and ice-distal isotopic records, in: C.R. Fielding, T.D. Frank, J.L. Isbell (Eds.), Geological Society of America Special Paper, vol. 441, 2008, pp. 331–342.

[53] A.G. Smith, K.T. Pickering, J. Geol. Soc. Lond. 160 (2003) 337–340.

[54] J.C. Zachos, 10 others, Science 308 (2005) 1611–1615.

[55] A.S. Cohen, A.L. Coe, D.B. Kemp, J. Geol. Soc. Lond. 164 (2007) 1093–1108.

[56] J.C. Zachos, G.R. Dickens, R.E. Zeebe, Nature 451 (2008) 279–283.

[57] D.J. Thomas, J.C. Zachos, T. Bralower, E. Thomas, S. Bohaty, Geology 30 (2002) 1067–1070.

[58] J.C. McElwain, J. Wade-Murphy, S.P. Hesselbo, Changes in carbon dioxide during an oceanic anoxic event linked to intrusion into Gondwana coals, Nature 435 (2005) 479–482.

[59] D.B. Kemp, A.L. Coe, A.S. Cohen, L. Schwark, Nature 437 (2005) 396–399.

[60] G.J. Bowen, 8 others, Nat. Geosci. 8 (2015) 44–47. (2014, online). http://dx.doi.org/10.1038/NGEO2316.

[61] M.E. Smith, A.R. Carroll, E.R. Mueller, Nat. Geosci. 1 (2008) 370–374.

[62] G.J. Bowen, J.C. Zachos, Nat. Geosci. 3 (2010) 866–869.

[63] J. Zachos, M. Pagani, L. Sloan, E. Thomas, K. Billups, Science 292 (2001) 686–693.

[64] C. Lear, T.R. Bailey, P.N. Pearson, H.K. Coxall, Y. Rosenthal, Geology 36 (2008) 251–254.

[65] A. Merico, T. Tyrrell, P.A. Wilson, Nature 452 (2008) 979–982.

[66] J.P. Kennett, J. Geophys. Res. 82 (1977) 3843–3859.

[67] P.N. Pearson, G.L. Foster, B.S. Wade, Nature 461 (2009) 1110–1113.

[68] D.J. Beerling, D.L. Royer, Nat. Geosci. 4 (2011) 418–420.

[69] A. Goldner, N. Herold, M. Huber, Nature 511 (2014) 574–577.

[70] G.J. Retallack, Palaeogeogr. Palaeoclim. Palaeoecol. 214 (2004) 97–123.

[71] A.R., Lewis and 12 others, PNAS 105 (2008) 10676–10680.

[72] B.P. Flower, J.P. Kennett, Relations between Monterey Formation deposition and middle Miocene global cooling; Naples Beach section, California, Geology 21 (1993) 877–880.

[73] A.E. Shevenell, J.P. Kennett, D.W. Lea, Science 305 (2004) 1766–1770.

[74] L. Liesecki, M.E. Raymo, Paleoceanography 20 (2005) 17. PA1003.

[75] P.L. Gibbard, 12 others, What status for the Quaternary? Boreas 34 (2005) 1–6.

[76] M. Mudelsee, M.E. Raymo, Slow dynamics of the Northern Hemisphere glaciation, Paleoceanography 20 (2005). PA-4022.

[77] J.A.I. Hennissen, M.J. Head, S. De Schepper, J. De Groeneveld, Palynological evidence for a southward shift of the North Atlantic Current at ~2.6 Ma during the intensification of late Cenozoic Northern Hemisphere glaciation, Paleoceanography 28 (2014). PA-002543.
[78] D.J. Lunt, G.L. Foster, A.M. Haywood, E.J. Stone, Nature 454 (2008) 1102–1105.
[79] J.M. Edmond, Science 258 (1992) 1594–1597.
[80] G.H. Haug, 10 others, Nature 433 (2005) 821–825.
[81] M. Milankovitch, Kanon der Erdbestrahlung und seine Anwendung auf das Eiszeiten-problem. Special Publication 132, R. Serbian Academy Belgrade, 1941.
[82] N. Shackleton, Nature 215 (1967) 15–17.
[83] N.J. Shackleton, N.D. Opdyke, Oxygen isotope and palaeomagnetic stratigraphy of equatorial Pacific core V28-238: oxygen isotope temperatures and ice volumes on a 105 and 106 year scale, Quatern. Res. 3 (1973) 39–55.
[84] A. Gale, J. Hardenbol, B. Hathway, W.J. Kennedy, J.R. Young, V. Phansalkar, Geology 30 (2002) 291–294.
[85] B.S. Wade, H. Pälike, Oligocene climate dynamics, Paleoceanography 19 (2004). PA-001042.
[86] P.U. Clark, R.B. Alley, D. Pollard, Science 286 (1999) 1104–1111.
[87] R. Bintanja, R.S. van der Wal, Nature 454 (2008) 869–872.
[88] M. Raymo, P. Huybers, Nature 451 (2008) 284–285.
[89] E.W. Wolff, Episodes 31 (2008) 219–221.
[90] J.R. Petit, 18 others, Nature 399 (1999) 429–436.
[91] NEEM community members, Nature 493 (2013) 489–494.
[92] Z.A. Zimov, E.A.G. Schuur, F.C. Chapin III, Science 312 (2006) 1612–1613.
[93] S.L. Jaccard, G.H. Haug, D.M. Sigman, T.F. Pedersen, H.R. Thierstein, U. Röhl, Science 308 (2005) 1003–1006.
[94] T.M. Marchitto, S.J. Lehman, J.D. Ortiz, J. Flüchtiger, A. van Gen, Science 316 (2007) 1456–1459.
[95] L.C. Skinner, S. Fallon, C. Waelbroeck, E. Michel, S. Barker, Science 328 (2010) 1147–1151.
[96] C. Huber, 8 others, Earth Planet. Sci. Lett. 245 (2006) 504–519.
[97] G. Bond, 9 others, Science 278 (1997) 1257–1266.
[98] G. Bond, 13 others, Nature 360 (1992) 245–249.
[99] EPICA community members, Nature 444 (2006) 195–198.
[100] E.J. Steig, Nature 444 (2006) 152–153.
[101] H. Braun, et al., Nature 438 (2005) 208–211.
[102] D.R. MacAyeal, Paleoceanography 8 (1993) 775–784.
[103] W.S. Broecker, Science 312 (2006) 1146–1147.
[104] R.B. Alley, Annu. Rev. Earth Planet. Sci. 35 (2007) 241–272.
[105] J.B. Murton, M.D. Bateman, S.R. Dallimore, J.T. Teller, Z. Yang, Nature 464 (2010) 740–743.
[106] T.F. Stocker, Science 297 (2002) 1814–1815.
[107] A.J. Weaver, O.A. Saenko, P.U. Clark, J.X. Mitrovica, Science 299 (2003) 1709–1713.
[108] EPICA community members, Nature 429 (2004) 623–628.
[109] W.F. Ruddiman, Clim. Change 61 (2003) 261–293.
[110] J.D. Marshall, 13 others, Geology 35 (2007) 639–642.
[111] K.-E. Behre, Boreas 36 (2007) 82–102.
[112] K. Lambeck, H. Rouby, A. Purcell, Y. Sun, M. Sambridge, PNAS 111 (43) (2014) 15296–15303.
[113] T.M. Rittenour, J. Brigham-Grett, M.E. Mann, Science 288 (2000) 1039–1042.
[114] W. Steffen, P.J. Crutzen, J.R. McNeill, Ambio 36 (2008) 614–621.
[115] W. Steffen, W. Broadgate, L. Deutsch, O. Gaffney, C. Ludwig, The Anthropocene Review 2 (2015) 81–98, http://dx.doi.org/10.1177/2053019614564785.
[116] P.J. Crutzen, Nature 415 (2002) 23.

[117] J. Zalasiewicz, 19 others, GSA Today 18 (2) (2008) 4–8.

[118] M. Williams, J. Zalasiewicz, A. Haywood, M. Ellis (Eds.), Philosophical Transactions of the Royal Society, vol. 369A, 2011, pp. 833–1112.

[119] C.W. Waters, J.A. Zalasiewicz, M. Williams, M. Ellis, A. Snelling (Eds.), Geological Society of London Special Publication, vol. 395, 2014.

[120] E. Monnin, A. Indermühle, A. Dällenbach, J. Flückiger, B. Stauffer, T.S. Stocker, D. Reynaud, J.-M. Barnola, Science 291 (2001) 112–114.

[121] L. Kump, T. Bralower, A. Ridgewell, Oceanography 22 (2009) 94–107.

[122] M. Huber, R. Knutti, Nature Geoscience 5 (2012) 31–36.

[123] IPCC (Intergovernmental Panel on Climate Change), Climate change 2014: Synthesis report. Summary for policy makers.

[124] W.S. Broecker, How to Build a Habitable Planet, Eldigio Press, New York, 1987, 291 pp.

[125] T. Tyrrell, J.G. Shepherd, S. Castle, Tellus B 59 (2007) 664–672.

INDICATORS OF CLIMATE CHANGE

2

GLOBAL SURFACE TEMPERATURES

2

Peter Thorne[1,2]

[1]*Nansen Environmental and Remote Sensing Centre, Bergen, Norway;*
[2]*Department of Geography, Maynooth University, Maynooth, Ireland*

CHAPTER OUTLINE

1. INTRODUCTION

Global surface temperatures are a broadly used indicator of climate change. There is not one single 'surface temperature' but rather a family of closely related but nonidentical temperatures: land surface air temperature (LSAT), land surface temperature (LST), marine air temperature, sea surface temperature (SST), ice surface temperature, lake surface temperature etc. [1]. Typical global surface temperature estimates are derived from combining LSAT arising from fixed meteorological sites with SST estimates arising from ships and, more latterly, buoys.

Analyses of global surface temperatures have a rich heritage [2], with the first estimate of a globally averaged surface temperature published over 75 years ago [3]. Over time, the methods used, the data sources and the computational capabilities have evolved and improved. In spite of these advances, the pioneering efforts at global surface temperature analyses stack up well against modern-day estimates [2].

The Working Group 1 contribution to the Fifth Assessment Report of the Intergovernmental Panel on Climate Change (IPCC) [4] concluded that

'The globally averaged combined land and ocean surface temperature data as calculated by a linear trend, show a warming of 0.85 [0.65 to 1.06] °C, over the period 1880 to 2012, when multiple independently produced datasets exist. The total increase between the average of the 1850–1900 period and the 2003–2012 period is 0.78 [0.72 to 0.85] °C, based on the single longest dataset available'.

Climate Change. http://dx.doi.org/10.1016/B978-0-444-63524-2.00002-6

The statement was made as a statement of fact, with no confidence or likelihood statement attached. Such factual statements were few and far between within the assessment and reserved only for the very few most certain aspects of the investigation. It is unequivocal that the world has warmed since the start of instrumental record taking. Much of the rest of this chapter outlines the observational and analytical basis that underlies this statement.

2. BASIC DATA AVAILABILITY

The basic building block of any analysis is the original observations. But observations never have been and never will be taken continuously across the globe at high resolution or specifically to monitor climate. Rather, observations have been taken where people live and for a myriad of reasons including shipping routing, aviation, agriculture, infrastructure, energy, or just personal or professional interest.

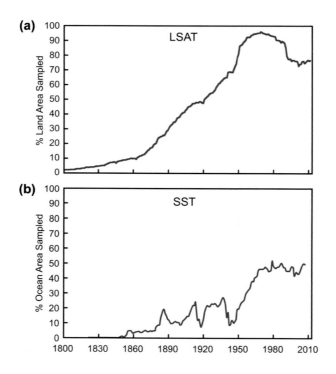

FIGURE 1

Change in percentage of possible sampled area for land records (top panel, a) and marine records (lower panel, b). Land data come from the Global Historical Climatology Network, version 3.2.0 [5] and marine data from the International Comprehensive Ocean-Atmosphere Data Set in situ record [6]. Coverage is defined as data present within a 5° grid box. *(Source: Hartmann et al. Appendix [7]; their Fig. 2. SM.2 p. 2SM-14. See http://www.climatechange2013.org/report/reports-graphic/.)* The graph has been reproduced with permission. Note, the *y*-axis should read: land area sampled/% and ocean area sampled/%.

Over land, measuring stations appear and are closed down frequently, and how the measurements have been taken has changed substantially through time. Over the ocean, with the exception of moored buoys or platforms, all measurements are vector measurements whereby the instruments themselves are constantly moving. Drifting buoys move with the currents while ships, and particularly commercial shipping, tend to follow well-worn shipping routes.

Substantial effort has been made to collate holdings of both LSAT and SST data in national, regional, and international holdings [5,6]. The percentage of both land and ocean observed has generally improved with time (Fig. 1). For LSAT data, historically available in many parts of the world is produced solely on a decadal basis as part of the World Weather Review process of the World Meteorological Organization Commission for Climatology. The last such compilation ended in 1990 and explains the drop-off in data availability at that time. The impact of the two World Wars on marine activity is clearly evident in the SST panel. The impact is twofold. First, there was a general reduction in shipping. Second, it was rather unwise to turn on a light at night to take a measurement as that might have alerted unfriendly forces and invited an attack.

3. ANALYSES OF LAND SURFACE AIR TEMPERATURES

The basic LSAT data holdings contain a myriad of data artefacts that arise from factors as diverse as:

- Station moves
- Instrument changes
- Observer changes
- Automation
- Time of observation biases
- Microclimate exposure changes
- Urbanization
- And so on and so forth.

By way of illustration, an example from Reno, Nevada, USA, is given in Fig. 2. Until the mid-1930s, the station was on the (probably white-painted) roof of the post office building in the town. In the 1930s, with the advent of aviation there was a need for observations at the airport, so the site was moved to the airport, which was a considerable distance from the town and its associated urban heat island (UHI). With time, the urban area has expanded and that UHI influence has led to a spurious multi-decadal warming trend since the 1970s at the site. Finally, in the late 1990s they moved the station from one end of the runway to the other and then back again. There is a clear thermal gradient along the length of the runway, which caused a temporary bias until they relocated the instrument back to its original location.

There exist numerous national, regional and global analyses that attempt to adjust for the non-climatic artefacts – a process termed *homogenization*. In general, the regional analyses are in concordance with the global analyses and so are not considered further here [7]. There are various approaches to the challenge of homogenization. Early techniques tended to consider the stations in isolation or their characteristics relative to some composite of series from neighbouring stations. Consideration of a station in isolation risks misdiagnosing a real change in climate system behaviour as a break in the series and so adjusting away the real climate signal. Consideration of a neighbour

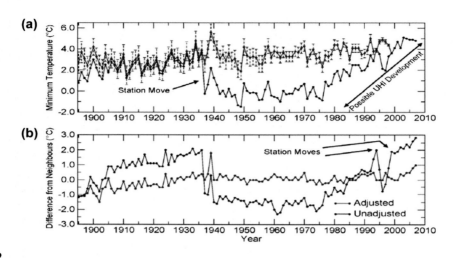

FIGURE 2

Example station series arising from Reno, Nevada, USA. Panel (a) is in absolute annual values. Panel (b) is in terms of differences from neighbours. The black line is the original series and the blue line denotes the homogenized series after Menne and Williams [8]. *(Modified with annotations from their Fig. 8.)* Note, the *y*-axis should read: minimum temperature/°C and difference in temperature from neighbours/°C.

composite has issues if the neighbours themselves contain biases. Such issues become critical when all or a substantial subset of the neighbours contain a common bias.

Most modern techniques utilize some form of pairwise neighbour comparison to identify the breaks [8,9]. These identify breaks in multiple candidate pairs of target minus reference station series and then seek to de-convolve the problem. For example, taking a network of 20 stations, it may be that a break is found to occur in 1950 in 15 of the pairwise comparisons. In 15 stations, this break is found once and in the 16th it occurs 15 times. In this case, it is this latter station that contains the real break. Once breaks have been attributed, they can either be adjusted [5] or the segments treated as effective stations for each homogeneous segment [10].

Several state-of-the-art homogenization techniques have been assessed against benchmark test cases [9,11]. Such test cases involve presenting to the dataset creators sets of data that have been synthetically produced and where the data originators know what the required data issues are to be found and what adjustments are necessary [12]. Benchmarking exercises undertaken to date show that modern techniques tend to improve the consistency and 'correctness' of the records but that no technique is perfect. Unsurprisingly, techniques tend to struggle when the data artefacts are small, numerous or both. It is therefore important to understand fully the likely impacts of common changes that have occurred across the global network. To this end a number of comparisons have been undertaken between modern and historical instrumentation at several sites [13,14].

There exist many possible ways to assess LSAT changes. Currently, there are four global analyses [5,10,15,16]. Each analysis takes a distinct approach to station selection, quality control and homogenization, interpolation and area averaging. The use of independent approaches serves to highlight the degree of sensitivity of resulting findings to methodological choices. The four different estimates are in broad agreement throughout the record, with differences becoming larger in the earlier period of record [7]. Differences between estimates in the global mean are substantively smaller than the long-term warming trend common to all estimates.

4. ANALYSES OF SEA SURFACE TEMPERATURES

The basic SST holdings have arisen from a broad range of measurement platforms using an array of measurement techniques that have changed substantially through time [17]. Biases in SST records are both larger and more systematic in nature than for LSAT and hence homogenization is essential. Measurements up until the 1940s were almost exclusively from buckets whereby a sample of the seawater from just below the surface would be hauled onto the ship deck and measured. Since World War II, there has been a preponderance for either engine room intake–based measurement or hull contact sensors. Since the 1990s, there has been an increasing ubiquity of drifting buoys so that today approximately 90% (by number but not coverage) of all measurements arise from this method. Each of these techniques has a distinct bias relative to the true SST, and failing to account for these effects would add substantial spurious multi-decadal variability to the records. Taking these in turn:

- Measurements based on buckets tended to be cold biased due to the effects of evaporative cooling that occurs between sampling the water and its subsequent measurement. Quite how cold biased depends upon the insulation efficiency of the bucket, the ship deck height, the delay between sampling and measurement, and the ambient weather conditions [18]. The effect is greatest when windy and when the atmosphere is substantially warmer or colder than the sea surface. Without accounting for these effects, pre-1942 measurements would be too cold by about 0.3°C globally averaged.
- Engine room intakes and hull contact sensors tend to sample water that has been warmed relative to the ambient temperature by the ship itself and therefore be warm biased.
- Drifting buoys exhibit little obvious bias and substantially smaller spread than ship-based measurements. They measure temperatures that are about 0.12°C–0.18°C colder than the modern ships, which are mainly engine room intake- or hull sensor-based measurements (Fig. 3).

Several global SST analyses have been undertaken that attempt to ascertain and adjust for biases either to some subset of the record or the entire record. Three analyses exist that consider global changes over the entire period of record [19–21]. These estimates take substantively different approaches to the problem. Despite this, they are closer to each other than they are to the original basic data on which they are based. The largest differences occur around the times of major transitions within the observing system (Fig. 3) or times when the observational record is dominated by ships sailing under a single flag (ship-based measurement protocols are broadly dictated on a national basis). For example, in World War II, most measures arise from the US fleet, which took measurements that were systematically warmer than most other nations. Having a mix of national measurements prior to 1939 and after 1945, but mainly US measurements in the middle, therefore yields a potential spurious SST maximum in the early 1940s in the raw data [22].

Overall, there is a greater sensitivity to choice of dataset construction methods in SST than there is in LSAT. Furthermore, the differences between the raw and adjusted data records are substantially

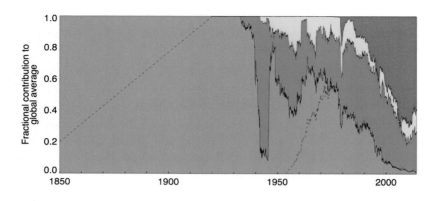

FIGURE 3

Best understanding of the changing mix of marine in situ observations since 1850. Blue is buoys with wooden to canvas bucket transition occurring between 1850 and 1920 (dashed line, approximate) and from 1954 to 1975 an uncertain switch from uninsulated to insulated buckets. Engine room intake (ERI)/hull contact measures are in green. Unknown measurement type is yellow and buoys are red. The *y*-axis is fractional contribution to global average and not an observation count. *(Modified from Kennedy et al. [19] and courtesy John Kennedy, UK Met Office.)*

larger at the global-mean scales. Whereas LSAT station biases tend to cancel somewhat regionally and globally, SST records afford no such luxury.

5. GLOBAL CHANGES

Global surface temperature datasets arise from combining underlying datasets of LSAT and SST. Choices are required in deciding:

- which underlying datasets are to be merged;
- how to undertake the merge;
- whether and, if so, how attempts are made to account for areas of missing data by interpolation.

The choice of whether to interpolate or not can have a significant impact, particularly on decadal timescale behaviour. Sampling is not uniform in space or in time and many regions have never been adequately sampled (deserts, rainforests, polar regions, regions of seasonal or perennial sea ice). If the temperatures in the unsampled regions are not well represented in the remainder of the sampled portion of the globe, then a biased estimate will result [23,24]. It appears that over the past 20 years in particular, interpolation has a distinct effect upon apparent global mean behaviour, with interpolated analyses showing greater warming in the global mean [24].

There are at least five datasets that estimate global average surface temperatures [10,16,24–26]. These products combine underlying SST and LSAT datasets in different ways and make distinct choices in how to calculate spatial and global averages. However, there are similarities that mean there are fewer true degrees of freedom than implied by having five estimates. For example, these five estimates are based upon solely two underlying SST datasets and three underlying LSAT datasets.

Indeed, two pairs of datasets differ solely in the post-processing of the underlying data, sharing the choice of both underlying LSAT and SST products. The different datasets broadly agree in their characterization of global mean changes at various space and timescales [7,27].

The global mean surface temperature has undoubtedly increased since the mid-nineteenth century (Fig. 4). The change has not been linear in nature. There exist several decade-plus stretches of either little change or even cooling. This includes the most recent 15 or so years. This period has been dubbed a 'hiatus' (a misnomer unless and until warming resumes) and elicited much scientific [28] and public interest. As is clear from Fig. 4, such periods are not atypical of the longer record [29]. Nevertheless, there is much debate around the nature, causes, and implications of the current hiatus [30–34 and numerous others]. This is understandable given that it raises legitimate questions around the short- to medium-term future trajectory of the system.

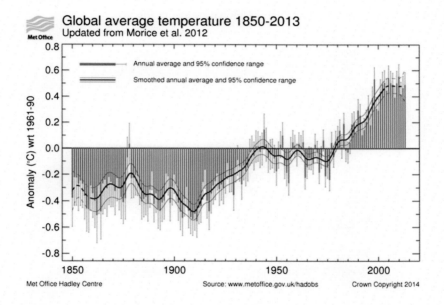

FIGURE 4

Global average surface temperature estimates and their uncertainties from the HadCRUT4 dataset [25]. Annual averages are shown as red bars with associated uncertainties. The blue trace and bounds denote a smoothed (using a 21-point binomial filter) estimate and associated uncertainties. *(Figure courtesy Colin Morice and sourced from www.metoffice.gov.uk/hadobs. Contains public sector information licenced under the Open Government Licence v1.0.)* Note, the y-axis should read: anomalous temperature wrt 1961–1990/°C.

In the same way that the global average change has not been linear, the global pattern of changes has not been uniform. Reasonable global spatial trend estimation has only been possible since the start of the twentieth century when the Southern Hemisphere sampling became sufficiently complete to estimate spatial anomaly patterns. It should be noted that estimation of a global average requires substantially fewer observations so long as they are well spaced. This is because anomalies in temperatures have large spatial scales. If it is unusually warm in London, the chances are that it is unusually warm also in Dublin, Edinburgh, Brussels and Paris. In reality, well-spaced sites of the order 150 would adequately characterize the global mean LSAT [35] and similar density would characterize SST. But, obviously, these would not provide local information. Figure 5 shows spatial trend estimates arising from three of the five datasets and two periods. The recent period contains more observations so all products are more complete. This is particularly marked for HadCRUT4, which is unique in not undertaking interpolation but choosing instead to account for sampling effects in its uncertainty estimates [25]. Longer-term trends tend to be spatially smoother as the effects of variability become less marked. Almost the whole globe has warmed since 1901. No region has experienced a statistically significant (distinguishable from a zero change) cooling. Since 1979, roughly 15% of the globe has cooled, and some regions do show significance, but most of the globe has warmed significantly. The period from 1979 exhibits a greatly enhanced rate of warming (also evident in the time series plot for HadCRUT4 global means in Fig. 4).

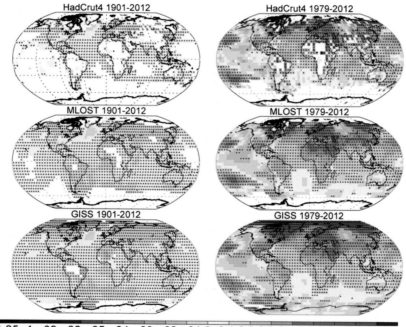

FIGURE 5

Spatial trend maps for the periods 1901–2012 and 1979–2012 for three global surface temperature estimates (HadRCUT4 [25]; MLOST [26] and GISS [16]). Trends have been calculated using ordinary least squares regression. Significance of trends has been assessed after accounting for AR(1) autocorrelation effects on the effective degrees of freedom [36]. *(The figure has been reproduced with permission.)* The label on the temperature code should read: Trend per decade/°C.

6. UNCERTAINTY QUANTIFICATION

Increasing attention is being paid to the quantification of uncertainties within surface temperature estimates. There exist several 'flavours' of uncertainty [37]. The most important are structural and parametric uncertainties. Structural uncertainties arise through choices of the overall method and can be quantified by comparing the estimates arising from different groups of analysts. Proper quantification would require a large ensemble of datasets that are produced independently. This is only partially true for surface temperature datasets (see earlier discussion). Parametric uncertainty involves the creation of an ensemble of estimates for a given dataset assessing sensitivity to uncertain choices within the methodology. For example, whether a break in a station series is assigned at 1%, 5% or 10% significance threshold for the breakpoint detection statistical test is not a choice with an a priori correct answer. There exist numerous such semi-subjective choices in all algorithms.

Several parametric uncertainty estimates have been constructed for SST [19,38] and LSAT [11,25] and combined [25]. These result in ensembles of possible realizations. Other approaches have also been applied to uncertainty quantification that do not result in ensembles [10,26]. However, ensembles are intuitively appealing because they allow the expression of uncertainties at various space and timescales to support users. Many users are interested in aspects of surface temperatures other than the global mean long-term trends.

Both structural and parametric uncertainties to the extent thus far quantified are an order of magnitude smaller than the estimated global-mean changes since the start of the instrumental record of surface temperatures. It would require a substantial hitherto unrecognized source of uncertainty to be discovered to call into question the conclusion that the globe has warmed on multi-decadal timescales. Further support for the conclusion that the world has warmed arises from our understanding of changes in a suite of co-related variables such as tropospheric temperatures, glacier volume, sea ice, surface humidity and ocean heat content. A variety of datasets produced by many scientific groups for each of these covariates conclude that they are changing in the manner that would be expected if the world is indeed warming [39]. It is a combination of the confidence in direct measurements and changes in these co-related variables that have led scientists, through the IPCC assessment process, to conclude that the warming is unequivocal.

7. CHARACTERIZATION OF EXTREMES AND VARIABILITY

Society experiences the weather and not the climate. Specifically, the major effects upon society relate to extremes of heat or cold or passing thresholds such as the need to heat or cool buildings or being able to grow certain types of crops. For such cases, global mean anomaly series discussed thus far in this chapter are of limited value. Several daily and sub-daily data holdings exist [40,41] and are used to monitor extremes and various societally relevant indices over land [42]. The Commission for Climatology Expert Team on Climate Change Detection and Indices (ETCDDI) has defined 27 core indices of which 16 are directly related to temperatures [43]. Several datasets have been created that indicate trends in these ETCDDI indices [42,44–46]. These provide useable and actionable information upon aspects such as the changing frequency of warm nights (Fig. 6). Overall, changes in warm extremes have tended to increase and changes in cold extremes decrease over the period of record. This is consistent with the observed warming in the mean climate. Changes to the tails of the distribution need not follow changes in the mean. Daily and sub-daily holdings generally do not go as far back as monthly holdings, and so information on these aspects only becomes really globally representative in the post-1950 period.

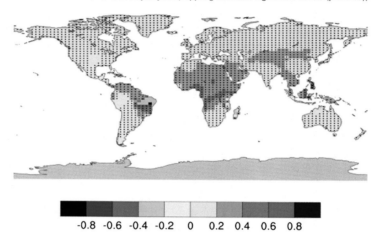

HADEX2 TN90p ANN Trend 1951-2010

unit: % of days / year (stippling indicates significant trends (p<=0.05))

-0.8 -0.6 -0.4 -0.2   0   0.2   0.4   0.6   0.8

copyright www.climdex.org, 2014-10-15

FIGURE 6

Changing frequency, in percent, of anomalously warm nights per year (the change in frequency of nights exceeding the local daily 90th percentile calculated over the 1961–1990 period) over 1951–2010 in the HadEX2 dataset [44]. *(Courtesy of the WMO/CLIVAR/JCOMM ETCCDI, Australian Research Council ARC Linkage Project LP100200690 and www.climdex.org (plotted at www.climdex.org). The figure has been reproduced with permission.)* Note, the label on the colour-coded ribbon should read: changing frequency/%.

8. FUTURE RESEARCH DIRECTIONS

In the latter part of the twentieth century, the pre-eminent questions were: Was the mean climate changing, how much was it changing, and to what extent were humans responsible for those changes? As we enter the twenty-first century, the demands as well as the expectations are different. Scientifically, it is unequivocal that the world has warmed and, it is virtually certain that humans are primarily responsible. Society, governments and industry require actionable information at the local level to make informed decisions. Actionable information begets openness and transparency, regional detail, and useable uncertainty estimates as well as information at daily and sub-daily timescales. The global science community is addressing this need through activities such as the ETCCDI and the International Surface Temperature Initiative [47].

The International Surface Temperature Initiative has created a new set of holdings by collating various national and international data collections and merging them. The first version released [48] consisted of monthly resolution holdings and represents a fourfold increase in available stations compared to the holdings used in most current LSAT records (Fig. 7). Accompanying these data will be a set of benchmark test datasets that mirror the databank sampling and formats [12]. It is then hoped that multiple pre-existing and new groups will analyze the data holdings and submit to the benchmarking exercise, leading to improved understanding of LSAT changes and their uncertainties.

Recent analyses of the effects of interpolation or not interpolating have reignited interest in research in this area with a focus on interpolation over the polar regions [24,49,50]. In part this is driven by the fact that in recent decades there is evidence that a majority of the unsampled regions have tended to warm faster than sampled regions [23,24]. As a result, there are likely to be several new and distinct attempts to create interpolated products that better account for regions that have not been directly measured.

There are also distinct data types that provide new tools to look at a minimum at recent period records. Satellite remote-sensing records starting in the late 1970s can tell us about sea and land skin temperatures. Most of such measures are in the infrared spectrum, which provides information only under clear-sky rather than all-sky conditions. However, the satellites do sense the skin temperatures over almost the entire globe including many regions where measurements have never been made using in situ measurement techniques. Thereby they provide the possibility of a truly global assessment. Analyses of satellite records of SST are mature and the SST records from in situ estimates and satellite records are in broad concurrence [7]. These records provide important spatial detail on SST variations. The analysis of land surface temperatures is somewhat less mature, and the clear-sky biases and differences in the diurnal cycle of LST and LSAT are likely to be more of a challenge. Several new projects such as the EUSTACE project aim to merge satellite and in situ records over land and oceans to create improved estimates are being implemented.

Reanalyses are data assimilation and forecast systems run in frozen mode that reanalyze a substantial fraction of the totality of the historical record. They are constrained by observed SST estimates, prescribed radiative forcings such as greenhouse gases, solar and volcanic, and many other aspects of the observed system (pressure readings, weather balloons, satellites etc.). Surface temperatures are not directly assimilated so the reanalyses fields are (almost, but not completely, in the subset of analyses that use them to indirectly nudge the soil boundary) independent of the actual LSAT measurements. They create geographically complete and physically internally consistent estimates of the atmospheric evolution, although this does not make them necessarily immune from in-homogeneities. Both a surface pressure–only reanalysis from the late nineteenth century onwards [51] and more traditional reanalyses since the latter part of the twentieth century, using weather balloons and satellites [23,52], have been shown to match the available direct observational estimates of LSAT. In the future, reanalyses will increasingly be useful in monitoring long-term changes. Haimberger et al. [53] have shown how the reanalyses can be utilized to infer the presence of breaks in the in situ balloon record. This can then be followed by adjustments. A similar approach may be possible for LSAT records.

In the longer term, there will be an increasing need to consider the homogeneity of daily or sub-daily data [54,55]. This is a much harder problem than homogenization of monthly time series. On monthly time scales, the effects of weather on the biases tend to cancel. For homogenization at daily and sub-daily time scales, it is likely that more physically based corrections are required. For example, it matters a lot more at these scales as to whether the day was sunny, rainy, windy or calm etc., and if so,

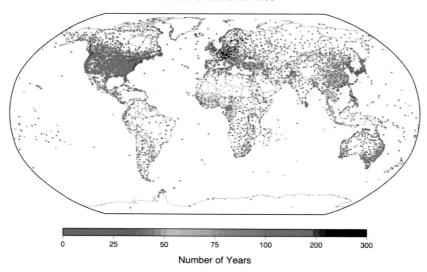

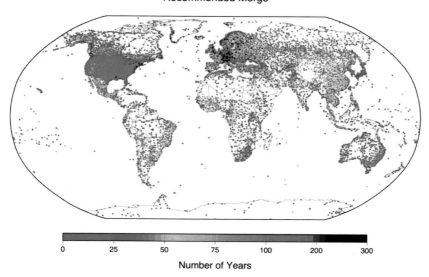

FIGURE 7

Station data innovations arising from the databank of the International Surface Temperature Initiative [48]. The top panel shows the LSAT station data used in several current products. The bottom panel shows newly available data holdings. *(Plots courtesy of Jared Rennie, CICS-NC.)*

how much the individual values in the series should be adjusted. Initial efforts are being made to build a database of parallel measurements where old and new measurements have been taken side by side to better understand these effects [13,14].

9. CONCLUSIONS

It is unequivocal that the global surface temperatures have warmed since the instigation of instrumental records. This change has not been linear and moreover has varied substantially over different geographical regions. Important uncertainties and challenges remain to be addressed regarding data availability, measurement understanding, merging in situ and satellite records, providing high temporal resolution data suitable for many applications and making robust uncertainty estimates that are useable. These challenges may alter important aspects of our understanding of surface temperatures but are very unlikely to affect the bottom-line conclusion that the world has warmed since the advent of global instrumental temperature records in the mid- to late-nineteenth century.

REFERENCES

[1] C.J. Merchant, S. Matthiesen, N.A. Rayner, J.J. Remedios, P.D. Jones, F. Olesen, B. Trewin, P.W. Thorne, R. Auchmann, G.K. Corlett, P.C. Guillevic, G.C. Hulley, Geosci. Instrum. Methods Data Syst. 2 (2013) 305–321, http://dx.doi.org/10.5194/gi-2-305-2013.

[2] E. Hawkins, P.D. Jones, Q. J. R. Meteorol. Soc. 139 (2013) 1961–1963.

[3] G.S. Callendar, Q. J. R. Meteorol. Soc. 64 (1938) 223–240, http://dx.doi.org/10.1002/qj.49706427503.

[4] IPCC, Summary for policymakers, in: T.F. Stocker, D. Qin, G.-K. Plattner, M. Tignor, S.K. Allen, J. Boschung, A. Nauels, Y. Xia, V. Bex, P.M. Midgley (Eds.), Climate Change 2013: The Physical Science Basis. Contribution of Working Group I to the Fifth Assessment Report of the Intergovernmental Panel on Climate Change, Cambridge University Press, Cambridge, UK and New York, NY, USA, 2013, pp. 1–30, http://dx.doi.org/10.1017/CBO9781107415324.004.

[5] J.H. Lawrimore, M.J. Menne, B.E. Gleason, C.N. Williams, D.B. Wuertz, R.S. Vose, J. Rennie, J. Geophys. Res. Atmos. 116 (2011) D19121.

[6] S.D. Woodruff, S.J. Worley, S.J. Lubker, Z. Ji, J.E. Freeman, D.I. Berry, P. Brohan, E.C. Kent, R.W. Reynolds, S.R. Smith, C. Wilkinson, Int. J. Climatol. 31 (2011) 951–967.

[7] D.L. Hartmann, A.M.G. Klein Tank, M. Rusticucci, L.V. Alexander, S. Brönnimann, Y. Charabi, F.J. Dentener, E.J. Dlugokencky, D.R. Easterling, A. Kaplan, B.J. Soden, P.W. Thorne, M. Wild, P.M. Zhai, Observations: atmosphere and surface, in: T.F. Stocker, D. Qin, G.-K. Plattner, M. Tignor, S.K. Allen, J. Boschung, A. Nauels, Y. Xia, V. Bex, P.M. Midgley (Eds.), Climate Change 2013: The Physical Science Basis. Contribution of Working Group I to the Fifth Assessment Report of the Intergovernmental Panel on Climate Change, Cambridge University Press, Cambridge, UK and New York, NY, USA, 2013.

[8] M.J. Menne, C.N. Williams, J. Clim. 22 (2009) 1700–1717.

[9] V.K.C. Venema, O. Mestre, E. Aguilar, I. Auer, J.A. Guijarro, P. Domonkos, G. Vertacnik, T. Szentimrey, P. Stepanek, P. Zahradnicek, J. Viarre, G. Müller-Westermeier, M. Lakatos, C.N. Williams, M.J. Menne, R. Lindau, D. Rasol, E. Rustemeier, K. Kolokythas, T. Marinova, L. Andresen, F. Acquaotta, S. Fratianni, S. Cheval, M. Klancar, M. Brunetti, C. Gruber, M. Prohom Duran, T. Likso, P. Esteban, T. Brandsma, Clim. Past 8 (2012) 89–115.

[10] R. Rohde, R. Muller, R. Jacobsen, S. Perlmutter, A. Rosenfeld, J. Wurtele, J. Curry, C. Wickham, S. Mosher, Geoinfor. Geostat. 1 (2013) 1–2.

[11] C.N. Williams, M.J. Menne, P.W. Thorne, J. Geophys. Res. Atmos. 117 (2012).

[12] K. Willett, C. Williams, I.T. Jolliffe, R. Lund, L.V. Alexander, S. Brönnimann, L.A. Vincent, S. Easterbrook, V.K.C. Venema, D. Berry, R.E. Warren, G. Lopardo, R. Auchmann, E. Aguilar, M.J. Menne, C. Gallagher, Z. Hausfather, T. Thorarinsdottir, P.W. Thorne, Geosci. Instrum. Methods Data Syst. 3 (2014) 187–200, http://dx.doi.org/10.5194/gi-3-187-2014.

[13] R. Bohm, P.D. Jones, J. Hiebl, D. Frank, M. Brunetti, M. Maugeri, Clim. Change 101 (2010) 41–67.

[14] M. Brunet, J. Asin, J. Sigro, M. Banon, F. Garcia, E. Aguilar, J.E. Palenzuela, T.C. Peterson, P.D. Jones, Int. J. Climatol. 31 (2011) 1879–1895.

[15] P.D. Jones, D.H. Lister, T.J. Osborn, C. Harpham, M. Salmon, C.P. Morice, J. Geophys. Res. Atmos. 117 (2012) D05127.

[16] J.R. Hansen, M. Ruedy, M. Sato, K. Lo, Rev. Geophys. 48 (2010) RG4004.

[17] J.J. Kennedy, Rev. Geophys. 52 (2014) 1–32, http://dx.doi.org/10.1002/2013RG000434.

[18] C.K. Folland, D.E. Parker, Q. J. R. Meteor. Soc. 121 (1995) 319–367.

[19] J.J. Kennedy, N.A. Rayner, R.O. Smith, D.E. Parker, M. Saunby, J. Geophys. Res. Atmos. 116 (2011) D14104.

[20] S. Hirahara, M. Ishii, Y. Fukuda, J. Clim. 27 (2014) 57–75, http://dx.doi.org/10.1175/JCLI-D-12-00837.1.

[21] B. Huang, et al., Extended Reconstructed Sea Surface Temperature version 4 (ERSST.v4), Part 1. Upgrades and Intercomparisons, J. Clim. 28 (2015) 911–930, http://dx.doi.org/10.1175/JCLI-D-14-00006.1.

[22] D.W.J. Thompson, J.J. Kennedy, J.M. Wallace, P.D. Jones, Nature 453 (2008) 646.

[23] A.J. Simmons, K.M. Willett, P.D. Jones, P.W. Thorne, D.P. Dee, J. Geophys. Res. Atmos. 115 (2010) D01110.

[24] K. Cowtan, R.G. Way, Q. J. R. Meteorol. Soc. 140 (2014) 1935–1944, http://dx.doi.org/10.1002/qj.2297.

[25] C.P. Morice, J.J. Kennedy, N.A. Rayner, P.D. Jones, J. Geophys. Res. Atmos. 117 (2012) 22.

[26] R.S. Vose, D. Arndt, V.F. Banzon, D.R. Easterling, B. Gleason, B. Huang, E. Kearns, J.H. Lawrimore, M.J. Menne, T.C. Peterson, R.W. Reynolds, T.M. Smith, C.N. Williams, D.B. Wuertz, Bull. Am. Meteor. Soc. 93 (2012) 1677–1685.

[27] Thorne, P.W. et al., Investigating the recent apparent hiatus in surface temperature increases: 2. Comparison of model ensembles to observational estimates. JGR, http://dx.doi.org/10.1002/2014JD022805.

[28] G. Flato, J. Marotzke, B. Abiodun, P. Braconnot, S.C. Chou, W. Collins, P. Cox, F. Driouech, S. Emori, V. Eyring, C. Forest, P. Gleckler, E. Guilyardi, C. Jakob, V. Kattsov, C. Reason, M. Rummukainen, Evaluation of climate models, in: T.F. Stocker, D. Qin, G.-K. Plattner, M. Tignor, S.K. Allen, J. Boschung, A. Nauels, Y. Xia, V. Bex, P.M. Midgley (Eds.), Climate Change 2013: The Physical Science Basis. Contribution of Working Group I to the Fifth Assessment Report of the Intergovernmental Panel on Climate Change, Cambridge University Press, Cambridge, UK and New York, NY, USA, 2013.

[29] B. Liebmann, R.M. Dole, C. Jones, I. Blade, D. Allured, Bull. Am. Meteor. Soc. 91 (2010) 1485–1491.

[30] J.L. Cohen, J.C. Furtado, M. Barlow, V.A. Alexeev, J.E. Cherry, Geophys. Res. Lett. 39 (2012) L04705.

[31] Y. Kosaka, S. Xie, Nature 501 (2013) 403–407, http://dx.doi.org/10.1038/nature12534.

[32] K.E. Trenberth, J.T. Fasullo, G. Branstator, A.S. Phillips, Nat. Clim. Change (2014), http://dx.doi.org/10.1038/NCLIMATE2341.

[33] G.A. Meehl, H. Teng, J.M. Arblaster, Nat. Clim. Change (2014), http://dx.doi.org/10.1038/NCLIMATE2357.

[34] J.S. Risbey, S. Lewandowsky, C. Langlais, D.P. Monselesan, T.J. O'Kane, N. Oreskes, Well-estimated global surface warming in climate projections selected for ENSO phase, Nat. Clim. Change (2014), http://dx.doi.org/10.1038/nclimate2310.

[35] P.D. Jones, Clim. Change 31 (1995) 545–558.

[36] B.D. Santer, P.W. Thorne, L. Haimberger, K.E. Taylor, T.M.L. Wigley, J.R. Lanzante, S. Solomon, M. Free, P.J. Gleckler, P.D. Jones, T.R. Karl, S.A. Klein, C.A. Mears, D. Nychka, G.A. Schmidt, S.C. Sherwood, F.J. Wentz, Int. J. Climatol. 28 (2008) 1703–1722.

[37] P.W. Thorne, D.E. Parker, J.R. Christy, C.A. Mears, Bull. Am. Meteor. Soc. 86 (2005) 1437.

[38] W. Liu, et al., Extended Reconstructed Sea Surface Temperature version 4 (ERSST.v4): Part II. Parametric and Structural Uncertainty Estimations, J. Clim 28 (2015) 931–951. http://dx.doi.org/10.1175/JCLI-D-14-00007.1.

[39] J.J. Kennedy, P.W. Thorne, T.C. Peterson, R.A. Ruedy, P.A. Stott, D.E. Parker, S.A. Good, H.A. Titchner, K.M. Willett, Bull. Am. Meteor. Soc. (2010) S26–S27.

[40] M.J. Menne, I. Durre, R.S. Vose, B.E. Gleason, T.G. Houston, J. Atmos. Oceanic Technol. 29 (2012) 897–910, http://dx.doi.org/10.1175/JTECH-D-11-00103.1.

[41] R.J.H. Dunn, K.M. Willett, P.W. Thorne, E.V. Woolley, I. Durre, A. Dai, D.E. Parker, R.S. Vose, Clim. Past 8 (2012) 1649–1679, http://dx.doi.org/10.5194/cp-8-1649-2012.

[42] M.G. Donat, L.V. Alexander, H. Yang, I. Durre, R. Vose, J. Caesar, Bull. Am. Meteor. Soc. 94 (2013) 997–1006.

[43] T.C. Peterson, C. Folland, G. Gruza, W. Hogg, A. Mokksit, N. Plummer, WMO, Rep. WCDMP-47, WMO-TD 1071, Geneve, Switzerland, 143 pp.

[44] M.G. Donat, L.V. Alexander, H. Yang, I. Durre, R. Vose, R.J.H. Dunn, K.M. Willett, E. Aguilar, M. Brunet, J. Caesar, B. Hewitson, C. Jack, A.M.G. Klein Tank, A.C. Kruger, J. Marengo, T.C. Peterson, M. Renom, C. Oria Rojas, M. Rusticucci, J. Salinger, A.S. Elrayah, S.S. Sekele, A.K. Srivastava, B. Trewin, C. Villarroel, L.A. Vincent, P. Zhai, X. Zhang, S. Kitching, J. Geophys. Res. Atmos. 118 (2013) 2098–2118.

[45] L.V. Alexander, X. Zhang, T.C. Peterson, J. Caesar, B. Gleason, A.M.G. Klein Tank, M. Haylock, D. Collins, B. Trewin, F. Rahimzadeh, A. Tagipour, R. Kumar Kolli, J.V. Revadekar, G. Griffiths, L. Vincent, D.B. Stephenson, J. Burn, E. Aguilar, M. Brunet, M. Taylor, M. New, P. Zhai, M. Rusticucci, J.L. Vazquez Aguirre, J. Geophys. Res. Atmos. 111 (2006) D05109.

[46] J. Caesar, L. Alexander, R. Vose, J. Geophys. Res. Atmos. 111 (2006) D05101.

[47] P.W. Thorne, K.M. Willett, R.J. Allan, S. Bojinski, J.R. Christy, N. Fox, S. Gilbert, I. Jolliffe, J.J. Kennedy, E. Kent, A. Klein Tank, J. Lawrimore, D.E. Parker, N. Rayner, A. Simmons, L. Song, P.A. Stott, B. Trewin, Bull. Am. Meteorol. Soc. (2011), http://dx.doi.org/10.1175/2011BAMS3124.1.

[48] J.J. Rennie, J.H. Lawrimore, B.E. Gleason, P.W. Thorne, C.P. Morice, M.J. Menne, C.N. Williams, W. Gambi de Almeida, J.R. Christy, M. Flannery, M. Ishihara, K. Kamiguchi, A.M.G. Klein-Tank, A. Mhanda, D.H. Lister, V. Razuvaev, M. Renom, M. Rusticucci, J. Tandy, S.J. Worley, V. Venema, W. Angel, M. Brunet, B. Dattore, H. Diamond, M.A. Lazzara, F. Le Blancq, J. Luterbacher, H. Mächel, J. Revadekar, R.S. Vose, X. Yin, The international surface temperature initiative global land surface databank: monthly temperature data release description and methods. Geosci. Data Journal (2014), http://dx.doi.org/10.1002/gdj3.8.

[49] E.M.A. Dodd, C.J. Merchant, N.A. Rayner, C.P. Morice, An investigation into the impact of using various techniques to estimate Arctic surface air temperature anomalies. Journal of Climate, 28 (5). pp. 1743–1763. ISSN 1520–0442, http://dx.doi.org/10.1175/JCLI-D-14-00250.1.

[50] E.J. Steig, D.P. Schneider, S.D. Rutherford, M.E. Mann, J.C. Comiso, D.T. Shindell, Nature 460 (2009) 766.

[51] G.P. Compo, P.D. Sardeshmukh, J.S. Whitaker, P. Brohan, P.D. Jones, C. McColl, Geophys. Res. Lett. 40 (2013) 3170–3174, http://dx.doi.org/10.1002/grl.50425.

[52] A.J. Simmons, P.D. Jones, V. da Costa Bechtold, A.C.M. Beljaars, P.W. Kållberg, S. Saarinen, S.M. Uppala, P. Viterbo, N. Wedi, J. Geophys. Res. 109 (2004) D24115, http://dx.doi.org/10.1029/2004JD005306.

[53] L. Haimberger, C. Tavolato, S. Sperka, J. Clim. 25 (2012) 8108–8131.

[54] B. Trewin, Int. J. Climatol. 33 (2012) 1510–1529.

[55] R.J.H. Dunn, K.M. Willett, C.P. Morice, D.E. Parker, Clim. Past 10 (2014) 1501–1522, http://dx.doi.org/10.5194/cp-10-1501-2014.

ARCTIC SEA ICE

3

David Schroeder

Centre for Polar Observation and Modelling, Department of Meteorology, University of Reading, Reading, UK

CHAPTER OUTLINE

1. INTRODUCTION

During winter, most parts of the Arctic Ocean are covered by a layer of sea ice with a thickness generally ranging between 0 m and 5 m. During March, the Arctic sea ice reaches its maximum area of around 13×10^6 km², more than 50 times the size of the UK. A significant part of the winter sea ice melts every summer resulting in a minimum ice area during September that varies between 3×10^6 and 7×10^6 km² during the last 36 years [1]. From September to March the sea ice grows in extent through freezing of sea water and thickens. While sea ice formation is primarily thermodynamically driven, sea ice changes are also caused by the ice drift in response to wind and ocean currents. Sea ice consists of individual pieces of ice, called floes, with diameters between 100 m and 5 km [2]. Divergence and shear of the ice drift create areas of open water, while convergence pushes ice floes together, stacking them on top of each other, in a process called ridging. During winter, the floes weld together to form a continuous cover. Unlike freshwater ice, sea ice is not just frozen water but contains sea salt, mostly in pockets of concentrated brine. Sea ice is a two-phase (solid, liquid), two-component (water, salt), reactive porous medium – a so-called mushy layer [3]. The salinity affects the porosity of sea ice, its mechanical strength and its thermal properties.

A crucial step in observing the sea ice state was the advent of satellite multichannel passive microwave imaging systems in late 1978 [4]. In the following section, our focus lies on studying the observed changes in the state of Arctic sea ice during the last 36 years. In Section 3, the reasons for the observed changes are discussed. Can these changes be attributed to climate change? Here, results from

Climate Change. http://dx.doi.org/10.1016/B978-0-444-63524-2.00003-8

physical models, which capture important feedback mechanisms, are included. Finally, a summary is provided in Section 4.

2. OBSERVED CHANGES IN THE STATE OF ARCTIC SEA ICE

2.1 SEA ICE EXTENT AND CONCENTRATION

Sea ice concentration can be derived from passive microwave imagery. The observed brightness temperature enables a distinction between ice covered and open ocean. Apart from a gap around the poles, complete daily data exist with a horizontal resolution of around 25 km from November 1978 to present. Several algorithms have been developed to retrieve sea ice concentration from the brightness temperature, and some important differences regarding absolute values, strengths and weaknesses exist [5–7]. To reduce the impact of the uncertainty in ice concentration, sea ice extent (defined as the areal coverage with at least 15% ice coverage) is widely used as a matrix for presenting time series of sea ice cover [8–10]. The time series of sea ice extent presented here in Fig. 1 are derived from two different algorithms: the NASA Team sea ice concentration algorithm [11,12] and the Bootstrap sea ice algorithm [13].

The Arctic sea ice extent did shrink during all seasons between 1979 and 2013. The declining trend is smallest in March (−2.5% per decade for the NASA Team algorithm and −2.3% per decade for the Bootstrap algorithm, Fig. 1(a)) and largest in September (−14% per decade for both algorithms, Fig. 1(c)). In spite of the interannual variability all trends are statistically highly significant (confidence level >99%). While the absolute values clearly differ between the two algorithms, in particular in August (Fig. 1(b)), with the larger extent values according to the NASA Team algorithm, trends and interannual variability are quite similar. The same is true for the other existing algorithms (not shown). Under melting conditions, the signal received by the satellite is modified by the liquid water (melt ponds, wet snow) on top of the underlying ice. Such effects limit the accuracy of satellite-retrieved passive microwave ice products [6]. The decline of Arctic sea ice took place in almost all regions of the Arctic with exception of the Bering Sea [14].

While the negative trend for Arctic sea ice extent is significant, 35 years is a short time period with respect to climate change. If we want to extend the time series of sea ice into the past, we must rely on proxy data from tree rings, ice cores and sediment cores. Several paleo records of ice conditions from locations around the Arctic have been derived during the last years [4,15–17]. Figure 2 presents results from a reconstruction of variability in sea ice cover in the Arctic over the past 1450 years based on a network of high-resolution climate proxy records from the circumpolar region [15]. The shown reconstruction is smoothed with a 40-year low-pass filter and represents the August sea ice extent, the month with most available observations. In spite of the increasing uncertainty with age (shaded in red in Fig. 2), the results demonstrate that the Arctic sea ice extent has never been lower during the last 1450 years than in the last decade and that the observed decrease during the last 35 years is unprecedented in both magnitude and rate [15]. According to the reconstruction, a pronounced sea ice decrease was observed already twice in the past: during 550–650 CE, marked as T3 in Fig. 2, and during 1500–1700 CE, marked as T2, but in both cases the total drop and the decline rate are much lower than what has been observed during the last 35 years.

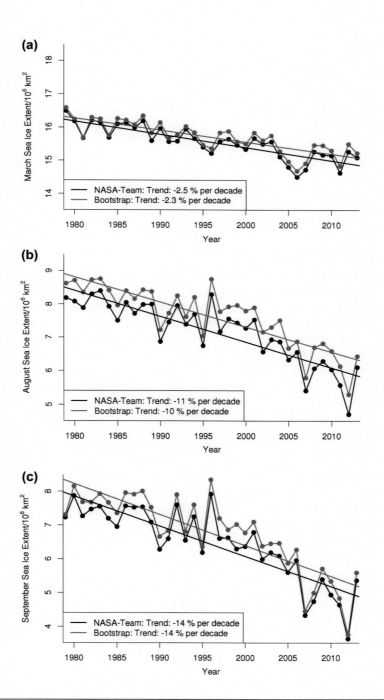

FIGURE 1

Time series of monthly average Arctic sea ice extent for March (a), August (b) and September (c) for 1979–2013 and linear trend. *(Data provided by NSIDC [11,13].)*

FIGURE 2

Paleo time series of Arctic sea ice extent in August (smoothed over 40 years with 95% confidence interval) from proxy data (561–1995, red) and observations (blue). T1–T3 highlights periods of reduced sea ice coverage. *(Data from Ref. [15].)*

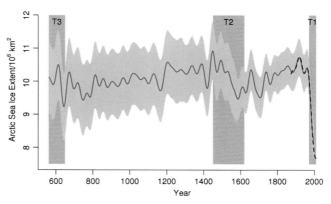

2.2 SEA ICE THICKNESS AND VOLUME

Observations of sea ice thickness are limited due to the sparse spatial and temporal coverage and uncertainties in measurements. Before 1992, most data were collected from submarines operating beneath the Arctic pack ice. Upward-looking sonars measure the submerged portion of the ice (draft), which is converted to thickness by making assumptions about the ice and snow density. The first evidence of 'basin-wide' decrease in ice thickness was reported in 1990 [18]. Between 1958 and 1977, the average ice draft decreased by about 42% [19] and a decline of 43% was found for the Eurasian Basin between 1976 and 1996 [20,21].

Based on 34 US Navy submarine 'basin-wide' time series of mean Arctic sea ice, thickness in spring and autumn could be derived over the period 1975 to 2000 ([22], Fig. 3). The gathered data cover an area of almost 38% of the Arctic Ocean, and the larger amount of data results in a higher representativeness in comparison to previous submarine data. Figure 3 shows that the mean spring ice thickness decreased from 3.6 m in 1980 to 2.4 m in 2000, while the mean autumn ice thickness decreased from 2.8 to 1.6 m during the same time period. The absolute value of thickness reduction (1.2 m) is clearly larger than the uncertainty of the measurement, which is about 0.5 m.

More recently, the cryospheric-focused altimeters Ice, Cloud, and land Elevation Satellite (ICESat) and CryoSat-2 allowed to determine the sea ice thickness distribution throughout the Arctic [23–25].

FIGURE 3

Sea ice thickness changes in spring (blue) and autumn (red) from submarine cruises and ICES at campaigns. For the submarine data, a multiple regression analysis (RA) has been employed to separate the interannual change, the annual cycle, and the spatial field. Blue error bars show residuals in the regression and the quality of data. *(Adapted from Ref. [24] (Fig. 2(b)) with permission.)* Note, the *y*-axis should read: thickness/m.

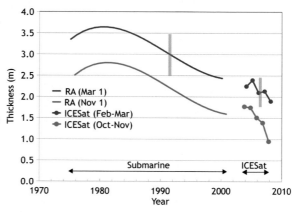

Similarly, to the derivation of sea ice thickness from ice draft using submarine data, the freeboard (the height of sea ice above the water surface) is measured from the satellites and can be converted to thickness, assuming an average density of ice and snow. The principal challenges in deriving an accurate ice thickness using satellite altimetry are in the discrimination of ice and open water, and the estimate the thickness of the snow cover [26].

The basin-wide ice thickness from ICESat is added in Fig. 3 for the years 2003–2007 representing the same area as the submarine data. In comparison to 2000, the mean ice thickness decreased by 44 cm in spring and by 55 cm in autumn resulting in mean values of 1.65 m in February/March 2007 and 1.15 m in October/November 2007. The autumn thickness is less than half the value observed during the 1970 and 1980s [24]. CryoSat-2 data indicate a continued thinning since ICESat measurements and a considerable loss of thick ice along the northern coast of Greenland and Canadian Archipelago [25]. In the same study, the mean Arctic Basin wide sea ice volume is calculated from ICESat and CryoSat-2 data. During October/November, the sea ice volume decreased by 4290 km^3 and by 1480 km^3 in February/March between the ICESat (2003–2008) and CryoSat-2 (2010–2012) periods. The decrease cannot be caused artificially by the difference of the measurement systems as the expected bias between ICESat and CryoSat-2 should not be larger than 700 km^3 [25].

2.3 SEA ICE DRIFT

Sea ice is not fixed at a certain location, but it drifts forced primarily by winds and ocean currents. Ice motion has an impact on the distribution of sea ice thickness in the Arctic through deformation, advection and export of ice. On short time scales, winds are mostly responsible for changes in the sea ice extent [21]. Sea ice drift can be measured from drifting buoys [27] and derived from passive microwave measurements obtained by the Special Sensor Microwave Imager. See Ref. [28] for details of the sea ice tracking procedure and an assessment of the data quality. The satellite-retrieved sea ice drift allows us to examine the basin-wide trends in sea ice circulation between 1982 and 2009. While the averaged ice drift speed did not change significantly from 1982 to 2000, a strong increase occurred between 2001 and 2009 with a trend of 24% per decade during winter and 18% per decade during summer [29]. These trends cannot be explained by the wind forcing because the changes in wind speed were marginal, therefore, the positive trend in ice drift speed suggests a thinning and reduction of multiyear ice [29]. The drift of sea ice is the result of the acting forces due to wind stress, ocean stress, internal ice stress, Coriolis stress and a stress caused by sea surface tilt [30]. Thus, ice drift depends not only on wind speed and ocean current but also on ice strength (thinner ice is weaker), top and bottom surface roughness, and ice concentration. The positive trends in drift speed are found in regions with reduced multiyear sea ice coverage giving confidence in the physical relationship.

2.4 SEA ICE AGE

Sea ice can be classified depending on its age: first-year ice is defined as sea ice that has not yet survived a whole melting season, and third-year ice is ice that has persisted throughout two melting seasons. From satellite data and drifting buoys, it is also possible to observe the formation, movement, persistence and disappearance of sea ice [31–33]. This history can then be used to estimate age [34].

While the oldest ice (four and more years old) is located on the Canadian side of the Arctic Basin and in the Bering Sea, most ice in the Shelves Seas (continental shelf), but also substantial parts of the Arctic Basin, consists of first-year ice in March 2014 (Fig. 4(a)). The 32-year record of estimated sea

Arctic Sea Ice Age, March 2014

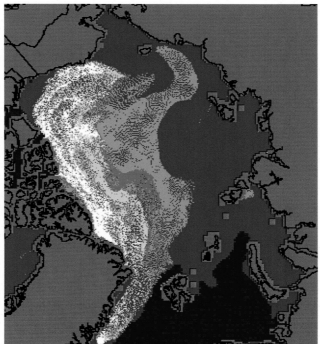

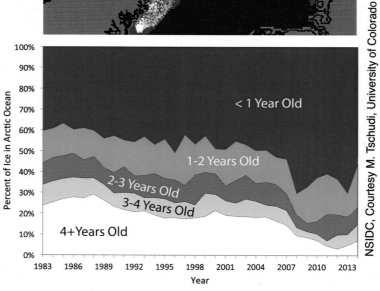

NSIDC, Courtesy M. Tschudi, University of Colorado

FIGURE 4

Ice age field for March 2014 (top) and time series of areal composition of March ice age types for 1983–2014. *(Data from Ref. [32]. Image from National Snow and Ice Data Center (NSIDC) Arctic Sea Ice News and Analysis, April 2014, http://nsidc.org/arcticseaicenews/. See also, Ref. [4].)*

ice shows an increase of first-year ice from around 40% in the 1980s to about 70% in 2007 and 2012 (Fig. 4(b)). This is accompanied by a net long-term decrease in multiyear ice extent. Multiyear ice is typically thicker than younger ice [32], hence the findings above indicate an overall thinning of the sea ice cover. The decline has been particularly strong in the past decade, with a decline in end-of-summer multiyear ice area from over 3×10^6 km^2 in 2000–2006 to less than 2×10^6 km^2 in 2008 based on scatterometer-derived estimates [35].

However, it is important to be aware of the substantial interannual variability of sea ice (Fig. 4(b)) [36]. Although a rapid reduction of second-year ice during 2007 caused by a combination of dynamic and thermodynamic reasons [37] was observed, the last years indicate some recovery of the area of older ice. This is consistent with the larger summer ice extent in 2013 and 2014.

2.5 LENGTH OF MELT SEASON

Daily passive microwave data allow to determine the duration of the ice-free season from ice concentration records [38,39] as well as the length of the melt season due to the clear changes in surface emissivity at the beginning of the melt season and the start of the freeze-up [40–42].

Between 1979 and 2011 most Arctic regions characterized by seasonal sea ice cover show significant trends toward an earlier retreat of sea ice in spring, a later advance in autumn and, consequently, a shorter duration of the ice-free season in summer [43]. The strongest changes occurred in the East Siberian Sea and in the western Beaufort Sea where the summer ice-free season increased by 28 days per decade. The total area with ice seasons shortened by at least 5 days per decade between 1979 and 2013 amounts to 12.4×10^6 km^2 [44].

For the entire Arctic, the length of the continuous melt season increased by 6.4 days per decade between 1979 and 2007 [45]. Consistent with results for the length of the summer ice-free season, the largest and most significant trends (at the 99% level) of more than 10 days per decade occurred for Hudson Bay, the East Greenland Sea, the Laptev/East Siberian Seas and the Chukchi/Beaufort Seas. The length of the melt season has an impact on the amount of solar energy absorbed by the ice cover and the ocean. In particular, the early melt season is decisive for the strength of the summer ice retreat [46]. The area of sea ice covered by melt ponds, pools of water that form on top of sea ice from snow and ice melt, is strongly correlated with the following September ice extent [47].

3. CLASSIFYING AND COMPREHENDING THE OBSERVED CHANGES

The observational record of sea ice is clear: the Arctic sea ice cover is in decline [48]. In order to understand the causes of the decline, the impact of changing dynamics (responsible for ice motion) and thermodynamics (causing ice growth and melt) must be assessed [48]. The years with minimum sea ice extent, 2007 and 2012, are linked to both natural variability and external forcing. In 2007, the ice cover was more vulnerable due to several preceding years of thinning and shrinking of ice cover [1,49]. A key driver causing the low ice extent was the very stable pressure pattern throughout the summer leading warm southerly winds in the Chukchi and East Siberian Seas [50,48]. As in 2007, the record low in 2012 was caused by preconditioning the sea ice state (low fraction of multiyear ice in spring 2012) [51] and anomalous atmospheric events (e.g. a rare Arctic storm in August 2012) [52,53].

Climate models can be used to quantify the relative impacts of natural variability and the impact of climate change (anthropogenic increase of atmospheric CO_2) on the observed decline of Arctic Sea ice. Using CMIP4 model output [54], it can be shown that the observed and modelled late-twentieth century Arctic sea ice loss cannot result from natural variability alone and that roughly 50% of the observed trend can be explained by internal variability of the system [55]. However, the results do depend on how well the key processes are represented in climate models. An observational study analysing the observed sea ice extent from 1950 to present confirms that internal variability alone cannot explain recent trend and magnitude of observed sea ice retreat [56]. There is a significant correlation between decreasing sea ice extent and increasing CO_2 concentration [56,57], but not with any other drivers, such as irradiance or A0-index [56]. Thus, it can be concluded that the observed sea ice decrease is most likely caused by increase in atmospheric CO_2 concentration. A strong physical argument can explain the relationship between sea ice extent and CO_2 concentration: the incoming long-wave radiation dominates annual mean surface heat balance of sea ice in the Arctic [58], and thus, an increase in CO_2 leads to an increase in long-wave radiation resulting in a decrease in sea ice [56].

Arctic sea ice is seen as a key indicator of the state of global climate due to its sensitivity to warming and its potential role in amplifying climate change [59]. The most notable positive feedback is the sea ice–albedo feedback: due to the much lower albedo of open water than sea ice, a decrease of sea ice leads to additional heating of the ocean, which will result in an increase of ice melt [60]. However, observations show a strong recovery from years with low and high ice extent [56], and model simulations show a complete recovery from total removal or strong increase of sea ice within a couple of years [61]. This is not only the case for present-day or past climate but also for any kind of model climate state in the twenty-first century [62]. This shows that the ice-albedo feedback, which could

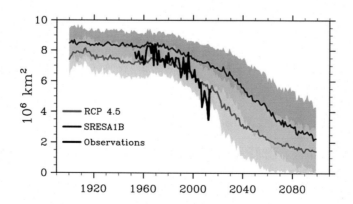

FIGURE 5

Observed September sea ice extent for 1952–2014 (bold black line) and extents for 1900–2100 from the CMIP3 models using the 'business as usual' SRESA1B greenhouse gas emissions scenario (the blue line averaging results from all of the model runs with the blue shading showing the ±1 standard deviation of the different model runs) and from the CMIP5 archive, using the moderate RCP 4.5 scenario (pink line and pink shading). The darker pink shading shows where the simulations from CMIP3 and CMIP5 overlap each other. *(Adapted from Ref. [9] (Fig. 2(a) inset) with permission and updated until 2014.)* Note, the *y*-axis to read: Sea ice extent/10^6 km^2.

potentially lead to a self-acceleration of sea ice retreat, must be compensated by negative feedbacks, such as stronger ice growth over thinner ice [63,64]. Knowing that sea ice cover would quickly regrow under a colder climate does increase our confidence that the observed decline of Arctic sea ice is an indicator of climate change.

Coupled global climate models not only provide estimates for the future of Arctic sea ice but also provide insight into the impact of anthropogenic forcing versus internal climate variability on the observed sea ice trends [9,55]. Figure 5 compares the observed September sea ice extent with the multimodel ensemble projections for Intergovernmental Panel on Climate Change (IPCC) assessment reports. All the sea ice components of the IPCC AR4 and Fifth Assessment Report (AR5) CMIP models [65,66] indicate a declining trend in Arctic sea ice extent over the coming century for the business-as-usual scenarios [21,54]. While the observed downward trend in September ice extent exceeds the simulated trends for the AR4 models (Fig. 5, blue line), as a group, the CMIP5 models (red line) are more consistent with observations over the satellite era [9].

4. CONCLUSIONS

Passive microwave imagery documents indicate that the Arctic sea ice extent declined significantly in all seasons from 1979 to present time. The strongest decline occurred during September with a negative trend of -14% per decade (Fig. 1). This is accompanied by a decrease in ice thickness (Fig. 3) and an increase of ice drift speed indicating that the Arctic sea ice has become thinner and less compact [29]. This is confirmed by a reduction of older ice and increase of first-year ice from around 40% in the 1980s to about 70% in 2007 and 2012 (Fig. 4). The reduction of summer sea ice comes along with an increase of the length of the melt season. In particular, the early onset of the melt season and the amount of spring melt ponds seem to be decisive for the strength of the summer ice retreat [46,47]. A multi-proxy reconstruction demonstrates that the Arctic summer sea ice extent has never been smaller during the last 1450 years than in the last decade and that the recent observed decrease is unprecedented (Fig. 2).

While there is no doubt about the decline of Arctic sea ice during the last decades, it is more difficult to identify and prove the causes for the observed decline. There is a strong physical argument that climate warming does cause a decrease of sea ice: an increase of atmospheric CO_2 leads to an increase in long-wave radiation, which results in an increase of melting ice [58]. However, other processes and feedback mechanisms could be dominating and natural variability can have a strong impact. It is worth noting that the sea ice in the Southern Ocean is not declining. An observational study reveals that the observed Arctic trend cannot be explained only by natural variability [56].

Climate model simulations allow us to quantify the contributions from internal variability and external forcing. A study based on CMIP4 model output reveals that 50% of observed sea ice decline can be explained by internal variability, but the other half is due to anthropogenic influence [55]. The reliability of such studies depends on how well key processes are captured in the climate model [9]. CMIP5 models generally simulate the observed sea ice decline better than the older CMIP3 models (Fig. 5) indicating that the improved representation of sea ice processes is important [67,9]. During past years several new physical processes have been added to the widely used open access Los Alamos sea ice model CICE [68] and they promise to improve the skill of future climate models [47,69–72].

Arctic sea ice responds quickly to changes of the global climate and is, therefore, a prominent indicator of climate change. This does not mean that the Arctic sea ice will decrease continuously in future. Several years of recovery, as observed in 2013 and 2014, or even a decade with sea ice increase are possible due to natural variability. But there is little doubt that the mean sea ice extent, for example the period 2050 to 2060, will be significantly lower than the mean observed in the last 10 years.

ACKNOWLEDGEMENTS

I thank Christophe Kinnard for providing the paleo data shown in Fig. 2, the National Snow and Ice Data Center (Boulder, Colorado, USA) for providing the sea ice extent data and Fig. 4. I also thank my colleagues Daniela Flocco and Michel Tsamados for proofreading the manuscript and giving useful feedback.

REFERENCES

[1] J.C. Stroeve, M. Serreze, M. Holland, J. Kay, J. Maslanik, A. Barrett, Clim. Change 110 (2012) 1005–1027, http://dx.doi.org/10.1007/s10584-011-0101-1.

[2] C. Lüpkes, V.M. Gryanik, J. Hartmann, E.L. Andreas, J. Geophys. Res. 117 (2012) D13112, http://dx.doi.org/10.1029/2012JD017630.

[3] D.L. Feltham, N. Untersteiner, J.S. Wettlaufer, M.G. Worster, J. Geophys. Res. Lett. 33 (2006) L14501, http://dx.doi.org/10.1029/2006GL026290.

[4] W.N. Meier, G.K. Hovelsrud, B.E.H. van Oort, J.R. Key, K.M. Kovacs, C. Michel, C. Haas, M.A. Granskog, S. Gerland, D.K. Perovich, A. Makshtas, J.D. Reist, Rev. Geophys. 51 (2014) 185–217, http://dx.doi.org/10.1002/2013RG000431.

[5] J.C. Comiso, D.J. Cavalieri, C.L. Parkinson, P. Gloersen, Remote Sens. Environ. 60 (1997) 357–384, http://dx.doi.org/10.1016/S0034-4257(96)00220-9.

[6] W.N. Meier, IEEE Trans. Geosci. Remote Sens 43 (6) (2005) 1324–1337.

[7] S. Andersen, R. Tonboe, L. Kaleschke, G. Heygster, L.T. Pedersen, J. Geophys. Res. 112 (C08004) (2007), http://dx.doi.org/10.1029/2006JC003543.

[8] H.J. Zwally, C.L. Parkinson, J.C. Comiso, Science 220 (1983) 1005–1012, http://dx.doi.org/10.1126/science.220.4601.1005.

[9] J.C. Stroeve, V. Kattsov, A. Barrett, M. Serreze, T. Pavlova, M. Holland, W.N. Meier, Geophys. Res. Lett. 39 (2012) L16502, http://dx.doi.org/10.1029/2012GL052676.

[10] F. Massonnet, T. Fichefet, H. Goosse, C.M. Bitz, G. Philippon-Berthier, M.M. Holland, P.-Y. Barriat, Cryosphere 6 (2012) 1383–1394, http://dx.doi.org/10.5194/tc-6-1383-2012.

[11] D.J. Cavalieri, C. Parkinson, P. Gloersen, H.J. Zwally, Sea Ice Concentrations from Nimbus-7 SMMR and DMSP SSM/ISSMIS Passive Microwave Data, 1979–2013, National Snow and Ice Data Center, Digital media, Boulder, Colorado, 1996, updated 2014.

[12] D.J. Cavalieri, C.L. Parkinson, P. Gloersen, J.C. Comiso, H.J. Zwally, J. Geophys. Res. 104 (C7) (1999) 15,803–15,814, http://dx.doi.org/10.1029/1999JC900081.

[13] J.C. Comiso, Bootstrap Sea Ice Concentrations from Nimbus-7 SMMR and DMSP SSM/I-SSMIS. Version 2, NASA DAAC at the National Snow and Ice Data Center, Boulder, Colorado, USA, 2000, updated 2014.

[14] D.J. Cavalieri, C.L. Parkinson, Cryosphere 6 (2012) 881–889, http://dx.doi.org/10.5194/tc-6-881-2012.

[15] C. Kinnard, C.M. Zdanowicz, D.A. Fisher, E. Esaksson, A. de Vernal, L.G. Thompson, Nature 24 (2011) 509–512, http://dx.doi.org/10.1038/nature10581.

[16] J. Halfar, W.H. Adey, A. Kronz, S. Hetzinger, E. Edinger, W.W. Fitzhugh, Proc. Natl. Acad. Sci. U.S.A 110 (49) (2013) 19737–19741.

[17] R. Stein, K. Fahl, Org. Geochem. 55 (2013) 98–102.

[18] P. Wadhams, Nature 345 (1990) 795–797.

[19] D.A. Rothrock, Y. Yu, G.A. Maykut, Geophys. Res. Lett. 26 (1999) 3469–3472.

[20] P. Wadhams, N.R. Davis, Geophys. Res. Lett. 27 (2000) 3973–3975.

[21] Intergovernmental Panel on Climate Change, Climate change 2013: the physical science basis, in: T.F. Stocker, et al. (Eds.), Contribution of Working Group I to the Fifth Assessment Report of the Intergovernmental Panel on Climate Change, Cambridge Univ. Press, Cambridge, UK, and New York, 2013, 1535 pp.

[22] D.A. Rothrock, D.B. Percival, M. Wensnahan, J. Geophys. Res. Oceans 113 (2008) C05003.

[23] K.A. Giles, S.W. Laxon, A.L. Ridout, Geophys. Res. Lett. 35 (2008) L22502.

[24] R. Kwok, D.A. Rothrock, Geophys. Res. Lett. 36 (2009) L15501.

[25] S.W. Laxon, et al., Geophys. Res. Lett. 40 (2013) 732–737.

[26] K.A. Giles, S.W. Laxon, D.J. Wingham, D.W. Wallis, W.B. Krabill, C.J. Leuschen, D. McAdoo, S.S. Manizade, R.K. Raney, Remote Sens. Environ. 111 (2007) 182–194, http://dx.doi.org/10.1016/j.rse.2007.02.037.

[27] P. Rampal, J. Weiss, D. Marsan, J. Geophys. Res. 114 (2009) C05013, http://dx.doi.org/10.1029/2008JC005066.

[28] R. Kwok, A. Schweiger, D.A. Rothrock, S. Pang, C. Kottmeier, J. Geophys. Res. 103 (1998) 8191–8214.

[29] R. Kwok, G. Spreen, S. Pang, J. Geophys. Res. Oceans 118 (2013) 2408–2425, http://dx.doi.org/10.1002/jgrc.20191.

[30] W.D. Hibler, J. Phys. Oceanogr. 9 (4) (1979) 815–846, http://dx.doi.org/10.1175/1520-0485.

[31] C. Fowler, W. Emery, J.A. Maslanik, IEEE Trans. Geosci. Remote Sens. Lett. 1 (2) (2003) 71–74.

[32] J.A. Maslanik, C. Fowler, J. Stroeve, S. Drobot, J. Zwally, D. Yi, W. Emery, Geophys. Res. Lett. 34 (2007) L24501, http://dx.doi.org/10.1029/2007GL032043.

[33] J. Maslanik, J. Stroeve, C. Fowler, W. Emery, Distribution and trends in Arctic sea ice age through spring 2011, Geophys. Res. Lett. 38 (2011) L13502, http://dx.doi.org/10.1029/2011GL047735.

[34] I.G. Rigor, J.M. Wallace, Geophys. Res. Lett. 31 (2004) L09401, http://dx.doi.org/10.1029/2004GL019492.

[35] R. Kwok, G.F. Cunningham, M. Wensnahan, I. Rigor, H.J. Zwally, D. Yi, J. Geophys. Res. 114 (2009) C07005, http://dx.doi.org/10.1029/2009JC005312.

[36] J.C. Comiso, J. Clim. 25 (2012) 1176–1193.

[37] S.V. Nghiem, I.G. Rigor, D.K. Perovich, P. Clemente-Colon, J.W. Weatherly, G. Neumann, Geophys. Res. Lett. 34 (6) (2007) L19504.

[38] C.L. Parkinson, Ann. Glaciol. 34 (2002) 435–440.

[39] S.E. Stammerjohn, D.G. Martinson, R.C. Smith, X. Yuan, D. Rind, J. Geophys. Res. Oceans 113 (2008) C03S90.

[40] D.M. Smith, Geophys. Res. Lett. 25 (1998) 655–658.

[41] S.D. Drobot, M.R. Anderson, J. Geophys. Res. Atmos. 106 (2001) 24033–24049.

[42] G.I. Belchansky, D.C. Douglas, N.G. Platonov, J. Clim. 17 (2004) 67–80.

[43] S.E. Stammerjohn, R. Massom, D. Rind, D. Martinson, Geophys. Res. Lett. 39 (2012) L06501.

[44] C.L. Parkinson, Geophys. Res. Lett. 41 (2014) 4316–4322, http://dx.doi.org/10.1002/2014GL060434.

[45] T. Markus, J.C. Stroeve, J. Miller, J. Geophys. Res. Oceans 114 (2009) C12024.

[46] D.K. Perovich, B. Light, H. Eicken, K.F. Jones, K. Runciman, S.V. Nghiem, Geophys. Res. Lett. 34 (2007) L19505.

[47] D. Schröder, D.L. Feltham, D. Flocco, M. Tsamados, Nat. Clim. Change 4 (5) (2014) 353–357, http://dx.doi.org/10.1038/NCLIMATE2203.

[48] D.K. Perovich, Oceanography 24 (3) (2011) 162–173.

[49] R.W. Lindsay, J. Zhang, A. Schweiger, M. Steele, H. Stern, J. Clim. (2009), http://dx.doi.org/10.1175/2008JCLI2521.1.

[50] J. Wang, J. Zhang, E. Watanab, M. Ikeda, K. Mizobata, J.E. Walsh, X. Bai, B. Wu, Geophys. Res. Lett. 36 (2009) L05706, http://dx.doi.org/10.1029/2008GL036706.

[51] C.L. Parkinson, J.C. Comiso, Geophys. Res. Lett. 40 (2013) 1356–1361, http://dx.doi.org/10.1002/grl.50349.

[52] I. Simmonds, I. Rudeva, Geophys. Res. Lett. 39 (2012) L23709, http://dx.doi.org/10.1029/2012GL054259.

[53] J. Zhang, R. Lindsay, A. Schweiger, I. Rigor, Geophys. Res. Lett. 39 (2012) L20503, http://dx.doi.org/10.1029/2012GL053545.

[54] Intergovernmental Panel on Climate Change, in: S. Solomon, et al. (Eds.), Contribution of Working Group I to the Fourth Intergovernmental Panel on Climate Change, Cambridge Univ. Press, Cambridge, UK, and New York, 2007.

[55] J.E. Kay, M.M. Holland, A. Jahn, Geophys. Res. Lett. 38 (2011) L15708, http://dx.doi.org/10.1029/2011GL048008.

[56] D. Notz, J. Marotzke, Geophys. Res. Lett. 39 (2012) L051094, http://dx.doi.org/10.1029/2012GL051094.

[57] O. Johannessen, Atmos. Oceanic Sci. Lett. 1 (1) (2008) 51–56.

[58] G.A. Maykut, N. Untersteiner, J. Geophys. Res. 76 (1971) 1550–1575.

[59] W. Maslowski, J.C. Kinney, M. Higgins, A. Roberts, Ann. Rev. Earth Planet. Sci. 40 (2012) 625–654, http://dx.doi.org/10.1146/annurev-earth-042711-105345.

[60] D.K. Perovich, C. Polashenski, Geophys. Res. Lett. 39 (2012) L08501, http://dx.doi.org/10.1029/2012GL051432.

[61] D. Schröder, W.M. Connolley, Geophys. Res. Lett. 34 (2007) L14502, http://dx.doi.org/10.1029/2007GL030253.

[62] S. Tietsche, D. Notz, J.H. Jungclaus, J. Marotzke, Geophys. Res. Lett. 38 (2011) L02707, http://dx.doi.org/10.1029/2010GL045698.

[63] I. Eisenman, J.S. Wettlaufer, Proc. Natl. Acad. Sci. U.S.A. 106 (2009) 28–32, http://dx.doi.org/10.1073/pnas.0806887106.

[64] M.M. Holland, C.M. Bitz, B. Tremblay, D.A. Bailey, in: E.T. DeWeaver, C.M. Bitz, L.B. Tremblay (Eds.), Geophys. Monogr. Ser., vol. 180, AGU, Washington, DC, 2008, pp. 133–150.

[65] S.J. Vavrus, M.M. Holland, A. Jahn, D. Bailey, B.A. Blazey, J. Clim. 25 (2012) 2696–2710, http://dx.doi.org/10.1175/JCLI-D-11-0020.1.

[66] A. Jahn, et al., J. Clim. 25 (2012) 1431–1452.

[67] J. Stroeve, M.M. Holland, W. Meier, T. Scambos, M. Serreze, Geophys. Res. Lett. 34 (2007) L09501, http://dx.doi.org/10.1029/2007GL029703.

[68] E.C. Hunke, W.H. Lipscomb, A.K. Turner, N. Jeffrey, S. Elliot, CICE Version 5.0. Tech. Rep. LA-CC-06–012, Los Alamos National Laboratory, 2013. Available at: http://climate.lanl.gov/Models/CICE.

[69] D. Flocco, D. Schröder, D.L. Feltham, E.C. Hunke, J. Geophys. Res. 117 (2012) C09032.

[70] M. Tsamados, D.L. Feltham, A. Wilchinsky, J. Geophys. Res. 118 (2013) 91–107.

[71] A.K. Turner, E.C. Hunke, C.M. Bitz, J. Geophys. Res. Oceans 118 (5) (2013) 2279–2294.

[72] M. Tsamados, D.L. Feltham, D. Schroeder, D. Flocco, S. Farrell, N. Kurtz, S. Laxon, S. Bacon, J. Phys. Oceanogr. 44 (2014) 1329–1353, http://dx.doi.org/10.1175/JPO-D-13-0215.1.

ANTARCTIC SEA ICE CHANGES AND THEIR IMPLICATIONS

4

Peter Wadhams

Department of Applied Mathematics and Theoretical Physics, University of Cambridge, Cambridge, UK

CHAPTER OUTLINE

1. INTRODUCTION

Within the cryosphere current attention is focused strongly on the Arctic Ocean because the rapid retreat of sea ice greatly exceeds the predictions of models and is already changing the face of the Arctic, replacing rugged multiyear ice by thinner, weaker first-year ice, and replacing ice of any kind in summer by open water. This is a massive and accelerating change and has many impacts on other global processes. Yet the sea ice in the Antarctic Ocean is also behaving in a way that defies the predictions of most models, since at a time of climatic warming the ice is actually advancing in area, albeit slowly.

This matters for at least two reasons. Firstly, the albedo feedback effect from snow and ice retreat must be computed for the planet as a whole if we are to assess the impact of cryospheric changes on global heating. (Albedo is the fraction of solar energy – shortwave radiation – reflected from the Earth back into space. It is a measure of the reflectivity of the earth's surface. Ice, especially with snow on top of it, has a high albedo; most sunlight hitting the surface bounces back toward space.) If Antarctic sea ice is advancing, then this helps to offset the global albedo reduction due to Arctic sea ice and snowline retreat. Secondly, if Antarctic sea ice is advancing at a time of warming, this is just as much of a challenge to global climatic models as is the rapid retreat of Arctic sea ice, and we must seek the reasons for it.

2. WHY ANTARCTIC ICE IS DIFFERENT

Antarctic sea ice is not like Arctic sea ice. Of course they are both made of frozen water. But Antarctic sea ice forms differently and has different properties and appearance from Arctic ice. The reason is that the Antarctic is a continent surrounded by a vast ocean. Sea ice starts to form early in the winter close to the Antarctic coast and advances northward into that ocean as winter progresses, experiencing all the power of the sea as it grows. The mechanism by which the ice is generated was not elucidated until the first expedition was able to work in the pack ice zone during early winter, the time of ice edge advance. This was in 1986, in the Winter Weddell Sea Project, using the German research ship F.S. 'Polarstern'. I was aboard that ship on her memorable winter cruise. As we traversed the ice margin region, we did a careful study of the ice conditions and characteristics, and identified what we called the **pancake-frazil cycle** (Fig. 1) as the source of much of the first-year sea ice seen further inside the pack [1–3]. A map of the Antarctic and its seas is given in Fig. 2.

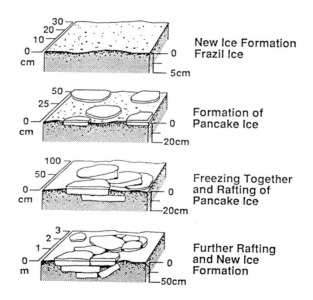

FIGURE 1

Schematic illustration of the frazil-pancake cycle. *(Reproduced with permission from Ref. [2] work done by Lange et al).*

   Ice growing on calm water forms an initial skim that solidifies into a thin transparent layer called nilas; water molecules freezing onto the bottom of a nilas sheet then extend the ice downward, with a selection factor favouring crystals with c-axes horizontal, to yield eventually a first-year ice sheet. Ice forming at the extreme Antarctic ice edge cannot grow directly into a continuous sheet of nilas like this because of the high energy and turbulence in the Southern Ocean wave field, which maintains the new ice as a dense suspension of frazil ice, a suspension of tiny ice crystals in water. This suspension undergoes cyclic compression because of the particle orbits in the wave field, and during the compression phase the crystals can freeze together to form small coherent cakes of slush that grow

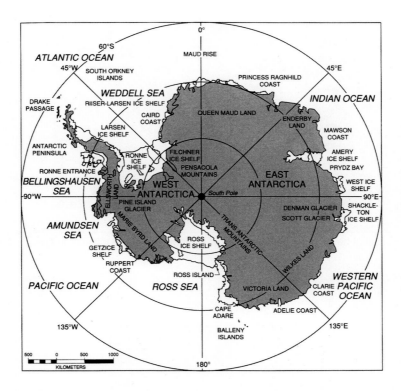

FIGURE 2

The seas around the Antarctic continent.

larger by accretion from the frazil ice and more solid through continued freezing between the crystals. This becomes known as **pancake ice** because collisions between the cakes pump frazil ice suspension onto the edges of the cakes, then the water drains away to leave a raised rim of frazil ice that gives each cake the appearance of a pancake. At the ice edge, the pancakes are only a few centimetres in diameter, but they gradually grow in diameter and thickness with increasing distance from the ice edge, until they may reach 3 m–5 m diameter and 50 cm–70 cm thickness. The surrounding frazil continues to grow and supply material to the growing pancakes since the water surface is not completely closed off by ice, and so a large ocean-atmosphere heat flux is still possible, which can dispose of latent heat.

At greater distances inside the ice edge, the pancakes begin to freeze together in groups (Fig. 3), but in the case of the Antarctic the wave field was found to be powerful enough to prevent overall freezing until a penetration of some 270 km is reached. Here the pancakes coalesce to form first large floes (Fig. 4) then finally a continuous sheet of first-year ice (Fig. 5). At this point, with the open water surface cut off, the growth rate drops to a very low level (estimated at 0.4 cm d^{-1} where d refers to day) [1] and the ultimate thickness reached by first-year ice is only a few centimetres more than the thickness attained at the time of consolidation of the pancakes [4].

First-year ice formed in this way is known as **consolidated pancake ice** and has a different bottom shape from Arctic ice. The pancakes at the time of consolidation are jumbled together and rafted over one another, and freeze together in this way with the frazil acting as 'glue'. The result is a very rough,

FIGURE 3

Pancakes at the stage of having acquired raised rims, beginning to freeze together in groups. *(Photo by author.)*

FIGURE 4

The pancake groups have frozen together into floes, 270 km from ice edge, with remaining swell energy sufficient to produce a regular pattern of leads. *(Photo by author.)*

FIGURE 5

The floes have frozen together into fully consolidated pancake ice, with the pancake rims still visible through a snow layer. *(Photo by author.)*

jagged bottom, with rafted cakes doubling or tripling the normal ice thickness, and with the edges of pancakes protruding upward to give a surface topography resembling a 'stony field'. The contrast between such ice and ice formed in calm conditions in the normal way is shown in Fig. 6, which depicts profiles generated by drilling holes at 1 m intervals.

The rafted bottom of consolidated pancake ice provides a large surface area per unit area of sea surface, providing an excellent substrate for algal growth and a refuge for krill. The thin ice permits much light to penetrate, and the result is a fertile winter ice ecosystem, which is thought to contribute about 30% to the total biological production of the Southern Ocean.

Even after another 29 years, comparatively few ships have worked in the Antarctic pack ice in midwinter; the Alfred Wegener Institute ran a Winter Weddell Gyre Study in 1989 [5], and there were more recent experiments in the Weddell Sea where the ship worked in conjunction with an ice camp – Ice Station Weddell-1 in 1992 [6] and ISPOL in 2004–2005 [7]. We do not yet have enough evidence to be sure of whether the frazil-pancake sequence of ice growth is followed around the entire periphery of the Antarctic, but if it is, then the area occupied by Antarctic pancake ice in early winter could be as great as 6×10^6 km^2, making it an important yet seldom-seen feature of the Earth's surface. It is quite extraordinary that this amazing landscape of heaving white pancakes should occupy such a vast area yet should so seldom be seen by humans. Probably fewer than a thousand people have seen it.

3. SNOW ON THE ICE

The annual snowfall in the Antarctic is much greater than in the Arctic because of the presence of the unconfined Southern Ocean, and in coastal regions snow is also blown onto the sea ice by katabatic winds off the tops of ice shelves. During the July–September 1986 'Polarstern' cruise in the eastern Weddell Sea, we found a mean snow thickness of 14 cm–16 cm on the surface of first-year ice. Since

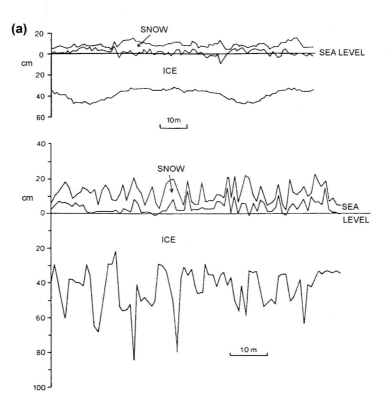

FIGURE 6

(a) Winter thickness profiles across Antarctic ice fields, from holes drilled at 1-m intervals [3] showing the difference between ice that has grown in calm conditions with a smooth underside (top) and consolidated pancake ice, where the pancakes have frozen together in a jumbled fashion to give a jagged underside with rafting to two or three pancake thicknesses (bottom). *(Reproduced from Ref. [3] with permission.)* The y-axes should read: Thickness/cm. (b) Winter thickness profiles across Antarctic ice fields, from holes drilled at 1-m intervals [3] showing the difference between first-year ice (top) and multiyear ice (bottom) in the western Weddell Sea, demonstrating the way in which the weight of the snow cover pushes the ice surface below sea level, especially for multiyear ice. *(Reproduced from Ref. [3] with permission.)*

the ice itself is so thin, this was sufficient to bring the ice surface below sea level in 15%–20% of holes drilled, leading to the infiltration of sea water into the overlying snow and the formation of either a wet slushy layer on top of the ice or, in the case of freezing, the formation of a 'snow-ice' layer between the unwetted snow and the original ice upper surface. In September–October 1989, the snow was even thicker, especially over multiyear ice in the western Weddell Sea, into which we ventured. This was sufficient to push the ice surface below sea level in almost every case. Figure 6 shows the contrast between the two types of ice cover in this respect [4]. The thick snow insulates the ice, and its slushy

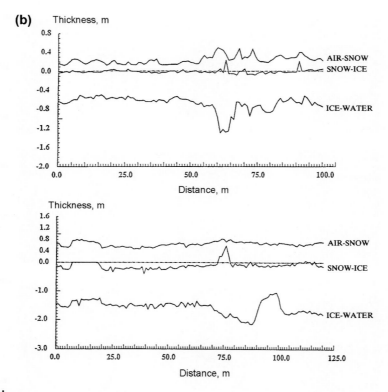

FIGURE 6 Cont'd

wetness means that satellite radar methods for mapping ice thickness do not work well because the radar beam is reflected by the wet snow. Snow on Antarctic sea ice has been reviewed by Massom et al. [8], while Eicken et al. [9] have studied the question of the contributions of snow and ice derived from snow by water infiltration (termed *meteoric ice*) to the overall mass balance.

4. THE ANNUAL ICE CYCLE AND ITS CHANGES

Awkwardly for climate change models, the extent of Antarctic sea ice has been observed to be slowly increasing in recent years, but these changes have high regional variability.

Figure 7 shows the 'normal' annual cycle of ice extent for a late-twentieth century epoch of 1978–1987 [10]. In summer there are only two substantial areas of sea ice remaining, in the western Weddell and Ross seas, so these are the only regions that can contain much multiyear ice (ice that has survived at least one summer melt period), an ice type that until recently was dominant in the Arctic. The minimum varies little from year to year. As winter sets in, new ice forms north of the ice edge and the sea ice limit advances northward until it reaches a limit at about 55°S–66°S by the end of winter (August–September), whereupon it retreats back to its start point. The northern limit is 55°S in the Indian

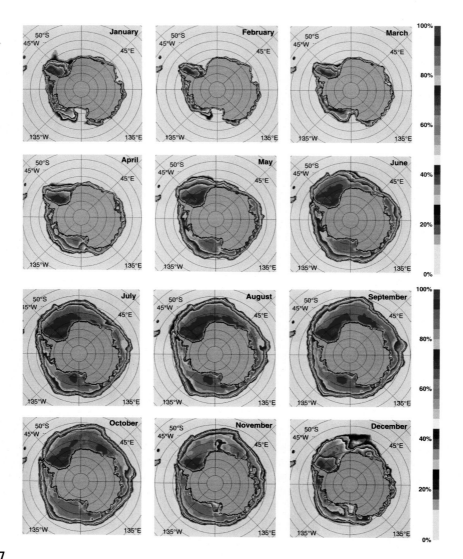

FIGURE 7

Average monthly Southern Hemisphere sea ice extent for the 1979–1987 period, from passive microwave data. *(Reproduced with permission from Ref. [10] from work by Gloersen et al.)*

Ocean sector at about 15°E, but lies at about 60°S around most of the rest of East Antarctica, then slips even further south to 65°S off the Ross Sea. The edge moves slightly north again to 62°S at 150°W, then again shifts southward to 66°S off the Amundsen Sea before moving north again to engulf the South Shetland and South Orkney Islands off the Antarctic Peninsula and complete the circle. The zonal variation in latitude of this winter maximum therefore amounts to some 11°, which is not negligible. The absolute limit of northward progression is the edge of the Antarctic Circumpolar Current, where surface water temperature changes abruptly in the Polar Front, or Antarctic Convergence. The ice seldom reaches this far, however, because its advance is limited both by mesoscale ocean processes (storms, eddies) that break up the ice, and by surface air temperatures; it was shown by Zwally et al. [11] that the winter advance of the ice edge follows closely the advance of the 271.2 K ($-2.0°C$) isotherm in surface air temperature (freezing point of seawater is $-2.0°C$) and almost coincides with this isotherm at the time of maximum advance. The cycle of this annual variation in extent (that is, area south of the main ice edge) can be easily measured by satellites, especially the useful passive microwave satellites SMMR, SSM/I and SSMIS, and Fig. 7 shows results obtained over the period 1978 to 2010 [12]. The average maximum and minimum ice extents for the period were 18.5×10^6 km^2 and 3.1×10^6 km^2.

As Fig. 8 shows, there is clearly a slow upward trend in the winter maximum extent for the Antarctic as a whole, amounting to $17\,100 \pm 2300$ km^2 a^{-1}, where 'a' refers to annum. This, however, conceals regional and seasonal variations in trend. The most rapid growth has been in the sector of the Ross Sea ($13\,700 \pm 1500$ km^2 a^{-1}) with lesser contributions from the Indian Ocean and eastern

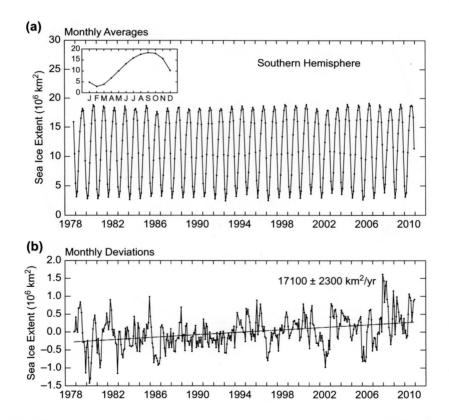

FIGURE 8

(a) Monthly averaged Southern Hemisphere sea ice extents for November 1978 through December 2010. The inset shows the average annual cycle. (b) Monthly deviations for the sea ice extents. *(The graphs have been reproduced from D.J. Cavalieri, C.L. Parkinson, Cryosphere 6 (2012) 871–880 with permission.)* The y-axis for graph (a) and for graph (b) should read: Sea Ice Extent/10^6 km^2.

Weddell Sea sectors, while the Bellingshausen/Amundsen seas of West Antarctica have experienced a retreat rate of 8200 ± 1200 km^2 a^{-1}. Steig [13] found that air temperatures over the Pacific sector of the Antarctic continent (Antarctic Peninsula to Ross Sea) have warmed twice as fast as over the rest of the continent, while an analysis of the temperature record at Byrd Station (120°W longitude) shows that from 1958 to 2010 it warmed by 1.6°C–3.2°C, a very large increase [14]. The fast warming over the Pacific sector of Antarctica (West Antarctica) is reflected in a decrease in the length of the ice-covered season (number of days per year in which a given location has an ice cover) by one to three days per year over much of this sector of the Southern Ocean between 1979 and 2010 (Fig. 9; Maksym et al.

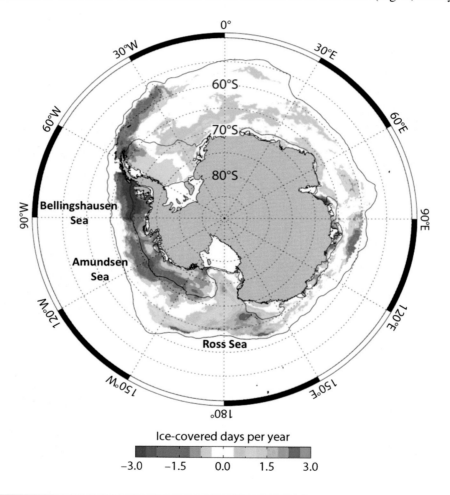

FIGURE 9

Trends in the lengths of the sea ice season (change in days per year) over 1979/80 to 2010/11. Blue and purple show where sea ice is advancing, and orange and red show where it is decreasing. The Bellingshausen/ Amundsen duration was decreasing at the same time the Ross Sea duration was increasing. *(Reproduced from Ref. [15] from work by Maksym et al.)* The units of the scale should read: d a^{-1}, where 'd' refers to days and 'a' to annum.

[15]), while the Atlantic-Indian Ocean sectors showed a slower increase in the ice-covered season. The message from the ice cover is clear – a wide swath of East Antarctica has an ice cover that is growing slowly, while a narrower swath of West Antarctica has an ice cover that is shrinking more rapidly, the net effect being a very slow growth.

There are other, more detailed, ice variations related to local topographic factors that are usually visible in spring or summer. In the sector off Enderby Land at 0°E–20°E, a large gulf opens up in December to join a coastal region of reduced ice concentration that opens in November. This is a much attenuated version of a winter polynya (an area of open water surrounded by sea ice), which was detected in the middle of the pack ice in this sector during 1974–1976 [11,16] but which has not been seen as a full open water feature since that date. It was known as the Weddell Polynya and lay over the Maud Rise, a plateau of reduced water depth. The area was investigated in winter 1986 by the Winter Weddell Sea Project cruise of the icebreaker FS 'Polarstern', and it was found that the region is already part of the Antarctic Divergence, where upwelling of warmer water can occur, and that additional circulating currents and the doming of isopycnals (layers of constant water density) over the Rise could allow enough heat to reach the surface to keep the region ice-free in winter [17,18]. Since this has not occurred since 1976, the region is presumably balanced on the edge of instability as far as its winter ice cover is concerned. The 1986 winter cover was of high concentration but was very thin [1]. The December distribution also shows an open water region appearing in the Ross Sea (see Fig. 9), the so-called Ross Sea Polynya, with ice still present to the north. In November and December, a series of small coastal polynyas can be seen to be actively opening along the East Antarctica coast, mostly driven by offshore (katabatic) winds driving ice away from the coast as fast as it can form.

5. **WHAT IS HAPPENING TO THE ICE?**

We know that much of the ice in the winter pack is of pancake origin and is quite thin. Given that the climate of much of the Antarctic is warming, why is the ice limit advancing rather than retreating, with a regional pattern described above?

A clue to a possible continent-wide mechanism came in a paper published in 2014 [19]. The key is the winds around Antarctic, the great circumpolar west wind belt, which recently has been increasing in intensity. The average wind speed is higher, and the wind is coming mainly from the west. Consider a typical ice floe. It is blown eastward by the direct force of the wind stress on its surface, but as it moves it experiences a force tending to turn it to the left, i.e. to give it some component of motion northward. This is the well-known Coriolis force due to the rotation of the Earth – it acts to the right in the northern hemisphere, to the left in the southern, and is zero at the Equator. Any moving object experiences this force. With a higher wind speed, the northward-acting Coriolis force on the ice floe increases, so the floe moves northward more rapidly, even as its main motion is eastward. Therefore, although it will reach a latitude where the warmer atmosphere will melt it, it will get further north before this happens. So the whole Antarctic pack ice zone is like a great merry-go-round driven by the wind, which is throwing its ice northward into warmer water. But, this may be a too simplistic view. Firstly, the mechanism would lead to an increase in ice *extent* but not necessarily of *area*, since it deals only with the dynamics of existing ice. Secondly, the increase is bound to be temporary as eventually global warming will win out over increased wind speeds and the ice will fail to reach the lower latitudes. However, the mechanism is rooted in simple physics and the observed fact that circumpolar wind speeds have indeed increased.

6. RESPONSE OF THE ANTARCTIC TO CHANGES ELSEWHERE

To explain the *regional* nature of the sea ice trends it is likely that we need a model that considers forcing from elsewhere that can cause regional effects on the Antarctic ice.

One obvious cause, although the effects will be long term, is the Antarctic ice sheet itself, which is starting to lose mass (e.g. Jacobs et al. [20]), although more slowly than the Greenland ice sheet. If Antarctic ice loss increases, it has been projected that the Filchner-Ronne and Ross ice shelves will disintegrate, allowing Antarctic glaciers (for instance, those in the Transantarctic Mountains) to debouch directly into the ocean. This will rapidly accelerate the rate of mass loss from Antarctica, leading to acceleration in global sea level rise, but will also impact Antarctic sea ice (if it still exists by then). Such changes are not expected to occur for a few centuries except for the possible case of a disintegration of the ice shelf around Pine Island Bay and a region of East Antarctica where the ice sheet is believed to be potentially unstable should a 'plug' of coastal ice decay [21].

For more immediate effects, which are determining the *present* regional variation in Antarctic sea ice advance or retreat, we need to look for teleconnections that can exist with the lower latitude oceans and atmosphere and even with northern latitudes extending up to the Arctic. There are many candidates for the linking mechanism. Peterson and White [22] considered the Antarctic Circumpolar Wave, a system of waves on the Antarctic Circumpolar Current that propagate slowly eastward (though westward relative to the current) and may interact with the tropical El Niño–Southern Oscillation (ENSO) system. The El Niño is a warm ocean current of variable intensity that develops after late December along the coast of Ecuador and Peru and sometimes causes catastrophic weather conditions. More recent work focuses on the Southern Annular Mode (SAM), e.g. Comiso et al. [23] for the Ross Sea ice. Rind et al. [24] suggested that an El Niño year leads to more sea ice in the Weddell Sea and less sea ice in the Pacific, with the opposite for a La Niña year, but the ENSO link is complicated by recent discoveries about Central Pacific (or 'Modoki') El Niño events (e.g. Wilson et al. [25]). La Niña refers to a cooling of the ocean surface off the western coast of South America that occurs periodically every 4–12 years and affects the Pacific and other weather patterns. Wider-ranging latitudinal teleconnections could relate to the link between Arctic warming and lower latitude weather extremes due to distortions in the jet stream [26], which might then involve onward links with tropical and Southern Hemisphere patterns.

Any complete explanation for why Antarctic and Arctic sea ice behaviour differ must also depend on the fundamental differences between Arctic and Antarctic sea ice. The Antarctic is bound to warm more slowly than the Arctic because of the greater area of ocean with its high heat capacity and the way in which the Antarctic Circumpolar Current insulates the continent from the warmer ocean to the north. The Antarctic sea ice limits are set in a different way from the Arctic – in summer the ice retreats to the land, leaving substantial ice mass only in awkwardly shaped bights like the Weddell Sea, while the winter limits are thermodynamic and set by conditions in the open ocean. In the Arctic, the situation is the opposite – the winter limit is set by surrounding land masses, while in summer the ice retreats to an ocean limit that is thermodynamically and dynamically set. Albedo feedback is also less important in the Antarctic than the Arctic, since at the time of maximum summer insolation the Antarctic sea ice has already retreated almost to the continent while the Arctic sea ice still has a long time to go to retreat to its September minimum and is thus susceptible to changes in forcing.

A final point relating to rate of warming is that the rapid Arctic warming itself creates feedbacks that lead to a further acceleration of warming, e.g. the albedo feedback mentioned above, which was estimated by Pistone et al. [27] as adding a quarter to the rate of warming induced by greenhouse gas growth alone; a further albedo feedback due to terrestrial snowline retreat; and the potentially very serious additional warming that may be created by unlocking methane from newly ice-free Arctic continental shelves [28]. Such feedbacks cannot occur to the same extent in the Antarctic – because of lack of shallow shelves and the inflexible area of terrestrial snow cover. The Arctic feedback and Arctic warming amplifications mean that, whatever the interactions between Antarctic sea ice and temperate oceans, it will always be the case over the next few decades that the Arctic will be a major determinant of the rate of global warming, and thus the overall warming that will be experienced by the Antarctic sea ice. In this sense, the Arctic is a driver and the Antarctic can be thought of as a passive trailer in the global warming feedback stakes.

REFERENCES

[1] P. Wadhams, M.A. Lange, S.F. Ackley, J. Geophys. Res. 92 (1987) 14535–14552.
[2] M.A. Lange, S.F. Ackley, P. Wadhams, G.S. Dieckmann, H. Eicken, Ann. Glaciol. 12 (1989) 92–96.
[3] P. Wadhams, Ice in the Ocean, Taylor and Francis, Abingdon, UK, 2000, pp. 368.
[4] A.P. Worby, C.A. Geiger, M.J. Paget, M.L. Van Woert, S.F. Ackley, T.L. DeLiberty, J. Geophys. Res. 113 (2008) C05S92, http://dx.doi.org/10.1029/2007JC004254.
[5] P. Wadhams, D.R. Crane, Polar Rec. 27 (1991) 29–38.
[6] S.F. Ackley, V.I. Lytle, B. Elder, D. Bell, Antarct. J. US 27 (1992) 111–113.
[7] H.H. Hellmer, M. Schröder, C. Haas, G.S. Dieckmann, M. Spindler, Deep-Sea Res. 55 (2008) 8–9.
[8] R.A. Massom, H. Eicken, C. Haas, M.O. Jeffries, M.R. Drinkwater, M. Sturm, A.P. Worby, X. Wu, V.I. Lytle, S. Ushio, K. Morris, P.A. Reid, S.G. Warren, I. Allison, Rev. Geophys. 39 (2001) 413–445.
[9] H. Eicken, M.A. Lange, H.-W Hubberten, P. Wadhams, Ann. Geophys. 12 (1994) 80–93.
[10] P. Gloersen, W.J. Campbell, D.J. Cavalieri, J.C. Comiso, C.L. Parkinson, H.J. Zwally, Arctic and Antarctic Sea Ice, 1978–1987: Satellite Passive-Microwave Observations and Analysis, National Aeronautics and Space Administration, Rep. NASA SP-511, 1992, pp. 290.
[11] H.J. Zwally, J.C. Comiso, C.L. Parkinson, W.J. Campbell, F.D. Carsey, P. Gloersen, Antarctic Sea Ice 1973–1976: Satellite Passive Microwave Observations, NASA, Washington, DC, 1983. Rept. SP-459.
[12] C.L. Parkinson, D.J. Cavalieri, Cryosphere 6 (2012) 871–880.
[13] E.J. Steig, D.P. Schneider, S.D. Rutherford, M.E. Mann, J.C. Comiso, D.T. Shindell, Nature 457 (2009) 459–462.
[14] D.H. Bromwich, J.P. Nicolas, A.J. Monaghan, M.A. Lazzara, L.M. Keller, G.A. Weidne, A.B. Wilson, Nat. Geosci. 6 (2013) 139–145.
[15] T. Maksym, S.E. Stammerjohn, S. Ackley, R. Massom, Oceanography 25 (2012) 140–151.
[16] H.J. Zwally, P. Gloersen, Polar Rec. 18 (1977) 431–450.
[17] N.V. Bagriantsev, A.L. Gordon, B.A. Huber, J. Geophys. Res. 94 (1989) 8331–8334.
[18] A.L. Gordon, B.A. Huber, J. Geophys. Res. 95 (1990) 11655–11672.
[19] J. Zhang, J. Clim. 27 (2014) 202–214, http://dx.doi.org/10.1175/JCLI-D-12-00139.1.
[20] S. Jacobs, A. Jenkins, H. Hellmer, C. Giulivi, F. Nitsche, B. Huber, R. Guerrero, Oceanography 25 (2012) 154–163, http://dx.doi.org/10.5670/oceanog.2012.90.
[21] M. Mengel, A. Levermann, Nature Climate Change 4 (2014) 451–455, http://dx.doi.org/10.1038/NCLIMATE2226.
[22] R.G. Peterson, W.B. White, J. Geophys. Res. 103 (1998) 24573–24583.

[23] J.C. Comiso, R. Kwok, S. Martin, A.L. Gordon, J. Geophys. Res. 116 (2011) C04021, http://dx.doi.org/10.1029/2010JC006391.

[24] D. Rind, M. Chandler, J. Lerner, D.G. Martinson, X. Yuan, J. Geophys. Res. 106 (2001) 20161–20173.

[25] A.B. Wilson, D.H. Bromwich, K.M. Hines, S.-H. Wang, J. Clim. 27 (2014) 8934–8955, http://dx.doi.org/10.1175/JCLI-D-14-00296.1.

[26] J.A. Francis, S.J. Vavrus, Geophys. Res. Lett. 39 (2012) L06801, http://dx.doi.org/10.1029/2012GL051000.

[27] K. Pistone, I. Eisenman, V. Ramanathan, Proc. Nat. Acad. Sci. U.S.A. 111 (2014) 3322–3326.

[28] G. Whiteman, Hope, P. Wadhams, Nature 499 (2013) 401–403.

LAND ICE: INDICATOR AND INTEGRATOR OF CLIMATE CHANGE

5

Jonathan Bamber

University of Bristol, School of Geographical Sciences, Bristol Glaciology Centre, Bristol, UK

CHAPTER OUTLINE

1. INTRODUCTION

Glaciers are frequently depicted as an iconic example of a warming planet. Whenever a climate change story emerges in the news, it is frequently accompanied by footage of a collapsing ice front crashing into the ocean, creating a myriad of icebergs in the process. In reality, such processes occur with or without a changing climate and are part of the normal process of mass turnover of a glacier that terminates in the ocean. This imagery is used, however, to imply that in a warming world, ice melts faster and, although this is broadly correct, the timescale for the response and its magnitude varies dramatically between the various different locations and size of ice mass that currently exist. In the case of Antarctica, an increase in air temperature may result in growth of the ice sheet in the short term [1], so the realities of how land ice responds to a changing climate are more complex than may first appear.

This chapter will introduce the concepts and processes that influence the gains and losses of land ice (termed the mass balance) and how climate change affects these various processes. It will discuss why understanding and measuring the mass balance of land ice is important and will present some of the latest observational evidence for changes to glacier volume and the ice sheets that cover Greenland and Antarctica. Before doing this, it is necessary to provide a brief definition of what is being discussed. We consider here all permanent ice that covers the land and, in the case of Antarctica and Greenland, additionally ice shelves that extend out from the coast into the ocean. The difference between a glacier and ice sheet is, predominantly, one of scale and morphology.

Glaciers (including here ice caps), range in size from a few kilometres in length up to around $10\,000$ km^2 in area for the larger ice caps in the Canadian Arctic. The ice sheets, on the other hand, are on a far grander scale: 1.8×10^6 and 13.9×10^6 km^2 for Greenland and Antarctica, respectively. In fact, they represent about 99.5% of the total volume of land ice on the planet. Their potential contribution to sea level rise, if completely melted, is 66 m, while glaciers would contribute around 0.5 m[1]. How fast land ice responds to a change in climate, however, is crudely a function of its size, and thus glaciers being much smaller than the ice sheets can potentially react more rapidly. The dynamic response time, t, of land ice (how quickly it reaches a new equilibrium after a change in climate) can be approximated by the relationship

$$t \approx H/a \tag{1}$$

where H is the maximum ice thickness and a is the surface melt rate [2]. If typical values for Greenland are used in Eqn (1), we find that it has a present-day response time of about 3000 years, while for a valley glacier t can be on the order of decades to a few centuries [3]. This relationship does not take into account, however, the possible rapid response of marine-terminating land glaciers that is discussed in Section 4. Given this caveat, land ice is, in general, an integrator of climate over these time scales, and this is an important attribute as it means it acts to smooth out short-term (interannual) fluctuations in temperature and precipitation that relate to variability due to 'weather' rather than climate.[2] The ice sheets, on the other hand, are still responding (to a small degree) to changes that took place at the end of the last Ice Age, around 12 ka BP [4]. It is important to note, however, that this is the equilibrium response of the ice sheets and that they are never truly in equilibrium with the climate at any given time. It is also important to note, as will be discussed in Section 4, that parts of the ice sheets can react rapidly to short-term perturbations in external forcing, such as, for example, tidal forcing by the oceans [5,6].

1.1 GEOGRAPHICAL CONTEXT

Land ice requires persistent temperatures below 0°C to be sustained and is found, therefore, at either high latitudes or altitudes. In most studies investigating global trends in glaciers, the coverage is grouped into 19 distinct regions, depicted in Fig. 1. Each region is considered to be generally influenced by the same climatic regime and trends. Over half (56%) of glaciers, by area, lie in the Arctic, a region that has been experiencing greater warming than the global average [7]. Outside of the polar regions, the next largest concentration of land ice lies in Central and South Asia (sectors 13–15 in Fig. 1), which is sometimes referred to as the Third Pole and includes the Himalaya, Karakoram and Tibetan Plateau. Around the margins of Greenland (sector 5) and parts of Antarctica (sector 19), there are glaciers that are disconnected from the adjacent ice sheets. These are considered not to be part of the ice sheet and can, and do, respond more rapidly than their much larger neighbours. In the Northern Hemisphere, the Greenland Ice Sheet (GrIS) is by far the largest ice mass, with a volume about 20 times that of all glaciers. It covers some 83% of the largest island in the world. Peripheral glaciers in

---

[1]The ratio of 0.5/66 implies 0.75%, but a substantial portion of the Antarctic ice sheet is grounded below sea level, and this volume of ice only has a small effect on sea level rise. Hence, in terms of total volume rather than sea level contribution, the ratio is closer to 0.5%.

[2]In meteorology, a 30-year mean is generally considered to be adequate to minimise interannual variability due to weather and to represent trends in climate. Thus glaciers are well placed to reflect the latter rather than the former.

Greenland (sector 5) cover an additional 4% of the land area. The Antarctic Ice Sheet (AIS) is usually split into three regions, with different glaciological and climatic characteristics: the West Antarctic Ice Sheet (WAIS), East (EAIS) and Antarctic Peninsula (Fig. 2). The WAIS is similar in volume to the GrIS, but a large fraction of it rests on bedrock substantially below sea level (at its deepest, around 2500 m below sea level). The bedrock also slopes downward inland and this configuration is believed to be inherently unstable, such that a small change in ice thickness can result in a rapid decay of the ice sheet [8,9]. This behaviour has been termed the Marine Ice Sheet Instability (MISI) hypothesis and will be discussed further in Section 4.

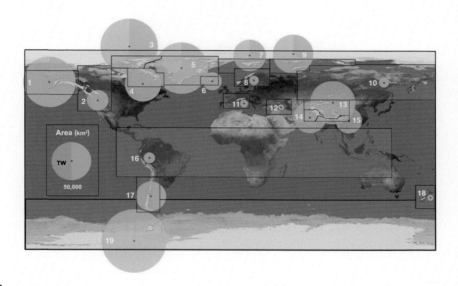

FIGURE 1

Distribution of glaciers around the world, divided into 19 standardized regions: (1) Alaska, (2) Western Canada and US, (3) Arctic Canada North, (4) Arctic Canada South, (5) Greenland, (6) Iceland, (7) Svalbard, (8) Scandinavia, (9) Russian Arctic, (10) North Asia, (11) Central Europe, (12) Caucasus and Middle East, (13) Central Asia, (14) South Asia (west), (15) South Asia (east), (16) Tropical glaciers, (17) Southern Andes, (18) New Zealand, (19) Antarctica and Sub-Antarctic. The size of the circles is proportional to the areal coverage and the percentage that is marine terminating is shown in blue (*From Ref. [10]*). Note that for Greenland and Antarctica, only those glaciers that are disconnected from the ice sheets are included here. *(This figure has been reproduced with permission.)*

1.2 ANATOMY OF AN ICE SHEET

To understand how land ice loses mass, particularly when it terminates in the ocean, it is useful to introduce some terms and concepts related to the flow of the ice from the interior toward the coast and to the interactions that take place on the way and when it reaches the ocean. Figure 3 is a schematic diagram showing the key components of an idealized ice sheet. These components exist for the Antarctic Peninsula (AP), WAIS, EAIS and GrIS. The ice sheet interior moves slowly toward the coast (*cf.* Fig. 2) at a rate of a few tens of m a^{-1} but at various locations it gets channelled into faster-flowing conduits known as ice streams (and also, confusingly, outlet glaciers). These 'arteries' of the ice sheet flow at rates of 10–100 times faster than the inland ice (Fig. 2) and eventually reach the ocean.

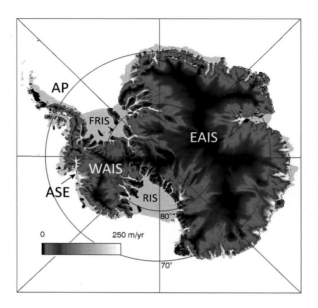

FIGURE 2

Balance velocities for the Antarctic ice sheet showing the steady-state ice flow required to keep the ice sheet in equilibrium with present-day snowfall [11]. Areas of faster flow, which are the ice streams and outlet glaciers, are shown in blue to white colours and drain into the fringing ice shelves that are shaded grey. FRIS: Filchner Ronne Ice Shelf; RIS: Ross Ice Shelf, ASE: Amundsen Sea Embayment. Within the figure, the 250 m a^{-1} is the correct way to abbreviate 250 m year^{-1}.

The point at which the base of the ice comes into contact with the ocean is known as the grounding line (Fig. 3). It is a triple junction between water, ice and rock and plays a critical control on the stability and behaviour of the inland ice. This is partly because it is the point at which the ocean can interact with ice through basal melting and also because beyond the grounding line, the ice forms floating shelves that are in hydrostatic equilibrium with the ocean and can, therefore, have no more impact on sea level. If an ice shelf melts, it has a negligible effect on sea level, but if ice is transferred from the land into the ocean it has a direct and significant impact. The ice shelves that form in Antarctica can be very large; the Filchner Ronne Ice Shelf, for example, is around 1000 km wide and 800 km deep and the thickest ice, close to the grounding line, exceeds 1200 m [12]. These floating expanses act to buttress the interior ice and changes in their size and thickness can, therefore, affect how rapidly the inland, grounded ice enters the ocean and also the stability of that inland ice [13,14].

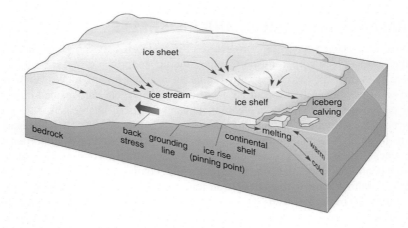

FIGURE 3

A schematic diagram of a cross-section through the margin of an idealized ice sheet, showing the key elements that influence flow into the ocean and the formation of ice shelves. *(Courtesy R. Bindschadler.)*

1.3 LAND ICE AND SEA LEVEL

Glaciers, especially those at lower latitudes in Asia and the Andes, are critical reservoirs of water and help maintain river flows during the drier summer months. They are, therefore, important for water resources. One of the most damaging consequences of climate change, however, is future sea level rise (SLR) [15]. Glaciers and ice sheets represent the largest potential source of SLR and also the most uncertain. The other main source is thermal expansion of the oceans as they warm. When land ice melts, it has a direct impact on sea level, as long as the ice was resting on bedrock above mean sea level. Any ice resting on bedrock below sea level (such as in parts of the WAIS, see Section 1.1) will be replaced by the equivalent volume of water when the ice melts, so its sea level effect is small (and relates to the difference between the density of ice and sea water, which is about 11%). Mass exchange between land ice and the ocean is usually expressed in units of Gt year^{-1}, where a Gt is equivalent to a cubic kilometre (1 km^3) of water (360 Gt of ice is sufficient to raise global sea level by 1 mm). Estimates of SLR for the twentieth century are 1.5 mm a^{-1}–1.8 mm a^{-1}, with about half of this coming from glaciers and the rest from thermal expansion [16]. Little is known about the mass balance of the ice sheets during most of this period but their contribution is believed to be relatively small [16]. During the last decade of the twentieth century and up to the present day, the rate of SLR has roughly doubled to 3.2 mm a^{-1} with a significantly increased contribution from both glaciers and the ice sheets. Since 1992, satellite observations of both the oceans and land ice have given us unprecedented detail about the sources of SLR and the mass changes of glaciers and ice sheets. Much of Section 4, on observations, relates, therefore, to this relatively short period where we have detailed and comprehensive information about ice mass trends.

2. MASS BALANCE OF GLACIERS AND ICE SHEETS

The mass balance of a glacier or ice sheet represents the trade-off between gains through, primarily, snowfall and the losses, which compromise melting at the upper surface of the ice (also called surface ablation), iceberg calving (in the case of marine-terminating land ice), bottom melting underneath floating ice shelves and basal melting at the ice/bedrock interface, which is generally a small term. Calving and bottom melting can be aggregated into a single term, which is the ice discharge across the grounding line (remember, once it has crossed this threshold it has made its contribution to SLR). For land-terminating glaciers, such as those in Asia and the European Alps, only surface melting is important – there is no discharge component. In the case of marine-terminating glaciers, such as those in most of the Arctic (see Fig. 1), discharge is an important, and sometimes dominant, component of the total mass removal. For Antarctica, surface melting is negligible because air temperatures, even in summer, are generally below freezing, except for parts of the Peninsula. Here, calving and bottom melting beneath ice shelves are the two key components of mass removal, with roughly equal magnitudes integrated over the whole continent but with large regional variations depending on the oceanic setting [17]. For Greenland, about 60% of the present-day mass removal is from ice discharge and 40% from surface ablation [18]. These two components have different sensitivities to external forcing and different response times and it is worth discussing, therefore, more detail of the factors that influence mass turnover.

2.1 SURFACE MASS BALANCE

This refers to all processes that relate to mass gain or loss at the upper surface of the glacier or ice sheet. As mentioned above, the two most important processes are snowfall and surface melting, also called runoff. Both snowfall and runoff are influenced by atmospheric drivers such as changes in circulation, moisture content and air temperature. The surface mass balance (SMB) adjusts almost instantaneously to changes in these atmospheric drivers, while changes in discharge are driven by ice flow, which responds, in general, more slowly and in a more complex way to changes in external forcing.

2.2 ICE DYNAMICS

Discharge is controlled by ice dynamics – the faster the ice flows across the grounding line, the larger the discharge will be (cf. Fig. 2). What then controls ice dynamics? The answer to this question could fill a chapter on its own. Gravity is the driving force that makes ice flow from the interior to the margin of an ice sheet. The magnitude of this gravitational driving force is proportional to surface slope, such that steeper sloping ice will flow faster than flatter regions. If surface melting increases at lower elevations, then this can result in a steepening of the glacier profile, which, will in turn result in a speed up of the ice. This speed-up is a gradual process, however, that takes decades to centuries to reach a new steady state. Changes in temperature (the thermodynamics) of the ice also influence how fast it flows but a step change in surface air temperature can take millennia to reach deep into the ice in Antarctica, which is up to 5-km thick. Changes in surface gradient and thermodynamics are, consequently, slow processes that, for the ice sheets at least, are believed to be relatively unimportant on a centennial timescale [19]. The basal boundary conditions (i.e. what the conditions are like at the ice/bed interface) are also important in controlling ice motion. In particular, if the bed is at the pressure melting temperature of ice, then water may be present, which acts as a lubricant for motion, termed basal sliding [2]. As mentioned, changes in temperature at the surface propagate to the bed very slowly. In the case of the GrIS, an alternative mechanism has been proposed, which involves the routing of surface melt water to the bed via englacial conduits [20]. It is likely that melt water does reach the bed, and in a warming climate, surface melting will increase, which means more melt water could reach the bed. It is less clear, however, how this might affect the flow rates of the ice because of the complex nature of the subglacial hydrological system [21]. The same mechanism for enhanced basal lubrication from surface melting exists for other Arctic glaciers, but the same caveat holds for whether it will have any significant impact on flow rates. The final factor that can influence ice dynamics on short timescales, which has been mentioned already, is the buttressing effect of ice shelves on the inland, grounded ice. After the collapse of the Larsen A ice shelf on the AP in 1995, it was observed that a number of the glaciers that fed into the shelf had accelerated dramatically, in some cases by as much as a factor 10, after the removal of the shelf [13,22]. Parts of West Antarctica (discussed further in Section 4) are believed to have speeded up by more than a factor two due to thinning of the adjacent ice shelves as a result of changes in ocean circulation [23,24]. Thus, the ocean can have an important (crucial) impact on the short-term dynamic response of glaciers when they are marine terminating. In summary, the atmosphere is responsible for changes in SMB while the ocean plays a central role in short-term changes in discharge, with only a weak influence from the atmosphere. This point is important to bear in mind when considering observations of land ice changes and their relationship to climate change.

3. LONG-TERM BEHAVIOUR

In the previous sections, the timescales for a response to external forcing were discussed in relation to the nature of the forcing. It was mentioned that part of the present-day dynamic signal in the ice sheets is due to changes in climate that took place at the end of the last glacial, around 12 ka BP, and it is worth discussing briefly how land ice cover has changed on millennial timescales in response to long-term secular changes in climate. The aim is to provide a context for recently observed changes, particularly during the satellite record.

3.1 GLACIAL-INTERGLACIAL CYCLES

Glacial-interglacial cycles are believed to be driven by changes in the orbital pattern of the Earth that has periods of about 20 ka, 40 ka and 100 ka [25]. During the last glacial cycle, an ice sheet covered most of North America, Eurasia, the Barents Sea and the northern half of the UK. At this time, both the GrIS and AIS were around 10% larger than present, extending out roughly to their continental margins [25]. At Last Glacial Maximum, about 22 ka BP, sea level was 120 m lower than present. It is interesting to note, however, that during the last interglacial, around 130 ka BP, a period termed the Eemian, temperatures were around 2°C higher than present and sea level was between 5 and 8 m higher [26]. A 2°C warming is at the lower end of projections for 2100 [27]. A SLR of this magnitude (5–8 m) would have the most profound consequences for humankind [28]. Less than a metre would have come from thermal expansion and the rest must have been from Antarctica and Greenland. The exact contribution from these two regions is a topic of ongoing research [29,30].

3.2 HOLOCENE VARIABILITY

The period since the end of the last glacial is known as the Holocene and has been characterized by a relatively stable climate compared, for example, to the last glacial and interglacial [31]. Nonetheless, there have been a number of periods of climate change, although of relatively small amplitude [32]. One period in particular is worth noting because of its impact on present-day and recent past land ice; it is termed the Little Ice Age (LIA). The LIA was a period of modest cooling of about 0.5°C that occurred from roughly 1550–1850 AD with regional variations in its start and end date. Most glacier regions, shown in Fig. 1, reached their Holocene maximum (or close to it) toward the end of the LIA [33]. As a consequence, most of these glacier regions have been experiencing recession since then. It is important, in the discussion that follows, to separate recession as a consequence of this from that due to recent (late twentieth and twenty-first century) climate warming. Partly because of their size, both the AIS and GrIS were less strongly influenced by these relatively small Holocene climate fluctuations including the LIA. For the AIS, colder (warmer) air temperatures can result in wastage (growth) of the ice sheet because of the reduced (increased) moisture carrying capacity of the atmosphere [1]. In other words, colder air temperature means less snowfall, and vice versa. For the GrIS, the reduced (increased) snowfall is outweighed by the reduction (increase) in surface melting in a cooling (warming) climate.

4. OBSERVATIONS OF RECENT CHANGES

The preceding sections provide an introduction to the main processes and factors that influence the mass balance and, therefore, the decay and growth of glaciers and ice sheets. They show that the response can be complex and is a function of the size of the ice mass, its proximity and contact with the ocean and whether a significant area of surface melting exists or not. This background is important for understanding the signals that have been seen in changing glacier and ice sheet volume.

4.1 GLACIERS AND ICE CAPS

The discussion of recent changes is split between glaciers and the ice sheets because the timescales and drivers of change are not the same, and we begin by considering glacier volume and extent changes since 1850. Some glacier length records extend back that far, whereas for other sectors, records are not available until well into the twentieth century [34,35], while the most complete estimates cover the last 50 years [10] (Fig. 4). Information about glacier mass changes comes from a variety of sources including ground-based field measurements of surface mass balance using time-consuming, labour-intensive survey techniques and for some regions also from repeat airborne and satellite geodetic surveys [36]. The latter involves measuring the surface elevation of the glacier at different dates and calculating the volume change from this. Airborne stereo photographs can be used for this purpose and extend back to the 1950s and earlier for several sectors [37]. More recent technology includes airborne [38] and satellite laser altimetry [39] as well as (since 2003) satellite gravimetry [40]. A complete and robust estimate of glacier mass trends from a combination of all these satellite and in situ observations has been produced for the period 2003–2009 [41]. It provides a benchmark for assessing other approaches and is shown in Fig. 4 in red. All 19 regions show a decline in area since 1850, as would be expected after emerging from the LIA [10]. This, then, is not a surprise. However, individual glaciers, in the same sector, can demonstrate markedly different magnitudes of response. This is due to differences in size, aspect (are they north or south facing, for example), altitude and morphology. To improve the interpretation, it is necessary, therefore, to take into account these factors and to increase the number of glaciers included. This was done in a study that turned the problem around: if glaciers integrate climate changes over decades, can they detect trends in the climate? Oerlemans [42] examined length records of 169 glaciers around the world and applied some simple principles of glacier dynamics to these to reconstruct the temperature record that their length variations implied. This resulted in a consistent, robust signal of twentieth-century warming of 0.5°C, in good agreement with land surface temperature reconstructions [43]. By inference, the glaciers must therefore be responding to the temperature changes that they are recording in their length changes in a way that is consistent with glaciological theory. An important development to this topic has been the attribution of glacier volume changes to anthropogenic and natural causes. As mentioned earlier, some mass wastage is expected as a consequence of changes in climate at the end of the LIA, but the rates of loss have generally been increasing in most of the 19 sectors over the last few decades, particularly for Central

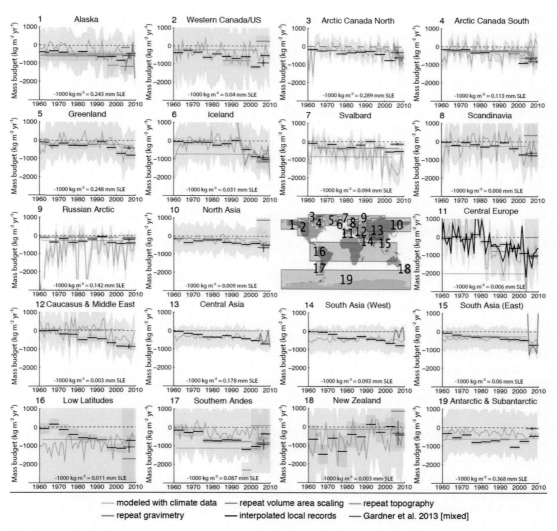

FIGURE 4

Glacier mass balance estimates for five-year intervals, over the period 1960–2010, from a range of different sources including modelling estimates, geodetic approaches, satellite gravimetry and field-based observations. Estimates are aggregates for all glaciers in each of the 19 sectors shown in Fig. 1 (*From Ref. [10]*). Note that the five-year averages shown by the solid black lines show negative mass balance for almost all periods and all regions. (*The figure has been reproduced with permission.*) Note: the *y*-axis should read: mass budget/(kg m^{-2} a^{-1}).

and South Asia (sectors 13, 14, 15), Greenland (sector 5), Central Europe (sector 11) and Iceland (sector 6) [10]. What proportion of these changes, particularly for the last few decades, is due to natural climate variability and what proportion to human interference with the climate? This question was addressed in a study that combined a suite of atmospheric GCM simulations for 1850–2010 with a glacier mass balance model [44]. During the entire period only 25% ± 35% of the retreat was attributed to anthropogenic sources, i.e. a statistically insignificant amount. Importantly, though, when they examined the most recent 20 years (1991–2010), the proportion was 69% ± 24%. In other words, since 1991, more than two-thirds of the wastage is attributed to human interference with the climate system. Using the same sorts of GCMs to simulate the climate through the twenty-first century suggests that a significant number of sectors shown in Fig. 1 could be largely ice free by 2100 [45]. These include Western Canada and the US, Iceland, North Asia, Central Europe, Caucasus and low latitudes. Such a dramatic change in the geography of these regions will have profound impacts on water resources and local to regional economies.

4.2 ICE SHEETS

The ice sheets (GrIS, WAIS and EAIS) are less important than glaciers in terms of water resources and local economies but are two orders of magnitude more important in terms of their potential contribution to global SLR. The GrIS has a sea level equivalent of 7.4 m [46], the WAIS and Antarctic Peninsula 4.8 m [8], and the EAIS 53.5 m [47]. These values take into account ice that is grounded below sea level, as discussed earlier, which only has a small effect [8]. As mentioned in Section 3.1, the ice sheets contributed multiple metres (around 4–7 m) to SLR during the last interglacial, when global temperatures were about 2°C warmer than present, which raises the interesting question of what contribution they might make in a similarly warmer world in the future. A mean global warming of 2°C above preindustrial lies at the lower end of projections for 2100 [27], and it is reasonable, therefore, to look at the Eemian as a potential analogue for the future. It is, however, not just the absolute sea level rise that is important but also the rate at which it takes place. This is because adaptation costs and options are closely tied to rates rather than the final number [15].

Because of their size and the severity of the environment, field-based measurements are not practical for providing ice sheet–wide estimates of mass trends. They are useful, for example, for validating models and observations [48] but only locally. The AIS covers an area greater than the lower 48 states of the United States, and extrapolating measurements over thousands of kilometres is of limited scientific value. As a consequence, robust and accurate estimates of the mass balance of the ice sheets, prior to the satellite era, are limited [10]. Estimates of thermal expansion and land hydrology contributions to sea level during the twentieth century provide a constraint on the contribution of glaciers and ice sheets as they will be the residual in the observed SLR measured by tide gauge data during this period [49]. This approach has been used to infer that the Antarctic ice sheet was fairly close to balance during the twentieth century and contributed somewhere between -0.2 mm a^{-1} and 0.5 mm a^{-1} [49]. A significant component of the spread in this range is due to the uncertainty in the contribution from the GrIS. There is evidence from examination of satellite imagery of glacier retreat rates that the GrIS was relatively stable from 1972 to 1985 with rates of retreat increasing from then and accelerating significantly from around 2000 [50]. These data provide an indirect indication of stability of the ice sheet but cannot give quantitative estimates of mass balance. That was not possible (with an acceptable level of accuracy) until the launch of a new era of Earth observation satellites beginning in 1991 with the European Space Agency's ERS-1 mission. This satellite carried a suite of instruments on board including two that have proved particularly valuable for ice sheet research. The first was a radar altimeter that allowed repeat observations of elevation changes over almost the entirety of Greenland and four-fifths of Antarctica [51]. The second instrument was a synthetic aperture radar (SAR) that could be used to measure, with high accuracy, the ice surface velocity over the ice sheets using a technique termed interferometric SAR, or InSAR [52,53]. If the ice thickness is known at the grounding line of outlet glaciers and ice streams (see Fig. 3), then, combined with the velocity, this provides an estimate of the flux of ice leaving the land and entering the ocean. An increase in velocity translates directly to an increase in flux. ERS-1 was succeeded by a suite of satellite missions, from both ESA and NASA, that included the first satellite laser mission, ICESat, and a gravity mission called GRACE. The data from these various missions have been used to determine, with unprecedented detail, the mass trends of the ice sheets since 1991 (e.g. Ref. [54]). Multiple studies have been published using various combinations of data and time periods, which were synthesized in the most recent IPCC report chapter on cryosphere observations [10]. The results of this synthesis suggest that both ice sheets were close to balance in the early 1990s (Fig. 5) but have experienced an accelerating mass loss since then, especially in the case of Greenland [55,56].

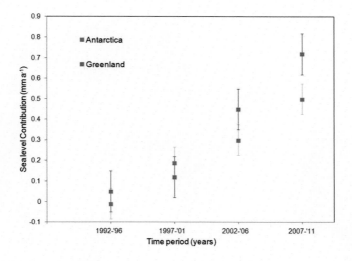

FIGURE 5

Five yearly mean estimates of mass balance of the Antarctic and Greenland ice sheets for the 20-year period of the high-quality satellite era. The estimates are a synthesis of different methods and data sets [10]. The label along the *y*-axis should appear as Sea Level Contribution/(mm a^{-1}) and that along the *x*-axis as Time period/years.

The spatial pattern of this mass wastage is not uniform across the ice sheets – it is concentrated around the margins of the GrIS and in the Amundsen Sea Embayment of the WAIS [57] (Fig. 6). Thinning in Greenland (at least over this relatively short interval) is concentrated, in particular, along the northwest and southeast margins of the ice sheet and are associated with an almost equal contribution from ice flow speed-up (dynamics) and increased surface melting [58]. Modest thickening in the interior is likely due to an increase in precipitation associated with warmer temperatures. This has not been sufficient, however, to offset the enhanced melting and dynamic mass loss that are also linked to increased atmospheric temperatures and intrusion of warm ocean waters [59–61]. Hence, mass loss has been increasing in Greenland at the same time as increases in air temperature [62]. It is important to note that what is being observed in the satellite record is a transient response to external forcing rather than the equilibrium response that was discussed in Section 1 and characterized by Eqn (1). This has a longer time constant than the years to decades of the instrumental record.

The other area of major thinning (and mass wastage) is in the Amundsen Sea Embayment of the WAIS. Recall that the WAIS and, in particular, this sector is vulnerable to the Marine Ice Sheet Instability mechanism discussed earlier. Bed elevations for this part of the WAIS reach as deep as

2500 m below sea level. Multiple studies have identified sustained grounding line retreat [24], ice flow speed-up [63], and increasing mass loss [64] since the mid-1990s. Recent assessments of this retreat and mass wastage suggest that it has exceeded a stable threshold and will continue for decades to centuries into the future [24,65]. With the potential to raise global mean sea level by up to 3.3 m through the Marine Ice Sheet Instability mechanism [8], this is a particularly worrying development. As mentioned previously, it is, however, not just the sea level commitment but also the rate of SLR that is important here. This depends on future, and uncertain, climate change that, in turn, depends critically on the actions of humans over the coming decades.

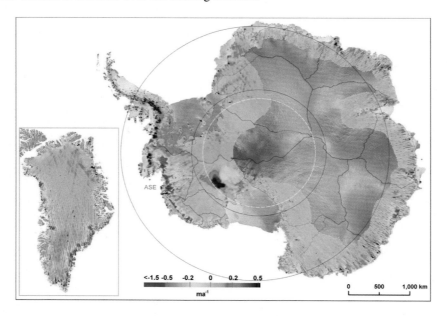

FIGURE 6

Rates of thickening (blues) and thinning (reds) for the period 2003–2007 obtained from the NASA satellite laser altimeter ICESat. *(Courtesy H. Pritchard, British Antarctic Survey.)* East Antarctic rates are truncated at an elevation of 2500 m elevation. ASE = Amundsen Sea Embayment. The white dashed circle marks the southern latitudinal limit of satellite altimetry data prior to the launch of ICESat in 2002.

The magnitude and speed of changes in ice dynamics for the ice sheets came as a surprise to many glaciologists when the results began to emerge from the satellite record. Prior to that time, the general consensus was that the ice sheets were relatively insensitive to changes in external forcing and that any signal of this forcing would take centuries to become significant [4]. This view arose from the fact that the first generation of numerical models of ice sheet evolution, developed during the 1980s, displayed a weak sensitivity to climate change that evolved slowly in time [4]. Observations of rapid, and large, changes in ice dynamics of major outlet glaciers in both Greenland [66] and Antarctica [67] were, therefore, unexpected and highlighted deficiencies in the first generation of numerical models. Since then, more sophisticated models have been developed that include more of the physics of the system and, in particular, a more realistic representation of grounding line behaviour [9]. These models have been more effective in reproducing observations and are being used to aid in projecting the response of the ice sheets to future climate change [68].

5. CONCLUDING REMARKS

Glaciers and ice sheets respond to climate change in a complex manner. For land-terminating glaciers, the relationship is relatively simple and, in general, as air temperatures rise, they lose more mass and retreat. When the ice is in contact with the ocean, the relationship is more complex because mass is also lost by iceberg calving and bottom melting beneath floating ice shelves. In this case, the oceans play a critical role in modulating the rate and characteristics of mass loss. This is particularly important for the ice sheets that cover Greenland and Antarctica where, in the case of the former, around 65% of mass removal is from ice discharge into the ocean while for the latter it is close to 100%. Despite these complexities in the different sensitivities of land ice to ocean and atmospheric forcing, the observations are unequivocal about the increasing rates of mass loss for most glacier regions, the Greenland Ice Sheet and the vulnerable West Antarctic Ice Sheet over the last two decades. These increased rates of loss coincide with a doubling, or potentially more [69], in the rate of sea level rise during the same period. Projecting the behaviour of land ice over the next century is a challenging task, but the signal of changes over the last 20 years does not bode well for the future of mountain glaciers or for the fate of low-lying nations at risk from sea level rise.

REFERENCES

[1] J.M. Gregory, P. Huybrechts, Phil. Trans. R. Soc. A Math. Phys. Eng. Sci. 364 (2006) 1709–1731.
[2] W.S.B. Paterson, The Physics of Glaciers, third ed., Pergamon, Oxford, 1994, pp. 480.
[3] J. Oerlemans, Science 264 (1994) 243–245.
[4] P. Huybrechts, J. de Wolde, J. Clim. 12 (1999) 2169–2188.
[5] J.L. Bamber, R.B. Alley, I. Joughin, Earth Planet. Sci. Lett. 257 (2007) 1–13.
[6] G.H. Gudmundsson, Nature 444 (2006) 1063–1064.
[7] M.C. Serreze, A.P. Barrett, J.C. Stroeve, D.N. Kindig, M.M. Holland, Cryosphere 3 (2009) 11–19.
[8] J.L. Bamber, R.E.M. Riva, B.L.A. Vermeersen, A.M. LeBrocq, Science 324 (2009) 901–903.
[9] C. Schoof, J. Geophys. Res. 112 (2007) F03S28.
[10] D.G. Vaughan, J.C. Comiso, I. Allison, J. Carrasco, G. Kaser, R. Kwok, et al., in: T.F. Stocker, et al. (Eds.), Climate Change 2013: The Physical Science Basis, Contribution of Working Group I to the Fifth Assessment Report of the Intergovernmental Panel on Climate Change, Cambridge University Press, Cambridge, UK and New York, NY, USA, 2013, pp. 317–382 (Chapter 4).
[11] J.L. Bamber, D.G. Vaughan, I. Joughin, Science 287 (2000) 1248–1250.
[12] J.A. Griggs, J.L. Bamber, J. Glaciol. 57 (2011) 485–498.
[13] H. De Angelis, P. Skvarca, Science 299 (2003) 1560–1562.
[14] H.H. Hellmer, F. Kauker, R. Timmermann, J. Determann, J. Rae, Nature 485 (2012) 225–228.
[15] R.J. Nicholls, R.S.J. Tol, Phil. Trans. R. Soc. A Math. Phys. Eng. Sci. 364 (2006) 1073–1095.
[16] J.A. Church, P.U. Clark, A. Cazenave, J.M. Gregory, S. Jevrejeva, A. Levermann, M.A. Merrifield, et al., in: T.F. Stocker, et al. (Eds.), Climate Change 2013: The Physical Science Basis, Contribution of Working Group I to the Fifth Assessment Report of the Intergovernmental Panel on Climate Change, Cambridge University Press, Cambridge, UK and New York, NY, USA, 2013, pp. 1137–1216 (Chapter 13).

[17] M.A. Depoorter, J.L. Bamber, J.A. Griggs, J.T.M. Lenaerts, S.R.M. Ligtenberg, M.R. van den Broeke, G. Moholdt, Nature 502 (2013) 89–92.
[18] J. Bamber, M. van den Broeke, J. Ettema, J. Lenaerts, E. Rignot, Geophys. Res. Lett. 39 (2012) L19501.
[19] H. Goelzer, P. Huybrechts, J.J. Furst, F.M. Nick, M.L. Andersen, T.L. Edwards, X. Fettweis, A.J. Payne, S. Shannon, J. Glaciol. 59 (2013) 733–749.
[20] H.J. Zwally, W. Abdalati, T. Herring, K. Larson, J. Saba, K. Steffen, Science 297 (2002) 218–220.
[21] R.S.W. van de Wal, W. Boot, M.R. van den Broeke, C. Smeets, C.H. Reijmer, J.J.A. Donker, J. Oerlemans, Science 321 (2008) 111–113.
[22] E. Rignot, G. Casassa, P. Gogineni, W. Krabill, A. Rivera, R. Thomas, Geophys. Res. Lett. 31 (2004). Art. No. L18401.
[23] I. Joughin, S. Tulaczyk, R. Bindschadler, S.F. Price, J. Geophys. Res. Solid Earth 107 (2002). Art. no.-2289.
[24] E. Rignot, J. Mouginot, M. Morlighem, H. Seroussi, B. Scheuchl, Geophys. Res. Lett. (2014), 2014GL060140.
[25] A.G. Dawson, Ice Age Earth: Late Quaternary Geology and Climate, Routledge, London, 1992, pp. 293.
[26] R.E. Kopp, F.J. Simons, J.X. Mitrovica, A.C. Maloof, M. Oppenheimer, Nature 462 (2009) 863–867.
[27] M. Collins, R. Knutti, J. Arblaster, J.-L. Dufresne, T. Fichefet, P. Friedlingstein, et al., in: T.F. Stocker, et al. (Eds.), Climate Change 2013: The Physical Science Basis, Contribution of Working Group I to the Fifth Assessment Report of the Intergovernmental Panel on Climate Change, Cambridge University Press, Cambridge, UK and New York, NY, USA, 2013, pp. 1029–1136 (Chapter 12).
[28] R.S.J. Tol, M. Bohn, T.E. Downing, M.L. Guillerminet, E. Hizsnyik, R. Kasperson, K. Lonsdale, C. Mays, R.J. Nicholls, A.A. Olsthoorn, G. Pfeifle, M. Poumadere, F.L. Toth, A.T. Vafeidis, P.E. Van der Werff, I.H. Yetkiner, J. Risk Res. 9 (2006) 467–482.
[29] K.M. Cuffey, S.J. Marshall, Nature 404 (2000) 591–594.
[30] D. Dahl-Jensen, M.R. Albert, A. Aldahan, N. Azuma, D. Balslev-Clausen, M. Baumgartner, et al., NEEM Community, Nature 493 (2013) 489–494.
[31] W. Dansgaard, S.J. Johnsen, H.B. Clausen, D. Dahl-Jensen, N. Gunderstrup, C.U. Hammer, Nature 364 (1993) 218–220.
[32] P.A. Mayewski, E.E. Rohling, J. Curt Stager, W. Karlén, K.A. Maasch, L. David Meeker, et al., Quatern. Res. 62 (2004) 243–255.
[33] S. Ivy-Ochs, H. Kerschner, M. Maisch, M. Christl, P.W. Kubik, C. Schlüchter, Quat. Sci. Rev. 28 (2009) 2137–2149.
[34] P.W. Leclercq, J. Oerlemans, Clim. Dynam. 38 (2012) 1065–1079.
[35] P.W. Leclercq, J. Oerlemans, J.G. Cogley, Surv. Geophys. 32 (2011) 519–535.
[36] J.G. Cogley, Ann. Glaciol. 50 (2009) 96–100.
[37] E. Thibert, R. Blanc, C. Vincent, N. Eckert, J. Glaciol. 54 (2008) 522–532.
[38] W. Krabill, R. Thomas, K. Jezek, K. Kuivinen, S. Manizade, Geophys. Res. Lett. 22 (1995) 2341–2344.
[39] A. Kaab, E. Berthier, C. Nuth, J. Gardelle, Y. Arnaud, Nature 488 (2012) 495–498.
[40] T. Jacob, J. Wahr, W.T. Pfeffer, S. Swenson, Nature XXX (2012) XXX.
[41] A.S. Gardner, G. Moholdt, J.G. Cogley, B. Wouters, A.A. Arendt, J. Wahr, et al., Science 340 (2013) 852–857.
[42] J. Oerlemans, Science 308 (2005) 675–677.
[43] M.E. Mann, P.D. Jones, Geophys. Res. Lett. 30 (2003).
[44] B. Marzeion, J.G. Cogley, K. Richter, D. Parkes, Science 345 (2014) 919–921.
[45] V. Radic, A. Bliss, A.C. Beedlow, R. Hock, E. Miles, J.G. Cogley, Clim. Dynam. 42 (2014) 37–58.
[46] J.L. Bamber, J.A. Griggs, R.T.W.L. Hurkmans, J.A. Dowdeswell, S.P. Gogineni, I. Howat, et al., Cryosphere 7 (2013) 499–510.

[47] P. Fretwell, H.D. Pritchard, D.G. Vaughan, J.L. Bamber, N.E. Barrand, R. Bell, et al., Cryosphere 7 (2013) 375–393.

[48] W. Greuell, J. Oerlemans, Ann. Glaciol. 42 (2005) 107–117.

[49] J.M. Gregory, N.J. White, J.A. Church, M.F.P. Bierkens, J.E. Box, M.R. den Broeke, et al., J. Clim. 26 (2013) 4476–4499.

[50] I.M. Howat, A. Eddy, J. Glaciol. 57 (2011) 389–396.

[51] C.G. Rapley, J.L. Bamber, J.G. Morley, ESA SP-359 1 (1993) 235–240.

[52] R.M. Goldstein, H.A. Zebker, C.L. Werner, Radio Sci. 23 (1988) 713–720.

[53] I.R. Joughin, D.P. Winebrenner, M.A. Fahnestock, Geophys. Res. Lett. 22 (1995) 571–574.

[54] E. Rignot, I. velicogna, M. van den Broeke, A. Monaghan, J. Lenaerts, Geophys. Res. Lett. (2011).

[55] E.M. Enderlin, I.M. Howat, S. Jeong, M.J. Noh, J.H. van Angelen, M.R. van den Broeke, Geophys. Res. Lett. 41 (2014) 866–872.

[56] R.T.W.L. Hurkmans, J.L. Bamber, C.H. Davis, I.R. Joughin, K.S. Khvorostovsky, B.S. Smith, N. Schoen, Cryosphere 8 (2014) 1725–1740.

[57] H.D. Pritchard, R.J. Arthern, D.G. Vaughan, L.A. Edwards, Nature 461 (2009) 971–975.

[58] M. van den Broeke, J. Bamber, J. Ettema, E. Rignot, E. Schrama, W.J. van de Berg, E. van Meijgaard, I. Velicogna, B. Wouters, Science 326 (2009) 984–986.

[59] D.M. Holland, R.H. Thomas, B. De Young, M.H. Ribergaard, B. Lyberth, Nat. Geosci. 1 (2008) 659–664.

[60] F. Straneo, G.S. Hamilton, D.A. Sutherland, L.A. Stearns, F. Davidson, M.O. Hammill, G.B. Stenson, A. Rosing-Asvid, Nat. Geosci. 3 (2010) 182–186.

[61] F. Straneo, R.G. Curry, D.A. Sutherland, G.S. Hamilton, C. Cenedese, K. Vage, L.A. Stearns, Nat. Geosci. 4 (2011) 322–327.

[62] R.B. Alley, J.T. Andrews, J. Brigham-Grette, G.K.C. Clarke, K.M. Cuffey, J.J. Fitzpatrick, et al., Quat. Sci. Rev. 29 (2010) 1728–1756.

[63] E. Rignot, Geophys. Res. Lett. 35 (2008).

[64] T.C. Sutterley, I. Velicogna, E. Rignot, J. Mouginot, T. Flament, M.R. van den Broeke, J.M. van Wessem, et al., Geophys. Res. Lett. (2014), 2014GL061940.

[65] I. Joughin, B.E. Smith, B. Medley, Science 344 (2014) 735–738.

[66] I. Joughin, W. Abdalati, M. Fahnestock, Nature 432 (2004) 608–610.

[67] E. Rignot, D.G. Vaughan, M. Schmeltz, T. Dupont, D. MacAyeal, Ann. Glaciol. 34 (2002) 189–194.

[68] F. Gillet-Chaulet, O. Gagliardini, H. Seddik, M. Nodet, G. Durand, C. Ritz, T. Zwinger, R. Greve, D.G. Vaughan, Cryosphere 6 (2012) 1561–1576.

[69] C.C. Hay, E. Morrow, R.E. Kopp, J.X. Mitrovica, Nature 517 (2015) 481–484.

CHAPTER

POLEWARD EXPANSION OF THE ATMOSPHERIC CIRCULATION

6

Thomas Reichler

Department of Atmospheric Sciences, University of Utah, Salt Lake City, UT, USA

CHAPTER OUTLINE

1. INTRODUCTION

The strength, direction, and steadiness of the prevailing winds are crucial for climate. Winds associated with the atmospheric circulation lead to transports of heat and moisture from remote areas and thereby modify the local characteristics of climate in important ways. Specific names, such as extratropical westerlies, tropical trades, and equatorial doldrums remind us of the significance of winds for the climate of a region and for the human societies living in it.

The purpose of this chapter is to discuss changes in the structure of the atmospheric circulation that have taken place over the past and that are expected to take place in the future with ongoing climate change. These changes are best described as poleward displacements of major wind belts and pressure systems throughout the global three-dimensional atmosphere. The associated trends are important indicators of climate change and are likely to have profound influences on our ecosystems and societies because they control the regional expression of global climate change.

Changes in regional precipitation are amongst the most important and least well-understood consequences of climate change. Such changes are largely determined by shifts in the large-scale circulation, which controls vertical motions, cloud development, and thus precipitation [1,2]. Unfortunately, our ability to understand and predict circulation-related change is low compared to temperature-related change [3,4]. Predictions of circulation-related aspects of climate differ widely amongst models, and even simulations from the same model are strongly affected by internal variability [5]. Better understanding circulation change and its relationship to other aspects of climate change is therefore a crucial piece in the puzzle of climate sciences. In fact, the question of how the position, strength, and variability of winds change and how this affects the formation of clouds are at the heart of one of the recently identified Grand Science Challenges of the World Climate Research Programme [6].

This review is focused on important aspects of recent change in the large-scale circulation: First, tropical circulation change related to a poleward expansion and reduced intensity of the Hadley cell (HC), and second, extratropical circulation change, as manifested by a poleward shift of the storm tracks and the zone of high westerly winds in the midlatitudes, also known as an enhanced positive phase of the annular modes (AMs). Although both changes are associated with similar poleward displacements, it still remains to be seen to what extent the two phenomena are connected.

Much progress has been made over the past decade in the study of the problem, and scientific focus is shifting away from finding evidence for circulation change and determining its magnitude to attributing it to specific causes and understanding the underlying dynamical mechanisms. This review is therefore focused on the causes and mechanisms of circulation change. The reader who is more interested in detection issues is referred to other excellent reviews on the widening phenomenon [7–9]. Further, this review gives about equal room to tropical and extratropical circulation change, as research suggests that both phenomena are important and they are related to each other.

As with most aspects of climate change, the circulation changes that occurred over the past are still relatively subtle, making it difficult to distinguish them from naturally occurring variations. The problem to reliably monitor the global circulation is an additional complication. Long-term records of the atmosphere exist at few locations only, and most regions of the Earth are not observed. The problem of sparse observations can be partly overcome by utilizing meteorological reanalyses, which represent a combination of numerical weather predictions and available observations. In the present context, however, reanalyses are only of limited use, since changes in the mix of used observations over time create spurious trends in the data [10].

Because of the difficulty of observing the atmospheric circulation and its long-term trends, most studies of circulation change not only rely on observation-based evidence but also include findings from general circulation models (GCMs). GCMs are not perfect, but they are extremely valuable in situations where observations alone are not giving sufficient information. For example, they allow producing consistent time series of virtually any length, location, and quantity. GCMs can be used to perform actual experiments of the Earth's climate system in its full complexity. This makes GCMs indispensable research tools, in particular for the search for human influences on climate.

At the beginning of this review, we will develop some basic understanding for the nature of the atmospheric circulation, which is necessary information for the remainder of this chapter. Next, I will give a brief overview of observation and model-based evidence of past and future circulation change. I will also clarify the connection between tropical and extratropical circulation change. Then, I will focus on the two most pressing questions regarding circulation change: what are the causes, and what are the physical mechanisms? As I will explain, there are still many unknowns regarding these questions, but continuous progress is being made and studies begin to converge toward similar answers. I will conclude by summarizing some outstanding research questions and by explaining the significance of circulation change for other aspects of climate change.

2. THE GENERAL CIRCULATION OF THE ATMOSPHERE

The general circulation of the atmosphere describes the global three-dimensional structure of atmospheric winds. Halley [11] was probably the first to realize that the sphericity of the Earth, and the resulting spatially nonuniform distribution of solar heating, are the basic drivers behind this circulation. The tropics absorb about twice the solar energy than the higher latitudes absorb, creating a meridional gradient in temperature and potential energy. Some of the potential energy is converted into kinetic energy [12], which is manifested as wind. The winds are then deflected under the influence of the rotating Earth, creating the complicated flow patterns of the general circulation.

Atmospheric flow leads to systematic transports and conversions of energy within the Earth climate system. The different forms of energy involved are sensible heat, latent heat, potential energy, and kinetic energy. Typically, the energy transports are directed against spatial gradients, thus reducing the contrasts between geographical regions. For example, the winds transport warm air from the tropics to the extratropics and cold air in the opposite direction, decreasing the temperature contrasts between low and high latitudes. Similarly, the general circulation re-distributes water from the oceans to the continents and supplies land surfaces with life-bringing precipitation. In other words, the atmospheric circulation exerts a moderating influence on climate and reduces the extremes in weather elements. The atmospheric winds also help drive the oceans, which in turn redistribute heat from low to high latitudes, nutrients from the ocean interior to the surface, and carbon from the atmosphere to the ocean. Because of its important role in redistributing properties within the climate system, the general circulation has also been dubbed the 'great communicator' [13].

The distinction into tropical and extratropical regimes is fundamental for the Earth atmosphere. In the extratropics, large-scale motions are governed by quasi-geostrophic theory, a simple framework related to the near perfect balance between the pressure gradient force and the Coriolis force. The extratropical circulation is dominated by cyclones, which are also called storms, eddies, or simply waves. These cyclones are the product of baroclinic instability, which develops particularly strongly during winter as a consequence of the intense pole-to-equator temperature gradient during that season. The storm-track regions over the western parts of the Pacific and Atlantic oceans are the preferred locations for the development of such systems.

In the tropics, the Coriolis force is weak, and other effects such as friction, and diabatic and latent heating, become important [14]. The resulting tropical circulation is very distinct from the extra-tropics. The HC [15] is the most prominent tropical circulation feature. It extends through the entire depth of the troposphere from the equator to the subtropics (*ca.*30° latitude) over both hemispheres (Fig. 1). The cell develops in response to intense solar heating in the Inter Tropical Convergence Zone (ITCZ) near the equator. The moist tropical air warms, becomes buoyant, and rises toward the upper troposphere. The rising air cools adiabatically, leading to condensation, release of latent heat, and production of clouds and intense precipitation. In the upper troposphere, the air then diverges toward the poles and descends in the subtropics. The air is now dry and warm since it lost its moisture but retained much of the latent heat gained while rising. Consequently, the climate under the descending branch of the HC is characterized by dry conditions and relatively high pressure. The HC is closed by the trade winds at the surface, which take up moisture from the oceans before they converge into the ITCZ.

The Walker circulation [16,17] is another important tropical circulation system, representing east-west oriented overturning of air across the equatorial Pacific. It is driven by low pressure and convection in the west and high pressure and subsidence in the east. The pressure differences across the Pacific are due to warm sea surface temperatures (SSTs) over the west and rather cool SSTs over the east. Variations in these SSTs and the Walker circulation are closely related to the El Niño Southern Oscillation (ENSO) phenomenon, a naturally occurring instability of the coupled atmosphere-ocean system that has worldwide climate impacts [18].

The meridional overturning associated with the HC is also important for the extratropical circulation. For example, the poleward moving air in the upper branch of the HC tends to conserve angular momentum, spinning up a region of high zonal winds over the subsiding branch of the HC. This is the subtropical jet (Fig. 1). The jet, however, is not entirely angular momentum conserving, mainly because of the stirring action of the midlatitudes storms [19]. The stirring creates net fluxes of zonal momentum out of the jet and into the midlatitudes, which are so-called eddy stresses or divergences and convergences of eddy momentum. The consequence of these fluxes is a slowing of the subtropical jet and the creation of another wind maximum poleward of the subtropical jet. This second zone of high wind speeds is the eddy-driven or polar-front jet [20]. This jet is often merged with the subtropical jet, giving the appearance of only one tropospheric jet centred at ∼30° latitude [21]. Only over the southern hemisphere (SH) and during winter are the two jet systems fairly well separated.

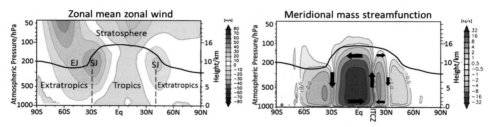

FIGURE 1

Zonal mean view of the general circulation during JJA. Vertical axis is atmospheric pressure (in hPa) and height (in km) and horizontal axis is latitude (in degrees). The continuous black line denotes the thermally defined tropopause. (Left) Zonal mean zonal winds (in m s^{-1}) derived from NCEP/NCAR reanalysis. SJ indicates the two subtropical jet cores, and EJ denotes approximate position of the eddy-driven jet. (Right) Mean meridional mass stream function (in kg s^{-1}), with arrows indicating the direction and strength of the zonal mean overturning associated with the Hadley cell, with a strong winter cell in the SH and a weak summer cell in the NH.

How does climate change impact the atmospheric circulation? Alterations of the radiative balance of the Earth due to climate change modify regional temperature and humidity structures. The winds respond to the resulting gradients and change the intensity and structure of the circulation. In the following sections, I will present evidence that such change is already happening, discuss the specific causes for it, and explain some of the underlying dynamical mechanisms.

3. EVIDENCE FOR CIRCULATION CHANGE
3.1 THE WIDENING TROPICS

It was about 10 years ago that scientists started to realize that shifts in the atmospheric circulation are amongst the many consequences of climate change. Using a variety of metrics and data, it was found that the edges of the HC have been moving poleward at both its northern and southern margins since the 1970s. The finding was significant because the edges of the HC also represent the boundaries of the tropics, regions with sharp meridional gradients in precipitation. There, even small changes in the location of the HC edge can create significant variations in precipitation, leading to either significantly drier or wetter conditions.

Rosenlof [22] was one of the first to mention potential trends in tropical width. She examined the latitudinal extent of the upwelling branch of the Brewer–Dobson circulation in the lower stratosphere, which can be regarded as one indicator for the width of the tropical belt. From reanalysis she found that the width increased by about 3° latitude per decade (1992–2001). From what we know today, this is an unrealistically large trend, which may be related to considerable observational uncertainty.

Later, Reichler and Held [23] investigated the structure of the global tropopause surface as another indicator of tropical width. This indictor is based on the well-known distinction between a high (*ca.*16 km) tropical and a low (*ca.*10 km) extratropical tropopause (Fig. 1). Using data from radio-sondes (Fig. 2) and reanalyses, it was shown that the tropics have been expanding by about 0.4° latitude per decade since 1979. Similar trends were found from measuring the separation between the cores of the northern and southern subtropical jets.

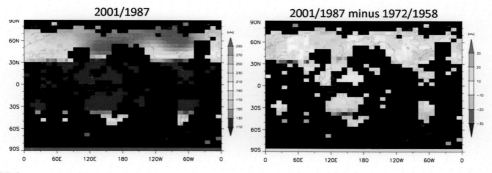

FIGURE 2

Changes in tropopause pressure during boreal winter (DJF) derived from gridded radiosonde data HADRT V2.1 [194]. (Left) Absolute tropopause pressure (in hPa) averaged over 1987–2001. (Right) Differences in tropopause pressure (in hPa) between the late period 1987–2001 and an early period 1958–1972. Bluish grid points indicate that tropopause pressure is decreasing and tropopause heights are increasing. The bluish-banded structures over southern Australia and southern Europe indicate a trend toward much lower pressure, indicative for tropical tropopause conditions, and thus a widening of the tropics. *(Adapted from Ref. [23].)*

The initial studies sparked a flurry of new research activity, aimed at detecting and understanding the new phenomenon and its underlying cause. For instance, Fu et al. [24] examined long-term data (1979–2005) from the satellite-borne microwave sounding unit (MSU) and found that the mid-tropospheric global warming signal was most pronounced in the subtropics (15°–45°). It was argued that the enhanced warming was caused by a poleward shift of the subtropical jets. Hudson et al. [25] defined the location of the tropical edges from the characteristic distribution of total ozone between the tropics and the extratropics. They examined long-term records of total ozone from the Total Ozone Mapping Spectrometer instruments and found that the area over the northern hemisphere (NH) occupied by low ozone concentrations, which is indicative for tropical regions, has increased over time. Seidel and

Randel [26] also used the tropopause criterion to distinguish between the tropics and extratropics and examined the bimodal distribution of tropopause heights in the subtropics. Applying this measure to radiosonde and reanalysis data, they again concluded that the tropics have been expanding. Additional satellite-derived indicators for tropical expansion are data for precipitation and clouds [27,28], outgoing long-wave radiation [29,30], and lower-stratospheric MSU temperatures [31].

Now, after one decade of research and after multiple independent studies using a variety of metrics were conducted, there is no longer doubt that the tropical belt has been widening and will continue to do so as a consequence of ongoing climate change. Some studies [29] suggest that the trends have a regional and seasonal structure, and that trends are strongest during summer of the respective hemisphere. Other studies [32] find that the hemispheric and seasonal differences in trends are not statistically significant.

3.2 INDICATORS OF TROPICAL WIDTH

A large variety of indicators have been used to determine the tropical edge and other circulation features. Having so many indicators is problematic because outcomes from the individual studies are difficult to compare with each other. The large number of indicators is a consequence of the lack of a unique and commonly accepted definition for the tropics. As shown in Fig. 3, atmospheric properties vary more or

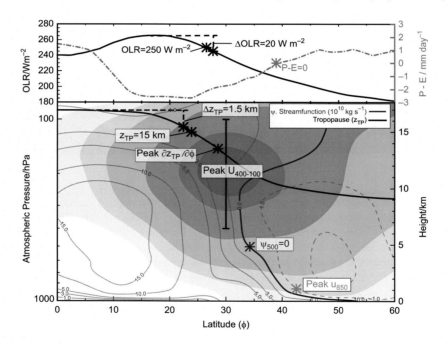

FIGURE 3

Indicators for circulation shifts. Shown are cross-sections of various atmospheric properties used to diagnose the location of the tropical edge and the eddy-driven jet. The quantities are zonal-mean zonal wind (reddish shading, 5 m s^{-1} contours), mean meridional stream function (blue contours), height of the tropopause (bold black line), difference between precipitation and evaporation (P-E, broken green line), and outgoing long-wave radiation (OLR) (top panel black line). The location of the individual diagnostics is denoted by asterisks. *(Adapted from Ref. [32].)*

less gradually between the tropics and the extratropics, and there is no easily identifiable boundary between the two zones. Some of the indicators involve arbitrary thresholds, which is problematic as changes in tropical width may interfere with other aspects of climate change [32]. For example, Birner [33] found that tropopause-derived widening trends are sensitive to the assumed tropopause height.

The indicators used so far can be roughly divided into two groups. The first group is based on dynamical definitions, focussing on specific aspects of the circulation at the outer edges of the tropics. Examples are the overturning stream function, the subtropical jet cores, or the latitude where low-level winds change from westerly to easterly. The second group is based on more physical properties of the atmosphere, such as the outgoing long-wave radiation, the concentration in stratospheric ozone, the height of the thermal or dynamical tropopause, the relative humidity, or the difference between precipitation and evaporation at the surface. Lucas et al. [9] give a detailed explanation of these and other indicators and provide an in-depth discussion of the differences and uncertainties associated with them.

3.3 THE DECREASING INTENSITY OF THE TROPICAL CIRCULATION

The intensity of tropical circulation systems may also have been changing during recent decades. Although there is no direct observational evidence for it, thermodynamic arguments suggest that global warming should weaken tropical circulation systems like the HC and the Walker cell [34,35]. This can be understood from the increasing moisture-holding capacity of warmer air, which is not followed along by an equivalent intensification of the hydrological cycle. A circulation slowdown is required to compensate for the difference.

Long-term observations of sea level pressure over the tropical Pacific indeed suggest a slowing of the Walker circulation [36,37], as suggested by the theory. Climate model simulations forced with anthropogenic factors are able to reproduce the past trend, indicating that the trend is related to warming SSTs [38]. Model simulations further suggest that mostly the Walker cell and to a much lesser degree the HC are affected by it and that the decrease will continue into the future [35,39–41]. Atmospheric reanalyses give a somewhat mixed picture of past changes, with some indicating intensification and others showing no change in the intensity of the HC [42–44].

3.4 EXTRATROPICAL CIRCULATION CHANGE

After realizing that climate change is connected to an expansion of the tropics, it was found that large-scale extratropical circulation features are also shifting poleward. For example, both the storm tracks and the eddy-driven jets have been moving toward higher latitudes [28] and are expected to continue to do so in the future [45,46]. The poleward movement of the eddy-driven jet is closely related to changes of the annular mode (AM) index [47], which is the dominant pattern of extratropical circulation variability [48]. Changes in the AM index are associated with shifts of atmospheric pressure (or mass) between the high and middle latitudes, and thus with changes in geopotential height, wind, and temperature.

Multiple lines of research indicate that the AMs underwent positive trends during recent decades [49–51]. The trends were unambiguous over the SH [24,50,52] but much less clear over the NH [53–58]. Simulations with climate models indicate that the past AM trends are caused by a warming climate associated with increased amounts of greenhouse gases and by stratospheric ozone depletion

[59–64]. Further, models indicate that under future increases of greenhouse gases the AMs tend to become even more positive [59,65,66], except over the SH during summer when the expected recovery of stratospheric ozone counteracts the trend from increasing greenhouse gas concentrations. Consistent with positive AM trends, the twenty-first-century simulations of the Coupled Model Intercomparison Project Phase 3 (CMIP3) [60], CMIP5 [66], and other models [67] show that the positions of the extratropical storm tracks and the zone of maximum surface westerlies move poleward and become more intense.

There is evidence that tropical and extratropical circulation trends are connected by similar causes and mechanisms. Kang and Polvani [68] found a significant positive correlation between interannual variations in the latitude of the eddy-driven jet and the edge of the HC—for every degree of poleward HC shift, the eddy-driven jet moves in the same direction by about twice that amount. Staten and Reichler [69] extended this analysis and confirmed that the tropical and the extratropical circulation move in certain proportions. However, the ratio between the two shifts changes by season, hemisphere, model, and even for the same model under different forcings. Kidston et al. [70] suggested that the separation distance between the HC and the eddy-driven jet is a good indicator for the ratio. In other words, the closer the two circulation systems are together, the tighter is the coupling between them.

Tropical and extratropical circulation systems are not only connected on interannual timescales. Analysis of the CMIP3 simulations shows that future upward trends in the AMs explain about half of the future expansion of the tropics [71]. Per standard deviation increase in the AM index, the HC is displaced poleward over the NH by $0.40°$ and over the SH by $0.26°$.

3.5 MAGNITUDE OF PAST AND FUTURE TRENDS

In terms of the magnitude of the past poleward shifts of the HC edge, earlier studies arrived at estimates that ranged between $0.3°$ and $3°$ latitude per decade [7]. The large range of outcomes can be explained from observational uncertainties, different data sets, different methodologies, and different start and end dates [32]. Some of the earlier estimates of tropical widening were probably too large because they would have been associated with extreme shifts in climate that were not observed.

Table 1 provides an overview of observational and model based poleward circulation shifts from some of the more recent studies. Compared to the earlier estimates, the trend numbers are now smaller and probably more realistic. Allen et al. [72] used a combination of observations and indicators to arrive at best estimates of past expansion of $0.35°$ per decade over the NH and $0.17°$ per decade over the SH, which is at the low end of the earlier studies. Another independent study by Fu and Lin [31], using satellite-derived estimates of lower-stratospheric temperatures, arrives at very similar estimates.

When climate models are driven by the observed history of forcings, the simulations also produce a widening. Most of the twentieth-century scenario integrations from the CMIP3 project reproduce the widening of the tropopause [23]. Using the mean meridional circulation as indicator for the tropical edge, the consensus estimate of all CMIP3 models is a total (NH and SH combined) widening of $0.2°$ latitude per decade over the period 1979–2005 [73]. The mean of the historical simulations from the latest CMIP5 models arrives at even smaller rates: $0.05°$ over the NH and $0.13°$ over the SH [72]. The numbers demonstrate that the model-simulated trends for the past are about 3–6 times smaller than the observed trends, in particular over the NH. This led some to speculate that perhaps model deficiencies are to blame for the discrepancies between models and observations [73]. Seidel et al. [7] raised the possibility that the poor representation stratospheric processes in the CMIP3 models [74,75]

Table 1 Annual mean expansion rates (in degrees latitude per decade) over NH and SH

Period	Metric	Data	NH	SH	Source
1979—2009	HC edge (combination of 5 methods)	Various reanalyses and observations	0.35°	0.17°	[72]
1979—2009	HC edge (stream function)	Various reanalyses	1° (NH + SH)		[32]
1979—2009	Subtropical jets	MSU lower-stratospheric temperatures	0.5°	0.5°	[31]
1979—2005	HC edge (stream function)	CMIP3 (historical simulations, all forcings)	0.2° (NH + SH)		[73]
1979—2009	HC edge	CMIP5 (historical simulations, all forcings)	0.05°	0.13°	[72]
2000—2100	Subtropical jets	CMIP3 A2	0.2° (NH + SH)		[41,73]
1993—2100	Low-level wind maximum (eddy-driven jet)	CMIP5 RCP8.5	0.09°	0.19°	[66]

may be in part responsible. However, there are also other reasonable explanations, such as the shortness of the observed data record, combined with observational uncertainty and internal climate variability. More recently, Allen et al. [72] found that when the CMIP5 models are forced with the observed history of SSTs, the simulated widening agrees much better with the observations than when the models predict their own SSTs. Allen and colleagues conclude that internal climate variability associated with the recent trend of the PDO is the most likely reason for the discrepancies. If correct, then there is no reason to believe that model-simulated circulation widening is unrealistic.

Given the relatively small past expansion rates of 0.2°–0.4° latitude per decade, one may ask how models respond to future stronger greenhouse gas increases. Kushner et al. [59] forced a fully coupled GCM with $\sim 1\%$ CO_2 increase per year and found over the SH a strong poleward shift of the westerly jet and of several related dynamical fields. Under the A2 scenario of the CMIP3 project, models reproduce robust poleward shifts of the jets [76], with an ensemble mean response of about 0.2° latitude per decade (NH and SH combined) over the period 2000–2100 [41,73]. The CMIP5 models under the RCP8.5 scenario predict very similar trends of 0.09° over the NH and 0.19° over the SH [66]. In addition, the speed of the eddy-driven jet is found to increase markedly, but only over the SH [66].

4. CAUSE FOR CIRCULATION CHANGE

Which factors are to blame for atmospheric circulation change? Numerous studies have tried to answer this question, using single-forcing experiments with models to isolate the role of the individual natural and anthropogenic factors. But finding conclusive answers turns out to be difficult, simply because climate change–related shifts of the general circulation are complex. Some of the complicating issues are the superposition of various, partly opposing effects, nonlinearities of the system, previously unknown or unrecognized factors, and unrealistic model responses to such factors.

Two types of models are commonly used in the study of circulation change. The most common are complex GCMs, which simulate as faithfully as possible most known climate processes. However, the complexity of these models limits the number and length of simulations one can afford. The second type are the relatively simple GCMs. They simulate all details of the atmospheric circulation but represent only crudely the physical processes that force the circulation. The simplicity of these models allows more and longer simulations. Therefore, simple GCMs are mostly used in theoretical studies of circulation change, in particular to study the dynamical mechanisms and the sensitivity to variations in the strength and structure of external forcings.

Studies with relatively simple GCMs use idealized and regionally localized thermal forcings to mimic certain aspects of anthropogenic climate change. For example, heating in the upper tropical troposphere, most commonly associated with the global warming response, produces an expansion of the circulation [77–79], but only if the heating is broad enough [79,80]. Wang et al. [78] find an abrupt and large poleward jump of the circulation when the warming in the tropical upper troposphere exceeds a certain threshold. Cooling in the polar stratosphere, representing the effects of ozone depletion, also drives poleward circulation shifts [77]. Surface warming over the pole, representing the effects of Arctic sea ice decline and the associated polar amplification effect, leads to an equatorward contraction of the circulation [77]. Studies with simple GCMs successfully identified some of the principal drivers for circulation change, but the simplicity of these models produces answers that are of limited practical use. In most cases complex GCMs are required to arrive at more realistic answers.

4.1 DIRECT VERSUS INDIRECT EFFECTS

From a theoretical perspective, circulation change is driven by both direct and indirect effects [38]. Direct effects are due to the radiative forcing from changing amounts of greenhouse gases, ozone, and aerosol. These forcings directly impact the atmosphere's temperature structure and this may in turn alter its circulation. However, the imbalance in the Earth's energy due to the radiative forcings also impacts global SSTs. The SSTs in turn further impact the atmosphere's temperature and circulation, leading to an indirect effect. It is still an open question whether the direct or the indirect effect is more important in creating circulation change. Various studies explicitly investigated the relative roles played by the direct and indirect effect. One study finds that SST forcing and direct atmospheric radiative forcing contribute equally to the circulation trends [38], another study finds that the widening can be attributed entirely to the direct radiative forcing [81], and the remaining studies find that the indirect SST effect is the main driver for circulation change [65,72,82]. Probable reasons for the opposing answers are differences in models and methodologies. For example, it depends on the type, structure, and strength of the forcings, whether ozone depletion is considered as a forcing or not, and on the period and hemisphere under investigation. Waugh and colleagues [83] argue that over the SH during the period 1979–1999 ozone depletion was the single most important driver for the poleward expansion of the HC, but that changes in SST dominated the circulation response after that period.

4.2 NATURAL AND ANTHROPOGENIC SEA SURFACE TEMPERATURE VARIATIONS

SSTs change due to natural and anthropogenic reasons. While warming SSTs due to anthropogenic forcings almost always result in a poleward expansion of the circulation, the oscillatory nature of natural SST modes can cause either contraction or expansion. Natural causes are linked to major modes of ocean variability, like ENSO, the Pacific Decadal Oscillation (PDO), and the Atlantic Multi-decadal Oscillation (AMO). Anthropogenic causes are related to radiative forcing from increasing concentrations of greenhouse gases and other constituents and have created a significant long-term warming trend in SSTs [84]. The observed trend over the tropical Pacific somewhat resembles the SST pattern that exists during the warm phase of ENSO, and it is also related to the climate transition from 1976/1977 and the associated upward swing of the PDO [85–87]. Aerosol emissions from human activity and from volcanic eruptions have a regional cooling effect on the oceans, in particular over the North Pacific [87] and North Atlantic [88,89].

4.3 TROPICAL SEA SURFACE TEMPERATURE VARIATIONS

Tropical SST variations, for example, those related to ENSO, impact tropospheric temperatures and circulation in important ways. Lu et al. [81] find that, when a GCM is forced by the observed history of SSTs, and when direct radiative effects from ozone and greenhouse gas changes are excluded, the tropical circulation contracts. This perhaps somewhat surprising result can be understood from the upward trending PDO during the 1980s and 1990s and the fact that a positive PDO is similar to an El Niño-like warming in the equatorial Pacific [57]. In other words, a positive PDO or a warm ENSO anomaly contracts the circulation because equatorial SSTs warm. Impacts of the AMO on the latitudinal extent of the circulation are less certain, but the tropical lobe of the AMO may have similar effects on the atmospheric circulation as the PDO. Indeed, there is modelling evidence [90] that the positive phase of the AMO leads to a negative NAO and thus to an equatorward shift of the eddy-driven jet. If this holds true, the recent trend of the AMO toward its positive phase [57] would counteract the widening of the circulation, at least over the Atlantic sector.

Allen et al. [72] investigated the role of SST variations in historical simulations of the CMIP5 project. They found that prescribing the observed evolution of SSTs in models yields NH expansion rates that are in very good agreement with the observations. However, when models predict SSTs, they tend to underestimate the recent widening trend, in particular, over the NH. This indicates that the past expansion over the NH is largely driven by natural multidecadal SST variability related to the PDO. The SST influence leads to a contraction when the PDO is in its positive (El Niño-like) phase and to an expansion when the PDO is in its negative (La Niña-like) phase. In other words, it is likely that the recent downward trend of the PDO toward its negative phase reinforces the human driven expansion. Since model-predicted SSTs and also the PDO are not in synch with the observations, this provides a good explanation for why models tend to underpredict the past NH widening. However, a somewhat similar investigation using CMIP3 models does not support this hypothesis since forcing these models with observed SSTs (the so-called AMIP simulations) does not improve model agreement with the observations [73].

4.4 EXTRATROPICAL SEA SURFACE TEMPERATURE VARIATIONS

SST variations over the extratropics also influence the atmosphere and its circulation, in particular, on longer timescales (decades and more). This is suggested not theory [91] and observations [92]. As with tropical SST anomalies, extratropical SST variations are related to both anthropogenic influences and natural modes of ocean variability (AMO and PDO). The SSTs anomalies force the atmosphere, modify its meridional temperature gradients, and influence its circulation. However, relatively little is known about the importance and effectiveness of extratropical SST anomalies in driving the circulation. Peings and Magnusdottir [90] found that in terms of the AMO, both tropical and extratropical influences are important, and that the extratropical effect is related to a reinforcement of eddy-mean flow interaction in the atmosphere by the SSTs. Other studies [45,93] also argue that interactive air-sea coupling is crucial in shaping the circulation response to climate change, in particular, over the extratropics. For example, Woolings et al. [45] report that the response of the midlatitude storm track and the eddy-driven jet over the North Atlantic crucially depends on ocean-atmosphere interaction and SST patterns.

4.5 STRUCTURE OF SEA SURFACE TEMPERATURE VARIATIONS

On longer climate change timescales, there is indication that most of the circulation change is the result of global mean warming, and that the exact pattern of warming SSTs has little impact on the response [93]. Frierson et al. [94] studied the response of the HC to global mean and pole-to-equator temperature SST gradients. They also find that the primary sensitivity of the circulation is with increase in mean temperatures. Such findings are also consistent with Deser et al. [5], who argue that the regional details of the circulation response are mostly the result of internal and thus unpredictable variability. However, Gastineau et al. [95] come to different conclusions about the importance of SST patterns in shaping the circulation response. In their model they find that the circulation response is sensitive to the zonal-mean meridional structure of SSTs, and even longitudinal SST variations play some role. Using more idealized models, several studies [79,80,96] also demonstrate that details of the zonal-mean meridional structure of SSTs are crucially important for the latitudinal change of the HC edge and eddy-driven jet. For example, narrow SST warming at the equator produces a contraction of the HC, while a broader, more global warming-like SST increase leads to an expansion.

4.6 ARCTIC TEMPERATURE CHANGE

The global warming single at the surface is particularly strong over the Arctic, a phenomenon commonly referred to as 'polar amplification'. Polar amplification is thought to be the result from positive feedbacks associated with the melt of snow and ice, as exemplified by the dramatic decline of Arctic sea ice during the past decade. In autumn and early winter the decline produces warmer near-surface temperatures, reduced static stability, and increased lower-tropospheric thicknesses [97]. The study by Butler et al. [77], using a simple GCM, demonstrated that polar warming results in an equatorward contraction of the circulation, presumably because of the decrease of baroclinicity in the atmosphere. Several other modelling studies find a negative NAO during early winter in response to sea ice decline [98–103], consistent with an equatorward movement of the eddy-driven jet. Thus, it is likely that polar amplification and Arctic sea ice decline, particularly that in the Kara Sea to the north of Europe [104], counteract to some extent the effect of most other known forcings in expanding the circulation poleward.

4.7 GREENHOUSE GAS INCREASES

Multiple studies identified the thermal forcing from changing concentrations of greenhouse gases, ozone, and aerosol as an important contributor to circulation change. The 2001 study by Kushner et al.

[59] was probably the first to recognize this. Many subsequent studies were conducted, using observations [105], specific climate models [61,65,67,81,106–112], or climate models from the CMIP3 [41,60,64,73,76,113–115] and CMIP5 [66,72,82,116,117] archives. Most studies arrive at the same two main conclusions: first, greenhouse gas–related global warming leads to an expansion of the circulation over both hemispheres; and second, historical simulations using predicted SSTs do not simulate the full extent of the widening seen in past observations [7,73]. Allen et al. [72] argue that the discrepancy between models and observations can be explained from the deficient simulation of the effects from natural SST variations and anthropogenic aerosol. There may also be issues with the prescribed ozone depletion that is too weak in some of the studies [83] and with insufficient vertical resolution in the stratosphere [118].

4.8 DEPLETION AND RECOVERY OF STRATOSPHERIC OZONE

Thompson and Solomon [50] were the first to make the connection between upward trend of the southern AM and stratospheric ozone depletion over Antarctica. The initial paper created lots of interest and was succeeded by a large number of modelling studies [62,65,74,109–111,119–123]. The studies confirmed that localized cooling in the lower polar stratosphere due to ozone depletion is an important driver for circulation shifts during late spring over the SH and that models that do not incorporate this effect fail to reproduce the observed trends of the southern AM. Later, it was found that ozone depletion not only affects the AM but also contributes to a poleward shift of the SH tropospheric midlatitude jet and to an expansion of the HC in summer. Further, it was shown that the magnitude of these circulation changes scales quasi-linearly with the stratospheric ozone changes and the associated cooling in the models [121]. Ozone has also been declining over the Arctic, but to a lesser extent than over the Antarctic [124]. The impacts of Arctic ozone depletion to NH tropospheric circulation change are not well understood.

The effects of Antarctic ozone changes for the SH circulation are mostly confined to the period of austral summer (~November–January). Past ozone trends caused a widening and future trends will reverse the previous expansion as ozone is expected to recover. Increasing greenhouse gas concentrations also cause an expansion, but in contrast to ozone, the effects from greenhouse gases are more seasonally uniform and have the same sign in the past and future. A recent observational study [105] finds that past ozone depletion contributed about 50% more than greenhouse gases toward the observed shifts in the SH jet during austral summer. This agrees well with results from several modelling studies [109,119,122], which also find that the ozone effect dominated past trends. In the future, ozone recovery and continued greenhouse gas increases will lead to opposing trends, more or less cancelling their effects on the SH circulation [109,110,119,125]. However, during seasons other than austral summer, the SH circulation is expected to expand because of the continued increase in greenhouse gas concentrations.

4.9 SOLAR VARIABILITY

Natural variability of the sun associated with the 11-year sun spot cycle has been repeatedly suggested as a driver for tropospheric circulation shifts. Observational and modelling studies find that solar maximum produces a poleward shifted jet [126] and a positive NAO [127], and that solar minimum leads to a negative AM response [128]. However, CMIP5 models do not produce significant AM responses in response to solar irradiance variations [117], which may be either related to the unrealistic stratospheric representation and solar forcings in these models or to solar influences that are too tiny to be detectable. Overall, the issue of solar influences is controversial, simply because the observational record is too short, irradiance changes associated with the solar cycle are small, and unknown amplifying mechanisms are required to produce a consistent tropospheric circulation response.

A so-called 'top-down' mechanism was suggested to explain solar forcing of the troposphere [129]. In this mechanism, enhanced UV radiation associated with the solar cycle heats the tropical strato-sphere and causes shifts in the tropospheric circulation by changing eddy momentum fluxes and reinforcing feedbacks between eddies and the circulation [60]. Another suggestion for solar influences is the so-called 'bottom-up' mechanism, in which the solar forcing warms the oceans [130] and perhaps induces an extratropical ocean-atmosphere feedback [127]. Meehl et al. [130] find that both the top-down and the bottom-up mechanisms are important in producing a consistent solar response in a model.

4.10 NATURAL AND ANTHROPOGENIC AEROSOL

Strong equatorial volcanic eruptions are observed to be followed by winters with an anomalously positive NAO [131]. It is suggested that this is related to the localized heating of the tropical strato-sphere due to the radiative effects from the volcanic aerosol, strengthening the equator-to-pole tem-perature gradient. However, the CMIP3 [132] and CMIP5 [117,118,133] models do not simulate significant and consistent AM changes in response to volcanic eruptions. This indicates that either the models' responses or the observational basis for volcanic effects are flawed.

A relatively small number of studies examined the effects of anthropogenic aerosol on the cir-culation. Arblaster and Meehl [61] found that past aerosol emissions from natural and also from anthropogenic sources led to a small negative trend of the southern AM. However, Cai and Cowan [134] found the opposite, that ocean cooling from past anthropogenic aerosol emissions shifted the AM over the SH toward a more positive state. Over the NH, the CMIP5 models suggest a small widening trend when forced with the observed history of aerosol [72]. This impact is assumed to be the consequence of an indirect cooling effect of the aerosol on the PDO.

Aerosol may also directly impact the circulation by cooling or warming the atmosphere. Chen et al. [135] found this effect to be relatively small. Later, Allen and Sherwood [136] found in a model a considerable poleward jet shift from modest heating due to prescribed black carbon and sulphate aerosol. They explained the shift from the heterogeneous nature of the aerosol, which impacts only certain regions, and which creates meridional gradients in warming. The differential nature of anthropogenic aerosol over the NH can be somewhat compared to Antarctic ozone depletion, which represents the main heterogeneous forcing agent over the SH. In a follow-up study, Allen and col-leagues [137] found that forcing from black carbon and tropospheric ozone is particularly important for the simulation of the observed widening trend, providing a nice explanation for why past model simulations were unable to replicate the magnitude of the observed NH widening.

4.11 LINEARITY OF THE RESPONSE

An interesting question is how additive the effects of individual forcings for the overall circulation response are in the presence of many nonlinear processes in the climate system. In other words, how meaningful are single forcing experiments with models, and does the sum of individual forcings lead to the same circulation response as that seen when using the combined forcings? There is unanimous agreement amongst studies with complex GCMs [65,38,109] that the response of the circulation to individual forcings is linearly additive, and that to a very good degree the sum of circulation responses to individual forcings equals the response when all forcings are applied simultaneously. Butler and colleagues [77], however, find in their simple GCM considerable nonlinearities in the responses to various thermal forcings. The reason for this discrepancy may be related to the more idealized character of the latter study.

5. EMERGING DYNAMICAL MECHANISMS

Understanding the dynamical mechanisms by which anthropogenic and natural factors modify the circulation is an area of active research. Climate change is related to anomalous heating of atmosphere and the surface. The regionally uneven distribution of the heating creates spatial temperature gradients, alters the baroclinicity, modifies the mean wind, and changes the structure and propagation of atmospheric waves. However, the importance of the individual events, the sequence in which and the region where they take place, how they are initiated, and how exactly they lead to an expansion of the circulation have yet to be fully unravelled. Past research into these issues led to a range of explanations, indicating that the problem is complex and that perhaps no single answer can provide a full explanation. It further indicates the need for additional research, since understanding the mechanisms will perhaps help reduce the prediction uncertainties of models.

5.1 RELATIONSHIP BETWEEN CIRCULATION, CLOUDS, AND RADIATION

Circulation change is closely linked to change in cloudiness and precipitation [41,123]. Lu et al. [41] demonstrated that the expansion of the HC is correlated with the poleward expansion of the subtropical dry zone. Changes in cloudiness lead to localized anomalies of latent and radiative heating, which in turn change baroclinicity and create feedbacks on the circulation. This raises the question whether and how changes in clouds, radiation, heating, and circulation are connected to each other. Several recent studies started to investigate these and similar issues. All studies highlight the problem of model uncertainties related to change in cloud properties in response to climate change. Grise and Polvani [138] show that cloud changes seen in some of the CMIP5 models in response to jet shifts are inconsistent with the observations, and they argue that this is relevant for future SH circulation change. Ceppi et al. [116] found in the CMIP5 models a tight relationship between changes in the meridional gradient of absorbed shortwave radiation and shifts in the SH jet. The authors argue that intermodel differences in jet shifts are related to model-specific change of cloud properties, which impact shortwave radiation and baroclinicity. Voigt and Shaw [139] also find that different regional circulation responses in a model are related to the way cloud radiative properties respond to global warming.

5.2 STRATOSPHERIC LINKAGES

Many studies have suggested links between climate change in the stratosphere and tropospheric circulation change over the SH. Observed increases in greenhouse gases and ozone-depleting substances over the past led to substantial cooling of the stratosphere, especially over the higher latitudes [140].

The resulting changes in the zonal winds and the subsequent dynamical effects on the troposphere [141,142] influence the tropospheric circulation [75]. Despite intensive research over the past decade, the exact mechanisms behind this influence are not well understood [143,144]. Possibilities include the balanced tropospheric response to stratospheric climate change [145], change on tropospheric eddies exerted by the stratosphere [146], or influences of lower stratospheric temperature perturbations on tropospheric baroclinic instability [147]. The lower stratospheric cooling from changing levels of ozone and greenhouse gases also increases the height of the extratropical tropopause, which may be linked to tropospheric circulation change.

Polar amplification and Arctic sea ice decline have been repeatedly associated with a tendency toward a negative NAO and equatorward circulation shifts [148]. The stratosphere may be involved in the response of the circulation to polar warming. This may be related to changes in the amplitude and position of planetary waves by the surface heating, enhancing the upward planetary wave flux and weakening the stratospheric polar vortex [149–153]. The dynamical coupling between the stratosphere and troposphere could explain the subsequent tendency toward a negative NAO.

5.3 TROPOPAUSE HEIGHTS

Analysis of radiosonde [154] and reanalysis data [155,156] shows that the height of the global tropopause has increased over the past decades, and GCM experiments indicate that anthropogenic climate change is likely responsible for this increase [106,157]. The increase is related to systematic temperature changes below and above the tropopause [158]. Temperatures have been warming in the troposphere and cooling in the stratosphere, both of which have been shown to be related to anthropogenic activity [159–161]. The pattern of warming and cooling also affects the zonal wind structure in the region of the subtropical upper troposphere and lower stratosphere (UTLS). This is related to the height structure of the tropopause. In the tropics, the tropopause is high and global warming reaches up to ∼16 km. In the extratropics, the tropopause is low and warming reaches only up to ∼12 km, followed by cooling in the stratosphere above. Thus, at intermediate heights of the UTLS region (∼12–16 km) the tropics warm and the extratropics cool, leading to an increase in meridional temperature gradients, and, by the thermal wind relationship, to an increase of zonal wind speeds above.

Various studies related the lifting of the tropopause to the poleward expansion of the circulation [41,76,162]. According to a theory from Held and Hou [163], the meridional extent of the HC varies proportionally with the square root of its vertical depth. However, applying this scaling to the past observed tropical tropopause height increase of about 200 m [154,157] leads to a tropical expansion of only 0.1° latitude per decade, which is much less than what is suggested by the observations and by most models. Analysis of idealized [164,165] and complex models [41,94] also demonstrates that the Held and Hou theory does not provide a full explanation for the parameter dependence of the meridional extent of the HC, indicating that additional mechanisms are at work.

Some model studies suggest that lifting of extratropical tropopause heights is connected to poleward shifts of the jet and the tropical edges [76,126,166]. However, some caution is required. Studies lift the tropopause by imposing external thermal forcings to their models. This not only affects the height of the tropopause but other aspects of the atmosphere as well, like the meridional temperature gradient, the zonal wind, and the vertical wind shear. The additional changes make it difficult to unequivocally assign the cause for the tropical widening to the lifting of the tropopause. In addition, lifting of the extratropical tropopause in itself still does not provide a satisfying dynamical interpretation for the poleward expansion. A study by Wu et al. [167] finds little evidence for the rise in tropopause leading to poleward jet shifts, arguing that an increase in the spatial scale of Rossby waves is responsible for the tropopause rise and poleward shifting jets.

5.4 STATIC STABILITY

Changes in the vertical temperature structure of the atmosphere provide another plausible explanation for the tropical widening. Vertical temperature changes are primarily related to the vertical nonuniformity of the tropospheric global warming signal. Observations as well as model experiments indicate that global warming in the upper troposphere is more pronounced than in the lower troposphere and that the signal maximizes in the tropical upper troposphere [168,169]. This so-called upper tropospheric amplification is a well-established consequence of the quasi-moist adiabatic adjustment of the atmosphere, which increases the static stability in both tropics [170,171] and extratropics [172–174].

The theory by Held [19] establishes a connection between static stability and tropical width. The theory assumes that the upper, poleward moving branch of the HC is angular momentum conserving. The poleward moving air increases its zonal wind speed until it becomes unstable and breaks down under the growing vertical wind shear. This marks the latitude of the outer boundary of the HC. Global warming–related increases in static stability postpone the point where the atmosphere becomes baroclinically unstable. As a consequence, the HC expands toward higher latitudes.

The original theory [19] was later refined, arguing that the poleward movement of the HC is intimately tied to the eddy-driven jet [175]. Global warming–related reductions of baroclinicity at the equatorward flank of the eddy-driven jet stabilize eddy growth and move the jet and the associated subsidence toward the poles. The HC follows along since in the subtropics both HC and eddy-driven jet are associated with subsidence.

Independent of which interpretation is best, studies with both idealized [94,164] and full models [41,162,175] confirm that the Held [19] theory holds reasonably well. For example, in idealized parameter sweep experiments, which were forced with prescribed SSTs, Frierson et al. [94] find that the global mean warming is the primary reason for the expansion of the HC and that increases in meridional temperature gradients play only a secondary role. It is also noteworthy that the global warming–related increase in static stability is expected to be particularly strong during summer and over the SH [173], which is consistent with the regional and seasonal patterns of the observed tropical widening [23,29].

5.5 TROPICAL PUSH

As the HC edge represents the tropical-extratropical boundary, one may ask whether there are factors that push the HC edge outward from the tropics and factors that pull it in from the extratropics. Indeed, past research finds that both factors play a role in shaping circulation change.

A good indication for tropical push factors is the observation that heating of the deep tropics is an effective driver for the extent of the global atmospheric circulation [176,177]. Concentrated heating around the equator leads to meridional contraction, including equatorward displacements of the jet, storm track, eddy momentum divergence, and edge of the HC [47,38,175,178–180]. Heating in the tropics can have several causes. Global warming projects on tropical SSTs [84] and leads to pronounced warming in the upper tropical troposphere. There are also natural reasons for warming of tropical SSTs, such as the warm phases of ENSO (El Niño), PDO, and AMO. Modelling studies [69,162] suggest that the warming tropics explain roughly one-third of the total climate change–related HC expansion.

Tropical SST anomalies can influence the extratropical circulation through the generation of Rossby waves [18]. However, this does not explain meridional circulation shifts. The contraction of the HC under warm equatorial SSTs can be better understood from the fact that the circulation of the HC tends to intensify in response to the warming SSTs [181]. The stronger HC leads to a westerly acceleration in its upper, poleward moving branch and thus to a strengthening of the subtropical jet. As discussed by various authors [164,175,178,179,182], the strengthening zonal winds allow extratropical Rossby waves, which tend to propagate toward the equator, to penetrate deeper into the tropics before they dissipate. In more technical terms, increasing zonal winds move the critical latitude for equatorward extratropical wave propagation in the subtropics toward the equator. Since the zone of dissipation marks the edge of the HC, the HC shifts equatorward as equatorial SSTs and zonal winds intensify. During neutral and cold ENSO conditions the opposite relationship holds.

An alternative explanation for the contraction of the tropics during El Niño is that the increased equatorial heating increases the pole-to-equator temperature gradient and draws the zone of maximum baroclinicity toward the equator. Consequently, the eddy-driven part of the circulation is shifted toward lower latitudes. There is also evidence that the increase in surface baroclinicity in association with El Niño impacts the type and number of wave breaking, which in turn may change the structure and position of the jet [180,183]. Rivière [147] used similar arguments to explain the global warming related expansion as a response to enhanced upper-tropospheric baroclinicity.

5.6 EXTRATROPICAL PULL

Past research demonstrated that extratropical processes are crucial for the circulation widening [123,175,184]. This can be demonstrated by the aforementioned close connection between shifts of the AMs, the eddy-driven jet, and the HC edge [71]. This influence is related to the action of dissipating extratropical Rossby waves and how extratropical climate change alters the propagation of the waves. There is also evidence that not only the width but also the strength of the HC is controlled by extratropical waves [182,185,186].

Rossby waves are large-scale atmospheric waves that transport momentum and other properties through the atmosphere. The propagation of Rossby waves follows laws similar to that of sound or electromagnetic waves. For example, Rossby wave propagation depends on the size and phase speed of the waves and on the state of the atmosphere in which the waves travel. This is somewhat analogous to the wavelength-dependent refraction of light passing through a prism. Rossby waves can also be reflected, very much like light hitting a mirror.

Interaction of Rossby waves with the background mean flow extracts momentum at one location and deposits it at another. More specifically, momentum is deposited where the waves are created and from which they propagate, forming the so-called eddy-driven jet, and momentum is extracted from locations where the waves dissipate, which is usually the subtropical jet. Anthropogenic climate change and other natural factors can alter the characteristics of the waves and of the mean flow. This alters how the waves propagate and transport momentum and leads to shifts in the latitudinal position of the jet and HC. This is the basic idea behind most expansion mechanisms proposed that involve extratropical processes. The different mechanisms differ only in the details of how climate change alters the eddies and their propagation.

Chen and Held [187] were the first to suggest a connection between climate change and the behaviour of extratropical Rossby waves. Rossby waves tend to move equatorward toward the subtropics until they approach critical latitudes, where their phase speed equals the speed of the background zonal flow. There, the waves grow in amplitude, break irreversibly, and decelerate the flow as a result of the absorbed wave activity. A key to understanding the expansion mechanism proposed by [187] is that the zonal wind in the UTLS region determines the eastward phase speed of extratropical tropospheric waves. Increased greenhouse gases and/or stratospheric ozone depletion lead to upper tropospheric warming and lower stratospheric cooling across the tropopause slope, which increases the UTLS winds and the phase speed of the waves. According to critical layer theory, the now faster waves cannot penetrate as far equatorward into the regions of decreasing zonal winds. This in turn confines the zone of the eddy-driven jet more poleward and leads to a more positive AM. At the same time, the eddy-driven subsidence in the subtropics is also shifting poleward, which helps to explain the correlation between AM and tropical width. Critical elements of the mechanism were identified by [187] in model simulations and in observations. Later, this mechanism was extended by arguing that the poleward shift of the eddy-driven subsidence in the subtropics not only affects the AM but also the edge of the HC [175].

Another extratropical mechanism is based on the model and observation-derived evidence [188] that Rossby waves tend to grow in size in a warmer climate. Kidston et al. [189] argue that the increase in length scale causes a decrease of the waves' phase speed relative to the mean flow. This allows the waves to travel further away from the jet before they reach their critical latitude and dissipate. Kidston et al. hypothesize that this leads to acceleration on the poleward flank of the jet and thus to a poleward movement of the jet.

Lorenz [190] focuses on the possibility that climate change alters the conditions for Rossby wave reflection on the poleward side of the jet. Climate change–related increases in wind speed create more favourable conditions for wave reflection, such that more poleward propagating waves are reflected back toward the equator. The result is that more waves dissipate on the equatorward side of the jet. Since wave dissipation is equivalent with momentum removal, this leads to a poleward-moving jet.

Both Wu et al. [167] and Staten et al. [191] used a large number of switch-on forcing experiments with climate models to prove or disprove some of the proposed mechanisms. Staten et al. found some evidence for the Chen and Held eddy phase speed mechanism but concluded that it more likely plays the role of a feedback and not that of an initiator. Using a different model, Wu et al. were unable to find evidence for the Chen and Held mechanism. They argued that changes in eddy phase speed are model dependent, citing results from other studies [147,192]. Staten et al. found some support for Kidston's length scale mechanism but concluded that it is probably the least important one. Likewise, experiments by Wu et al. could not validate the eddy length scale mechanism as initiator for poleward jet shifts. Staten et al. found solid evidence for the Lorenz mechanism and concluded that Rossby wave reflection provides the most plausible explanation for the poleward expansion of the circulation.

---

6. SUMMARY, OUTSTANDING PROBLEMS, AND CONCLUSIONS

After one decade of intense research there is now overwhelming evidence that key elements of the atmospheric circulation have been moving poleward. Detection studies found that the past trends have been about 0.2° degree latitude per decade and per hemisphere. Climate model simulated future expansion rates are somewhat smaller as such estimates do not include the effects from natural climate variability. Based on climate models, by the end of the twenty-first century, the width of the tropical belt is expected to expand by about 200 km–300 km. Experiments with climate models also indicate that long-term circulation trends can be attributed to anthropogenic activity, in particular, to changes in greenhouse gases, stratospheric ozone, and aerosol. On shorter, decade-long timescales, natural climate variability associated with major modes of ocean variability is also important.

There has also been a marked improvement in our theoretical understanding for circulation change. For example, there is strong indication that circulation change is largely driven by extratropical processes in association with changing Rossby wave behaviour. Several plausible dynamical mechanisms have been proposed so far, but still it is not entirely clear which, if any, is correct. Since there are different causes for circulation change, it is conceivable that not one single mechanism can explain the entire changes. The still existing lack of understanding for the dynamical mechanism indicates the need for additional research. This includes the role of change in stratospheric ozone and the dynamical coupling between the stratosphere and the troposphere.

What are the consequences of circulation change for climate? So far, the changes in position and intensity have been modest in magnitude. But even small shifts in the location of the HC, jets, and storm tracks can have important implications for regional climates by modifying patterns of storminess, temperature, and precipitation [67,193]. Particularly sensitive are regions with large spatial gradients in their normal distribution of precipitation, like the subtropical dry zones. There, even small trends decide whether there is a surplus or a deficit in overall rain. For example, the expansion of the HC may cause drier conditions over the subtropical semiarid regions, including the Mediterranean, the southwestern United States, southern Australia, and southern Africa [7], and it was speculated that this process is already under way [193]. Atmospheric circulation change may also alter ocean currents. Because oceans are important regulators of climate, this may induce complicated and unexpected feedbacks, which either amplify or diminish the original cause.

Studies agree that circulation change is closely linked to regional aspects of climate change, in particular to those related to precipitation. Improved regional climate information would have tremendous societal value, but such information is still severely lacking from most circulation studies. In part, this is related to the little confidence one has in the regional details of model simulated circulation change. Past studies of circulation change were almost exclusively focused on zonal means, since at this level there is acceptable agreement amongst the different models. However, as our understanding for cause and effect of circulation change grows and as model predictions converge, it is to hope that regional predictions will improve as well. Several forcings known to impact the circulation have a spatially heterogeneous character, which may create local circulation changes that have not been studied yet. It is also conceivable that some forcings, like a cooling PDO and a warming AMO, have opposing but regionally limited circulation impacts, which may lead to complex regional circulation responses. These issues need to be addressed in future research.

ACKNOWLEDGEMENTS

I gratefully acknowledge financial support from the National Science Foundation under grant ATM 1446292. I also thank the German Aerospace Centre (DLR), where I wrote part of this review, and the University of Utah for their support of my sabbatical leave.

LIST OF ABBREVIATIONS

AM	Annular mode
AMO	Atlantic Multidecadal Oscillation
CMIP3	Coupled Model Intercomparison Project Phase 3
CMIP5	Coupled Model Intercomparison Project Phase 5
ENSO	El Niño Southern Oscillation
GCM	General circulation model
HC	Hadley cell
ITCZ	Inter tropical convergence zone
NAO	North Atlantic Oscillation
NH	Northern hemisphere
PDO	Pacific decadal oscillation
SH	Southern hemisphere
SSTs	Sea surface temperatures
UTLS	Upper troposphere/lower stratosphere

REFERENCES

[1] J. Scheff, D. Frierson, J. Clim. 25 (2012) 4330.

[2] N.P. Gillett, T.D. Kell, P.D. Jones, Geophys. Res. Lett. 33 (2006).

[3] K. Marvel, C. Bonfils, Proc. Natl. Acad. Sci. U.S.A. 110 (2013) 19301.

[4] T.G. Shepherd, Nat. Geosci. 7 (2014) 703.

[5] C. Deser, A. Phillips, V. Bourdette, H. Teng, Clim. Dynam. 38 (2012) 527.

[6] WCRP Joint Scientific Committee, vol. 2015, 2015.

[7] D.J. Seidel, Q. Fu, W.J. Randel, T. Reichler, Nat. Geosci. 1 (2008) 21.

[8] K. Rosenlof, L. Terray, C. Deser, A. Clement, H. Goosse, S. Davis, in: G.R. Asrar, J.W. Hurrell (Eds.), Climate Science for Serving Society, Springer, Netherlands, 2013.

[9] C. Lucas, B. Timbal, H. Nguyen, Wiley Interdiscip. Rev. Clim. Change 5 (2014) 89.

[10] K.E. Trenberth, D.P. Stepaniak, J.W. Hurrell, M. Fiorino, J. Clim. 14 (2001) 1499.

[11] E. Halley, Philos. Trans. 16 (1686) 153.

[12] E.N. Lorenz, Tellus 7 (1955) 157.

[13] D.L. Hartmann, Global Physical Climatology, vol. 56, International Geophysics Series Academic Press, San Diego, 1994.

[14] T. Reichler, J.O. Roads, J. Clim. 18 (2005) 619.

[15] G. Hadley, Philos. Trans. 39 (1735) 58.

[16] G.T. Walker, Mem. Indian Meteor. Dept. 24 (1924) 275.

[17] P.R. Julian, R.M. Chervin, Mon. Wea. Rev. 106 (1978) 1433.

[18] K.E. Trenberth, G.W. Branstator, D. Karoly, A. Kumar, N.C. Lau, C. Ropelewski, J. Geophys. Res. Oceans 103 (1998) 14291.

[19] I.M. Held (Ed.), Woods Hole Geophysical Fluid Dynamics Program, Woods Hole Oceanographic Institute, Woods Hole, MA, 2000, pp. 54.

[20] G.K. Vallis, Atmospheric and Oceanic Fluid Dynamics: Fundamentals and Large-Scale Circulation, Cambridge University Press, Cambridge, 2006.

[21] S. Lee, H.-K. Kim, J. Atmos. Sci. 60 (2003) 1490.

[22] K.H. Rosenlof, J. Meteorol. Soc. Jpn. 80 (2002) 831.
[23] T. Reichler, I. Held (Eds.), Paper Presented at the AMS Conference on Climate Variability and Change, Cambridge, MA, 2005.
[24] Q. Fu, C. Johanson, J. Wallace, T. Reichler, Science 312 (2006) 1179.
[25] R.D. Hudson, M.F. Andrade, M.B. Follette, A.D. Frolov, Atmos. Chem. Phys. 6 (2006) 5183.
[26] D.J. Seidel, W.J. Randel, J. Geophys. Res. 112 (2007), http://dx.doi.org/10.1029/2007JD008861.
[27] Y.P. Zhou, K.-M. Xu, Y.C. Sud, A.K. Betts, J. Geophys. Res. Atmos. 116 (2011) D09101.
[28] F.-M. Bender, V. Ramanathan, G. Tselioudis, Clim. Dynam. 38 (2012) 2037.
[29] Y. Hu, Q. Fu, Atmos. Chem. Phys. 7 (2007) 5229.
[30] Y. Hu, C. Zhou, J. Liu, Adv. Atmos. Sci. 28 (2011) 33.
[31] Q. Fu, P. Lin, J. Clim. 24 (2011) 5597.
[32] S.M. Davis, K.H. Rosenlof, J. Clim. 25 (2012) 1061.
[33] T. Birner, J. Geophys. Res. Atmos. 115 (2010) D23109.
[34] T.R. Knutson, S. Manabe, J. Clim. 8 (1995) 2181.
[35] I.M. Held, B.J. Soden, J. Clim. 19 (2006) 5686.
[36] G.A. Vecchi, B.J. Soden, A.T. Wittenberg, I.M. Held, A. Leetmaa, M.J. Harrison, Nature 441 (2006) 73.
[37] M. Zhang, H. Song, Geophys. Res. Lett. 33 (2006), http://dx.doi.org/10.1029/2006GL025942.
[38] C. Deser, A.S. Phillips, J. Climate, 22 (2009) 396.
[39] G.A. Vecchi, B.J. Soden, J. Clim. 20 (2007) 4316.
[40] H.L. Tanaka, N. Ishizaki, A. Kitoh, Tellus A 56 (2004) 250.
[41] J. Lu, G.A. Vecchi, T. Reichler, Geophys. Res. Lett. 34 (2007), http://dx.doi.org/10.1029/2006GL028443.
[42] C. Mitas, A. Clement, Geophys. Res. Lett. 32 (2005), http://dx.doi.org/10.1029/2004GL021765.
[43] C.M. Mitas, A. Clement, Geophys. Res. Lett. 33 (2006), http://dx.doi.org/10.1029/2005GL024406.
[44] C. Kobayashi, S. Maeda, Geophys. Res. Lett. 33 (2006) L22703.
[45] T. Woollings, J.M. Gregory, J.G. Pinto, M. Reyers, D.J. Brayshaw, Nat. Geosci. 5 (2012) 313.
[46] T. Woollings, M. Blackburn, J. Clim. 25 (2011) 886.
[47] W.A. Robinson, in: T. Schneider, A. Sobel (Eds.), The Global Circulation of the Atmosphere, Princeton University Press, Pasadena, 2007.
[48] D.W.J. Thompson, J.M. Wallace, J. Clim. 13 (2000) 1000.
[49] D.W.J. Thompson, J.M. Wallace, G.C. Hegerl, J. Clim. 13 (2000) 1018.
[50] D.W.J. Thompson, S. Solomon, Science 296 (2002) 895.
[51] J.M. Jones, R.L. Fogt, M. Widmann, G.J. Marshall, P.D. Jones, M. Visbeck, J. Clim. 22 (2009) 5319.
[52] G. Marshall, J. Clim. 16 (2003) 4134.
[53] J. Marshall, Y. Kushnir, D. Battisti, P. Chang, A. Czaja, R. Dickson, J. Hurrell, M. McCartney, R. Saravanan, M. Visbeck, Int. J. Climatol. 21 (2001) 1863.
[54] J. Cohen, M. Barlow, J. Clim. 18 (2005) 4498.
[55] J.E. Overland, M. Wang, Geophys. Res. Lett. 32 (2005), http://dx.doi.org/10.1029/2004GL021752.
[56] J.W. Hurrell, Y. Kushnir, M. Visbeck, Science 291 (2001) 603.
[57] K.E. Trenberth, J.T. Fasullo, Earth's Future 1 (2013) 19.
[58] T. Reichler, J. Kim, E. Manzini, J. Kroger, Nat. Geosci. 5 (2012) 783.
[59] P.J. Kushner, I.M. Held, T.L. Delworth, J. Clim. 14 (2001) 2238.
[60] J.H. Yin, Geophys. Res. Lett. 32 (2005) L18701.
[61] J.M. Arblaster, G.A. Meehl, J. Clim. 19 (2006) 2896.
[62] N.P. Gillett, D.W.J. Thompson, Science 302 (2003) 273.
[63] N.P. Gillett, M.R. Allen, K.D. Williams, Q. J. R. Meteorol. Soc. 129 (2003) 947.
[64] W. Cai, T. Cowan, J. Clim. 20 (2007) 681.
[65] P.W. Staten, J.J. Rutz, T. Reichler, J. Lu, Clim. Dynam. 39 (2012) 2361.
[66] E.A. Barnes, L. Polvani, J. Clim. 26 (2013) 7117.

[67] L. Bengtsson, K.I. Hodges, E. Roeckner, J. Clim. 19 (2006) 3518.

[68] S.M. Kang, L.M. Polvani, J. Clim. 24 (2010) 563.

[69] P.W. Staten, T. Reichler, Clim. Dynam. 42 (2014) 1229.

[70] J. Kidston, C.W. Cairns, P. Paga, Geophys. Res. Lett. 40 (2013) 2328.

[71] M. Previdi, B.G. Liepert, Geophys. Res. Lett. 34 (2007), http://dx.doi.org/10.1029/2007GL031243.

[72] R.J. Allen, J.R. Norris, M. Kovilakam, Nat. Geosci. 7 (2014) 270.

[73] C.M. Johanson, Q. Fu, J. Clim. 22 (2009) 2713.

[74] S.-W. Son, L.M. Polvani, D.W. Waugh, H. Akiyoshi, R. Garcia, D. Kinnison, S. Pawson, E. Rozanov, T.G. Shepherd, K. Shibata, Science 320 (2008) 1486.

[75] M.P. Baldwin, M. Dameris, T.G. Shepherd, Science 316 (2007) 1576.

[76] D.J. Lorenz, E.T. DeWeaver, J. Geophys. Res. 112 (2007) D10119.

[77] A.H. Butler, D.W.J. Thompson, R. Heikes, J. Clim. 23 (2010) 3474.

[78] S. Wang, E.P. Gerber, L.M. Polvani, J. Clim. 25 (2012) 4097.

[79] N.F. Tandon, E.P. Gerber, A.H. Sobel, L.M. Polvani, J. Clim. 26 (2012) 4304.

[80] G. Chen, R.A. Plumb, J. Lu, Geophys. Res. Lett. 37 (2010) L12701.

[81] J. Lu, C. Deser, T. Reichler, Geophys. Res. Lett. 36 (2009) L03803.

[82] K.M. Grise, L.M. Polvani, Geophys. Res. Lett. 41 (2014), 2014GL061638.

[83] D.W. Waugh, C.I. Garfinkel, L.M. Polvani, J. Clim., in press.

[84] T.P. Barnett, D.W. Pierce, K.M. AchutaRao, P.J. Gleckler, B.D. Santer, J.M. Gregory, W.M. Washington, Science 309 (2005) 284.

[85] C. Deser, A.S. Phillips, J.W. Hurrell, J. Clim. 17 (2004) 3109.

[86] Y. Zhang, J.M. Wallace, D.S. Battisti, J. Clim. 10 (1997) 1004.

[87] J. Lu, Nat. Geosci. 7 (2014) 250.

[88] M.E. Mann, K.A. Emanuel, Eos, Trans. Am. Geophys. Union 87 (2006) 233.

[89] G. Villarini, G.A. Vecchi, J. Clim. 26 (2012) 3231.

[90] Y. Peings, G. Magnusdottir, J. Clim., in press.

[91] J. Bjerknes, in: H.E. Landsberg, J.V. Mieghem (Eds.), Advances in Geophysics, vol. 10, Academic Press, New York, 1964.

[92] S.K. Gulev, M. Latif, N. Keenlyside, W. Park, K.P. Koltermann, Nature 499 (2013) 464.

[93] J. He, B.J. Soden, B. Kirtman, Geophys. Res. Lett. 41 (2014), 2014GL059435.

[94] D. Frierson, J. Lu, G. Chen, Geophys. Res. Lett. 34 (2007), http://dx.doi.org/10.1029/2007GL031115.

[95] G. Gastineau, L. Li, H. Le Treut, J. Clim. 22 (2009) 3993.

[96] C. Michel, G. Rivière, J. Atmos. Sci. 71 (2013) 349.

[97] J.A. Screen, I. Simmonds, Nature 464 (2010) 1334.

[98] G. Magnusdottir, C. Deser, R. Saravanan, J. Clim. 17 (2004) 857.

[99] M.A. Alexander, U.S. Bhatt, J.E. Walsh, M.S. Timlin, J.S. Miller, J.D. Scott, J. Clim. 17 (2004) 890.

[100] I.A. Seierstad, J. Bader, Clim. Dynam. 33 (2009) 937.

[101] C. Deser, R. Tomas, M. Alexander, D. Lawrence, J. Clim. 23 (2010) 333.

[102] J.A. Screen, I. Simmonds, C. Deser, R. Tomas, J. Clim. 26 (2013) 1230.

[103] Y.J. Orsolini, R. Senan, R.E. Benestad, A. Melsom, Clim. Dynam. 38 (2012) 2437.

[104] S. Yang, J.H. Christensen, Geophys. Res. Lett. 39 (2012) L20707.

[105] S. Lee, S.B. Feldstein, Science 339 (2013) 563.

[106] S.M. Kang, C. Deser, L.M. Polvani, J. Clim. 26 (2013) 7541.

[107] W.-T. Chen, H. Liao, J.H. Seinfeld, J. Geophys. Res. Atmos. 112 (2007) D14209.

[108] A.Y. Karpechko, N.P. Gillett, L.J. Gray, M. Dall'Amico, J. Geophys. Res. Atmos. 115 (2010) D22117.

[109] C. McLandress, T.G. Shepherd, J.F. Scinocca, D.A. Plummer, M. Sigmond, A.I. Jonsson, M.C. Reader, J. Clim. 24 (2011) 1850.

[110] L.M. Polvani, M. Previdi, C. Deser, Geophys. Res. Lett. 38 (2011) L04707.

[111] D.T. Shindell, G.A. Schmidt, Geophys. Res. Lett. 31 (2004).

[112] X.-W. Quan, M.P. Hoerling, J. Perlwitz, H.F. Diaz, T. Xu, J. Clim. 27 (2013) 1999.

[113] J.M. Arblaster, G.A. Meehl, D.J. Karoly, Geophys. Res. Lett. 38 (2011) L02701.

[114] R.L. Miller, G.A. Schmidt, D.T. Shindell, J. Geophys. Res. 111 (2006), http://dx.doi.org/10.1029/2005JD006323.

[115] E.P. Salathé, Geophys. Res. Lett. 33 (2006) L19820.

[116] P. Ceppi, M.D. Zelinka, D.L. Hartmann, Geophys. Res. Lett. 41 (2014) 3244.

[117] N.P. Gillett, J.C. Fyfe, Geophys. Res. Lett. 40 (2013) 1189.

[118] A.J. Charlton-Perez, M.P. Baldwin, T. Birner, R.X. Black, A.H. Butler, N. Calvo, et al., J. Geophys. Res. Atmos. 118 (2013) 2494.

[119] J. Perlwitz, S. Pawson, R.L. Fogt, J.E. Nielsen, W.D. Neff, Geophys. Res. Lett. 35 (2008), http://dx.doi.org/10.1029/2008GL033317.

[120] S.-W. Son, N.F. Tandon, L.M. Polvani, D.W. Waugh, Ozone hole and Southern Hemisphere climate change, Geophys. Res. Lett. 36 (2009) L15705, http://dx.doi.org/10.1029/2009GL038671.

[121] S.W. Son, E.P. Gerber, J. Perlwitz, L.M. Polvani, N.P. Gillett, K.H. Seo, et al., J. Geophys. Res. Atmos. 115 (2010) D00M07.

[122] L.M. Polvani, D.W. Waugh, G.J.P. Correa, S.-W. Son, J. Clim. 24 (2010) 795.

[123] S.M. Kang, L.M. Polvani, J.C. Fyfe, M. Sigmond, Science 332 (2011) 951.

[124] G.L. Manney, M.L. Santee, M. Rex, N.J. Livesey, M.C. Pitts, P. Veefkind, et al., Nature 478 (2011) 469.

[125] J. Perlwitz, Nat. Clim. Change 1 (2011) 29.

[126] J.D. Haigh, M. Blackburn, R. Day, J. Clim. 18 (2005) 3672.

[127] L.J. Gray, A.A. Scaife, D.M. Mitchell, S. Osprey, S. Ineson, S. Hardiman, N. Butchart, J. Knight, R. Sutton, K. Kodera, J. Geophys. Res. Atmos. 118 (2013), 2013JD020062.

[128] S. Ineson, A.A. Scaife, J.R. Knight, J.C. Manners, N.J. Dunstone, L.J. Gray, J.D. Haigh, Nat. Geosci. 4 (2011) 753.

[129] J.D. Haigh, Science 272 (1996) 981.

[130] G.A. Meehl, J.M. Arblaster, K. Matthes, F. Sassi, H. van Loon, Science 325 (2009) 1114.

[131] G. Stenchikov, A. Robock, V. Ramaswamy, M.D. Schwarzkopf, K. Hamilton, S. Ramachandran, J. Geophys. Res. Atmos. 107 (2002) 4803.

[132] G. Stenchikov, K. Hamilton, R.J. Stouffer, A. Robock, V. Ramaswamy, B. Santer, H.-F. Graf, J. Geophys. Res. Atmos. 111 (2006) D07107.

[133] S. Driscoll, A. Bozzo, L.J. Gray, A. Robock, G. Stenchikov, J. Geophys. Res. Atmos. 117 (2012) D17105.

[134] W. Cai, T. Cowan, Geophys. Res. Lett. 34 (2007).

[135] W.-T. Chen, H. Liao, J.H. Seinfeld, J. Geophys. Res. 112 (2007).

[136] R.J. Allen, S.C. Sherwood, Clim. Dynam. 36 (2011) 1959.

[137] R.J. Allen, S.C. Sherwood, J.R. Norris, C.S. Zender, Nature 485 (2012) 350.

[138] K.M. Grise, L.M. Polvani, J. Clim. 27 (2014) 6074.

[139] A. Voigt, T.A. Shaw, Nat. Geosci., advance online publication (2015).

[140] WMO, Scientific Assessment of Ozone Depletion: 2010, World Meteorological Organization, Geneva, Switzerland, 2011.

[141] M.P. Baldwin, T.J. Dunkerton, J. Geophys. Res. Atmos. 104 (1999) 30937.

[142] L.M. Polvani, P.J. Kushner, Geophys. Res. Lett. 29 (2002), http://dx.doi.org/10.129/2001GL014284.

[143] E.P. Gerber, A. Butler, N. Calvo, A. Charlton-Perez, M. Giorgetta, E. Manzini, et al., Bull. Am. Meteorol. Soc. 93 (2012) 845.

[144] D.W.J. Thompson, S. Solomon, P.J. Kushner, M.H. England, K.M. Grise, D.J. Karoly, Nat. Geosci. 4 (2011) 741.

[145] D.W.J. Thompson, J.C. Furtado, T.G. Shepherd, J. Atmos. Sci. 63 (2006) 2616.

[146] P.J. Kushner, L.M. Polvani, J. Clim. 17 (2004) 629.

[147] G. Rivière, J. Atmos. Sci. 68 (2011) 1253.

[148] J. Liu, J.A. Curry, H. Wang, M. Song, R.M. Horton, Proc. Natl. Acad. Sci. U.S.A. 109 (2012) 4074.

[149] R. Jaiser, K. Dethloff, D. Handorf, Tellus Ser. A Dyn. Meteorol. Oceanogr. 65 (2013) 1.

[150] J. Cohen, M. Barlow, K. Saito, J. Clim. 22 (2009) 4418.

[151] J. Cohen, J.A. Screen, J.C. Furtado, M. Barlow, D. Whittleston, D. Coumou, J. Francis, K. Dethloff, D. Entekhabi, J. Overland, J. Jones, Nat. Geosci. 7 (2014) 627.

[152] J. Overland, K. Wood, M. Wang, Polar Res. (2011), http://dx.doi.org/10.3402/polar.v30i0.15787.

[153] B.-M. Kim, S.-W. Son, S.-K. Min, J.-H. Jeong, S.-J. Kim, X. Zhang, T. Shim, J.-H. Yoon, Nat. Commun. 5 (2014).

[154] D.J. Seidel, W.J. Randel, J. Geophys. Res. 111 (2006), http://dx.doi.org/10.1029/2006JD007363.

[155] W. Randel, F. Wu, D. Gaffen, J. Geophys. Res. 105 (D12) (2000) 15509.

[156] R. Sausen, B.D. Santer, Meteorol. Z. 12 (2003) 131.

[157] B.D. Santer, M.F. Wehner, T.M.L. Wigley, R. Sausen, G.A. Meehl, K.E. Taylor, C. Ammann, J. Arblaster, W.M. Washington, J.S. Boyle, W. Brüggemann, Science 301 (2003) 479.

[158] J. Austin, T. Reichler, J. Geophys. Res. 113 (2008) D00B10.

[159] IPCC, Climate change 2007: the physical science basis, in: S. Solomon, et al. (Eds.), Contribution of Working Group I to the Fourth Assessment Report of the Intergovernmental Panel on Climate Change, Cambridge University Press, Cambridge, 2007.

[160] WMO, Scientific Assessment of Ozone Depletion: 2006, World Meteorological Organization, Geneva, Switzerland, 2007.

[161] V. Ramaswamy, M.D. Schwarzkopf, W.J. Randel, B.D. Santer, B.J. Soden, G.L. Stenchikov, Science 311 (2006) 1138.

[162] S.M. Kang, J. Lu, J. Clim. 25 (2012) 8387.

[163] I.M. Held, A.Y. Hou, J. Atmos. Sci. 37 (1980) 515.

[164] C.C. Walker, T. Schneider, J. Atmos. Sci. 63 (2006) 3333.

[165] T. Schneider, Annu. Rev. Earth Planet. Sci. 34 (2006) 655.

[166] G.P. Williams, J. Atmos. Sci. 63 (2006) 1954.

[167] Y. Wu, R. Seager, T.A. Shaw, M. Ting, N. Naik, J. Clim. 26 (2012) 918.

[168] T.R. Karl, S.J. Hassol, C.D. Miller, W.L. Murray, Temperature Trends in the Lower Atmosphere: Steps for Understanding and Reconciling Differences U.S. Climate Change Science Program and the Subcommittee on Global Change Research, National Oceanic and Atmospheric Administration, National Climatic Data Center, Asheville, NC, 2006.

[169] B.D. Santer, P.W. Thorne, L. Haimberger, K.E. Taylor, T.M.L. Wigley, J.R. Lanzante, et al., Int. J. Climatol. (2008), http://dx.doi.org/10.1002/joc.1756.

[170] K.-M. Xu, K.A. Emanuel, Mon. Wea. Rev. 117 (1989) 1471.

[171] R.J. Allen, S.C. Sherwood, Nat. Geosci. 1 (2008) 399.

[172] M.N. Juckes, J. Atmos. Sci. 57 (2000) 3050.

[173] D.M.W. Frierson, Geophys. Res. Lett. 33 (2006), http://dx.doi.org/10.1029/2006GL027504.

[174] D.M.W. Frierson, J. Atmos. Sci. 65 (2008) 1049.

[175] J. Lu, G. Chen, D.M.W. Frierson, J. Clim. 21 (2008) 5835.

[176] N.-C. Lau, A. Leetmaa, M.J. Nath, J. Clim. 19 (2006) 3607.

[177] M. Hoerling, A. Kumar, Science 299 (2003) 691.

[178] R. Seager, N. Harnik, Y. Kushnir, W. Robinson, J. Miller, J. Clim. 16 (2003) 2960.

[179] W.A. Robinson, Geophys. Res. Lett. 29 (2002) 1190.

[180] I. Orlanski, J. Atmos. Sci. 62 (2005) 1367.

[181] E. Yulaeva, J.M. Wallace, J. Clim. 7 (1994) 1719.

[182] S. Bordoni, T. Schneider, J. Atmos. Sci. 67 (2009) 1643.
[183] J.T. Abatzoglou, G. Magnusdottir, J. Clim. 19 (2006) 6139.
[184] I.R. Simpson, T.A. Shaw, R. Seager, J. Atmos. Sci. 71 (2014) 2489.
[185] R. Caballero, Geophys. Res. Lett. 34 (2007) L22705.
[186] T. Schneider, S. Bordoni, J. Atmos. Sci. 65 (2008) 915.
[187] G. Chen, I.M. Held, Geophys. Res. Lett. 34 (2007), http://dx.doi.org/10.1029/2007GL031200.
[188] J. Kidston, S.M. Dean, J.A. Renwick, G.K. Vallis, Geophys. Res. Lett. 37 (2010) L03806.
[189] J. Kidston, G.K. Vallis, S.M. Dean, J.A. Renwick, J. Clim. 24 (2011) 3764.
[190] D.J. Lorenz, J. Atmos. Sci. 71 (2014) 2370.
[191] P.W. Staten, T. Reichler, J. Lu, J. Clim. 27 (2014) 9323.
[192] I.R. Simpson, M. Blackburn, J.D. Haigh, J. Atmos. Sci. 66 (2009) 1347.
[193] R. Seager, M. Ting, I. Held, Y. Kushnir, J. Lu, G. Vecchi, H.-P. Huang, N. Harnik, A. Leetmaa, N.-C. Lau, C. Li, J. Velez, N. Naik, Science 316 (2007) 1181.
[194] D.E. Parker, M. Gordon, D. Cullum, D. Sexton, C. Folland, N. Rayner, Geophys. Res. Lett. 24 (1997) 1499.

WEATHER PATTERN CHANGES IN THE TROPICS AND MID-LATITUDES

Ricardo M. Trigo[1], Luis Gimeno[2]

[1]*Faculdade de Ciências, Instituto Dom Luiz (IDL), Universidade de Lisboa, Lisbon, Portugal;* [2]*Departamento de Física Aplicada, Faculty of Sciences, University of Vigo, Ourense, Spain*

CHAPTER OUTLINE

1. INTRODUCTION

The overwhelmingly majority of the scientific community supports the existence of a causal link between increasing levels of anthropogenic greenhouse gas concentrations and the observed global warming trend of the climate system, particularly since the 1960s. Successive Assessment Reports (ARs) of the Intergovernmental Panel on Climate Change (IPCC) published in 1996 (AR2), 2001 (AR3), 2007 (AR4) and 2013 (AR5) have only consolidated that understanding [1]. However, in recent years this wide support has been partially undermined by the impact in the media and the general public of two phenomena, namely: (1) the existence of the so-called hiatus in the global temperature trend [2] and (2) the occurrence of several unusually cold winters in Europe and in the USA associated with the meandering of the polar jetstream [3]. A number of high impact publications have focused on both topics and obtained results that provide useful new insights. Thus, in relation to the first issue, several works have shown that since the end of the 1990s most of the excessive heat retained in the climate system has been stored in the ocean depths of either the Pacific [4] or the Atlantic [5] oceans. Additionally, and perhaps more important from a societal perspective, it has been showed that the occurrence of hot extreme events in the atmospheric component has maintained the same positive trend as before the hiatus [2]. On the other hand, the occurrence of extremely cold winters in both Europe and the USA in recent years has been associated with the strong decrease in Arctic ice cover [6] although other authors point toward the role played by the tropical warmer waters [3]. In any case, both issues are useful to underline the difficulty in ascribing

Climate Change. http://dx.doi.org/10.1016/B978-0-444-63524-2.00007-5

monotonic global trends simultaneously to all components of the climate system (atmosphere, hydrosphere, cryosphere, lithosphere) at global and regional scales. The climate system is too complex, with fluxes of energy and mass between different components operating over a wide range of temporal and spatial scales, often plagued by nonlinear interactions [7]. The concerns stated above are particularly relevant within the scope of this chapter, which focuses on the assessment of weather pattern changes as an indicator of global changes. Thus, the note stressed in the previous edition of this book is still valid, i.e. it is likely that recent (and future) trends in temperature, precipitation and other climatic variables are amplified or partially offset at the regional scale, as a consequence of changes of major patterns of atmospheric circulation [8].

2. OBSERVED CHANGES IN SEA LEVEL PRESSURE

Since the publication of the AR4, several updated observational and reanalyzed datasets have been disseminated, allowing for an improved assessment of the potential changes of atmospheric circulation based on sea level pressure (SLP), winds or geopotential height [1]. A long-term assessment of SLP between 1949 and 2009 has found an increase in the tropics and subtropics throughout the year and a decrease at higher latitudes in both hemispheres [9]. However, these SLP trends are not consistent over the most recent decades, for which we have access to higher resolution and more robust data. Thus, a shorter SLP and geopotential height trend assessment between 1979 and 2012 was discussed in the recent AR5 [1], and the results are depicted in Fig. 1. These results show a positive and statistically

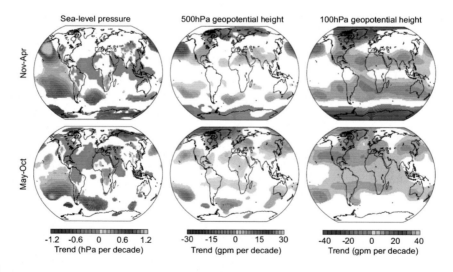

FIGURE 1

(Left) Trends in sea level pressure, SLP; (middle) 500 hPa geopotential height (gph); and (right) 100 gph: in (top) November to April and (bottom) May to October. Geopotential height is a vertical coordinate referenced to the Earth's mean sea level and is roughly the height (in units of metres) above sea level of a pressure level. All the data refer to the time between 1979 and 2012 from ERA-Interim data. Trends are only represented in statistically significant (i.e. within the 90% confidence) intervals (from IPCC, 2013). *(The figure is reproduced with permission. Cambridge Press (IPCC report, 2013).)*

significant SLP trend over the southern Atlantic and large sectors of the northern and southern Pacific and negative trends in the tropical and northern subtropical Atlantic during the entire year and also in northern Russia during the boreal summer.

The spatial distribution of SLP is particularly relevant, being directly associated with the distribution of atmospheric mass, which is the surface imprint of the atmospheric circulation [7]. In this regard it can be particularly useful to evaluate any major change in the location and strength of semipermanent pressure centres for both hemispheres as depicted in Fig. 2. Despite the lack of evidence for long-term trends since the mid-twentieth century, results do show significant decadal variability.

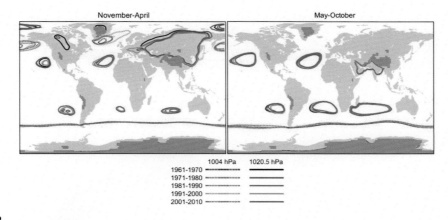

FIGURE 2

Decadal averages of sea level pressure (SLP) from the twentieth-century reanalysis (20CR) for (left) November to April and (right) May to October shown through selected contours: 1004 hPa (dashed lines) and 1020.5 hPa (solid lines) (from IPCC 2013). *(The figure is reproduced with permission. Cambridge Press (IPCC report, 2013).)*

Due to their impact on the northern and southern European climate, it is worth mentioning that both the Azores high and the Icelandic low in boreal winter were smaller in the 1960s and 1970s, and increased in the 1980s and 1990s, particularly the Icelandic low. During the winter half there appears to be a clear eastward shift of the Aleutian low from the 1970s to the 2000s, in agreement with previous results [10]. Similarly, the Atlantic and Pacific high-pressure systems show an extended elongation westward, although restricted to the 1960s and 1970s, as captured by their corresponding high SLP contours. Overall, decadal variability during the boreal winter was larger than in boreal summer, although during the latter both Atlantic and Pacific high-pressure systems tend to be smaller in recent decades.

3. OBSERVED CHANGES IN EXTRATROPICAL PATTERNS

Since the late 1990s, an increasing number of atmospheric science researchers have studied changes that have occurred on the frequency, phase and amplitude of the most important circulation patterns that modulate the climate on a regional and even on a continental scale. Indices of teleconnections can be very useful when describing the state of a climate system as associated to individual modes of climate variability [1]. These indices and the associated spatial patterns often encapsulate a large share of spatiotemporal climate variability.

Different approaches can be used to derive the main atmospheric circulation patterns that characterize the large-scale circulation over the entire northern hemisphere [11,12]. Here we focused the analysis of the extra tropical patterns in those affecting most of the two large NH landmasses, the North Atlantic Oscillation (NAO) and the Pacific North American (PNA), as obtained from the U.S. National Oceanographic and Atmospheric Administration (NOAA) Climate Prediction Centre (CPC). Spatial patterns of the NAO and PNA can be seen in Figs 3 and 4, respectively, and represent

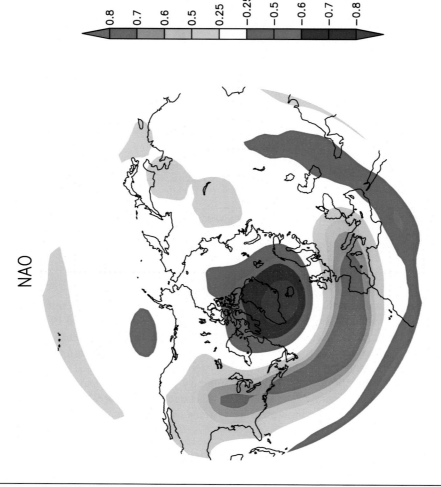

FIGURE 3

Spatial pattern of the NAO as given by the temporal correlation between the Winter (DJFM) monthly standardized 50 kPa geopotential height anomalies at each point and the monthly teleconnection pattern time series from 1960 to 2000 (from Letcher [8]). The numbers 0.8–0.8 are correlation values and have no units. *(The figure is reproduced with permission. Elsevier Press (Letcher, 2009).)*

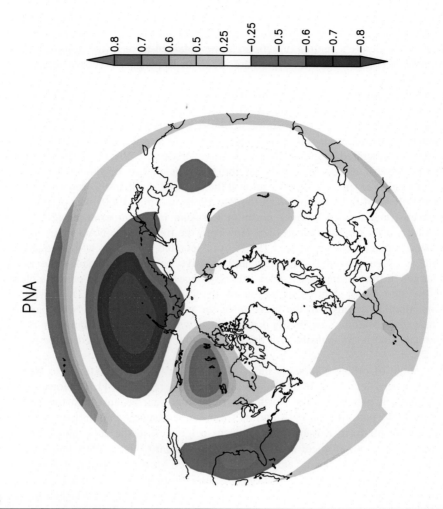

FIGURE 4

Spatial pattern of PNA as given by the temporal correlation between the winter (DJFM) monthly standardized 50 kPa geopotential height anomalies at each point and the monthly teleconnection pattern time series from 1960 to 2000 (from Letcher [8]). The numbers 0.8−0.8 are correlation values and have no units. *(The figure is reproduced with permission. Elsevier Press (Letcher, 2009).)*

the temporal correlation between the monthly standardized geopotential height anomalies at each point and the monthly teleconnection pattern time series from 1960 to 2000.

The North Atlantic Oscillation was first identified in the 1920s [13] but has been used more extensively since the availability of longer datasets and computers resources in the 1980s, coupled with appropriate statistical methods [11,12]. Historically, the NAO has been defined as a simple index that measures the difference in surface pressure between a southern station (e.g. Lisbon, Gibraltar or Ponta Delgada in the Azores) and some northern station in Iceland [14,15]. However, a more objective determination of the dipole's centres of action can be obtained through the application of principal component analysis (PCA) to sea level pressure (SLP) or tropospheric geopotential height [12] as performed by NOAA.

The winter NAO pattern is the most important mode of atmospheric variability in the northern hemisphere, explaining a large fraction of SLP and impinging a strong impact on the surface climate of the Atlantic/European sector in particular (e.g. Refs [14,16,17]). To a certain extent, the strong influence of this pattern is related to major changes of storm tracks in the North Atlantic and European sectors associated to negative and positive phases of NAO [18,19]. Moreover, major decadal fluctuations (or long-term trends) of the NAO index can be reflected in corresponding oscillations (or trends) of specific climate fields at the regional scale. In this regard, the trend toward a more positive phase of the NAO between the late 1960s and mid-1990s was partially responsible for the observed trend toward warmer northern Eurasian land temperatures [20,21] and also a tendency toward a drier Mediterranean [22]. However, the trend of the NAO index toward positive values has been replaced since the mid-1990s by a levelled evolution with frequent alternation between winters characterized by negative and positive values [1]. Besides the assessment of potential changes in time of the NAO index it is also necessary to evaluate possible changes in the location of the two major centres of action of the NAO pattern (Fig. 3). Previous work has identified considerable changes in the location of the main centres of the NAO pattern, with the northern centre (the Icelandic low) being displaced eastward toward Scandinavia in recent decades [23,24] — a result that is compatible with the SLP decadal changes depicted in the northern hemisphere (Fig. 2).

The wide range of climatic impacts induced by the NAO pattern extends to North America, particularly to eastern sectors of both Canada and the USA (e.g. Ref. [16]). However, the NAO is not the most relevant pattern for the North American continent because that top rank position is occupied by the PNA pattern. The PNA and its southern hemisphere counterpart, the Pacific-South American (PSA) result from Rossby wave patterns originated in the subtropical western Pacific, associated with anomalous tropical heating [25]. The standard winter PNA pattern involves four centres of action (with decreasing amplitude) that extend from subtropical Pacific and North America (Fig. 4). Similarly for the NAO for Europe, the PNA holds strong impact for North America, namely through the steering of the Aleutian Low [26] or the frequency of Alaskan blocking events and associated cold air outbreaks over the western USA in winter [27]. While some multi-decadal variability of both PNA and PSA have been detected, probably modulated by the ENSO signal, no systematic trend in their frequency of amplitude has been detected with statistical significance [1].

4. CHANGES IN TROPICAL PATTERNS
4.1 EL NIÑO SOUTHERN OSCILLATION

Unlike the NAO, PNA, blocking and other atmospheric circulation patterns, the ENSO is a truly coupled ocean-atmosphere oscillation mode with impacts on a global scale [28], including Europe

[29]. El Niño events are characterized by anomalous warming of tropical waters in central and eastern Pacific associated with the decrease in strength of the trade winds. This overall pattern leads to an increase of precipitation in the central and eastern tropical Pacific accompanied by a decrease in the western Pacific [28]. The opposite phase (La Niña) is characterized by below-average temperatures in central and eastern tropical waters and produces reverse impacts to those of El Niño. The frequency and strength of ENSO has varied over time in a 2- to 8-year cycle but also in decadal and centennial intervals [30]. In recent decades there has been a resurgence of strong (e.g. 1982/1983 and 1997/1998) and prolonged (1991–1995) El Niño episodes. However, the last decade has been characterized by two important La Niña events with global impacts on climate, water resources and vegetation dynamics [31].

In recent years, the scientific community has embraced new assessments of the ENSO phenomenon partially driven by two novel aspects of this large-scale coupled mode. These are the recent (and likely future) changes in: (1) the frequency of extreme ENSO events; and (2) the position of the El Niño Sea Surface Temperature (SST) warming centre. The first issue is mostly addressed through assessments of the likelihood of changes in the frequency of extreme ENSO events within a warmer future climate. This topic is extremely relevant in terms of socioeconomic consequences because such extreme episodes of ENSO can cause widespread impacts. Such was the case during the 1997/1998 event that caused damages estimated at US 35×10^9 (35 billion dollars) worldwide [32]. This outstanding El Niño episode was the largest on record, contributing significantly to the highest global average temperature recorded in 1998. The current generation of coupled ocean atmosphere climate models are capable of reproducing El Niño events and their impact relatively well [1]. When forced with distinct climate change scenarios, these very same models predict continued ENSO interannual variability. Despite these achievements it is still difficult to infer conclusively on future changes in the frequency of extreme El Niño events. However, it is possible to look for changes in conditions that favour the occurrence of such as extremes, as shown by Cai et al. [33]. These authors have found a spatial asymmetry in the warming of the Pacific Ocean with a multi-model ensemble, being higher in the eastern equatorial than in the surroundings. This pattern has an impact on atmospheric convection, which is paramount in El Niño extreme events. According to these estimations, the frequency of extreme El Niño events could be doubled in the late twenty-first century.

The second issue, on possible changes in the position of the El Niño SST warming centre, has fuelled an intense debate about the importance of the ENSO phenomena [34]. The classical pattern of El Niño consists of an SST warming centre located along the eastern Pacific (hereafter EP-El Niño) and spreading north and south close to the American coast. However, in the last few decades it has been observed that this pattern has been less frequent, evolving toward a different configuration with the SST warming centre placed further in the central Pacific (Fig. 5). This new configuration has been given various names, including: Central Pacific El Niño (hereafter CP-El Niño), El Niño Modoki, the Warm Pool El Niño or dateline El Niño [1]. Due to the different pattern of CP versus classical ENSO, there are many differences in the structure, evolution and teleconnection patterns between them [35] (Fig. 6). In the last three decades, the CP-El Niño configuration has been frequently observed and projections with global warming scenarios point toward an increased frequency of CP-El Niño compared to EP-El Niño [36]. Although it is difficult to conclude which of the two types of El Niño will occur more frequently in a future warmer climate, the majority of models tend to agree that CP-El Niño exhibits larger variability than the Ep-one [37].

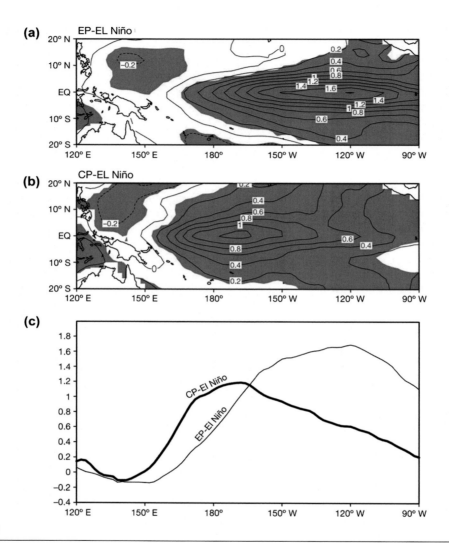

FIGURE 5

(After Yeh et al. [36].) Deviations of mean sea surface temperature for the two characteristics of El Niño from the 1854–2006 climatology: (a) the EP-El Niño; (b) the CP-El Niño, here the contour interval is 0.2°C and the shading denotes a statistical confidence at 95% confidence level based on a student's *t*-test; and (c) the zonal structure for the composite EP-El Niño (thin line) and CP-El Niño (thick line) averaged over 2°N to 2°S The label along the *y*-axis of figures (a), (b) and (c) should read Deviation Temperature/°C. *(The figure is reproduced with permission. Nature Publishing Group.)*

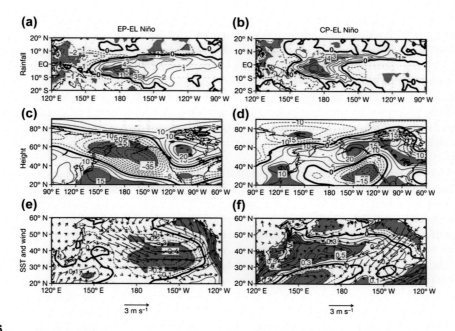

FIGURE 6

(After Yeh et al. [36].) Deviations for the two characteristics of El Niño from their climatology conditions: (a, b), the deviation of mean rainfall for the EP-El Niño (a) and the CP-El Niño (b), here the contour interval is 1 mm d^{-1} (where d refers to day), the *y*-axis label should read as 'Deviation of mean rainfall/mm d^{-1}'; (c, d), the mean geopotential height at 500 hPa, here the contour interval is 5 m, the *y*-axis label should read as 'Geopotential height at 500 hPa'; (e, f), the mean winds at 925 hPa (arrows, see scale arrow below, 3 m s^{-1}) and the mean SST (line), here the solid (dotted) line denotes positive (negative) deviations from the mean and the contour interval is 0.1°C, with the shading in all panels indicating the region exceeding 95% significance based on a *t*-test and the zero line is denoted by the thick line for (e, f) the label along the *y*-axis should read as 'SST/°C and Wind Speed/m s^{-1}'. The climate periods are 1979–2006 (for rainfall), 1950–2006 (for geopotential height and winds) and 1854–2006 (for SST), respectively. *(The figure is reproduced with permission. Nature Publishing Group.)*

In this regard, the increase of the CP-El Niño in recent decades and the possible increase in future climate projections could be mainly due to climate variability [37]. The last assessment report from IPCC acknowledges that it is currently not possible to determine whether the recent change in the mean climate state is driving this change in the distribution of frequency between the two types of El Niño events [1].

4.2 TROPICAL CYCLONES

Tropical cyclones (TC) are among the most destructive natural hazards in the world [38]. Consequently, any trends or large decadal fluctuations in tropical cyclone activity are particularly relevant to society, primarily in those coastal areas where populations are affected. The dependence of TC on high SST values has opened the debate of a possible increase in the frequency and intensity of tropical cyclones in a warmer climate. The public and media awareness has been captivated by this possibility in the follow-up of a number of high-impact TC events in the last decade in both the developed (e.g. Katrina in 2005 and Sandy 2011 in USA) and developing (e.g. Haiyan in 2013 in the Philippines) world [39]. Several authors have suggested that the recent rise in SST, partially prompted by anthropogenic global warming, has already led to a greater number of intense tropical cyclones in recent decades [40,41]. However, other assessments do not agree and state that part of these trends is simply an artefact of the short length and nonhomogeneity of records [42].

The evidence of significant changes, either in frequency or intensity in TCs, is thin because these changes are within natural interannual and inter-decadal variability, reducing the significance attributable to long-term trends [43]. In particular, studies of TC variability in the north Atlantic basis (the most studied one) reveal large interannual and inter-decadal swings in storm frequency that have been linked to different large-scale climate phenomena such as the ENSO, the NAO, the stratospheric quasi-biennial oscillation and multi-decadal oscillations in the North Atlantic region [1].

Since the publication of the AR4 there have been two additional aspects of TCs that deserved further attention, namely, (1) the poleward migration in the average location of TCs, and (2) the impact of climate change on the frequency of the most intense and damaging TCs. The most recent analysis of the former question [44] shows a poleward migration of TC in both hemispheres in the last 30 years of about one degree of latitude per decade (Fig. 7), consistent with the expansion in the tropics [45]. These results are coherent with the increased/decreased shear/potential intensity in the deep tropics and decreased/increased shear/potential intensity in the subtropics, i.e. two of the key factors controlling TC occurrence by inhibit/enhance storm development. The migration occurred in an asymmetric way for the ocean basins, with the western North Pacific the largest contributor to the global trend for the northern hemisphere and the North Atlantic the smallest (no significant trend). In the case of the southern hemisphere, the South Pacific exhibited a more pronounced trend than the South Indian Ocean. If climate change induces further poleward migration of TC average latitudinal location, there will be obvious consequences for life and property, especially in coastal cities outside the tropics.

The average global annual economic loss from TCs is about US 26×10^9 [46] and it is expected to increase in a warming climate [1]. This increment is partially anthropogenic and does not necessarily imply an increased number of TCs making landfall, as it is expected that there will be an increase in population (and income) along coastlines in the future and consequently damage will increase, even if the frequency of extreme events remains similar. Models that do quantify damages have to deal with estimations of changes in income and population but also in frequency, intensity and location of tropical cyclones in the different oceanic basins [47]. Recent sophisticated approaches find that the damage associated to tropical cyclones could double by 2100, with almost all the damage being concentrated in North American, East Asian and the Caribbean regions [48].

4.3 MONSOONS

Over certain continental tropical areas there is intense heating inducing strong ascent rise of air that is balanced with adiabatic cooling, producing a global atmospheric overturning monsoon [49]. On a more regional scale there are several well-established monsoon systems, namely: Africa,

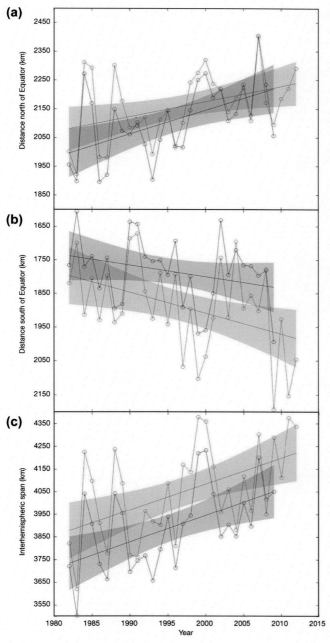

(a)

(b)

(c)

FIGURE 7

(After Kossin et al. [44].) (a, b) The time series of annual mean latitude of tropical cyclone LMI calculated from the best-track historical data (red) and the ADT-HURSAT reanalysis (blue) in the northern (a) and southern (b) hemispheres; (c) relates to the annual mean difference between a and b and shows the global migration of the latitude of LMI away from the tropics. Linear trend lines are shown with their 95% two-sided confidence intervals (shaded). Note that the *y*-axis in b increases downward. Note also that the *y*-axis labels should read: (a) distance north of the equator/km; (b) distance south of the equator/km; (c) inter-hemispheric span/km and that the decimal comma mark ',' should be replaced by a decimal point, '.'.
(The figure is reproduced with permission. Nature Publishing Group.)

Australia–Asia, North America, South America, and the Pacific and Atlantic oceans. The life of about one-sixth of the world population is affected by the south Asian summer monsoon rainfall, so the negative trend of the observed summer rainfall since 1950 is a topic of intense interest not only for scientific reasons but for socioeconomic ones, with large populations depending on the rainfall for basic human needs. The acceleration of the hydrological cycle due to climate change results in an

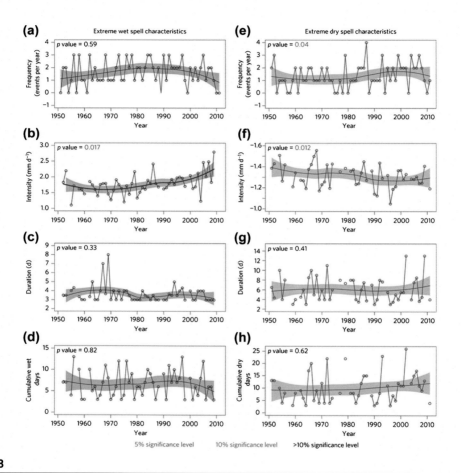

FIGURE 8

(After Singh et al. [52].) (a–h) Time series of wet (blue) and dry (red) spell frequency, duration, intensity and cumulative days over the core monsoon domain. Missing links in the time series are years with no wet or dry spells. Trend lines are estimated using the nonparametric LOESS regression technique; shading represents the 90% confidence intervals of the estimated trends; p values are obtained from testing the difference in means of the distributions of each variable 1951–1980 and 1981–2011 using the nonparametric moving block bootstrap test. Colours indicate the significance level of the p values. Note: the units for the y-axis should be: Fig. 8(b) and (f) intensity/mm d^{-1}; (c) and (g) duration/d. *(The figure is reproduced with permission. Nature Publishing Group.)*

increase of available moisture, low-level moisture convergence and convective available energy in the Indian subcontinent during summer. However, and in spite of this, summer rainfall declined between 1950 and 2000 for the region as a whole, with the decline mainly in the monsoon core region of central and northern India [50]. Several mechanisms appear to be responsible for this apparent contradiction; however, the role of aerosols seems to be the most important. The increasing industrialization of India in the second half of the twentieth century has resulted in an increase in atmospheric aerosols containing black carbon and sulphate resulting in a strong radiative effect due to their scattering and absorbing properties. This has resulted in a reduction of the solar radiation reaching the surface, provoking a relative cooling of the northern Indian Ocean thus weakening the thermal gradient characteristics along the meridian of the monsoon and reducing the monsoon rainfall [51].

Another interesting aspect of the monsoon summer precipitation is to know whether this significant decrease in peak season precipitation is associated with changes in frequency of either wet spells (periods of heavy precipitation) or dry spells (periods of low precipitation). An analysis of the daily scale precipitation variability [52] has recently shown that there is a significant increase in the frequency of both wet and dry spells (defined as events of at least three consecutive days with precipitation anomalies higher than one standard deviation of daily precipitation) in the core monsoon region (18° 28°N and 73° 82°E). Based on this evidence, the authors derived a typical range for the frequency of wet spells of between 0 and 3 per year, with a positive trend from 1950 to 1980 and negative trend thereafter. Likewise, the number of dry spells per year ranged from 0 to 4, with the number of dry spells in the period 1981–2011 being double the number in the period 1951–1980 (Fig. 8). These results point toward an increment in the risks related to extremes in the Indian subcontinent with clear relevance to infrastructures, agriculture or water resources.

Projections of future monsoon rainfall obtained from models, as a consequence of thermodynamic forcing, continue to show a positive increase. The increase in precipitation of water using the specific humidity scale (together with the Clausius–Clapeyron equation) is about 6.5% K^{-1} [53]. This indicates that there will be increasing moisture content over the warmer Indian ocean, an increment in the integrated moisture fluxes over the subcontinent with an increase in precipitation. However, the complex structure of sources of moisture for precipitation (up to six different sources) [54,55], and the complex dynamical feedbacks in the region force us to be prudent.

5. CONCLUSIONS

There is no doubt that the unrelenting growth in the concentration of greenhouse gases in the atmosphere and oceans will further increase the global average temperature by the end of this century, most likely within the range of 2°C–4°C, depending on the emission scenarios [1]. In fact, the psychological barrier of 400×10^{-6} CO_2 equivalent concentration was reached for the first time in 2014. If the current pattern is sustained, this temperature increase will be considerably larger in the polar regions compared to the tropical belt, as a consequence of positive ice-albedo-temperature feedback loop. Nevertheless, these trends will be regionally dependent, as some of the large-scale driving patterns described in this chapter may enhance (or partially offset) the warming at the regional scale. In this regard, it is important to evaluate recent trends as well as potential future changes of these tropical and extratropical modes under global warming. In particular it is worth mentioning recent changes on the location or extent of important mid-latitude features such as the semipermanent anticyclones or higher latitude preferred sectors of low pressure systems. Such decadal changes can be identified more clearly in the boreal winter, for example for the Azores high and the Icelandic low that were both smaller in extension during the 1960s and 1970s, and increased in the 1980s and 1990s, or the eastward shift of the Aleutian low from the 1970s to the 2000s. In relation to the tropical patterns considered here, the most likely future perspectives point toward a maintenance of ENSO as the dominant mode of interannual global variability, with a strong influence on regional precipitation changes, a strengthening of global monsoon precipitation and a global frequency of tropical cyclones unchanged but with higher global rainfall and increased wind [1].

All considered, there is a continuous need to maintain the international effort of determining to what extent climate models yield a realistic picture of the variability in the present climate and evaluating the fraction of future regional climate change that can be attributed to future trends of both tropical and extratropical circulation patterns, since these patterns will probably be partially responsible for regional differences in climate in the future.

REFERENCES

[1] IPCC, Climate change 2013: the physical science basis, in: T.F. Stocker, D. Qin, G.K. Planttner, M. Tignor, S.K. Allen, J. Boschung, A. Nauels, Y. Xia, V. Bex, P.M. Midgley (Eds.), Contribution of Working Group 1 to the Fifth Assessment Report of the Intergovernmental Panel on Climate Change, Cambridge University Press, Cambridge, UK and New York, NY, USA, 2013.
[2] S.I. Seneviratne, M.G. Donat, B. Mueller, L.V. Alexander, Nat. Clim. Change 4 (2014) 161, http://dx.doi.org/10.1038/nclimate2145.
[3] T. Palmer, Science 344 (2014) 803–804.
[4] Y. Kosaka, S.P. Xie, Nature 501 (2013) 403–407.
[5] X. Chen, K.K. Tung, Science 345 (2014) 897, http://dx.doi.org/10.1126/science.1254937.
[6] Q. Tang, X. Zhang, J.A. Francis, Nat. Clim. Change 4 (2014) 45–50, http://dx.doi.org/10.1038/nclimate2065.
[7] J.P. Peixoto, A.H. Oort, Physics of Climate, American Institute of Physics, New York, USA, 1992, 520 pp.
[8] T.M. Letcher (Ed.), Climate Change: Observed Impacts on Planet Earth, Elsevier, Oxford, 2009, 25 chapters, pp. 444.
[9] N.P. Gillett, P.A. Stott, Geophys. Res. Lett. 36 (2009) L23709, http://dx.doi.org/10.1029/2009GL041269.
[10] A. Favre, A. Gershunov, Clim. Dynam. 26 (2006) 617–629, http://dx.doi.org/10.1007/s00382-005-0101-9.
[11] J.M. Wallace, D.S. Gutzler, Mon. Wea. Rev. 109 (1981) 784–812.
[12] A.G. Barnston, R.E. Livezey, Mon. Wea. Rev. 115 (1987) 1083–1126.
[13] G.T. Walker, Mem. Indian Meteorol. Dep. 24 (1924) 225–232.
[14] J.W. Hurrell, Science 269 (1995) 676–679.
[15] P.D. Jones, T. Jonsson, D. Wheeler, Int. J. Climatol. 17 (1997) 1433–1450.
[16] T.J. Osborn, K.R. Briffa, S.F.B. Tett, P.D. Jones, R.M. Trigo, Clim. Dynam. 15 (1999) 685–702.
[17] R.M. Trigo, T.J. Osborn, J.M. Corte-Real, Clim. Res. 20 (2002) 9–17.
[18] U. Ulbrich, M. Christoph, Clim. Dynam. 15 (1999) 551–559.
[19] R.M. Trigo, I.F. Trigo, C.C. DaCamara, T.J. Osborn, Clim. Dynam. 23 (2004) 17–28.
[20] J.W. Hurrell, Geophys. Res. Lett. 23 (1996) 665–668.
[21] L. Gimeno, L. de la Torre, R. Nieto, R. García, E. Hernández, P. Ribera, Earth. Planet. Sci. Lett. 206 (2003) 15–20.
[22] P. Sousa, R.M. Trigo, P. Aizpurua, R. Nieto, L. Gimeno, R. Garcia-Herrera, Nat. Hazards Earth Syst. Sci. 11 (2011) 33–51, http://dx.doi.org/10.5194/nhess-11-33-2011.

[23] J. Lu, R.J. Greatbatch, Geophys. Res. Lett. 29 (2002), http://dx.doi.org/10.1029/2001GLO14052.

[24] T. Jung, M. Hilmer, E. Ruprecht, S. Kleppek, S.K. Gulev, O. Zolina, J. Clim. 16 (2003) 3371–3382.

[25] J.D. Horel, J.M. Wallace, Mon. Wea. Rev. 109 (1981) 813–829.

[26] D.M. Straus, J. Shukla, J. Clim. 15 (2002) 2340–2358.

[27] G.P. Compo, P.D. Sardeshmukh, J. Clim. 17 (2004) 3701–3720.

[28] S.G. Philander, El Niño, La Nina, and the Southern Oscillation, Elsevier, New York, 1989.

[29] S. Brönnimann, Rev. Geophys. 45 (2007) (RG3003).

[30] R.D. D'Arrigo, E.R. Cook, R.J. Wilson, R. Allan, M.E. Mann, Geophys. Res. Lett. 32 (2005) L03711, http://dx.doi.org/10.1029/2004GL022055.

[31] A. Bastos, S.W. Running, C.M. Gouveia, R.M. Trigo, J. Geophys. Res. Biogeosci. 118 (2012) 1–9.

[32] K. Sponberg, Compendium of Climatological Impacts, University Corporation for Atmospheric Research, vol. 1, National Oceanic and Atmospheric Administration, Office of Global Programs, 1999.

[33] W. Cai, S. Borlace, M. Lengaigne, P. van Rensch, M. Collins, G. Vecchi, A. Timmermann, A. Santoso, M.J. McPhaden, L. Wu, M.H. England, G. Wang, E. Guilyardi, F.-F. Jin, Nat. Clim. Change 4 (2014) 111–116.

[34] J.S. Kug, F.-F. Jin, S.-I. An, J. Clim. 22 (2009) 1499–1515.

[35] H.-Y. Kao, J.-Y. Yu, J. Clim. 22 (2009) 615–632.

[36] S.W. Yeh, J.-S. Kug, B. Dewitte, M.-H. Kwon, B.P. Kirtman, F.-F. Jin, Nature 461 (2009) 511–514.

[37] S.W. Yeh, B.P. Kirtman, J.-S. Kug, W. Park, M. Latif, Geophys. Res. Lett. 38 (2011) L02704, http://dx.doi.org/10.1029/2010GL-045886.

[38] R. Munich, Natural Catastrophes 2013 Analyses, Assessments, Positions, 2013.

[39] S.C. Herring, M.P. Hoerling, T.C. Peterson, P.A. Stott (Eds.), Explaining Extreme Events of 2013 from a Climate Perspective, Bull. Am. Meteorol. Soc. 95 (9) (2014) S1–S96.

[40] K. Emanuel, Nature 436 (2005) 686–688.

[41] P.J. Webster, G.J. Holland, J.A. Curry, H.R. Chang, Science 309 (2005) 1844–1846.

[42] G.J. Holland, P.J. Webster, Phil. Trans. R. Soc. A. 365 (2007) 2695–2716.

[43] C.W. Landsea, Nature 438 (2005) E11–E13, http://dx.doi.org/10.1038/nature04477.

[44] J.P. Kossin, K.A. Emanuel, G.,A. Vecchi, Nature 509 (7500) (2014) 349–352.

[45] C. Lucas, B. Timbal, H. Nguyen, WIREs Clim. Change 5 (2014) 89–112.

[46] EMDAT, The OFDA/CRED International Disaster Database, Universite Catholique de Louvain, 2009. www.emdat.be.

[47] R.A. Pielke, Phil. Trans. R. Soc. 365 (2007) 1–13.

[48] R. Mendelsohn, K. Emanuel, S. Chonabayashi, L. Bakkensen, Nat. Clim. Change 2 (2012) 205–209, http://dx.doi.org/10.1038/nclimate1357.

[49] K.E. Trenberth, D.P. Stepaniak, J.M. Caron, J. Clim. 13 (2000) 3969–3993.

[50] A.G. Turner, H. Annamalai, Nat. Clim. Change 2 (2012) 587–595.

[51] V. Ramanathan, et al., Atmospheric brown clouds: impacts on South Asian climate and hydrological cycle, Proc. Natl. Acad. Sci. U.S.A. 102 (2005) 5326–5333.

[52] D. Singh, et al., Nat. Clim. Change 4 (2014) 456–461.

[53] I.M. Held, B.J. Soden, J. Clim. 19 (2006) 5686–5699.

[54] L. Gimeno, A. Drumond, R. Nieto, R.M. Trigo, A. Stohl, Geophys. Res. Lett. 37 (2010), http://dx.doi.org/10.1029/2010GL043712.

[55] L. Gimeno, R. Nieto, A. Drumond, R. Castillo, R.M. Trigo, Geophys. Res. Lett. 40 (2013) 1443–1450, http://dx.doi.org/10.1002/grl.50338.

BIRD ECOLOGY

8

Wolfgang Fiedler[1,2]

[1]*Department of Migration and Immuno-Ecology, Max Planck Institute for Ornithology, Radolfzell, Germany;*
[2]*Department of Biology, University of Konstanz, Konstanz, Germany*

CHAPTER OUTLINE

1. INTRODUCTION

Birds are highly mobile and easy to observe. They are relatively easy to recognise and their occurrence and habits are noted by millions of passionate birdwatchers or just interested persons. It is not surprising that changes in abundance or behaviour of birds are among the best-documented changes known in the animal world. Changes in the arrival of migrating birds at their breeding grounds and their disappearance in autumn have been used as cues to forecast weather in many cultures for centuries. Modern biology understands bird behaviour not as a result of miraculous wisdom of individuals but as a result of the action of evolution through mutation, selection and reproduction.

Since a central goal in evolution is adaptation to the environment, climate change, as well as global change in a wider sense, will change selection pressures and reproductive success of various behavioural types. This is, indeed, what is presently being observed and birds show us that we are already in the middle of massive changes.

However, it is important to note that not all changes in bird behaviour, as they are currently observed, can be attributed to climate change. Other factors, such as changes in land use, can influence the migration behaviour of birds. Changes in agriculture, in industrial activities or in human behaviour may offer or destroy suitable wintering sites. Examples include a new food source for European cranes (*Grus grus*) in fields of winter weed in northern France, ice-free waters for coots (*Fulica atra*) due to power plant cooling in Lithuania or bird feeders for blackcaps (*Sylvia atricapilla*) wintering in Great Britain [1]. Effects can be accelerated or attenuated by climate change and in some cases it will not be possible to identify the primary source of change that affects a certain behavioural modification. Nevertheless, all of the environmental changes currently experienced, that top the list in terms of speed and extent, are very likely a result of human activity and thus share a common source. Also it became

Climate Change. http://dx.doi.org/10.1016/B978-0-444-63524-2.00008-7

121

clear that birds as higher trophic level consumers and homeothermic animals with many behavioural options rarely are considerably affected directly by abiotic factors (like temperature) but rather by biotic mechanisms, namely altered species interactions such as predation or food availability [2].

2. INDICATORS OF CHANGE

2.1 RANGE

2.1.1 Size and Position of Breeding Ranges

Changes in the distribution of birds, especially in their breeding range, were one of the earliest topics discussed among ornithologists in the context of climate change [3]. These observations meanwhile were supported by many studies showing that range boundaries are moving poleward or upward in altitude as the climate gets warmer. This even is the case for Antarctic gentoo penguins (*Pygoscelis papua*) expanding their ranges southward [4].

Sound data comes from comparing standardized breeding bird surveys. Such comparisons have been done, among others, on data from the United Kingdom and Ireland [5], from Finland [6] and from the United States [7] and are summarized in Table 1.

Changes in breeding distribution registered, so far, are likely to be the first indications of rather severe ecological shifts and species rearrangements in some areas. Based on museum material of 1179 bird species and some mammal and butterfly species occurring in Mexico, ecological niche models have been developed with a genetic algorithm and were projected onto two predicted climate surfaces (conservative and liberal) for 2055. While extinctions and dramatic range losses were expected to be few, the turnover in some regions was predicted to reach more than 40% of the species [8].

The studies mentioned above lead to the expectation that in the first instance species richness should increase in areas with incoming southern species and northern species that have not left. Indeed, the Lake Constance area in Central Europe could be an example for this. Based on 2×2 km grid cells, breeding birds in an area of 1212 km^2 were counted in a semi-quantitative way in the periods 1980–1981, 1990–1992, 2000–2002 and 2010–2013. During this time species numbers increased from 141 to 157. In the second last decade a significant increase of species with a southern centre of distribution has been shown in Ref. [9].

A flagship species for a northward range extension of a southerly distributed bird species in Europe is the bee-eater (*Merops apiaster*). Distributed mainly in warmer areas such as the Mediterranean, the species starts breeding in higher European latitudes as soon as there are periods with warmer temperatures. This has been shown by a comparison between bee-eater records in Central Europe and the size of growth rings in oak wood. The rings give evidence of warmer periods with higher annual growth rates in the oaks from the sixteenth century onward [10]. Currently, bee-eaters are showing significant population increases (Fig. 1) and now breed as far north as Poland and Scandinavia.

In an analysis of the distribution areas of 116 British terrestrial breeding birds, Bradshaw et al. [11] distinguished 74 range contractors and 42 range expanders. Largest area changes in northerly distributed species were found in warmer and drier areas whereas largest changes in southerly distributed species occurred in colder and wetter environments. Slow life history (larger body size) and lower geographical fragmentation of the ranges decreased expansion probability. Surprisingly, this was also the case for species with greater natal dispersal. In this analysis, climate was shown to be the dominant driver of changes but the direction of the effects depended on geographical context.

Table 1 Changes in borders of the range of bird species as revealed from breeding bird surveys

Region and source	Period 1	Period 2	Distribution within region	Number of species	Distance and type of shift	Shift/year
Great Britain and Ireland [3]	1968–1972	1988–1991	Southerly distribution	59	Northern border moved 18.9 km northward	*ca.* 1 km a^{-1}
			Northerly distribution	42	Southern border did not show systematic movements	
Finland [4]	1974–1979	1986–1989	Southerly distribution	119	Northern border moved 18.8 km northward	*ca.* 1.7 km a^{-1}
			Northerly distribution	34	Southern border did not show systematic movements	
North America [5]	1967–1971	1998–2002	Southerly distribution	26	Northern border moved 72.9 km northward	*ca.* 2.4 km a^{-1}
			Northerly distribution	29	Southern border did not show systematic movements	

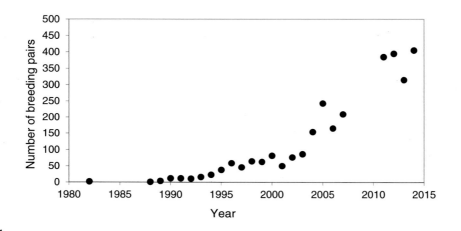

FIGURE 1

Number of breeding pairs of the Bee-eater (*Merops apiaster*) in the country of Baden-Württemberg (southwestern Germany) 1982–2013. (*Data from Boschert, Rupp, and Todte.*) Data 2011–2014 show the mean of a range estimate of ±15 pairs.

From the conservation point of view range expansions or reductions and, mostly related, population increases or decreases are the focus of interest. Siegel et al. [12] identified species associated with alpine/subalpine and aquatic habitats of California's Sierra Nevada to be most vulnerable to climate change. Together with birds breeding in coniferous woodland, again species of alpine habitats and wetlands in Switzerland were identified to be most vulnerable to climate and land use changes [13].

On a European scale, climatic variables in the actual breeding ranges of bird species have been used to forecast the future distributions based on the assumptions of first-generation climate change models in the Climate Atlas of Breeding Birds in Europe [14]. This modelling leads to the prediction that between the two periods 1960–1990 and 2070–2100 breeding distribution areas of European birds shall move on average 550 km northward and that many species shall suffer from area losses. This predicted value of a northward movement rate of 5.5 km a^{-1} is higher than has actually been found in the studies cited in Table 1.

One major problem with this approach is the negligence of changes in human land use that may result in habitat loss also in regions that might show suitable climate conditions in the future [15]. New generation models considerably improve the quality of prediction by including, among others, dispersal behaviour and conservation strategies [16], but the most realistic prediction of future regional land use as well as the selection of the most realistic climate change scenario for the prediction models remain to be a major challenge.

2.1.2 Ranges During the Nonbreeding Season

Besides breeding ranges, winter distributions are also changing. This is obviously the case where migrating birds can stay closer to their breeding grounds when closer areas become more suitable wintering areas for them, or when closer wintering areas become less suitable (e.g. drier) and thus birds are forced to migrate longer distances [1].

Data from the Christmas Bird Count in North America between 1975 and 2004 showed a mean northward movement of the northern border of wintering ranges of migratory bird species of 1.5 km a^{-1}. At the same time, winter distributions of nonmigrants also moved northward during that time period [17].

In Europe, the effect of the drying up of the Sahel belt (a dry savannah area south of the Sahara desert where many Palaearctic long-distance migrants have their wintering areas) has been considered as one of the main reasons for population declines in the common whitethroat (*Sylvia communis*), the sedge warbler (*Acrocephalus schoenobaenus*) and in many other species [18]. This indicates that the potential for rather simple latitudinal shifts of wintering areas, corresponding to changes in climate, is limited and might not be an option for all species.

2.2 MIGRATION

2.2.1 Timing of Migration to the Breeding Grounds

Despite its complexity and genetic component, bird migration behaviour appears to be highly flexible and changeable in many species. Changes in migration behaviour have been and still are the subject of numerous publications.

Lehikoinen et al. [19] showed in an analysis of 21 long-term studies of 10 European countries a consistent advancement of arrival times at the breeding grounds for sand martin (*Riparia riparia*), blackcap (*Sylvia atricapilla*), chiffchaff (*Phylloscopus collybita*), wagtail (*Motacilla alba*), barn swallow (*Hirundo rustica*), pied flycatcher, sedge warbler, tree pipit (*Anthus trivialis*) and house martin (*Delichon urbica*). In contrast, whinchat (*Saxicola rubetra*), spotted flycatcher (*Muscicapa striata*) and cuckoo (*Cuculus canorus*) did not advance their arrival times in half of the reported studies.

More examples are given in the first edition of this book [20]. In a very large compilation of different studies in Eurasia, Sparks et al. [21] calculated an advance in arrival times of 2.5 d K^{-1}–3.3 d K^{-1} warmer mean temperature.

The evidence for earlier arrival of birds at their breeding grounds in concordance with the warming up of the climate is enormous and corresponds well with the finding of a consistent global advancement of phenological events in spring between two and five days per decade [22]. However, some species react stronger than others to the advancement of spring phenology in certain regions and a few species seem unable to follow the changes. It was generally found that among 56 species in Lithuania, those species that arrive early in spring advance their arrival dates more than those species arriving later in the spring [22]. There is, however, a variation in response on the individual level, but in general the first birds to arrive at their breeding grounds advance their arrival by four days per decade, while the mean arrival date (average arrival date of a population) advances only by one day per decade [22].

From an evolutionary point of view, this indicates that some birds might benefit from an earlier arrival at their breeding grounds and thus show a strong response to changed environmental or climate conditions, while others change their timing at a much slower rate. One reason for the variation in the rate of changes within populations can be explained by the proximate factors driving the advancement of arrival times. The earlier arrival of pied flycatchers in recent years in southern Finland correlates well with higher temperatures in the winter quarters and along the homeward migration routes. However, the last birds to arrive did not advance their arrival dates, and late spring temperatures did not change [23].

In Europe, many studies used the North Atlantic Oscillation (NAO) [24] as a measure of climatic conditions. Almost all bird species in those parts of Europe influenced by the NAO can adjust their homeward migration timing to rising temperatures. This seems to be true for long-distance migrants (migration routes from Europe to at least sub-Saharan Africa) as well as for short-distance migrants (migration between Europe north of the Alps and the Mediterranean). However, it seems to be necessary that birds experience the warmer temperatures not only after arrival at the breeding grounds but along migration routes and also in their wintering quarters. In Europe and North America, birds do not arrive early at their breeding grounds if temperatures in these breeding areas rise but do not rise along the migration routes [25]. However, correlations between arrival times and temperature in the breeding areas have been found. For example, in passing migrants, over the Courish Spit (Southern Baltic), a strong negative correlation between April temperatures and passage times of 20 songbird species has been observed [26].

Positive NAO values in Europe can mean not only warmer temperatures but poor conditions in the Mediterranean and Sahel zone. For example, barn swallows in Italy arrive later in years when there are poor conditions in Africa [27]. In Spain, an increasing delay in the spring arrival of migrants in the 1970s and a current return to the level of the 1940s has been found [28] despite increasing local temperatures. It has been assumed that this is an effect of poor conditions in northern Africa (mainly due to low precipitation), resulting in a poor food supply, which in turn means a delay of fat deposition and consequently a later takeoff to the breeding grounds [29].

2.2.2 Timing of Migration from the Breeding Grounds

In contrast to the fairly consistent patterns of more or less pronounced advancements of spring arrival at the breeding grounds, when mean temperatures rise, the post-breeding migration timing shows a very different picture.

From a 42-year dataset of 65 migrating bird species, passing the Swiss alpine pass Col de Bretolet, the autumn passage of migrants wintering south of the Sahara has advanced in recent years, while migrants wintering north of the Sahara have delayed their autumn passage [30]. This advancement of post-breeding migration timing in long-distance migrants might be seen under the light of a selection pressure to cross the Sahel before its seasonal dry period. Species with shorter migration routes might benefit from a less constrained time schedule for breeding and moulting during summer when autumns are warmer and the risk of bad weather during autumn is reduced.

This assumption is supported by the additional finding that species with a variable rather than a fixed number of broods per year also delay their passage, possibly because they are free to attempt more broods [30,31].

This picture of advancements and delays in post-breeding migration timing, being dependent on the species, seems to be consistent (at least) all over Europe, but the assumption of a rather simple division between advancing long-distance migrants and delayed short-distance migrants is not supported generally at other places [32]. While in most European Studies more species show a delay in post-breeding migration timing [32,33], some studies clearly showed different trends at different time periods [34], and at the autumn passage on the Kola peninsula in Northern Russia the number of advances was much the same as the number of delays [35].

Despite the self-evident assumption that those birds advancing their autumn departure might benefit from an earlier arrival, an earlier onset of breeding and an earlier onset of post-breeding moult [36], no marked relationship between timing in autumn and timing in the preceding spring has consistently been found [32].

2.2.3 Migration Routes and Wintering Areas

Results gained over more than a century of bird ringing enable us, at least in some regions with sufficient data, to detect possible changes in the migration routes and in the position of the wintering quarters. Birds marked with a small coded ring at the breeding grounds and recovered later outside the breeding season enable insights into the position of various areas used by the birds through the year as well as insights into the changes of the positions of these areas. Presumably, wintering grounds and other areas used by birds during the nonbreeding season like moulting areas or stopover sites during migration, will change in the same way as changes of the breeding range have been described above. Generally, it can be expected that in regions with less severe winters migration routes will be shortened or that migration behaviour even will be reduced to zero. There is much evidence for a selection pressure toward earlier arrival at the breeding grounds for many bird species. Besides that, positions of wintering areas will also change when areas become unsuitable due to environmental changes. This may be true especially for birds wintering in areas endangered by desertification such as the Sahel Belt in Africa or parts of the Mediterranean Basin.

Studies available so far support these assumptions. Among 30 bird species investigated in Germany, 13 showed evidence of shorter migration routes, 11 showed evidence of a northward move of mean wintering latitude and 9 species showed increased numbers of winter recoveries within 100 km around the breeding place. Only a few species showed the opposite trend [37]. On a larger dataset of 66 species from the United Kingdom and Ireland it was found that 27 species showed increasingly northern wintering areas and 11 showed a northward move of the mean wintering latitude [38].

However, global warming might also lead to longer migration routes when breeding ranges are extended into higher latitudes and at the same time the wintering areas do not change much. For example, the European bee-eaters showed a range expansion northward and increased the intra-European part of their migration routes by up to 1000 km, but still winter south of the Sahara. Also the black-winged stilt (*Himantopus himantopus*) expanded its breeding areas from the Mediterranean northward into France, Ukraine and Russia but still winters south of 40° latitude [1]. Evidence for increasing migratory activity also comes from white-rumped and little swifts (*Apus caffer*) and (*A. affinis*), which colonize the Mediterranean area from the south, leaving these areas during nonbreeding periods while they are resident in almost all of the rest of their African breeding ranges [39,40].

2.2.4 Partial Migration

Partial migration describes the widespread phenomenon of some birds of a population migrating, while others do not. This situation has been described as the turntable of migratory and sedentary behaviour that enables selection to favour either more migratory or more sedentary behaviour according to environmental conditions [41]. Increasing numbers of winter records of otherwise migratory bird species give evidence of the development of partial migratory populations in Europe and North America and presumably elsewhere [1]. The Central European blackbird (*Turdus merula*) is a well-known example of this phenomenon. It was once considered as a migrating thrush of European woodlands, but in the early twentieth century it successfully started colonizing human settlements and reduced migration to become the first entirely sedentary populations in recent decades [41,42].

2.2.5 Eruptions

The mass movements of parts of local populations, which may be directed but seldom are reversible, are commonly called eruptions or evasions. In less migratory species, with highly variable population sizes, living under highly variable food conditions, such as tits in forest habitats and other boreal seedeaters, these eruptions occur repeatedly every few years. In a German blue tit (*Parus caeruleus*) population it has been shown that along with rising environmental temperatures, the numbers of eruptions have decreased remarkably [43]. While population size did not drop significantly, this observation (which might be a common phenomenon), may indicate a constant and improved food supply, making it unnecessary for parts of the population to emigrate.

2.3 REPRODUCTION

2.3.1 Onset of Breeding Period

The reproduction of birds is influenced by weather and thus by climate change in many ways. It is known that temperature, precipitation and resulting food supply can trigger the start of breeding [44]. An analysis of the relationship between ambient temperature and time of the first egg laid showed that 45 out of 57 bird species advanced the time of the first egg, when temperatures were high. Therefore, under current global warming it is not surprising that there are numerous studies indicating advancements in the onset of breeding in many species.

With respect to migrating species, the general advancement of arrival times in breeding areas has been mentioned above. Only a few studies report a delay in breeding phenology like an analysis of 47 datasets of southern hemisphere seabirds [45] where 67% of all significant trends showed this delay. For these species oceanographic conditions and sea ice obviously are more important than (terrestrial) spring arrival.

Early breeding depends on fitness, which stems from the availability of food insects, which in turn depends on early leafing and flowering of plants under elevated spring temperatures [44,46]. Based on data from the British nest record scheme, for the period 1971–1995, Crick et al. [46] found significant trends toward earlier laying dates for 20 of 65 species analyzed, with only one species having a delayed breeding date. The shift of the 20 species advancing their laying dates averaged 8.8 d (0.37 d a^{-1}). These species could neither be assigned to distinct migration strategies nor to ecological or taxonomic groups and comprise early and late breeders as well as long-distance migrants and residents. Among arctic bird species the advancement of laying date was reported to be 0.4 d a^{-1}–0.8 d a^{-1} [47]. The same order of magnitude was found in the tree swallows (*Tachycineta bicolor*) throughout North America with up to nine days between 1959 and 1991 [48] and for German great tits (*Parus major*) and blue tits with six and nine days between 1970 and 1995 [49]. Based on Danish bird ringing data of Arctic terns (*Sterna paradisaea*), A. P. Møller and colleagues reported an advancement of the ringing dates of chicks by 18 d during a 70 a period. This was explained by an increase in mean temperatures in April and May [50]. These are only a few examples out of a long list of reports that in most cases indicated homologous trends. Dunn and Møller [51] found that species with multiple broods per season that occur in nonagricultural habitats had the largest advances of the laying date.

2.3.2 Length of Breeding Period

As discussed briefly above, not only is the earlier onset of breeding beneficial but it also may lead to an extension of the breeding period. In species with high nest predation rates, longer breeding periods can offer more time for replacement clutches or species might successfully raise more than one brood per season. Calculated durations of the stay of 20 migrating bird species at their breeding grounds, from passage data on the German island of Heligoland showed an average increase over a decade of 2.2 d [32]. A prolongation of the breeding period has also been shown in reed warblers (*Acrocephalus scirpaceus*) in Poland [52]. Between 1970 and 2006 the peak of egg laying advanced 18 d but the end of the breeding season did not change. Replacement clutches, in cases of nest failure, were produced in early years by 15% of breeding pairs while in recent years 35% of failing pairs started a second, third or up to a fifth laying attempt. For example, evidence for an increase in second broods (a brood following a successful brood in the same season) comes from German swifts – during the past few years swifts have arrived at their breeding grounds earlier than before, have delayed post-breeding migration [53,54] and have increased the number of second broods [55]. Also, correlations between weather, food availability and multiple broods per season have been shown in a series of studies on various bird species [56–59]. As a more general figure, from an analysis of studies on European and North American birds Dunn and Møller [51] reported an increase in the length of the breeding season for birds with multiple broods per season and a decrease for single-brooded species.

2.3.3 Breeding Success

Earlier arrival at breeding sites and earlier onset of egg laying in many bird species means also larger clutch sizes since there is a link between the length of daylight and the clutch size with clutches produced earlier often containing more eggs [60]. In a 30-year study of reed warblers breeding in Southern Germany, the median of the date of the first egg advanced 15 d and the mean clutch size increased by about 0.5 eggs [61]. A similar relationship between onset of breeding, mean clutch size and breeding success can be found in Southern German collared flycatchers (*Ficedula albicollis*) (Fig. 2). However, reduced post-fledging survival may prevent those populations from growing even when more young are produced. Capercaillies (*Tetrao urogallus*) in Scotland advanced the onset of breeding but suffer from a drop in breeding success, presumably due to seasonal changes in the insect supply for the chicks [62].

In extensive data analysis on temporal change in laying date and clutch size of European and North American birds it was found that laying date advanced significantly across studies, while clutch size did not [51]. However, within populations, changes in laying date and clutch size were positively correlated but these changes were not related to changes in population size.

Optimal food supply of the young in the nest is crucial for reproductive success. Since timing of breeding as well as of moulting and migration is always a trade-off between multiple environmental and physiological requirements, phenological processes as induced by global warming may desynchronize. Visser et al. [63] and Both et al. [64] presented a 'textbook example' for this with pied flycatchers and great tits in Europe. In nine Dutch study areas, rising spring temperatures over the last 40 a, were connected with an advance of leafing and of the spring development of caterpillars of an abundant moth species (*Operophtera brumata*). These caterpillars form the most important food for nestlings of pied flycatchers and great tits and the birds aim to synchronize their breeding in a way that the caterpillar peak matches the time of highest food requirement for the nestlings. This is the time shortly before fledging, when large chicks have to be fed by the adults. Both bird species advanced laying dates in recent years but for the pied flycatcher (a long-distance migrant wintering south of the Sahara and spending two-thirds of its lifetime outside Central Europe), other factors seem to prevent them from advancing the breeding period to match the advancing hatching times of the caterpillars.

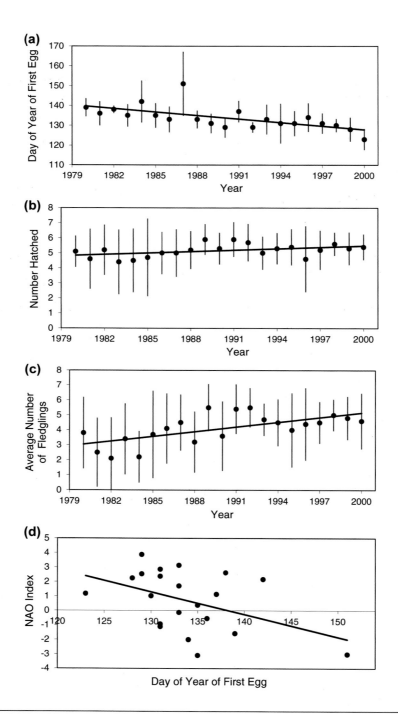

FIGURE 2

Date of first egg (a), hatching success (b) and fledging success (c) of Collared Flycatchers (*Ficedula albicollis*) in a southwestern German study area. R square and ANOVA probabilities > F: (a) −0.40, <0.003; (b) −0.21, <0.04; (c) −0.42, <0.002 (*Data from Renz, Dallmann and Braun*). (d) Data show the correlation between the date of the first egg and the NAO Station based annual index (*NAO Index Data provided by the Climate Analysis Section, NCAR, Boulder, USA, [24]*); R square = 0.19, ANOVA probability > F is 0.047.

As a consequence, nestlings miss the caterpillar peak and breeding success decreases. In areas where caterpillars hatch very early, pied flycatcher populations dropped by up to 90% while in areas with less advancing caterpillar timing, decreases only reached up to 10%.

2.3.4 Sexual Selection

In the large majority of migrating bird species, pairs do not migrate together and males arrive some time earlier at the breeding grounds than females. This phenomenon, called protandry, has been assumed to be affected by sexual selection because males emerging first at breeding grounds can occupy better territories and hence enjoy a mating advantage [65–67]. However, arriving too early at a breeding ground is a risk because food supply and weather conditions might not yet be suitable. If warmer spring temperatures reduce the risk of arriving too early at a breeding site, changes in the relation of costs and benefits of early arrival should have a greater effect on the sex arriving first, which in the majority of cases is the male. Indeed, in Danish barn swallows during 1971–2003, males advanced their arrival significantly while females did not [67]. It has also been shown that species with stronger female choice showed greatest advancements in arrival times, which is in accordance with the assumption that early arrival of males is favoured by female choice [68,69].

In blackcaps breeding in southwestern Germany and wintering either 1800 km southwest in Portugal and Spain or 1000 km northwest in the United Kingdom and Ireland, it has been shown that earlier arrival is not only related to a higher breeding success but also drives assortative mating among mates with comparable timing, which drives evolution especially rapidly in one direction [70]. Birds wintering at higher latitudes not only face shorter distances to return to the breeding grounds but also experience a daylight-night regime that triggers their circannual rhythms and accelerates pre-breeding migration, gonadal development and the onset of breeding [71,72].

2.4 MORPHOLOGY

It has been hypothesized that birds may change body size or other morphological traits as a response to climate change. This could be linked to thermoregulatory issues (for discussion about Bergmann's rule, see Ref. [73]), to improving or deteriorating feeding conditions or to changes in the requirements on the flight apparatus in the light of reduced migratory activity. Although negative correlations between body size and climate warming in concordance to Bergmann's rule have been found in 5 out of 12 passerine species in Southern Germany [73], this could not be shown for the other species and also not in a series of other studies, e.g. [74]. In a study on Southern German museum skins dating from a range of 120 years, Salewski et al. [75] found decreasing wing length in the majority of species but no consistent effect on wing pointedness. The latter is expected to decrease when migration activity decreases [76]. Although these studies show a considerable amount of time-dependent variation of morphological traits, the trends seem to be generally nonconsistent and thus indicate that fluctuations are mainly influenced by factors other than temperature.

2.5 DISEASES

It has been suggested that climate change and human movements and trade play a role in recent range expansions of avian pathogens [77]. Since parasites and pathogens are heterogeneously distributed in space [78], immune systems have a local component and thus changes in the distributions of both, hosts and parasites, as they are likely in the context of climate change, should be related to fitness costs of the host resulting from a less well-adapted immune system. Pérez-Rodríguez et al. [79] forecasted the distribution of three genera of avian blood parasites present in Spanish Blackcaps based on climate and parasite distribution data. They predict an area reduction for the abundant genera *Hemoproteus* and *Leucocytozoon* and an expansion of the more virulent *Plasmodium*. Although consequences for the affected birds are difficult to estimate, this at least predicts a modification of host-parasite relationships along with climate change.

Another effect of changing climate and the range of Avian malaria is expected for the Hawaiian Island of Kaua'I – increasing mean temperatures could lead to expansion of malaria into habitats where cooler temperatures currently limit transmission to highly susceptible endemic bird species [80]. Indeed, prevalences of infections and vector populations (mosquitos) increased in these areas between 1994 and 2013 along with increasing mean air temperature and declining precipitation.

3. CONCLUSIONS

The ecology of birds can clearly serve as an indicator of climate and global change. Almost all aspects in the life cycle of birds that have been studied so far show recent changes that can be linked to environmental changes. It is not surprising that birds show a high potential to adapt even complex behaviour such as breeding or migration to changing environments – either through evolutionary mechanisms acting on the genetic basis of behaviour or through available phenotypic plasticity. Ever since very early bird species evolved on earth 200 Ma ago, birds have had to cope with floating continents, rising and eroding mountains, ice ages and other massive environmental changes. A high degree of agility and mobility might have helped birds to adapt better to new conditions than other organisms might have done.

This is not to say that there is no conservation concern behind the reactions of birds to climate change. Some of the studies presented above clearly give evidence of problems that birds might face when they need to adapt their behaviour to rapid environmental and climatic changes. It is very likely that among bird species there will be winners and losers resulting from the current climate and global change and it might also be that the rate of losers will be high and extinctions of bird species will reach a level exceeding extinction rates seen in earlier times in birds' evolution. Since birds are easy to observe, are present in all parts of the world and are objects of interest to many people, they are ideal flagships to observe the consequences and the impacts of future environmental changes on organisms and on ecosystems.

REFERENCES

[1] W. Fiedler, in: P. Berthold, E. Gwinner, E. Sonnenschein (Eds.), Avian Migration, Springer, Berlin, 2003, pp. 21–38.
[2] N. Ockendon, D.J. Baker, J.A. Carr, E.V.C. White, R.E.A. Almond, T. Amano, E. Bertram, R.B. Bradbury, C. Bradley, S.H.M. Butchart, N. Doswald, W. Foden, D.J.C. Gill, R.E. Green, W.J. Sutherland, Glob. Change Biol. 20 (2014) 2221–2229.
[3] J.F. Burton, Birds and Climate Change, Helm, London, 1995.
[4] G.V. Glucas, M.J. Dunn, G. Dyke, S.D. Emslie, H. Levy, R. Naveen, M.J. Polito, O.G. Pybus, A.D. Rogers, T. Hart, Art.no. 5024, Sci. Rep. 4 (2014), http://dx.doi.org/10.1038/srep05024.

[5] C.D. Thomas, J.J. Lennon, Nature 399 (1999) 213.

[6] M. Luoto, R. Virkkala, R.K. Heikkinen, Glob. Ecol. Biogeogr. 16 (2007) 34–42.

[7] D.I. Leech, H.Q.P. Crick, Ibis 249 (Suppl. 2) (2007) 138–145.

[8] A. Townsend Peterson, M.A. Ortega-Huerta, J. Bartley, V. Sánchez-Cordero, J. Soberón, R.H. Buddemeier, D.R.B. Stockwell, Nature 416 (2002) 626–628.

[9] N. Lemoine, H.-G. Bauer, M. Peintinger, K. Böhning-Gaese, Conserv. Biol. 21 (2007) 495–503.

[10] R. Kinzelbach, B. Nicolai, R. Schlenker, J. Ornithol. 138 (1997) 297–308.

[11] C.J.A. Bradshaw, B.W. Brook, S. Delean, D.A. Fordham, S. Herrando-Perez, P. Cassey, R. Early, C.H. Sekercioglu, M.B. Araujo, Proc. R. Soc. Lond. B 28 (2014) 1471–2954.

[12] R.B. Siegel, P. Pyle, J.H. Thorne, A.J. Holguin, C.A. Howell, S. Stock, M.W. Tingley, Avian Conserv. Ecol. 9 (2014), http://dx.doi.org/10.5751/ACE-00658-090107.

[13] R. Maggini, A. Lehmann, N. Zbinden, N.E. Zimmermann, J. Bolliger, B. Schröder, R. Foppen, H. Schmid, M. Beniston, L. Jenni, Diversity Distrib. 6 (2014), http://dx.doi.org/10.1111/ddi.12207.

[14] B. Huntley, R.E. Green, Y.C. Collingham, S.G. Willis, Climate Atlas of Breeding Birds in Europe, Lynx Editions, Barcelona, 2007.

[15] M.B. Araúja, C. Rahbek, Science 313 (2006) 1396–1397.

[16] J.M.J. Travis, M. Delgado, G. Bocedi, M. Baguette, K. Barton, D. Bonte, I. Boulangeat, J.A. Hodgson, A. Kubisch, V. Penteriani, M. Saastamoinen, V.M. Stevens, J.M. Bullock, Oikos 122 (2013) 1532–1540.

[17] F.A. La Sorte, F.R. Thompson, Ecology 88 (2007) 1803–1812.

[18] P. Berthold, Naturwiss. Rundsch. 51 (1998) 337–346.

[19] E. Lehikoinen, T.H. Sparks, M. Zalakevicius, in: A.P. Møller, W. Fiedler, P. Berthold (Eds.), Birds and Climate Change, Elsevier Science, London, 2004.

[20] W. Fiedler, in: T.M. Letcher (Ed.), Climate Change: Observed Impacts on Planet Earth, Elsevier, Oxford, 2009.

[21] T.H. Sparks, F. Bairlein, J.G. Bojarinova, O. Hüppop, E.A. Lehikoinen, K. Rainio, L.V. Sokolov, D. Walker, Glob. Change Biol. 11 (2005) 22–30.

[22] C. Parmesan, Glob. Change Biol. 13 (2007) 1860–1872.

[23] M. Ahola, T. Laaksonen, K. Sippola, T. Eeva, K. Rainio, E. Lehikoinen, Glob. Change Biol. 10 (2004) 1610–1617.

[24] J.W. Hurrell, Y. Kushnir, M. Visbeck, Science 291 (2001) 603–605.

[25] P.P. Marra, C.M. Francis, R.S. Mulvihill, F.R. Moore, Oecologia 142 (2005) 307–315.

[26] L.V. Sokolov, M.Y. Markovets, A.P. Shapoval, G.Y. Morozov, Zool. Zhurnal 78 (1999) 1102–1109.

[27] N. Saino, T. Szep, M. Romano, D. Rubolini, F. Spina, A.P. Moller, Ecol. Lett. 7 (2004) 21–25.

[28] O. Gordo, J.J. Sanz, Glob. Change Biol. 12 (2006) 1993–2004.

[29] O. Gordo, L. Brotons, X. Ferrer, P. Comas, Change Biol. 11 (2005) 12–21.

[30] L. Jenni, M. Kery, Proc. R. Soc. Lond. Ser. B Biol. Sci. 270 (2003) 1467–1471.

[31] P.A. Cotton, Proc. Nat. Acad. Sci. 100 (2003) 12219–12222.

[32] O. Hüppop, K. Hüppop, Proc. R. Soc. Lond. B 270 (2003) 233–240.

[33] T.H. Sparks, C.F. Mason, Ibis 146 (2004) 57–60.

[34] L.V. Sokolov, M.Y. Markovets, Y.G. Morozov, Avian Ecol. Behav. 2 (1999) 1–18.

[35] A. Gilyazov, T.H. Sparks, Avian Ecol. Behav. 8 (2002) 35–47.

[36] H.Q.P. Crick, T.H. Sparks, Nature 399 (1999) 423–424.

[37] W. Fiedler, U. Köppen, F. Bairlein, in: A.P. Møller, W. Fiedler, P. Berthold (Eds.), Birds and Climate Change, Elsevier Science, London, 2004.

[38] A. Soutullo, Climate Change and Shifts in Winter Distribution of European Breeding Birds, MSc dissertation, University of East Anglia, 2004.

[39] D.W. Snow, C.M. Perrins, Birds of the Western Palearctic, Oxford University Press, 1998.

[40] J.E. del Hoyo, A. Elliott, J. Sargatal (Eds.), Handbook of the Birds of the World, Lynx Editions, Barcelona, 1999.

[41] P. Berthold, Bird Migration: A General Survey, Oxford University Press, 2001.
[42] H. Schwabl, J. Ornithol 124 (1983) 101–116.
[43] W. Winkel, M. Frantzen, J. Ornithol 132 (1991) 81–96.
[44] C.M. Perrins, Acta XIX Congr. Int. Ornithol. 1 (1988) 892–899.
[45] L.E. Chambers, P. Dann, B. Cannell, E.J. Woehler, Int. J. Biometeorol. 58 (2014) 603–612.
[46] H.Q.P. Crick, C. Dudley, D.E. Glue, D.L. Thomson, Nature 388 (1997) 526.
[47] J.R. Liebezeit, K.E.B. Gurney, M. Budde, S. Zack, D. Ward, Polar Biol. (2014) 1309–1320.
[48] P.O. Dunn, D.W. Winkler, Proc. R. Soc. Lond. B 266 (1999) 2487–2490.
[49] W. Winkel, H. Hudde, J. Avian Biol. 28 (1997) 187–190.
[50] A.P. Møller, E. Flensted-Jensen, W. Mardal, J. Anim. Ecol. 75 (2006) 657–665.
[51] P.O. Dunn, A.P. Møller, J. Anim. Ecol. 83 (2014) 729–739.
[52] L. Halupka, A. Dyrcz, M. Borowiec, J. Avian Biol. 39 (2008) 95–100.
[53] M. Peintinger, S. Schuster, Vogelwarte 43 (2005) 161–169.
[54] W. Gatter, Vogelzug und Vogelbestände in Mitteleuropa, Aula, Wiebelsheim, 2000.
[55] E. Kaiser, Vogelwelt 125 (2004) 113–115.
[56] M. Gucco, G. Malacarne, G. Orecchia, G. Boano, Ecography 15 (1992) 184–189.
[57] J. Valencia, C. De la Cruz, J. Carranza, Etología 8 (2000) 25–28.
[58] S. Verhulst, J.M. Tinbergen, S. Daan, Funct. Ecol. 11 (2003) 714–722.
[59] S.F. Eden, A.G. Horn, M.L. Leonhard, Ibis 131 (2008) 429–432.
[60] L. von Haartman, Proc. Int. Ornithol. Congr. 14 (1967) 155–164.
[61] T. Schaefer, G. Ledebur, J. Beier, B. Leisler, J. Ornithol. 147 (2006) 47–56.
[62] R. Moss, J. Oswald, D. Baines, J. Anim. Ecol. 70 (2001) 47–61.
[63] M. Visser, A.J. van Noordwijk, J.M. Tinbergen, C.M. Lessells, Proc. R. Soc. Lond. B 265 (1998) 1867–1870.
[64] C. Both, S. Bouwhuis, C.M. Lessells, M.E. Visser, Nature 441 (2006) 81–83.
[65] R. Thornbill, J. Alcock, The Evolution of Insect Mating Systems, Harvard University Press, Cambridge, 1984.
[66] M. Andersson, Sexual Selection, Princeton University Press, Princeton, 1994.
[67] A.P. Møller, Glob. Change Biol. 10 (2004) 2028–2035.
[68] D. Rubolini, F. Spina, N. Saino, Behav. Ecol. 15 (2004) 592–601.
[69] C.N. Spottiswoode, A.P. Tøttrup, T. Coppack, Proc. R. Soc. Lond. B 273 (2006) 3023–3029.
[70] S. Bearhop, W. Fiedler, R.W. Furness, S.C. Votier, S. Waldron, J. Newton, G.J. Bowen, P. Berthold, K. Farnsworth, Science 310 (2005) 502–504.
[71] P. Berthold, S.B. Terill, Ringing Migr. 9 (1988) 153–159.
[72] T. Coppack, F. Pulido, M. Czisch, D.P. Auer, P. Berthold, Proc. R. Soc. Lond. B 270 (Suppl. 1) (2003) 43–46.
[73] V. Salewski, W.M. Hochachka, W. Fiedler, Oecologia 162 (2010) 247–260.
[74] J.L. Gardner, T. Amano, B.G. Mackey, W.J. Sutherland, M. Clayton, A. Peters, Glob. Change Biol. 20 (2014) 2062–2075.
[75] V. Salewski, K.H. Siebenrock, W.M. Hochachka, F. Woog, W. Fiedler, PLoS One 9 (2014), http://dx.doi.org/10.1371/journal.pone.0101927.
[76] W. Fiedler, Ann. N.Y. Acad. Sci. 1046 (2005) 253–263.
[77] T. Fuller, S. Bensch, I. Müller, J. Novembre, J. Pérez-Tris, R.E. Ricklefs, T.B. Smith, J. Waldenström, Ecohealth 9 (2012), http://dx.doi.org/10.1007/s10393-012-0750-1.
[78] A.T. Peterson, Naturwissenschaften 95 (2008) 483–491.
[79] A. Pérez-Rodríguez, I. de la Hera, S. Farnández-González, J. Pérez-Tris, Glob. Change Biol. 20 (2014) 2406–2416.
[80] C.T. Atkinson, R.B. Utzurrum, D.A. Lapointe, R.J. Camp, L.H. Crampton, J.T. Foster, T.W. Giambelluca, Glob. Change Biol. 20 (2014) 2426–2436.

MAMMAL ECOLOGY

Jeremy R. Brammer, Murray M. Humphries

Department of Natural Resource Sciences, McGill University, Montreal, QC, Canada

CHAPTER OUTLINE

1. INTRODUCTION: HOW DOES CLIMATE IMPACT MAMMALS?

On January 12, 2002, the temperature in New South Wales, Australia, exceeded 42°C. On this day more than 3500 flying foxes (fruit bats of the genus *Pteropus*) succumbed to hyperthermia. A similar die-off of 5000 flying foxes also occurred in 2014 [1]. Mass die-offs of flying foxes associated with heat waves occurred three times before 1990, three times between 1990 and 2000, and 13 times between 2000 and 2007 [2]. Flying foxes, like other mammals, use a combination of physiological and behavioural mechanisms to regulate their body temperature [3]. Thermoregulation is a defining feature of endotherms; it allows them to decouple their core body temperature from air temperature. Thus, despite exposure to very cold or very hot air temperatures, thermoregulatory responses ensure that mammal core body temperature never varies by more than a few degrees centigrade between birth and death [4]. Even mammals that express torpor (i.e. decreased physiological activity to conserve energy) do not abandon thermoregulation but rather lower their thermoregulatory set point [5]. For all endotherms, the abandonment of thermoregulation is fatal.

Thermoregulation requires the use of metabolic energy to regulate body temperature [3]. When air temperatures are moderate, thermoregulation requires no additional heat production or dissipation because small, energetically insignificant adjustments in conductance (e.g. vasodilation, piloerection, and postural changes) are sufficient to maintain a constant body temperature [3]. This is an animal's **thermoneutral zone**. When air temperatures drop below a critical threshold, metabolic rates increase to produce heat and maintain a constant body temperature (Fig. 1(a)) [3]. When air temperatures rise above a critical threshold, endotherms actively dissipate heat through panting, perspiration, saliva spreading, and in the case of bats, wing fanning [3]. Because these responses increase heat production, contributing to the problem that it solves, the slope of the increase in metabolism at warm temperatures

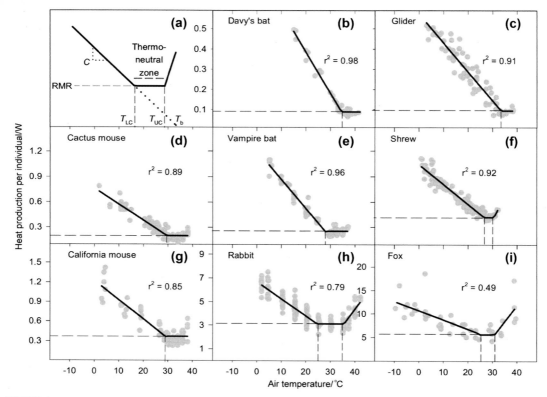

FIGURE 1

Metabolic rate as a function of air temperature: interpretive framework and examples. (a) Theoretical metabolic rate as a function of air temperature. The thermoneutral zone indicates temperatures where metabolic rate is independent of air temperature. Above the upper critical temperature (T_{UC}), and below the lower critical temperature (T_{LC}), metabolic rate increases due to thermoregulation. (b)–(i) Metabolic rate as a function of air temperature in eight species: (b) *Pteronotus davyi*; (c) *Acrobates pygmaeus*; (d) *Peromyscus eremicus* (Nevada); (e) *Desmodus rotundus*; (f) *Blarina brevicauda*; (g) *Peromyscus californicus* (*parasiticus*); (h) *Sylvilagus audubonii* (winter); (i) *Vulpes macrotis* (winter). *(Adapted from Ref. [7].)*

is always steeper than the slope at cold temperatures (Fig. 1(f) and (i)). Due to the inefficiency of heat dissipation, endotherms are particularly vulnerable to heat stress and occupy microenvironments to avoid it [6]. These elaborate adaptations to heat and cold stress might be expected to enable a wide degree of thermal independence. But 3500 dead flying foxes suggests that is incorrect [2]. To understand why, we must expand our consideration beyond heat and cold stress to a broader set of **direct** and **indirect** climate impacts on mammals.

Climate, in the form of temperature, precipitation, and humidity, can **directly** impact a mammal's metabolic rate, thus influencing its resource requirements and survival [8]. Mammals require more resources to stay alive when they metabolically increase heat production or heat loss in response to cold or hot temperatures. Precipitation and humidity mediate this heat loss through evaporative cooling, conduction, and convection [9]. As well, although difficult to isolate, precipitation can impact resource requirements by facilitating or inhibiting movement (e.g. snow depth [10]). Ultimately, heat production and loss are not without limits. Exposure to extreme temperatures can exceed thermoregulatory capacity, first leading to hypo- or hyperthermia, then death. Exposure leading to mortality is increasingly likely during extreme weather events like heat waves [2], icing events [11], winter storms [12], cyclones [13], and hurricanes [14]. Thus, through exposure to changes in air temperature, precipitation, and humidity, climate has direct impacts on the metabolism, resource requirements, and survival of mammals.

Climate exerts additional, **indirect** effects on mammals through their resources, competitors, and predators. Temperature has a fundamental effect on all biological processes [15]; thus climate should profoundly affect most interactions between a mammal and its environment. These indirect effects are likely to be so strong that they will frequently supersede direct effects of climate. It is true that the mass die-off of flying foxes provides a potent example of a direct effect of climate [2]. But even here, it is likely that indirect climate effects played a role. For example, although 1453 flying foxes from the Dallis Park colony succumbed to hyperthermia, more than 25 000 individuals exposed to the same thermal conditions survived [2]. Many factors determine thermoregulatory capacity: body size, age, social rank, reproductive condition, body composition, and aerobic capacity [2,3]. Most of these are determined by an individual's lifetime experience with resources, competitors, and predators. As we will see below, climate impacts on mammals are most commonly a combination of both **indirect** and **direct** effects on energetics, resources, competitors, and predators [11,16,17].

2. CLIMATE IMPACTS SCALE FROM LOCAL TO GLOBAL, DAILY TO GEOLOGICAL

What impacts of climate change are documented is strongly influenced by the duration and area of study, or more specifically the **temporal** and **spatial scales** of research. When data are analyzed across large areas, or **spatial extents** (e.g. continents and hemispheres), phenomena like diversity [18], abundance [19], body size [20], and metabolism [21] are often correlated with climate. Of course, climate incorporates many related variables, and those with the highest predictive power vary by region and taxon. In contrast at local and regional spatial extents, interspecific interactions frequently emerge as more powerful predictors of ecological variation [22,23]. This is perhaps overly simplistic; at all spatial scales it should be clear that ecological variation is driven by a complex relationship between climate, biotic interactions, physical geography, and history. But importantly, our capacity to quantify and detect those influences varies with spatial scale, particularly the extent and the resolution of what we observe.

Similar to space, documenting mammalian responses to climate change depends on the time period of observation [24]. Data collected over limited **temporal extents** (e.g. weeks to decades) document mammal responses to weather or short-term climate variation. At these extents, the responses observed are frequently behavioural and other forms of phenotypic plasticity [25,26]. But these observations reveal less about the impacts of long-term climate change, particularly when this will involve climate conditions outside of what was observed historically. Paleontological observations of larger temporal extent (e.g. millions of years) incorporate more of this variation but, due to their coarser resolution, can obscure the significance of events occurring over short time periods. For example, Gingerich [27] argues that the failure to detect a relationship between climate and diversity over the 65 Ma Cenozoic time period resulted from the averaging of climate and diversity over 1 Ma (million years) intervals when most significant changes occurred over 1000 a. Generally paleontological studies are, due to coarser temporal resolution and the nature of past climatic change, confined to studying changes that are much more gradual than current climate change. Thus, there is a continuum of temporal extents available to study the effects of climate change on mammals—paleontological time series represent one end, ecological time series the other. Most studies are situated at either end of this continuum;

unfortunately, to understand how mammals will respond to climate change, we need paleontological time series detailing brief periods of rapid climate change at finer resolutions, together with ecological time series of longer durations. We must remain conscious of this continuum of temporal extents, together with its spatial parallel, in the following discussion of climate impacts on mammals.

3. DEMONSTRATED IMPACTS OF CLIMATE CHANGE ON MAMMALS

Demonstrated impacts of climate change on mammals is a broad topic, in evolutionary time, geographic scope, and taxonomic diversity, which cannot be covered comprehensively in a short chapter. We refer the reader to many excellent reviews regarding climate impacts on mammal paleoecology [28,29], mammal population dynamics [30], animal phenology and distribution [31], mammal morphology [32,33], mammal demographics [34], Arctic marine mammals [35,36], Arctic mammal molecular ecology [37], Australian fauna [38], migratory fauna [39], and bats [40]. In the following sections, we will proceed from lower to higher levels of organization in the following order: mammal metabolism (Section 3.1); morphology (Section 3.2); phenology (Section 3.3); population dynamics (Section 3.4); range limits (Section 3.5); and community composition and diversity (Section 3.6).

3.1 CLIMATE IMPACTS ON MAMMAL METABOLISM

Mammals at lower latitudes are metabolically slower with lower body temperatures than their higher latitude counterparts (Fig. 2(a) and (b)) [21,41]. Classic work demonstrating that cold environments

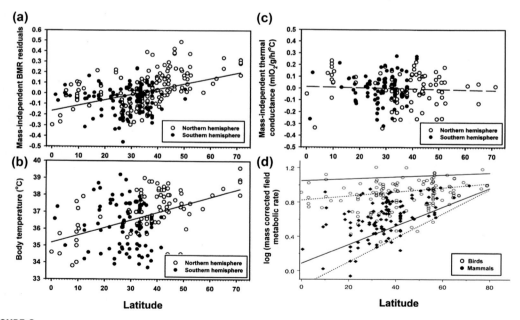

FIGURE 2

(a) Mass independent $\log_{10}$ basal metabolic rate (BMR); (b) body temperature; (c) mass independent thermal conductance; and (d) mass corrected log field metabolic rate (FMR) as a function of latitude. In (d) the field metabolic rate of birds and mammals tends to decrease with increasing air temperature but also tends to be more variable in warmer environments. The label along the *y*-axis of the figure in the lower left quadrant should read body temperature/°C. *(Adapted from Refs [21,45].)*

require increased heat production and reduced heat loss [42,43] have been expanded using phylogenetically informed analyses [21,44]. In addition to the impacts of climate averages, global correlative studies [21,41] have shown that species inhabiting regions with less predictable climates (usually lower latitudes) have lower resting metabolic rates than species inhabiting more predictable climates (usually higher latitudes) (Fig. 2(a–c)). Similarly, the field metabolic rate of mammals tends to decrease with increasing air temperature but also tends to be more variable in warmer environments (Fig. 2(d)) [45]; see Ref. [8]. This supports the metabolic niche hypothesis that the range of energetically feasible lifestyles becomes restricted in cold environments [45,46]. In highly variable environments of any temperature, torpor expression during seasonal hibernation is a common adaptation by mammals to reduce metabolic rates during resource shortages [5]. While hibernation may become more common at high latitudes under warming [47], at lower latitudes warmer temperatures could limit the energy savings of torpor [9,48]. Indeed, interspecific differences in thermoregulatory characteristics should mediate warming impacts on mammals. For example, it is suggested that thermal generalists who are more capable of functioning over a greater range of body temperatures should cope better with extreme temperature events than thermal specialists [48]. It is an ongoing challenge to document the thermoregulatory characteristics of species, populations, and individuals and how these interact with weather, to understand how directional climate change will impact mammal metabolism.

3.2 CLIMATE IMPACTS ON MAMMAL MORPHOLOGY

Similar to metabolism, mammals in warmer environments tend to be smaller (Fig. 3(a)). This tendency, known as Bergmann's rule, is well supported by studies of mammals distributed along latitudinal and climatic gradients [20,49,50]. There are many notable exceptions, but in general, more than two-thirds of mammals conform to this trend [49,50]. Similarly, several multiyear to multiepoch

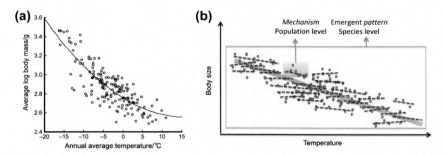

FIGURE 3

Empirical and conceptual representations of the relationship between mammalian body size and temperature. (a) Average log body mass (g) of North American mammals in 48 400 km^2 grid cells as a function of average annual temperature. (b) Conceptual representation of the mammal body size–temperature relationship demonstrating how population level responses can be contrary to Bergmann's rule due to the smaller temperature gradient experienced. *(Panels adapted from Refs [20,55].)*

studies have documented declines in mammal body size that correlate with warming climate [29,32,51,52]. However, other decadal studies have documented the opposite trend [53–55]. Overall, evidence of any impacts of current warming on body size is mixed [33,55], likely due to large discrepancy between temperature gradients observed globally and paleontologically, and those observed in decadal time series (Fig. 3(b); Δ12.29°C vs Δ0.93°C in Ref. [55]). In addition, uncertainty regarding the mechanisms driving Bergmann's rule further complicates the prediction of climate impacts on body size. The original explanation for Bergmann's rule, increased heat conservation in cold environments, is no longer widely accepted because the trend occurs in both large and small endotherms and in many ectotherms [49,50,56]. A more likely explanation is resource accessibility that is mediated by climate, productivity, competition, and predation [33,56,57]. Teasing apart these mechanisms is context dependent; it will require longer time series to identify how contemporary climate change will impact mammal body size.

3.3 CLIMATE IMPACTS ON MAMMAL PHENOLOGY

Changes in phenology, such as the annual timing of reproduction or dormancy, are among the best documented impacts of climate change on animals and plants [31,58,59]. Examples among mammals are fewer [26,59], but they provide important lessons. In Colorado, USA, the date of weaning of yellow-bellied marmots (*Marmota flaviventris*) advanced 5.6 d (days) over 30 a, while their date of emergence from hibernation advanced by 38 d over 25 a [60,61] (Fig. 4(b)). These phenological

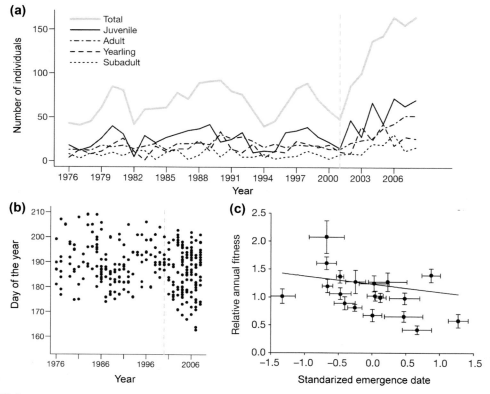

FIGURE 4

How phenology can both positively (a–b) and negatively (c) impact fitness. Increasing abundance (a) and advancing weaning date (b) of yellow-bellied marmots (*Marmota flaviventris*) over a 30-year warming period in Colorado, USA; and (c) the mean relative annual fitness (winter survival + 0.5 * reproductive success) of Colombian ground squirrels (*Urocitellus columbianus*) as a function of emergence date. Vertical lines indicate phases of population dynamics. Error bars represent s.e.m. *(Adapted from Refs [61,62].)*

changes had demographic consequences: marmot body mass, survival, and abundance increased (Fig. 4(a)) [61]. But phenological shifts do not always increase fitness. In Alberta, Canada, Columbia ground squirrels (*Urocitellus columbianus*) have delayed their emergence from hibernation 9.4 d over 20 a, resulting in decreased winter survival and reproductive success (Fig. 4(d)) [62]. This appears to be driven not by a temperature trend but by delayed snowmelt due to an increase in late winter storms [62]. Whether these phenological changes reduce fitness through a shortened active season or through a trophic mismatch remains unclear. The latter has been documented in bighorn sheep (*Ovis canadensis*), alpine ibex (*Capra ibex*), and caribou (*Rangifer tarandus*) [63,64]. Documenting these links between climate and mammal phenology, demography, and fitness is rare, but it is necessary to understand the mechanisms linking climate to mammals [65]. Currently, the majority of documented changes in mammal phenology represent phenotypic not evolutionary responses [26]. Phenotypic plasticity was behind two-thirds of the advancing parturition date of red squirrels (*Tamiasciurus hudsonicus*) in Yukon, Canada, whereas one-third involved microevolution [66,67]. This study represents one of the few demonstrations of mammalian evolutionary capacity to adapt to contemporary climate change, but the taxonomic generality and long-term sustainability of this evolutionary capacity is unclear [25,26].

3.4 CLIMATE IMPACTS ON MAMMAL POPULATION DYNAMICS

Rigorous demonstration of population-level impacts of climate requires long-term monitoring of mammal abundance alongside climatic and biotic drivers [30]. Several case studies provide compelling evidence of climate impacts on mammals. Polar bears (*Ursus maritimus*) play a prominent role in the discourse on climate change because they rely on sea ice that is projected to decline [68]. The strongest evidence of climate impacts on polar bears comes from a 20 a time series of the Western Hudson Bay subpopulation [69]. There, juvenile, subadult, and senescent-adult survival declined 6%, leading to population declines of 22% over a period when ice breakup advanced by three weeks (Fig. 5(a–e)) [69]. Similarly, in a shorter 5 a time series from the Southern Beaufort Sea subpopulation, breeding female survival declined 20% during an increase in the ice-free season of 34 days (Fig. 5(f)) [70]. Comparable data are generally lacking for other populations [71]. Still, if warming and sea ice decline continues at projected rates, Stirling and Derocher [71] predict polar bears will disappear from their southern ranges and will have an uncertain outlook overall [71,72].

Evidence of climate impacts on the population dynamics of other mammals arise from the influence of large-scale climate oscillations on population dynamics in: muskox (*Ovibos moschatus*) and caribou (*Rangifer tarandus*; [73]), soay sheep (*Ovis aries*) and red deer (*Cervus elaphus*; [74]), wolves (*Canis lupus*) and moose (*Alces alces*; [16]), ibex (*Capra ibex*; [75]) grey-sided voles (*Myodes rufocanus*; [76]), lynx (*Lynx canadensis*; [77]), collared and American pikas (*Ochotona collaris* and *Ochotona princeps*; [78,79]), South American leaf-eared mice (*Phyllotis darwini*; [80]), southern elephant seal (*Mirounga leonine*; [81]) and Savanna ungulates [82]. While the generality of this single-species approach has been questioned [83], longer time series analyses have frequently identified the demographic basis of population-level weather effects. However, these studies provide a weaker basis to predict future climate impacts as this requires extrapolation beyond historical conditions. In addition, the absence of predator and prey data [84], as well as landscape modifications, clouds interpretation of the mechanistic links between climate and mammal population dynamics (e.g. Ref. [85]).

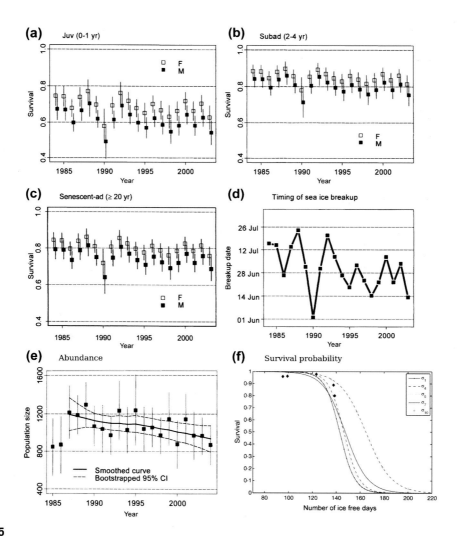

FIGURE 5

Survival of western Hudson's Bay polar bear (a) juveniles, (b) subadults, (c) and senescent adults; (d) timing of sea ice breakup; (e) polar bear abundance; and (f) survival as a function of ice-free days in the southern Beaufort Sea. Error bars represent 95% confidence intervals. Survival curves are plotted for five life stages (subadult females to adult males) and diamonds represent model-averaged survival estimates. (*Adapted from Refs [69,70].*)

Additionally, climate change has been blamed for the disappearance or dampening of many of the renowned mammal population cycles [86]. Large-scale spatial variation in cycle amplitude and period, including the pattern of cyclic dynamics at northern latitudes transitioning to stable dynamics at lower latitudes, has been attributed to variation in seasonality [86]. In Europe, lower winter growth rates appear to be driving the decline in vole cyclicity [87]. But the hypothesis that cold winters drive cycles may be too simplistic. Finnish voles cycle in regions with long, cold, snowy winters (Fig. 6(a)), and in regions with warm summers (Fig. 6(d)), but not in regions that are intermediate (Fig. 6(c)) [88]. While

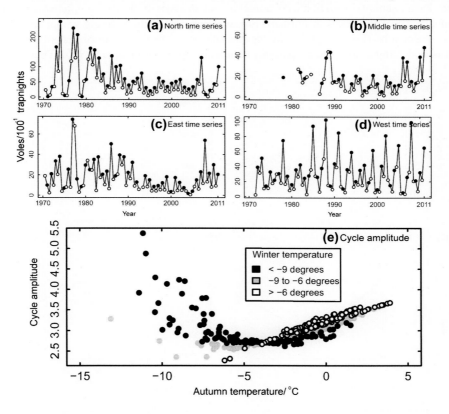

FIGURE 6

Population cycles in vole density in different regions of Finland. Vole density time series are presented in panels (a–d), with filled circles representing fall density and open circles representing spring density. Panel (e) presents estimated cycle amplitude as a function of autumn and winter temperatures. Note how amplitude declines with rising autumn temperatures at sites with cold winter temperatures, while the reverse occurs at sites with warmer winter temperatures. *(Adapted from Ref. [88].)*

warming has been associated with a decline in cyclicity in colder regions (Fig. 6(a) and (e) black points), in some regions we may see an increase in cyclicity (Fig. 6(d) and (e) white points).

3.5 CLIMATE IMPACTS ON MAMMAL RANGE LIMITS

Range shifts are a commonly documented impact of climate change on mammals; but understanding the mechanisms driving this phenomenon is an ongoing challenge [28,89,90]. Climatic conditions do correlate with the range boundaries of many mammals. For example, the breeding distribution of grey seals (*Halichoerus grypus*) may be limited by cold air temperatures on the fasting endurance of recently weaned pups [91]. The winter distribution of little brown bats (*Myotis lucifugus*) is limited to where hibernacula are warm enough so the energetic costs of hibernation do not exceed fat reserves [92]. The northward range limit of nine-banded armadillos (*Dasypus novemcinctus*) and Virginia opossums (*Didelphis virginiana*) seems to be constrained by long bouts of cold winter weather [93,94]. Relatively few studies have examined variation in mammal abundance across climatic gradients with their range, but the abundance of red foxes in northern Eurasia decreases with declining winter temperature and increasing seasonality (Fig. 7) [95], and the abundance of beaver in northeastern North America decreases with declining potential evapotranspiration and spring temperatures [96]. There is a need for

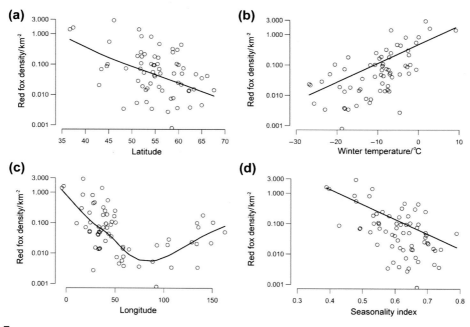

FIGURE 7

Red fox density in Eurasia as a function of (a) latitude; (b) winter temperature; (c) longitude; and (d) seasonality. Seasonality is calculated as the yearly variance of photosynthetically active radiation estimated using MODIS imagery. Lines represent smoothing splines in panels (a) and (c) (df = 3 and 5, respectively), and predictions from a GLM in panels (b) and (d) using winter temperature and seasonality as explanatory variables. *(Adapted from Ref. [95].)*

more studies of mammal abundance across species' ranges, because how abundance varies near range limits will dictate whether climate change impacts will be greater at range edges or cores [96].

The distribution of mammal abundance within their range, together with different life histories, and other forms of environmental change, complicate attempts to project range shifts under climate change. A classic example is the southern range contraction of arctic foxes (*Vulpes lagopus*) concomitant with the northern range expansion of red foxes (*Vulpes vulpes*), which is mediated by climate-driven changes in primary productivity and prey base [97]. This hypothesis is partially supported by comparing sites abandoned and occupied by arctic foxes, but there are many site-specific contingencies and alternative explanations [98]. In North America, many of the mammals with poleward range expansions are species that are affiliated with human-modified habitats, including red foxes, Virginia's opossum (*Didelphis virginiana*) [93], raccoons (*Procyon lotor*) [99], white-tailed deer (*Odocoileus virginianus*) [100], and coyotes (*Canis latrans*) [101]. Similar patterns have been observed in range changes along altitudinal gradients [102]. To further complicate matters, a review of 452 elevation range shifts in Californian flora and fauna found no trend of range movement up elevation gradients due to a combination of life history differences, land use changes, trophic interactions, and other climate impacts [90]. In addition, in many regions of the world little is known about how much plants and animals actually vary across regional climate gradients (i.e. across hundreds to thousands of kilometres), much less how and why this variation is correlated to climate. Still, climate change is contributing to the poleward range shifts of many mammals and this will be an important impact over the next century [31,58,103,104]. Species of particular concern are those whose dispersal ability is limited, and whose poleward range limit is constrained by barriers like coastlines, mountain ranges, or human land, leaving them nowhere to go as they shift toward these barriers [105].

3.6 CLIMATE IMPACTS ON MAMMAL COMMUNITIES AND DIVERSITY

Latitudinal gradients in species diversity are among the strongest and most general patterns in ecology [106,107]. Warm productive regions of the globe host a greater diversity of mammals. In North America, measures of local mammal diversity vary from 178 species in tropical Central America to 20 species in arctic Canada ([18]; Fig. 8). Almost 90% of this variation can be explained by five variables,

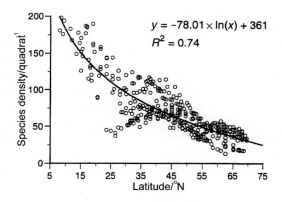

$$y = -78.01 \times \ln(x) + 361$$
$$R^2 = 0.74$$

FIGURE 8

Latitudinal variation in the diversity of North American mammals. The diversity measure is the number of terrestrial mammal species present in 58 275 km^2 quadrats distributed across the continent. Five environmental variables, representing seasonal extremes of temperature, annual energy and moisture, and elevation, predicted 88% of continental variation in this variable. *(Adapted from Ref. [18].)*

representing seasonal temperature, precipitation, evapotranspiration, and elevation [18]. A global analysis found that three environmental variables, mostly the actual evapotranspiration, could explain 63% of the spatial variation in mammal diversity, but that this relationship varies by region [108].

North America appears to be the major exception, which Davies et al. suggest could be due to the megafaunal extinction of the late Pleistocene that decoupled local mammal diversity from productivity [108]. Irrespective of this regional variation, climate variables usually outperform land-cover variables as predictors of diversity [109,110], particularly when spatial grains are large [111]. But it is unclear whether this is because climate is the more important mechanistic driver of diversity or because it is difficult to classify land cover appropriately. Regardless, the generality and strength of climate–diversity correlations across biogeographical space has caused them to occupy a central role in predicting the impacts of climate change on mammals.

Similar to biogeographic patterns, paleontological studies demonstrate that climate has pronounced effects on mammal communities [24,29,112]. Climate has been observed to impact mammal communities over time scales of thousands of years [24,113,114] to millions of years [29,115]. For example, the megafaunal extinctions of the late Pleistocene were driven by, together with human impacts, climate forcing (see Fig. 2 in Ref. [114]) [29,114]. In North America, mammal communities were significantly altered, with roughly 70 of the largest 220 mammals becoming extinct [113,114]. Research using ancient DNA to reconstruct abundance trends of Beringian steppe bison (*Bison* spp.) has established that preextinction population declines coincided with periods of climate warming and forest expansion [116]. Related genetics research suggests a similar pattern of climate-driven population declines in Pleistocene bears, horses, and mammoths [117–119], but the relative contributions of human and climate forcing to megafaunal extinctions continues to be debated [114,120]. Like megafauna, small mammal communities also responded to climatic changes; northern Californian small mammal communities declined in evenness and richness over this same period [113]. Over a longer period of 22 Ma, fossil rodent communities from central Spain responded to climate impacts with pulses of species turnover at 1–2 Ma intervals coinciding with astronomical climate forcing associated with Milankovitch oscillations [115].

The strong associations between climate and faunal composition observed across paleontological timescales suggest the maintenance of some degree of thermal niche conservatism over prolonged periods of climate change. Direct evidence of this niche conservatism is provided by the demonstration by Martinez-Meyer et al. [121] that 23 extant North American mammals with fossil records have spatially tracked consistent climate profiles for the last 18 000 a. Associations between paleoclimate reconstructions and the contemporary diversity and composition of fossil mammal assemblages are sufficiently strong that some have advocated using fossil mammal assemblages to reconstruct paleoclimates (e.g. Refs [122,123]). This thermal niche conservatism is likely driven by ecological niche conservatism, as mammals track particular habitat during climatic change [29].

This niche conservatism leads to a remarkable, albeit counterintuitive, observation of mammal community stability over the past million years [28,29]. While individual species at particular localities may change with climate (tracking their particular niche), the trophic structure of the mammal community remains similar through a shuffling of community composition and relative abundances [28,29]. As a result, with the exception of the Pleistocene-Holocene megafaunal extinction, paleontological analyses of climate change have not found the large losses of mammal biodiversity projected by some (up to 37% of species by 2050 [124]) even during periods of similarly rapid warming [28]. Instead, these studies highlight the importance of range shifts, microrefugia, and life history responses that reduce the biodiversity impacts of climate change [28,29]. The notable exception, the megafaunal extinctions at the end of the Pleistocene, appears to be the consequence of a combination of climate change and human activities. It is this combination of swift climate change, and anthropogenic harvest and land use changes, that pose the greatest threat to mammal diversity [114,120].

4. CONCLUSION: LINKING TIME AND SPACE IN MAMMAL RESPONSES TO CLIMATE

Here we have briefly reviewed observed climate impacts on mammal metabolism, morphology, phenology, population dynamics, range limits, and diversity. However, a great deal of uncertainty remains regarding impacts of future rapid climate change. A better understanding of the impacts of this change on mammals awaits integration of temporal and spatial approaches, with careful attention paid to the scales of comparison. We end with a brief review of two approaches with considerable promise in this regard.

Phylogeography, as its name implies, integrates spatial and temporal approaches by examining historical influences on geographical distributions. Much of phylogeography is based on genetic approaches, but palaeontology spread across geographical gradients provides similar insights [112]. For example, comparing the distribution of woolly mammoths (*Mammuthus primigenius*) and reconstructed climates at different time intervals during the Pleistocene provides insight into the roles of climate change and human hunting in this species' range contraction and extinction [125]. Genetic estimates of the timing of population and species divergence linked with paleoclimate and habitat reconstructions can provide critical insight into the influence of past environmental change on contemporary diversity. The divergence of Antarctic minke whales (*Balaenoptera bonaerensis*) and common minke whales (*Balaenoptera acutorostrata*) is estimated to have occurred during an extended warming period in the Pliocene, when elevated ocean temperatures would have disrupted the spatial homogeneity of oceanic upwelling and promoted allopatric speciation [126].

A contemporary analogue to phylogeography is provided by spatially extensive, long-term monitoring of species diversity, population abundance, and individual traits (e.g. Refs [127,128]). Unfortunately, the best of these programs, such as breeding and winter bird surveys conducted for many decades over continental spatial scales [129], involve nonmammalian taxa. Nevertheless, these programs can serve as a model for the type and spatial extent of long-term data needed to document the effects of anthropogenic climate change on mammals. They also provide a warning of the difficulty in disentangling the effects of climate change from other forms of environmental change, even with fantastic data [130]. As anthropogenic climate change accelerates, we need to continue and expand the few global examples of spatially extensive, long-term mammal monitoring programs (e.g. Refs [131,132]); for the many mammal taxa and regions currently excluded, we need to initiate rigorous monitoring programs before it is too late.

REFERENCES

[1] M. Godfrey, Heatwave Hell as 5000 Dead Bats Drop from Trees in Casino, Northern New South Wales, The Daily Telegraph, November 19, 2014.
[2] J.A. Welbergen, S.M. Klose, N. Markus, P. Eby, Proc. Biol. Sci. 275 (2008) 419.
[3] B.K. McNab, The Physiological Ecology of Vertebrates: A View from Energetics, Cornell University Press, Ithaca, 2002.

[4] B. Heinrich, Am. Nat. 111 (1977) 623.

[5] F. Geiser, Annu. Rev. Physiol. 66 (2004) 239.

[6] C.R. Vispo, G.S. Bakken, Ecology 74 (1993) 377.

[7] R.W. Hill, T.E. Muich, M.M. Humphries, PLoS One 8 (2013) e76238.

[8] M.M. Humphries, S. Boutin, D.W. Thomas, J.D. Ryan, C. Selman, A.G. McAdam, D. Berteaux, J.R. Speakman, Ecol. Lett. 8 (2005) 1326.

[9] B.G. Lovegrove, C. Canale, D. Levesque, G. Fluch, M. Reháková-Petrů, T. Ruf, Physiol. Biochem. Zool. 87 (2013) 30.

[10] A. Loison, R. Langvatn, E.J. Solberg, Ecography (Cop.) 22 (1999) 20.

[11] K.-S. Chan, A. Mysterud, N.A. Øritsland, T. Severinsen, N.C. Stenseth, Oecologia 145 (2005) 556.

[12] T.B. Hallett, T. Coulson, J.G. Pilkington, J.M. Pemberton, B.T. Grenfell, Nature 430 (2004) 71.

[13] G. Jones, D. Jacobs, T. Kunz, M. Willig, P. Racey, Endanger. Species Res. 8 (2009) 93.

[14] C.A. Langtimm, C.A. Beck, Ecol. Appl. 13 (2003) 257.

[15] J.F. Gillooly, J.H. Brown, G.B. West, V.M. Savage, E.L. Charnov, Science 293 (2001) 2248.

[16] E. Post, R.O. Peterson, N.C. Stenseth, B.E. Mclaren, Nature 88 (1999) 905.

[17] S.A.R. Mduma, A.R.E. Sinclair, R. Hilborn, J. Anim. Ecol. 68 (1999) 1101.

[18] C. Badgley, D.L. Fox, J. Biogeogr. 27 (2000) 1437.

[19] E. Post, N.C. Stenseth, Ecology 80 (1999) 1322.

[20] T.M. Blackburn, B.A. Hawkins, Ecography (Cop.) 27 (2004) 715.

[21] B.G. Lovegrove, J. Comp. Physiol. B 173 (2003) 87.

[22] S.E. Williams, H. Marsh, J. Winter Ecol. 83 (2002) 1317.

[23] P.L. Meserve, D.a. Kelt, W.B. Milstead, J.R. Gutiérrez, Bioscience 53 (2003) 633.

[24] A.D. Barnosky, E.A. Hadly, C.J. Bell, J. Mammal. 84 (2003) 354.

[25] D. Berteaux, D. Réale, A.G. McAdam, S. Boutin, Integr. Comp. Biol. 44 (2004) 140.

[26] S. Boutin, J.E. Lane, Evol. Appl. 7 (2014) 29.

[27] P.D. Gingerich, Trends Ecol. Evol. 21 (2006) 246.

[28] C. Moritz, R. Agudo, Science 341 (2013) 504.

[29] J.L. Blois, E.A. Hadly, Annu. Rev. Earth Planet. Sci. 37 (2009) 181.

[30] M.C. Forchhammer, N.M. Schmidt, T.T. Høye, T.B. Berg, D.K. Hendrichsen, Population dynamical responses to climate change, in: H. Meltofte, T.R. Christensen, B. Elberling, M.C. Forchhammer, M. Rasch (Eds.), Advances in Ecological Research, fortieth ed., Elsevier Ltd, London, 2008, pp. 391–419.

[31] C. Parmesan, Annu. Rev. Ecol. Evol. Syst. 37 (2006) 637.

[32] V. Millien, S. Kathleen Lyons, L. Olson, F.a. Smith, A.B. Wilson, Y. Yom-Tov, Ecol. Lett. 9 (2006) 853.

[33] J.L. Gardner, A. Peters, M.R. Kearney, L. Joseph, R. Heinsohn, Trends Ecol. Evol. 26 (2011) 285.

[34] V. Grosbois, O. Gimenez, J.M. Gaillard, R. Pradel, C. Barbraud, J. Clobert, A.P. Møller, H. Weimerskirch, Biol. Rev. Camb. Philos. Soc. 83 (2008) 357.

[35] S.E. Moore, H.P. Huntington, Ecol. Appl. 18 (2008) S157.

[36] K.L. Laidre, I. Stirling, L.F. Lowry, Ø. Wiig, M.P. Heide, S.H. Ferguson, Ecol. Appl. 18 (2008) S97.

[37] G. O'Corry-Crowe, Ecol. Appl. 18 (2008) S56.

[38] L. Hughes, Austral Ecol. 28 (2003) 423.

[39] R. Robinson, H. Crick, J. Learmonth, I. Maclean, C. Thomas, F. Bairlein, M. Forchhammer, C. Francis, J. Gill, B. Godley, J. Harwood, G. Hays, B. Huntley, A. Hutson, G. Pierce, M. Rehfisch, D. Sims, B. Santos, T. Sparks, D. Stroud, M. Visser, Endanger. Species Res. 7 (2009) 87.

[40] H.a. Sherwin, W.I. Montgomery, M.G. Lundy, Mamm. Rev. 43 (2013) 171.

[41] B. Lovegrove, Am. Nat. 156 (2000) 201.

[42] P.F. Scholander, V. Walters, R. Hock, L. Irving, Biol. Bull. 99 (1950) 225.

[43] P.F. Scholander, R. Hock, V. Walters, L. Irving, Biol. Bull. 99 (1950) 259.

[44] B.G. Lovegrove, J. Comp. Physiol. B 175 (2005) 231.
[45] K.J. Anderson, W. Jetz, Ecol. Lett. 8 (2005) 310.
[46] A. Clarke, K.J. Gaston, Proc. Biol. Sci. 273 (2006) 2257.
[47] M.M. Humphries, J. Umbanhowar, K.S. McCann, Integr. Comp. Biol. 44 (2004) 152.
[48] J.G. Boyles, F. Seebacher, B. Smit, A.E. McKechnie, Integr. Comp. Biol. 51 (2011) 676.
[49] K.G. Ashton, M.C. Tracy, A. De Queiroz, Am. Nat. 156 (2000) 390.
[50] S. Meiri, T. Dayan, J. Biogeogr. 30 (2003) 331.
[51] F.A. Smith, J.L. Betancourt, J.H. Brown, Science 270 (1995) 2012.
[52] F.A. Smith, H. Browning, U.L. Shepherd, Ecography (Cop.) 21 (1998) 140.
[53] Y. Yom-Tov, S. Yom-Tov, Biol. J. Linn. Soc. 82 (2004) 263.
[54] Y. Yom-Tov, J. Yom-Tov, J. Anim. Ecol. 74 (2005) 803.
[55] C. Teplitsky, V. Millien, Evol. Appl. 7 (2014) 156.
[56] B.K. McNab, Oecologia 164 (2010) 13.
[57] Y. Yom-Tov, E. Geffen, Biol. Rev. Camb. Philos. Soc. 86 (2011) 531.
[58] G. Walther, E. Post, P. Convey, A. Menzel, C. Parmesan, T.J.C. Beebee, J. Fromentin, O. Hoegh-Guldberg, F. Bairlein, Nature 416 (2002) 389.
[59] C. Parmesan, G. Yohe, Nature 421 (2003) 37.
[60] D.W. Inouye, B. Barr, K.B. Armitage, B.D. Inouye, Proc. Natl. Acad. Sci. U.S.A. 97 (2000) 1630.
[61] A. Ozgul, D.Z. Childs, M.K. Oli, K.B. Armitage, D.T. Blumstein, L.E. Olson, S. Tuljapurkar, T. Coulson, Nature 466 (2010) 482.
[62] J.E. Lane, L.E.B. Kruuk, A. Charmantier, J.O. Murie, F.S. Dobson, Nature 489 (2012) 554.
[63] N. Pettorelli, F. Pelletier, A. Von Hardenberg, M. Festa-bianchet, S.D. Côté, Ecology 88 (2007) 381.
[64] E. Post, M.C. Forchhammer, Philos. Trans. R. Soc. Lond. B Biol. Sci. 363 (2008) 2369.
[65] M.E. Visser, Nature 466 (2010) 445.
[66] D. Réale, A.G. McAdam, S. Boutin, D. Berteaux, Proc. R. Soc. B Biol. Sci. 270 (2003) 591.
[67] D. Réale, D. Berteaux, A.G. Mcadam, S. Boutin, Evolution (N.Y.) 57 (2003) 2416.
[68] A.E. Derocher, N.J. Lunn, I. Stirling, Integr. Comp. Biol. 44 (2004) 163.
[69] E.V. Regehr, N.J. Lunn, S.C. Amstrup, I. Stirling, J. Wildl. Manage. 71 (2007) 2673.
[70] E.V. Regehr, C.M. Hunter, H. Caswell, S.C. Amstrup, I. Stirling, J. Anim. Ecol. 79 (2010) 117.
[71] I. Stirling, A.E. Derocher, Glob. Change Biol. 18 (2012) 2694.
[72] COSEWIC, COSEWIC Assessment and Update Status Report on the Polar Bear *Ursus maritimus* in Canada, 2008.
[73] M. Forchhammer, E. Post, Popul. Ecol. 46 (2004) 1.
[74] J. Hone, T.H. Clutton-Brock, J. Anim. Ecol. 76 (2007) 361.
[75] V. Grøtan, B.-E. Saether, F. Filli, S. Engen, Glob. Change Biol. 14 (2007) 218.
[76] T. Saitoh, B. Cazelles, J.O. Vik, H. Viljugrein, N.C. Stenseth, Clim. Res. 32 (2006) 109.
[77] N.C. Stenseth, A. Shabbar, K.-S. Chan, S. Boutin, E.K. Rueness, D. Ehrich, J.W. Hurrell, O.C. Lingjaerde, K.S. Jakobsen, Proc. Natl. Acad. Sci. U.S.A. 101 (2004) 10632.
[78] S.F. Morrison, D.S. Hik, J. Anim. Ecol. 76 (2007) 899.
[79] E.A. Beever, S.Z. Dobrowski, J. Long, A.R. Mynsberge, N.B. Piekielek, Ecology 94 (2013) 1563.
[80] M. Lima, N.C. Stenseth, F.M. Jaksic, Proc. Biol. Sci. 269 (2002) 2579.
[81] C.R. McMahon, M.N. Bester, M.A. Hindell, B.W. Brook, C.J.A. Bradshaw, Oecologia 159 (2009) 69.
[82] J.O. Ogutu, N. Owen-Smith, Ecol. Lett. 6 (2003) 412.
[83] J. Knape, P. de Valpine, Proc. Biol. Sci. 278 (2011) 985.
[84] C.C. Wilmers, E. Post, A. Hastings, J. Anim. Ecol. 76 (2007) 1037.
[85] A. Middleton, M. Kauffman, D. McWhirter, J.G. Cook, R.C. Cook, A.A. Nelson, M.D. Jimenez, R.W. Klaver, Ecology 94 (2013) 1245.

[86] R.A. Ims, J.-A. Henden, S.T. Killengreen, Trends Ecol. Evol. 23 (2008) 79.

[87] T. Cornulier, N.G. Yoccoz, V. Bretagnolle, J.E. Brommer, A. Butet, F. Ecke, D.A. Elston, E. Framstad, H. Henttonen, B. Hörnfeldt, O. Huitu, C. Imholt, R.A. Ims, J. Jacob, B. Jedrzejewska, A. Millon, S.J. Petty, H. Pietiäinen, E. Tkadlec, K. Zub, X. Lambin, Science 340 (2013) 63.

[88] K. Korpela, M. Delgado, H. Henttonen, E. Korpimäki, E. Koskela, O. Ovaskainen, H. Pietiäinen, J. Sundell, N.G. Yoccoz, O. Huitu, Glob. Change Biol. 19 (2013) 697.

[89] A.R. Hof, R. Jansson, C. Nilsson, Ecol. Modell. 246 (2012) 86.

[90] G. Rapacciuolo, S.P. Maher, A.C. Schneider, T.T. Hammond, M.D. Jabis, R.E. Walsh, K.J. Iknayan, G.K. Walden, M.F. Oldfather, D.D. Ackerly, S.R. Beissinger, Glob. Change Biol. 20 (2014) 2841.

[91] S. Hansen, D.M. Lavigne, Physiol. Zool. 70 (1997) 85.

[92] M.M. Humphries, D.W. Thomas, J.R. Speakman, Nature 418 (2002) 313.

[93] L.L. Kanda, Ecography (Cop.) 28 (2005) 731.

[94] J.F. Taulman, L.W. Robbins, J. Biogeogr. 23 (1996) 635.

[95] K.A. Bartoń, A. Zalewski, Glob. Ecol. Biogeogr. 16 (2007) 281.

[96] S.I. Jarema, J. Samson, B.J. Mcgill, M.M. Humphries, Glob. Change Biol. 15 (2009) 508.

[97] P. Hersteinsson, D.W. Macdonald, Oikos 64 (1992) 505.

[98] S.T. Killengreen, R.a. Ims, N.G. Yoccoz, K.A. Bråthen, J.A. Henden, T. Schott, Biol. Conserv. 135 (2007) 459.

[99] S. Lariviere, Wildl. Soc. Bull. 32 (2004) 955.

[100] A.M. Veitch, Can. F. Nat. 115 (2001) 172.

[101] D.H. Cluff, Can. F. Nat. 120 (2006) 67.

[102] R.J. Rowe, J.A. Finarelli, E.A. Rickart, Glob. Change Biol. 16 (2009) 2930.

[103] T.L. Root, J.T. Price, K.R. Hall, S.H. Schneider, C. Rosenzweig, J.A. Pounds, Nature 421 (2003) 57.

[104] R. Warren, J. VanDerWal, J. Price, J.A. Welbergen, I. Atkinson, J. Ramirez-Villegas, T.J. Osborn, A. Jarvis, L.P. Shoo, S.E. Williams, J. Lowe, Nat. Clim. Change 3 (2013) 678.

[105] C.A. Schloss, T.A. Nuñez, J.J. Lawler, Proc. Natl. Acad. Sci. U.S.A. 109 (2012) 8606.

[106] D.J. Currie, Am. Nat. 137 (1991) 27.

[107] K.J. Gaston, Nature 405 (2000) 220.

[108] T.J. Davies, L.B. Buckley, R. Grenyer, J.L. Gittleman, Philos. Trans. R. Soc. Lond. B Biol. Sci. 366 (2011) 2526.

[109] W. Thuiller, M.B. Araujo, S. Lavorel, J. Biogeogr. 31 (2004) 353.

[110] B. Huntley, R.E. Green, Y.C. Collingham, J.K. Hill, S.G. Willis, P.J. Bartlein, W. Cramer, W.J.M. Hagemeijer, C.J. Thomas, Ecol. Lett. 7 (2004) 417.

[111] W. Thuiller, L. Brotons, M.B. Araújo, S. Lavorel, Ecography (Cop.) 27 (2004) 165.

[112] G.M. MacDonald, K.D. Bennett, S.T. Jackson, L. Parducci, F.A. Smith, J.P. Smol, K.J. Willis, Prog. Phys. Geogr. 32 (2008) 139.

[113] J.L. Blois, J.L. Mcguire, E.A. Hadly, Nature 465 (2010) 771.

[114] P.L. Koch, A.D. Barnosky, Annu. Rev. Ecol. Evol. Syst. 37 (2006) 215.

[115] J.A. van Dam, H. Abdul Aziz, M.A. Alvarez Sierra, F.J. Hilgen, L.W. van den Hoek Ostende, L.J. Lourens, P. Mein, A.J. van der Meulen, P. Pelaez-Campomanes, Nature 443 (2006) 687.

[116] B. Shapiro, A.J. Drummond, A. Rambaut, M.C. Wilson, P.E. Matheus, A. V Sher, O.G. Pybus, M.T.P. Gilbert, I. Barnes, J. Binladen, E. Willerslev, A.J. Hansen, G.F. Baryshnikov, J.A. Burns, S. Davydov, J.C. Driver, D.G. Froese, C.R. Harington, G. Keddie, P. Kosintsev, M.L. Kunz, L.D. Martin, R.O. Stephenson, J. Storer, R. Tedford, S. Zimov, A. Cooper, Science 306 (2004) 1561.

[117] R. Debruyne, G. Chu, C.E. King, K. Bos, M. Kuch, C. Schwarz, P. Szpak, D.R. Gröcke, P. Matheus, G. Zazula, D. Guthrie, D. Froese, B. Buigues, C. de Marliave, C. Flemming, D. Poinar, D. Fisher, J. Southon, A.N. Tikhonov, R.D.E. MacPhee, H.N. Poinar, Curr. Biol. 18 (2008) 1320.

[118] R.D. Guthrie, Nature 426 (2003) 169.

[119] I. Barnes, P. Matheus, B. Shapiro, D. Jensen, A Cooper, Science 295 (2002) 2267.

[120] D.A. Burney, T.F. Flannery, Trends Ecol. Evol. 20 (2005) 395.

[121] E. Martínez-meyer, A.T. Peterson, W.W. Hargrove, Glob. Ecol. Biogeogr. 13 (2004) 305.

[122] S. Legendre, S. Montuire, O. Maridet, G. Escarguel, Earth Planet. Sci. Lett. 235 (2005) 408.

[123] M.H. Fernández, P. Peláez-Campomanes, Glob. Ecol. Biogeogr. 14 (2005), 39.

[124] C.D. Thomas, A. Cameron, R.E. Green, M. Bakkenes, L.J. Beaumont, Y.C. Collingham, B.F.N. Erasmus, M.F. De Siqueira, A. Grainger, L. Hannah, L. Hughes, B. Huntley, A.S. Van Jaarsveld, G.F. Midgley, L. Miles, M.A. Ortega-Huerta, A.T. Peterson, O.L. Phillips, S.E. Williams, Nature 427 (2004) 145.

[125] D. Nogués-Bravo, J. Rodriguez, J. Hortal, P. Batra, M.B. Araujo, PLoS Biol. 6 (2008) 0685.

[126] L.A. Pastene, M. Goto, N. Kanda, A.N. Zerbini, D. Kerem, K. Watanabe, Y. Bessho, M. Hasegawa, R. Nielsen, F. Larsen, P.J. Palsbøll, Mol. Ecol. 16 (2007) 1481.

[127] M.S. Douglas, J.P. Smol, W. Blake Jr., Science 266 (1994) 416.

[128] J. Esper, U. Büntgen, D.C. Frank, D. Nievergelt, A. Liebhold, Proc. Biol. Sci. 274 (2007) 671.

[129] A.H. Hurlbert, E.P. White, Ecol. Lett. 8 (2005) 319.

[130] C. Rahbek, G.R. Graves, Proc. Natl. Acad. Sci. U.S.A. 98 (2001) 4534.

[131] J. Pellikka, H. Rita, H. Lindén, Ann. Zool. Fennici 42 (2005) 123.

[132] S. Harris, D.W. Yalden, Mamm. Rev. 34 (2004) 157.

INSECT COMMUNITIES

10

Audrey M. Maran, Shannon L. Pelini

Department of Biological Sciences, Bowling Green State University, Bowling Green, OH, USA

CHAPTER OUTLINE

1. INTRODUCTION

Insects make up over half of known species on Earth and are present in every trophic level, i.e. food chain link, in both terrestrial and aquatic systems [1]. Recognizing the current and future effects of climate change on insects is important not only because they provide ecosystem services (e.g. pollination, nutrient cycling, seed dispersal), but also because their activity directly affects human beings (e.g. crop pests and disease vectors) [1]. The ubiquity of insects, along with their dependence on climactic conditions, makes them excellent indicators of ecosystem responses to climate change. Insects are poikilothermic ectotherms, meaning that their internal temperature varies with the environmental temperature [2]. Because of this dependence on environmental conditions, climatic warming will affect insect development time, voltinism (number of generations per year), foraging behaviour, emergence time, survivorship, and geographic range [3,4]. Changes in these factors could significantly alter insect community dynamics by affecting where species occur (range changes, see Section 2), when species can interact (phenology, see Section 3), and which species encounter one another (physiology, see Section 4). Furthermore, these dynamics are affected by other atmospheric conditions such as levels of CO_2 and precipitation (see Section 5). Indeed, the diversity and composition of insect communities have responded to recent climatic changes (see Section 6).

This chapter provides a thorough, but not exhaustive, overview of the effects of climate change on insects. We will focus on aspects that are most well supported in the literature, first describing the direct effect and then discussing the consequences of these changes. The first three sections are

Climate Change. http://dx.doi.org/10.1016/B978-0-444-63524-2.00010-5

categorized by the direct effect under discussion (i.e. geographical range shift, phenological shift, effects on physiology), followed by a section separately describing responses to CO_2 levels and precipitation, and closing with a section describing how climate change will change insect communities as a whole. Although we do not include a separate section on evolution of insects in response to climate change, we do present examples during the discussion of other topics.

2. RANGE CHANGES

Due to the fact that insects are dependent on environmental conditions, many species respond to rising temperatures and changing precipitation by expanding, contracting, or shifting where they are found geographically (geographic range) [5–13]. There are several pressures that drive range changes. As temperature increases, species may no longer be able to inhabit the warmer, equatorial areas of their ranges, causing them to contract or shift their geographic location to survive [11,14,15,16]. In contrast, at the historically cooler, poleward edges of ranges, species will be able to expand into areas previously too cold for them to inhabit. This has several benefits, including access to new resources and escape from natural enemies for species that are able to disperse more readily than their predators [17]. Researchers have long expected that species will track climate changes and shift or expand their ranges in a poleward direction or to higher altitudes. While there is a large body of evidence supporting this [5–13,18–22], it has become apparent that predicting future species distributions is more complex than simple poleward shifts. Success of range change in response to climate change depends on a variety of factors, such as life history, heterogeneity of the environment, resource availability, thermal tolerance, and physical traits (e.g. dispersal ability, body size). We will look more closely at each of these factors in the following sections.

2.1 RANGE CHANGES – LIFE HISTORY

One of the most important factors determining species' responses to climate change is their life history characteristics [3]. Growth rate and voltinism, in particular, affect how a species will alter its range. Organisms that can grow more quickly and have shorter generation times are better able to respond quickly to changes in the environment [23]. In addition, they spend less time in a vulnerable larval stage [24], which allows them to escape predation and expand their ranges more successfully than those with longer generation times. Taken together, these benefits mean that fast-growing, multivoltine (i.e. production of multiple generations per year) species will be better able not only to extend their range but also remain successful in their current range through adaptation [3,25]. In contrast, slow-growing, univoltine (i.e. production of one generation per year) species are unable to adapt as easily to change and may need to move poleward if their current range becomes too warm. For example, in an area of Japan where previously two species of green stink bug (*Nezara spp.*) co-occurred, a multivoltine species was able to outcompete and replace a univoltine species under climate change (reviewed in Ref. [25]). While many factors were important in determining which species outcompeted the other, one of strongest factors was voltinism. Bale et al. [3] considered these effects of growth rate and voltinism, and adding diapause requirements, have suggested that future range shifts in response to climate change can be predicted more broadly. Diapause, the state during which insects arrest development, is an important part of this model because many insects must diapause over winter to complete development successfully [26]. Bale et al. [3] predict that fast-growing insects that are either nondiapausing or do not require a low temperature to trigger diapause will likely expand their ranges. However, fast-growing insects that require a low temperature to either induce or terminate diapause will likely contract their range, since warming temperatures in parts of their range could interrupt diapause from occurring. Slow-growing insects that require a low temperature for successful diapause will likely not be able to expand their ranges and experience range contraction.

2.2 RANGE CHANGES – ENVIRONMENTAL FACTORS AND RESOURCE AVAILABILITY

In addition to the influence of the species' life history traits, changes in a species' distribution are dependent on whether or not resources necessary for survival and reproduction are present in areas to which the range may expand [25,27]. In some cases, a nearby habitat may contain the resources, but not suitable temperatures, to support a species [25]. To populations at the poleward edges of a species' range, this previously uninhabited area could allow the population to expand as the climate warms. For example, significant pine pests, mountain pine beetles (*Dendroctonus* spp.), are responding to warmer temperatures by expanding their ranges into pine forests that were previously too cold [28]. As with invasive species, pests may be particularly successful in new areas where hosts have not adapted resistance. In addition, expanding species may be more successful when they enter an enemy-free space (area not inhabited by their predators or competitors) [17]. These enemy-free zones occur when predators are not adapted to feed on new prey species. Conversely, in a species' historical range, predators may seek cooler temperatures due to the increased sensitivity of higher trophic levels to warming [29–32], thereby relieving pressure on prey. This subsequently results in population increases of the prey. Barton [33] demonstrated this in a grassland system. Predatory nursery web spiders (*Pisaurina mira*) and their prey, red-legged grasshoppers (*Melanoplus femurrubrum*), normally have a large degree of spatial overlap, but under warming spiders moved out of the canopy and closer to the soil, presumably in search of cooler temperatures. This reduced spatial overlap resulted in lower predation on grasshoppers, which do not need to seek the cooler temperatures and continue to forage in the canopy of the grassland.

Specialist species, or those that have restrictive resource demands, are more limited in range expansion if other species they depend on are not present [34,35]. While it is not uncommon for *generalists* to switch their preferred food source in order to expand their range, it is much rarer in 'specialists' [36,37]. However, there is some evidence for specialists adapting to new host plants. For example, the brown argus butterfly (*Aricia agestis)* was able to expand its range by using dovesfoot geranium (*Geranium molle*) plants as hosts in areas with warmer climate where its normal host, rockrose (*Helianthemum nummularium*), was unavailable. In the case of the brown argus butterfly, there may have been additional pressure to expand; butterflies on the new hosts had lower risks of acquiring parasites that are typically associated with rockrose [17,38].

The plight of the 'specialist' may seem intuitive, but other restrictions to range expansion may not be so obvious. For example, extreme weather events in newly colonized locales could reverse range expansions [39]. Furthermore, presence of competitors or predators[6] and environmental features required for reproduction can affect the success of the species attempting to colonize. Capinha, Rocha, and Sousa et al. [40] demonstrated how environmental features can determine range by modelling the current distribution of the yellow fever mosquito (*Aedes aegypti*). They were able to predict the distribution of the mosquito in the Americas with considerable accuracy based on climate occupied in other regions, including climates that would have been unsuitable if not for human sources of water or shelter from temperature extremes. Their model had better predictive ability than previous models that relied solely on climate as a predictor and did not consider human sources of suitable habitat in otherwise uninhabitable areas. This demonstrates the importance of factors aside from climate that determine the habitability of a region for species.

2.3 RANGE CHANGES AND ADAPTATION

Range expansion or contraction can result in adaptation of populations in the newly colonized area. Several studies have found that species at range edges are better adapted for flight (e.g. larger wingspans, changes in flight metabolism, and larger thoraxes (reviewed in Ref. [22])). For example, bush crickets (*Metrioptera roeselii*) have two wing morphs: small rudimentary wings or long wings. Long wing morphs are normally not common, but at range edges they are dominant [41,42]. In species that do not have multiple wing morphs, there are other indications of adaptation; larger thorax size, which includes muscles required for flight, is common at range edges [43]. This phenomenon is most apparent in species that can move over long distances and establish new populations. In species where small groups can move a long distance and establish new populations, there is less genetic mixing than those that steadily move poleward [22,44]. The low genetic variation at newly colonized sites a great distance from the rest of the species' range experience a population bottleneck, which results in the founder populations having individuals with traits that are advantageous for dispersal but less able to adapt in the future [22]. More recent studies on the brown argus butterfly (discussed in the previous section) have shown that the butterflies adapted to use dovesfoot geranium as a host plant show reduced egg laying on rockrose when moved back to their home range [45,46]. Higher latitudes are dominated by rockrose, which suggests that the butterflies adapted to dovesfoot geranium may be unsuccessful in expanding their range further. It is worth noting, however, that adaptation is limited by lack of underlying genetic variation.

3. CHANGES IN PHENOLOGY

In addition to range shifts, changes in phenology as a result of climate change are well documented [21]. Changes in phenology arise because insect development rates are affected by many factors, including climatic variables such as minimum winter temperatures, average and maximum summer temperatures, precipitation, and aridity [20,47]. Increases in temperature have been found to increase growth rate, generations per year, and survivorship both over winter and during vulnerable life stages such as during larval development [3,20,48,49]. Consequently, climate change has been associated with early emergence in many insects, including butterflies, flies, and beetles, but this is likely because research has focused on these groups rather than because other insects are not exhibiting similar changes (reviewed in Ref. [21]).

3.1 PHENOLOGY CHANGES – MISMATCHES

Despite the widespread use of similar cues to determine phenology (e.g. photoperiod, temperature, precipitation), changes in phenology are not necessarily synchronous across all species [50]. In fact, studies have demonstrated that there are not just differences in the way that plants and insects alter phenology but also differences among insect species [51,52]. These varying phenological responses can disrupt community dynamics by causing a mismatch between interacting species [53,54]. For example, several studies have found mismatches between different species of plants and their bee pollinators [55–57], which resulted in reduced reproductive capability for the plants [45,46]. Mismatches have also been well documented in pest species, which are commonly studied because of their economic importance. Researchers have found that climate-driven phenological mismatches can mediate pest outbreaks [58,59]. For example, Schwartzberg et al. [60] demonstrated that under warming, the forest tent caterpillar (*Malacosoma disstria*) fell out of sync with its host trees, aspen (*Populus tremuloides*) and birch (*Betula papyrifera*). Warming caused both bud break of the trees and egg hatch of the caterpillars to happen earlier, but the shift of bud break was greater. Since leaves are not nutritive enough for high caterpillar survival right after bud break and leaves later in the season have more defenses, shifts in herbivore timing can reduce the damage done to host plants (reviewed in Ref. [50]). Although pest damage may be reduced by a mismatch, the pest may also find a new host that is synchronous and cause an equal amount of damage on the new host [61]. In an artificial warming experiment, Liu et al. [62] found that larva of the noctuid moth (*Melanchra pisi*) reached peak abundance later in the season, while a nearby plant the moth does not typically feed on, gentian (*Gentiana formosa*), flowered earlier in the season. These shifts in phenology aligned the larva temporally with the gentian. As a result, there was more damage to gentian in the artificially warmed environment due to the strengthened trophic relationship of the larva and plant, which were previously not strongly linked. Even with the large body of evidence demonstrating mismatches caused by climate change, it is possible that the mismatches may be short lived. In populations where there are individuals emerging at different times, those that align with their food source will be selected for, resulting in an insect population that is shifting in time with its partner [57,61].

Mismatches between species dependent on one another have been hypothesized to have less of an effect on ecosystem stability in areas with high diversity. This assertion is based on the biodiversity insurance hypothesis, which posits that biodiversity buffers ecosystems from losses of species during significant disturbances such as climate change [63,64]. Analysis of long-term datasets of apple trees and their bee pollinators suggests that while different species of bees shifted their phenology faster or slower than apple trees, the average phenological shift of bees was still synchronous with apple bloom. Because bee diversity was high enough, bees active during the apple bloom were able to pollinate trees, resulting in no negative change to the apple tree populations [57]. In order to help us predict the effect climate change will have on insect communities, it is useful to recognize the importance of biodiversity, which will reduce net variation in an ecosystem by providing many species that can fulfil functions necessary for ecosystem stability (e.g. pollination, predation) [63,64].

3.2 PHENOLOGY CHANGES – GENERATIONS AND ABUNDANCE

Warmer temperatures will lengthen insect active seasons because suitable temperatures are reached both earlier in the spring and later into the fall. This larger temporal window will allow some species to complete more generations than they have historically, an effect that has already been observed across several taxa [39,49,65–67]. However, temperature alone is not enough to keep an organism alive; resource availability is imperative to survival of future generations [68]. Altermatt [68] found that diet can be a good predictor of shifts in phenology. More specifically, lepidopterans that feed on woody plants, which only produce new unprotected leaves early in the season [69], were limited in their ability to increase generations. Those lepidopterans that feed on herbaceous plants, which produce new unprotected leaves throughout the season, were much more likely to have a second generation. An increase in voltinism may have far-reaching effects, contributing to population growth and possibly yielding more frequent or severe pest outbreaks [59,70,71]. In addition, since the shift in voltinism is food dependent, there may be uneven changes between forests and herb-dominated landscapes [68]. Intuitively, potential for multiple generations seems beneficial to insects, but there may also be drawbacks to extended reproductive seasons. When turnip moths (*Agrotis segetum*) complete a second generation, they have reduced population sizes the following year. This is likely because the second generation is not reaching a size at which it can successfully overwinter due to being more meta-bolically active on a smaller energy budget [67,72]. However, as temperatures continue to warm and the growing season extends even further, the turnip moth may reach a point where it can successfully complete two generations. Nevertheless, transitions between the numbers of generations a species can complete may result in reduced population sizes.

3.3 PHENOLOGY AND ADAPTATION

We have discussed shifts in phenology as a result of phenotypic and physiological responses to warming temperatures and longer growing seasons, but there may be a heritable component that leads to phenological adaptation. While some of the examples mentioned in the previous sections may have genetic components, in this section we will present research that explicitly addresses adaptation resulting in shifted phenology. Temperature and photoperiod (day length) are both important for insect emergence and development [73,74], and there is evidence of evolution involving both of these factors to cope with climate change [75]. As suggested by the studies we have presented previously, insects often deal with warming by altering their response without a genetic change [75]. However, there are examples of genetic changes associated with altered climate. For example, winter moths (*Operophtera brumata*) are advancing egg hatch faster than bud break in their oak (*Quercus robur*) hosts, leading them to be asynchronous [76]. Van Asch et al. [77] found evidence of selection for delayed egg hatch in winter moths, which restores synchrony with the oaks. For events in the life cycle that are more strongly influenced by photoperiod, such as diapause, insects may adapt to climate change by altering critical photoperiod (day length that cues the beginning of diapause) [23,78]. Indeed, under long-term experimental warming, several taxa showed an adaptive response: critical photoperiods became similar to those in populations where the growing season is longer [79–82]. This allows the insects to take advantage of the longer growing season by waiting to commence diapause until later in the fall [23].

4. PHYSIOLOGY

Shifts in range and phenology have been well studied in insects, but physiological impacts of climate change have received much less attention; only 14% of insect climate change studies investigate physiology, while nearly 30% of studies focus on phenology and geographical range [21]. Physiological effects of climate change on insects are manifested when temperatures exceed the organism's critical thermal maximum (temperature at which it would likely die due to a loss of coordination) [2]. Since individuals across species generally have a similar critical thermal maximum, these effects may vary across the species' geographical range. At lower latitudes in the species' range, individuals will be living closer to their thermal maximum than those in the poleward areas of the range [2,83]; therefore, warming in lower latitudes will be more likely to harm a species than the same degree of warming in the higher latitudes.

The trend of climatic changes having more severe effects on insects near the equator can be generalized not just within species but also across species. Species living in areas with greater seasonality are expected to be less sensitive to changes in temperature than those that live in areas with stable climate, since species in seasonal areas are already adapted to fluctuating temperatures [2,84,85]. Populations that live nearer to the equator, which has a more stable climate, are generally thought to be more vulnerable to warming since they are not adapted to cope with changes in temperature that will occur with climate change. In addition, since insect metabolic rate increases exponentially, not linearly, with temperature [86], tropical insects will also be receiving relatively larger boosts in metabolic rate than equal levels of warming in naturally cooler climates [87]. A body of evidence supports the trend of warming tolerance decreasing toward the equator in several different taxa [2,83,88,89]. However, some data suggest that using latitude to predict how vulnerable insect species will be to warming temperatures is too narrow in scope. Researchers have also found evidence that heterogeneity in the environment reduces the latitudinal trend [90] – some species can adapt quickly, and behavioural plasticity, dispersal ability, and acclimation can aid species in dealing with warming [2]. We have discussed studies that address many of these factors in other sections. In addition, climate change involves more than average warming across the globe, and taking into consideration other effects such as drought, higher CO_2 levels, and extreme weather events is crucial to understanding how insects will be affected by climate change. Kingsolver et al. [91] found that although insects in higher latitudes will have longer growing seasons as the climate warms, conferring physiological benefits such as increased number of generations per year, insects in temperate locations will also experience more extreme temperature events than those found near the tropics. These extreme events could ultimately result in more negative effects at mid-high latitudes rather than at the tropics, where research has been focused recently [91].

5. RESPONSES TO OTHER CLIMACTIC VARIABLES

Most of the examples we presented in preceding sections were attributed to increasing temperatures, which is a reflection of insect–climate change literature [21]. However, there is also a large body of evidence supporting insect responses to changes in precipitation, CO_2 levels, ozone levels, and snow cover, among other factors [21]. While these factors receive less attention, they are important to insect activity. Precipitation largely determines moisture of the environment, which directly impacts insects by causing varied physiological and, potentially, behavioural responses and indirectly impacts insects through their interactions with plants [58,92,93]. The CO_2 studies investigate insect responses due to their consumption of plants, which change in nutritive quality as a result of greater carbon availability [94]. Following, we will briefly summarize insect responses to some of these other factors.

5.1 RESPONSES TO PRECIPITATION

Altered precipitation can have large and varied effects on insect communities. Zhu et al. [93] investigated the increased precipitation in a semiarid grassland where water was a limiting factor. They found that insect community structure, i.e. species richness, changed the year following increased precipitation. However, there was not a consistent increase in abundance or species richness as precipitation increased; the response varied across insect taxa. The researchers believe that several factors contributed to the change in community structure. The cause may have been a combined effect of a shift in the plant community and high soil moisture. Raised soil moisture is often beneficial to insects, but levels too high can negatively affect insects reliant on soil for development [95] or dilute the nutrient pool, ultimately changing plant physiology [93,96]. Precipitation is also important to ecosystems less limited by water. Using a long-term dataset, Haynes, Allstadt, and Klimetzek [58] found that outbreaks of a pine pest, pine tree lappet moth (*Dendrolimus pini*), followed years of drought. The moths likely have larger outbreaks resulting from combined effects of moisture on their food source and a parasitic fungus. Drought stresses plants, often making them both more susceptible to herbivory [97] and reducing the survival of parasitic fungi that normally control pine tree lappet moth populations [98,99]. Direct effects of precipitation on insect survival, as discussed in the previous examples, are important, but precipitation may induce direct behavioural changes in insects that can indirectly affect survival. For example, during times of higher precipitation, ants (*Messor* spp.) altered their foraging behaviour by putting more effort into searching out food items matching their individual size [100]. Depending on the environment, this can either increase efficiency of the colony or decrease efficiency if too much time is spent foraging for food the 'right size', resulting in too little food returned to the nest [100]. The studies we presented above suggest that responses to changes in precipitation are significant but dependent on ecosystem. Therefore, predicting responses of insects to climate change would benefit from a better understanding of how precipitation affects various ecosystems.

5.2 RESPONSES TO CO_2 LEVELS

Carbon dioxide levels will continue to rise [101], and there is evidence that CO_2 concentrations may have appreciable effects on insect communities. Higher atmospheric CO_2 concentrations increase carbon availability for plants, which results in the plants having higher carbon-to-nitrogen ratios, essentially diluting nitrogen in the foliage [94]. Higher CO_2 concentrations, ozone levels, and warming have been shown to increase volatile organic compounds [102–105], which are used in both positive (e.g. pollination) and negative (e.g. plant defenses) interactions with insects [102,106]. The increase of the ratio of carbon to nitrogen in the atmosphere, together with chemical changes, has directly affected herbivores and their behaviour. A metanalysis of elevated CO_2 studies [107] found an overall decline in insect abundance, with an increase in consumption rates, development time, growth rate, and weight. The quality of food for herbivores can have effects reaching beyond the insects themselves. For example, herbivore frass is a significant source of labile carbon and nitrogen for decomposers, and its quality and quantity are important to nutrient cycling [108–110]. Couture and Lindroth [111] found that insect frass quality changes in a similar way to foliar chemistry, 'mirroring' the changes in the plant, though differing by insect species. This suggests that elevated CO_2 levels could be reflected in the quality of frass that will eventually be decomposed.

5.3 INTERACTIVE EFFECTS OF WARMING, PRECIPITATION, AND CO_2

While it is important to consider effects of individual aspects of climate change on insects, in order to predict future response, we need to understand the interactive effects of the components. Studies looking at multiple factors are not common, and those that have been done have conflicting results, with some showing additive and others mediating effects [112–114]. For example, an experiment investigating heather beetle (*Lochmaea suturalis*) responses to drought, temperature, and CO_2 found that elevated CO_2, drought, and warming (to a lesser degree) had negative effects on heather beetle weight and survival. Furthermore, when these treatments were applied simultaneously, detrimental effects to the heather beetle increased with each added treatment [112]. However, earlier data from the same long-term experiment revealed dampening effects between the different treatments when looking at plants and belowground invertebrates [114]. These conflicting results may suggest that multiple climatic factors interact to produce either additive or dampening effects depending on the response organisms, which reveals a need for more studies combining multiple climate variables.

6. INSECT COMMUNITIES UNDER CLIMATE CHANGE

We have discussed changes on insect populations and species geographical range, phenology, physiology, evolution, and effects of other climactic variables separately. Climactic variables can also interact to affect insects at the community level. Through a combination of evidence from current experiments and from past warming events as documented in the fossil record, it is possible to make some predictions about the impacts of climate change on ecological communities [115]. In general, communities are expected to become less diverse and more homogenous (similar) to other communities [116–119]. Historically, changes in climate resulted in communities dominated by *generalists* [120,121]. This is possibly because *specialists* have a narrow range of environmental conditions that they can tolerate and cannot easily switch food sources, therefore they are more sensitive to changes in the environment or decline of their food source [45,122]. There are recent examples of climate change causing a shift toward generalist-dominated communities. Menendez et al. [123] found that specialist butterflies are less able to expand their range in newly climatically suitable habitats because they are restricted by lack of resources, ultimately resulting in communities where the less restricted generalists have become dominant [124].

Shifts to more generalist-dominated communities are not the only way that community structure (occurrence and abundance of species in a community) is altered by climate change. Community structure can be altered in a variety of ways and does not necessarily require a change in diversity. For example, a study done by de Sassi, Lewis, and Tylianakis [118] found changes in species present in a community, which did not lead to a decrease in the number of species present (species richness) per se. They found that in response to increased temperature and nitrogen concentrations, the herbivore community shifts in which species were most abundant (dominant) and which species were present, but not a significant change in species richness. The lack of change in species richness, but a change in which species were present, suggests there was species replacement (a new species entering the community and taking the place of one no longer present) [118]. Additionally, they found evidence of the warmed communities becoming more homogenous, so even without a decrease in species richness within the plots themselves, there was a decrease in diversity *among* the communities under warming [118]. Changes to insect community structure may be a delayed response to climate change. Suttle, Thomsen, and Power [125] found reduced invertebrate species richness (by 20%) when they increased the rainy season in a grassland with a Mediterranean climate to mimic predicted future climate; however, the result was delayed, with the indirect community effects only becoming apparent after five years. The researchers hypothesize that a dramatic increase in grasses relative to forbs over the preceding years reduced the food and habitat quality for invertebrate herbivores, ultimately resulting in declines in species richness at higher trophic levels. In addition to considering larger temporal scales, there is evidence that community effects may differ depending on location. Pelini et al. [126] conducted warming experiments in deciduous forests that demonstrate the importance of spatial scale in community research. After 2.5 years of experimental warming in a high and a low latitude site, they found differences in species richness and community composition at the low latitude site but not at the high latitude site. They suspect that this may be a result of species living closer to their critical thermal maximum at lower latitudes (see Section 3.1).

Warming may alter community structure if responses vary by trophic level [127]. Predicting how warming will indirectly affect an organism within a community can become complicated because there can be indirect effects from both top-down (predator effect on lower trophic levels) and bottom-up (producer effect on higher trophic levels) interactions [127]. We have discussed examples of how bottom-up interactions can shape the invertebrate community in detail in previous sections, so we will focus mainly on top-down community effects. Higher trophic levels are expected to be disproportionately negatively affected by climate change both because they are more sensitive to abiotic factors [31,128] and because of the combined effect of climate change on lower trophic levels [127]. Additionally, climate change has been shown to strengthen top-down interactions [129,130], leading to changes in community structure [130]. One example of how top-down interactions could affect community structure under climate change is demonstrated by a three-trophic-level food web involving spiders (*Pisaurina mira*), grasshoppers (*Melanoplus femurrubrum*), and various herbaceous plants [130]. The presence of spiders has been shown to alter grasshopper behaviour, causing them to preferentially feed on forbs rather than grasses [131]. Under warming, this relationship is strengthened, causing the presence of spiders to positively affect grasses with 30% more strength and negatively affect forbs with 40% more strength per 1°C of warming [130]. The effect of warming in this case would result in a community shifted more heavily toward grasses than forbs. Considering previous examples we have presented that demonstrate bottom-up control of the invertebrate community, this shift in plant community would likely have far-reaching effects and may change the structure of the invertebrate community. In this section, we have presented several ways in which community structure may be altered due to climate change.

7. CONCLUSION

In this chapter we discussed insect responses to climate change, with examples of effects that are already happening in situ and those that may occur as predicted by experimental manipulations or modelling. These changes are varied, with different magnitudes and directions, and may be spatial or temporal, direct or indirect. Insects may respond to climate change by shifting their geographic range, either to leave areas where the climate is no longer suitable, or enter areas where climate has become suitable. Insects that do not shift their range may adapt to changes in environmental conditions through altered phenology, such as emergence happening earlier or later, life stages occurring at different times, or increasing the number of generations per year. These shifts can lead to novel interactions and changes in community structure. Climate change will affect insects because of their dependence on environmental temperature to control their physiological functions. Individuals living closer to their critical thermal maximum will either have to shift spatially, shift temporally, or adapt to tolerate higher temperatures. Insects may evolve to combat climate change, although studies explicitly addressing evolution are not as common in the insect–climate change literature. However, that may be due to a focus on plastic responses, which are comparatively short term and easier to study than genetic responses [75,132]. In fact, due to short generation times and fast growth, insects may evolve rapidly in response to climate change [23,132,133]. It worth noting that even with the ability to reproduce rapidly, evolution can be limited by several factors, including low underlying variation or one genetic trait being linked with another trait that prevents the first from change [134]. Despite the dearth of evolutionary studies, we included a few compelling examples that demonstrated increased frequencies of dispersal characteristics, poleward populations changing their critical photoperiod to match those of equatorial populations, and specialists evolving to exploit new hosts. Considering the complex response of insects to climate change, we will likely see very different insect communities in the future.

REFERENCES

[1] G.C. McGavin, Essential entomology: An order by order introduction, Oxford University Press, New York, 2001, p. 318.

[2] C.A. Deutsch, J.J. Tewksbury, R.B. Huey, K.S. Sheldon, C.K. Ghalambor, D.C. Haak, P.R. Martin, Proc. Natl. Acad. Sci. U.S.A. 105 (2008) 6668–6672.

[3] J.S. Bale, G.J. Masters, I.D. Hodkinson, C. Awmack, T.M. Bezemer, V.K. Brown, J. Butterfield, A. Buse, J.C. Coulson, J. Farrar, J.E.G. Good, R. Harrington, S. Hartley, T.H. Jones, R.L. Lindroth, M.C. Press, I. Symrnioudis, A.D. Watt, J.B. Whittaker, Global Change Biol. 8 (2002) 1–16.

[4] R. Menendez, Tijdschr. Entomol. 150 (2007) 355–365.

[5] C. Parmesan, N. Ryrholm, C. Stefanescu, J.K. Hill, C.D. Thomas, H. Descimon, B. Huntley, L. Kaila, J. Kullberg, T. Tammaru, W.J. Tennent, J.A. Thomas, M. Warren, Nature 399 (1999) 579–583.

[6] A. Battisti, M. Stastny, E. Buffo, S. Larsson, Global Change Biol. 12 (2006) 662–671.

[7] A. Battisti, M. Stastny, S. Netherer, C. Robinet, A. Schopf, A. Roques, S. Larsson, Ecol. Appl. 15 (2005) 2084–2096.

[8] C.V. Baxter, K.D. Fausch, W.C. Saunders, Freshw. Biol. 50 (2005) 201–220.

[9] L. Crozier, Oecologia 135 (2003) 648–656.

[10] J.K. Hill, Y.C. Collingham, C.D. Thomas, D.S. Blakeley, R. Fox, D. Moss, B. Huntley, Ecol. Lett. 4 (2001) 313–321.

[11] J.K. Hill, C.D. Thomas, R. Fox, M.G. Telfer, S.G. Willis, J. Asher, B. Huntley, Proc. R. Soc. B Biol. Sci. 269 (2002) 2163–2171.

[12] R. Karban, S.Y. Strauss, Ecol. Entomol. 29 (2004) 251–254.

[13] C.D. Thomas, E.J. Bodsworth, R.J. Wilson, A.D. Simmons, Z.G. Davies, M. Musche, L. Conradt, Nature 411 (2001) 577–581.

[14] A.M.A. Franco, J.K. Hill, C. Kitschke, Y.C. Collingham, D.B. Roy, R. Fox, B. Huntley, C.D. Thomas, Global Change Biol. 12 (2006) 1545–1553.

[15] C.D. Thomas, A.M.A. Franco, J.K. Hill, Trends Ecol. Evol. 21 (2006) 415–416.

[16] R.J. Wilson, D. Gutierrez, J. Gutierrez, D. Martinez, R. Agudo, V.J. Monserrat, Ecol. Lett. 8 (2005) 1138–1146.

[17] R. Menendez, A. Gonzalez-Megias, O.T. Lewis, M.R. Shaw, C.D. Thomas, Ecol. Entomol. 33 (2008) 413–421.

[18] I.C. Chen, J.K. Hill, R. Ohlemuller, D.B. Roy, C.D. Thomas, Science 333 (2011) 1024–1026.

[19] R. Hickling, D.B. Roy, J.K. Hill, R. Fox, C.D. Thomas, Global Change Biol. 12 (2006) 450–455.

[20] T.M. Letcher (Ed.), Climate Change: Observed Impacts on Planet Earth, Elsevier, Oxford, 2009, pp. 215–229.

[21] N.R. Andrew, S.J. Hill, M. Binns, M.H. Bahar, E.V. Ridley, M.-P. Jung, C. Fyfe, M. Yates, M. Khusro, PeerJ 1 (2013) e11.

[22] J.K. Hill, H.M. Griffiths, C.D. Thomas, in: M.R. Berenbaum, R.T. Carde, G.E. Robinson (Eds.), Annual Review of Entomology, vol. 56, 2011. Annual Reviews: Palo Alto, pp. 143–159.

[23] R. Stoks, A.N. Geerts, L. De Meester, Evol. Appl. 7 (2014) 42–55.

[24] K.M. Clancy, P.W. Price, Ecology 68 (1987) 733–737.

[25] D.L. Musolin, Global Change Biol. 13 (2007) 1565–1585.

[26] S.F. Gilbert, Developmental Biology, Sinauer Associates, Inc., MA, 2013 xviii+894pp.

[27] M.B. Araujo, M. Luoto, Global Ecol. Biogeogr. 16 (2007) 743–753.

[28] B.J. Bentz, J. Regniere, C.J. Fettig, E.M. Hansen, J.L. Hayes, J.A. Hicke, R.G. Kelsey, J.F. Negron, S.J. Seybold, Bioscience 60 (2010) 602–613.

[29] W. Voigt, J. Perner, A.J. Davis, T. Eggers, J. Schumacher, R. Bahrmann, B. Fabian, W. Heinrich, G. Kohler, D. Lichter, R. Marstaller, F.W. Sander, Ecology 84 (2003) 2444–2453.

[30] E.L. Preisser, D.R. Strong, Am. Nat. 163 (2004) 754–762.

[31] B.A. Menge, J.P. Sutherland, Am. Nat. 130 (1987) 730–757.

[32] M. Marquis, I. Del Toro, S.L. Pelini, Ecology 95 (2014) 9–13.

[33] B.T. Barton, Ecology 91 (2010) 2811–2818.

[34] O. Schweiger, J. Settele, O. Kudrna, S. Klotz, I. Kuhn, Ecology 89 (2008) 3472–3479.

[35] S.E. Gilman, M.C. Urban, J. Tewksbury, G.W. Gilchrist, R.D. Holt, Trends Ecol. Evol. 25 (2010) 325–331.

[36] B. Braschler, J.K. Hill, J. Anim. Ecol. 76 (2007) 415–423.

[37] A. Colles, L.H. Liow, A. Prinzing, Ecol. Lett. 12 (2009) 849–863.

[38] R.M. Pateman, J.K. Hill, D.B. Roy, R. Fox, C.D. Thomas, Science 336 (2012) 1028–1030.

[39] J.S. Bale, S.A.L. Hayward, J. Exp. Biol. 213 (2010) 980–994.

[40] C. Capinha, J. Rocha, C.A. Sousa, Ecohealth 11 (2014) 420–428.

[41] A.D. Simmons, C.D. Thomas, Am. Nat. 164 (2004) 378–395.

[42] D. Poniatowski, S. Heinze, T. Fartmann, Evol. Ecol. 26 (2012) 759–770.

[43] C.L. Hughes, C. Dytham, J.K. Hill, Ecol. Entomol. 32 (2007) 437–445.

[44] K.M. Ibrahim, R.A. Nichols, G.M. Hewitt, Heredity 77 (1996) 282–291.

[45] J. Buckley, J.R. Bridle, Ecol. Lett. 17 (2014) 1316–1325.

[46] J. Buckley, R.K. Butlin, J.R. Bridle, Mol. Ecol. 21 (2012) 267–280.

[47] H.V. Daly, J.T. Doyen, A.H. Purcell, Introduction to Insect Biology and Diversity, Oxford University Press, Oxford, 1998.

[48] H. Audusseau, S. Nylin, N. Janz, Ecol. Evol. 3 (2013) 3021–3029.

[49] F. Altermatt, Proc. R. Soc. B Biol. Sci. 277 (2010) 1281–1287.

[50] M.J. Solga, J.P. Harmon, A.C. Ganguli, Nat. Areas J. 34 (2014) 227–233.

[51] C. Parmesan, Global Change Biol. 13 (2007) 1860–1872.

[52] T.T. Hoye, A. Eskildsen, R.R. Hansen, J.J. Bowden, N.M. Schmidt, W.D. Kissling, Curr. Zool. 60 (2014) 243–251.

[53] T.L. Root, J.T. Price, K.R. Hall, S.H. Schneider, C. Rosenzweig, J.A. Pounds, Nature 421 (2003) 57–60.

[54] M.E. Visser, C. Both, Proc. R. Soc. B Biol. Sci. 272 (2005) 2561–2569.

[55] J.D. Thomson, Philos. Trans. R. Soc. B Biol. Sci. 365 (2010) 3187–3199.

[56] G. Kudo, Y. Nishikawa, T. Kasagi, S. Kosuge, Ecol. Res. 19 (2004) 255–259.

[57] I. Bartomeus, M.G. Park, J. Gibbs, B.N. Danforth, A.N. Lakso, R. Winfree, Ecol. Lett. 16 (2013) 1331–1338.

[58] K.J. Haynes, A.J. Allstadt, D. Klimetzek, Global Change Biol. 20 (2014) 2004–2018.

[59] E.H. DeLucia, P.D. Nabity, J.A. Zavala, M.R. Berenbaum, Plant Physiol. 160 (2012) 1677–1685.

[60] E.G. Schwartzberg, M.A. Jamieson, K.F. Raffa, P.B. Reich, R.A. Montgomery, R.L. Lindroth, Oecologia 175 (2014) 1041–1049.

[61] R.T. Gilman, N.S. Fabina, K.C. Abbott, N.E. Rafferty, Evol. Appl. 5 (2012) 2–16.

[62] S. Liu, B. Li, L. Meng, X. Zhang, G. Pan, Biodivers. Sci. 22 (2014) 502–507.

[63] S. Naeem, S.B. Li, Nature 390 (1997) 507–509.

[64] M. Loreau, S. Naeem, P. Inchausti, J. Bengtsson, J.P. Grime, A. Hector, D.U. Hooper, M.A. Huston, D. Raffaelli, B. Schmid, D. Tilman, D.A. Wardle, Science 294 (2001) 804–808.

[65] E. Braune, O. Richter, D. Sondgerath, F. Suhling, Global Change Biol. 14 (2008) 470–482.

[66] A.M. Jonsson, G. Appelberg, S. Harding, L. Barring, Global Change Biol. 15 (2009) 486–499.

[67] P. Esbjerg, L. Sigsgaard, Crop Prot. 62 (2014) 64–71.

[68] F. Altermatt, Ecol. Lett. 13 (2010) 1475–1484.

[69] P. Feeny, Ecology 51 (1970) 565.

[70] M. van Asch, P.H. Tienderen, L.J.M. Holleman, M.E. Visser, Global Change Biol. 13 (2007) 1596–1604.

[71] M.J. Steinbauer, D.J. Kriticos, Z. Lukacs, A.R. Clarke, For. Ecol. Manage. 198 (2004) 117–131.

[72] D.A. Hahn, D.L. Denlinger, Annu. Rev. Entomol. 56 (2011) 103–121.

[73] M.V. Nunes, J. Theor. Biol. 146 (1990) 369–378.

[74] D.S. Saunders, Entomol. Sci. 17 (2014) 25–40.

[75] A. Donnelly, A. Caffarra, C.T. Kelleher, B.F. O'Neill, E. Diskin, A. Pletsers, H. Proctor, R. Stirnemann, J. O'Halloran, J. Penuelas, T.R. Hodkinson, T.H. Sparks, Clim. Res. 53 (2012) 245–262.

[76] M.E. Visser, L.J.M. Holleman, Proc. R. Soc. B Biol. Sci. 268 (2001) 289–294.

[77] M. van Asch, L. Salis, L.J.M. Holleman, B. van Lith, M.E. Visser, Nat. Clim. Change 3 (2013) 244–248.

[78] W.E. Bradshaw, C.M. Holzapfel, Science 312 (2006) 1477–1478.

[79] W.E. Bradshaw, C.M. Holzapfel, Mol. Ecol. 17 (2008) 157–166.

[80] W.E. Bradshaw, C.M. Holzapfel, Proc. Natl. Acad. Sci. U.S.A. 98 (2001) 14509–14511.

[81] T. Harada, S. Takenaka, S. Maihara, K. Ito, T. Tamura, Physiol. Entomol. 36 (2011) 309–316.

[82] J. Urbanski, M. Mogi, D. O'Donnell, M. DeCotiis, T. Toma, P. Armbruster, Am. Nat. 179 (2012) 490–500.

[83] S.E. Diamond, D.M. Sorger, J. Hulcr, S.L. Pelini, I. Del Toro, C. Hirsch, E. Oberg, R.R. Dunn, Global Change Biol. 18 (2012) 448–456.

[84] D.H. Janzen, Am. Nat. 101 (1967) 233.

[85] K.J. Gaston, S.L. Chown, Oikos 84 (1999) 309–312.

[86] J.F. Gillooly, J.H. Brown, G.B. West, V.M. Savage, E.L. Charnov, Science 293 (2001) 2248–2251.

[87] M.E. Dillon, G. Wang, R.B. Huey, Nature 467 (2010), 704–U88.

[88] J.A. Zeh, M.M. Bonilla, E.J. Su, M.V. Padua, R.V. Anderson, D. Kaur, D.S. Yang, D.W. Zeh, Global Change Biol. 18 (2012) 1833–1842.

[89] J.M. Sunday, A.E. Bates, N.K. Dulvy, Proc. R. Soc. B Biol. Sci. 278 (2011) 1823–1830.

[90] T.C. Bonebrake, C.A. Deutsch, Ecology 93 (2012) 449–455.

[91] J.G. Kingsolver, S.E. Diamond, L.B. Buckley, Funct. Ecol. 27 (2013) 1415–1423.

[92] M.A. Jamieson, A.M. Trowbridge, K.F. Raffa, R.L. Lindroth, Plant Physiol. 160 (2012) 1719–1727.

[93] H. Zhu, D.L. Wang, L. Wang, J. Fang, W. Sun, B.Z. Ren, Ecol. Entomol. 39 (2014) 453–461.

[94] D.E. Lincoln, D. Couvet, N. Sionit, Oecologia 69 (1986) 556–560.

[95] G.J. Masters, V.K. Brown, I.P. Clarke, J.B. Whittaker, J.A. Hollier, Ecol. Entomol. 23 (1998) 45–52.

[96] Y. Wang, S. Yu, J. Wang, Ecol. Lett. 10 (2007) 401–410.

[97] B.J. Hicks, A.D. Watt, J. Appl. Entomol. 127 (2003) 553–559.

[98] N.W. Siegert, D.G. McCullough, R.C. Venette, A.E. Hajek, J.A. Andresen, Can. J. For. Res. Revue Canadienne De Recherche Forestiere 39 (2009) 1958–1970.

[99] D.R. Smitley, L.S. Bauer, A.E. Hajek, F.J. Sapio, R.A. Humber, Environ. Entomol. 24 (1995) 1685–1695.

[100] U. Segev, K. Tielborger, Y. Lubin, J. Kigel, Ecol. Entomol. 39 (2014) 427–435.

[101] L.E. Rustad, Sci. Total Environ. 404 (2008) 222–235.

[102] R.L. Lindroth, J. Chem. Ecol. 36 (2010) 2–21.

[103] R.L. Lindroth, in: G.R. Iason, M. Dicke, S.E. Hartley (Eds.), Ecology of Plant Secondary Metabolites: From Genes to Global Processes, Cambridge Univ Press, the Pitt Building, Trumpington St, Cambridge Cb2 1rp, Cambs, UK, 2012, pp. 120–153.

[104] J.K. Holopainen, J. Gershenzon, Trends Plant Sci. 15 (2010) 176–184.

[105] J. Penuelas, M. Staudt, Trends Plant Sci. 15 (2010) 133–144.

[106] T. Cornelissen, Neotrop. Entomol. 40 (2011) 155–163.

[107] P. stiling, T. Cornelissen, Global Change Biol. 13 (2007) 1823–1842.

[108] M.D. Madritch, J.R. Donaldson, R.L. Lindroth, Soil Biol. Biochem. 39 (2007) 1192–1201.

[109] H.L. Throop, M.T. Lerdau, Ecosystems 7 (2004) 109–133.

[110] G.M. Lovett, A.E. Ruesink, Oecologia 104 (1995) 133–138.

[111] J.J. Couture, R.L. Lindroth, Arthropod Plant Interact. 8 (2014) 33–47.

[112] C. Scherber, D.J. Gladbach, K. Stevnbak, R.J. Karsten, I.K. Schmidt, A. Michelsen, K.R. Albert, K.S. Larsen, T.N. Mikkelsen, C. Beier, S. Christensen, Ecol. Evol. 3 (2013) 1449–1460.

[113] S. Leuzinger, Y.Q. Luo, C. Beier, W. Dieleman, S. Vicca, C. Korner, Trends Ecol. Evol. 26 (2011) 236–241.

[114] K.S. Larsen, L.C. Andresen, C. Beier, S. Jonasson, K.R. Albert, P. Ambus, M.F. Arndal, M.S. Carter, S. Christensen, M. Holmstrup, A. Ibrom, J. Kongstad, L. van der Linden, K. Maraldo, A. Michelsen, T.N. Mikkelsen, K. Pilegaard, A. Prieme, H. Ro-Poulsen, I.K. Schmidt, M.B. Selsted, K. Stevnbak, Global Change Biol. 17 (2011) 1884–1899.

[115] J.L. Blois, P.L. Zarnetske, M.C. Fitzpatrick, S. Finnegan, Science 341 (2013) 499–504.

[116] O.E. Sala, F.S. Chapin, J.J. Armesto, E. Berlow, J. Bloomfield, R. Dirzo, E. Huber-Sanwald, L.F. Huenneke, R.B. Jackson, A. Kinzig, R. Leemans, D.M. Lodge, H.A. Mooney, M. Oesterheld, N.L. Poff, M.T. Sykes, B.H. Walker, M. Walker, D.H. Wall, Science 287 (2000) 1770–1774.

[117] R.J. Wilson, D. Gutierrez, J. Gutierrez, V.J. Monserrat, Global Change Biol. 13 (2007) 1873–1887.

[118] C. de Sassi, O.T. Lewis, J.M. Tylianakis, Ecology 93 (2012) 1892–1901.

[119] J. Resasco, S.L. Pelini, K.L. Stuble, N.J. Sanders, R.R. Dunn, S.E. Diamond, A.M. Ellison, N.J. Gotelli, D.J. Levey, Plos One 9 (2014).

[120] S. Sahney, M.J. Benton, Proc. R. Soc. B Biol. Sci. 275 (2008) 759–765.

[121] J.K. Schubert, D.J. Bottjer, Palaeogeogr. Palaeoclimatol. Palaeoecol. 116 (1995) 1–39.

[122] R.R. Dunn, N.C. Harris, R.K. Colwell, L.P. Koh, N.S. Sodhi, Proc. R. Soc. B Biol. Sci. 276 (2009) 3037–3045.

[123] R. Menendez, A.G. Megias, J.K. Hill, B. Braschler, S.G. Willis, Y. Collingham, R. Fox, D.B. Roy, C.D. Thomas, Proc. R. Soc. B Biol. Sci. 273 (2006) 1465–1470.

[124] R. Menendez, A. Gonzalez-Megias, Y. Collingham, R. Fox, D.B. Roy, R. Ohlemuller, C.D. Thomas, Ecology 88 (2007) 605–611.

[125] K.B. Suttle, M.A. Thomsen, M.E. Power, Science 315 (2007) 640–642.

[126] S.L. Pelini, S.E. Diamond, L.M. Nichols, K.L. Stuble, A.M. Ellison, N.J. Sanders, R.R. Dunn, N.J. Gotelli, Ecosphere 5 (2014) 125.

[127] C. Both, M. van Asch, R.G. Bijlsma, A.B. van den Burg, M.E. Visser, J. Anim. Ecol. 78 (2009) 73–83.

[128] L. Cagnolo, S.I. Molina, G.R. Valladares, Biodivers. Conserv. 11 (2002) 407–420.

[129] M. Jochum, F.D. Schneider, T.P. Crowe, U. Brose, E.J. O'Gorman, Philos. Trans. R. Soc. B Biol. Sci. 367 (2012) 2962–2970.

[130] B.T. Barton, A.P. Beckerman, O.J. Schmitz, Ecology 90 (2009) 2346–2351.

[131] O.J. Schmitz, Ecol. Stud. 173 (2004) 277–302.

[132] C. Parmesan, Annual Review of Ecology Evolution and Systematics, vol. 37, 2006. Annual Reviews: Palo Alto, pp. 637–669.

[133] J.A. Thomas, Philos. Trans. R. Soc. B Biol. Sci. 360 (2005) 339–357.

[134] M.W. Blows, A.A. Hoffmann, Ecology 86 (2005) 1371–1384.

SEA LIFE (PELAGIC ECOSYSTEMS)

11

Martin Edwards[1,2]

[1]*Sir Alister Hardy Foundation for Ocean Science, Plymouth, UK;* [2]*Marine Institute, University of Plymouth, Plymouth, UK*

CHAPTER OUTLINE

1. PELAGIC AND PLANKTONIC ECOSYSTEMS

The marine pelagic realm is the largest ecological system on the planet, occupying 71% of the planetary surface and a major part of the Earth's overall biosphere. As a consequence of this, pelagic ecosystems play a fundamental role in modulating the global environment via its regulatory effects on the Earth's climate and its role in biogeochemical cycling. Changes caused by increased warming on marine pelagic communities are likely to have important consequences on ecological structure and function thereby leading to significant feedbacks on the Earth's climate system.

Apart from discussing the effects of climate on higher trophic organisms, particularly pelagic fish, the overall emphasis of this chapter is focused on the planktonic community. More specifically, the chapter will concentrate on observational evidence from contemporary plankton indicators over the past multi-decadal period rather than paleoplanktonic indicators. The free-floating photosynthesizing life of the oceans (algal phytoplankton, bacteria and other photosynthesizing protists), at the base of the marine food web, provides food for the animal plankton (zooplankton), which in turn provide food for many other marine organisms ranging from the microscopic to whales. The carrying capacity of pelagic ecosystems in terms of the size of fish resources and recruitment to individual stocks as well as the abundance of marine wildlife (e.g. seabirds and marine mammals) is highly dependent on variations in the abundance, seasonal timing and composition of the plankton.

Phytoplankton also comprise approximately half of total global primary production and play a crucial role in climate change through biogeochemical cycling and the export of the greenhouse gas to the deep ocean by carbon sequestration in what is known as the 'biological pump'. Phytoplankton have thus already helped mitigate some of the climate effects of elevated CO_2 observed over the last

200 years, with the oceans taking up to about 40% of anthropogenic CO_2 [1]. In terms of feedback mechanisms on Earth's climate, it is speculated that these biological pumps will be less efficient in a warmer world due to changes in the phytoplankton composition. These changes favour small flagellates [2] and less overall nutrient mixing due to increased stratification (see Section 1.1). It is also predicted that warmer temperatures will shift the metabolic balance between production and respiration in the world's oceans toward an increase in respiration, thus reducing the capacity of the oceans to capture CO_2 [3]. Apart from playing a fundamental role in the Earth's climate system and in marine food webs, plankton are also highly sensitive contemporary and paleoindicators of environmental change and provide rapid information on the 'ecological health' of our oceans. There is some evidence that suggests that plankton are more sensitive indicators than environmental variables themselves and can amplify weak environmental signals due to their nonlinear responses [4]. A plankton species, defined by its abiotic envelope, in effect has the capacity to simultaneously represent an integrated ecological, chemical and physical variable.

1.1 SENSITIVITY OF PELAGIC AND PLANKTONIC ECOSYSTEMS TO CLIMATE AND GLOBAL CHANGE

Temperature is a key driver of marine ecosystems, and in particular its effects on pelagic populations are manifested very rapidly [5–7]. This is hardly surprising when more than 99% of pelagic and planktonic organisms are ectothermal; this makes them highly sensitive to fluctuations in temperature [8]. The rapidity of the planktonic response is predominantly due to their short life cycles and in their mainly passive response to advective changes. For example, phytoplankton fix as much CO_2 per year as all terrestrial plants, but due to being unicellular they represent at any one time only 1% of the Earth's biomass. This means the rate of turnover in the world's oceans is huge, and on average the global phytoplankton population is consumed in days to weeks [9]. This all makes plankton tightly coupled to fluctuations in the marine environment and highly sensitive indicators of environmental change such as nutrient availability, ocean current changes and climate variability.

In the marine environment the effect of short-term climate variability and interannual variability on populations of higher trophic levels such as seabirds and whales can to a degree be somewhat buffered due to their longer life cycles. In the long term, their ability to undergo large geographical migrations may also help them to mitigate some of the effects of global change; however, this hypothesis has not been investigated. In both cases, this is not applicable to planktonic organisms. Biologically speaking, changes in temperature have direct consequences on many physiological processes (e.g. oxygen metabolism, adult mortality, reproduction, respiration, reproductive development) and control virtually all life processes from the molecular to the cellular to whole regional ecosystem level and biogeographical provinces. Ecologically speaking, temperature also modulates both directly and indirectly species interactions (e.g. competition, prey–predator interactions and food web structures). Ultimately, changes in temperatures can lead to impacts on the biodiversity, size structure and functioning of the whole pelagic ecosystem [10–12].

While temperature has direct consequences on many biological and ecological traits, it also modifies the marine environment by influencing oceanic circulation and by enhancing the stability of the water column and hence nutrient availability. The amount of nutrients available in surface waters directly dictates phytoplankton growth and is the key determinant of the plankton size, community and food web structure. In terms of nutrient availability, warming of the surface layers increases water column stability, enhancing stratification and requiring more energy to mix deep, nutrient-rich waters into surface layers. Particularly warm winters will also limit the degree of deep convective mixing and thereby limit nutrient replenishment necessary for the following spring phytoplankton bloom. In summary, climatic warming of surface waters will increase the density contrast between the surface layer and the underlying nutrient-rich waters. The availability of one of the principal nutrients (nitrate) that limits phytoplankton growth has therefore been found to be negatively related to temperatures globally [13,14]. Similarly, a global analysis of satellite-derived chlorophyll data shows a strong inverse relationship between sea surface temperatures (SST) and chlorophyll concentration [9]. Furthermore, other abiotic variables like oxygen concentration (important to organism size and metabolism [15]), nitrate metabolism [16] and the viscosity of sea water (important for the maintenance of buoyancy for plankton) are also directly linked to temperature. So unlike terrestrial environments, where precipitation plays a key role, the chemical and upper-ocean temperature regime in open oceans and its consequent biological composition are inexorably entwined.

1.2 MARINE AND TERRESTRIAL BIOLOGICAL RESPONSES TO CLIMATE AND GLOBAL CHANGE

Many planktonic organisms live in narrow temperature ranges (stenothermal) and often undergo a much more rapidly observed change due to temperature, be it biogeographically or phenologically [10,11], in comparison to their terrestrial counterparts [17]. Apart from this and the fact that planktonic organisms having shorter life cycles, already mentioned previously, there are a number of distinct reasons why the speed of the response to climate and global change of pelagic organisms differs from those of terrestrial organisms. There are a number of primary reasons for this. Firstly, due to the high specific heat index of water in open ocean systems, many planktonic organisms are largely buffered against extremes in daily and seasonal temperature fluctuations. Daily and seasonal variations in temperature are therefore less variable in comparison to the terrestrial domain, allowing marine species to become firmly embedded in their optimum thermal envelope. Secondly, and unlike terrestrial environments, many planktonic organisms can quickly track evolving bioclimatic envelopes by being largely free of geographical barriers hindering their dispersal range and do not need a large amount of energy expenditure to do so, being primarily passively advected. Ocean currents, therefore, provide an ideal mechanism for dispersal over large distances and this is seemingly why a vast many of marine organisms have evolved at least a portion of their life cycles as planktonic entities. Thirdly, many terrestrial organisms are geographically and ecologically bound by their habitat type, mainly dictated by the vegetative composition. In terrestrial systems the development of these vegetative types can be particularly slow moving (e.g. forest ecosystems) and hence organisms that rely on this habitat will be restricted in terms of their geographical spread. This is not the case for phytoplankton, which have extremely short life cycles in comparison, allowing rapid temporal and spatial spread of planktonic herbivores and associated communities. Furthermore, the presence of inimitably terrestrial anthropogenic pressures, such as habitat fragmentation and habitat loss, which clearly limits the geographical spread of organisms in the terrestrial environment, is seemingly absent from open ocean systems [18].

1.3 OCEAN ACIDIFICATION AND OTHER ANTHROPOGENIC INFLUENCES ON PELAGIC AND PLANKTONIC ECOSYSTEMS

While temperature, light and nutrients are probably the most important physical variables structuring marine ecosystems, the pelagic realm will also have to contend with, apart from global climate change, the impact of anthropogenic CO_2 directly influencing the pH of the oceans [19,20]. Evidence collected and modelled to date indicates that rising CO_2 has led to chemical changes in the ocean, which has led to the oceans becoming more acidic. Ocean acidification has the potential to affect the process of calcification and therefore certain planktonic organisms (e.g. coccolithophores, foraminifera, pelagic molluscs) may be particularly vulnerable to future CO_2 emissions. Apart from climate warming, potentially chemical changes to the oceans and its effect on the biology of the oceans could further reduce the ocean's ability to absorb additional CO_2 from the atmosphere, which in turn could affect the rate and scale of global warming. Other anthropogenic driving forces of change that are operative in pelagic ecosystems are predominantly overfishing and its effect on modifying marine pelagic food webs [21], and in coastal regions nutrient input from terrestrial sources leading in some cases to enhanced biological production and harmful algal blooms and other general chemical and inorganic contaminants. The impacts of atmospheric-derived anthropogenic nitrogen on the open ocean have only been recently investigated but may also play a significant role in annual new marine biological production [22,23].

2. OBSERVED IMPACTS ON PELAGIC AND PLANKTONIC ECOSYSTEMS

There is a large body of observed evidence to suggest that many pelagic ecosystems, both physically and biologically, are responding to changes in regional climate caused predominately by the warming of air and SSTs and to a lesser extent by the modification of precipitation regimes and wind patterns. The biological manifestations of rising SSTs have variously taken the form of biogeographical, phenological, biodiversity, physiological, species abundance changes and whole ecological regime shifts. Any observational change in the marine environment associated with climate change, however, should be considered against the background of natural variation on a variety of spatial and temporal scales. Recently, long-term decadal observational studies have focused on known natural modes of climatic oscillations at similar temporal scales such as the El Niño Southern Oscillation (ENSO) in the Pacific and the North Atlantic Oscillation (NAO) in the North Atlantic in relation to pelagic ecosystem changes (see reviews [24,25]). Many of the biological responses observed have been associated with rising temperatures. However, approximating the effects of climate change embedded in natural modes of variability, particularly multi-decadal oscillations like the Atlantic multi-decadal oscillation (AMO) [26,27], is extremely difficult, and therefore observed evidence of planktonic changes directly attributable to anthropogenic climate and global change must be treated with a degree of scientific caution.

Evidence for observed pelagic changes is also biased toward regions, particularly seas around Europe and North America, which have had some form of biological monitoring in place over a consistently long period [28]. Apart from a number of important long-term coastal research stations sampling plankton (e.g. Helgoland Roads time series in the southern North Sea [29]), there are only a few long-term biological surveys that sample the open ocean. For this reason, some of the strongest evidence detected for observed changes in open ocean ecosystems comes from the North Atlantic where an extensive spatial and long-term biological survey exists in the form of the Continuous Plankton Recorder (CPR) survey. The CPR survey has been in operation in the North Sea and North Atlantic since 1931 and has systematically sampled up to 500 planktonic taxa from the major regions of the North Atlantic at a monthly resolution [30]. Important multi-decadal evidence from the Pacific is mainly derived from the Californian Cooperative Oceanic Fisheries Investigations (CalCOFI) survey operating off the coast of Californian since 1949 [28].

2.1 BIOGEOGRAPHICAL CHANGES AND NORTHWARD SHIFTS

Some of the strongest evidence of large-scale biogeographical changes observed in our oceans comes from the CPR survey. A study encompassing the whole of the northeast Atlantic Ocean over a 50-year period, by Beaugrand et al. [10,31], highlighted the rapid northerly movements of the biodiversity of a key zooplankton group (calanoid copepods). During the last 50 years there has been a northerly movement of warmer-water plankton by 10° latitude in the northeast Atlantic and a similar retreat of colder-water plankton to the north (a mean poleward movement of 200 km–250 km per decade) (Fig. 1).

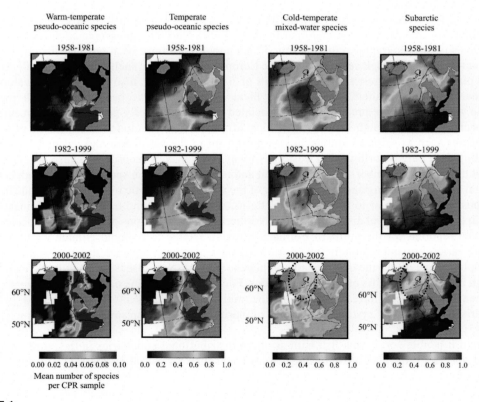

FIGURE 1

Changes in the geographical distribution of four different plankton assemblages over a multi-decadal period. There has been a rapid northerly movement of warm-temperate species and a subsequent decline in subarctic species over 40 years. Particularly rapid movement is observed along the European Continental Shelf. *(Data derived from the Continuous Plankton Recorder survey. Updated from Ref. [10].)*

This geographical movement is much more pronounced than any documented terrestrial study, mainly due to advective processes and in particular the shelf-edge current running north along the northern European continental shelf. The rapid movement of plankton northward is only seen along the continental shelf, where deeper water is warming much more rapidly. Further along the shelf, plankton are upwelled from this deeper water to make an appearance in the surface plankton community. Hence the plankton have moved 10° latitude northward via mainly deep water advective processes not seen in the movement of surface isotherms. In other areas in the northeast Atlantic the plankton shifts were more moderate and varied between 90 km and 200 km per decade, still faster than any other documented terrestrial study, which has a metanalytic average of 6 km per decade [17]. Similar to the North Atlantic, the northeast Pacific has seen a general increase in the frequency of southern species moving northward [32]. Interestingly, in the northwest Atlantic, pelagic organisms have been moving southward [33]. This initially seems to contradict general thinking of homogenous global climate warming throughout the world's oceans. However, this movement has been linked to the strengthening of the Labrador Current, which has spread colder water southward over the last decade, carrying pelagic organisms with cold-water affinities as far south as Georges Bank.

These large-scale biogeographical shifts observed in the plankton have also seen paralleled latitudinal movements of fish species distribution [34–38]. Northerly geographical range extensions or changes in the geographical distribution of fish populations have been recently documented for European Continental shelf seas and along the European Continental shelf edge [35,38–40]. Similar to the plankton, the largest movements north of fish species have also been observed along the European Continental shelf. These geographical movements have been related with regional climate warming and are predominantly associated with the northerly geographical movement of fish species with more southern biogeographical affinities. These include the movement of pelagic fish species such as sardines and anchovies northward in the North Sea and red mullet and bass extending their ranges northward to western Norway [35,39]. New records were also observed over the last decade for a number of Mediterranean and northwest African species on the south coast of Portugal [35]. The cooling and the freshening of the northwest Atlantic over the last decade has had an opposite effect similar to the plankton patterns, with some ground fish species moving further south in their geographical distribution [41]. Northerly range extensions of pelagic fish species have also been reported for the northern Bering Sea region related to regional climate warming [42]. Climate variability and regional climate warming have also been associated with variations in the geographic range of marine diseases and pathogens [43,44]. New diseases typically have emerged through host or range shifts of known pathogens. For example, over the past few decades pathogens detrimental to oysters have spread from the mid-Atlantic states into New England [43]. In comparison to terrestrial systems, epidemics of marine pathogens can spread at an extremely rapid rate [45].

Again it is noteworthy that fish with northern distributional boundaries in the North Sea have shifted northward at rates of up to three times faster than terrestrial species [17,36]. One of the largest biogeographical shifts ever observed for fish species is the dramatic increase and subsequent geographical spread northward of the snake pipefish (*Entelurus aequoreus*). Once confined mainly to the south and west of the British Isles before 2003, it can now be found as far north as the

Barents Sea and Spitzbergen [46]. While this present discussion has described surface geographical changes in epipelagic organisms, it is worth remembering the three-dimensional nature of the pelagic environment. Research has observed not just changes in fish bio-geography but also changes in fish species depth (toward deeper waters) in response to climate warming [47]. This change can be seen as analogous with the upward altitudinal movement of terrestrial organisms in alpine environments. All these studies highlight the consistency of pelagic organisms undergoing large-scale distributional changes in response to hydroclimatic variability.

2.2 LIFE CYCLE EVENTS AND PELAGIC PHENOLOGY

Phenology, or repeated seasonal life cycle events, such as annual migrations or spawning, are highly sensitive indicators of climate warming. This is because many terrestrial and marine organisms, apart from photoperiod, are dependent on temperature as a trigger for seasonal behaviour. In the terrestrial realm, phenology events such as bird migrations, egg laying, butterfly emergence and flowering of certain plants are all getting earlier in response to milder spring weather [17]. In terms of the pelagic phenological response to climate warming, many plankton taxa have also been found to be moving forward in seasonal cycles [11]. In some cases a shift in seasonal cycles of over six weeks was detected, again a far larger shift than observed for terrestrial-based observations. Summarizing a terrestrial study of phenology using over 172 species of plants, birds, insects and amphibians, Parmesan and Yorke [17] calculated a mean phenological change of 2.3 days. It is thought that temperate pelagic environments are particularly vulnerable to phenological changes caused by climatic warming because the recruitment success of higher trophic levels is highly dependent on synchronization with pulsed planktonic production [11]. Furthermore, in the marine environment, and just as important, was that the response to regional climate warming varied between different functional groups and trophic levels, leading to mismatch in timing between trophic levels (Fig. 2). For example, while the spring bloom has remained relatively stable in seasonal timing over five decades (mainly due to light limitation and photoperiod rather than temperature dictating seasonality [11,48]) many zooplankton organisms as well as fish larvae have moved rapidly forward in their seasonal cycles.

These changes, seen in the North Sea, have the potential to be detrimental to commercial fish stocks via trophic mismatch. For example, regional climate warming in the North Sea has affected cod recruitment via changes at the base of the food web [49]. Cod, like many other fish species, are highly dependent on the availability of planktonic food during their pelagic larval stages. Key changes in the planktonic assemblage and phenology, significantly correlated with the warming of the North Sea over the last few decades, have resulted in a poor food environment for cod larvae and hence an eventual decline in overall recruitment success. The rapid changes in plankton communities observed over the last few decades in the North Atlantic and European regional seas, related to regional climate changes, have enormous consequences for other trophic levels and biogeochemical processes. Similarly, other pelagic phenology changes have been observed in the North Sea [29], the Mediterranean [50] and the Pacific [51,52].

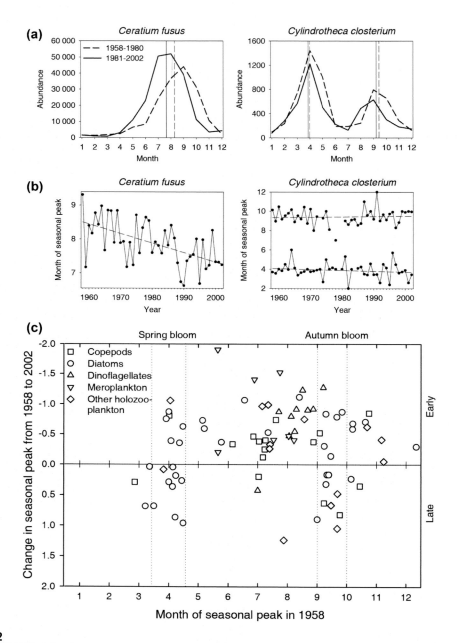

FIGURE 2

(a) Seasonal cycles for two phytoplankton for the periods 1958–1980 and 1981–2002: the dinoflagellate *Ceratium fusus* and the diatom *Cylindrotheca closterium*. (b) Interannual variability of the seasonal peak for the above two species from 1958 to 2002. (c) The change in the timing of the seasonal peaks (in months) for the 66 taxa over the 45-year period from 1958 to 2002 plotted against the timing of their seasonal peak in 1958. For each taxon, the linear regression in (b) was used to estimate the difference between the seasonal peak in 1958 and 2002. A negative difference between 1958 and 2002 indicates seasonal cycles are becoming earlier. Standard linear regression was considered appropriate because there was minimal auto-correlation (determined by the Durbin–Watson statistic) in the phenology time series. *(From Ref. [11].)*

2.3 PLANKTON ABUNDANCE AND PELAGIC PRODUCTIVITY

Contemporary observations of satellite in situ blended ocean chlorophyll records indicate that global ocean net primary production has declined over the last decade [9]. Although this time series is only 10 years long, it does show a strong negative relationship between primary production and SST and is evidence of the closely coupled relationship between ocean productivity and climate variability at a global scale. In the North Atlantic and over multi-decadal periods, both changes in phytoplankton and zooplankton species and communities have been associated with northern hemisphere temperature trends and variations in the NAO index. These have included changes in species distributions and abundance, the occurrence of subtropical species in temperate waters, changes in overall phytoplankton biomass and seasonal length (Fig. 3), changes in the ecosystem functioning and productivity of the North Atlantic [7,10–12,53–59]. The increase in overall phytoplankton biomass in the North Sea has been associated with an increase in smaller flagellates, which favour warmer and stratified conditions [53,54]. Over the whole northeastern Atlantic there has been an increase in phytoplankton biomass with increasing temperatures in cooler regions but a decrease in phytoplankton biomass in

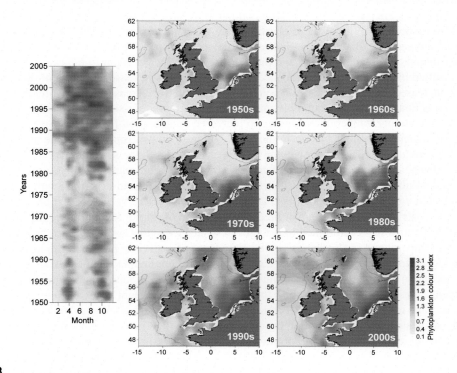

FIGURE 3

Spatial-temporal maps of the changes in the abundance of phytoplankton colour (an index of total phytoplankton biomass) for the NE Atlantic averaged per decade from the 1950s to the present. The contour plot shows monthly mean values from 1950 to 2005 of phytoplankton colour averaged for the North Sea. Large increases in phytoplankton colour are observed toward the end of the 1980s and have continued since. The increase in colour has been associated with a regime shift in the North Sea. *(Updated from Ref. [54].)*

warmer regions [60]. Presumably this is a trade-off between increased phytoplankton metabolic rates caused by temperature in cooler regions but a decrease in nutrient supply in warmer regions. Regional climate warming in the North Sea has also been associated with an increase in certain harmful algal blooms in some areas of the North Sea [61,62]. A recent link has been established between the changes in the plankton in the North Sea to sand eels and eventual seabird breeding success (encompassing four trophic levels) [63]. In the North Sea the population of the previously dominant and ecologically important zooplankton species, the cold-water species *Calanus finmarchicus*, has declined in biomass by 70% since the 1960s [64]. Species with warmer-water affinities are moving northward to replace this species, but these species are not as numerically abundant or nutritionally beneficial to higher trophic levels. This has had inevitably important ramifications for the overall carrying capacity of the North Sea ecosystem.

The ecological changes that have occurred in the North Sea since the late 1980s (predominately driven by change in temperature regime and more warmer winters) have also been documented for the Baltic Sea for zooplankton and fish stocks [65,66]. The related changes that have taken place in these northern European waters are sufficiently abrupt and persistent to be termed as 'regime shifts' [67]. Similarly in the Mediterranean, zooplankton communities have also been linked to regional warming and the NAO index [68,69]. All these observed changes appear to be closely correlated to climate-driven sea temperature fluctuations. Indirectly the progressive freshening of the Labrador Sea region, attributed to climate warming and the increase in freshwater input to the ocean from melting ice, has resulted in the increasing abundance, blooms and shifts in seasonal cycles of dinoflagellates due to the increased stability of the water column [70]. Similarly, increases in coccolithophore blooms in the Barents Sea and harmful algal blooms in the North Sea are associated with negative salinity anomalies and warmer temperatures leading to increased stratification [61,71].

In the Benguela upwelling system in the South Atlantic, long-term trends in the abundance and community structure of coastal zooplankton have been related to large-scale climatic influences [72]. Similarly, changes in mesozooplankton abundance have also been related to large-scale climate influences in the Californian upwelling system [73]. The progressive warming in the Southern Ocean has been associated with a decline in krill [74] and an associated decline in the population sizes of many seabirds and seals monitored on several breeding sites [75,76]. In the Southern Ocean the long-term decline in krill stock has been linked to changes in winter ice extent, which in turn has been related to warming temperatures [74]. Changes in the abundance of krill have profound implications for the Southern Ocean food web. The progressive warming of the Southern Ocean has also been associated with the decline in the population sizes of many seabirds and seals monitored on several breeding sites [76]. Recent investigations of planktonic foraminifera, from sediment cores encompassing the last 1400 years, have revealed anomalous changes in the community structure over the last few decades. The study suggests that ocean warming has already exceeded the range of natural variability [77]. A recent major ecosystem shift in the northern Bering Sea has been attributed to regional climate warming and trends in the Arctic oscillation [42]. Decadal changes in zooplankton related to climatic variability in the west subarctic North Pacific [78] and in the Japan/East Sea [79] have also been observed.

Many changes in abundance of marine commercial fish stocks have been observed over the last few decades in the Atlantic and Pacific oceans, but it is extremely difficult to separate, in terms of changes in population densities and recruitment, regional climate effects from direct anthropogenic influences like fishing. Geographical range extensions mentioned earlier or changes in the geographical distribution of fish populations, however, can be more confidently linked to hydroclimatic variation and regional climate warming. Similar to the observed changes in marine planktonic systems, many long-term changes in pelagic fish populations have been associated with known natural modes of climatic oscillations such as ENSO and the Pacific Decadal Oscillation (PDA) in the Pacific and the NAO/AMO in the North Atlantic (see reviews: [5,24,25,27,32]). For example, variations in SST driven by NAO fluctuations have been linked to fluctuations in cod recruitment both off Labrador and Newfoundland and in the Barents Sea [80]. Populations of herring, sardine, salmon and tuna have also been related to fluctuations in the NAO/AMO index [5,7,27]. Warm events related to El Nino episodes and climate-induced ecological regime shifts in the Pacific have been related to the disruption of many commercial fisheries [24,32,81]. These changes highlight the sensitivity of fish populations to environmental change. Direct evidence of biological impacts of anthropogenic climate change is, however, difficult to discern due to the background of natural variation on a variety of spatial and temporal scales and in particular natural oscillations in climate. A recent study based on a 50-year larval fish time series from CalCOFI showed that exploited fish species were more vulnerable to the impacts of climate change than nonexploited species. The authors suggest that the enhanced response to environmental change of exploited species was due to a reduced spatial heterogeneity caused by fishery-induced age truncation and a restriction of geographic distribution that had accompanied fishing pressure [82].

2.4 PELAGIC BIODIVERSITY AND INVASIVE SPECIES

At the ocean basin scale, pelagic biodiversity is closely related to temperature with a higher diversity in the subtropical and tropical zones of the oceans. Over the last few decades the increase in warming in many other ocean regions has also increased the diversity [12,83–85]. In particular, increases in diversity are seen when a previously low diversity system like Arctic and cold-boreal provinces undergo prolonged warming events. The overall diversity patterns of pelagic organisms, peaking between 20° and 30° north or south, follow temperature gradients in the world's oceans [86]. Similarly, phytoplankton show a relationship between temperature and diversity, which is linked to the phytoplankton community having a higher diversity but an overall smaller size fraction and a more complex food web structure (i.e. microbial-based vs diatom-based production) in warmer, more-stratified environments. Climate warming will therefore increase planktonic diversity throughout the cooler regions of the world's oceans as temperature isotherms shift poleward. However, the relationship between temperature and pelagic fish diversity is far more complex due to other anthropogenic pressures such as overfishing apparently playing a significant role in diversity patterns [21].

Climate warming will open up new thermally defined habitats for previously denied nonindigenous species (e.g. subtropical species in the North Sea) and invasive species, allowing them to establish viable populations in areas that were once environmentally unsuitable. Apart from these thermal boundaries limits moving progressively poleward and in some cases expanding, the rapid climate change observed in the Arctic may have even larger consequences for the establishment of invasive species and the biodiversity of the North Atlantic. The thickness and areal coverage of summer ice in the Arctic have been melting at an increasingly rapid rate over the last two decades to reach the lowest-ever recorded extent in September 2007. In the spring following the unusually large ice-free period in 1998, large numbers of a Pacific diatom *Neodenticula seminae* were found in samples taken by the CPR survey in the Labrador Sea in the North Atlantic. *Neodenticula seminae* is an abundant member of the phytoplankton in the subpolar North Pacific and has a well-defined paleohistory based on deep sea cores. According to the paleoevidence and modern surface sampling in the North Atlantic since 1948, this was the first record of this species in the North Atlantic for at least 800 000 years. The reappearance of *N. seminae* in the North Atlantic, and its subsequent spread southward and eastward to other areas in the North Atlantic, after such a long gap, could be an indicator of the scale and speed of changes that are taking place in the Arctic and North Atlantic oceans as a consequence of climate warming [87]. The diatom species could be the first evidence of a trans-Arctic migration in modern times and a harbinger of a potential inundation of new organisms into the North Atlantic. The consequences of such a change to the function, climatic feedbacks and biodiversity of Arctic systems are at present unknown.

3. CONCLUSION AND SUMMARY OF KEY INDICATORS

The case studies highlighted in this review collectively indicate that there is substantial observational evidence that many pelagic ecosystems, both physically and biologically, are responding to changes in regional climate. These changes are caused predominately by the warming of sea surface temperatures, ocean current changes and to a lesser extent by the modification of precipitation regimes and wind patterns. The biological manifestations of climatic variability have rapidly taken the form of biogeographical, phenological, biodiversity, physiological, species abundance changes, community structural shifts and whole ecological regime shifts. Some of the most convincing evidence for the biological response to regional climate variability comes from the bottom of the marine pelagic food web, especially from phytoplankton and zooplankton communities. Many other responses associated with climate warming on higher trophic levels are also indirectly associated with changes in the plankton and imply bottom-up control of the marine pelagic environment. It is therefore assumed that one of the ways in which populations respond to climate is in part determined by changes in the food web structure where the population is embedded, with synchrony between predator and prey (match-mismatch) playing an important role.

At the species level, some of the first consequences of climate warming and global change are often seen in a species phenology (i.e. timing of annual occurring life cycle events) and in species geographical distribution responses. This is mainly because temperature continually impacts on the life cycle of the species. As a result the population will naturally respond over time, providing it is not biotically restrained or spatially restricted to its optimum position within its bioclimatic envelope. Whether this is within a temporal niche as in seasonal succession (observed as a phenological response) or in its overall biogeographical distribution (observed as a geographical movement in a population). These biological changes as well as those changes observed in biodiversity and planktonic abundance and productivity are perhaps the key indicators signifying the large-scale changes occurring in our world's oceans as a consequence of climate and global change.

Summarizing the observed case studies, what particularly stands out in this review is the rapidity of the pelagic and planktonic response, be it biogeographically or phenologically, to climate warming and global change compared to their terrestrial counterparts. For example, plankton shifts of up to 200 km per decade [10] have been observed in the northeast Atlantic compared with a metanalytic terrestrial average of 6 km per decade [17]. Similarly, changes in phenology of up to six weeks have been observed in pelagic ecosystems [11] compared with a mean phenological change of 2.3 collectively observed for 172 species of plants, birds, insects and amphibians [17]. Of the myriad differences between the terrestrial and marine realm (see Section 1.2), the rapidity of the planktonic response is predominantly due to their short life cycles and in their mainly passive response to advective changes. These changes highlighted in this review are set to continue into the future following current climate warming projections. It is therefore thought that the currently observed and future warming have and are likely to continue altering the geographical distribution of primary and secondary planktonic production [60], affecting marine ecosystem services such as oxygen production, carbon sequestration and biogeochemical cycling and placing additional stress on already depleted fish and mammal populations.

In terms of feedback mechanisms on Earth's climate, it is thought that these biological pumps will be less efficient in a warmer world due to changes in the phytoplankton composition (floristic shifts) and less overall nutrient mixing (reduced bulk properties) due to increased stratification of the world's oceans. In particular, this will affect large areas of the tropical oceans that are permanently stratified [9]. There also exists a strong negative relationship between ocean productivity and SST (linked through nutrient availability) at a global scale [9,13,14]. Although climate change and its spatially heterogeneous effect on surface wind patterns, wind strength, upwelling and deep-water mixing makes many regional predictions beset with uncertainty. It is also worth noting that potential habitat expansion for pelagic organisms in the northern hemisphere due to the melting of Arctic ice will be severely restricted by light limitations dictating seasonal phytoplankton production. However, many of these scenarios are still at their infancy stage; while it is relatively simpler to predict changing ocean physics under climate forcing, understanding the biological response due to the underlying complexity of biological communities and their quite often nonlinear responses to environmental change makes predicting floristic changes fraught with uncertainty. Investigating the importance of biological nitrogen fixation and the production of dimethylsulfide by certain phytoplankton is currently needed to understand the biological consequences of increased stratification on nitrogen cycles [88,89] and biological feedbacks [90,91].

Ecologically speaking and on a planetary scale, plankton and pelagic ecosystems as a metaphorical collective entity are perhaps some of the most sensitive organisms to environmental change and include some very important biological communities. They are responsible for the overwhelming majority of marine biological production that fuel marine food webs and nutrient cycling as well as contributing to approximately half of the world's oxygen production and carbon sequestration. Virtually all the biological observations highlighted in this review result from financially fragile multi-decadal monitoring programmes. Future biological monitoring of these ecosystems, through an integrated and sustained observational approach, will be essential in understanding the continuing impacts of climate and global change on our planetary system [28]. This in turn may allow us through international collaboration to mitigate and adaptively manage some of their more detrimental impacts.

REFERENCES

[1] R.E. Zeebe, J.C. Zachos, K. Caldeira, T. Tyrrell, Science 321 (2008) 51–52.
[2] L. Bopp, O. Aumont, P. Cadule, S. Alvain, M. Gehlen, Geophys. Res. Lett. 32 (2005).
[3] A. Lopez-Urrutia, E. San Martin, R.P. Harris, X. Irigoien, Proc. Natl. Acad. Sci. U.S.A. 103 (2006) 8739–8744.
[4] A.H. Taylor, J.I. Allen, P.A. Clark, Nature 416 (2002) 629–632.

[5] K.F. Drinkwater, A. Belgrano, A. Borja, A. Conversi, M. Edwards, C.H. Greene, G. Ottersen, A.J. Pershing, H. Walker, Geophys. Monogr. 134 (2003).
[6] G. Beaugrand, F. Ibañez, J.A. Lindley, Mar. Ecol. Prog. Ser. 219 (2001) 205–219.
[7] G. Beaugrand, P.C. Reid, Glob. Change Biol. 9 (2003) 801–817.
[8] D. Atkinson, R.M. Sibly, Trends Ecol. Evol. 12 (1997) 235–239.
[9] M.J. Behrenfeld, R.T. O'Malley, D.A. Siegel, C.R. McClain, J.L. Sarmiento, G.C. Feldman, A.J. Milligan, P.G. Falkowski, R.M. Letelier, E.S. Boss, Nature 444 (2006) 752–755.
[10] G. Beaugrand, P.C. Reid, F. Ibanez, J.A. Lindley, M. Edwards, Science 296 (2002) 1692–1694.
[11] M. Edwards, A.J. Richardson, Nature 430 (2004) 881–884.
[12] G. Beaugrand, M. Edwards, L. Legendre, Proc. Natl. Acad. Sci. U.S.A. 107 (2010) 10120–10124.
[13] D. Kamykowski, S.J. Zentara, Deep Sea Res. (I Oceanogr. Res. Pap.) 33 (1986) 89–105.
[14] D. Kamykowski, S.J. Zentara, Deep Sea Res. (I Oceanogr. Res. Pap.) 52 (2005) 1719–1744.
[15] H.O. Portner, R. Knust, Science 315 (2007) 95–97.
[16] J.A. Berges, D.E. Varela, P.J. Harrison, Mar. Ecol. Prog. Ser. 225 (2002) 139–146.
[17] C. Parmesan, G. Yohe, Nature 421 (2003) 37–42.
[18] L. Hannah, T.E. Lovejoy, S.H. Schneider, Biodiversity and climate change in context, in: T.E. Lovejoy, L. Hannah (Eds.), Climate Change and Biodiversity, Yale University Press, New Haven, CT, USA, 2005, pp. 3–14.
[19] R.A. Feely, C.L. Sabine, K. Lee, W. Berelson, J. Kleypas, V.J. Fabry, F.J. Millero, Science 305 (2004) 362–366.
[20] G. Beaugrand, A. McQuatters-Gollop, M. Edwards, E. Goberville, Nat. Clim. Change 3 (2012) 263–267.
[21] B. Worm, E.B. Barbier, N. Beaumont, J.E. Duffy, C. Folke, B.S. Halpern, J.B.C. Jackson, H.K. Lotze, F. Micheli, S.R. Palumbi, E. Sala, K.A. Selkoe, J.J. Stachowicz, R. Watson, Science 314 (2006) 787–790.
[22] R.A. Duce, J. LaRoche, K. Altieri, K.R. Arrigo, A.R. Baker, D.G. Capone, S. Cornell, F. Dentener, J. Galloway, R.S. Ganeshram, R.J. Geider, T. Jickells, M.M. Kuypers, R. Langlois, P.S. Liss, S.M. Liu, J.J. Middelburg, C.M. Moore, S. Nickovic, A. Oschlies, T. Pedersen, J. Prospero, R. Schlitzer, S. Seitzinger, L.L. Sorensen, M. Uematsu, O. Ulloa, M. Voss, B. Ward, L. Zamora, Science 320 (2008) 893–897.
[23] I.-N. Kim, K. Lee, N. Gruber, D.M. Karl, J.L. Bullister, S. Yang, T.-W. Kim, Science 346 (2014) 1102–1106.
[24] J. Overland, S. Rodionov, S. Minobe, N. Bond, Prog. Ocean 77 (2008) 92–102.
[25] N.C. Stenseth, G. Ottersen, J.W. Hurrell, A. Mysterud, M. Lima, K.S. Chan, N.G. Yoccoz, B. Aadlandsvik, Proc. R. Soc. Lond. Ser. B: Biol. Sci. 270 (2003) 2087–2096.
[26] R.T. Sutton, D.L.R. Hodson, Science 309 (2005) 115–118.
[27] M. Edwards, G. Beaugrand, P. Helaouët, J. Alheit, S. Coombs, PLoS One 8 (2013) e57212.
[28] M. Edwards, G. Beaugrand, G.C. Hays, J.A. Koslow, A.J. Richardson, Trends Ecol. Evol. 25 (2010) 602–610.
[29] W. Greve, S. Prinage, H. Zidowitz, J. Nast, F. Reiners, ICES J. Mar. Sci. 62 (2005) 1216–1223.
[30] P.C. Reid, J.M. Colebrook, J.B.L. Matthews, J. Aiken, Prog. Oceanogr. 58 (2003) 117–173.
[31] G. Beaugrand, C. Luczak, M. Edwards, Glob. Change Biol. 15 (2009) 1790–1803.
[32] J.A. McGowan, D.R. Cayan, L.M. Dorman, Science 281 (1998) 210–217.
[33] D.G. Johns, M. Edwards, S.D. Batten, Can. J. Fish Aquat. Sci. 58 (2001) 2121–2124.
[34] J.-C. Quero, M.-H. Du Buit, J.-J. Vayne, Oceanol. Acta 21 (1998) 345–351.
[35] K. Brander, G. Blom, M.F. Borges, K. Erzini, G. Henderson, B.R. MacKensie, H. Mendes, J. Ribeiro, A.M.P. Santos, R. Toresen, ICES Mar. Sci. Symp. 219 (2003) 261–270.
[36] A.L. Perry, P.J. Low, J.R. Ellis, J.D. Reynolds, Science 308 (2005) 1912–1915.
[37] K. Brander, J. Mar. Syst. 79 (2010) 389–402.
[38] I. Montero-Serra, M. Edwards, M.J. Genner, Glob. Change Biol. 21 (2014) 1–10.
[39] D. Beare, F. Burns, E. Jones, K. Peach, E. Portilla, T. Greig, E. McKenzie, D. Reid, Glob. Change Biol. 10 (2004) 1209–1213.

[40] M.J. Genner, D.W. Sims, V.J. Wearmouth, E.J. Southall, A.J. Southward, P.A. Henderson, S.J. Hawkins, Proc. R. Soc. Lond. Ser. B: Biol. Sci. 271 (2004) 655–661.

[41] G.C. Rose, R.L. O'Driscoll, ICES J. Mar. Sci. 59 (2002) 1018–1026.

[42] J.M. Grebmeier, J.E. Overland, S.E. Moore, E.V. Farley, E.C. Carmack, L.W. Cooper, K.E. Frey, J.H. Helle, F.A. McLaughlin, S.L. McNutt, Science 311 (2006) 1461–1464.

[43] C.D. Harvell, K. Kim, J.M. Burkholder, R.R. Colwell, P.R. Epstein, D.J. Grimes, E.E. Hofmann, E.K. Lipp, A. Osterhaus, R.M. Overstreet, J.W. Porter, G.W. Smith, G.R. Vasta, Science 285 (Suppl. 5433) (1999) 1505–1510.

[44] L. Vezzulli, I. Brettar, E. Pezzati, P.C. Reid, R.R. Colwell, M.G. Höfle, C. Pruzzo, ISME J. 6 (2012) 21–30.

[45] H. McCallum, D. Harvell, A. Dobson, Ecol. Lett. 6 (2003) 1062–1067.

[46] M.P. Harris, D. Beare, R. Toresen, L. Nottestad, M. Kloppmann, H. Dorner, K. Peach, D.R.A. Rushton, J. Foster-Smith, S. Wanless, Mar. Biol. 151 (2007) 973–983.

[47] N.K. Dulvy, S.I. Rogers, S. Jennings, V. Stelzenmuller, S.R. Dye, H.R. Skjodal, J. Appl. Ecol. (2008), http://dx.doi.org/10.1111/j.1365-2664.2008.01488.x.

[48] U. Sommer, K. Lengfellner, Glob. Change Biol. 14 (2008) 1199–1208.

[49] G. Beaugrand, K.M. Brander, J.A. Lindley, S. Souissi, P.C. Reid, Nature 426 (2003) 661–664.

[50] J.C. Molinero, F. Ibanez, S. Souissi, M. Chifflet, P. Nival, Oecologia 145 (2005) 640–649.

[51] D.L. Mackas, R. Goldblatt, A.G. Lewis, Can. J. Fish Aquat. Sci. 55 (1998) 1878–1893.

[52] S. Chiba, M.N. Aita, K. Tadokoro, T. Saino, H. Sugisaki, K. Nakata, Prog. Ocean 77 (2008) 112–126.

[53] M. Edwards, G. Beaugrand, P.C. Reid, A.A. Rowden, M.B. Jones, Mar. Ecol. Prog. Ser. 239 (2002) 1–10.

[54] M. Edwards, P. Reid, B. Planque, ICES J. Mar. Sci. 58 (2001) 39–49.

[55] J.M. Fromentin, B. Planque, Mar. Ecol. Prog. Ser. 134 (1996) 111–118.

[56] R.R. Kirby, G. Beaugrand, J.A. Lindley, A.J. Richardson, M. Edwards, P.C. Reid, Mar. Ecol. Prog. Ser. 330 (2007) 31–38.

[57] C.P. Lynam, S.J. Hay, A.S. Brierley, Limnol. Oceanogr. 49 (2004) 637–643.

[58] P.C. Reid, M. Edwards, Senck. Marit. 31 (2001) 107–115.

[59] P.C. Reid, M. Edwards, H.G. Hunt, A.J. Warner, Nature 391 (1998) 546.

[60] A.J. Richardson, D.S. Schoeman, Science 305 (2004) 1609–1612.

[61] M. Edwards, D.G. Johns, S.C. Leterme, E. Svendsen, A.J. Richardson, Limnol. Oceanogr. 51 (2006) 820–829.

[62] S.L. Hinder, G.C. Hays, M. Edwards, E.C. Roberts, A.W. Walne, M.B. Gravenor, Nat. Clim. Change 2 (2012) 271–275.

[63] M. Frederiksen, M. Edwards, A.J. Richardson, N.C. Halliday, S. Wanless, J. Anim. Ecol. 75 (2006) 1259–1268.

[64] M. Edwards, P. Licandro, D.G. Johns, A.W.G. John, D.P. Stevens, SAHFOS Tech. Rep. 3 (2006) 1–8.

[65] F.W. Koster, C. Mollmann, H.H. Hinrichsen, K. Wieland, J. Tomkiewicz, G. Kraus, R. Voss, A. Makarchouk, B.R. MacKenzie, M.A. St John, D. Schnack, N. Rohlf, T. Linkowski, J.E. Beyer, ICES J. Mar. Sci. 62 (2005) 1408–1425.

[66] J. Alheit, C. Mollmann, J. Dutz, G. Kornilovs, P. Loewe, V. Mohrholz, N. Wasmund, ICES J. Mar. Sci. 62 (2005) 1205–1215.

[67] G. Beaugrand, Prog. Ocean 60 (2004) 245–262.

[68] J.C. Molinero, F. Ibanez, P. Nival, E. Buecher, S. Souissi, Limnol. Oceanogr. 50 (2005) 1213–1220.

[69] A. Conversi, S. Fonda Umani, T. Peluso, J.C. Molinero, A. Santojanni, M. Edwards, S.F. Umani, PLoS One 5 (2010) e10633.

[70] D.G. Johns, M. Edwards, A. Richardson, J.I. Spicer, Mar. Ecol. Prog. Ser. 265 (2003) 283–287.

[71] T.J. Smyth, T. Tyrrell, B. Tarrant, Geophys. Res. Lett. 31 (2004).

[72] H.M. Verheye, S. Afr. J. Mar. Sci./S-Afr Tydskr Seewet 19 (1998) 317–332.

[73] B.E. Lavaniegos, M.D. Ohman, Deep Sea Res. (II Top Stud. Oceanogr.) 50 (2003) 2473–2498.

[74] A. Atkinson, V. Siegel, E. Pakhomov, P. Rothery, Nature 432 (2004) 100–103.

[75] C. Barbraud, H. Weimerskirch, Nature 411 (2001) 183–186.

[76] H. Weimerskirch, P. Inchausti, C. Guinet, C. Barbraud, Antarct. Sci. 15 (2003) 249–256.

[77] D.B. Field, T.R. Baumgarter, C.D. Charles, V. Ferreira-Bartrina, M.D. Ohman, Science 311 (2006) 63–66.

[78] S. Chiba, K. Tadokoro, H. Sugisaki, T. Saino, Glob. Change Biol. 12 (2006) 907–920.

[79] S. Chiba, T. Saino, Prog. Oceanogr. 57 (2003) 317–339.

[80] N.C. Stenseth, A. Mysterud, G. Ottersen, J.W. Hurrell, K.S. Chan, M. Lima, Science 297 (2002) 1292–1296.

[81] F.P. Chavez, J. Ryan, S.E. Lluch-Cota, M. Niquen, Science 299 (2003) 217–221.

[82] C.H. Hsieh, C.S. Reiss, R.P. Hewitt, G. Sugihara, Can. J. Fish Aquat. Sci. 65 (2008) 947–961.

[83] G. Beaugrand, F. Ibanez, Mar. Ecol. Prog. Ser. 232 (2002) 197–211.

[84] G. Beaugrand, F. Ibanez, J.A. Lindley, Mar. Ecol. Prog. Ser. 219 (2001) 189–203.

[85] G. Beaugrand, F. Ibanez, J.A. Lindley, P.C. Reid, Mar. Ecol. Prog. Ser. 232 (2002) 179–195.

[86] S. Rutherford, S. D'Hondt, W. Prell, Nature 400 (1999) 749–753.

[87] P.C. Reid, D.G. Johns, M. Edwards, M. Starr, M. Poulins, P. Snoeijs, Glob. Change Biol. 13 (2007) 1910–1921.

[88] K.R. Arrigo, Nature 437 (2005) 349–355.

[89] D.G. Capone, J.A. Burns, J.P. Montoya, A. Subramaniam, C. Mahaffey, T. Gunderson, A.F. Michaels, E.J. Carpenter, Glob. Biogeochem. Cycles 19 (2005) 1–17.

[90] L. Bopp, O. Boucher, O. Aumont, S. Belviso, J.L. Dufresne, M. Pham, P. Monfray, Can. J. Fish Aquat. Sci. 61 (2004) 826–835.

[91] S.L. Strom, Science 320 (2008) 1043–1045.

CHANGES IN CORAL REEF ECOSYSTEMS

12

Martin J. Attrill, Nicola L. Foster

Marine Institute, Plymouth University, Plymouth, UK

CHAPTER OUTLINE

1. INTRODUCTION

In comparison with terrestrial systems and other components of the marine environment, relatively little is known about how climate and global change has been affecting the organisms of the seabed, with one exception: tropical coral reefs. Recent reviews [1,2] and metanalyses [2,3] have pulled together all existing knowledge on how climate is impacting, for example, range shifts and phenology [3] of the world's species, but, corals aside, few, if any, examples within these overviews are from the marine seabed. In her excellent review, Parmesan [4] devotes a section to marine community shifts, but the majority of examples are from either the pelagic (Chapter 11) or intertidal (Chapter 14) zones; only two studies on fish [5,6] are associated with the subtidal seabed. Terrestrial and freshwater examples dominate these studies. A sizeable proportion of research assessing climate impact on marine systems has investigated the response to the major climate cycles, such as the El Niño Southern Oscillation (ENSO) and North Atlantic Oscillation (NAO). These provide information on how systems respond to cooling and warming trends across the extremes of these cycles, and thus provide a model of how potentially organisms and systems may respond to climate warming [7], particularly as, for example, the occurrence and severity of ENSO events is predicted to increase under warming scenarios [8,9]. This future relationship between atmospheric dynamics and temperature change is, however, uncertain, so there also needs to be some caution about how past responses of biological systems to these climatic cycles reflect ongoing and future climate change [4]. Nevertheless, such studies provide much of the information available on climate responses of marine systems.

In contrast to other subtidal benthic systems, the impact of global warming–related issues on coral reefs has had one of the highest profiles in recent years, particularly following the worldwide impact of the 1997–1998 extreme El Niño event [10], the extensive public concern about this ecosystem and

Climate Change. http://dx.doi.org/10.1016/B978-0-444-63524-2.00012-9

recent reports predicting widespread losses of reef and extinction of species [11–13]. Climate change can potentially impact coral reefs through several key mechanisms, in particular increasing sea surface temperatures, ocean acidification, increasing storminess and sea level rise [14,15]. The latter three mechanisms are dealt with specifically in other chapters, so this chapter will focus on the impact of rising sea temperatures on coral reef ecosystems, particularly the effect of mass bleaching of the corals themselves.

2. TROPICAL CORAL REEF ECOSYSTEMS

Concern about the human impact on coral reefs has existed for decades and, until comparatively recently, the major threats to the integrity of reef systems have been considered to be overfishing and pollution [14,16]. Such impacts can be potentially managed at the local level, but any such management will be unsuccessful when put into the context of more recent recognized effects of global climate change [14]. Similarly, climate impacts may be exacerbated by the additional effect of these other local anthropogenic factors, which have made coral reefs systems in some areas of the world more susceptible to damage. The link between climate change and region-scale mass bleaching of corals is now incontrovertible [14,15], in particular the direct link between bleaching and sea surface temperature (SST) anomalies [17,18]. There are no records of mass bleaching prior to the 1980s, though it is unclear how extensive such bleaching was earlier in the twentieth century before widespread reporting [19]; it is unlikely, however, that bleaching of the scale seen in recent years would have gone unnoticed. In the Great Barrier Reef, for example, bleaching events have become more widespread since the 1980s (Fig. 1(a)), coinciding with a decline in coral cover since this time [20]; globally, mass coral bleaching has become more frequent and intense in recent decades [11,21].

Bleaching is due to a whitening of the corals following the expulsion of the symbiotic zooxanthellae, the algae providing most of the coral's pigment. Loss of the zooxanthellae therefore leaves comparatively colourless coral tissue plus the white calcareous reef skeleton. The process is often considered as a response to increasing ambient temperatures above a threshold, approximately 0.8°C–1°C above summer average temperatures for at least four weeks [22,23]. Bleaching thresholds across coral species are likely to represent a broad spectrum of responses (Fig. 1(b)), however, and susceptibilities will change over time following phenotypic and genetic responses of the corals [14,24]. It is clear that many coral species exist over a wide biogeographical range of temperatures, and individuals have subsequently different bleaching thresholds in terms of absolute temperature, indicating adaptive ability within species [25]. The key driver for bleaching therefore appears to be temperature increases above those generally experienced by corals in any given location.

It has been hypothesized that the bleaching response is an adaptive process [26,27], the corals expelling susceptible symbionts and taking up more resistant ones [28–30]. While a shift in the symbiont community within corals following bleaching was initially described as a physiological or stress response [31,32], more recent work has demonstrated that a shift in symbiont types can occur prior to the bleaching event itself [33], providing support to the theory that bleaching is an adaptive response by the host corals. Furthermore, some coral species have been found to be flexible in their associations with symbionts, hosting background populations of more stress-tolerant symbiont types and reflecting a potential mechanism for corals to respond to thermally stressful events [34,35]. Evidence has also been presented for increasing thermal tolerance of corals following major and/or

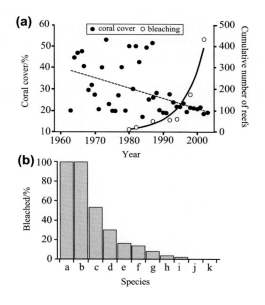

FIGURE 1

(a) Trends in coral cover and number of reefs with mass bleaching on the Great Barrier Reef, Australia. *(Adapted from Ref. [20].)* (b) Differential bleaching responses of nine species of corals in Raiatea, French Polynesia, during May 2002. *(Redrawn from Ref. [14].)* (a, *Acropora anthoceris*; b, *Acropora retusa*; c, *Montipora tuberculosa*; d, *Pocillopora verrucosa*; e, *Montipora caliculata*; f, *Leptastrea transversa*; g, *Pocillopora eydouxi*; h, *Pocillopora meandrina*; i, *Leptastrea bewickensis*; j, *Porites lobata*; k, *Leptastrea purpurea*.)

repeated bleaching events, suggesting adaptation or acclimation of the coral to thermally stressful conditions. Corals that have experienced bleaching in previous years and survived have been observed to be more resilient of bleaching in future years [36–39]. A number of possible mechanisms have been suggested for increased thermal tolerance, including shifts in the symbiont community, as described above [40], selective removal of susceptible coral phenotypes from the population through mortality [41], maintenance of high concentrations of energy reserves by the host coral [40], or changes in gene expression that may increase thermal tolerance of the holobiont [42,43]. However, the ability of corals to adapt or acclimate to more frequent bleaching events is species specific, and annual bleaching events (predicted to occur by 2040 under current CO_2 emissions) may lead to the selective loss of coral diversity from reefs [40].

What remains uncontested, however, is that major bleaching events can severely impact coral reefs in the long term – if bleaching is prolonged or exceeds 2°C above seasonal maxima, corals can die [15]. Major bleaching events were observed in 1982–1983, 1987–1988, 1994–1995, 1997–1998 [15], 2002 (GBR [44]) and 2005 (Caribbean [18,23]) and have often, but not always, been associated with intense El Niño events, which enhance global sea temperatures. The 1997–1998 event was the most extreme El Niño on record [10] and resulted in extensive bleaching recorded across the world's coral reefs [45,46]. An estimated 16% of the world's coral was lost in this one event, in particular within the Indian Ocean/SE Asia [45,46] (Table 1), with only partial recovery evident. Overall, approximately

Table 1 Summary of status of coral reefs in 17 regions of the world as of 2004 from Ref. [45], indicating proportion of coral reefs in each region that have been destroyed (i.e. 90% coral lost and unlikely to recover), plus proportion lost and recovered following the 1997−1998 El Niño event.

Region	Coral reef area/ km^2	Reefs destroyed/ (%)	Reefs destroyed in 1998/(%)	Reefs recovered/ (%)
The Gulfs	3800	65	15	2
South Asia	19 210	45	65	13
SE Asia	91 700	38	18	8
SW Indian Ocean	5270	22	41	20
US Caribbean	3040	16	NA	NA
S Tropical America	5120	15	NA	NA
E & N Asia	5400	14	10	3
East Africa	6800	12	31	22
East Antilles	1920	12	NA	NA
Central America	4630	10	NA	NA
Micronesian Islands	12 700	8	2	1
North Caribbean	9800	5	4	3
Red Sea	17 640	4	4	2
SW Pacific Islands	27 060	3	10	8
Australia & PNG	62 800	2	3	1
Polynesian Islands	6733	2	1	1
Hawaiian Islands	1180	1	NA	NA
Total	284 803	20	16	6.4

NA = not applicable as no losses recorded in 1998.

three-quarters of the reefs affected in 1998 have recovered [46]. The 2005 event, however, occurred without an El Niño and has provided evidence of the impact of the underlying increasing trend in global sea water temperatures (Chapter 2); this has been related to anthropogenic forcing in the Atlantic since the 1970s [23].

Anomalously warm temperatures were recorded across the Caribbean and tropical Atlantic resulting in exceptional levels of bleaching [18,23,47]. In the British Virgin Islands, 90% of coral cover experienced bleaching [23] and over 80% of corals were bleached at many other sites across the region, with 20%–50% mortality of corals at many sites [18,47]. Analysis of local temperature anomalies revealed that SSTs were higher than the expected annual maxima for longer than had been previously recorded in the last 150 years [18,23] (Fig. 2), resulting in the exceptional bleaching observed. It is also notable that such maxima have been exceeded every year since 1995 (Fig. 2); prior to this, such extremes were rare. Extensive bleaching was also apparent in the southern Caribbean in 2010, when the second highest value occurred [48,49] and in the Caribbean in 1998, when the third highest value occurred [50]. Following strong coral recovery in the Indian Ocean and Western Pacific after the 1998 bleaching event, the total area of the world's coral reefs effectively lost has reduced from

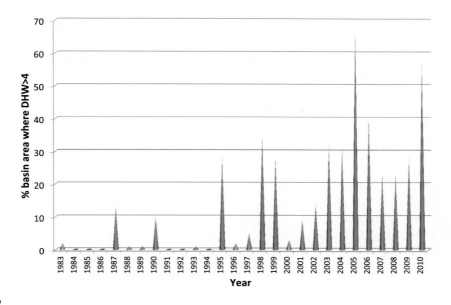

FIGURE 2

Acute thermal stress events on Caribbean coral reefs since 1983, identified when degree heating weeks (DHW) were over a specific threshold (4°C wk^{-1}) where wk refers to week. DHWs are a measure of accumulated thermal stress by summing temperatures 1°C above the usual summertime maximum over a 12-week period [85]. DHWs above four degree weeks have been related to bleaching and associated mortality. *(Data from Ref. [85].)*

20% in 2004 (Table 1; [45]) to 19% in 2008 (Table 2; [46]). However, the area of reef effectively lost in the Caribbean (US and north) and Central America has increased from 31% in 2004 (Table 1; [45]) to 47% in 2008 (Table 2), largely due to mortality following the 2005 bleaching event [46].

The only long-term data available on the impact of climate on coral species is from the geological record, as reviewed by Hughes et al. [14]. Many extant species of coral can be traced back in time to the Pliocene (1.8–5.3 million years ago), so have experienced periods of extensive and rapid warming and cooling during the Pleistocene and Holocene prior to human impact [16,51]. In response to climatic changes, there is evidence species underwent large shifts in their distributional range [52]; for corals this involved poleward range expansions and equatorial range retractions [51], with distributions extending up to 500 km further south in Australia, for example [14]. Until very recently there was no evidence of such shifts in response to modern climate change, but, following increasing temperatures a number of cases have been documented. Two species of *Acropora* have re-expanded their ranges 50 km northward along the Florida Peninsula into areas where they have not been recorded for 6000 years [53]. The coral *Goniopora norfolkensis* has extended its geographical distribution 1250 km southward along the coast of Western Australia in recent years [54]. In Japan, four coral species have also been observed to undergo poleward range expansion into temperate areas within the last 80 years, at speeds of up to 14 km a^{-1} (where 'a' refers to annum for year) in some cases, which is greater than that observed in other marine benthic species [55]. Furthermore, adult colonies in the newly colonized

Table 2 Summary of status of coral reefs in 17 regions of the world as of 2008 from Ref. [46].

Region	Coral reef area/km^{2a}	Effectively lost reefs/ (%)[b]	Reefs at critical stage/(%)[c]	Reefs at threatened stage/(%)[d]	Reefs at low threat level/ (%)[e]
Red Sea	17 640	4	4	10	82
The Gulfs	3800	70	15	12	3
Eastern Africa	6800	15	22	28	35
SW Indian Ocean	5270	9	24	39	29
South Asia	19 210	25	20	25	30
SE Asia	91 700	40	20	25	15
E & N Asia	5400	20	22	28	40
Australia, PNG	62 800	3	4	10	83
SW Pacific Islands	27 060	4	17	35	44
Polynesian Islands	6733	3	2	5	90
Micronesian Islands	12 700	8	7	15	70
Hawaiian Islands	1180	2	4	8	86
US Caribbean	3040	21	31	19	29
North Caribbean	9800	12	13	30	45
Central America	4630	14	24	22	40
Lesser Antilles	1920	13	31	22	34
S Tropical America	5120	13	40	17	30
Total	**284 803**	**19**	**15**	**20**	**45**

Estimates were determined using considerable coral reef monitoring data, some anecdotal reports and the expert opinion of hundreds of people associated with the Global Coral Reef Monitoring Network. Assessments should be regarded as indicative as there is insufficient data for many regions to make definitive statements and predictions.
[a]*Coral reef area from the* World Atlas of Coral Reefs *[84].*
[b]*Reefs 'effectively lost' with 90% of corals lost and unlikely to recover soon.*
[c]*Reefs at a critical stage with 50%–90% loss of corals and likely to join category 2 in 10 years–20 years.*
[d]*Reefs threatened with moderate signs of damage: 20%–50% loss of corals and likely to join category 2 in 20 years–40 years.*
[e]*Reefs under no immediate threat of significant losses (except for global climate change).*

regions exhibited spawning, indicating that the corals have the potential to reproduce and further expand their range [55]. The most striking example of range expansion is that of Tosa Bay in southern Japan, which has seen a dramatic change in its benthic assemblage as water temperatures have increased [56]. Within the last 30 years, there has been a shift from temperate kelp to reef building corals, with over 136 coral species from 50 genera now documented to occur within the bay [56].

Much evidence therefore exists that increasing sea-surface temperatures are impacting coral reefs through extensive bleaching, and subsequent mortality, particularly during exceptional years where average maximum temperatures are exceeded. Corals will also be impacted through changes in seawater pH, affecting their ability to produce calcareous reef skeletons as covered in Chapter 18 (see also, Ref. [11]), and potentially further affected by severe storms [57] and sea level rise [58], and other consequences of climate change [15]. Unlike past climate changes, however, coral reefs are now also markedly influenced by the synergistic effect of other anthropogenic activities, such as fishing and pollution, making them much more susceptible to changes associated with current climate warming [14].

3. **THE ASSOCIATED FAUNA OF CORAL REEFS**

Unlike the corals themselves, comparatively little research has been focused on how climate and global change may be influencing the vast biodiversity associated with coral reefs (about a quarter of all known marine species [15,59]), perhaps due to the logistic difficulties and expense of constructing long-term data sets on these organisms. Additionally, studies on changes to fish populations and their immediate prey are greatly influenced by other stresses, particularly fishing pressure [15], making it more difficult to identify signals of climate change. However, although over 4000 species of fish are associated with coral reefs [60], until recently, remarkably few studies had addressed the impact of climate change on this group.

Recent studies, however, have demonstrated both direct and indirect impacts of climate change on reef fish populations, with some evidence suggesting that coral reef fish may be impacted more severely by the indirect effects of mass coral bleaching [61–63]. In a study of Seychelles reefs following the 1997–1998 El Niño [64], intense bleaching resulted in decreased fish abundance and a shift within the assemblage from corallivores to species feeding on invertebrates. The size structure of fish also changed with an increase in large fish [65], a possible time-lag response due to a reduction in coral structural complexity affecting fish recruitment and thus the number of juveniles. However, a minimal effect of this 1997–1998 event was noted on the diversity and abundance of cryptobenthic fish species on the Great Barrier Reef [66], suggesting coral reef fishes may be comparatively resilient to short-term perturbations [66] as long as reef structure is sustained [63]. However, as corals worldwide continue to experience mortality due to the impacts of climate change, one unavoidable consequence of this is the erosion of their skeletons and a subsequent loss of structural complexity. Reef complexity influences predator–prey interactions and competition in reef fish populations through the provision of prey refugia [67], and it has been suggested that a loss of complexity may result in declines in the abundance, diversity and productivity of reef fish [67–69].

The more direct impacts of climate change on reef fish populations have also been investigated. Increases in temperature have been observed to reduce swimming speeds of some fish species, being more pronounced in smaller individuals [70,71]. A reduction in swimming speed has implications for the fitness and survival of both individuals and populations through changes in prey encounter rates, prey capture ability, predator avoidance, foraging ability, energy intake and ability to swim against currents commonly found in their habitats [70,71]. Furthermore, increases in temperature have been associated with higher oxygen needs of fish, reduced aerobic scope and a reduction in the tolerance of hypoxic conditions, which may have negative effects on their capacity for feeding, growth and reproduction [72]. However, variations in response to temperature differ among species, demonstrating that some species will be more susceptible to rising temperature than others [70,72]. Despite the negative implications of temperature rises on coral reef fish, transgenerational acclimation to elevated temperatures has been observed. Using the damselfish, *Acathochromis polyacanthus*, Donelson and Munday [73] demonstrated that fish reared at 3°C above the present day average water temperature had reduced resting oxygen consumption compared with fish reared at present day temperatures and tested at elevated temperatures. Furthermore, Donelson and Munday [74] demonstrated that when both parents and offspring were reared throughout their lives at the elevated temperature, the offspring were able to completely restore their aerobic scope compared with fish not acclimated. These studies demonstrate the potential for some reef fish to acclimate to changing environmental conditions, albeit over multiple generations. However, evidence of distributional shifts in coral reef fish is limited, despite being predicted by climate impact reviews [60]. This may be due to a lack of habitat availability unless coral reefs also extend their range [53]. Nevertheless, there are a small number of cases where tropical fishes have been reported on temperate reefs and are causing shifts in the dominant benthic community [75]. For example, in Tosa Bay there has been a shift from temperate kelp to reef-building corals in response to increasing water temperature, and despite the temperate latitude of the bay, tropical reef fish now dominate temperate reef fish communities [56].

The most extensive study on the impact of a climate event on coral-associated invertebrates (>500 species) has been undertaken in Bahia, Brazil, by Kelmo and colleagues [76–79]. The 1997–1998 El Niño resulted in anomalous high temperatures and a reduction in the usual high turbidity of the area allowing more UV light to reach the reefs. All groups (except sponges) studied showed extensive reductions in diversity for several years following El Niño (Fig. 3 in Ref. [80]), including the local extinction of one species of coral (*Porites astreoides*); this was most likely due to extensive neoplastic tumours on the corals following UV damage [81]. The density of the majority of species also decreased dramatically (Fig. 3 in Ref. [80]), with only the urchin *Diadema antillarum* showing an opportunistic response to the changing conditions and disappearance of competitive taxa [82]. What is remarkable about this data set is the recovery of groups after 2001, with diversity returning to, or exceeding, levels prior to El Niño (Fig. 3 in Ref. [80]). Reef assemblages clearly have the ability to recover from extreme climate events, but only if no further such events subsequently occur; in Bahia, no major El Niño or bleaching event was evident after 1998. Models suggest ENSO events are likely to be more frequent and severe in the future [8,9], so this recovery ability of coral reef communities may be compromised by climate change.

4. CONCLUSION

There is clear, unequivocal evidence that climate change is affecting coral reef systems [14], with a particular concern about mass coral bleaching due to rising temperatures and the subsequent effects this will have on coral survival and thus the associated organisms. Whilst corals have an innate ability to acclimatize to such change [83], as evidenced in the past [14], severe and regular El Niño events coupled with the modern synergistic impact of other human activities such as fishing, pollution (including comparatively fast ocean acidification [11]) and tourism make coral reefs amongst the most vulnerable of the world's ecosystems under current scenarios of future climate change.

REFERENCES

[1] G.R. Walther, E. Post, P. Convey, A. Menzel, C. Parmesan, T.J.C. Beebee, J.M. Fromentin, O. Hoegh-Guldberg, F. Bairlein, Nature 416 (2002) 389–395.
[2] C. Parmesan, Glob. Change Biol. 13 (2007) 1860–1872.
[3] C. Parmesan, G. Yohe, Nature 421 (2003) 37–42.
[4] C. Parmesan, Annu. Rev. Ecol. Evol. Syst. 37 (2006) 637–669.
[5] S.J. Holbrook, R.J. Schmitt, J.S. Stephens, Ecol. Appl. 7 (1997) 1299–1310.
[6] A.L. Perry, P.J. Low, J.R. Ellis, J.D. Reynolds, Science 308 (2005) 1912–1915.
[7] C.D.G. Harley, A.R. Hughes, K.M. Hultgren, B.G. Miner, C.J.B. Sorte, C.S. Thornber, L.F. Rodriguez, L. Tomanek, S.L. Williams, Ecol. Lett. 9 (2006) 228–241.
[8] A. Timmermann, J. Oberhuber, A. Bacher, M. Esch, M. Latif, E. Roeckner, Nature 398 (1999) 694–697.

[9] W.J. Cai, S. Borlace, M. Lengaigne, P. van Rensch, M. Collins, G. Vecchi, A. Timmermann, A. Santoso, M.J. McPhaden, L.X. Wu, M.H. England, G.J. Wang, E. Guilyardi, F.F. Jin, Nat. Clim. Change 4 (2014) 111–116.

[10] M.J. McPhaden, Nature 398 (1999) 559–562.

[11] O. Hoegh-Guldberg, P.J. Mumby, A.J. Hooten, R.S. Steneck, P. Greenfield, E. Gomez, C.D. Harvell, P.F. Sale, A.J. Edwards, K. Caldeira, N. Knowlton, C.M. Eakin, R. Iglesias-Prieto, N. Muthiga, R.H. Bradbury, A. Dubi, M.E. Hatziolos, Science 318 (2007) 1737–1742.

[12] K.E. Carpenter, M. Abrar, G. Aeby, R.B. Aronson, S. Banks, A. Bruckner, A. Chiriboga, J. Cortes, J.C. Delbeek, L. DeVantier, G.J. Edgar, A.J. Edwards, D. Fenner, H.M. Guzman, B.W. Hoeksema, G. Hodgson, O. Johan, W.Y. Licuanan, S.R. Livingstone, E.R. Lovell, J.A. Moore, D.O. Obura, D. Ochavillo, B.A. Polidoro, W.F. Precht, M.C. Quibilan, C. Reboton, Z.T. Richards, A.D. Rogers, J. Sanciangco, A. Sheppard, C. Sheppard, J. Smith, S. Stuart, E. Turak, J.E.N. Veron, C. Wallace, E. Weil, E. Wood, Science 321 (2008) 560–563.

[13] K. Frieler, M. Meinshausen, A. Golly, M. Mengel, K. Lebek, S.D. Donner, O. Hoegh-Guldberg, Nat. Clim. Change 3 (2013) 165–170.

[14] T.P. Hughes, A.H. Baird, D.R. Bellwood, M. Card, S.R. Connolly, C. Folke, R. Grosberg, O. Hoegh-Guldberg, J.B.C. Jackson, J. Kleypas, J.M. Lough, P. Marshall, M. Nystrom, S.R. Palumbi, J.M. Pandolfi, B. Rosen, J. Roughgarden, Science 301 (2003) 929–933.

[15] M.L. Parry, O.F. Canziani, J.P. Palutikof, P.J. van der Linden, C.E. Hanson, Cross-chapter case study, in: Climate Change 2007: Impacts, Adaptation and Vulnerability. Contribution of Working Group II to the Fourth Assessment Report of the IPCC, Cambridge University Press, Cambridge, UK, 2007, pp. 843–868.

[16] J.B.C. Jackson, M.X. Kirby, W.H. Berger, K.A. Bjorndal, L.W. Botsford, B.J. Bourque, R.H. Bradbury, R. Cooke, J. Erlandson, J.A. Estes, T.P. Hughes, S. Kidwell, C.B. Lange, H.S. Lenihan, J.M. Pandolfi, C.H. Peterson, R.S. Steneck, M.J. Tegner, R.R. Warner, Science 293 (2001) 629–638.

[17] J.P. McWilliams, I.M. Cote, J.A. Gill, W.J. Sutherland, A.R. Watkinson, Ecology 86 (2005) 2055–2060.

[18] C.M. Eakin, J.A. Morgan, S.F. Heron, T.B. Smith, G. Liu, L. Alvarez-Filip, B. Baca, E. Bartels, C. Bastidas, C. Bouchon, M. Brandt, A.W. Bruckner, L. Bunkley-Williams, A. Cameron, B.D. Causey, M. Chiappone, T.R.L. Christensen, M.J.C. Crabbe, O. Day, E. de la Guardia, G. Díaz-Pulido, D. DiResta, D.L. Gil-Agudelo, D.S. Gilliam, R.N. Ginsburg, S. Gore, H.M. Guzmán, J.C. Hendee, E.A. Hernández-Delgado, E. Husain, C.F.G. Jeffrey, R.J. Jones, E. Jordán-Dahlgren, L.S. Kaufman, D.I. Kline, P.A. Kramer, J.C. Lang, D. Lirman, J. Mallela, C. Manfrino, J.-P. Maréchal, K. Marks, J. Mihaly, W.J. Miller, E.M. Mueller, E.M. Muller, C.A. Orozco Toro, H.A. Oxenford, D. Ponce-Taylor, N. Quinn, K.B. Ritchie, S. Rodríguez, A.R. Ramírez, S. Romano, J.F. Samhouri, J.A. Sánchez, G.P. Schmahl, B.V. Shank, W.J. Skirving, S.C.C. Steiner, E. Villamizar, S.M. Walsh, C. Walter, E. Weil, E.H. Williams, K.W. Roberson, Y. Yusuf, PLoS One 5 (2010) e13969.

[19] A.D. Barton, K.S. Casey, Coral Reefs 24 (2005) 536–554.

[20] D.R. Bellwood, T.P. Hughes, C. Folke, M. Nystrom, Nature 429 (2004) 827–833.

[21] A.C. Baker, P.W. Glynn, B. Riegl, Estuar. Coast. Shelf Sci. 80 (2008) 435–471.

[22] O. Hoegh-Guldberg, Mar. Freshwater Res. 50 (1999) 839–866.

[23] S.D. Donner, T.R. Knutson, M. Oppenheimer, Proc. Natl. Acad. Sci. U.S.A. 104 (2007) 5483–5488.

[24] J.M. Pandolfi, S.R. Connolly, D.J. Marshall, A.L. Cohen, Science 333 (2011) 418–422.

[25] S.D. Donner, W.J. Skirving, C.M. Little, M. Oppenheimer, O. Hoegh-Guldberg, Glob. Change Biol. 11 (2005) 2251–2265.

[26] R.W. Buddemeier, D.G. Fautin, Bioscience 43 (1993) 320–326.

[27] A.C. Baker, C.J. Starger, T.R. McClanahan, P.W. Glynn, Nature 430 (2004) 741.

[28] R. Rowan, N. Knowlton, A. Baker, J. Jara, Nature 388 (1997) 265–269.

[29] A.C. Baker, Nature 411 (2001) 765–766.

[30] A.C. Baker, Annu. Rev. Ecol. Evol. Syst. 34 (2003) 661–689.

[31] A.H. Baird, P.A. Marshall, Mar. Ecol. Prog. Ser. 237 (2002) 133–141.

[32] A.E. Douglas, Mar. Pollut. Bull. 46 (2003) 385–392.

[33] T.C. LaJeunesse, R.T. Smith, J. Finney, H. Oxenford, Proc. R. Soc. B Biol. Sci. 276 (2009) 4139–4148.

[34] R.N. Silverstein, A.M.S. Correa, A.C. Baker, Proc. R. Soc. B Biol. Sci. 279 (2012) 2609–2618.

[35] E.V. Kennedy, N.L. Foster, P.J. Mumby, J.R. Stevens, Coral Reefs 34 (2015) 519–531.

[36] J.A. Maynard, K.R.N. Anthony, P.A. Marshall, I. Masiri, Mar. Biol. 155 (2008) 173–182.

[37] D.M. Thompson, R. van Woesik, Proc. R. Soc. B Biol. Sci. 276 (2009) 2893–2901.

[38] J. Carilli, S.D. Donner, A.C. Hartmann, PLoS One 7 (2012) e34418.

[39] J.R. Guest, A.H. Baird, J.A. Maynard, E. Muttaqin, A.J. Edwards, S.J. Campbell, K. Yewdall, Y.A. Affendi, L.M. Chou, PLoS One 7 (2012) e33353.

[40] A.G. Grottoli, M.E. Warner, S.J. Levas, M.D. Aschaffenburg, V. Schoepf, M. McGinley, J. Baumann, Y. Matsui, Glob. Change Biol. 20 (2014) 3823–3833.

[41] M.S. Pratchett, D. McCowan, J.A. Maynard, S.F. Heron, PLoS One 8 (2013).

[42] C.D. Kenkel, E. Meyer, M.V. Matz, Mol. Ecol. 22 (2013) 4322–4334.

[43] S.R. Palumbi, D.J. Barshis, N. Traylor-Knowles, R.A. Bay, Science 344 (2014) 895–898.

[44] R. Berkelmans, G. De'ath, S. Kininmonth, W.J. Skirving, Coral Reefs 23 (2004) 74–83.

[45] C. Wilkinson (Ed.), Status of Coral Reefs of the World: 2004, vol. 1, AIMS, Townsville, Australia, 2004, p. 316.

[46] C. Wilkinson (Ed.), Status of Coral Reefs of the World: 2008, Global Coral Reef Monitoring Network and Reef and Rainforest Research Centre, Townsville, Australia, 2008, p. 296.

[47] C. Wilkinson, D. Souter, Status of Caribbean Coral Reefs after Bleaching and Hurricanes in 2005, Global Coral Reef Monitoring Network, and Reef and Rainforest Research Centre, Townsville, 2008.

[48] J.B.I. Alemu, Y. Clement, PLoS One 9 (2014) e83829.

[49] C. Bastidas, D. Bone, A. Croquer, D. Debrot, E. Garcia, A. Humanes, R. Ramos, S. Rodriguez, Rev. Biol. Trop. 60 (2012) 29–37.

[50] R.B. Aronson, W.F. Precht, M.A. Toscano, K.H. Koltes, Mar. Biol. 141 (2002) 435–447.

[51] W. Kiessling, C. Simpson, B. Beck, H. Mewis, J.M. Pandolfi, Proc. Natl. Acad. Sci. U.S.A. 109 (2012) 21378–21383.

[52] K. Roy, D. Jablonski, J.W. Valentine, Ecol. Lett. 4 (2001) 366–370.

[53] W.F. Precht, R.B. Aronson, Front. Ecol. Environ. 2 (2004) 307–314.

[54] D. Thomson, J. R. Soc. West. Aust. 93 (2010) 81–83.

[55] H. Yamano, K. Sugihara, K. Nomura, Geophys. Res. Lett. 38 (2011).

[56] Y. Nakamura, D.A. Feary, M. Kanda, K. Yamaoka, PLoS One 8 (2013) e81107.

[57] T.A. Gardner, I.M. Cote, J.A. Gill, A. Grant, A.R. Watkinson, Ecology 86 (2005) 174–184.

[58] N. Knowlton, Proc. Natl. Acad. Sci. U.S.A. 98 (2001) 5419–5425.

[59] C.M. Roberts, C.J. McClean, J.E.N. Veron, J.P. Hawkins, G.R. Allen, D.E. McAllister, C.G. Mittermeier, F.W. Schueler, M. Spalding, F. Wells, C. Vynne, T.B. Werner, Science 295 (2002) 1280–1284.

[60] P.L. Munday, G.P. Jones, M.S. Pratchett, A.J. Williams, Fish Fish. 9 (2008) 261–285.

[61] D.J. Booth, G.A. Beretta, Mar. Ecol. Prog. Ser. 245 (2002) 205–212.

[62] K.C. Garpe, S.A.S. Yahya, U. Lindahl, M.C. Ohman, Mar. Ecol. Prog. Ser. 315 (2006) 237–247.

[63] U. Lindahl, M.C. Ohman, C.K. Schelten, Mar. Pollut. Bull. 42 (2001) 127–131.

[64] M.D. Spalding, G.E. Jarvis, Mar. Pollut. Bull. 44 (2002) 309–321.

[65] N.A.J. Graham, S.K. Wilson, S. Jennings, N.V.C. Polunin, J. Robinson, J.P. Bijoux, T.M. Daw, Conserv. Biol. 21 (2007) 1291–1300.

[66] D.R. Bellwood, A.S. Hoey, J.L. Ackerman, M. Depczynski, Glob. Change Biol. 12 (2006) 1587–1594.

[67] A. Rogers, J.L. Blanchard, P.J. Mumby, Curr. Biol. 24 (2014) 1000–1005.

[68] M.S. Pratchett, A.S. Hoey, S.K. Wilson, V. Messmer, N.A.J. Graham, Diversity 3 (2011) 424–452.

[69] M.S. Pratchett, P.L. Munday, S.K. Wilson, N.A.J. Graham, J.E. Cinner, D.R. Bellwood, G.P. Jones, N.V.C. Polunin, T.R. McClanahan, Effects of climate-induced coral bleaching on coral-reef fishes – ecological and economic consequences, in: R.N. Gibson, R.J.A. Atkinson, J.D.M. Gordon (Eds.), Oceanography and Marine Biology: An Annual Review, vol. 46, 2008, pp. 251–296.

[70] J.L. Johansen, G.P. Jones, Glob. Change Biol. 17 (2011) 2971–2979.

[71] J.L. Johansen, V. Messmer, D.J. Coker, A.S. Hoey, M.S. Pratchett, Glob. Change Biol. 20 (2014) 1067–1074.

[72] G.E. Nilsson, S. Ostlund-Nilsson, P.L. Munday, Comp. Biochem. Physiol. A Mol. Integr. Physiol. 156 (2010) 389–393.

[73] J.M. Donelson, P.L. Munday, M.I. McCormick, G.E. Nilsson, Glob. Change Biol. 17 (2011) 1712–1719.

[74] J.M. Donelson, P.L. Munday, M.I. McCormick, C.R. Pitcher, Nat. Clim. Change 2 (2012) 30–32.

[75] A. Vergés, P.D. Steinberg, M.E. Hay, A.G.B. Poore, A.H. Campbell, E. Ballesteros, K.L. Heck, D.J. Booth, M.A. Coleman, D.A. Feary, W. Figueira, T. Langlois, E.M. Marzinelli, T. Mizerek, P.J. Mumby, Y. Nakamura, M. Roughan, E. van Sebille, A.S. Gupta, D.A. Smale, F. Tomas, T. Wernberg, S.K. Wilson, Proc. R. Soc. B Biol. Sci. 281 (2014).

[76] M.J. Attrill, F. Kelmo, M.B. Jones, Clim. Res. 26 (2004) 151–158.

[77] F. Kelmo, M.J. Attrill, R.C.T. Gomes, M.B. Jones, Biol. Conserv. 118 (2004) 609–617.

[78] F. Kelmo, J.J. Bell, M.J. Attrill, PLoS One 8 (2013).

[79] F. Kelmo, J.J. Bell, S.S. Moraes, R.D.T. Gomes, E. Mariano-Neto, M.J. Attrill, PLoS One 9 (2014).

[80] M.J. Attrill, Changes in coral reef ecosystems as an indication of climate and global change, in: T.M. Letcher (Ed.), Climate Change: Observed Impacts on Planet Earth, Elsevier, Oxford, 2009, pp. 253–261.

[81] F. Kelmo, Ecological Consequences of the 1997-98 El Niño Southern Oscillation on the Major Coral Reef Communities from Northern Bahia, Brazil, University of Plymouth, Plymouth, 2002, 245.

[82] M.J. Attrill, F. Kelmo, Estuar. Coast. Shelf Sci. 73 (2007) 243–248.

[83] P.J. Edmunds, R.D. Gates, Mar. Ecol. Prog. Ser. 361 (2008) 307–310.

[84] M.D. Spalding, C. Ravilious, E.P. Green, World Atlas of Coral Reefs. Prepared at the UNEP World Conservation Monitoring Centre, University of California Press, Berkeley, USA, 2001.

[85] I. Chollett, P.J. Mumby, Biol. Conserv. 167 (2013) 179–186.

MARINE BIODIVERSITY AND CLIMATE CHANGE

13

Boris Worm, Heike K. Lotze

Department of Biology, Dalhousie University, Halifax, NS, Canada

CHAPTER OUTLINE

1. INTRODUCTION

Our planet has a number of features that make it unique, namely the presence of large oceans and the evolution of life forms therein. Biodiversity, commonly defined as the variability among living organisms [1], likely originated in the oceans, and most of the larger taxonomic groups still reside there today. Over evolutionary timescales, there have been massive changes to the ocean's biodiversity, including several mass extinctions [2–4] that have shaped the diversity of life over millions of years [5,6]. Some, if not most, of these events are thought to correlate with large-scale climate change that perturbed ocean temperature, circulation, chemistry and productivity [6,7]. In general, observed patterns of biodiversity change are increasingly being understood in relation to variation in temperature, both over time [8] and space [9].

Today, we are living through another episode of rapid climate change [10], which is causing global changes in weather patterns, temperature and ice cover [11,12] that affect the sea level, thermal stratification regime, ocean circulation patterns and productivity [13–19]. Most attempts to trace the ecological effects of climate change, whether on land or in the sea, have concentrated on individual species [20–23], as discussed elsewhere in this volume. It is only quite recently that community metrics such as species composition and diversity have been studied in direct relation to climate change [24–29]. Here we attempt to summarize this emerging literature, to detect common patterns in the effects of climate change on marine biodiversity. Biodiversity has three main components: diversity contained within species, between species, and of ecosystems or habitats [1]. We will discuss

Climate Change. http://dx.doi.org/10.1016/B978-0-444-63524-2.00013-0

195

changes in all three components, but note that studies to date have mostly focused on species diversity, namely patterns and changes in species richness (the number of species in a given area), likely because it represents the most easily quantifiable aspect of biodiversity.

Despite its taxonomic prominence, marine biodiversity is sometimes overlooked in the climate change discussion, undoubtedly because much of it is less understood than its terrestrial counterpart. Yet, marine biodiversity needs to be accounted for, not just because of its different taxonomic composition and large geographic extent (oceans comprise >70% of the planet's surface and >90% of the living biosphere by volume), but also as it provides important ecosystem goods and services such as fishery yields, shoreline protection, carbon and nutrient cycling, climate regulation, among others [30–33]. The ocean's biodiversity should therefore be carefully studied in order to understand and project how it will change with climate change and what the consequences may be for human well-being [6,32,34,35].

In this chapter, we first discuss both observed and predicted changes in biodiversity at various scales and how they relate directly to warming and other climate-related factors. Then we outline some indirect effects of climate change that arise from complex interactions with biotic and abiotic factors, and the cumulative effects of climate and other global changes. Finally, we highlight the importance of biodiversity for maintaining ecosystem resilience in the face of climate change. We do not pretend to give a complete overview but instead discuss some prominent patterns by example, largely focusing on the effects of increasing temperature. Herein we shall focus on documented changes from the published literature and highlight how these effects are projected to develop into the future. The primary question we are asking is whether diversity, here defined as the number of genotypes, species or habitats in a location, changes in some predictable way with climate change. A secondary question is how climate effects on marine biodiversity are modified by and interact with other, co-occurring aspects of global change.

2. CLIMATE CHANGE IN THE OCEANS

Climate change has a range of effects on the abiotic marine environment, which are documented in detail elsewhere in this volume. From a biodiversity perspective, the prominent physical changes include ocean warming [11,36,37], increased climatic variability leading to more frequent extreme events [38,39] and changes in sea level, sea ice extent, thermal stratification and ocean circulation [13,40,41]. In addition, both warming and altered ocean circulation act to reduce subsurface oxygen (O_2) concentrations [42]. Carbon dioxide emissions, which in large part drive anthropogenic climate change, also cause ocean acidification (see Chapter 18, this volume for details). All of these processes can act on biodiversity directly (for example when local temperature exceeds individual species' physiological tolerances) or indirectly (for example by altering habitat availability, species interactions, or productivity). Furthermore, potentially complex interactions between climate change and other aspects of global change, notably those due to fishing, eutrophication, habitat destruction, invasions and disease can be important [40,43–46] and are briefly highlighted in this review. This latter point suggests an important difference between the current and previous episodes of climate change in Earth's history – recent changes in climate are superimposed on other stressors that have already compromised biodiversity in many places [6,30]. From a scientific perspective, this added complexity can make it more difficult to clearly attribute observed changes in diversity to a single factor. From a conservation perspective, the loss of biodiversity already diminishes adaptive capacity and diversity of biotic responses to climate change [47].

3. EFFECTS OF CLIMATE CHANGE ON MARINE BIODIVERSITY

What are the recently observed changes in biodiversity, and how do they relate to climate? In the following Sections 3.1–3.3, we first review evidence for the effects of climate warming that are emerging at increasing scales, from local (<10 km) to regional (10 km–1000 km) and global (>1000 km), respectively. In Section 3.4, we discuss factors other than changes in temperature. Observed effects are summarized in Table 1.

Table 1 Some observed effects of climate change on marine biodiversity

Cause	Effect	Net effect on diversity	Selected references
Temperature increase (tropical regions)	Local extinction of heat-sensitive species;	↓	[28,72,86]
	Coral bleaching and associated habitat change	↓	[53,54,57,137]
Temperature increase (temperate regions)	Warm-adapted species replace cold-adapted ones	↑	[26,49,69,71,77]
	Shifts of some habitat-forming species poleward and deeper	↓	[52,68,122]
Temperature increase (polar regions)	Invasion of subpolar species	↑	[64,66,67]
	Decline of endemics and sea ice dependents	↓	[60−63,66]
Increased climate variability (heat waves, storms)	Increased rates of disturbance	↓	[111,112]
Increased upwelling intensity	Mid-water hypoxia	↓	[106−108]
Increasing water column stratification	Lower nutrient supply and productivity	?	[13,16,17,110]
Sea level rise	Erosion of coastal habitats	↓	[115]
Changes in ocean circulation	Changes in larval transport	?	[138]

3.1 LOCAL SCALE

Changes in biodiversity at the local scale are often driven by the interplay of local and regional abiotic and biotic factors. The effects of a regional change in sea surface temperature (SST), for example, may be mediated or exacerbated by local factors such as wave exposure, tidal mixing, upwelling, and species composition. Nevertheless, some common patterns have been observed.

In temperate locations, slow changes in species composition often lead to an overall net increase in species richness. This was first shown by Southward and colleagues in their classic long-term studies in the English Channel [48]. Both intertidal and pelagic communities changed predictably during periods of climate warming, with warm-adapted species increasing in abundance and cold-adapted species decreasing, leading to overall increases in diversity. Reverse patterns were observed during periods of cooling [48]. Similar changes occurred in the northwest Pacific (Monterey Bay, California) where eight of nine southern species of intertidal invertebrates increased between the 1930s and 1990s, while five out of eight northern species decreased [49]. This change tracked observed increases in both mean and maximum temperature and led to an overall increase in invertebrate species richness by 7%, due to three species newly invading from the south [49]. A similar pattern was documented for a temperate reef fish community in southern California [50]. In this case, however, sudden warming in the 1970s also led to a decline in productivity, 80% loss of large zooplankton biomass and recruitment failure of many reef fish. This may explain why total biomass declined significantly, and total species richness also declined by 15%–25% at the two study sites [50]. These two contrasting examples illustrate that predictions based on temperature alone can be misleading, if concomitant changes in productivity or other overriding factors are involved. Moreover, it has been shown that local differences in tidal exposure rendered some northern sites more thermally stressful than southern sites, counteracting the poleward shift of southern species discussed above and possibly causing localized extinctions [51]. In addition to latitudinal shifts, species may also move into deeper, colder waters, as observed for example in fish [29] and seaweeds [52].

In tropical locations, warming can lead to species loss and a decline in diversity, as maximum temperature tolerances are exceeded. So far, this has been particularly documented in tropical coral reefs that are affected by bleaching events (reviewed for example by Refs [45,53,54] and in Chapter 13 of this volume). Poised near their upper thermal limits, coral reefs have experienced mass bleaching where sea temperatures have exceeded long-term summer averages by more than 1°C for several weeks [53,55]. The loss of sensitive coral species causes secondary changes of reef-associated fauna and flora [56]. For reef fish specifically, available studies so far indicate large changes in species composition after bleaching events, and a decline in species diversity that is linearly related to disturbance intensity [57] (see Ref. [58] for an exception).

Polar marine ecosystems are also thought to be particularly sensitive to climate change. In some parts of the Arctic and Antarctic atmospheric temperatures are rising at rates more than double the global average, and sea ice extent is shrinking rapidly, particularly in the Arctic [10]. Partly spurred by warmer climate and more open waters, there are other growing human impacts on polar regions such as pollution, exploitation and development. Therefore, the rate of change in species abundances and composition can be rapid [59]. While sea ice–dependent species such as polar bears [60], krill [61] and some penguins [62,63] have sharply decreased in abundance at some locations, there are signs of increasing invasion of subpolar and ice-independent species in other places [62]. For example, killer whale (*Orcinus orca*) sightings are increasingly almost exponentially in the Canadian Arctic, as sea ice retreats [64]. As these powerful consumers can set off trophic cascades [65], the resulting changes to the food web could be complex. However, only a few observations on net changes in species richness are available so far [66]; for example, surveys of Arctic macrobenthos suggest slow increases in species numbers at sites that are accessible to larval advection from southern locations [67]. Large uncertainties remain due in part to low sampling effort [66], and in part due to the complexity of this highly seasonal environment and the compounding effect of changes in sea ice, salinity, stratification, runoff and acidity [59].

3.2 **REGIONAL SCALE**

A growing number of studies have examined changes in species composition and diversity at regional scales. Much of this work was conducted using fisheries or plankton monitoring data, but recent work has extended to other species groups, such as seaweeds [68]. Again, a dominant observation is the replacement of cold-adapted by warm-adapted species. This appears to occur simultaneously at various levels in the food web, for example in North Atlantic zooplankton [69,70], as well as fish communities [27,71]. These changes are not necessarily synchronized; Beaugrand and colleagues documented a growing mismatch between warming-related changes in North Sea zooplankton communities since the 1980s and the emergence of cod larvae and juveniles. Cod populations were directly affected by changes in temperature but also indirectly by changes in their planktonic prey that compromised growth and survival of cod larvae. Perry et al. observed that larger species with slower life histories (such as cod) adapted their range much more slowly to changing conditions compared to fast-growing species [71]. This finding has implications for fisheries, as species with slower life histories are already more vulnerable to overexploitation [72] and may also be less able to compensate for warming through rapid demographic responses. Constraints to range shifts, however, appear to be less important than on the land. In the North Sea, the average rate of northward change was 2.2 km a^{-1} (where 'a' refers to annum), which is more than three times faster than observed range shifts in terrestrial environments, which reportedly average 0.6 km a^{-1} [21]. Likewise, a metanalysis of species range shifts showed that marine species fill their thermal niches more fully and move more readily at both cold and warm range boundaries compared to terrestrial species [73]. These findings may not be surprising, given the absence of hard physical boundaries in marine, and particularly pelagic, environments.

The net effect of such temperature-induced compositional changes on species richness can be surprisingly large – an almost 50% increase in the number of species recorded per year in North Sea bottom trawl surveys was documented between 1985 and 2006 [26]. This change correlated tightly with increasing water temperatures [26]. Similar trends have been found in the Bristol Channel, UK, where fish species richness increased by 39% from 1982 to 1998 [74]. In both cases, increases in richness were mainly driven by invasion of small-bodied southern species. It is noteworthy that similar regional changes have been observed nearby on land, where the species richness of British butterflies [25] and epiphytic lichen in the Netherlands [75] have increased with warming over time.

Such decadal changes in species richness and diversity are superimposed on significant year-to-year variation in temperature and other climatic factors. For example, in the NW Atlantic there is a well-documented latitudinal gradient in fish species richness that covaries with temperature [76]. This latitudinal gradient in diversity was previously treated as static. Recently it has been shown how temperature variability readjusts diversity gradients year by year [77]. Temperature variability is linked to large-scale pressure differences across the North Atlantic, known as the North Atlantic Oscillation (NAO) [78]. Positive NAO anomalies cause temperature gradients in the NW Atlantic to steepen, which leads to rapid adjustments in species diversity – northern areas decline and southern areas increase in diversity [77]. During NAO-negative years, the gradients flattens – northern areas increase and southern areas decrease in diversity. Although the north-south trend of increasing fish diversity does not reverse, there are substantial differences in its slope. This dynamic pattern is mostly driven by expansions and contractions of species at their northern or southern range limits [77]. Again, warming waters increase overall diversity in temperate regions; cooling waters have the opposite effect.

Similar mechanisms have been shown to affect pelagic fish diversity across the tropical to temperate Pacific Ocean. Here, pressure differences in the central Pacific lead to periodic warming and cooling of surface waters in the eastern tropical Pacific, the well-known El Niño Southern Oscillation (ENSO) that affects weather patterns around the planet [79]. Positive ENSO years are characterized by regional warming of the eastern tropical Pacific and an increase in species diversity in the following year [24]. Regional cooling leads to decreases in diversity. Single species such as blue marlin [24] or skipjack tuna [80] appear to readjust their distribution year by year in response to these temperature changes. These studies show how species diversity not only serves as an indicator of long-term climate change but accurately tracks short-term variability in climate as well. A caveat for many fish populations is of course that intense exploitation can override climate signals on diversity. For example, in the Atlantic and Indian Oceans there has been a long-term decline in tuna and billfish species richness, which is most likely explained by fishing [24]. In the Pacific, however, a similar decline is counteracted by increasing warming after 1977 [24].

In contrast to marine fish, plankton communities are not affected by exploitation, except maybe indirectly through trophic cascades [81]. For both phyto- and zooplankton phenological changes (for example the timing of the spring bloom), range shifts and changes in species composition have been shown to track changes in climate quite well [14,69,82]. For example, long-term Continuous Plankton Recorder data in central and European waters show that zooplankton communities there are gradually shifting away from cold-temperate and subarctic species to more species-rich warm-temperate and mixed assemblages [69]. Thus plankton communities are increasingly used as indicators of recent climate change [82].

Another approach used for macroalgae entailed the compilation of >20 000 herbarium records collected in Australia since the 1940s [68]. The study shows shifts in species distribution along both the coasts of the Indian and Pacific oceans consistent with rapid warming over the past five decades. If these waters continue to warm, hundreds of species could be pushed out of suitable habitats along the Australian coastline, with no place further south to go. These species (and others that depend on particular seaweed habitats) could face regional or even global extinction [68].

Recently, regional climate change has been discussed in the context of climate velocity, which is defined as the speed and direction of climate change across the landscape [83]. The advantage of this concept is that it tracks the spatial dimension of climate change and appears to predict very well the observed movements of surveyed fish populations around North America [29]. Interestingly, there was much variability in species responses in this region, with many populations moving northward but others tracking in the opposite direction, moving deeper, or further offshore. Climate velocity, however, was able to capture much of this variation and explained the recent shift in the distribution of individual species and entire communities better than individual species characteristics [29].

3.3 GLOBAL SCALE

There are a growing number of global-scale studies analyzing the effects of global climate variability and change on marine biodiversity. The argument has been made on land, albeit controversially, that a large number of extinctions could be caused by climate change by compressing species thermal habitats, particularly for species of restricted ranges [84]. Whether to expect global marine extinctions due to climate change is yet unclear, although much concern is focusing on coral reefs that are simultaneously threatened by warming and acidifying waters [45,85]. Dulvy and co-workers [72] note the possible global extinction of two coral species due to bleaching (*Siderastrea glynni*, *Millepora boschmai*), both of which have limited geographic ranges in the Eastern Pacific. Another study calculates that one-third of reef-building corals are faced with elevated extinction risk from climate change and other impacts [86]. Some coral-associated fish have also disappeared over the course of recent bleaching events [72]. Other habitat-forming species, such as sea grasses, mangroves and some seaweeds, also face elevated extinction risk due to warming and sea level rise [52,87,88], with consequences for communities dependent on these habitats.

Harnik et al. have more broadly compared the patterns and drivers of extinction across both the fossil record and the recent Anthropocene [6]. They found a fundamental difference in drivers of extinction, with climatic changes, perturbations in ocean chemistry, acidity and oxygen content explaining most marine extinctions across the fossil record, while combinations of human-driven overharvesting, habitat loss and pollution tended to explain more recent (last 400 years) extinctions. The authors speculated, however, that climate change is poised to recreate some of the conditions that led to previous mass extinctions, and hence may again become a dominant driver of extinction in the future, in combination with other cumulative stressors [6].

Although the question of projected extinctions due to climate change is contentious [89,90], there is little doubt that temperature is a major driver of marine diversity at the global scale. Global diversity patterns have so far been synthesized for 13 major taxa ranging from zooplankton to whales [9]. Two fundamental patterns emerged. For coastal species such as corals, sea grasses, coastal fish, among others, global reef diversity peaked at tropical latitudes in the Western Pacific [91], with a secondary hotspot in the Caribbean [9]. Pelagic plankton, fish and mammals, however, all peaked at intermediate latitudes, around 20°–30° North or South [24,92–94]. These patterns were most parsimoniously explained by spatial variation in sea surface temperature, which explained 45%–90% of the variation in species diversity for these groups (Fig. 1, showing examples of 6 out of 13 taxa analyzed). As mentioned above, variation in sea surface temperature well explains not just the broad spatial patterns but also much of the interannual variation in tuna and billfish richness in the Pacific [24] as well as seasonal variation in cetacean diversity in the Atlantic [94]. Moreover, the global richness pattern of tuna and billfish could be independently reconstructed from individual species' temperature tolerances [95]. Thus, it appears that temperature might be a powerful and general predictor of species richness at global scales. A prominent ecological theory supports this observation: the metabolic theory of ecology [96] predicts well-known relationships between body size, temperature and metabolic rate, which also correlates with the rate of mutation and speciation across long time horizons [97,98]. Taken together, these processes may explain higher diversity in the tropics but fail to account for the different pattern of diversity in pelagic species.

The empirical relationships between SST, species distributions and species richness can be used to derive hypotheses about the potential effects of future warming on species richness. For example, deep-water cetaceans show a unimodal relationship with SST, which under future warming scenarios projects substantial diversity increases at high latitudes but decreases in the tropical ocean [94]. This prediction was verified by a more detailed species-by-species habitat model that tracks preferences for temperature, sea ice and depth across all species of marine mammals to predict a synthetic diversity pattern [99]. In a warming ocean, the highest latitudes increased dramatically in species richness, but tropical locations showed fewer marine mammal species. Substantial turnover in species composition was predicted for temperate locations, but the effect on net richness remained small.

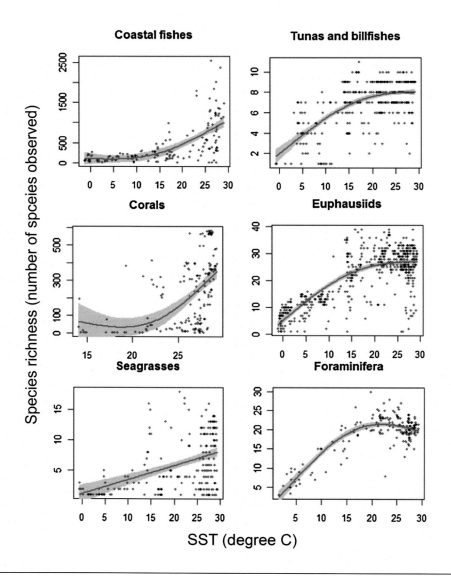

FIGURE 1

Temperature is a dominant environmental predictor of marine biodiversity. Shown are relationships as richness for six species with global coverage plotted against sea surface temperature (SST/°C). Species richness is the number of species present per grid cell across a global 880 × 880 km equal-area grid. *(Redrawn from data in Ref. [9] with permission.)*

These patterns for mammals broadly agree with similar predictions for 1066 exploited fishes and invertebrates, which were also modelled in relation to their temperature, salinity, depth and sea ice preferences, as well as their association with habitat features such as coastal upwelling, coral reef, estuaries and seamounts [28,100]. The authors mapped predicted changes in species invasions (Fig. 2(a)), extinctions (Fig. 2(b)) and turnover (Fig. 2(c)) across the world oceans. Invasions and turnover of species composition were predicted to occur particularly frequently at high latitudes, and

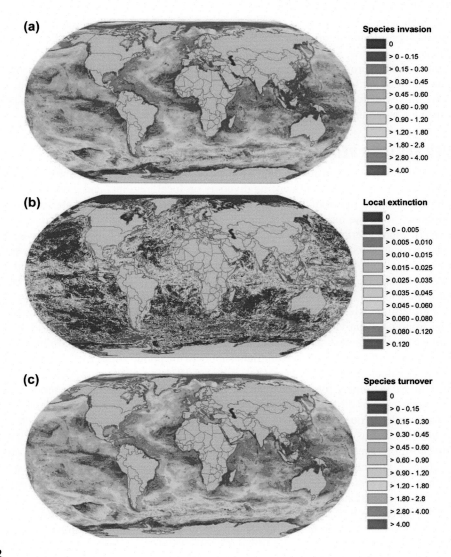

FIGURE 2

Predicted impact of ocean warming on biodiversity of 1066 species of fish and invertebrates. Predicted changes are expressed in terms of: (a) invasion intensity; (b) local extinction intensity; and (c) species turnover in 2050 relative to the mean of 2001–2005 (high-range climate change scenario). Intensity is expressed proportional to the initial species richness in each 0.5° × 0.5° cell. *(Reprinted with permission from Ref. [28].)*

extinctions were especially common in the tropics, the North Atlantic and the Southern Ocean. Again, net loss of species richness was most commonly predicted at tropical latitudes.

Finally, a global modelling study of plankton communities again reached similar conclusions, with large predicted losses of diversity in warm waters (assuming the lack of a rapid evolutionary response) and gains at high latitudes [101]. It is striking how plankton, fish and whales show similar current patterns in diversity relating to global temperature fields [9], and also similar predicted responses to climate change.

So far, most of these projections exclude the possibility of significant evolutionary change or other adaptive changes altering temperature or other habitats preferences for species faced with significant warming. Yet, the evidence for both short-term acclimatory and longer-term adaptive acquisition of climate resistance is mounting, for example in corals [102] and phytoplankton [103].

3.4 OTHER FACTORS RELATING TO CLIMATE CHANGE

In addition to the strong observed and predicted effects of temperature on biodiversity, there are clearly other factors that are important in influencing diversity on local, regional and global scales. For tuna and billfish, for example, the availability of thermal fronts that act to concentrate food supply is of great importance, as is the availability of sufficient oxygen concentrations (>2 mL L^{-1} at 100 m depth) [24,104]. Many marine animals may also concentrate in areas of high productivity [93], and export productivity appears to be a major correlate of deep-sea species richness [105]. However, all of these factors (fronts, oxygen, surface-ocean and export productivity) are both directly and indirectly affected by climate change (Table 1). For example, increased variability in wind stress has been shown to affect the intensity of upwelling, leading to periodic hypoxia and death of marine organisms [106–108]. Furthermore, climate change is implicated in the observed shallowing of oxygen minimum zones in the tropical ocean [109], which is likely compromising local biodiversity at intermediate depths. Surface-ocean biomass and productivity is also affected by global warming, particularly through increased stratification and lower nutrient supply to the photic zone [13,14,16,18,19]. This will most likely have direct effects on surface diversity [82,101] but also indirect effects on deep sea ecosystems that are extremely food limited and depend almost entirely on export production [110].

Even in the absence of directional trends in mean climate variables, increasing climate variability can affect biodiversity through extreme events, such as intense storms or heat waves, which can lead to large-scale die-offs such as in shallow-water corals or sea grass meadows [53,54,111,112]. Particularly when habitat-forming species such as sea grasses and corals are affected, such events can lead to substantial changes in local diversity, including genetic, species and habitat diversity, at least on short to intermediate timescales. Similarly, ocean acidification is occurring as a direct consequence of increasing CO_2 emissions, independent of trends in other climate variables. Observations at naturally acidified sites [113] as well as experiments [114] so far suggest that acidification forces substantive shifts in community composition but not necessarily large reductions in the number of species.

Finally, climate change leads to sea level rise (Chapter 16, this volume) and changes in ocean circulation (Chapter 17, this volume). Sea level rise in concert with increasing climate variability and more frequent storms can lead to increasing coastal erosion and the loss of coastal habitats (Chapter 20, this volume), which may affect the diversity of coastal species that depend on wetlands, salt marshes or mangroves [87,115]. Shorelines are increasingly fortified against rising water levels thereby preventing the adaptive inland movement of wetlands and upward movement of intertidal habitats, which decline or disappear over time together with their associated flora and fauna [52,115]. Further away from the coastlines, ocean circulation patterns can be sensitive to changes in temperature, precipitation, runoff, salinity and wind. So far, the effects of changing circulation and currents on ocean diversity have not been studied in any detail, with the exception of upwelling studies mentioned above [106–108].

We conclude that climate change leads to a range of physical changes, many of which are known to have effects on species diversity not directly mediated by changes in temperature. These effects may further interact with other aspects of global change that are unrelated to climate. Such complexities are discussed in more detail below.

4. CUMULATIVE IMPACTS AND INDIRECT EFFECTS OF CLIMATE CHANGE

A major challenge in ecological research is the disentanglement of multiple factors that are driving ecological change [46]. Up to this point we have reviewed the direct effects of increasing temperature and climate variability, and resulting changes in upwelling, stratification, sea level, oxygen and currents (Table 1). In reality, however, these processes are likely interacting with other impacts on biodiversity, such as exploitation, eutrophication, disease, and physical disturbance, among others. Species composition and abundance are also influenced to a large degree by local species interactions, such as predation, competition and facilitation. Through changing species interactions, and by interacting with other drivers, climate change can have a number of indirect effects that are sometimes surprising and difficult to predict. Here we highlight such indirect effects, giving some well-documented examples for illustration.

Consider the classic example of a keystone predator, the starfish *Pisaster ochraceus*, which maintains intertidal diversity by feeding on competitively dominant mussels *Mytilus californianus* [116]. This interaction, however, is temperature dependent – increases in upwelling lead to colder waters, lower predation rates and higher mussel cover [117]. Therefore, possible effects of climate change on diversity are mediated by a strong interaction between a predator and a competitively dominant prey.

Another well-documented complexity concerns the interaction between warming temperatures and disease. There is good evidence that climate warming can increase pathogen development and survival, disease transmission and host susceptibility [44,118]. This has become evident both in the sea and on land following large-scale warming events associated with ENSO, which are implicated in several coral diseases, oyster pathogens, crop pathogens, Rift Valley fever and human cholera [44,118]. These effects occurred both in tropical and temperate locations, with some documented range shifts of pathogens toward higher latitudes. In some cases, such temperature-mediated disease outbreaks may contribute to observed shifts in the latitudinal distribution of species. The range centroid of American lobster, for example, has shifted northward by about 2° latitude since 1970 [29]. In all likelihood, this is not caused by the northward migration of individuals but by a shifting balance of growth and mortality at the fringes of the present range. In the case of lobster populations, both paramoebic [119] and fungal diseases [120] have recently contributed to the collapse of populations at the southern range limit, and may be moving north, as temperatures continue to warm.

Climate change can also affect the interaction between our own species and marine biodiversity. Over the past two centuries humans have already had a marked impact on marine biodiversity, including a number of local, regional and global extinctions [6,72]. To date, exploitation and habitat destruction have probably had the most severe effects [121]. The existing rate of coastal habitat destruction will likely be accelerated by climate-driven losses due to warming, sea level rise, acidification and bleaching [45,87,88,115,122]. Similarly, the effects of exploitation are likely exacerbated by climate change. This is because most fisheries effectively truncate the age structure and size structure of target fish, by preferentially removing larger, older individuals. The fishery then becomes increasingly dependent on the recruitment of new individuals into the fishery. Recruitment however is strongly affected by climate variability [123] and change [124]. Removing the older age classes increases susceptibility both of the stock and the fishery to climatically induced fluctuations [34]. In addition, documented changes in temperature and plankton concentration have already affected the capacity of major fish stocks to produce new recruits [124]. Reducing fishing mortality and rebuilding spawning stock biomass may be the only feasible means of mitigating the impacts of climate change on recruitment [34,124,125].

Taken together, these results have implications for priority setting in global conservation initiatives. Tropical regions are predicted and observed to suffer most from species loss due to climate change, while often lacking resources and capacity for mitigation and adaptation [126]. At the same time, these regions are hotspots of fisheries-induced pressures, coincident with low management capacity [127]. Four out of five global fisheries–conservation hotspots identified in a recent metanalysis [127] coincide with areas of high predicted extinction intensity to climate change (Fig. 2), namely the Red Sea, Canary Current, Gulf of California and Indonesian Sea large marine ecosystems. The combination of multiple and intensifying threats, low capacity to meet these threats and high reliance of coastal populations on seafood and other marine goods and services motivate a strong management and conservation focus on tropical developing countries. Polar ecosystems also range high on the priority list due to the extraordinary rapid rate of change observed there, the possible extinction of polar specialist and ice-dependent species, and the growing pressures from development [59].

5. BIODIVERSITY AS INSURANCE AGAINST CLIMATE CHANGE IMPACTS

There is now good evidence that in addition to being a response variable to changes in temperature and climate, biodiversity may also provide resilience against climate change. This is because high genetic, species and habitat variation enhance the diversity of possible responses and adaptive capacity in the face of environmental variation [47,128,129]. For example, in a study on sea grass loss after the 2003 European heat wave, high genetic diversity (manipulated experimentally) led to faster recovery of damaged habitat [111]. This was driven both by selection of heat-adapted genotypes and by some form of facilitation that led to increased survival [111]. This observation was verified by laboratory experiments that manipulated temperature and genetic diversity in a controlled environment [112]. Another field study documented that high genetic diversity in sea grass also increased resilience to physical disturbance from overgrazing [130]. Theoretical studies have come to similar conclusions. For example, Yachi and Loreau [131] showed two major insurance effects of species richness on ecosystem productivity: (1) a reduction in the temporal variance of productivity, and (2) an increase in the temporal mean of productivity despite stochastic disturbances.

From these mechanistic studies follows the prediction that a loss in biodiversity should lead to a loss in productivity and resilience, which could magnify any effect of climate change (or other disturbances) on marine systems. An increase in biodiversity should have the opposite effect.

Broad evidence in support of this prediction comes from a series of metanalyses examining local experiments, regional time series and global fisheries data [32]. The vulnerability to climate change in particular was examined by regional studies of Alaskan salmon fisheries that have been carefully managed to avoid loss of stock diversity [129,132]. These stock complexes show a remarkable resilience to climatic change due to a large number of local life-history adaptations that are preserved within the stock complex. As environmental conditions changed, overall productivity was maintained by different sub-stocks that were adapted to thrive under those conditions [132]. This 'portfolio effect', which is analogous to the effects of asset diversity on the stability of financial portfolios, was recognized as an important insurance against climate-driven fluctuation in exploited fish stocks [129]. Maintaining population, species and habitat diversity is now generally seen to be of critical importance in stabilizing ecosystem services in a variable and changing world [30,32,133].

6. CONCLUSIONS

In this short review, we examined the relationship between marine biodiversity and climate change. It appears that temperature changes (both warming and cooling) produce predictable changes in marine biodiversity, both on ecological [27,29] and evolutionary timescales [6]. This effect is particularly visible at large spatial scales where diversity patterns are strongly linked to temperature [9]. On a global scale, it appears that as oceans warm, the tropics lose diversity, temperate regions show species turnover and sometimes increases in net diversity, whereas polar environments so far mostly show declines in ice-dependent species and some invasion of subpolar taxa. Underlying these dynamic patterns is a redistribution of species ranges, with range expansions of warm-adapted and range contractions of cold-adapted species toward the poles, as well as local extirpations and new invasions that are driven by local climate velocity. On local scales, however, other factors may modify the effects of temperature change depending on local context. As a result, species communities and food webs on all scales reorganize. Sometimes this involves the loss of particular habitats and their associated communities, the decoupling of predator populations from their prey, or other changes in species interactions due to shifts in phenology and physiology. Little is known about how entire communities or food webs reassemble with climate change; this should be a germane topic for further research.

From a biodiversity management perspective, concerns about climate change as a threat to biodiversity focus on tropical and polar regions because of the documented and anticipated species losses there, often exacerbated by growing population pressures, cumulative impacts and scarce management capacity. But even in regions with well-developed science and management capacity, little can be done to control the shifting of species ranges and the reorganization of ecosystems. There is some debate about assisted migration, for example in helping warm-adapted coral species to spread more widely [134], but this strategy is controversial [135]. It is generally accepted, however, to maintain as much as possible the existing diversity both within and between species and habitats that is evidently so important for adaptation and resilience. This can be achieved by controlling the impacts of other factors that may compromise biodiversity and by minimizing cumulative impacts [32,136]. In an era of rapid climate change, however, complex and surprising effects on biodiversity are to be expected, and any form of management must necessarily be highly adaptive and precautionary.

ACKNOWLEDGEMENTS

The authors wish to thank H. Whitehead and D. P. Tittensor for discussion and insight, as well as NSERC for financial support.

REFERENCES

[1] UN, Convention on Biological Diversity, United Nations, New York, 1992. http://www.biodiv.org/.
[2] D.M. Raup, J.J.J. Sepkoski, Science 215 (1982) 1501–1503.
[3] A.I. Miller, Science 281 (1998) 1157–1160.
[4] M.J. Benton, R.J. Twitchett, Trends Ecol. Evol. 18 (2003) 358–365.
[5] J.W. Kirchner, A. Weil, Nature 404 (2000) 177–180.
[6] P.G. Harnik, H.K. Lotze, S.C. Anderson, Z.V. Finkel, S. Finnegan, D.R. Lindberg, L.H. Liow, R. Lockwood, C.R. McClain, J.L. McGuire, Trends Ecol. Evol. 27 (2012) 608–617.
[7] G.J. Vermeij, Evol. Ecol. Res. 6 (2004) 315–337.
[8] P.D. Mannion, P. Upchurch, R.B. Benson, A. Goswami, Trends Ecol. Evol. 29 (2014) 42–50.
[9] D.P. Tittensor, C. Mora, W. Jetz, H.K. Lotze, D. Ricard, E.V. Berghe, B. Worm, Nature 466 (2010) 1098–1101.
[10] M.R. Allen, V.R. Barros, J. Broome, W. Cramer, R. Christ, J.A. Church, L. Clarke, Q. Dahe, P. Dasgupta, N.K. Dubash, IPCC Fifth Assessment Synthesis Report-Climate Change 2014 Synthesis Report, Intergovernmental Panel on Climate Change, Geneva, Switzerland, 2014.
[11] T.P. Barnett, D.W. Pierce, R. Schnur, Science 292 (2001) 270–273.
[12] C. Mora, A.G. Frazier, R.J. Longman, R.S. Dacks, M.M. Walton, E.J. Tong, J.J. Sanchez, L.R. Kaiser, Y.O. Stender, J.M. Anderson, C.M. Ambrosino, I. Fernandez-Silva, L.M. Giuseffi, T.W. Giambelluca, Nature 502 (2013) 183–187.
[13] J.L. Sarmiento, R. Slater, R. Barber, L. Bopp, S.C. Doney, A.C. Hirst, J. Kleypas, R. Matear, U. Mikolajewicz, P. Monfray, V. Soldatov, S.A. Spall, R. Stouffer, Global Biogeochem. Cycles 18 (2004), http://dx.doi.org/10.1029/2003GB002134.
[14] A.J. Richardson, D.S. Schoeman, Science 305 (2004) 1609–1612.
[15] H.L. Bryden, H.R. Longworth, S.A. Cunningham, Nature 438 (2005) 655–657.
[16] M.J. Behrenfeld, R.T. O'Malley, D.A. Siegel, C.R. McClain, J.L. Sarmiento, G.C. Feldman, A.J. Milligan, P.G. Falkowski, R.M. Letelier, E.S. Boss, Nature 444 (2006) 752–755.
[17] J.J. Polovina, E.A. Howell, M. Abecassis, Geophys. Res. Lett. 35 (2008) L03618.
[18] A.M. Lewandowska, D.G. Boyce, M. Hofmann, B. Matthiessen, U. Sommer, B. Worm, Ecol. Lett. 17 (2014) 614–623.
[19] D. Boyce, M. Lewis, B. Worm, Nature 466 (2010) 591–596.
[20] T.L. Root, J.T. Price, K.R. Hall, S.H. Schneider, C. Rosenzweig, J.A. Pounds, Nature 421 (2003) 57–60.
[21] C. Parmesan, G. Yohe, Nature 421 (2003) 37–42.
[22] G.-R. Walther, E. Post, P. Convey, A. Menzel, C. Parmesan, T.J.C. Beebee, J.-M. Fromentin, O. Hoegh-Guldberg, F. Bairlein, Nature 416 (2002) 351–460.
[23] C. Parmesan, Annu. Rev. Ecol. Evol. Syst. 37 (2006) 637–669.
[24] B. Worm, M. Sandow, A. Oschlies, H.K. Lotze, R.A. Myers, Science 309 (2005) 1365–1369.
[25] R. Menéndez, A.G. Megías, J.K. Hill, B. Braschler, S.G. Willis, Y. Collingham, R. Fox, D.B. Roy, C.D. Thomas, Proc. R. Soc. B 273 (2006) 1465–1470.
[26] J.G. Hiddink, R. ter Hofstede, Glob. Change Biol. (2008) 453–460.
[27] W.W. Cheung, R. Watson, D. Pauly, Nature 497 (2013) 365–368.
[28] W.W.L. Cheung, V.W.Y. Lam, J.L. Sarmiento, K. Kearney, R. Watson, D. Pauly, Fish Fish. 10 (2009) 235–251.

[29] M.L. Pinsky, B. Worm, M.J. Fogarty, J.L. Sarmiento, S.A. Levin, Science 341 (2013) 1239–1242.

[30] MEA, Millenium Ecosytem Assessment, Ecosystems and Human Well-Being: Synthesis, Island Press, Washington, DC, USA, 2005.

[31] H.K. Lotze, M. Glaser, in: E.R.J. Urban, B. Sundby, P. Malanotte-Rizzoli, J. Melillo (Eds.), Watersheds, Bays and Bounded Seas, Island Press, Washington, DC, USA, 2009, pp. 227–249.

[32] B. Worm, E.B. Barbier, N. Beaumont, J.E. Duffy, C. Folke, B.S. Halpern, J.B.C. Jackson, H.K. Lotze, F. Micheli, S.R. Palumbi, E. Sala, K. Selkoe, J.J. Stachowicz, R. Watson, Science 314 (2006) 787–790.

[33] O. Hoegh-Guldberg, J.F. Bruno, Science 328 (2010) 1523–1528.

[34] K.M. Brander, Proc. Natl. Acad. Sci. U.S.A. 104 (2007) 19709–19714.

[35] S.C. Doney, M. Ruckelshaus, J.E. Duffy, J.P. Barry, F. Chan, C.A. English, H.M. Galindo, J.M. Grebmeier, A.B. Hollowed, N. Knowlton, J. Polovina, N.N. Rabalais, W.J. Sydeman, L.D. Talley, Ann. Rev. Mar. Sci. 4 (2012) 11–37.

[36] S. Levitus, J.I. Antonov, T.P. Boyer, C. Stephens, Science 287 (2000) 2225–2229.

[37] A.B. Hollowed, S. Sundby, Science 344 (2014) 1084–1085.

[38] C. Schär, P.L. Vidale, D. Lüthi, C. Frei, C. Häberli, M.A. Liniger, C. Appenzeller, Nature 427 (2004) 332–336.

[39] S. Rahmstorf, D. Coumou, Proc. Natl. Acad. Sci. U.S.A. 108 (2011) 17905–17909.

[40] C.D.G. Harley, A.R. Hughes, K.M. Hultgren, B.G. Miner, C.J.B. Sorte, C.S. Thornber, L.F. Rodriguez, L. Tomanek, S.L. Williams, Ecol. Lett. 9 (2006) 228–241.

[41] A. Schmittner, Nature 434 (2005) 628–633.

[42] R.F. Keeling, A. Körtzinger, N. Gruber, Ann. Rev. Mar. Sci. 2 (2010) 199–229.

[43] H.K. Lotze, B. Worm, Limnol. Oceanogr. 47 (2002) 1734–1741.

[44] C.D. Harvell, C.E. Mitchell, J.R. Ward, S. Altizer, A.P. Dobson, R.S. Ostfeld, M.D. Samuel, Science 296 (2002) 2158–2162.

[45] O. Hoegh-Guldberg, P.J. Mumby, A.J. Hooten, R.S. Steneck, P. Greenfield, E. Gomez, C.D. Harvell, P.F. Sale, A.J. Edwards, K. Caldeira, N. Knowlton, C.M. Eakin, R. Iglesias-Prieto, N. Muthiga, R.H. Bradbury, A. Dubi, M.E. Hatziolos, Science 318 (2007) 1737–1742.

[46] C.M. Crain, K. Kroeker, B.S. Halpern, Ecol. Lett. 11 (2008) 1304–1315.

[47] T. Elmqvist, C. Folke, M. Nyström, G. Peterson, J. Bengtsson, B. Walker, J. Norberg, Front. Ecol. Environ. 1 (2003) 488–494.

[48] A.J. Southward, S.J. Hawkins, M.T. Burrows, J. Therm. Biol. 20 (1995) 127–155.

[49] J.P. Barry, C.H. Baxter, R.D. Sagarin, S.E. Gilman, Science 267 (1995) 672–675.

[50] S.J. Holbrook, R.J. Schmitt, J.S. Stephens Jr., Ecol. Appl. 7 (1997) 1299–1310.

[51] B. Helmuth, C.D.G. Harley, P.M. Halpin, M. O'Donnell, G.E. Hofmann, C.A. Blanchette, Science 298 (2002) 1015–1017.

[52] C.D.G. Harley, K.M. Anderson, K.W. Demes, J.P. Jorve, R.L. Kordas, T.A. Coyle, M.H. Graham, J. Phycol. 48 (2012) 1064–1078.

[53] O. Hoegh-Guldberg, Mar. Freshwater Res. 50 (1999) 839–866.

[54] T.P. Hughes, A.H. Baird, D.R. Bellwood, M. Card, S.R. Connolly, C. Folke, R. Grosberg, O. Hoegh-Guldberg, J.B.C. Jackson, J. Kleypas, J.M. Lough, P. Marshall, M. Nystrom, S.R. Palumbi, J.M. Pandolfi, B. Rosen, J. Roughgarden, Science 301 (2003) 929–933.

[55] S.D. Donner, W.J. Skirving, C.M. Little, M. Oppenheimer, O.V.E. Hoegh-Guldberg, Glob. Change Biol. 11 (2005) 2251–2265.

[56] T. McClanahan, N. Muthiga, S. Mangi, Coral Reefs 19 (2001) 380–391.

[57] S.K. Wilson, N.A.J. Graham, M.S. Pratchett, G.P. Jones, N.V.C. Polunin, Glob. Change Biol. 12 (2006) 2220–2234.

[58] D.R. Bellwood, A.S. Hoey, J.L. Ackerman, M. Depczynski, Glob. Change Biol. 12 (2006) 1587–1594.

[59] C. Michel, B. Bluhm, V. Gallucci, A.J. Gaston, F.J.L. Gordillo, R. Gradinger, R. Hopcroft, N. Jensen, T. Mustonen, A. Niemi, T.G. Nielsen, Biodiversity 13 (2012) 200–214.

[60] I. Stirling, N.J. Lunn, J.L. Iacozza, Arctic 52 (1999) 294–306.
[61] A. Atkinson, V. Siegel, E. Pakhomov, P. Rothery, Nature 432 (2004) 100–103.
[62] S.D. Emslie, W. Fraser, R.C. Smith, W. Walker, Antarct. Sci. 10 (1998) 257–268.
[63] C. Barbraud, H. Weimerskirch, Nature 411 (2001) 183–186.
[64] J.W. Higdon, D.D. Hauser, S.H. Ferguson, Mar. Mammal Sci. 28 (2012) E93–E109.
[65] J.A. Estes, M.T. Tinker, T.M. Williams, D.F. Doak, Science 282 (1998) 473–476.
[66] P. Wassmann, C.M. Duarte, S. Agusti, M.K. Sejr, Glob. Change Biol. 17 (2011) 1235–1249.
[67] J.M. Węsławski, M.A. Kendall, M. Włodarska-Kowalczuk, K. Iken, M. Kędra, J. Legezynska, M.K. Sejr, Mar. Biodivers. 41 (2011) 71–85.
[68] T. Wernberg, B.D. Russell, M.S. Thomsen, C.F.D. Gurgel, C.J.A. Bradshaw, E.S. Poloczanska, S.D. Connell, Curr. Biol. 21 (2011) 1828–1832.
[69] G. Beaugrand, P.C. Reid, F. Ibañez, J.A. Lindley, M. Edwards, Science 296 (2002) 1692–1694.
[70] M. Edwards, A.J. Richardson, Nature 430 (2004) 881–884.
[71] A.L. Perry, P.J. Low, J.R. Ellis, J.D. Reynolds, Science 308 (2005) 1912–1915.
[72] N.K. Dulvy, Y. Sadovy, J.D. Reynolds, Fish Fish. 4 (2003) 25–64.
[73] J.M. Sunday, A.E. Bates, N.K. Dulvy, Nat. Clim. Change 2 (2012) 686–690.
[74] P.A. Henderson, J. Mar. Biol. Assoc. U.K. 87 (2007) 589–598.
[75] C.M. van Herk, A. Aptroot, H.F. van Dobben, Lichenologist 34 (2002) 141–154.
[76] K.T. Frank, B. Petrie, N.L. Shackell, Trends Ecol. Evol. 22 (2007) 236–242.
[77] J.A.D. Fisher, K.T. Frank, B. Petrie, W.C. Leggett, N.L. Shackell, Ecol. Lett. 11 (2008) 883–897.
[78] J.W. Hurrell, Science 269 (1995) 676–679.
[79] M.J. McPhaden, S.E. Zebiak, M.H. Glantz, Science 314 (2006) 1740–1745.
[80] P. Lehodey, M. Bertignac, J. Hampton, A. Lewis, J. Picaut, Nature 389 (1997) 715–718.
[81] K.T. Frank, B. Petrie, J.S. Choi, W.C. Leggett, Science 308 (2005) 1621–1623.
[82] G.C. Hays, A.J. Richardson, C. Robinson, Trends Ecol. Evol. 20 (2005) 337–344.
[83] S.R. Loarie, P.B. Duffy, H. Hamilton, G.P. Asner, C.B. Field, D.D. Ackerly, Nature 462 (2009) 1052–1055.
[84] C.D. Thomas, A. Cameron, R.E. Green, M. Bakkenes, L.J. Beaumont, Y.C. Collingham, B.F.N. Erasmus, M.F. de Siqueira, A. Grainger, L. Hannah, L. Hughes, B. Huntley, A.S. van Jaarsveld, G.F. Midgley, L. Miles, M.A. Ortega-Huerta, A.T. Peterson, O.L. Phillips, S.E. Williams, Nature 427 (2004) 145–148.
[85] J.M. Pandolfi, S.R. Connolly, D.J. Marshall, A.L. Cohen, Science 333 (2011) 418–422.
[86] K.E. Carpenter, M. Abrar, G. Aeby, R.B. Aronson, S. Banks, A. Bruckner, A. Chiriboga, J. Cortes, J.C. Delbeek, L. DeVantier, G.J. Edgar, A.J. Edwards, D. Fenner, H.M. Guzman, B.W. Hoeksema, G. Hodgson, O. Johan, W.Y. Licuanan, S.R. Livingstone, E.R. Lovell, J.A. Moore, D.O. Obura, D. Ochavillo, B.A. Polidoro, W.F. Precht, M.C. Quibilan, C. Reboton, Z.T. Richards, A.D. Rogers, J. Sanciangco, A. Sheppard, C. Sheppard, J. Smith, S. Stuart, E. Turak, J.E.N. Veron, C. Wallace, E. Weil, E. Wood, Science 321 (2008) 560–563.
[87] B.A. Polidoro, K.E. Carpenter, L. Collins, N.C. Duke, A.M. Ellison, J.C. Ellison, E.J. Farnsworth, E.S. Fernando, K. Kathiresan, N.E. Koedam, S.R. Livingstone, T. Miyagi, G.E. Moore, V. Ngoc Nam, J.E. Ong, J.H. Primavera, S.G. Salmo, J.C. Sanciangco, S. Sukardjo, Y. Wang, J.W.H. Yong, PLoS One 5 (2010) e10095.
[88] F.T. Short, B. Polidoro, S.R. Livingstone, K.E. Carpenter, S. Bandeira, J.S. Bujang, H.P. Calumpong, T.J.B. Carruthers, R.G. Coles, W.C. Dennison, P.L.A. Erftemeijer, M.D. Fortes, A.S. Freeman, T.G. Jagtap, A.H.M. Kamal, G.A. Kendrick, W. Judson Kenworthy, Y.A. La Nafie, I.M. Nasution, R.J. Orth, A. Prathep, J.C. Sanciangco, B.v. Tussenbroek, S.G. Vergara, M. Waycott, J.C. Zieman, Biol. Conserv. 144 (2011) 1961–1971.
[89] W. Thuiller, M.B. Araújo, R.G. Pearson, R.J. Whittaker, L. Brotons, S. Lavorel, Nature 430 (2004), http://dx.doi.org/10.1038/nature02716.

[90] D.B. Botkin, H. Saxe, M.B. Araujo, R. Betts, R.H.W. Bradshaw, T. Cedhagen, P. Chesson, T.P. Dawson, J.R. Etterson, D.P. Faith, S. Ferrier, A. Guisan, A.S. Hansen, D.W. Hilbert, C. Loehle, C. Margules, M. New, M.J. Sobel, D.R.B. Stockwell, BioScience 57 (2007) 227–236.

[91] C.M. Roberts, C.J. McClean, J.E.N. Veron, J.P. Hawkins, G.R. Allen, D.E. McAllister, C.G. Mittermeier, F.W. Schueler, M. Spalding, F. Wells, C. Vynne, T.B. Werner, Science 295 (2002) 1280–1284.

[92] S. Rutherford, S. D'Hondt, W. Prell, Nature 400 (1999) 749–753.

[93] J. Schipper, J.S. Chanson, F. Chiozza, N.A. Cox, M. Hoffmann, V. Katariya, J. Lamoreux, A.S.L. Rodrigues, S.N. Stuart, H.J. Temple, J. Baillie, L. Boitani, T.E. Lacher Jr., R.A. Mittermeier, A.T. Smith, D. Absolon, J.M. Aguiar, G. Amori, N. Bakkour, R. Baldi, R.J. Berridge, J. Bielby, P.A. Black, J.J. Blanc, T.M. Brooks, J.A. Burton, T.M. Butynski, G. Catullo, R. Chapman, Z. Cokeliss, B. Collen, J. Conroy, J.G. Cooke, G.A.B. da Fonseca, A.E. Derocher, H.T. Dublin, J.W. Duckworth, L. Emmons, R.H. Emslie, M. Festa-Bianchet, M. Foster, S. Foster, D.L. Garshelis, C. Gates, M. Gimenez-Dixon, S. Gonzalez, J.F. Gonzalez-Maya, T.C. Good, G. Hammerson, P.S. Hammond, D. Happold, M. Happold, J. Hare, R.B. Harris, C.E. Hawkins, M. Haywood, L.R. Heaney, S. Hedges, K.M. Helgen, C. Hilton-Taylor, S.A. Hussain, N. Ishii, T.A. Jefferson, R.K.B. Jenkins, C.H. Johnston, M. Keith, J. Kingdon, D.H. Knox, K.M. Kovacs, P. Langhammer, K. Leus, R. Lewison, G. Lichtenstein, L.F. Lowry, Z. Macavoy, G.M. Mace, D.P. Mallon, M. Masi, M.W. McKnight, R.A. Medellin, P. Medici, G. Mills, P.D. Moehlman, S. Molur, A. Mora, K. Nowell, J.F. Oates, W. Olech, W.R.L. Oliver, M. Oprea, B.D. Patterson, W.F. Perrin, B.A. Polidoro, C. Pollock, A. Powel, Y. Protas, P. Racey, J. Ragle, P. Ramani, G. Rathbun, R.R. Reeves, S.B. Reilly, J.E. Reynolds III, C. Rondinini, R.G. Rosell-Ambal, M. Rulli, A.B. Rylands, S. Savini, C.J. Schank, W. Sechrest, C. Self-Sullivan, A. Shoemaker, C. Sillero-Zubiri, N. De Silva, D.E. Smith, C. Srinivasulu, P.J. Stephenson, N. van Strien, B.K. Talukdar, B.L. Taylor, R. Timmins, D.G. Tirira, M.F. Tognelli, K. Tsytsulina, L.M. Veiga, J.-C. Vie, E.A. Williamson, S.A. Wyatt, Y. Xie, B.E. Young, Science 322 (2008) 225–230.

[94] H. Whitehead, B. McGill, B. Worm, Ecol. Lett. 11 (2008) 1198–1207.

[95] D.G. Boyce, D.P. Tittensor, B. Worm, Mar. Ecol. Prog. Ser. 355 (2008) 267–276.

[96] J.H. Brown, J.F. Gillooly, A.P. Allen, V.M. Savage, G.B. West, Ecology 85 (2004) 1771–1789.

[97] A.P. Allen, J.H. Brown, J.F. Gillooly, Science 297 (2002) 1545–1548.

[98] A.P. Allen, J.F. Gillooly, V.M. Savage, J.H. Brown, Proc. Natl. Acad. Sci. U.S.A. 103 (2006) 9130–9135.

[99] K. Kaschner, D. Tittensor, J. Ready, T. Gerrodette, B. Worm, PLoS One 6 (2011) e19653.

[100] W.W.L. Cheung, C. Close, V. Lam, R. Watson, D. Pauly, Mar. Ecol. Prog. Ser. 365 (2008) 187–197.

[101] M.K. Thomas, C.T. Kremer, C.A. Klausmeier, E. Litchman, Science 338 (2012) 1085–1088.

[102] S.R. Palumbi, D.J. Barshis, N. Traylor-Knowles, R.A. Bay, Science 344 (2014) 895–898.

[103] L. Schlüter, K.T. Lohbeck, M.A. Gutowska, J.P. Gröger, U. Riebesell, T.B. Reusch, Nat. Clim. Change 4 (2014) 1024–1030.

[104] P.N. Sund, M. Blackburn, F. Williams, Oceanogr. Mar. Biol. Ann. Rev. 19 (1981) 443–512.

[105] D.P. Tittensor, M.A. Rex, C.T. Stuart, C.R. McClain, C.R. Smith, Biol. Lett. 7 (2011) rsbl20101174.

[106] A. Bakun, S.J. Weeks, Ecol. Lett. 7 (2004) 1015–1023.

[107] B.A. Grantham, F. Chan, K.J. Nielsen, D.S. Fox, J.A. Barth, A. Huyer, J. Lubchenco, B.A. Menge, Nature 429 (2004) 749–754.

[108] F. Chan, J.A. Barth, J. Lubchenco, A. Kirincich, H. Weeks, W.T. Peterson, B.A. Menge, Science 319 (2008) 920.

[109] L. Stramma, G.C. Johnson, J. Sprintall, V. Mohrholz, Science 320 (2008) 655–658.

[110] C.R. Smith, F.C. De Leo, A.F. Bernardino, A.K. Sweetman, P.M. Arbizu, Trends Ecol. Evol. 23 (2008) 518–528.

[111] T.B.H. Reusch, A. Ehlers, A. Hämmerli, B. Worm, Proc. Natl. Acad. Sci. U.S.A. 102 (2005) 2826–2831.

[112] A. Ehlers, B. Worm, T.B.H. Reusch, Mar. Ecol. Prog. Ser. 355 (2008) 1–7.

[113] J.M. Hall-Spencer, R. Rodolfo-Metalpa, S. Martin, E. Ransome, M. Fine, S.M. Turner, S.J. Rowley, D. Tedesco, M.-C. Buia, Nature 454 (2008) 96–99.

[114] I.E. Hendriks, C.M. Duarte, M. Álvarez, Estuar. Coast. Shelf Sci. 86 (2010) 157–164.

[115] D. Scavia, J.C. Field, D.F. Boesch, R.W. Buddemeier, V. Burkett, D.R. Cayan, M.J. Fogarty, M.A. Harwell, R.W. Howarth, C. Mason, D.J. Reed, T.C. Royer, A.H. Sallenger, J.G. Titus, Estuaries 25 (2002) 149–164.

[116] R.T. Paine, Am. Nat. 100 (1966) 65–76.

[117] E. Sanford, Science 283 (1999) 2095–2097.

[118] C.D. Harvell, K. Kim, J.M. Burkholder, R.R. Colwell, P.R. Epstein, D.J. Grimes, E.E. Hofmann, E.K. Lipp, A.D.M.E. Osterhaus, R.M. Overstreet, J.W. Porter, G.W. Smith, G.R. Vasta, Science 285 (1999) 1505–1510.

[119] T.E. Mullen, S. Russell, M.T. Tucker, J.L. Maratea, C. Koerting, L. Hinckley, S. De Guise, S. Frasca Jr., R.A. French, T.G. Burrage, J. Aquat. Anim. Health 16 (2004) 29–38.

[120] M.F. Tlusty, H.O. Halvorson, R. Smolowitz, U. Sharma (Eds.), State of Lobster Science: Shell Disease Workshop, Aquatic Forum Series 05–01, The New England Aquarium, 2005.

[121] H.K. Lotze, H.S. Lenihan, B.J. Bourque, R. Bradbury, R.G. Cooke, M.C. Kay, S.M. Kidwell, M.X. Kirby, C.H. Peterson, J.B.C. Jackson, Science 312 (2006) 1806–1809.

[122] A. Jueterbock, L. Tyberghein, H. Verbruggen, J.A. Coyer, J.L. Olsen, G. Hoarau, Ecol. Evol. 3 (2013) 1356–1373.

[123] N.C. Stenseth, A. Mysterud, G. Ottersen, J.W. Hurrell, K.-S. Chan, M. Lima, Science 297 (2002) 1292–1296.

[124] G.L. Britten, M. Dowd, B. Worm, Proc. Natl. Acad. Sci. U.S.A., in revision (2015).

[125] B. Worm, R.A. Myers, Nature 429 (2004) 15.

[126] T.R. McClanahan, J.E. Cinner, J. Maina, N.A.J. Graham, T.M. Daw, S.M. Stead, A. Wamukota, K. Brown, M. Ateweberhan, V. Venus, N.V.C. Polunin, Conserv. Lett. 1 (2008) 53–59.

[127] B. Worm, T.A. Branch, Trends Ecol. Evol. 27 (2012) 594–599.

[128] J.E. Duffy, Mar. Eco. Prog. Ser. 311 (2006) 233–250.

[129] D.E. Schindler, R. Hilborn, B. Chasco, C.P. Boatright, T.P. Quinn, L.A. Rogers, M.S. Webster, Nature 465 (2010) 609–613.

[130] A.R. Hughes, J.J. Stachowicz, Proc. Natl. Acad. Sci. U.S.A. 101 (2004) 8998–9002.

[131] S. Yachi, M. Loreau, Proc. Natl. Acad. Sci. U.S.A. 96 (1999) 1463–1468.

[132] R. Hilborn, T.P. Quinn, D.E. Schindler, D.E. Rogers, Proc. Natl. Acad. Sci. U.S.A. 100 (2003) 6564–6568.

[133] M. Loreau, Philos. Trans. R. Soc. B Biol. Sci. 365 (2010) 49–60.

[134] B.M. Riegl, S.J. Purkis, A.S. Al-Cibahy, M.A. Abdel-Moati, O. Hoegh-Guldberg, PLoS One 6 (2011) e24802.

[135] N. Hewitt, N. Klenk, A. Smith, D. Bazely, N. Yan, S. Wood, J. MacLellan, C. Lipsig-Mumme, I. Henriques, Biol. Conserv. 144 (2011) 2560–2572.

[136] T.P. Hughes, D.R. Bellwood, C. Folke, R.S. Steneck, J. Wilson, Trends Ecol. Evol. 20 (2005) 380–386.

[137] M.S. Pratchett, A.S. Hoey, S.K. Wilson, V. Messmer, N.A. Graham, Diversity 3 (2011) 424–452.

[138] C.J. Svensson, S.R. Jenkins, S.J. Hawkins, P. Aberg, Oecologia 142 (2005) 117–126.

INTERTIDAL INDICATORS OF CLIMATE AND GLOBAL CHANGE

14

Nova Mieszkowska[1,2]

[1]*The Marine Biological Association of the UK, Plymouth, UK;* [2]*School of Environmental Sciences, University of Liverpool, Liverpool, UK*

CHAPTER OUTLINE

1. INTRODUCTION

The rocky intertidal zone spans the region of the coastline from the highest vertical level reached at high water during spring tides (with associated wave splash) to the lowest level exposed to the air during low water springs. A wide variety of taxa inhabit the rocky intertidal zone, including algae, molluscs, echinoderms, cnidarians and crustaceans. Owing to the accessibility of rocky shores, intertidal species have been studied extensively throughout the nineteenth and twentieth centuries by amateur naturalists [1,2] and professional researchers as model systems for the development of ecological and biological theory [3–15].

Intertidal invertebrates and macroalgae are ectotherms of marine evolutionary origin, but due to the daily regime of emersion and immersion they must contend with both marine and terrestrial regimes. Therefore, they provide a unique insight into the impacts of changes in both aquatic and terrestrial climatic environments. Diurnal tidal cycles and seasonal fluctuations in both sea and air temperature mean that intertidal organisms are subject to extremes of temperature with resultant fluctuations in body temperature of over 30°C frequently experienced even in temperate regions [16]. Additional stressors such as desiccation [17], current and wave forces [18], rapid fluctuations in salinity [19–21] oxygen availability [22,23] and nutrient levels [24] mean that organisms are often living close to their physiological tolerance limits [25–32].

Marine ectothermic species respond faster than terrestrial species to environmental change; the typically short life spans [33] and sessile or sedentary nature of the adult and juvenile stages prevent

Climate Change. http://dx.doi.org/10.1016/B978-0-444-63524-2.00014-2

escape from changing environmental regimes. The larval stage of most intertidal species is planktonic and therefore also provides an indication of the impacts of environmental change in the pelagic zone. Changes in distribution and abundance are therefore likely to be driven by the direct response of organisms to changes in the environment. Intertidal invertebrates and marine macroalgae are from low trophic levels and thus would be expected to respond quicker to alterations in local conditions than species at higher trophic levels, often showing the first response in a cascade of effects up the food chain to tertiary and apex predators [34,35]. Variation in the abundance of keystone structural or functional species can alter the composition and dynamics of entire rocky communities [5,36,37] and these small changes in environmental conditions can lead to major alterations in community structure and functioning [38,39]. Taking all of the above factors into account, the rocky intertidal ecosystem is likely to be one of the most sensitive natural systems to climate and show some of the earliest responses to climate change [40,41].

The geographical ranges of intertidal species are essentially one-dimensional, as they occupy a narrow strip of coastline between the low and high tide levels [42]. In combination with the highly stressful and fluctuating environment in which these species live, the rocky intertidal zone is thus an ideal model system in which to study the effects of climate change. Rocky intertidal ecosystems occur all over the globe and thus facilitate spatial and temporal comparisons of the relative impacts of global environmental change. Responses to environmental change can be divided into two categories: (1) proximate ecological responses that depend upon relationships between abiotic factors and organismal-level processes, population dynamics and community structure [43], and (2) direct impacts on individual performance during various life stages through changes in physiology, morphology and behaviour. These impacts scale up to the population level response, which can be additionally affected by climate-driven changes in hydrographic processes that affect dispersal of the pelagic larval life stages and recruitment. All lead to alterations in distributions, biodiversity, productivity and micro-evolutionary processes.

2. CLIMATE CHANGE AND BIOGEOGRAPHY

Biogeographical studies were first introduced by Tournefort [44] in the 1700s, and work undertaken in the early 1900s [24,45–51] is used as the basis for ecological climate change research today. The major marine biogeographic provinces have been defined primarily on where clusters of biogeographic distributional limits occur for taxa of interest. Sea temperature has been assumed to ultimately set biogeographical ranges of marine species (see [24,52] for reviews), although for intertidal species, air temperature will also be a major factor. Low and high latitude biogeographic limits have been associated with August and February sea surface isotherms, respectively, for fauna and flora across a diverse range of taxonomic groups including marine algae [53–58], cirripedes [59] and molluscs [60,61]. The relationship between species' distributions and climate is not simple, however, and biogeographic studies are often complicated by covarying environmental parameters, species-specific effects and responses to local environmental changes that prevent cause-and-effect relationships from being fully understood.

Suitable habitat exists beyond the distributional limits of many species of marine invertebrates [62–65], but the unsuitability of environmental conditions currently prevents their colonization and therefore the ranges are assumed to be limited by climate. This principle is termed the 'climate envelope' of a species and is the basis for many bioclimatic models in use today [66,67] (but see Refs

[68,69]). In locations where environmental conditions alter to fall within the physiological tolerance range of a species, range extensions of the leading edge of the distribution are predicted as organisms are able to colonize new sites of suitable habitat, whilst range contractions may occur at the trailing edge of the range where climate becomes unsuitable for continued persistence. At large spatial scales, environmental tolerances of individual species broadly match changes in biogeographic distributions with changing climatic conditions, however, the actual range edges may lie some distance inside the fundamental niche 'envelope'. Interactions between species and between organisms and environmental factors, and local influences such as a lack of suitable habitat, poor dispersal and connectivity of suitable habitat space act to set this realized niche for each species [70–72]. The effects of climatic variability on the distributions of algae, ectothermic and endothermic marine animals and their interactions must therefore be measured across a range of spatial and temporal scales in order to understand and ultimately forecast changes in marine ecosystems [73,74].

2.1 USING LONG-TERM DATA SETS TO DETECT CLIMATE CHANGE

Some of the most spatially and temporally extensive data sets in the world exist for the distribution and abundance of intertidal invertebrates and macroalgae along the coastline of the northeast Atlantic [75]. Intensive and wide-ranging surveys were made in the 1930s, 1940s and 1950s by Fischer-Piette [76–79] along the Atlantic coastlines of France, Spain, Portugal and North Africa. Crisp and Southward made similar surveys around the coastlines of Britain and Ireland during the 1950s [80,81,175]. These data sets are particularly valuable within the context of climate change monitoring as they provide extensive baselines from which to measure the rate and extent of changes in distribution and abundance of intertidal species during periods of warming and cooling over the past 70 years [82]. Time series data for abundance and population structures for barnacles, trochids and limpets also exist for British shores dating back to the 1950s, 1970s and 1980s, respectively [63,80–83].

The Marine Biodiversity and Climate Change Project 'MarClim' was established by the Marine Biological Association of the UK in 2001 to assess and forecast the influence of climate change on rocky intertidal biodiversity in Britain and Ireland. It combined historical data with contemporary resurveys at over 400 rocky shores (Fig. 1) to provide evidence of changes in abundance, population structure and geographical distribution of intertidal species in relation to recent climate change [64]. MarClim survey protocols were the same as those used in the original surveys made in the 1950s [80,81] to map the distribution and range limits of over 50 species of invertebrates and macroalgae of both cold and warm water origins. In addition, quantitative data on the abundance and population dynamics of key species of barnacles, limpets and trochids were collected. These surveys were carried out at locations spanning sites from the range edges to locations closer to the centres of distribution. These combined data sets have been used to track the changes in abundances and relative dominance of warm- and cold-water species on shores where they coexist in response to fluctuating climatic conditions throughout the twentieth century.

A similar survey network, PISCO, has been set up to survey the rocky intertidal zone along the Pacific coastline of the United States since 1999. PISCO is a network of universities that has combined studies of variability in ocean climate with simultaneous multispecies experimental studies of larval and adult populations over most of the US west coast [85,86].

2.2 BIOGEOGRAPHIC RESPONSES OF INTERTIDAL BIOTA

Contractions and expansions of geographic range edges due to global environmental change are resulting in species both being lost from and introduced to intertidal assemblages around the world. Such changes are initially being recorded at the periphery of the geographic range of a species, where organisms are often already experiencing temperatures close to their thermal limits [26]. However, there can also be local or regional heterogeneity within the geographic range of a species as evidenced by environmental hot spots [31,40] or cold spots [84] occurring far from the distributional limits of sessile invertebrates. Such changes in turn influence the outcomes of species interactions, for example competition, facilitation and predation, ultimately altering the structure of communities and marine ecosystem processes [41,43,87–89].

2.2.1 Europe

Alterations in distributional limits of a wide range of intertidal taxa have already occurred in Britain since rapid warming of the climate began in the mid-1980s. Northern and eastern range edges of warm-water trochid gastropods such as *Phorcus lineatus* and *Gibbula umbilicalis*, barnacles including *Chthamalus montagui*, *Chthamalus stellatus* and *Perforatus (Balanus) perforatus* and the brown macroalga *Bifurcaria bifurcata* have extended between 85 km and 180 km since previous records in the twentieth century [64,65,90] and at rates of up to 50 km per decade. This rate is much faster than the average movement of 6.1 km per decade for terrestrial species' ranges [91] and is concordant with northward shifts in plankton recorded in British coastal waters (see Chapter 11). The limpet *Patella rustica* has recently bridged a historical gap in its distribution in north Portugal during a period of weakened upwelling in coastal waters [84]. Several factors may be responsible, including increase in sea surface temperature (SST), decrease in upwelling strength in the southern Bay of Biscay and an increase in the strength of the Western Iberian Shelf Current, all of which are driven by the global climate.

Saccharina latissima (*Laminaria saccharina*) has disappeared from large areas across northern Europe during the last decade, with significant losses of populations along the coasts of Scandinavia, Helgoland and southern Brittany [87,92]. No such decline has been recorded in populations of this macroalga on British coastlines, however [64], suggesting that the causal factor may not be climatic but potentially disease related.

In the Mediterranean, influxes of warmer water and propagules of tropical species from the Atlantic, combined with lessepsian migration of species from the Red Sea via the Suez Canal and human introduction of exotic species are altering ecosystem structure with potential impacts on the trophic web [93]. Marine caves have also been impacted by increases in warm-water mysids and severe declines in an endemic cold-water congener [94].

Species of cold-water origin including the barnacle *Semibalanus balanoides* [95], tortoiseshell limpet *Tectura testudinalis* and the brown macroalga *Alaria esculenta* [65,96] have shown retractions in the southern range limits and decreases in abundance in Britain and Europe during the last few decades.

Surveys of macroalgal distributions along the coast of Portugal during the 1950s, 1960s [97,98] and the 2000s [99] have identified approximately 120 conspicuous species that have shown significant alterations in the location of a range edge between these periods. Warm-water species have all shifted their high latitude range limits further north, with significant correlations between distributional movement and mean annual inshore sea surface temperature (SST) since 1941 [100]. Species classified as cold water in origin displayed both north and south shifts with no significant change when considered as a group. This alternative response may be in part due to the grouping of these species for this study, some of which are non-native and others that are widely considered to have a cosmopolitan distribution throughout Europe rather than a warm or cold affinity [82].

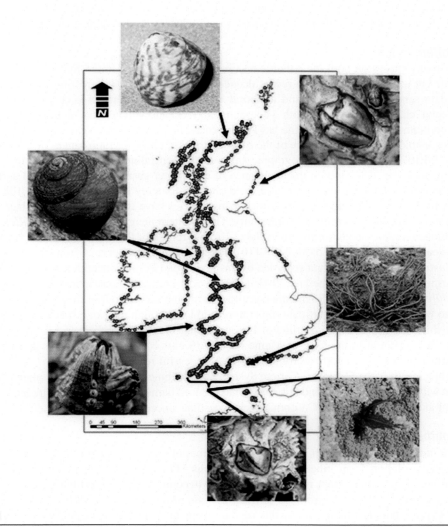

FIGURE 1

MarClim long-term survey sites (red circles) around the UK. Locations of leading range extensions marked by arrows clockwise from left: the barnacle Perforatus perforatus, topshells Phorcus lineatus and Gibbula umbilicalis, and brown alga Bifurcaria bifurcata. Trailing edge retractions for the brown alga Alaria esculenta and the barnacle Semibalanus balanoides.

2.2.2 Arctic

The blue mussel *Mytilus edulis* extended its distribution from the Norwegian mainland north by 500 km to Bear Island on the Svalbard archipelago between 1977 and 1994 [101], and was found on the Arctic island of Svalbard itself for the first time in 1000 years after a period of anomalously warm Atlantic seawater influx between 2002 and 2004 [102,103]. This reappearance represented a huge poleward shift in distributional limits of approximately 1000 km, probably due to transport of pelagic larvae north in the warm-water current. However, it is not known whether these populations are sustainable or if the prevailing climate is still too cold for this species to reproduce and survive at such high latitudes.

2.2.3 United States

Studies of rocky shores on both the Atlantic and Pacific coasts of the United States date back to the early 1900s but are mostly limited to recent decades, stemming from the growing awareness of the need for data sets of broad spatial and temporal coverage to track and predict impacts of global environmental change [104,105]. Both geography and oceanography have a large influence on intertidal community structure along the US Pacific coast. In warm-regime years, some species from the Californian biogeographic province of the east Pacific have extended their ranges north into higher latitudes [106,107]. Additional biogeographic shifts have been inferred from increases in the abundance of warm versus decreases in the abundance of cold-water gastropods, limpets and anthozoans between the early 1930s and the 1990s at a fixed site in Monterey Bay, California [38,108]. These alterations in the relative dominance of co-occurring species have changed the assemblage composition to a more typically warm-water community. The southern neogastropod *Kelletia kelletii* shifted its entire biogeographic range northward in the California region of the northeast Pacific between the late 1970s/early 1980s and the 2000s [109,110]. Fossil records and surveys from the 1830s to the present illustrate that this is the first recorded extension beyond Point Conception and coincided with strong warming of SST during the late 1900s [103]. A 350 km range contraction of the blue mussel *Mytilus edulis* occurred on the US Atlantic coastline during the second half of the twentieth century in response to warming summer intertidal temperatures [111], although the same species has not shown any range shifts on the Atlantic coastline of Europe during this time. *Codium fragile*, a warm-water green alga has appeared in the Gulf of Maine on the US east coast since the 1970s as summer sea temperatures have increased sufficiently to facilitate successful reproduction [112].

Upwelling events in the California Current System have become less frequent but stronger and longer in duration with climate warming, and this change has reduced barnacle recruitment but increased mussel recruitment along the Oregon and California rocky coastline since the mid-1960s [87].

2.2.4 Southern Hemisphere

Both Australia and New Zealand have a very high number of endemics due to their extensive history of geographic isolation from other temperature regions [113,114]. In such regions, reductions in abundance and geographic extent may lead to species becoming globally extinct. In Australia, new research programmes have been initiated to track the rate of biogeographic shifts of rocky intertidal species [115], but data is still sparse for this region. Tropical species of rock pool fishes are already being found at temperate latitude locations on the east coast as coastal water temperatures have increased [116]. Temperate species of kelp that form a dense zone from the low intertidal to shallow subtidal in Tasmania have been continually declining due to the direct impact of rising sea temperatures [117,118]. The decline has been exacerbated by intensive grazing from the spread of the warm-water long-spined sea urchin *Centrostephanus rodgersii* from the mainland since the 1960s [119] and increases in abundance of the warm-water urchin *Janus edwardsii* and the abalone *Haliotis rubra* [115]. The western rock lobster *Parulirus cygmus* is the most important single species fishery in Australia [120]. Fisheries landings correlate strongly with the strength of the Leeuwin Current, which drives cross-shelf transport of larvae and hence productivity of the stock [120]. The Leeuwin Current strength is highly correlated with ENSO events, with a weakening during El Niño years. Since the 1970s, El Niño events have become more common [121], resulting in more frequent depressions in the size of the lobster fishery.

New Zealand has little quantitative data for intertidal species covering the entire coastline, although extensive time series exist for individual locations [122,176]. Research began in 2008 to quantitatively survey the New Zealand rocky coastline to establish a baseline from which future climate-induced shifts in species distributions and ecosystem-level responses can be measured [123]. Experiments show that the mussel *Perna canaliculus* is less tolerant to warm air temperatures than the co-occurring *Mytilus* species on New Zealand shores [124] and exhibits reduced growth and reproductive effort when transplanted to higher intertidal elevations, indicating potentially deleterious effects of climate change. This species typically inhabits the low shore and damp, shaded regions of the mid-shore and thus may not be subjected to as severe exposures as under experimentally manipulated conditions.

2.2.5 South Africa

Little climate-related research has been published from the African continent. Range extensions and population-level changes were reported for warm-water rocky intertidal species of limpets (*Patella longicosta*, *Patella oculus*) and winkle (*Oxystele variegata*). Recruitment failure was observed in the cold-water limpet *Patella granatina* in South Africa in response to the unusually warm surface temperatures in the southern Benguela Current (around South Africa) in 1982/1983 [125] probably connected to the strong 1982/1983 El Niño event in the Pacific [126]. Whilst there is plenty of evidence for species expanding their northern range limits, there is less for contraction of southern limits. This issue stems in part from the lack of knowledge of both past and present locations of southern biological limits of many intertidal species, and the paucity of data collected from southern limit populations, which tend to lie in African or South American coastal waters.

2.2.6 Asia

Few long-term or large-scale studies of intertidal systems have been carried out in Asia, and for many countries the biodiversity and biogeographies of rocky intertidal ecosystems are not fully known. Surveys and specimen collections were documented along the coastline of China in the 1950s and the 1980s, although many were in soft-sediment systems with no information on long-term changes for rocky reef systems [127].

2.3 EXTREME WEATHER EVENTS

The extreme cold winter of 1962/1963 in Britain substantially cut back northern range limits of many intertidal species by hundreds of kilometres as a direct result of exposure to subzero temperatures. Communities in north Wales were particularly severely impacted due to low water spring tides occurring in the early hours of the morning, when air temperatures are at their lowest. Populations were less affected further south in Wales and England due to low water occurring around midday, and northern limits were relocated in these regions [128]. The northern range edge of *P. lineatus* in north Wales did not show much recovery from the retraction caused by the extreme cold winter of 1962/1963 until the 1980s. In the intervening two decades the range has re-extended around the coastline and multi-age, breeding populations have become established beyond previous limits [129] and the leading range edge has expanded throughout the 2000s. After the cold winter of 2009/2010, species with colder thermal preferences again showed a temporary increase in recruitment success and small range extensions along the Atlantic coastline of Europe, however, this reversal of the long-term trend of increased success and distribution of species with warmer thermal windows did not continue, demonstrating that biogeographic change is punctuated by population responses to extreme events [130]. This cold winter also caused widespread mortality of invasive mussels on the US Atlantic coastline due to prolonged exposure to low air temperatures [131].

Heatwave events are becoming more common during the current period of rapid global environmental change and have also caused catastrophic mortalities of intertidal species such as the blue mussel *Mytilus edulis* on the US Atlantic coastline [132] and the tropical limpet *Cellana toreuma* [133]. These widespread observations demonstrate the role of weather in driving acute changes to species range limits, but these are set against the pervasive long-term poleward shift of species distributions in response to climate warming.

2.4 INTERACTIONS

Increasing numbers of species from warm climatic regions are beginning to replace those with colder climate affinities in temporal regions, leading to alterations in the composition of local assemblages [82,108,109,134,135]. These local-scale changes will also facilitate the poleward spread of species by altering the ratio of extinction to colonization events within range-edge populations. The Boreal barnacle *S. balanoides* and Boreal limpet *Patella vulgata* have also declined in abundance relative to Lusitanian barnacles (*Chthamalus* species) that have warmer thermal tolerance windows, and limpets (*Patella depressa*) around the coastline of Britain and northern Europe [82,136,137]. Models built and tested using the long-term barnacle time series for Britain show that this rate is increasing [138], and *S. balanoides* are predicted to be completely replaced by *Chthamalus* spp. between 2060 and 2080 [139]. Climate change is also altering facilitative interactions. The cold-water limpet *P. vulgata* preferentially locates its home scar under the shade of the macroalgae *Fucus vesiculosus*. When *F. vesiculosus* is experimentally removed from shores to simulate the impacts of climate warming, significant mortality is observed in *P. vulgata*, with surviving individuals relocating their home scars. In contrast, the warm-water congener *P. depressa* does not suffer increased mortality and does not alter its location [78,139].

In addition to competitive interactions, climate change is also altering interspecific interactions including predation. Studies along the Pacific coast of Canada and the United States combining field experiments and spatiotemporal comparisons of field surveys have highlighted the importance of climate-driven alterations in predator–prey interactions between the starfish *Pisaster ochraceus* and keystone biogenic species including the blue mussels *Mytilus trossulus* [71] and *Mytilus californianus* [140,141]. This was primarily due to differential responses of species to the local thermal climate, which set vertical location of *M. trossulus* and *M. californianus* on shores, whereas for *P. ochraceus* the relationship between vertical location and body temperature varied between locations, showing a tight link at some sites but no relationship at others, with no discernible latitudinal or thermal gradient underlying these differential responses. *Pisaster ochraceus* has been shown to adjust thermal inertia by modifying the volume of coelomic fluid, which acts as an internal coolant, thus providing a thermoregulatory 'backup' allowing increased predation time on intertidal *Mytilus* beds under warmer air temperatures at low tide [140,141].

3. MECHANISMS AND MICROCLIMATE

While correlational biogeographic studies can be used to obtain probabilistic maps of species occurrence and generate testable hypotheses, they cannot provide information on cause and effect [91,142]. In order to accurately predict the rate and extent of future biogeographic shifts in species distributions, the biological mechanisms driving these changes need to be better understood for both algae and animals [80,143]. Physical, ecological, evolutionary and physiological factors acting on the processes of reproduction, birth, dispersal, recruitment and mortality are all involved in shaping species' ranges [25,144–147] and must also be considered when studying the effects of a changing environment. Processes occurring at the level of the genome, cell and organism scale up to affect fitness, abundance and ultimately the distribution of species. Mechanistic responses to climate warming have been detected from the cellular and molecular to the organismal and population levels of biological organization, highlighting the need to understand how small-scale environmental changes act on individual organisms, and the resultant impacts at the population and species levels. It is now understood that as well as the large-scale latitudinal thermal gradient from the equator to the poles, microclimatic variation exists on rocky shores [148]. The location of an organism at the microclimate scale is likely to also have a significant effect on performance and survival, as well as the location of the organism within the larger scale of the species biogeographical range [149]. An integrative approach is required to understand the roles of environmental and physiological variance in driving ecological responses to climate change [74,150].

An example of species responses to long-term climate change being linked to changes in local environmental temperatures comes from a temporal study of co-occurring barnacles of differing thermal origins at a time series site in southwest England initiated in the 1950s. Annual surveys of abundances show that over the period of half a century, the chthamalid species that have a thermo-tolerance range encompassing higher temperatures increase in abundance and relative dominance in warmer climatic periods, whereas the Boreal *S. balanoides* that has thermal tolerance limits at lower temperatures has lower recruitment success in warmer years but has increased in abundance after colder winters [138]. These changes are not linked to large-scale climatic indices such as the Atlantic Multidecadal Oscillation or North Atlantic Oscillation but are tightly synchronized with local sea surface temperature.

3.1 PHYSIOLOGY

Advances in the understanding of the mechanisms by which animals thermally adapt to changes in local climate, and the resultant implications for performance, population-level success and ultimately distributional responses to climate change have been made over the past decade via the study of intertidal ectothermic invertebrates and development of novel technology and mechanistic models. Biomechanical methods including the determination of lethal and deleterious thermal tolerance thresholds have been developed for rocky intertidal gastropods that include measurement of heart rates [151] and body temperatures [152,153]. These field-based investigations have provided organismal-level response data to changing local environmental regimes that can be modelled to more accurately determine current, and predict future distributions in response to a changing climate [154]. Physiological measurements have been used to assess thermal performance and determine an organism's ability to adapt to changes in local thermal regimes. Studies on intertidal winkles and limpets have revealed how energetics underpins the ability of ectotherms to thermally adapt [155,156].

3.2 REPRODUCTION AND RECRUITMENT

Variations in sea and air temperatures have also long been known to influence the physiological performance and reproductive success of marine species [14,45,157–161]. Synchronous increases in abundance have been recorded in populations of southern trochids throughout Britain and northern France since the mid-1980s. These increases in abundance are linked to warming in the regional climate since the mid-1908s due to increased frequency of annual recruitment success [65]. The mechanisms behind these changes are earlier onset of annual reproductive cycles of southern trochid gastropods *Phorcus lineatus* and *Gibbula umbilicalis* in response to milder winters and warmer springs, coupled with increased survival of newly settled recruits (often the most sensitive life stage to environmental stress) exposed to milder, shorter winters on the shore [129]. The annual reproductive cycles of southern limpets are also starting earlier and lasting longer in southwest Britain. In contrast, less than 20% of the population of northern limpets is reaching gonad development stages at which spawning can occur [139]. Recruitment can also be influenced by oceanographic features, which can control dispersal of the pelagic larval phase. Large-scale surveys of the US Pacific coast have demonstrated a tight correlation between SST and recruitment, and large intraspecific differences in recruitment rate along the biogeographic range of keystone barnacle and mussel species [105] demonstrating the potential for alterations in recruitment success in a warming climate.

3.3 MODELLING

New mechanistic modelling approaches such as the dynamic energy budget approach, developed for blue mussels [162,163] and now being applied to other intertidal species including *P. ochraceus* is incorporating experimentally derived knowledge on the role that environmental temperature plays in setting the rate and efficiency of essential processes such as feeding, assimilation of food resources, and the resultant outputs via survival, growth and reproduction [164,165]. Mechanistic species distribution models based on thermal performance data are now being developed to link physiological processes to geographical climate patterns to determine the current and predict the future distribution of intertidal species under changing climatic conditions [154].

4. ADDITIONAL IMPACTS OF GLOBAL CHANGE
4.1 OCEAN ACIDIFICATION

Dealt with in Chapter 18, ocean acidification is predicted to impact upon calcifying rocky shore species such as barnacles, limpets and top-shells during the second half of the twenty-first century. Experiments on algae show concerning effects of decreased seawater pH, including regime shifts from kelp forests to turfing algal communities, with associated loss of biodiversity [165]. Mesocosm studies highlight the role of ocean acidification in structuring rocky shore biofilm communities, and subsequent effects of reduced grazing by intertidal gastropods [166].

Potential synergistic effects of warming seas and decreases in oceanic pH are still largely unexplored, and may also vary depending on the position of the organism within the latitudinal distribution of the species. For macroalgae, it is likely to be a complex issue with increases in CO_2 promoting photosynthesis, temperature changes driving latitudinal range shifts and interactions either positively or negatively impacting performance, reproduction and survival [167]. Long-term mesocosm studies are beginning to be carried out to determine realistic chronic impacts of climate change and ocean acidification, with negative implications of reduced performance scaling up to potential reductions in distributional coverage for intertidal gastropods such as the Boreal dogwhelk *Nucella lapillus* [168].

4.2 COASTAL ZONE DEVELOPMENT

Other facets of global environmental change are rising sea levels and extreme weather events [169]. Many areas of low-lying coastline are facing greater risk of flooding around the world. This risk has led to proliferation of coastal defences to protect property, agricultural land and infrastructure such as roads and railways. Localized defences can scale up to whole coastlines when multiple structures are built along large sections of the land–sea interface. This has occurred in the Mediterranean [170] and on the coasts of the northern Irish Sea and eastern English Channel and North Sea. These developments can have severe impacts with loss of sedimentary habitats and their replacement with artificial rocky shores with an impoverished biota [170,171]. Such large-scale coastal modification can also influence biogeographic processes. Recent range extensions of rocky shore species in the eastern English Channel are probably the result of a combination of increased reproductive success and the provision of artificial habitat (sea defences, marinas, seaside piers) as stepping stones for advance across patches of unfavourable habitat [172].

4.3 CLIMATE CHANGE AND NON-NATIVE SPECIES

Introductions of non-native species are increasing globally due to increases in maritime shipping, aquaculture, mariculture, the aquarium trade and imported live bait. Ballast water, hull fouling and intensive culture practices have been identified as high-risk vectors for the introduction of invasive non-native species into the marine environment. There is also evidence that such introductions are more likely in a warmer world [174]. The importation of shellfish species for aquaculture and mariculture businesses in the UK (e.g. the Pacific oyster *Crassostrea gigas*, the Mediterranean blue mussel *Mytilus galloprovinicialis*) have facilitated the spread of these species into natural ecosystems in Britain, Europe and the USA. Inshore sea temperatures are now warm enough to allow successful reproduction of these introduced species, resulting in aggressive colonization of areas outside their site of cultivation, often to the detriment of native congeners, which are outcompeted by the non-native species. Once non-natives become established in the natural environment, there are few barriers to prevent further spread. The introduction of non-native species from different biogeographical realms can be considered as a facet of global environmental change with the potential for global homogenization of biotas and hence decreased biodiversity.

5. CONCLUSIONS

Intertidal invertebrates and algae are already responding to global climate warming, with shifts in biogeographic distributions away from warmer low latitude regions toward the cooler poles. In general, the rate of recession of northern species is not as fast as the rate of advance of southern species. The rate and extent of change is also species specific, leading to alterations in community composition with knock-on effects for ecosystem structure and functioning and likely declines in biodiversity in temperate-tropical regions. These shifts are ultimately being driven by physiological responses to temperature, with additional indirect impacts from species interactions, oceanographic processes, coastal zone development and habitat availability. There is still a paucity of data with long-temporal and wide-spatial coverage, which is hampering the detection of environmentally driven changes in coastal ecosystems, especially in the southern hemisphere. It is of vital importance that research focuses on the combination of maintaining these valuable data sets with the development of experimental research to determine the mechanisms underpinning the observed responses, so that we may be better placed to predict the future impacts on intertidal systems from our rapidly changing environment.

ACKNOWLEDGEMENTS

The author would like to thank Natural England and Natural Resources Wales for continuing funding of the MarClim project http://www.marclim.com. Thanks to K. Richardson for comments and advice, which helped to improve this chapter. The INSHORE global intertidal network http://rockyinshore.org/ were instrumental in suppling findings and expertise from research projects around the world.

REFERENCES

[1] P.H. Gosse, Tenby: A Seaside Holiday, John van Voorst, Paternoster Row, London, 1856, pp. 1–397.
[2] C. Kingsley, Glaucus; or, the Wonders of the Shore, Macmillan & Company, Cambridge, 1856, pp. 1–168.
[3] J.H. Orton, Nature 123 (1929) 14–15.
[4] M.S. Doty, Ecology 27 (1946) 315–328.
[5] J.H. Connell, Ecology 42 (1961) 710–723.
[6] R.T. Paine, Am. Nat. 100 (1966) 65–75.
[7] R.T. Paine, Oecologia 15 (1974) 93–120.
[8] T. Carefoot, Pacific Seashores, J.J. Douglas, Vancouver, 1977, pp. 1–208.
[9] R.T. Paine, J.M. Levine, Ecol. Monogr. 51 (1981) 145–178.
[10] J.R. Lewis, The Ecology of Rocky Shores, English Universities Press, London, 1964, pp. 1–323.
[11] P.K. Paine, Ecol. Monogr. 41 (1971) 351–389.
[12] B.A. Menge, Ecol. Monogr. 46 (1976) 355–393.
[13] A.J. Southward, Nature 175 (1955) 1124–1125.
[14] A.J. Southward, J. Mar. Biol. Assoc. U.K. 36 (1957) 323–334.
[15] R.T. Paine, JAE 64 (1994) 425–427.
[16] B.T. Helmuth, M.W. Denny, Limnol. Oceanogr. 48 (2003) 1338–1345.
[17] B.A. Foster, Mar. Biol. 8 (1971) 12–29.
[18] J.A. Kitzes, M.W. Denny, Biol. Bull. 8 (2005) 114–119.
[19] J. Davenport, H. Macalister, J. Exp. Mar. Biol. Ecol. 76 (1996) 985–1002.

[20] R. Li, S.H. Brawley, Mar. Biol. 144 (2004) 205–213.

[21] L.E. Burnett, Am. Zool. 37 (1997) 633–640.

[22] R.F. Service, Science 305 (2004) 1099.

[23] E.P. Dahlhoff, B.A. Menge, Mar. Ecol. Prog. Ser. 144 (1996) 97–107.

[24] L.W. Hutchins, Ecol. Monogr. 17 (3) (1947) 325–335.

[25] R.M. MacArthur, Geographical Ecology, Harper & Row, 1972.

[26] J.R. Lewis, Hydrobiologia 142 (1986) 1–13.

[27] J.H. Brown, G.C. Stevens, D.M. Kaufman, Annu. Rev. Ecol. Syst. 27 (1996) 597–623.

[28] J.H. Brown, Am. Nat. 124 (1984) 255–279.

[29] J.H. Stillman, G.N. Somero, JEB 199 (1996) 1845–1855.

[30] A.A. Hoffmann, P.A. Parsons, Extreme Environmental Change and Evolution, Cambridge University Press, Cambridge, 1997, pp. 235.

[31] B. Helmuth, C.D. Harley, P.M. Halpin, M. O'Donnell, G.E. Hofmann, C.A. Blanchette, Science 298 (2002) 1015–1017.

[32] J.H. Stillman, Integr. Comp. Biol. 42 (2002) 790–796.

[33] M.H. Carr, J.E. Neigel, J.A. Estes, S. Andelman, R.R. Warner, J.L. Largier, Ecol. App 13 (2003) S90–S107.

[34] P.E. Smith, Can. J. Fish Aquat. Sci. 42 (1985) 69–82.

[35] S. Jenouvrier, C. Barbraud, H. Weimerskirch, JAE 72 (2003) 576–587.

[36] J.R. Lewis, Mar. Biol. Ann. Rev. 14 (1976) 371–390.

[37] S.J. Hawkins, R.G. Hartnoll, J. Exp. Mar. Biol. Ecol. 62 (1982) 271–283.

[38] J.P. Barry, C.H. Baxter, R.D. Sagarin, S.E. Gilman, Science 267 (1995) 672–674.

[39] J.R. Lewis, R.S. Bowman, M.A. Kendall, P. Williamson, Neth. J. Sea Res. 16 (1982) 18–28.

[40] B. Helmuth, N. Mieszkowska, P. Moore, S.J. Hawkins, Annu. Rev. Ecol. Evol. Syst. 37 (2006) 373–404.

[41] M.D. Bertness, G.H. Leonard, J.M. Levine, J.F. Bruno, Oecologia 120 (1999) 446–450.

[42] R.D. Sagarin, S.D. Gaines, J. Biogeogr. 29 (2002) 985–997.

[43] C.D.G. Harley, A.R. Hughes, K. Hultgren, B.G. Miner, C.J.B. Sorte, C.S. Thornber, L.F. Rodriguez, L. Tomanek, S.L. Williams, Ecol. Lett. 9 (2006) 228–241.

[44] J.P. de Tournefort, Relation d'un voyage du Levant, Paris, 1717, pp. 288.

[45] J. Grinnell, Am. Nat. 51 (1917) 115–128.

[46] J.H. Orton, J. Mar. Biol. Assoc. U.K. 2 (1920) 299–366.

[47] W.A. Setchell, Science (1920) 187–190.

[48] S.A. Cain, Foundations of Plant Geography, Harper Brothers, New York, London, 1944.

[49] L.W. Hutchins, Ecol. Monogr. 17 (1947) 325–335.

[50] E.V. Wulff, An Introduction to Historical Plant Geography (Translated from the Russian by E. Brissenden), Waltham, Massachusetts, 1950.

[51] H.G. Andrewartha, L.C. Birch, The Distribution and Abundance of Animals, University of Chicago Press, Chicago, 1954.

[52] E.C. Pielou, Biogeography, Wiley-Interscience, Chichester, 1979.

[53] V.D.C. Hoek, Biol. J. Linn. Soc. 18 (1982) 81–144.

[54] G. Michanek, Bot. Mar. 22 (1979) 375–391.

[55] K. Luhning, Seaweeds Their Environment, Biogeography and Ecophysiology, John Wiley and Sons, London, 1990, pp. 1–61.

[56] G.M. Voskoboinikov, A.M. Breeman, C. van den Hoek, V.N. Makarov, E.V. Shoshina, Bot. Mar. 39 (1996) 341–346.

[57] F.J. Molenaar, A.M. Breeman, J. Phycol. 33 (1997) 330–343.

[58] S. Orfanidis, A.M. Breeman, J. Phycol. 35 (1999) 919–930.

[59] A.J. Southward, Nature 165 (1950) 408.

[60] D.R. Franz, A.S. Merrill, Malacologia 19 (1980) 209–225.

[61] G.J. Vermeij, Evolution 36 (1982) 561–580.

[62] J.R. Lewis, The Ecology of Rocky Shores, English Universities Press, London, 1964.

[63] M.A. Kendall, J.R. Lewis, Hydrobiologia 142 (1986) 15–22.

[64] N. Mieszkowska, R. Leaper, P. Moore, M.A. Kendall, M.T. Burrows, D. Lear, E. Poloczanska, K. Hiscock, P.S. Moschella, R.C. Thompson, R.J. Herbert, D. Laffoley, J. Baxter, A.J. Southward, S.J. Hawkins, J. Mar. Biol. Assoc. U.K. 20 (2005) 1–55.

[65] N. Mieszkowska, M.A. Kendall, S.J. Hawkins, R. Leaper, P. Williamson, N.J. Hardman-Mountford, A.J. Southward, Hydrobiologia 555 (2006) 241–251.

[66] R.G. Pearson, T.P. Dawson, Glob. Ecol. Biogeogr. 12 (2003) 361–371.

[67] G.R. Walther, E. Post, P. Convey, A. Menzel, C. Parmesan, T.J.C. Beebee, T.J.C. Fromentin, O.H. Guldberg, F. Bairlein, Nature 416 (2002) 389–395.

[68] A.J. Davies, L.S. Jenkinson, J.H. Lawton, B. Shorrocks, S. Wood, Nature 391 (1998) 783–786.

[69] R.W. Brooker, R.W. Travis, E.J. Clark, C. Dytham, J. Therm. Biol. 245 (2007) 59–65.

[70] M.B. Araujo, A. Guisan, J. Biogeogr. 33 (2006) 1677–1688.

[71] C.D.G. Harley, Science 334 (2011) 1124–1127.

[72] S.A. Woodin, T.J. Hilbish, B. Helmuth, S.J. Jones, D.S. Wethey, Ecol. Evol. (2013) 3334–3346.

[73] C.J. Monaco, B. Helmuth, Adv. Mar. Biol. 60 (2011) 123–162.

[74] B. Helmuth, B.D. Russell, Y. Dong, C.D.G. Harley, F.P. Lima, G. Sara, G.A. Williams, N. Mieszkowska, Clim. Change Responses 1 (2014) 1–6, http://dx.doi.org/10.1186/s40665-014-0006-0.

[75] N. Mieszkowska, H.E. Sugden, L.B. Firth, S.J. Hawkins, Phil. Trans. R. Soc. A (2014), http://dx.doi.org/10.1098/rsta.2013.0339.

[76] E. Fischer-Piette, J. Conch. Paris 79 (1935) 5–66.

[77] E. Fischer-Piette, J. Linn. Soc. Zool. 40 (1936) 181–272.

[78] E. Fischer-Piette, Ann. Inst. Ocean. Monaco 31 (1955) 37–124.

[79] D.J. Crisp, E. Fischer-Piette, Ann. Inst. Ocean. Monaco 36 (1959) 276–381.

[80] A.J. Southward, D.J. Crisp, Proc. Roy. Irish Acad. 57 (1954) 1–29.

[81] D.J. Crisp, A.J. Southward, J. Mar. Biol. Assoc. U.K. 37 (1958) 157–208.

[82] A.J. Southward, S.J. Hawkins, M.T. Burrows, J. Therm. Biol. 20 (1995) 127–155.

[83] M.A. Kendall, J. Moll. Stud. 53 (1987) 213–222.

[84] F.P. Lima, N. Queiroz, P.A. Ribeiro, S.J. Hawkins, A.M. Santos, J. Biogeogr. 33 (2006) 812–822.

[85] A.C. Iles, T.C. Gouhier, B.A. Menge, J.S. Stewart, A.J. Haupt, M.C. Lynch, Glob. Change Biol. 18 (2) (2012) 783–796.

[86] B.A. Menge, S.D. Hacker, T. Freidenburg, J. Lubchenco, R. Craig, G. Rilov, M. Noble, E. Richmond, Ecol. Monogr. 81 (3) (2011) 493–509.

[87] B. Helmuth, J.G. Kingsolver, E. Carrington, Biophysics, physiological ecology and climate change: does mechanism matter? Ann. Rev. Phys. 67 (2005) 177–201.

[88] C. Parmesan, S. Gaines, L. Gonzales, D.M. Kaufman, J. Kingsolver, A.T. Peterson, R. Sagarin, Empirical perspective on species borders: from traditional biogeography to global change, Oikos 108 (2005) 58–75.

[89] P. Moore, S.J. Hawkins, R.C. Thompson, Mar. Ecol. Prog. Ser. 334 (2007) 11–19.

[90] R.J.H. Herbert, S.J. Hawkins, M. Sheader, A.J. Southward, J. Mar. Biol. Assoc. U.K. 83 (2003) 73–82.

[91] C. Parmesan, G. Yohe, Nature 421 (2003) 37–42.

[92] I. Bartsch, C. Wiencke, K. Bischof, C.M. Buchholz, B.H. Buck, A. Eggert, P. Feuerpfeil, D. Hanelt, S. Jacobsen, R. Karez, U. Karsten, M. Molis, M.Y. Roleda, H. Schubert, R. Schumann, K. Valentin, F. Weinberger, J. Wiese, Eur. J. Phycol. 43 (2008) 1–86.

[93] C.N. Bianchi, Hydrobiologia 580 (2007) 7–21.

[94] P. Chevaldonné, C. Lejeusne, Ecol. Lett. 6 (2003) 371–379.

[95] D.S. Wethey, S.A. Woodin, Hydrobiologia 606 (2008) 139–151.

[96] T. Vance, MRES Thesis, University of Plymouth, 2005.

[97] F. Ardré, Port. Acta Biol. 10 (1970) 1–423.

[98] F. Ardré, Bull. Cent. Etud. Rech. Sci. Biarritz 8 (1971) 359–574.

[99] F.P. Lima, P.A. Ribeiro, N. Queiroz, S.J. Hawkins, A.M. Santos, Glob. Change Biol. 13 (12) (2007) 2592–2604.

[100] R.T. Lemos, H.O. Pires, Int. J. Climatol. 24 (2004) 511–524.

[101] J.M. Wesławski, M. Zajączkowski, J. Wiktor, M. Szymelfenig, Polar Biol. 18 (1997) 45–52.

[102] O. Salvigsen, Norsk Geogr. Tidsskr-Den 56 (2002) 56–61.

[103] J. Berge, G. Johnsen, F. Nilsen, B. Gulliksen, D. Slagstad, Mar. Ecol. Prog. Ser. 303 (2005) 167–175.

[104] C.A. Blanchette, C.M. Miner, P.T. Raimondi, D. Lohse, K.E.K. Heady, B.R. Broitman, J. Biogeogr 35 (2008) 1593–1607.

[105] B.R. Broitman, C.A. Blanchette, B.A. Menge, J. Lubchenco, P.A. Raimondi, C. Krenz, M. Foley, D. Lohse, S.D. Gaines, Ecol. Monogr 78 (2008) 403–421.

[106] R.S. Burton, Evolution 52 (1998) 734–745.

[107] M.E. Hellberg, D.P. Balch, K. Roy, Science 292 (2001) 1707–1710.

[108] J.P. Barry, C.H. Baxter, R.D. Sagarin, S.E. Gilman, Science 267 (1995) 672–674.

[109] T.J. Herrlinger, Veliger 24 (1981) 78.

[110] D. Zacherl, S.D. Gaines, S.I. Lonhart, J. Biogeogr. 30 (2003) 913–924.

[111] S.G. Jones, F.P. Lima, D.S. Wethey, J. Biogeogr. 37 (12) (2010) 2243–2259.

[112] L.G. Harris, M.C. Tyrrell, Biol. Invasions 3 (2001) 9–21.

[113] G.C.B. Poore, The state of the marine environment report for Australia technical annex: 1, Mar. Environ. (2001) 75–84.

[114] R. Tsuchi, S. Nishimura, A.G. Beu, Tectonophysics 281 (1997) 83–97.

[115] E.S. Poloczanska, R.C. Bobcock, A. Butler, A.J. Hobday, O. Hoegh-Guldberg, T.J. Kunz, R. Matear, D.A. Milton, T.A. Okey, A.J. Richardson, Ocenogr. Mar. Biol. Ann. Rev. 45 (2008) 407–478.

[116] S.P. Griffiths, Estuar. Coast. Shelf Sci. 58 (2003) 173–186.

[117] K.S. Edyvane, Final Report for Environment Australia, Department of Primary Industries, Water and Environment, Hobart, 2003.

[118] G.J. Edgar, C.R. Samson, N.S. Barrett, Conserv. Biol. 19 (2005) 1294–1300.

[119] C. Johnson, S. Ling, J. Ross, S. Shepherd, K. Miller, Tasmanian Aquaculture and Fisheries Institute, FRDC, Australia, 2005. Project No. 2001/004.

[120] N. Caputi, C. Chubb, A. Pearce, Mar. Fresh. Res. 52 (2001) 1167–1174.

[121] K.E. Trenberth, Bull. Am. Meterol. Soc. 78 (12) (1997) 2771–2777.

[122] J.E. Morton, V.J. Chapman, Rocky Shore Ecology of the Leigh Area, North Auckland, University of Auckland Press, Auckland, 1968, 44 pp.

[123] N. Mieszkowska, C. Lundquist, Biogeographical patterns in limpet abundance and assemblage composition in New Zealand. J. Exp. Mar. Biol. Ecol. 400 (2011) 155–166.

[124] L.E. Petes, B.A. Menge, G.D. Murphy, J. Exp. Mar. Biol. Ecol. 351 (2007) 83–91.

[125] G.M. Branch, S. Afr. J. Sci. 80 (1984) 61–65.

[126] R.T. Barber, F.P. Chavez, Science 222 (1983) 1203–2110.

[127] J.Y. Liu, PLoS One (2013) e50719.

[128] D.J. Crisp, J. Anim. Ecol. 33 (1964) 165–210.

[129] N. Mieszkowska, S.J. Hawkins, M.T. Burrows, M.A. Kendall, J. Mar. Biol. Assoc. U.K. 87 (2007) 537–545.

[130] D.S. Wethey, S.A. Woodin, T.J. Hilbish, S.J. Jones, F.P. Lima, P.M. Brannock, J. Exp. Mar. Biol. Ecol. 400 (2011) 132–144.

[131] L.B. Firth, A.M. Knights, S.S. Bell, J. Exp. Mar. Biol. Ecol. 400 (2011) 250–256.

[132] S.J. Jones, N. Mieszkowska, D.S. Wethey, Biol. Bull. 217 (2009) 73–85.

[133] L.B. Firth, G.A. Williams, J. Exp. Mar. Biol. Ecol. 375 (2009) 70–75.

[134] J.A. McGowan, D.B. Chelton, A. Conversi, CalCOFI Rep. 37 (1996).

[135] S.J. Holbrook, R.J. Schmitt, J.S.J. Stevens, Ecol. Appl. (1997) 7.

[136] A.J. Southward, J. Mar. Biol. Assoc. U.K. 71 (1991) 495–513.

[137] S.J. Hawkins, P. Moore, M. Burrows, E. Poloczanska, N. Mieszkowska, S.R. Jenkins, R.C. Thompson, M. Genner, A.J. Southward, Clim. Change Res. 37 (2008) 123–133.

[138] N. Mieszkowska, M.T. Burrows, S.J. Hawkins, J. Sea Res. (2014), http://dx.doi.org/10.1016/j.jmarsys.2012.11.008.

[139] E.S. Poloczanska, S.J. Hawkins, A.J. Southward, M.T. Burrows, Ecology (2008) 3138–3149.

[140] P. Moore, R.C. Thompson, S.J. Hawkins, J. Exp. Mar. Biol. Ecol. 344 (2007) 170–180.

[141] B.R. Broitman, P.L. Szathmary, K.A.S. Mislan, C.A. Blanchette, B. Helmuth, Oikos 118 (2009) 219–224.

[142] S. Pincebourde, E. Sanford, B. Helmuth, Am. Nat. 174 (2009) 890–897.

[143] M.J. Fortin, T.H. Keitt, B.A. Maurer, M.L. Taper, D.M. Kaufman, T.M. Blackburn, Oikos 108 (2005) 7–17.

[144] C.D. Harley, K.M. Anderson, K.W. Demes, J.P. Jorve, R.L. Kordas, T.A. Coyle, M.H. Graham, J. Phycol. 48 (2012) 1064–1078.

[145] R.N. Carter, S.D. Prince, Nature 293 (1981) 644–645.

[146] D.M. Lodge, in: P.M. Karieva, J.G. Kingsolver, R.B. Huey (Eds.), Biotic Interactions and Global Change, Sinauer, Sunderland, MA, 1993, pp. 367–387.

[147] J.J. Lennon, J.R.G. Turner, D. Connell, Oikos 78 (1997) 486–502.

[148] B.D. Russell, C.D.G. Harley, T. Wernberg, N. Mieszkowska, S. Widdicombe, J.M. Hall-Spencer, S.D. Connell, Biol. Lett. 8 (2011) 164–166.

[149] B.T. Helmuth, G.E. Hofmann, Biol. Bull. 201 (2001) 371–381.

[150] K.A. Mislan, B. Helmuth, D.S. Wethey, Glob. Ecol. Biogeogr. 23 (2014) 744–756.

[151] B. Helmuth, J. Exp. Biol. 212 (2009) 753–760.

[152] N.P. Burnett, R. Seabra, M. de Pirro, D.S. Wethey, S.A. Woodin, B. Helmuth, F.P. Lima, Limnol. Oceanogr. Methods 11 (2013) 91–100.

[153] M.W. Denny, L.P. Miller, C.D. Harley, J. Exp. Biol. 209 (2006) 2420–2431.

[154] F.P. Lima, D.S. Wethey, Limnol. Oceanogr. Methods 7 (2009) 347–353.

[155] M.A. Kearney, A. Matzelle, B. Helmuth, J. Exp. Biol. 215 (2012) 922–933.

[156] D.J. Marshall, Y.W. Dong, C.D. McQuaid, G.A. Williams, J. Exp. Biol. 214 (2011) 3649–3657.

[157] G.D. Han, S. Zhang, D.J. Marshall, C.H. Ke, Y.W. Dong, J. Exp. Biol. 216 (2013) 3273–3282.

[158] R.A. Boolootian, in: R.A. Boolootian (Ed.), Physiology of Echinodermata, Wiley, New York, 1966, pp. 561–614.

[159] B. Cocanour, K. Allen, Comp. Biochem. Physiol. 20 (1967) 327–331.

[160] R.E. Stephens, Biol. Bull. 142 (93) (1972) 489–504.

[161] P.W. Frank, Mar. Biol. 31 (1975) 181–192.

[162] G. Sarà, M.A. Kearney, B. Helmuth, Chem. Ecol. 27 (2011) 135–145.

[163] G. Sarà, A. Rinaldi, V. Montalto, Mar. Ecol. (2014), http://dx.doi.org/10.1111/maec.12106.

[164] A. Matzelle, V. Montalto, G. Sarà, M. Zippay, B. Helmuth, Dynamic Energy Budget model parameter estimation for the bivalve Mytilus californianus: Application of the covariation method. J. Sea Res. 94 (2014) 105–110.

[165] C.J. Monaco, D.S. Wethey, B. Helmuth, PLoS One 9 (2014), http://dx.doi.org/10.1371/journal.pone.0104658.

[166] S.D. Connell, B.D. Russell, Proc. R. Soc. B (2010) rspb20092069.

[167] B.D. Russell, S.D. Connell, H.S. Findlay, K. Tait, S. Widdicombe, N. Mieszkowska, Phil. Trans. R. Soc. B 368 (2013) 20120438.

[168] J. Brodie, C.J. Williamson, D.A. Smale, N.A. Kamenos, N. Mieszkowska, R. Santos, M. Cunliffe, J.M. Hall-Spencer, Ecol. Evol. 4 (2014) 2787–2798.

[169] A. Querios, J.A. Fernandez, S. Faulwetter, J. Nunes, S.P.S. Rastrick, N. Mieszkowska, Y. Aritoli, P. Calosi, C. Arvanitidis, H.S. Findlay, M. Brange, W.W.L. Cheung, S. Widdicombe, Glob. Change Biol. (2014), http://dx.doi.org/10.1111/gcb.12675.

[170] IPCC, Summary for policymakers, in: Climate Change 2013: The Physical Science Basis. Contribution of Working Group I to the Fifth Assessment Report of the Intergovernmental Panel on Climate Change, 2013, pp. 1–28.

[171] L. Airoldi, M. Abbiati, M.W. Beck, S.J. Hawkins, P.R. Jonsson, D. Martin, P.S. Moschella, A. Sundelöf, R.C. Thompson, P. Åberg, Coastal Eng. 52 (2005) 1073–1087.

[172] P.M. Moschella, M. Abbiati, P. Åberg, L. Airoldi, J. Anderson, J.M. Bacchiocchi, F. Bulleri, G.E. Dinesen, M. Frost, E. Gacia, L. Granhag, P.R. Jonssonn, M.P. Satta, A. Sundelöf, R.C. Thompson, S.J. Hawkins, Coastal Eng. 52 (2005) 1053–1071.

[173] S.J. Hawkins, N. Mieszkowska, P. Moschella, unpublished data.

[174] J.J. Stachowicz, J.R. Terwin, R.B. Whitlatch, R.W. Osman, PNAS 99 (2002) 15497–15500.

[175] Southward, Crisp, unpublished.

[176] Ballantine, unpublished.

PLANT ECOLOGY

15

Mike D. Morecroft[1], Sally A. Keith[2]
[1]*Natural England, Worcester, UK;* [2]*Center for Macroecology, Evolution and Climate,*
Natural History Museum of Denmark, Copenhagen, Denmark

CHAPTER OUTLINE

1. INTRODUCTION

The distribution of types of vegetation around the world is clearly related to climate. Different combinations of temperature, rainfall and seasonality produce the global variety of biomes, from rainforest to tundra, that we take for granted. At a finer scale we can see changes in vegetation with more localized changes in climate such as on a mountain as conditions become cooler with altitude [1,2]. Individual species also have distribution patterns whose boundaries, at a global scale, are largely defined by climate. These distribution patterns reflect the influence of climate on plant survival, physiology and growth, together with climatic effects on ecological interactions, such as competition, pollination and herbivory. Different types of plants are adapted to different climatic conditions, from cold-tolerant, but slow-growing, alpine plants, to fast-growing trees in the wet tropics. It is therefore reasonable to expect that changes in climate would lead to a change in species distributions and community composition. Evidence of such changes has been accumulating at an accelerating rate in recent decades [3–5]. However, before we come to evaluate this evidence, we should consider some general principles.

To identify the ecological impacts of anthropogenic climate change with confidence, it is necessary both to be able to detect a change in an ecosystem and to reliably attribute it to a change in climate [5–7]. *Detection* of any change in an environmental variable requires a reliable data set with repeated measurements over a period of time. Good instrumental records of climate itself go back over 100 years in many countries, but very few biological data sets extend this far. In many cases climate

Climate Change. http://dx.doi.org/10.1016/B978-0-444-63524-2.00015-4

change impacts must be inferred from resurveys of early work carried out for quite different purposes. *Attribution* of impacts to climate change requires a relationship between climate and impact variables to be established and other potential causes of change ruled out. The effects of climate change on plants are complex (Fig. 1) and the presence or absence of a species from a particular location does not solely depend on its ability to tolerate physical conditions. In many cases climatic limits are determined by the influence of climate on a plant's ability to compete with other species [8]. Climate change may also disturb interactions between plants and their pollinators, mycorrhizae, herbivores or pathogens. Rising temperatures are the best understood aspects of climate change, but in the longer term, changes in precipitation or one-off extreme events, which are harder to predict, may be more important. Changing atmospheric composition, including carbon dioxide concentration, can also have effects on plant performance and interactions [9a,9b].

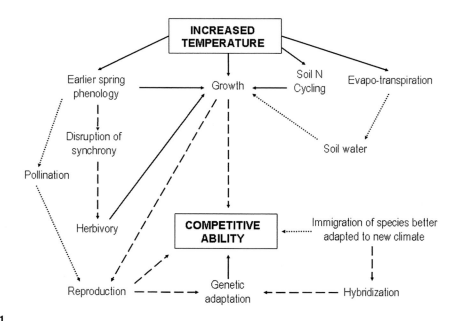

FIGURE 1

An example of complexity in plant responses to climate change. Factors influencing the effects of temperature on the competitive ability of a temperate, insect-pollinated plant. Note that this is a simplified diagram and does not take account of all factors, interactions or the role of other climate change factors such as changes in precipitation or extreme events. Solid arrows indicate positive effects, dotted arrows indicate negative effects and dashed arrows indicate effects that could be positive or negative.

A further issue is that many plant communities are composed of long-lived, slow dispersing species and only change slowly in response to incremental changes in climate [10,11]. This contrasts with many invertebrate species – most of these species have short generation times and in many cases high mobility leading to more rapid changes in distribution and community structure [12].

2. CHANGES IN PHENOLOGY

The recording of phenology, which is the seasonal timing of biological events such as leafing and flowering, provides several examples of unusually long-term data sets. A particularly good example is the Marsham family records for 'indications of spring' concerning over 20 plant and animal species for 200 a (years) in Norfolk, UK [13]. Analyses of these data, particularly correlations with equally lengthy climate data, have provided important information of past effects of climate on phenological events, which in turn have been used to anticipate future responses of these species to climate change.

The Marsham data formed a component of a much larger, European-wide meta-analysis of the relationship between phenology and temperature [14]. The meta-analysis included, inter alia, phenological trend data for 542 plant species from 21 countries. There was a clear correlation between warmer temperatures and the earlier onset of spring phenology (leaf opening and flowering) in 78% of plants (31% significantly). In contrast, autumn onset indicators were more ambiguous, showing no overall pattern of correlation with temperature, although some individual events did correlate with temperature. The paper demonstrated a mean advance in spring and summer phenology of 2.5 days per decade in the latter part of the twentieth century. Cook et al. [15] looked at two large, independent data sets for the northern hemisphere; they concluded that the patterns were broadly similar in showing advancing spring phenology but noted that the strength of the response was lower in locations with higher mean temperatures (for example lower latitudes).

At a global scale, ground-based observations of phenology tend to be concentrated in the northern hemisphere and temperate regions and close to centres of population. However, remote sensing allows a larger-scale assessment. Indices based on the spectral composition of light reflected from the surface of the earth, such as the normalized difference vegetation index, can identify the development and loss of a green canopy through the growing season across the temperate zones. These techniques have corroborated surface-based findings of a lengthening of the growing season through an earlier onset of leafing and delayed senescence [16]. Recent studies have added details about regional and temporal variations to this general pattern [17,18].

Phenology therefore provides a clear indicator of climate change impacts on plants. In itself, a change in phenology is arguably not a major issue if the species continues to persist in a current location. However, there is evidence that the changing phenology of a species can have important ecological consequences for pollinators [19], herbivores [20] and competitors [21]. Thackeray et al. [22] looked at a large number of taxa from the UK, representing different trophic levels and covering marine, terrestrial and freshwater habitats. This also showed a broad consistency in that the majority of spring and summer events were advanced. There were, however, differences in the extent of the advance in timing; in particular, secondary consumers tended to show less advance than lower trophic levels. There is, therefore, a risk that interactions within food webs will break down, for example animal species not reproducing when food supply is at its maximum. This may in turn have complex impacts for whole ecosystems.

3. CHANGES IN DISTRIBUTION

Alongside phenology, the most frequently reported changes in plant ecology in response to climate change are changes in species' geographical distributions. The mapping of distributions of species and vegetation types, whether local, national or international in scale, predates contemporary interest in climate change by several decades or more. Resurveys of distributions provide an opportunity to test whether changes consistent with the impacts of climate change are taking place [12,23,24]. Studies of this sort have been an important component of the impacts reported in the Intergovernmental Panel on Climate Change (IPCC) assessments [5,7].

Good examples of changes in distribution can be seen in altitudinal studies. Temperatures typically fall with altitude by approximately $6.5°C$ km^{-1} [25], although this varies with other factors such as humidity. Plant communities consequently change markedly with altitude. The clearest example of this is the presence of tree lines, beyond which trees do not grow. Many explanations for the occurrence of tree lines have been offered, but plants are thought to respond to combinations of temperature change, atmospheric CO_2 concentration, nutrient availability and solar radiation [26]. Regardless of the exact mechanism, which may vary between situations, natural tree lines (those not changed by forest management) are determined primarily by climate, particularly temperature. A warming of climate would therefore be expected to lead to tree lines shifting to higher altitudes. Evidence of this has been found with tree lines shifting at rates of 0.01 m a^{-1}–7.5 m a^{-1}, depending on the species of tree involved and the type of climatic forcing [27]. The length of data collection is also likely to affect the mean shift each year (in this and other variables) because longer data sets will be subject to a smoothing of the trend through natural variation and sign switching. Latitudinal tree line shifts have also been observed, correlated with warmer summer temperatures [26].

Tree line shifts are subject to time lags in their response to environmental change because of their long generation time, therefore, changes in nonwoody plants and dwarf shrubs might be expected to be more sensitive. Evidence of changes in altitudinal distribution have been found for alpine plants [23]. In a resurvey of vascular plants in the Alps of northern Italy, 52 of the 93 monitored vascular plants were found at a higher altitude than in the 1950s, moving upward at a rate of 23.9 m per decade [28]. The largest change in species richness was at an altitude that had experienced melting of permafrost, associated with increasing air temperature. Whilst data from the Alps were some of the earliest and clearest examples of this sort of trend, similar results have been reported more widely in Europe [29] and for other mountainous regions, including, for example, the Himalayas [30].

4. COMMUNITY COMPOSITION

Changes in distribution patterns are dependent on local extinctions and colonizations at species range margins. As this is where the effects of climate change are most likely to be seen first, they provide a sensitive early indicator of climate change impacts.

The appearance of a few individuals of species typical of warmer locations does not necessarily indicate a shift in the character of a whole community. Evidence of community scale changes has lagged behind changes in distribution but the literature is beginning to grow. It can also be challenging to separate the effects of climate from those of other variables. Reliable detection of a change in the balance between different species in unmanipulated communities can most reliably be achieved through long-term monitoring in permanent sample plots. Many studies of this sort have only recently started to have long enough time series to be detected. However, there are some longer-term examples from repeat recording of earlier surveys. Mountain and upland plant communities again provide some of the clearest examples of change. A good example is that of Britton et al. [31] who showed that vegetation of the Scottish Highlands changed between 1963 and 1987 with specialist northern and alpine species decreasing and lowland generalist species increasing. The net effect was a homogenization with vegetation becoming more similar across the altitudinal range. Mountain plant communities have been observed to change in ways consistent with the effects of warming over less than a decade in recent years. Gottfried et al. [32] showed a shift toward species typical of warmer climates in plots at high altitude in a network of monitoring plots across Europe, with effects on the overall species richness of the vegetation. In another study on the same data sets, Pauli et al. [3] showed that diversity was tending to increase in boreal temperate regions as a result of new colonizations but decreasing in Mediterranean regions as species were lost, with no new colonists to replace them. This sort of difference between community responses in different locations is an important aspect of climate change impacts and highlights the need to avoid oversimplistic generalizations.

Mountain areas also show some of the clearest effects in forest communities. Bertrand et al. [11] showed that in a large data set from France, change in highland (>500 m altitude) forests communities from 1986 to 2008 was equivalent to a difference of 0.54°C in terms of the mean distribution of species. In contrast, lowland forest changes were equivalent to only 0.02°C change in temperature. As the temperature rise in this period was just over 1°C, there is a lag in both highland and lowland communities but the lag is much greater in the lowlands. This reflects the larger distance that species need to disperse over in the lowlands to track their shifting climatic niches.

Changes in precipitation are likely with climate change but are subject to greater uncertainties than rising temperature and large local and regional variation. There are well-documented examples of the sensitivity of vegetation to precipitation. One particular area of concern is the potential impact of more frequent and/or intense droughts in the Amazon region. The Amazon Forest Inventory Network (RAINFOR) coordinates the monitoring of long-term forest plots throughout the Amazon rainforest [33]. Butt et al. [34] have recently shown a shift in the composition of these forest plots toward an increase in species associated with drier conditions; this may increase the chances of forest persistence in the face of drier conditions in the future. The RAINFOR plots have also provided evidence for a change in community composition of old-growth Amazonian forest, whereby slow-growing tree genera are decreasing and fast-growing tree genera are increasing in dominance or density [33,35]. There has also been an increase in density and dominance of lianas within these forests. These changes have been attributed with relative confidence to an increase in atmospheric CO_2 concentration [33].

Evidence of the responsiveness of vegetation to changing precipitation patterns has also been found in temperate ecosystems. In the UK, the Environmental Change Network is a good example of a programme in which plant community composition is monitored in permanently marked quadrats [36] (www.ecn.ac.uk). In this network vegetation quadrats form part of a larger ecosystem monitoring programme in which animal populations are also monitored, together with climate, soil nutrients and water content and other potential causes of change such as air pollution. This demonstrated a change in species composition of grasslands, specifically an increase in ruderal species in response to drought [37,38]. Ruderal plants are those that grow and reproduce quickly – they colonized gaps that opened up in the grassland in response to drought, before being excluded by competitors as wetter conditions returned. Drought can alter the course of forest dynamics for decades, as a study of Lady Park Wood in the UK has shown [39].

As we have noted, changes in plant communities detected in long-term monitoring can be hard to link to specific causes. Field experiments in which the climate is directly manipulated provide an opportunity to test the effects of different variables. One of the longest-running examples is an experiment in the subalpine zone of the Rocky Mountains (USA), where vegetation has been warmed using infrared lamps since 1990. The shrub, *Artemisia tridentata* (sagebrush), has increased in response to this treatment and herbaceous species have declined [40–43]. In this case, the effect of warming is mediated through a reduction in summer water availability as a result of earlier snowmelt. Another long-running field experiment at Buxton in Derbyshire, UK, has demonstrated that an old calcareous grassland community is very resistant to change, even after many years of increased temperature and drought treatment [10]. This contrasts with the much greater sensitivity of a more recent calcareous grassland [44,45]. Fridley et al. [46] showed that the large heterogeneity of soil depth on the old grassland site allowed small-scale species turnover, with species shifting between different microsites according to where conditions were most suitable for them. The factors that promote resilience in different communities are the subject of ongoing discussion and research [47].

Communities may shift from one recognizable type to another or to a degraded version of the present one. However, in the longer term novel combinations of species are likely to develop as species respond to changing climate at different rates and with complex interactions within the community and with other variables [48]. Paleoecology provides evidence of this happening during previous climate change events, indicating the formation of non-analogous communities that were of a different composition from anything currently recognized [49]. Where species are able to establish and spread outside their historic range, this can lead to radically different communities. A striking example of this is the establishment of palm trees within some forests in Switzerland [50]. This will have important implications for the functioning of communities and ecosystems and present challenges where current conservation policy is based on defined, historical communities.

5. PLANT GROWTH

Any change in species distribution or community composition is likely to be preceded by a change in plant growth. Plant growth may therefore be a sensitive indicator of climate change impacts. It is also of interest in its own right as it drives the production of food, materials and fuel and is responsible for the sequestration of carbon. The two main categories of plants whose growth is measured are crops and trees, and we will consider tree growth here.

The growth response of trees, as well as other plants, to climate differs between species, depending on their ecophysiology and life history characteristics. For example, Morecroft et al. [51] showed that the growth of sycamore (*Acer pseudoplatanus*) was adversely affected by drought to a greater extent than pedunculate oak (*Quercus robur*) and ash (*Fraxinus excelsior*). This was associated with reduced photosynthetic rates in dry soil conditions and may reflect relatively shallow rooting. Broadmeadow et al. [52] modelled broadleaved tree species' growth responses to future climate using a model based on empirical data for species-specific growth rates and their correlations with aspects of climate. They found that water limitation in southern England was likely to lead to reductions in growth and increased mortality, with beech (*Fagus sylvatica*) the worst affected.

One of the areas in which tree growth rates has been a particular subject of research interest has been the Amazon rainforest, with the RAINFOR network of old-growth forest plots again providing long-term observational evidence of changes [33]. The plots have shown evidence of an increase in growth rates and biomass in recent decades. More importantly when considering the carbon sink function of the Amazon forest, there is also an increase in turnover of tropical forest trees that is thought to be a function of increased mortality following more rapid growth. These responses are, like the changing communities' composition, most parsimoniously explained as a response to higher CO_2 concentrations, possibly combined with nutrient enrichment resulting from ash deposition from an increasing number of forest fires. Lewis et al. [53] reviewed results from long-term monitoring plot studies throughout the tropics and found a general increase in productivity and turnover of individual trees, which was best explained by rising CO_2 concentrations. There is also now good evidence that drought in the Amazon basin reduces carbon sequestration [54].

The trunks of most temperate and some tropical trees have annual rings, reflecting seasonal differences in growth rates. These provide a particularly valuable historical record of growth rates and are often used as proxies for the estimation of past climates. Tree ring data are useful indicators of climate change because they provide a 'self-kept' record of climate response over the lifetime of an individual tree, thereby circumventing the challenge of obtaining long-term monitoring data. Width of tree rings can be correlated with environmental data. In addition to assessing general trends in tree growth to trends in climate, tree rings are very useful for examining the response of trees to extreme climatic events. A reduction in productivity demonstrated by reduced tree ring width of old beech forests in Italy has been attributed to drought during the growing season [55].

A further strength of the use of tree ring data as an indicator of climate change is that changes can be explored in the context of a longer timeframe, potentially increasing our understanding of current trends. Touchan et al. [56] analyzed tree ring records from North West Africa for approximately the last 600 a to ascertain the influence of drought and found that drought between 1999 and 2002 was probably the most severe since the fifteenth century and consistent with projections from global circulation models.

6. CONCLUSIONS

For some years there has been clear evidence that plants are responding to climate change through changing phenology and distribution patterns, with species tending to disperse toward cooler areas. More far-reaching changes in community composition are now also emerging. Responses to temperature have been clearest to date at a global scale, but in the long term, local changes in precipitation or extreme events may be more important than the global trend in temperatures. There are also likely to be complex interactions within ecosystems and with other pressures, which we need to understand and model if attempts to mitigate climate change and adapt to it are to be successful.

REFERENCES

[1] B. Sieg, F.J.A. Daniels, Phytocoenologia 35 (2005) 887–908.

[2] A. Hemp, Plant Ecol. 184 (2006) 27–42.

[3] H. Pauli, M. Gottfried, S. Dullinger, O. Abdaladze, M. Akhalkatsi, J.L. Benito Alonso, G. Coldea, J. Dick, B. Erschbamer, R. Fernandez Calzado, D. Ghosn, J.I. Holten, R. Kanka, G. Kazakis, J. Kollar, P. Larsson, P. Moiseev, D. Moiseev, U. Molau, J. Molero Mesa, L. Nagy, G. Pelino, M. Puscas, G. Rossi, A. Stanisci, A.O. Syverhuset, J.-P. Theurillat, M. Tomaselli, P. Unterluggauer, L. Villar, P. Vittoz, G. Grabherr, Science 336 (2012) 353–355.

[4] G.-R. Walther, Phil. Trans. R. Soc. B Biol. Sci. 365 (2010) 2019–2024.

[5] J. Settele, R. Scholes, R. Betts, S.E. Bunn, P. Leadley, D. Nepstad, J.T. Overpeck, M.A. Taboada, in: C.B. Field, V.R. Barros, D.J. Dokken, K.J. Mach, M.D. Mastrandrea, T.C. Bilir, M. Chatterjee, K.L. Ebi, Y.O. Estrada, R.C. Genova, B. Girma, E.S. Kissel, A.N. Levy, S. MacCracken, P.R. Mastrandrea, L.L. White (Eds.), Climate Change 2014: Impacts, Adaptation, and Vulnerability. Part A: Global and Sectoral Aspects. Contribution of Working Group II to the Fifth Assessment Report of the Intergovernmental Panel of Climate Change, Cambridge University Press, Cambridge, UK and New York, NY, USA, 2014, pp. 271–359.

[6] G.F. Midgley, S.L. Chown, B.S. Kgope, S. Afr. J. Sci. 103 (2007) 282–286.

[7] C. Rosenzweig, G. Casassa, D.J. Karoly, A. Imeson, C. Liu, A. Menzel, S. Rawlins, T.L. Root, B. Seguin, P. Tryjanowski, Clim. Change (2007) 79–131.

[8] M. Morecroft, J. Paterson, Plant Growth Clim. Change (2006) 146–164.

[9] [a] A.D.B. Leakey, E.A. Ainsworth, C.J. Bernacchi, A. Rogers, S.P. Long, D.R. Ort, J. Exp. Bot. 60 (2009) 2859–2876.
[b] S.D. Smith, T.E. Huxman, S.F. Zitzer, T.N. Charlet, D.C. Housman, J.S. Coleman, L.K. Fenstermaker, J.R. Seemann, R.S. Nowak, Nature 408 (2000) 79–82.

[10] J.P. Grime, J.D. Fridley, A.P. Askew, K. Thompson, J.G. Hodgson, C.R. Bennett, Proc. Natl. Acad. Sci. U.S.A. 105 (2008) 10028–10032.

[11] R. Bertrand, J. Lenoir, C. Piedallu, G. Riofrio-Dillon, P. de Ruffray, C. Vidal, J.-C. Pierrat, J.-C. Gegout, Nature 479 (2011) 517–520.

[12] I.C. Chen, J.K. Hill, R. Ohlemueller, D.B. Roy, C.D. Thomas, Science 333 (2011) 1024–1026.

[13] T.H. Sparks, P.D. Carey, J. Ecol. 83 (1995) 321–329.

[14] A. Menzel, T.H. Sparks, N. Estrella, E. Koch, A. Aasa, R. Ahas, K. Alm-Kuebler, P. Bissolli, O.g. Braslavska, A. Briede, F.M. Chmielewski, Z. Crepinsek, Y. Curnel, A. Dahl, C. Defila, A. Donnelly, Y. Filella, K. Jatcza, F. Mage, A. Mestre, O. Nordli, J. Penuelas, P. Pirinen, V. Remisova, H. Scheifinger, M. Striz, A. Susnik, A.J.H. Van Vliet, F.-E. Wielgolaski, S. Zach, A. Zust, Glob. Change Biol. 12 (2006) 1969–1976.

[15] B.I. Cook, E.M. Wolkovich, T.J. Davies, T.R. Ault, J.L. Betancourt, J.M. Allen, K. Bolmgren, E.E. Cleland, T.M. Crimmins, N.J.B. Kraft, L.T. Lancaster, S.J. Mazer, G.J. McCabe, B.J. McGill, C. Parmesan, S. Pau, J. Regetz, N. Salamin, M.D. Schwartz, S.E. Travers, Ecosystems 15 (2012) 1283–1294.

[16] H.W. Linderholm, Agric. Forest Meteorol. 137 (2006) 1–14.

[17] S.J. Jeong, C.H. Ho, H.J. Gim, M.E. Brown, Glob. Change Biol. 17 (2011) 2385–2399.

[18] C. Jeganathan, J. Dash, P.M. Atkinson, Remote Sens. Environ. 143 (2014) 154–170.

[19] J. Memmott, P.G. Craze, N.M. Waser, M.V. Price, Ecol. Lett. 10 (2007) 710–717.

[20] E. Post, M.C. Forchhammer, Phil. Trans. Roy. Soc. B Biol. Sci. 363 (2008) 2369–2375.

[21] E.E. Cleland, N.R. Chiariello, S.R. Loarie, H.A. Mooney, C.B. Field, Proc. Natl. Acad. Sci. U.S.A. 103 (2006) 13740–13744.

[22] S.J. Thackeray, T.H. Sparks, M. Frederiksen, S. Burthe, P.J. Bacon, J.R. Bell, M.S. Botham, T.M. Brereton, P.W. Bright, L. Carvalho, T. Clutton-Brock, A. Dawson, M. Edwards, J.M. Elliott, R. Harrington, D. Johns, I.D. Jones, J.T. Jones, D.I. Leech, D.B. Roy, W.A. Scott, M. Smith, R.J. Smithers, I.J. Winfield, S. Wanless, Glob. Change Biol. 16 (2010) 3304–3313.

[23] G.R. Walther, S. Beissner, C.A. Burga, J. Vegetat. Sci. 16 (2005) 541–548.

[24] A.E. Kelly, M.L. Goulden, Proc. Natl. Acad. Sci. U.S.A. 105 (2008) 11823–11826.

[25] C.D. Whiteman, Mountain Meteorology: Fundamentals and Applications, Pacific Northwest National Laboratory, Richland, WA, USA, 2000.

[26] J. Grace, F. Berninger, L. Nagy, Ann. Bot. 90 (2002) 537–544.

[27] G.R. Walther, Perspect. Plant Ecol. Evol. Syst. 6 (2004) 169–185.

[28] G. Parolo, G. Rossi, Basic Appl. Ecol. 9 (2008) 100–107.

[29] J. Lenoir, J.C. Gegout, P.A. Marquet, P. de Ruffray, H. Brisse, Science 320 (2008) 1768–1771.

[30] Y. Telwala, B.W. Brook, K. Manish, M.K. Pandit, PLoS One 8 (2013).

[31] A.J. Britton, C.M. Beale, W. Towers, R.L. Hewison, Biol. Conserv. 142 (2009) 1728–1739.

[32] M. Gottfried, H. Pauli, A. Futschik, M. Akhalkatsi, P. Barancok, J.L. Benito Alonso, G. Coldea, J. Dick, B. Erschbamer, M.R. Fernandez Calzado, G. Kazakis, J. Krajci, P. Larsson, M. Mallaun, O. Michelsen, D. Moiseev, P. Moiseev, U. Molau, A. Merzouki, L. Nagy, G. Nakhutsrishvili, B. Pedersen, G. Pelino, M. Puscas, G. Rossi, A. Stanisci, J.-P. Theurillat, M. Tomaselli, L. Villar, P. Vittoz, I. Vogiatzakis, G. Grabherr, Nat. Clim. Change 2 (2012) 111–115.

[33] O.L. Phillips, S.L. Lewis, T.R. Baker, K.-J. Chao, N. Higuchi, Phil. Trans. R. Soc. B Biol. Sci. 363 (2008) 1819–1827.

[34] N. Butt, Y. Malhi, M. New, M.J. Macia, S.L. Lewis, G. Lopez-Gonzalez, W.F. Laurance, S. Laurance, R. Luizao, A. Andrade, T.R. Baker, S. Almeida, O.L. Phillips, Plant Ecol. Divers. 7 (2014) 267–279.

[35] W.F. Laurance, A.A. Oliveira, S.G. Laurance, R. Condit, H.E.M. Nascimento, A.C. Sanchez-Thorin, T.E. Lovejoy, A. Andrade, S. D'Angelo, J.E. Ribeiro, C.W. Dick, Nature 428 (2004) 171–175.

[36] J. Sykes, A. Lane, The United Kingdom Environmental Change Network: Protocols for Standard Measurements at Terrestrial Sites, The Stationery Office, 1996.

[37] M.D. Morecroft, C.E. Bealey, E. Howells, S. Rennie, I.P. Woiwod, Glob. Ecol. Biogeogr. 11 (2002) 7–22.

[38] M.D. Morecroft, C.E. Bealey, D.A. Beaumont, S. Benham, D.R. Brooks, T.P. Burt, C.N.R. Critchley, J. Dick, N.A. Littlewood, D.T. Monteith, W.A. Scott, R.I. Smith, C. Walmsey, H. Watson, Biol. Conserv. 142 (2009) 2814–2832.

[39] L. Cavin, E.P. Mountford, G.F. Peterken, A.S. Jump, Funct. Ecol. 27 (2013) 1424–1435.

[40] J. Harte, R. Shaw, Science 267 (1995) 876–880.

[41] J. Harte, Bioscience 51 (2001) 332–333.

[42] T. Perfors, J. Harte, S.E. Alter, Glob. Change Biol. 9 (2003) 736–742.

[43] F. Saavedra, D.W. Inouye, M.V. Price, J. Harte, Glob. Change Biol. 9 (2003) 885–894.

[44] M. Sternberg, V.K. Brown, G.J. Masters, I.P. Clarke, Plant Ecol. 143 (1999) 29–37.

[45] M.D. Morecroft, G.J. Masters, V.K. Brown, I.P. Clarke, M.E. Taylor, A.T. Whitehouse, Funct. Ecol. 18 (2004) 648–655.

[46] J.D. Fridley, J.P. Grime, A.P. Askew, B. Moser, C.J. Stevens, Glob. Change Biol. 17 (2002–2011).

[47] M.D. Morecroft, H.Q.P. Crick, S.J. Duffield, N.A. Macgregor, J. Appl. Ecol. 49 (2011) 547–551.

[48] S.A. Keith, A.C. Newton, R.J.H. Herbert, M.D. Morecroft, C.E. Bealey, J. Nat. Conserv. 17 (2009) 228–235.

[49] J.W. Williams, S.T. Jackson, Front. Ecol. Environ. 5 (2007) 475–482.

[50] G.-R. Walther, E.S. Gritti, S. Berger, T. Hickler, Z. Tang, M.T. Sykes, Glob. Ecol. Biogeogr. 16 (2007) 801–809.

[51] M.D. Morecroft, V.J. Stokes, M.E. Taylor, J.I.L. Morison, Forestry 81 (2008) 59–74.

[52] M.S.J. Broadmeadow, D. Ray, C.J.A. Samuel, Forestry 78 (2005) 145–161.

[53] S.L. Lewis, J. Lloyd, S. Sitch, E.T.A. Mitchard, W.F. Laurance, Annu. Rev. Ecol. Evol. Syst. 40 (2009) 529–549.

[54] O.L. Phillips, L.E.O.C. Aragao, S.L. Lewis, J.B. Fisher, J. Lloyd, G. Lopez-Gonzalez, Y. Malhi, A. Monteagudo, J. Peacock, C.A. Quesada, G. van der Heijden, S. Almeida, I. Amaral, L. Arroyo, G. Aymard, T.R. Baker, O. Banki, L. Blanc, D. Bonal, P. Brando, J. Chave, A.C. Alves de Oliveira, N.D. Cardozo, C.I. Czimczik, T.R. Feldpausch, M.A. Freitas, E. Gloor, N. Higuchi, E. Jimenez, G. Lloyd, P. Meir, C. Mendoza, A. Morel, D.A. Neill, D. Nepstad, S. Patino, M. Cristina Penuela, A. Prieto, F. Ramirez, M. Schwarz, J. Silva, M. Silveira, A.S. Thomas, H. ter Steege, J. Stropp, R. Vasquez, P. Zelazowski, E. Alvarez Davila, S. Andelman, A. Andrade, K.-J. Chao, T. Erwin, A. Di Fiore, E. Honorio C, H. Keeling, T.J. Killeen, W.F. Laurance, A. Pena Cruz, N.C.A. Pitman, P. Nunez Vargas, H. Ramirez-Angulo, A. Rudas, R. Salamao, N. Silva, J. Terborgh, A. Torres-Lezama, Science 323 (2009) 1344–1347.

[55] G. Piovesan, F. Biondi, A. Di Filippo, A. Alessandrini, M. Maugeri, Glob. Change Biol. 14 (2008) 1265–1281.

[56] R. Touchan, K.J. Anchukaitis, D.M. Meko, S. Attalah, C. Baisan, A. Aloui, Geophys. Res. Lett. 35 (2008) L13705.

RISING SEA LEVELS

16

Roland Gehrels

Department of Environment, University of York, York, UK

CHAPTER OUTLINE

1. INTRODUCTION

The Fifth Assessment Report (AR5) of the Intergovernmental Panel on Climate Change (IPCC) [1] represent a significant step forward for sea level studies since the previous Fourth Assessment Report (AR4) [2]. One of the major shortcomings of AR4, the inability to adequately model dynamical ice sheet processes, has been addressed by studies that are summarized in the new report. In addition, there has been an improvement in the understanding of the physical processes that drive sea level changes. Process-based models increasingly agree with observations of sea level change, so that the sea level 'budget' (i.e. the sum of the individual contributions explained by individual processes compared to the total observed global sea level change) appears to be closed within uncertainty, at least over the past 25 years; this gives us more confidence in sea level predictions.

2. IS SEA LEVEL RISING?

Although this question appears almost rhetorical and can at first glance be answered with a resounding 'yes', the direction of sea level changes, positive or negative, depends on the timescale of observations and the spatial scale under consideration. This is obvious on very small timescales, e.g. tidal fluctuations. However, sea level changes are also highly variable in time and space on decadal timescales. Satellite observations since the early 1990s have revealed the complex regional patterns of sea level changes (Fig. 1(a)). Linear trends over the period 1993–2014 show that global sea level is rising *on average* at a rate of about 3.2 mm a^{-1}. However, regional variability is very significant. Some parts of

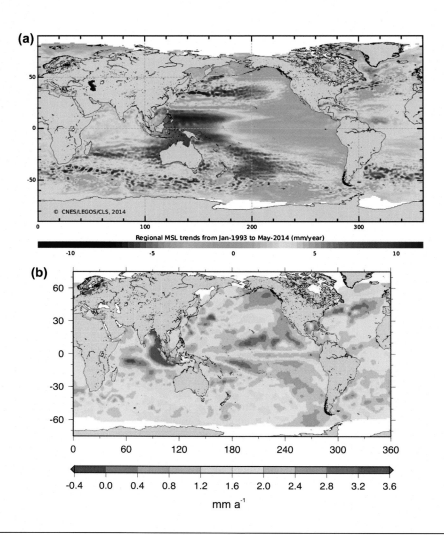

FIGURE 1

(a) Regional variations in mean sea level change/(mm a^{-1}) plotted along the y-axis from 1993 to 2014 based on satellite altimetry data from Aviso (http://www.aviso.oceanobs.com). (b) Regional variations in sea level changes from 1955 to 2003/(in mm a^{-1}) plotted along the y-axis. *(Reproduced from Fig. 5.16 in Ref. [26]. Updated from Ref. [5].)*

the world's oceans have experienced rates of sea level rise that far exceed the global average (>10 mm a^{-1} in places), while in others sea level has fallen. When averaged over the period 1955–2003 (Fig. 1(b)), linear rates are an order of magnitude smaller. The pattern, which is derived from tide-gauges, satellite observations and modelled reconstructions, is highly complex. Many areas have been subjected to sea level rise, but there are places, most notably in the Indian Ocean and the tropical Pacific, where sea level rise has been well below the global average, and in a few places sea level has fallen, albeit by a small amount. This picture does not take account of land-level movements and displacement of the ocean floor, which in many coastal locations need to be added to, or subtracted from, the mean sea level change to derive a figure that represents the relative change at a coastline. This is, after all, the number that is of most practical value from a coastal management perspective.

To answer the question adequately it is clear that sea level observations need to be averaged in some way. Since the first attempt by Gutenberg [3], many scientists have derived a global value of average sea level rise by using a range of statistical techniques and various data sets of measurements, many of which have been corrected for vertical movements of the coastlines to which tide gauges are attached. The IPCC consensus is that global sea level has risen during the twentieth century by about 17 cm [1,2,4]. Since the early 1990s, the rate of sea level rise has been about 3.2 mm a^{-1} [1], but it is too early to conclude that this change represents a true deviation from the twentieth-century global trend. This becomes evident when the global rate is analyzed at decadal timescales (Fig. 2). Rates of sea level

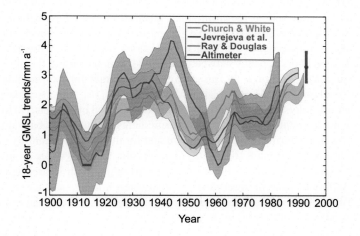

FIGURE 2

Global rates of sea level change since 1900, averaged over 18-year periods. *(Reproduced from Fig. 3.14 in Ref. [58]. Church and White data from Refs [27,57]. Jevrejeva et al. data from Ref. [54]. Ray and Douglas data from Ref. [59]. Satellite altimeter data are also shown.)*

rise, similar to those of the past two decades, have occurred previously, for example in the periods 1930–1950. Other studies show that maximum decadal rates have increased from ∼2.5 mm a^{-1} in the decade centred on 1970 to ∼6 mm a^{-1} in the late 1990s [5,6]. The question is, of course, whether the lengths of the records limit us in the conclusions we can draw. Are measurements during three decadal cycles sufficient to conclude that the rates of sea level rise are on the increase (i.e. that sea level rise is accelerating)? To complicate matters, apparent slowdowns of sea level rise, for example in the first decade of the twenty-first century, are due to interannual variability caused by, for example, the El Niño Southern Oscillation, and these need to be filtered out when assessing long-term trends [7]. Detecting changes in long-term trends is the field of statisticians. They have estimated that the global sea level signal forced by greenhouse gases will exceed natural sea level variability before the exceedance of natural variability is detectable in surface air temperature records [8]. To detect accelerations in sea level rise, interannual and multi-decadal variability should be removed and it could therefore be several decades before statistically significant accelerations show up in individual tide-gauge records [9]. Longer timescales of observations are also necessary to provide the appropriate context to identify possible accelerations in the rate of sea level rise. Long tide-gauge records and reconstructions based on proxy data are therefore crucial to inform the climate change debate (see Section 4).

3. WHY IS SEA LEVEL RISING?

Many factors contribute to the changes in sea level that the globe is experiencing today. Some cause sea levels to rise, others make a negative contribution. Not all factors are well constrained, and herein lies one of the bigger challenges of sea level science: can we explain the sea level rise we are observing?

According to the IPCC [1], the main contributors to sea level rise since 1993 are thermal expansion of the oceans (1.1 mm a^{-1}) and melting ice from small glaciers and ice caps (0.9 mm a^{-1}), with small amounts from the Greenland Ice Sheet (0.3 mm a^{-1}) and the Antarctic Ice Sheet (0.3 mm a^{-1}). When uncertainties are taken into account, the total contributions amount is between 2.3 mm a^{-1} and 3.4 mm a^{-1}. There is a small discrepancy between these modelled values and the observed global mean sea level rise of 2.8 mm a^{-1}–3.6 mm a^{-1}. This discrepancy is larger for the pre-altimetry era and has been called the 'sea level enigma' [4] or the 'attribution problem' [10]. The discrepancy implies one of three things: (1) either the individual contributions are underestimated by models, (2) there are sources of sea level rise that are not accounted for, or (3) the measurements produce a global value that is too high.

One of the most uncertain terms in the sea level budget for the twentieth century is the contribution of terrestrial water sources. Although the filling of reservoirs extracts water from the hydrological cycle and causes sea level to drop [11], other human interference with hydrological processes (e.g. wetland drainage, sedimentation in reservoirs, groundwater mining, surface water consumption, deforestation) contribute positively to sea level rise. It has been argued that the transfer of terrestrial water sources to the ocean could represent a 'missing' term in the sea level budget of the twentieth century [12]. Land water storage due to human interference with the hydrological cycle results in short-term variability that can mask the global long-term sea level trend [7].

Others have argued that the ice melt term has been underestimated [13]. The contribution of small glaciers and ice caps is reasonably well constrained, but accurate monitoring of mass-balance changes in land-based ice sheets is a relatively new activity so that the volumes of discharge of the polar ice sheets are not well known before the 1970s. It is now clear that dynamical changes in the outlet glaciers of the Greenland and Antarctic ice sheets can lead to large sea level contributions. Large outlet glaciers in Greenland, such as the Jakobshavn Isbrae glacier, have contributed about 0.1 mm a^{-1} of sea level rise since the mid-1990s [14]. The Amundsen Sea glaciers in West Antarctica, including the Pine Island Glacier, produced sea level rise contributions of 0.15 mm a^{-1} in the 1990s [15] and possibly as much as 0.24 mm a^{-1} between 2002 and 2005 [16]. Between 1992 and 2011, Antarctica as a whole is estimated to have contributed 0.19 mm $a^{-1} \pm$ 0.15 mm a^{-1} to global sea level rise [17], while for 2010–2013 the estimate is 0.45 mm $a^{-1} \pm$ 0.14 mm a^{-1} [18]. Combined ice losses from Greenland and Antarctica generated 11.2 mm $\pm$ 3.8 mm of sea level rise between 1992 and 2011 [17].

One way to tackle the ice mass term in the sea level budget is to map the 'sea level fingerprint' that would have been left by a melting ice mass. A shrinking ice mass produces a diminishing gravitational pull on the ocean surface and perturbs the sea surface as far as thousands of kilometres away, so that sea level rise is not globally uniform. Near the melt source sea level may be even falling, while far away from the melt source the sea level rise exceeds the global average [19,20]. In theory, therefore, it should be possible to determine from the patterns of sea level change measured by tide-gauges and the ice-mass contribution to sea level rise. For example, this method has been used to estimate a 1.0 mm a^{-1} ± 0.6 mm a^{-1} contribution of melting of the Greenland ice sheet to global sea level rise since 1960 [21]. Although other attempts to find a systematic pattern in tide-gauge measurements have been less successful [22], possibly due to steric (density) and isostatic overprints, estimates of ice loss between 2000 and 2008 demonstrate a distinct and measurable geographical sea level fingerprint [23]. It is clear that sea level fingerprinting has wide-ranging applications in sea level research [24,25].

Another study [4] concluded that the contribution of thermal expansion in past estimates has been underestimated because of biases in the way the expansion was calculated from observational data (Fig. 3). With the revised steric estimates, the sum of the contributions for the second half of

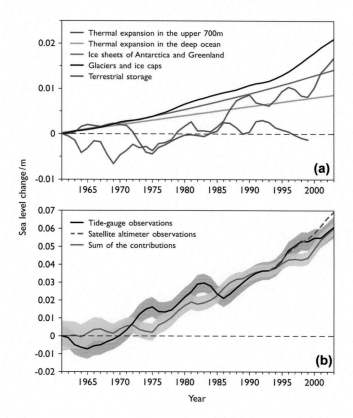

FIGURE 3

(a) Contributions to global sea level change since 1960, including thermal expansion in the upper 700 m of the oceans, thermal expansion in the deep ocean, polar ice sheets, glaciers and ice caps, and terrestrial water storage. (b) Sea level change estimated from global measurements and the sum of the contributions in (a). One standard deviation errors are also shown. *(From Ref. [4].)*

the twentieth century is close to what has been measured. A better understanding of the past contributions to sea level rise will enable modellers to improve their predictions of future sea level rise.

Could differences between observed and explained sea level rise be due to the various ways in which sea level rise has been measured? The measurements for the twentieth century are based on tide-gauge records, which only measure sea level change along coasts. For the past decade, the period for which satellite measurements are available, the sea level budget is almost closed [26]. At face value, this could highlight that satellites produce more accurate measurements of global sea level rise than tide gauges. However, when global averages of sea level rise based on satellite and tide-gauge measurements are compared, they tend to agree quite well [27].

A recent study [28] concluded that the twentieth-century rate of sea level rise has previously been overestimated. Using probabilistic techniques the authors calculated a rate of $1.2 \text{ mm a}^{-1} \pm 0.2 \text{ mm a}^{-1}$ for the period 1901–1990 and a rate of $3.0 \text{ mm a}^{-1} \pm 0.7 \text{ mm a}^{-1}$ for the period 1990–2010. The latter is in close agreement with previous estimates and, importantly, the lower twentieth-century estimate closes the sea level budget and resolves the 'enigma'. However, the lower twentieth-century trend implies that there has been a greater acceleration in sea level rise at the end of the twentieth century than previously thought.

Humans have had a measurable impact on sea level rise since about 1900 (Fig. 4). Model experiments demonstrate the influence of greenhouse gas emissions, producing an increased thermal expansion and greater glacier melt [29]. Volcanic eruptions have reduced the rate of mean sea level

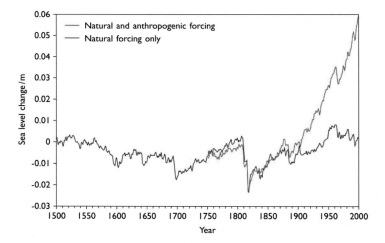

FIGURE 4

A model simulation for the past 500 years of natural and anthropogenically forced sea level change (red) and sea level changed forced by natural factors only (blue). Although the model simulations fail to reproduce the magnitude of both the observed long-term sea level trend and interannual and decadal variability, the onset of twentieth-century sea level rise appears to be controlled by anthropogenic forcing. Sea level rise in these model runs started after the eruption of Mount Tambora in 1815 but was driven by natural factors during the nineteenth century. *(From Ref. [29].)*

rise, and some of the twentieth-century rise in sea level was delayed by the eruptions of Krakatoa in 1886 and Pinatubo in 1991 [30], temporarily masking the impact of anthropogenic effects on sea level rise. A recent study [31] concluded that, on both regional and global scales, sea level rise is now beyond the limits of natural internal variability. At least 1 mm a^{-1} of global sea level rise during the twentieth century could be attributed to anthropogenic forcing. There are found to be large regional differences, but more than half the sea level rise measured in several tide-gauge stations in the Atlantic, Pacific and Indian Oceans, as well as the Mediterranean Sea, is statistically shown to be the direct result of human activities (i.e. greenhouse gas emissions).

The relationship between sea level change and greenhouse gas concentrations is well known on geological timescales. For example, when CO_2 concentrations were higher than 1000 ppm, i.e. 1000×10^{-6} (in this chapter ppm refers to μmol mol^{-1}) around 70 Ma, ice was absent from the planet and sea level was 73 m higher than today [14]. Figure 5 shows the Red Sea sea level record during the past 400 000 years [32] and a comparison with CO_2 concentrations measured in the Vostok ice core [33]. Although chronologies of both records have inherent uncertainties, and CO_2 fluctuations may in fact lead temperature change by several centuries [34], the correlation between CO_2 and sea level change is obvious – higher CO_2 levels correspond with increased sea levels. If the trend, shown in Fig. 5(c), persists and CO_2 concentrations exceed 300 ppm, i.e. 300×10^{-6}, it implies that sea levels will continue to rise significantly in the future. In 2014, CO_2 levels in the Earth's atmosphere reached 400 ppm, i.e. 500×10^{-6}. If the relationship in Fig. 5(c) were to be valid for higher than present concentrations of CO_2, all ice in Greenland and Antarctica would have to melt eventually. Climate models predict that with stabilization of CO_2 levels by the year 2100 at 550 ppm, i.e. 550×10^{-6} [14], or 700 ppm, i.e. 700×10^{-6} [2], sea level will continue to rise for another 1000 years. If CO_2 levels reach 1000 ppm, i.e. 1000×10^{-6}, the Greenland ice sheet will disappear in 3000 years, raising sea level by about 7 m [14].

4. ARE CONTEMPORARY RATES OF SEA LEVEL RISE UNUSUAL?

It is a well-known fact that rates of sea level rise in the past have been much higher than the ones we are experiencing today. For example, during the last deglaciation around 14,000 years ago, rapid melting of ice sheets during Meltwater Pulse 1A produced rates of sea level rise in excess of 40 mm a^{-1} [35,36]. However, the world was then emerging from an ice age, and many ice sheets contained unstable marine components that have now largely disappeared. The only remaining marine-based ice sheet is in West Antarctica, situated in one of the coldest regions of our planet. A comparison with the late glacial sea level history, therefore, does not provide a suitable analogue for modern (or future) conditions. Instead, it is more instructive to examine periods in the Earth's history when the cryosphere contained roughly the same volume of ice as today (or slightly less) and temperatures were similar (or slightly higher) than today's temperatures. Periods often cited as useful analogues include the Last Interglacial, the middle Holocene and the Medieval Climatic Optimum.

The position of sea level during the Last Interglacial is estimated at 6 m–9 m above present [37], but the exact height is difficult to determine due to uncertainties about land movements that have occurred since the Last Interglacial. Most evidence points at sea levels close to the present level, or slightly higher, for the time interval 128 ka–116 ka [38,39]. Only one study on sediment cores from the

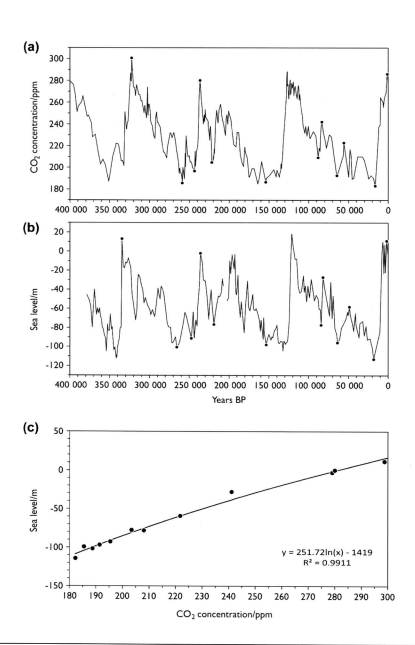

FIGURE 5

(a) The CO_2 concentrations (here ppm is $\times 10^{-6}$ and also refers to μmol mol^{-1}) during the past 400 ka as measured in the Vostok ice core from Antarctica [33]. (b) Sea level changes in the Red Sea during the past 400 ka [32]. (c) Relationship between CO_2 concentrations and sea level, assuming minima and maxima in (a) and (b) are of similar age. Points in (c) correspond to dots on the curves in (a) and (b). Note that the Red Sea maximum sea level for the Last Interglacial (+17 m) is considered to be unrealistically high and was left out of (c).

Red Sea provides a detailed assessment of the rates of sea level rise during the Last Interglacial [40]. It is estimated that the 'full potential range' of rates of sea level rise was between 0.6 m and 2.5 m per century. It is interesting to note that this estimate is within the same ballpark as some predictions made for the twenty-first century [41], although it is higher than those provided by the IPCC. About 2 m of the sea level rise during the Last Interglacial is thought to have come from Greenland, with the remaining contribution from Antarctica [42,43].

Evidence from many parts around the world suggests that temperatures in the current interglacial reached their maximum in the middle Holocene, although the Holocene 'thermal optimum' is spatially variable and not globally synchronous. There are many published sea level studies that argue for sea level fluctuations, some up to several metres in amplitude, during the Holocene, but these are almost always based on data with large vertical and age uncertainties and often use a 'connect-the-dot' approach that ignores these uncertainties [44]. Many Holocene sea level histories are only resolved on millennial timescales. The more robust sea level reconstructions that contain evidence for century-scale sea level oscillations are arguably from micro-atolls in Australia and have recorded rates of 0.1 m–0.2 m per century in the middle Holocene [45]. Even during the 8.2 ka event, which was caused by the final draining of a huge glacial lake (Lake Agassiz-Ojibway) and thus is not truly representative as a modern analogue, sea level rise may have only amounted to 0.4 m [46], although field evidence from the Netherlands points to a potential sea level jump of 2 m–4 m in just 200 years [47].

Sea level changes during the Medieval Climatic Optimum were small and have not been clearly resolved in paleo records [48]. Salt marshes in eastern North America provide evidence that rates did not exceed 0.2 m per century during the past millennium before the twentieth century [49,50]. In the North Atlantic Ocean [51] and in the Southwest Pacific [52], the recent acceleration of sea level rise started about 100 years ago [53]. These findings are based on microfossil evidence and high-precision dating of salt marsh sediments and are supported by some analyses of long tide-gauge records [54]. However, tide-gauge measurements that extend back into the eighteenth century are only available for a few stations in western Europe (Amsterdam since 1700, Stockholm since 1774, Liverpool since 1768 [55]). The earliest global acceleration of sea level rise that can be clearly demonstrated in instrumental sea level data [56] and in global reconstructions based on tide-gauge data [57] occurred in the 1930s.

5. CONCLUSION

Sea level rise is a major indicator of ongoing global change. Sea level has been rising at a rate of 3.0 mm a^{-1} or 3.2 mm a^{-1} since the early 1990s. This represents a substantial increase in rate compared to the period 1901–1990 for which various studies estimate a mean rate of 1.2 mm a^{-1}–1.7 mm a^{-1}, but it is too early to determine if this trend change represents a new global climate state. Model experiments show that twentieth-century sea level rise cannot be explained by natural processes alone. Anthropogenic forcing by greenhouse gases has become a dominant cause of sea level rise, generating thermal expansion of ocean waters and melting of land-based ice. The geological record of the past three glacial–interglacial cycles shows a strong positive relationship between atmospheric CO_2 concentrations and sea level. Modern rates of sea level rise started about 100 years ago, and the rate of twentieth century sea level rise appears to be faster than rates reconstructed for the warm intervals of the Medieval Climatic Optimum and the middle Holocene. However, during the Last Interglacial rates of sea level rise were possibly higher and may have been similar to those predicted in some future climate change scenarios.

REFERENCES

[1] J.A. Church, P.U. Clark, A. Cazenave, J.M. Gregory, S. Jevrejeva, A. Levermann, M.A. Merrifield, G.A. Milne, R.S. Nerem, P.D. Nunn, A.J. Payne, W.T. Pfeffer, D. Stammer, A.S. Unnikrishnan, Sea level change, in: T.F. Stocker, D. Qin, G.-K. Plattner, M. Tignor, S.K. Allen, J. Boschung, A. Nauels, Y. Xia, V. Bex, P.M. Midgley (Eds.), Climate Change 2013: The Physical Science Basis. Contribution of Working Group I to the Fifth Assessment Report of the Intergovernmental Panel on Climate Change, Cambridge University Press, Cambridge, UK and New York, NY, USA (2013).

[2] G.A. Meehl, T.F. Stocker, W.D. Collins, P. Friedlingstein, A.T. Gaye, J.M. Gregory, A. Kitoh, R. Knutti, J.M. Murphy, A. Noda, S.C.B. Raper, I.G. Watterson, A.J. Weaver, Z.-C. Zhao, in: S. Solomon, D. Qin, M. Manning, Z. Chen, M. Marquis, K.B. Averyt, M. Tignor, H.L. Miller (Eds.), Climate Change 2007: The Physical Science Basis. Contribution of Working Group I to the Fourth Assessment Report of the Intergovernmental Panel on Climate Change, Cambridge University Press, Cambridge, UK and New York, NY, USA (2007), pp. 747–845.

[3] B. Gutenberg, Geol. Soc. Am. Bull. 52 (1941) 721–772.

[4] C.M. Domingues, J.A. Church, N.J. White, P.J. Gleckler, S.E. Wijffels, P.M. Barker, J.R. Dunn, Nature 453 (2008) 1090–1094, http://dx.doi.org/10.1038/nature07080.

[5] J.A. Church, N.J. White, R. Coleman, K. Lambeck, J.X. Mitrovica, J. Clim. 17 (2004) 2609–2625.

[6] S.J. Holgate, P.L. Woodworth, Geophys. Res. Lett. 31 (2004), http://dx.doi.org/10.1029/2004GL019626.

[7] A. Cazenave, D. Habib-Boubacar, B. Meyssignac, K. von Schuckmann, B. Decharme, E. Berthier, Nat. Clim. Change 4 (2014) 358–361.

[8] K. Lyu, X. Zhang, J.A. Church, A.B.A. Slangen, J. Hu, Nat. Clim. Change 4 (2014) 1006–1010.

[9] I.D. Haigh, T. Wahl, E.J. Rohling, R.M. Price, C.B. Pattiaratchi, F.M. Calafat, S. Dangendorf, Nat. Commun. 5 (2014) 3635, http://dx.doi.org/10.1038/ncomms4635.

[10] W. Munk, Proc. Natl. Acad. Sci. U.S.A. 99 (2002) 6550–6555.

[11] B.F. Chao, Y.H. Wu, Y.S. Li, Science 320 (2008) 212–214, http://dx.doi.org/10.1126/science.1154580.

[12] T.G. Huntingdon, Hydr. Proc. 22 (2008) 717–723.

[13] L. Miller, B.C. Douglas, Nature 428 (2004) 406–409.

[14] R.B. Alley, P.U. Clark, P. Huybrechts, I. Joughin, Science 310 (2005) 456–460.

[15] A. Shepherd, D. Wingham, Science 315 (2007) 1529–1532.

[16] I. Velicogna, J. Wahr, Science 311 (2006) 1754–1756.

[17] A. Shepherd, E.R. Ivins, G.A. Valentina, R. Barletta, M.J. Bentley, S. Bettadpur, K.H. Briggs, D.H. Bromwich, R. Forsberg, N. Galin, M. Horwath, S. Jacobs, I. Joughin, M.A. King, J.T.M. Lenaerts, J. Li, S.R.M. Ligtenberg, A. Luckman, S.B. Luthcke, M. McMillan, R. Meister, G. Milne, J. Mouginot, A. Muir, J.P. Nicolas, J. Paden, A.J. Payne, H. Pritchard, E. Rignot, H. Rott, L. Sandberg Sørensen, T.A. Scambos, B. Scheuchl, E.J.O. Schrama, B. Smith, A.V. Sundal, J.H. van Angelen, W.J. van de Berg, M.R. van den Broeke, D.G. Vaughan, I. Velicogna, J. Wahr, P.L. Whitehouse, D.J. Wingham, D. Yi, D. Young, H.J. Zwally, Science 338 (2012) 1183–1189.

[18] M. McMillan, A. Shepherd, A. Sundal, K. Briggs, A. Muir, A. Ridout, A. Hogg, D. Wingham, Geophys. Res. Lett. 41 (2014) 3899–3905, http://dx.doi.org/10.1002/2014GL060111.

[19] J.A. Clark, J.A. Primus, in: M.J. Tooley, I. Shennan (Eds.), Sea-Level Changes, Institute of British Geographers, London (1987), pp. 356–370.

[20] J.X. Mitrovica, M.E. Tamisiea, J.L. Davis, G.A. Milne, Nature 409 (2001) 1026–1029.

[21] M. Marcos, M.N. Tsimplis, Geophys. Res. Lett. 34 (2008), http://dx.doi.org/10.1029/2007GL030641.

[22] B.C. Douglas, J. Coastal Res. 24 (2008) 218–227.

[23] J. Bamber, R. Riva, Cryosphere 4 (2010) 621–627.

[24] M.E. Tamisiea, J.X. Mitrovica, J.L. Davis, Earth Planet. Sci. Lett. 213 (2003) 447–485.

[25] H.-P. Plag, Philos. Trans. R. Soc. A 364 (2006) 821–844.

[26] N.L. Bindoff, J. Willebrand, V. Artale, A. Cazenave, J. Gregory, S. Gulev, K. Hanawa, C. Le Quéré, S. Levitus, Y. Nojiri, C.K. Shum, L.D. Talley, A. Unnikrishnan, in: S. Solomon, D. Qin, M. Manning, Z. Chen, M. Marquis, K.B. Averyt, M. Tignor, H.L. Miller (Eds.), Climate Change 2007: The Physical Science Basis. Contribution of Working Group I to the Fourth Assessment Report of the Intergovernmental Panel on Climate Change, Cambridge University Press, Cambridge, UK and New York, NY, USA (2007), pp. 385–432.

[27] J.A. Church, N.J. White, Surv. Geophys. 32 (2011) 585–602.

[28] C.C. Hay, E. Morrow, R.E. Kopp, J.X. Mitrovica, Nature (2015), http://dx.doi.org/10.1038/Nature14093.

[29] J.M. Gregory, J.A. Lowe, S.F.B. Tett, J. Clim. 19 (2006) 4576–4591.

[30] J.A. Church, N.J. White, J.M. Arblaster, Nature 438 (2005) 74–77.

[31] M. Becker, M. Karpytchev, S. Lennartz-Sassinek, Geophys. Res. Lett. 41 (2014) 5571–5580, http://dx.doi.org/10.1002/2014GL061027.

[32] M. Siddall, E.J. Rohling, A. Almogi-Labin, Ch Hemleben, D. Meischner, I. Schmelzer, D.A. Smeed, Nature 423 (2003) 853–858.

[33] J.R. Petit, J. Jouzel, D. Raynaud, N.I. Barkov, J.M. Barnola, I. Basile, M. Bender, J. Chappellaz, J. Davis, G. Delaygue, M. Delmotte, V.M. Kotlyakov, M. Legrand, V. Lipenkov, C. Lorius, L. Pépin, C. Ritz, E. Saltzman, M. Stievenard, Nature 399 (1999) 429–436.

[34] J. Ahn, E.J. Brook, Geophys. Res. Lett. 34 (2007) L10703, http://dx.doi.org/10.1029/2007GL029551.

[35] J.D. Stanford, E.J. Rohling, S.E. Hunter, A.P. Roberts, S.O. Rasmussen, E. Bard, J. McManus, R.G. Fairbanks, Paleoceanography 21 (2006), http://dx.doi.org/10.1029/2006PA001340.

[36] P. Deschamps, N. Durand, E. Bard, B. Hamelin, G. Camoin, A.L. Thomas, G.M. Henderson, J. Okuno, Y. Yokoyama, Nature 483 (2012) 559–564.

[37] R.E. Kopp, F.J. Simons, J.X. Mitrovica, A.C. Maloof, M. Oppenheimer, Nature 462 (2009) 863–867.

[38] D.R. Muhs, Quat. Res. 58 (2002) 36–40.

[39] M.J. O'Leary, P.J. Hearty, W.G. Thompson, M.E. Raymo, J.X. Mitrovica, J.M. Webster, Nat. Geosci. 6 (2013), http://dx.doi.org/10.1038/NGEO1890.

[40] E.J. Rohling, K. Grant, Ch. Hemleben, M. Siddall, B.A.A. Hoogakker, M. Bolshaw, M. Kucera, Nat. Geosci. 1 (2008) 38–42, http://dx.doi.org/10.1038/ngeo.2007.8.

[41] S. Rahmstorf, Science 315 (2007) 368–370, http://dx.doi.org/10.1126/science1135456.

[42] NEEM community members, Nature 493 (2013) 489–494.

[43] A. Dutton, J.M. Webster, D. Zwartz, K. Lambeck, B. Wohlfarth, Quat. Sci. Rev. 107 (2015) 182–196.

[44] W.R. Gehrels, J. Coastal Res. 17 (2000) 244–245.

[45] S.E. Lewis, R.A.J. Wüst, J.M. Webster, G.A. Shields, Terra Nova 20 (2008) 74–81, http://dx.doi.org/10.1111/j.1365-3121.2007.00789.x.

[46] R.A. Kendall, J.X. Mitrovica, G.A. Milne, T. Törnqvist, Y. Li, Geology 36 (2008) 423–426, http://dx.doi.org/10.1130/G24550A.1.

[47] M.P. Hijma, K.M. Cohen, Geology 38 (2010) 275–278.

[48] R. Gehrels, B. Horton, A. Kemp, E. Toker, D. Sivan, Eos 92 (2011) 289–296.

[49] O. van de Plassche, K. van der Borg, A.F.M. de Jong, Geology 26 (1998) 319–322.

[50] A.C. Kemp, B.P. Horton, C.H. Vane, C.E. Bernhardt, D.R. Corbett, S.E. Engelhart, S.C. Anisfield, A.C. Parnell, N. Cahill, Quat. Sci. Rev. 81 (2013) 90–104.

[51] W.R. Gehrels, J.R. Kirby, A. Prokoph, R.M. Newnham, E.P. Achterberg, E.H. Evans, S. Black, D.B. Scott, Quat. Sci. Rev. 24 (2005) 2083–2100.
[52] W.R. Gehrels, B.W. Hayward, R.M. Newnham, K.E. Southall, Geophys. Res. Lett. 35 (2008), http://dx.doi.org/10.1029/2007GL032632.
[53] W.R. Gehrels, P.L. Woodworth, Glob. Planet. Change 100 (2013) 263–277.
[54] S. Jevrejeva, J.C. Moore, A. Grinsted, P.L. Woodworth, Geophys. Res. Lett. 35 (2008), http://dx.doi.org/10.1029/2008GL033611.
[55] P.L. Woodworth, Geophys. Res. Lett. 26 (1999) 1589–1592.
[56] P.L. Woodworth, N.J. White, S. Jevrejeva, S.J. Holgate, J.A. Church, W.R. Gehrels, Int. J. Clim. 29 (2009) 777–789.
[57] J.A. Church, N.J. White, Geophys. Res. Lett. 33 (2006) L01602, http://dx.doi.org/10.1029/2005GL024826.
[58] M. Rhein, S.R. Rintoul, S. Aoki, E. Campos, D. Chambers, R.A. Feely, S. Gulev, G.C. Johnson, S.A. Josey, A. Kostianoy, C. Mauritzen, D. Roemmich, L.D. Talley, F. Wang, Observations: ocean, in: T.F. Stocker, D. Qin, G.-K. Plattner, M. Tignor, S.K. Allen, J. Boschung, A. Nauels, Y. Xia, V. Bex, P.M. Midgley (Eds.), Climate Change 2013: The Physical Science Basis. Contribution of Working Group I to the Fifth Assessment Report of the Intergovernmental Panel on Climate Change, Cambridge University Press, Cambridge, UK and New York, NY, USA (2013).
[59] R.D. Ray, B.C. Douglas, Prog. Oceanogr. 91 (2011) 496–515.

OCEAN CURRENT CHANGES

17

Torsten Kanzow[1,2], Martin Visbeck[3], Uta Krebs-Kanzow[1]

[1]*Fachbereich Klimawissenschaften, Alfred-Wegener-Institut für Polar- und Meeresforschung, Bremerhaven, Germany;*
[2]*Universität Bremen, Germany;* [3]*Fachbereich Ozeanzirkulation und Klimadynamik, GEOMAR Helmholtz-Zentrum für Ozeanforschung Kiel, Germany*

CHAPTER OUTLINE

1. INTRODUCTION

The high heat capacity of seawater and the relatively slow ocean circulation allow the oceans to provide significant 'memory' for the climate system. Bodies of water that descend from the sea surface may reside in the ocean interior for decades and centuries, while preserving their temperature and salinity signature, before they surface again to interact with the overlying atmosphere. In contrast to that, the residence time of water in the atmosphere is about 10 days and the persistence of dynamical states of the atmospheric circulation may last up to a few weeks. Thus, on long timescales ocean dynamics become important for climate, which implies that climate variations and climate change can only partially be understood without consideration of ocean dynamics and the intricate ocean-atmosphere interaction. The El Niño Southern Oscillation phenomenon in the tropical Pacific is a prominent example of tightly coupled ocean-atmosphere dynamics on interannual timescales; other more weakly coupled interactions exist throughout the system.

The oceans' role in climate and climate change is manifold. Ocean circulation transports large amounts of heat and freshwater on hemispheric space scales that have significant impacts on regional climate in the ocean itself but also noticeable consequences via atmospheric teleconnections on land. What is well known for the seasonal cycle with only moderate temperature changes between summer and winter in marine climates compared with much larger swings within the continents is also true on decadal timescales. Since 1960, the heat uptake of the oceans has been 20 times larger than that of the atmosphere. Thus, the oceans have been able to reduce the otherwise much more pronounced temperature rise in the atmospheric climate. Also, over the last 200 a (where 'a' refers to annum) the oceans have absorbed about half of the CO_2 release into the atmosphere by human activities (fossil fuel combustion, deforestation, cement production), thereby reducing the direct effect of greenhouse gases on atmospheric temperatures.

2. THE VARIABLE OCEAN

Bodies of water circulate throughout the oceans – both horizontally and vertically – as a consequence of physical forces exerted on them according to Newton's law. The oceanic circulation is not steady in time. Rather motions of water bodies in the ocean are known to vary on a broad range of spatial and temporal scales. The following four examples serve to highlight natural variations of large-scale circulation patterns:

1. Seasonal variations of the strength of the North Atlantic subtropical gyre at 26°N have amplitudes of 25 Sv[1] (peak to peak). This range is comparable to the time mean strength of the wind-driven, anti-cyclonic basin-scale gyre at this latitude [1].

2. The Pacific subtropical cells (STCs) form a meridional, upper-ocean pattern of circulation that links the subtropical subduction regions north and south of the equator to the equatorial thermocline. The STC has seen a decline of 11 Sv or 30% from the 1950s to the 1990s, however, displaying decadal variations of the same order of magnitude [2]. More recent findings indicate a substantial strengthening of the STC over the last 20 years [3]. The observed strong decadal and multi-decadal variations in sea surface temperature (SST) in the equatorial Pacific have been shown to be related to changes in STC strength.

3. The cyclonic circulation of the North Atlantic subpolar gyre has possibly weakened by 25% and shrunk in size since the mid-1990s [4,5]. This has been attributed to the transition of the North Atlantic Oscillation[2] (NAO, [6]) from a comparably strong phase between 1960 and 1995 (manifesting itself in stronger-than-average westerly winds at midlatitudes) to a significantly weaker one after 1995. The gyre's weakening and westward retreat has allowed large quantities of saline subtropical upper-ocean waters to flow northward past its eastern flank, as a consequence of which a drastic increase in salinities in the Nordic Seas[3] has been observed [7]. This is thought to have an impact on the sinking of waters as part of the Atlantic Meridional Overturning Circulation (AMOC), the latter being the primary focus of this chapter.

4. Although not having been observed directly, it is commonly thought that temporal changes in the strength of the AMOC, which is a basin-wide meridional circulation pattern that links upper-ocean net northward flow of warm, saline waters with cold southward return flow below roughly 1000 m throughout the Atlantic, explain large parts of the observed multi-decadal North Atlantic SST changes [8]. A recent summary and discussion of climate variability and its predictability in the Atlantic sector [9] provides a perspective on the difficulties one has to distinguish decadal variability from long-term, possibly anthropogenic-induced trends.

---

[1]In this chapter the Sverdrup is defined as: 1 Sv = 1×10^6 m^3 s^{-1} (unit for volumetric transport, named after Harald Ulrik Sverdrup). For comparison, the Amazon River discharge in the Atlantic is about 0.2 Sv.

[2]The NAO is the dominant mode of (winter) climate variability in the North Atlantic region ranging from central North America to Europe and into Northern Asia.

[3]Nordic Seas is used as collective term for Greenland Sea, Norwegian Sea and Iceland Sea.

The reason why the different components of the ocean circulation have the potential to change substantially over time is a consequence of the complex forcing at the sea surface (exchange of momentum, heat and freshwater between ocean and atmosphere) on the one hand and internal ocean dynamics on the other. Examples of internal ocean dynamics include advection of water of anomalous density by the mean large-scale ocean circulation, westward energy transfer by off-equatorial planetary waves, the equatorial wave guide, horizontal mixing by mesoscale eddies, deep-water formation due to convection or small-scale vertical mixing, acting to push the cold waters of the oceans' abyss upward. The large variations that basin-scale circulation patterns may exhibit have the potential to delay the detectability of climate change–related shifts in the flow field.

3. OCEANOGRAPHER'S TOOLS

Oceanographers have developed direct and indirect techniques for the observation of ocean currents in order to document and analyse the strength of the interior ocean circulation and its changes in space and time. Direct current measurements can be divided into two classes: Eulerian and Lagrangian. The Eulerian approach measures current velocity and direction at a fixed location. Current meters are typically used in this context that can be mounted on a stationary platform or 'mooring' [10], installed in a vessel's hull [11] or lowered from a vessel [12]. The scope of Lagrangian observations aims at deriving the streamlines of regional or ocean basin-scale flow fields. For this, drifters are used that move passively with the flow at a specified depth horizon and whose displacements are monitored over time [13].

The equation of motion for a fluid (i.e. Newton's law applied to a fluid) requires the various different physical forces acting on a unit mass of water to balance one another. The two probably most widely used indirect methods to study the flow field in the open ocean, viz. (1) the Ekman balance and (2) the geostrophic balance, rely on well-founded simplifications of the equation of motion that are valid on timescales longer than one or two days and spatial scales in excess of several tens of kilometres.

1. Within roughly the top 50 m of the water column (near-surface Ekman layer), the flow field results from a balance between the stress that the atmospheric wind field exerts onto the sea surface and the Coriolis force [14]. To the first order, the horizontal flow in the Ekman layer depends on strength of the wind speed but moves at right angles to the direction of the wind as a consequence of the Earth's rotation.[4] The availability of daily wind fields over the global ocean from space-borne measurements makes the Ekman balance a very powerful diagnostic tool of large-scale near-surface flows.

2. In the vast ocean interior below the Ekman layer, the horizontal movement results in long surfaces of constant pressure as a result of the near geostrophic balance between the horizontal pressures gradient force and the Coriolis force.[5] This is analogous to atmospheric conditions as depicted in weather charts, where (to first order) wind flows *around* cells of high or low pressure (and not from high to low pressure), again as a consequence of the Earth's rotation. As a result, the strength of the flow across a section between two points is approximately proportional to the difference in pressure

---

[4]The net flow in the Ekman layer is to the right of the direction of the wind stress in the Northern Hemisphere and to the left in the Southern Hemisphere.
[5]The geostrophic flow is directed such that the pressure increases to the right in the Northern Hemisphere and to the left in the Southern Hemisphere.

at the section's two end points (both in the ocean and in the atmosphere). The geostrophic balance is therefore an effective tool to diagnose the net strength of basin-scale ocean circulation patterns. All that is required are measurements of the pressure field at the section end points, while the actual horizontal structure of the flow in between does not need to be resolved. Practically, the pressure field of the ocean cannot be measured directly in the water column to derive reliable estimates of the strength of the flow. However, profiles of water density allow the computation of the ocean's pressure and velocity field relative to reference pressure level. This has served oceanographers for many decades to study strength and vertical structure of the ocean circulation [15].

Besides observing ocean currents, the numerical simulation of the circulation using so-called ocean general circulation models (OGCMs) has emerged as an important discipline in physical oceanography. To this end the equation of motion (if applicable, also the heat and salt conservation equations) are solved numerically on a predefined spatial model grid sequentially for each time step. This is commonly referred to as model 'integration'. For each time step, so-called boundary conditions have to be prescribed, which drive the model ocean. The boundary conditions mainly include observed fluxes of momentum, heat and freshwater between the ocean and the atmosphere. Typical horizontal resolutions of currently used OGCMs range between 10 and 100 km. The finer the resolution (and the larger the model region) that is chosen, the more computationally expensive it becomes to run the model, such that the integration periods that can be reached become shorter.

Typical integration periods are up to several decades. Besides OGCMs, the class of climate models has found widespread use in oceanography. Here, the equations of motion for the ocean and the atmosphere (and for ice sheets, if applicable) are solved simultaneously, and both model components are coupled at the air-sea interface. In climate models the ocean is not passively forced but instead can feed back on the atmosphere. Climate models are driven by orbital forcing (i.e. insolation at the top of the atmosphere). As climate models are typically integrated over climate-relevant timescales (centuries and longer), their horizontal resolution is often sparse compared to OGCMs, typically between 100 and 500 km.

The main advantage of numerical model simulations is that self-consistent estimates of the ocean circulation can be obtained for the entire spatial and temporal domain of interest. The degree of their shortcomings, however, is difficult to evaluate. One problem common to all numerical models is their finite resolution. That means that physical processes that take place on spatial scales smaller than the grid size in the real ocean are not included in the model physics and have to be parameterized. Other uncertainties derive from errors in the boundary conditions and the numerical integration itself.

4. THE ATLANTIC MERIDIONAL OVERTURNING CIRCULATION
4.1 MOTIVATION

For most of this chapter we will limit our attention to the aforementioned AMOC, which represents a circulation pattern most relevant for marine and terrestrial climates in many ways:

1. The AMOC represents a mechanism of long-term 'memory' in the climate system.
2. The AMOC is the most important oceanic flow component for meridional redistribution of heat.
3. The AMOC is an important pathway for the oceans' uptake of anthropogenic greenhouse gases and for the ventilation of the deep ocean interior.
4. The vigour of the AMOC and the associated heat transport are thought to experience a reduction of 30%–50% over the next century as a consequence of global warming.

Taken together, highly possible long-term changes in the AMOC are thought both to be indicative of climate change and to contribute to climate change. Because of this, the AMOC represents a subject of active ongoing research involving observations and numerical modelling. In the following we outline the AMOC's relevance for climate, climate variations and climate change. This is preceded by a description of the underlying pattern of circulation and its potential driving mechanisms. The authors intend to convey that although our knowledge about the AMOC has advanced dramatically over the last few decades, many uncertainties yet remain to be solved.

4.2 CIRCULATION, DRIVING MECHANISMS

A striking feature of the temperature distribution in the oceans – Fig. 1 displays a section of temperature along the meridional extent of the Atlantic – is the strong vertical contrast in temperatures at low and midlatitudes, with warm upper-ocean waters floating on top of cold deep and abyssal waters.

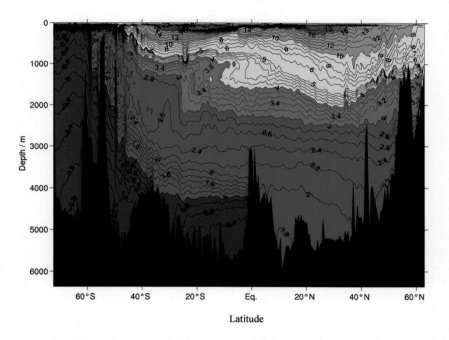

FIGURE 1

Section of potential temperature along the meridional extent of the Atlantic. For temperatures less than 5°C and greater than 5°C the black contours have a spacing of 0.2°C and 1°C, respectively. The red, Indian red, salmon, cyan, light blue and dark blue areas denote temperatures above 16°C, from 10°C–16°C, from 4°C–10°C, from 3°C–4°C, from 1°C–3°C and below 1°C, respectively. Lowered temperature measurements acquired during three research expeditions – aboard *RV Ronald H. Brown* in 2003 (Section A16N, PI: Bullister (PMEL)) and in 2005 (Section A16S; PIs: Wanninkhof (NOAA)/Doney (WHOI)) and aboard *RV James Clark Ross* in 1995 (Section A23; PIs: Heywood/King (NOCS)) – were joined together to compile this figure. *(Data source: Clivar and Carbon hydrographic data office (http://whpo.ucsd.edu/atlantic.htm. Adapted from a figure of Lynne D. Talley (http://sam.ucsd.edu/vertical_sections/Atlantic.html#a16a23).)*

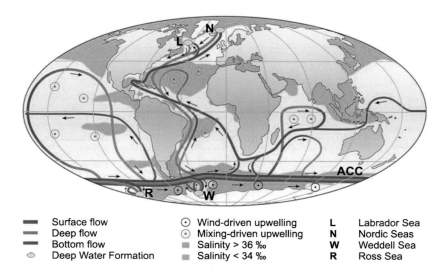

Surface flow
Deep flow
Bottom flow
Deep Water Formation

⊙ Wind-driven upwelling
⊙ Mixing-driven upwelling
▪ Salinity > 36 ‰
▪ Salinity < 34 ‰

L   Labrador Sea
N   Nordic Seas
W   Weddell Sea
R   Ross Sea

FIGURE 2

Strongly simplified sketch of the global overturning circulation system. In the Atlantic, warm and saline waters flow northward all the way from the Southern Ocean into the Labrador and Nordic Seas. By contrast, there is no deep-water formation in the North Pacific, and its surface waters are fresher. Deep waters formed in the Southern Ocean are denser and thus spread in deeper levels than those from the North Atlantic. Note the strongly localized deep-water formation areas in comparison with the widespread zones of mixing-driven upwelling. Wind-driven upwelling occurs along the Antarctic Circumpolar Current (ACC). *(This figure has been published by Kuhlbrodt et al. [18] and has been reproduced with permission.)*

The vertical layering of waters of different temperatures (densities) is referred to as stratification. It was already recognized as early as 1798 by Count Rumford that in the absence of any deep-ocean heat sinks at low latitudes, those cold waters had to originate from high latitudes propagating equatorward at depth.[6] Today it is well established that the observed temperature distribution is a consequence of the AMOC, that moves roughly 19 Sv of warm saline waters northward throughout the Atlantic and the same amount of cold water back south at depth ([17,18]; Fig. 2). Carried northward within the Gulf Stream/North Atlantic Current system, the near-surface waters release heat to the atmosphere and thus become gradually denser. The waters eventually reach the Nordic Seas and the Labrador Sea. Here, deep-reaching wintertime convection (i.e. vertical mixing throughout the upper 2000 m of water column) can occur [5,19,20], when the vertical stratification has eroded after periods of excessive heat loss (Fig. 2). The bulk of the newly formed deep waters that are subject to overflow and entrainment processes constitute the North Atlantic Deep Water (NADW). The NADW is subsequently exported southward, partly confined to the deep western boundary current (DWBC) along the Americas below roughly 1000 m. The intensity of the strongly localized, buoyancy-loss-induced formation of NADW

---

[6]Longworth and Bryden [16] give an exciting account of the history of the recovery of the Atlantic Meridional Overturning Circulation.

at high latitudes (Fig. 2) 'pushing' surface waters downward has long been thought to control the strength of the AMOC.

To close the circulation, the dense NADW needs to eventually return to the upper ocean. This is assumed to be accomplished mainly by two processes. The first process relates to winds and tides that represent the major sources of mechanical energy input into the ocean [21]. Ultimately this energy input is balanced by dissipation into small-scale motions, a process by which turbulent mixing occurs. Dissipation and mixing are ubiquitous in the open ocean. However, they seem most active in the vicinity of rough bathymetry such as exhibited by mid-oceanic ridges [22]. As a consequence, deeper (denser) waters from below are mixed with overlying warmer (less dense) waters, thus making deep waters gradually lighter. This allows them to rise and to return to the upper ocean. The fact that as a direct consequence of vertical mixing even the deep ocean below 1000 m exhibits a notable stable stratification (i.e. water becoming denser with depth, as shown in Fig. 1) has been used to argue that dissipation-induced vertical mixing 'pulling' deep water upward might ultimately have a stronger control on the vigour of the AMOC than the downward 'pushing' at high latitudes. To move waters vertically across surfaces of constant density, vertical mixing is required and this cannot be generated by high latitude buoyancy forcing [23].

The second potentially powerful mechanism to 'pull' deep water back to the upper ocean to close the overturning circulation can be motivated by a careful inspection of Fig. 1. The NADW flows southward away from the regions of its formation and eventually partly reaches the Southern Ocean. While north of 40°S the deep surfaces of constant temperature show only a weak upward slope toward the south, the situation changes dramatically south of 40°S. This is a direct result of 70% of the global wind energy input into the ocean taking place in this area. Due to the Ekman balance, the strong westerly winds over the Southern Ocean push large amounts of near-surface waters northward, which are then replaced by waters being sucked upward from the deep ocean. The manifestation of this process is the drastic increase in the upward tilt of the deep temperature surfaces toward the south (Fig. 1). In this scenario the transition from cold to warm waters (mixing across density surfaces) occurs near the sea surface in the Southern Ocean as suggested from model findings by Toggweiler and Samuels [24]. This scenario appears to have more and more support in the ongoing scientific debate [25]. Overall, the Southern Ocean's control on the upwelling branch of the AMOC represents an important element in the changing climate system, as reservoirs of heat and carbon may re-enter the surface layer of the ocean, where they may interact with the Antarctic ice shelves and the atmosphere, respectively [25].

Besides the AMOC, a second major pattern of meridional overturning exists. This involves formation of deep waters by means of convection around Antarctica. The Antarctic Bottom Water (AABW) spreads northward and represents the coldest and therefore deepest water mass in the Atlantic, Pacific and Indian Oceans (roughly represented by the dark-blue shaded part of the temperature field in Fig. 1).

In the Atlantic the waters gradually mix into the lower parts of the overlying NADW, and eventually return southward. Even though the volume of NADW flowing southward and AABW moving northward are comparable in size [26] or possibly larger for the southern cell [27], the contribution of the AABW-related meridional overturning cell to meridional heat transport is negligible, as the vertical temperature contrast between its upper and low branches is very small [28]. Nonetheless, the observed multi-decadal contraction in the coldest class of AABW corresponds to a net gain of heat in the deep ocean, resulting in a steric sea level rise south of 30°S of roughly 4 mm per decade [29].

4.3 THE AMOC'S ROLE IN HEAT TRANSPORT, OCEANIC UPTAKE OF CARBON AND VENTILATION OF THE DEEP OCEAN

The Earth's surface takes up heat by absorbing solar short-wave radiation. On a global average, this is almost exactly balanced by the Earth's emission of long-wave radiation. Regional budgets of radiative energy fluxes, however, are unbalanced, as they show pronounced heat gain at low latitudes in contrast to heat loss at high latitudes [30]. One-quarter of the 5 Pw[7] of maximum global heat transport, which the coupled ocean-atmosphere system is required to transport poleward in the northern hemisphere to balance approximately regional energy budgets, is carried by the AMOC in the subtropical North Atlantic [30]. While most of the remaining three-quarters of the heat transport is accomplished by the atmosphere, the AMOC is by far the most important oceanic component of meridional heat transport.

The steady increase in atmospheric CO_2 is widely regarded as one of the main drivers of the presently ongoing global warming. Exceeding the atmosphere in terms of carbon storage by more than a factor of 50 [31], the oceans exchange gases with the atmosphere. The CO_2 solubility in seawater increases with decreasing temperatures. The NADW formation at high latitudes, acting to increase carbon concentrations at depth, is considered a major element in the ocean's carbon uptake ('solubility pump'). Changes in the strength and spatial structure of the AMOC might affect atmospheric CO_2 concentrations and thus global temperatures. The Southern Ocean has historically been considered the most important area for uptake of atmospheric CO_2. This sink is undergoing a multidecadal decline, however, with the latitude band south of 50°S having developed from a sink into a source of atmospheric CO_2 [32].

The flow of well-oxygenated near-surface water to the deep ocean that goes along with the NADW formation at high latitudes and its subsequent export to the world ocean help to maintain the deep ocean basins as habitats for a diverse biota. This is probably the least-known component of biodiversity on Earth. Substantial changes in the rate of deep-water ventilation by the AMOC are thus expected to have consequences for deep-ocean habitats.

4.4 SIMULTANEOUS CHANGES OF THE AMOC AND ATLANTIC CLIMATE IN THE PAST

Especially in the northern hemisphere the amount of heat carried poleward by the oceans is very much tied to the strength of the AMOC. Analysis of ice cores from Greenland revealed more than 20 so-called 'Dansgaard-Oeschger events' during the last ice age (100 000 BCE to 10 000 BCE), over the course of which Greenland temperatures jumped by roughly 10°C within a few decades subsequently followed by a gradual cooling on a millennium timescale [33,34]. Based on the reconstructions of water mass properties [35,36] and of deep water residence times [37,38], this millennial-scale climate variability is thought to be linked to abrupt changes in the deep-ocean circulation in the North Atlantic The observations are in qualitative agreement with numerical model simulations that associate the climate variations with temporal changes in the vigour of the AMOC [39,40]. The North Atlantic cold phases are generally thought to be linked with a very weak (or inactive) state of the AMOC that goes along with a near cessation of NADW formation and northward heat transport in the North Atlantic. Warm phases are expected to coincide with a strong state of the AMOC. During the last ice age, some of the periods of reduced NADW formation rates are likely to have been enhanced by events of

---

[7]1 Pw = 1×10^{15} W. Note: 5 Pw correspond to the output of 5×10^6 average power stations.

massive input of freshwater from the Laurentide ice sheet (covering Canada) into the North Atlantic [41]. The subsequently fresher and thus less dense subpolar upper-ocean waters stabilized the vertical stratification of the water column, hence, heavily impeding deep-water formation.

To simulate the effect of freshwater input on deep-water formation, Vellinga and Wood [42] carried out a 'water hosing' experiment using a numerical model. They added a sufficient quantity of freshwater to the northern North Atlantic to cause the AMOC to switch off. This resulted in a strong cooling over the North Atlantic peaking at a temperature reduction of 8°C around Greenland, standing out from patterns of moderate cooling over the entire Northern Hemisphere and warming over the Southern Hemisphere (Fig. 3). Thus, besides its strong importance for climate over the North Atlantic section, the AMOC may also have a moderate impact on global climate patterns.

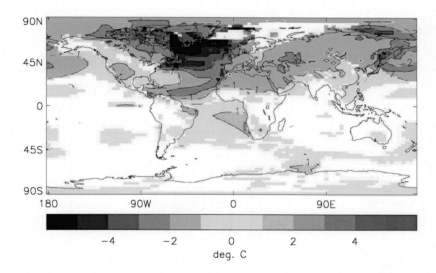

FIGURE 3

Change in surface air temperature during years 20–30 after the collapse of the AMOC in a water hosing experiment using the HadCM3 climate model. Areas where the anomaly is not significant have been masked. This figure has been published by Vellinga et al. [42] and has been reproduced with permission. The *x*-axis should read: temperature/°C.

Additionally, paleoclimate records suggest that changes in the global circulation involving the AMOC during the early last deglacial period (19 000 a–14 500 a) went along with a significant net transfer of CO_2 from the ocean to the atmosphere, leading atmospheric CO_2 concentration to rise by about 50 μmol mol^{-1} (50 × 10^{-6} vol) [43].

In a synthesis of climate reconstructions of the last deglaciations, Clark et al. [44] identify a large-scale climate mode strongly associated with variations of the AMOC. Together with a dominant mode of global warming, associated with the rising greenhouse gas concentrations, this 'AMOC mode' explains much of the variability in regional and global climate during the last deglaciation. Ritz et al. [45] demonstrate that on long timescales it might be possible to quantitatively deduce the strength of the AMOC from reconstructed climate patterns.

4.5 WHY SHOULD THE AMOC CHANGE AS PART OF ONGOING CLIMATE CHANGE?

In their fifth comprehensive climate assessment, the Intergovernmental Panel of Climate Change (IPCC) considers it 'very likely' that the AMOC will weaken over the twenty-first century relative to preindustrial values as a consequence of the greenhouse climate [46]. Climate model projections (CMIP5) based on the greenhouse gas emission scenarios yield best estimates and ranges for the reduction of 11% (1%–24%) in RCP2.6 and 34% (12%–54%) in RCP8.5 [46].

A complete (and possibly irreversible) AMOC shutdown is considered 'unlikely' over this time period. The future evolution of the AMOC in several selected climate model projections is shown in Fig. 4.

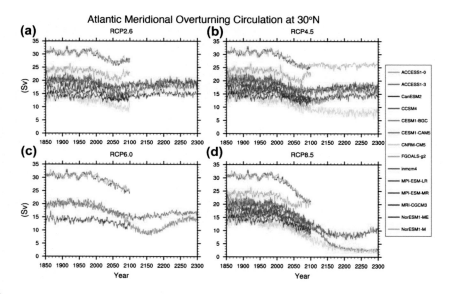

FIGURE 4

Evolution of the AMOC as defined by the maximum overturning at 30°N for the period 1850–2300 in different climate models (see legend) forced with different assumed future RCP forcing scenarios (panels a, b, c and d). RCP6.0 represents a radiative forcing of 60 W m^{-2} in 2100 relative to 1850. *(This figure was published in the fifth assessment of the IPCC [46] and has been reproduced with permission. The y-axis should read: transport/Sv.)*

Future greenhouse gas emission scenarios carry a high level of uncertainty as they depend on parameters such as economic and population growth, technology development and basic political and social conditions, all of which are difficult to predict. Also, none of the present-day climate models has sufficiently fine spatial resolution to resolve the processes that govern either the sinking or the rising, and has to rely on parameterizations instead. Both aspects may significantly add to the uncertainty in the prediction of the long-term AMOC evolution.

Reasons for a long-term greenhouse gas–induced reduction of the strength of the AMOC include straightforward effects, such as warming of surface waters, melting of continental ice sheets acting to reduce high-latitude salinity (a mechanism not included in many climate models), and intensification of the hydrological cycle [47]. All of these act to impede deep-water formation.

More complex feedbacks (that either stabilize or destabilize the AMOC) also involve wind field changes in the deep-water formation regions leading to buoyancy flux anomalies [48] and oceanic teleconnections driven by changes in the freshwater budget of the tropical Atlantic and South Atlantic [42,49–52].

Huang et al. [53] found a significant increase in wind stress and its energy input into the Southern Ocean between 1950 and 2000, which may have been caused by decreasing stratospheric ozone concentrations [54]. Using a climate model, Shindall and Schmidt [55] predict that the positive wind stress trend over the Southern Ocean will prevail until 2100 as a consequence of anthropogenic greenhouse gas–induced global warming. While oceanographers have not yet been able to establish a relationship between the multi-decadal wind stress increase and changes in the ocean circulation, which may partly be a consequence of unavailability of a sufficient number of suitable observations over the last 50 years, one might speculate that in the future, beyond the end of this century, in a then different climate 'pulling', the AMOC in the Southern Ocean might gain in importance relative to 'pushing' it in the North Atlantic [56].

4.6 CAN WE DETECT CHANGES IN THE AMOC? IS THE AMOC CHANGING ALREADY?

Previously direct estimates of the vigour of the AMOC have been obtained from transatlantic hydrographic (density profile) sections, assuming the geostrophic balance to hold on the ocean interior. Five such sections have been carried out along 24.5°N in the Atlantic over the last 50 years [50]. Taken together the five snapshots, each of which is assumed to be representative of the annual mean strength of the AMOC of the year in which they were taken (i.e. intra-seasonal variations assumed to be small), implies an AMOC slowdown of 30% (or 8 Sv) since 1957 [57]. Other measurements, focusing on single components of the AMOC, indeed show year-to-year changes, yet do not seem to support an ongoing long-term decline of the AMOC. Using a combination of direct and indirect transport measurement techniques, a gradual 1 Sv–2 Sv decrease in the amount of cold, dense overflow of deep waters from the Nordic Seas through the Faroe Bank Channel (feeding the NADW) has been found since 1970 [58], implying a long-term AMOC weakening. However, the continuation of the direct measurements showed an increase over the last few years back to the levels of the mid-1990s [59,60]. Overall, the volume transports that have been observed in the major deep-water overflow branches in Faroe Bank Channel, Denmark Strait have been stable between the mid-1990s and 2011 [61,62].

At the same time, measurements in the Deep Labrador Current, which represents a major pathway for the export of NADW from the deep-water formation regions, seem to have strengthened by 15% when comparing the 1996–1999 to 2000–2005 periods [63]. However, no significant long-term change was detected over the period between 1996 and 2009 [64]. After leaving the Labrador Sea it is uncertain how representative the strength of the DWBC is for the basin-wide AMOC. Hydrographic measurements in the mid- and high-latitude North Atlantic suggest that a substantial part of the southward export of NADW might be accomplished along a pathway in the ocean interior that feeds into the DWBC only in the subtropical North Atlantic [65]. Kanzow et al. [66] showed from observations and model simulations that fluctuations in the strength of the DWBC may not be a good indicator of AMOC changes in the tropical North Atlantic either, due to the presence of time-variable deep offshore recirculations.

A pilot system to measure the strength of the full AMOC (as opposed to individual components) continuously at 26.5°N (i.e. the zonally integrated meridional transport profile between Florida and Morocco) has been operating since April 2004 [10,17,67]. Figure 5 shows an eight-year-long time series of the AMOC between 2004 and 2012, exhibiting a time mean of 17.5 Sv [68]. The observed pronounced seasonal variability raises concerns whether the hypothesized 30% slowdown of the AMOC [50] may represent aliasing effects (as a consequence of not resolving the large intra-seasonal variations) rather than a sustained change of the ocean circulation [17,69]. The AMOC at 26.5°N was on average 2.7 Sv stronger during the 2004–2007 than during 2008–2011 (Fig. 5, [68]), suggesting a significant decline over the period of observations. This decline, however, is thought to be part of decadal, regional, wind-driven variations of the AMOC [70] rather than induced climate change. Decadal-scale variations in the transport of NADW (i.e. the deep limb of the AMOC) have also been found from continuous observations since 2000 in the tropical North Atlantic at 16°N [71], the cause of which has not been clarified. Indirect estimates of the AMOC based on coastal tide-gauges along the coastline of the USA revealed interannual variations yet no significant long-term trend over an almost 50-year-long period between 1960 and 2007 [72]. In summary, considerable progress has been made in the continuous observations of both the full and components of the AMOC over the past decade. The time series reveal temporal variations ranging from seasonal to multi-annual timescales, which at this point are thought to be part of natural fluctuations of the ocean-atmosphere system.

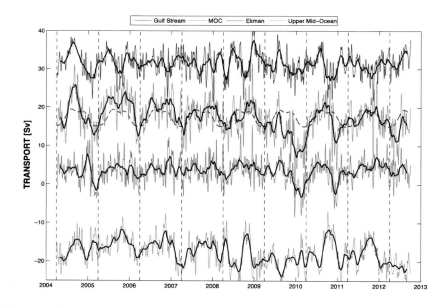

FIGURE 5

Ten-day (colours) and low-pass filtered time series of Gulf Stream transport (blue), Ekman transport (green), upper mid-ocean transport (magenta), and overturning transport (red) for the period 1 April 2004 to 1 October 2012. A dashed black line shows the mean annual cycle for the AMOC. Positive transports correspond to northward flow. *(This figure was published by Smeed et al. (2014, [69]) and has been reproduced with permission. The y-axis should read: Transport/Sv.)*

While oceanographers have not yet been able to document a statistically significant multi-decadal in the strength of the AMOC, it is worth asking how much time it would take to detect a possible long-term trend from continuous measurements at 26.5°N. Making assumptions about the short-term noise level of the AMOC, Baehr et al. [73] concluded from the analysis of an AMOC future projection that a 0.75 Sv per decade decline could be detected after three decades. A more abrupt (than currently expected) AMOC change would be detectable earlier. The detectability could most likely be shortened significantly if several continuously observing AMOC monitoring systems were operated simultaneously at different latitudes.

5. CONCLUSION

Observations have revealed that patterns of present-day regional and large-scale ocean circulation may display strong changes on intra-seasonal to multi-decadal timescales. Physical oceanographers have developed a variety of tools to quantify circulation changes, which involve direct and indirect measurement techniques and numerical simulations. While most of the documented present-day circulation changes are believed to fall within the class of natural (ocean climate) variability even at decadal and longer timescales, it is a nontrivial task to disentangle climate variability from presently possibly ongoing climate shifts.

The Atlantic Meridional Overturning Circulation has been in the focus of climate change research. The interpretation of paleoclimate records in the light of findings from numerical climate models reveals that the AMOC has undergone large changes in the Earth's past and that these went along with climate shifts in the North Atlantic sector and beyond. In the present-day climate, the AMOC represents the major oceanic mechanism of meridional heat transport. The AMOC moves volumes of cold waters (having sunk at high latitudes) southward throughout the Atlantic at depth and keeps them out of contact with the atmosphere for centuries, until the waters eventually rise to the upper ocean. Thereby the AMOC ventilates the deep ocean with oxygen rich waters. The sinking of waters in the Nordic Seas and the Labrador Sea (push) and their eventual rising (pull) are necessary ingredients for the existence of the AMOC, both of which are thought to change in a changing climate.

Model projections imply that the AMOC might slow down between 1% and 54% by the end of the twenty-first century relative to preindustrial values. This is thought to be due to an increase in vertical density stratification at high latitudes (both due to warming and freshening of surface waters) as a result of global warming. However, none of the present-day climate models has sufficiently fine spatial resolution to resolve the processes that govern either the sinking or the rising, and has to rely on parametrizations instead. Additionally, the climate model projections that produced the range of 0%–50% in the AMOC decline were based on different greenhouse gas forcing scenarios. The future climate forcing itself is therefore part of the uncertainty of the projections. There is clearly a need to monitor the state of the AMOC continuously over coming decades, in order to reduce the unsatisfactorily large uncertainties of the future projections.

To date there is no credible evidence that the AMOC has started to decrease in strength as a result of anthropogenic climate forcing. This is partly because the full AMOC has only been the subject of successful, continuous monitoring efforts for about one decade. These efforts represent a huge step forward because we can now start to use the time series to develop AMOC indices that go back in time significantly longer. Still, we argue that a reliable index of the strength of the AMOC spanning the last 50 years (or so) does not exist. The recent continuous measurements at 26.5°N suggest that the amplitudes of intra-seasonal, seasonal and interannual variations of the strength of AMOC are larger than previously thought. This makes it a challenging task to infer long-term AMOC changes from the few sporadic attempts to estimate the strength of the AMOC prior to the 1990s. Measurements that have focused on the observation of one particular component rather than the full AMOC (such as the strength of the deep western boundary current or the deep Nordic Sea inflow into the Atlantic) do not show clear, uniform trends either. In addition, it is difficult to assess how a change in one of the components translates into a change of the full AMOC. However, even with the recently started suitable continuous AMOC observations now well under way, the detection of a possible ongoing, global-warming-induced decline in the vigour of the AMOC may still be decades away. This will strongly depend on how fast the decline actually comes about, if at all.

The major difficulty for scientists to document ocean circulation changes on climate relevant timescales arises from the sparseness of historical in situ observations both in space and time. Over the last decades it has been (and partly still is) a technical, logistical and financial challenge to maintain ocean observatories at key locations continuously for more than a few years and/or to repeat measurement campaigns at a frequency that is sufficient to detect trends with a high level of confidence. However, the awareness that understanding the processes that govern ocean circulation changes may be vital for present and future societies has triggered dedicated, internationally coordinated field programmes and along with them technical developments (such as autonomous in situ profilers or advances in remote sensing and data-model syntheses). As a consequence, physical oceanography is currently undergoing a step change in capacity, capability and understanding, from which future generations will certainly profit.

REFERENCES

[1] T.N. Lee, W.E. Johns, R. Zantopp, E.R. Fillenbaum, J. Phys. Oceanogr. 26 (1996) 962–963.
[2] D. Zhang, M.J. McPhaden, Ocean Model. 15 (2006) 250–273.
[3] M. Feng, M.J. McPhaden, T. Lee, Geophys. Res. Lett. 37 (2010) L09606, http://dx.doi.org/10.1029/2010GL042796.
[4] S. Häkkinen, P. Rhines, Science 304 (2004) 555–559.
[5] M. Bersch, I. Yashayaev, K.P. Koltermann, Ocean Dyn. 57 (2007), http://dx.doi.org/10.1007/s10236-007-0104-7.
[6] J.W. Hurrell, Y. Kushnir, G. Ottersen, M. Visbeck, in: J.W. Hurrell, Y. Kushnir, G. Ottersen, M. Visbeck (Eds.), The North Atlantic Oscillation, Geophys. Monogr. Series, vol. 134, 2003, pp. 1–36.
[7] H. Hàtùn, A.B. Sandø, H. Drange, B. Hansen, H. Valdimarsson, Science 309 (2005) 1841–1844.
[8] J.R. Knight, R.J. Allan, C.K. Folland, M. Vellinga, M.E. Mann, Geophys. Res. Lett. 32 (2005) L20708, http://dx.doi.org/10.1029/2005GL024233.
[9] J.W. Hurrell, M. Visbeck, A. Busalacchi, R.A. Clarke, T.L. Delworth, R.R. Dickson, W.E. Johns, K.P. Koltermann, Y. Kushnir, D. Marshall, C. Mauritzen, M.S. McCartney, A. Piola, C. Reason, G. Reverdin, F. Schott, R. Sutton, I. Wainer, D. Wright, J. Clim. 19 (2006) 5100–5121.
[10] W.E. Johns, L.M. Beal, M.O. Baringer, J.R. Molina, S.A. Cunningham, T. Kanzow, D. Rayner, J. Phys. Oceanogr. 38 (2008) 605–623.
[11] T.M. Joyce, D.S. Bitterman, K.E. Prada, Deep Sea Res. 29 (1992) 903–913.
[12] J. Fischer, M. Visbeck, J. Atm. Ocean. Tech. 10 (1993) 764–773.
[13] R. Davis, W. Zenk, in: G. Siedler, J. Church, J. Gould (Eds.), Ocean Circulation and Climate, Academic Press, London, 2001.

[14] V.W. Ekman, Arch. Math. Astron. Phys. 2 (1905) 1–53.

[15] G. Wüst, Wissenschaftliche Ergebnisse der deutschen atlantischen expedition. "Meteor," 1925–1927, in: W.J. Emery (Ed.), English, the Stratosphere of the Atlantic Ocean, 1978, vol. 6, Amerind, 1935, pp. 109–288, 112 pp.

[16] H. Longworth, H.L. Bryden, in: A. Schmittner, J.C.H. Chiang, S.R. Hemmings (Eds.), Ocean Circulation: Mechanisms and Impacts – Past and Future Changes of Meridional Overturning, Geophys. Monogr. Series, vol. 173, American Geophysical Union, 2007.

[17] S.A. Cunningham, T. Kanzow, D. Rayner, M.O. Barringer, W.E. Johns, J. Marotzke, H.R. Longworth, E.M. Grant, J.J.-M. Hirschi, L.M. Beal, C.S. Meinen, H.L. Bryden, Science 317 (2007) 935–938.

[18] T. Kuhlbrodt, A. Griesel, M. Montoya, A. Levermann, M. Hofmann, S. Rahmstorf, Rev. Geophys. 45 (2007) RG2001, http://dx.doi.org/10.1029/2004RG000166.

[19] J.M. Lilly, P.B. Rhines, M. Visbeck, R. Davis, J.R.N. Lazier, F. Schott, D. Farmer, J. Phys. Oceanogr. 29 (1999) 2065–2098.

[20] J. Marshall, F. Schott, Rev. Geophys. 37 (1999) 1–64.

[21] C. Wunsch, R. Ferrari, Ann. Rev. Fluid Mech. 36 (2004) 281–314.

[22] K.L. Polzin, J.M. Toole, J.R. Ledwell, R.W. Schmitt, Science 276 (1997) 93–96.

[23] F. Paparella, W.R. Young, J. Fluid Mech. 466 (2002) 205–214.

[24] J.R. Toggweiler, B. Samuels, J. Phys. Oceanogr. 28 (1998) 1832–1852.

[25] J. Marshall, K. Speer, Nat. Geosci. 5 (2012) 171–180, http://dx.doi.org/10.1038/ngeo1391.

[26] A.H. Orsi, W.M. Smethie, J.L. Bullister, J. Geophys. Res. 107 (2002) 3122, http://dx.doi.org/10.1029/2001JC000976.

[27] G.C. Johnson, J. Geophys. Res. 113 (2008) C05027, http://dx.doi.org/10.1029/2007JC004477.

[28] G. Boccaletti, R. Ferrari, A. Adcroft, D. Ferreira, J. Marshall, Geophys. Res. Lett. 32 (2005) L10603, http://dx.doi.org/10.1029/2005GL022474.

[29] S.G. Purkey, G.C. Johnson, J. Clim. 26 (2013) 6105–6122, http://dx.doi.org/10.1175/JCLI-D-12-00834.1.

[30] H.L. Bryden, S. Imawaki, in: G. Siedler, J. Church, J. Gould (Eds.), Ocean Circulation and Climate, Academic Press, London, 2001, pp. 455–474.

[31] J.L. Sarmiento, N. Gruber, Ocean Biogeochemical Dynamics, Princeton University Press, Princeton, 2006.

[32] T. Takahashi, C. Sweeney, B. Hales, D.W. Chipman, T. Newberger, J.G. Goddard, R.A. Iannuzzi, S.C. Sutherland, Oceanography 25 (2012) 26–37, http://dx.doi.org/10.5670/oceanog.2012.71.

[33] S.J. Johnson, H.B. Clause, W. Dansgaard, K. Fuhrer, N. Gundestrup, C.H. Hammer, P. Iversen, J. Jouzel, B. Stauffer, J.P. Steensen, Nature 311 (1992) 313.

[34] W. Dansgaard, 10 others, Nature 364 (1993) 218–220.

[35] L. Vidal, L. Labeyrie, E. Cortijo, M. Arnold, J.C. Duplessy, E. Michel, S. Becqué, T.C.E. van Weering, Earth Planet. Sci. Lett. 146 (1997) 13–27.

[36] P.U. Clark, N.G. Pisias, T.F. Stocker, Nature 415 (2002) 863–869.

[37] J.F. McManus, R. Francois, J.-M. Gherardi, L.D. Keigwin, Nature 428 (2004) 834–837.

[38] J.M. Gherardi, L. Labeyrie, S. Nave, R. Francois, J.F. McManus, E. Cortijo, Paleoceanography 24 (2009) 2204.

[39] A. Ganopolski, S. Rahmsdorf, V. Petoukhov, M. Claussen, Nature 391 (1998) 351–356.

[40] A. Ganopolski, S. Rahmsdorf, Nature 409 (2001) 153–158.

[41] S. Hemming, Rev. Geophys. 42 (2004) 1005, http://dx.doi.org/10.1029/2003RG000128.

[42] M. Vellinga, R.A. Wood, J.M. Gregory, J. Clim. 15 (2002) 764–780.

[43] M. Sarnthein, P.M. Grootes, J.P. Kennett, M.-J. Nadeau, in: A. Schmittner, J. Chiang, S. Hemming (Eds.), Ocean Circulation: Mechanisms and Impacts – Past and Future Changes of the Meridional Overturning, Geophys. Monogr. Series, vol. 173, American Geophysical Union, New York, 2007, pp. 175–196.

[44] P.U. Clark, J.D. Shakun, P.A. Baker, P.J. Bartlein, S. Brewer, E. Brook, A.E. Carlson, H. Cheng, A. Darrell, S. Kaufman, Z. Liu, T.M. Marchitto, A.C. Mix, C. Morrill, B.L. Otto-Bliesner, K. Pahnke, J.M. Russell,

C. Whitlock, J.F. Adkins, J.L. Blois, J. Clark, S.M. Colman, W.B. Curry, B.P. Flower, F.g He, T.C. Johnson, J. Lynch-Stieglitz, V. Markgraf, J. McManus, J.X. Mitrovica, P.I. Moreno, J.W. Williams, Proc. Natl. Acad. Sci. U.S.A. 109 (2012) 1134–1142, http://dx.doi.org/10.1073/pnas.1116619109.

[45] S.P. Ritz, T.F. Stocker, J.O. Grimalt, L. Menviel, A. Timmermann, Nat. Geosci. 6 (2013) 208–214, http://dx.doi.org/10.1038/ngeo1723.

[46] M. Collins, R. Knutti, J. Arblaster, J.-L. Dufresne, T. Fichefet, P. Friedlingstein, X. Gao, W.J. Gutowski, T. Johns, G. Krinner, M. Shongwe, C. Tebaldi, A.J. Weaver, M. Wehner, Long-term climate change: projections, com- mitments and irreversibility, in: T.F. Stocker, D. Qin, G.-K. Plattner, M. Tignor, S.K. Allen, J. Boschung, A. Nauels, Y. Xia, V. Bex, P.M. Midgley (Eds.), Climate Change 2013: The Physical Science Basis. Contribution of Working Group I to the Fifth Assessment Report of the Intergovernmental Panel on Climate Change, Cambridge University Press, Cambridge, United Kingdom and New York, NY, USA, 2013.

[47] J.M. Gregory, K.W. Dixon, R.J. Stouffer, A.J. Weaver, E. Driesschaert, M. Eby, T. Fichefet, H. Hasumi, S. Hu, J.H. Jungclaus, I.V. Kamenkovich, A. Levermann, M. Montoya, S. Murakami, S. Nawrath, A. Oka, A.P. Sokolov, R.B. Thorpe, Geophys. Res. Lett. 32 (2005) L12703, http://dx.doi.org/10.1029/2005GL023209.

[48] P. Gent, Geophys. Res. Lett. 28 (2001) 1023–1026.

[49] M. Latif, E. Roeckner, U. Mikolajewicz, R. Voss, J. Clim. 13 (2000) 1809–1813.

[50] R.B. Thorpe, J.M. Gregory, T.C. Johns, R.A. Wood, J.F.B. Mitchell, J. Clim. 14 (2001) 3102–3116.

[51] A.X. Hu, G.A. Meehl, W.M. Washington, A. Dai, J. Clim. 17 (2004) 4267–4279.

[52] U. Krebs, A. Timmermann, J. Clim. 20 (2007) 4940–4956.

[53] R.X. Huang, W. Wang, L.L. Liu, Deep-Sea Res. 53 (2006) 31–41.

[54] D.W. Thompson, S. Solomon, Science 296 (2002) 895–899.

[55] D.T. Shindell, G.A. Schmidt Geophys, Res. Lett. 31 (2004) L18209, http://dx.doi.org/10.1029/2004GL020724.

[56] M. Visbeck, Nature 447 (2007) 383.

[57] H.L. Bryden, H.L. Longworth, S.A. Cunningham, Nature 438 (2005) 655–657.

[58] B. Hansen, W.R. Turrell, S. Østerhus, Nature 411 (2001) 927–929.

[59] B. Hansen, S. Østerhus, Progr. Oceanogr. 75 (2007) 865–871, http://dx.doi.org/10.1016/j.pocean.2007.09.004.

[60] S. Østerhus, T. Sherwin, D. Quadfasel, B. Hansen, in: R.R. ickson, et al. (Eds.), Arctic-subarctic Ocean Fluxes, Springer Science + Business Media B.V, 2008, pp. 427–441.

[61] K. Jochumsen, D. Quadfasel, H. Valdimarsson, S. Jónsson, J. Geophys. Res. 117 (2012) C12003, http://dx.doi.org/10.1029/2012JC008244.

[62] B. Berx, B. Hansen, S. Østerhus, K.M. Larsen, T. Sherwin, K. Jochumsen, Ocean Sci. 9 (2013) 639–654, http://dx.doi.org/10.5194/os-9-639-2013.

[63] M. Dengler, J. Fischer, F.A. Schott, R. Zantopp, Geophys. Res. Lett. 33 (2006) L21S06, http://dx.doi.org/10.1029/2006GL026702.

[64] J. Fischer, M. Visbeck, R. Zantopp, N. Nunes, Geophys. Res. Lett. 37 (2010) L24610, http://dx.doi.org/10.1029/2010GL045321.

[65] M.S. Lozier, Science 328 (2010) 1507–1511.

[66] T. Kanzow, U. Send, M. McCartney, Deep Sea Res. 1 (2008) 1601–1623, http://dx.doi.org/10.1016/j.dsr.2008.07.011.

[67] T. Kanzow, S.A. Cunningham, D. Rayner, J.J.-M. Hirschi, W.E. Johns, M.O. Baringer, H.L. Bryden, L.M. Beal, C.S. Meinen, J. Marotzke, Science 317 (2007) 938–941.

[68] T. Kanzow, S.A. Cunningham, W.E. Johns, J.J.-M. Hirschi, J. Marotzke, M.O. Baringer, C.S. Meinen, M.P. Chidichimo, C. Atkinson, L.M. Beal, H.L. Bryden, J. Collins, J. Clim. 23 (2010) 5678–5698.

[69] D.A. Smeed, G.D. McCarthy, S.A. Cunningham, E. Frajka-Williams, D. Rayner, W.E. Johns, C.S. Meinen, M.O. Baringer, B.I. Moat, A. Duchez, H.L. Bryden, Ocean Sci. 10 (2014) 29–38, http://dx.doi.org/10.5194/os-10-29-2014, 2014.

[70] J. Zhao, W.E. Johns, J. Geophys, Res. Oceans 119 (2014) 2403–2419, http://dx.doi.org/10.1002/2013JC009407.

[71] U. Send, M. Lankhorst, T. Kanzow, Geophys. Res. Lett. 38 (2011) L24606, http://dx.doi.org/10.1029/2011GL049801.

[72] R.J. Bingham, C.W. Hughes, Geophys. Res. Lett. 36 (2009) L02603, http://dx.doi.org/10.1029/2008GL036215.

[73] J. Baehr, H. Haak, S. Alderson, S.A. Cunningham, J.H. Jungclaus, J. Marotzke, J. Clim. 20 (2007) 5827–5841.

OCEAN ACIDIFICATION

18

Carol Turley, Helen S. Findlay
Plymouth Marine Laboratory, Plymouth, Devon, UK

CHAPTER OUTLINE

1. INTRODUCTION
1.1 CARBONATE CHEMISTRY

The ocean has the capacity to absorb large amounts of carbon dioxide (CO_2) because CO_2 dissolves and reacts in seawater to form bicarbonate (HCO_3^-) and protons (H^+). Between one-quarter to one-third of the CO_2 emitted into the atmosphere from the burning of fossil fuels, cement manufacturing and land use changes has been absorbed by the ocean [1]. Over thousands of years, the changes in pH have been buffered by bases, such as carbonate ions (CO_3^{2-}). However, the rate at which CO_2 is currently being absorbed into the ocean is too rapid to be buffered sufficiently to prevent substantial changes in ocean pH. As a consequence, the relative seawater concentrations of CO_2, HCO_3^-, CO_3^{2-} and pH have been altered. Since preindustrial times, ocean pH has decreased on a global average by 0.1 (compare Fig. 1(a) and (b)). The Intergovernmental Panel on Climate Change (IPCC) [2] projects a global increase in ocean acidification for all representative concentration pathway (RCP) scenarios, with decreases in mean surface ocean pH by the end of the twenty-first century in the range of 0.06–0.07 (RCP2.6), 0.14–0.15 (RCP4.5), 0.20–0.21 (RCP6.0) and 0.30–0.32 (RCP8.5). The RCP8.5 outcome represents an increase in acidity of >150% compared with preindustrial values [2,3] (Fig. 1(c)). By year 2300, surface ocean pH is projected to decrease by 0.7 [4]. It will take tens of thousands of years for these changes in ocean chemistry to be buffered through neutralization by calcium carbonate sediments and the level at which the ocean pH will eventually stabilize will be lower than it currently is [5].

Climate Change. http://dx.doi.org/10.1016/B978-0-444-63524-2.00018-X

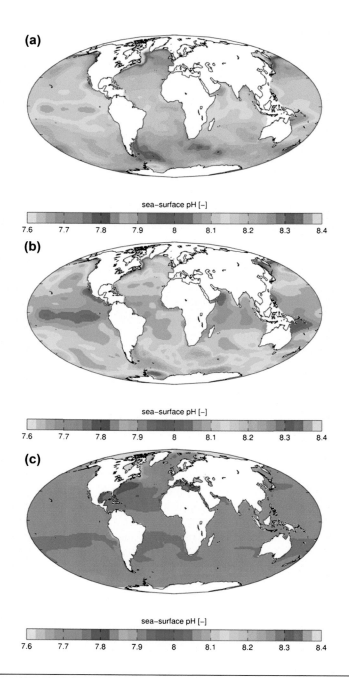

FIGURE 1

Modelled sea-surface pH in the (a) pre-industrial era (1700s), (b) present era (1990s) and (c) future (year 2100). Images show multi-model means using the RCP8.5 scenario (models are AR5 models: CanESM2, GFDL-ESM2G, IPSL-CM5A-LR, IPSL-CM5A-MR, MIROC-ESM-CHEM, MIROC-ESM, MPI-ESM-LR and MPI-ESM-MR).*(Courtesy of Andrew Yool (National Oceanography Centre, Southampton).)*

The $CO_3{}^{2-}$ concentration directly influences the saturation, and consequently the rate of dissolution, of calcium carbonate ($CaCO_3$) minerals in the ocean. The saturation state (Omega; Ω) is used to express the degree of $CaCO_3$ saturation in seawater:

$$\Omega = \left[Ca^{2+}\right]\left[CO_3^{2-}\right]\Big/K_{sp}^*$$

where K_{sp}^* is the solubility product for $CaCO_3$ and $[Ca^{2+}]$ and $[CO_3{}^{2-}]$ are the in situ calcium and carbonate concentrations, respectively. When $\Omega > 1$, seawater is supersaturated with respect to mineral $CaCO_3$ and the larger this value, the more suitable the environment will be for organisms that produce $CaCO_3$ (shells, liths and skeletons). When $\Omega < 1$, seawater is undersaturated and is corrosive to $CaCO_3$. Presently, the vast majority of the surface ocean is supersaturated with respect to $CaCO_3$. The depth at which $\Omega = 1$ is known as the saturation horizon. The three main mineral forms of $CaCO_3$, in order of least soluble to most soluble, are calcite, aragonite and high magnesium-calcite. Each form of the mineral has different saturation state profiles and saturation horizons, with the aragonite saturation horizon (ASH) being shallower than the calcite saturation horizon (CSH). Due to differences in ocean properties (salinity, temperature and pressure) both horizons vary with latitude and ocean basin. The depth of the ASH is 600 m or less in the North Pacific but can be over 2000 m deep in the North Atlantic. Increasing atmospheric CO_2 will cause Ω to decrease, as has already been occurring since preindustrial times [6].

1.2 COMBINED IMPACTS OF OCEAN ACIDIFICATION AND CLIMATE CHANGE

Changes in climate resulting from anthropogenic influences will synchronously alter environmental conditions such as temperature, pH, salinity, wind strength and oxygen levels [2]. Ocean warming dominates the increase in energy stored in the climate system, accounting for more than 90% of the energy accumulated between 1971 and 2010. On a global scale, the ocean warming is largest near the surface, and the upper 75 m warmed by $0.11°C$ per decade over the period 1971 to 2010 [7]. While seawater pH is sensitive to temperature, it is only a small contributing factor, such that the predicted range for future temperatures will not make a significant difference to the pH decline [8]. However, organisms' responses may be different with increasing temperature depending on the level at which they adapted [9]. The pH is also sensitive to changes in salinity, as a result of changes in total alkalinity and dissolved inorganic carbon, so organisms in coastal waters with riverine input can experience larger variability in pH than in the open ocean [10]. Both increasing temperature and decreasing salinity will also act to increase ocean stratification, which in turn will alter the nutrient supply that fuels primary production. A change in wind strength is also an important consideration for ocean acidification for two reasons. Firstly, wind strength determines the flux of CO_2 between the ocean and the atmosphere, so may reduce the ocean CO_2 sinks [11]. Secondly, wind strength drives ocean currents, mixes nutrients into the productive upper ocean and is particularly important for generating upwelling areas [12]. Upwelling areas, although rich in nutrients, are also rich in CO_2 and are therefore areas of natural low pH [13]. In parallel to warming, oxygen concentrations have decreased in coastal waters and in the open ocean thermocline in many ocean regions since the 1960s, with a likely expansion of tropical oxygen minimum zones in recent decades [7]. Additionally, an increase in nutrient load (e.g. caused by increased land run-off) can substantially increase biological productivity, and subsequent microbial decomposition of this excess productivity consumes large amounts of O_2 and releases CO_2 through respiration, causing hypoxia and low pH.

2. EVIDENCE FROM OBSERVATIONS
2.1 EVIDENCE FROM GEOLOGICAL AND ICE CORE RECORDS

Ice cores provide high resolution and accurate records of atmospheric CO_2 concentrations over the last 650 000 a and together with marine paleo-proxies (e.g. boron isotopes) serve to arrive at a reasonable estimate of ocean carbonate chemistry over millions of years [14] (Fig. 2).

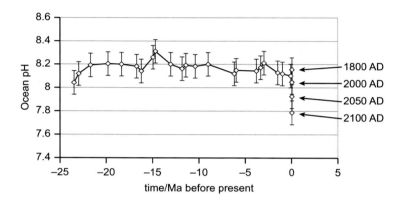

FIGURE 2

Past (white diamonds, data from Pearson and Palmer [14]) and contemporary variability of surface ocean pH (diamonds with dates). Future predictions are model-derived values based on IPCC 2007 mean scenarios. *(Adapted from Turley et al. [148].)*

The current rate of ocean acidification is unprecedented within the last 65 million years, (65 Ma) possibly the last 300 million years [15–17] and it will take tens of thousands of years for future ocean pH to return to near preindustrial conditions [18].

Observations that CO_2 variations in the glacial and interglacial periods of the last 50 000 a correlated with the shell weights of fossil planktonic foraminifers [19] indicate that marine calcifiers are influenced by small fluctuations in atmospheric CO_2 values and those effects are likely to progressively intensify with increasing CO_2.

2.2 EVIDENCE FROM LONG-TERM OCEANOGRAPHIC TIME SERIES

Direct measurements on ocean time series stations in the North Atlantic and North Pacific record increasing seawater pCO_2 as atmospheric CO_2 increases (Fig. 3(a)) and decreasing pH with rates ranging between -0.0014 a^{-1} and -0.0024 a^{-1} (Fig. 3(b)) [20–26]. Carbonate ions as well as pH also decreased over this period (Fig. 3(c)). These time series data show that the Pacific and the subtropical gyre at both sites on the North Atlantic are becoming more acidic as predicted by ocean general-circulation models (OGCMs) (see Section 2.3).

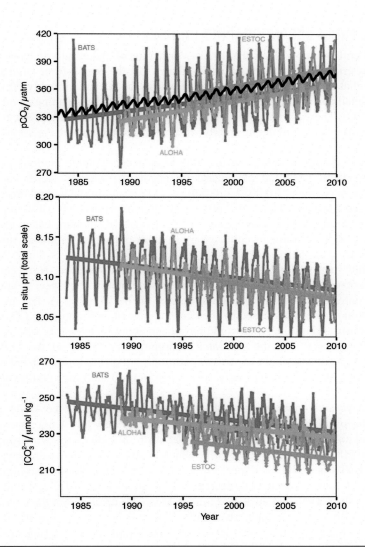

FIGURE 3

Long-term trends of surface seawater pCO_2 (top), pH (middle) and carbonate ion (bottom) concentration at three subtropical ocean time series in the North Atlantic and North Pacific Oceans, including the Bermuda Atlantic Time-series Study (BATS, 31°40′N, 64°10′W; green) and Hydrostation S (32°10′, 64°30′W) from 1983 to present *(Updated from Ref. [21].)*; the Hawaii Ocean Time-series (HOT) at Station ALOHA (A Long-term Oligotrophic Habitat Assessment; 22°45′N, 158°00′W; orange) from 1988 to present *(Updated from Ref. [24].)*; and the European Station for Time series in the Ocean (ESTOC, 29°10′N, 15°30′W; blue) from 1994 to present *(Updated from Ref. [26].)*. Atmospheric pCO_2 (black) from the Mauna Loa Observatory Hawaii is shown in the top panel [20].

2.3 EVIDENCE FROM OCEANOGRAPHIC CRUISES

An international collaboration since the 1990s, consisting of thousands of hydrographic stations collected on many cruises in different oceans has shown that the ocean stores large amounts of CO_2. These data show that the reservoir of inorganic carbon in the ocean is roughly 50 times that of the atmosphere [1]. Small changes in the ocean reservoir can therefore have an impact on the atmospheric concentration of CO_2. These data also show that the ocean is also an important sink for anthropogenic CO_2, which is currently accumulating nearly 30% of the total human emissions of CO_2 to the atmosphere [11].

This rate of pH change observed in the long-term oceanographic stations is also consistent with repeat transects of CO_2 and pH measurements in the western North Pacific (winter: -0.0018 ± 0.0002 a^{-1}; summer: -0.0013 ± 0.0005 a^{-1}) [27] and in the central North Pacific Ocean between Hawaii and Alaska (-0.0017 ± 0.001 a^{-1} decline in pH between 1991 and 2006) [28]. The paucity of long-term time series observations of pH changes in southern ocean surface waters are less certain, but pCO_2 measurements collected by ships-of-opportunity indicate similar rates of pH decrease there [29].

A hydrographic survey along the western coast of North America, from central Canada to northern Mexico, revealed upwelling of seawater undersaturated with respect to aragonite and with low pH (<7.75) onto large portions of the continental shelf [13] (Fig. 4). The areal extent of this natural phenomenon has been increased by the ocean uptake of anthropogenic CO_2. They estimated that during preindustrial days the ASH would have been about 50 m deeper with no undersaturated waters reaching the surface. With the additional anthropogenic CO_2 signal, the ASH has shoaled by around 1 m a^{-1} bringing increasingly corrosive conditions with pH as low as 7.6 not just to the deeper benthic communities but also, increasingly, to the productive, shallower continental shelf ecosystems [30].

Evidence from research cruises suggests that polar waters, particularly the Arctic, are experiencing the most rapid rates of ocean acidification, with waters already very low in aragonite saturation states [31]. Olafsson and colleagues [25] demonstrated that north of Iceland, the wintertime pH decreased between 1994 and 2008 at a rate of 0.0024 a^{-1}, which is 50% faster than average yearly rates at the tropical stations BATS and HOTS. Repeat cruises in the Beaufort Sea in the high Arctic suggest the surface aragonite saturation state has decreased from about 1.4 in 1997 to 1.0 in 2008 [32], with the majority of the decrease being attributed to increased CO_2 uptake from the atmosphere, but with additional impacts from sea ice melt, river influences and upwelling onto the shelf [33]. Local upwelling events, driving high CO_2 deep waters onto the Arctic shelves during storms, additionally exacerbates the issues of ocean acidification [34].

3. MODEL PREDICTIONS OF FUTURE CHANGE

OGCMs have been used to reconstruct, as well as predict, changes in climate. Forced by the physical dynamics of the ocean and atmosphere, and coupled together with biological models, OGCMs are able to reproduce biogeochemical cycling within the ocean that closely represents present and past observations. Using the IPCC RCP8.5 scenario, these models project a global average decrease in pH of 0.4 (equivalent to an increase in acidity >150%) by the year 2100 compared with preindustrial values (Fig. 1(c); [35–37]).

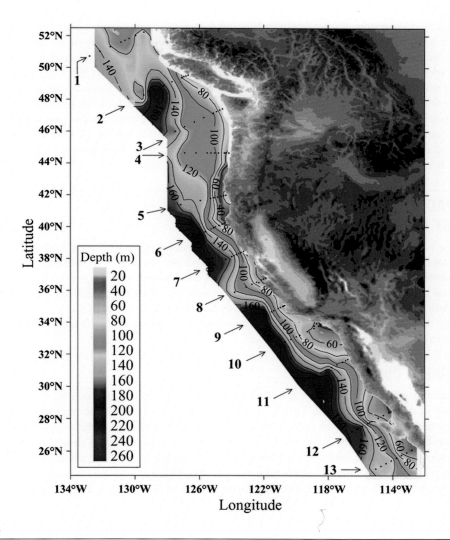

FIGURE 4

Distribution of the depths of the undersaturated water (aragonite saturation <1.0; pH < 7.7.5) on the continental shelf of western North America from Queen Charlotte Sound, Canada, to San Gregorio Baja California Sur, Mexico. On transect line 5 the corrosive water reaches all the way to the surface in the inshore waters near the coast. The black dots represent station locations. *(From Feely et al. [13].)*

Introducing changes in temperature, weathering and sedimentation into these simulations only reduced this maximum decline by 10% [3]. More detailed predictions of both carbonate ion and CO_2 concentration for different ocean regions and across latitudinal gradients strongly imply that the polar and subpolar waters are particularly vulnerable to ocean acidification [6,35]. The carbonate ion concentration is already much lower in these regions so they are particularly vulnerable to a reduction in pH [38]. Surface waters are projected to become seasonally corrosive to aragonite in parts of the Arctic and in some coastal upwelling systems within a decade, and in parts of the Southern Ocean within one to three decades in most scenarios. Aragonite under saturation becomes widespread in these regions at atmospheric CO_2 levels of (500–600) μmol mol^{-1} (500–600 ppm) [39]. Regional models are now being developed to assess the spatial and temporal variability in pH; for example, the future pH of the North Sea is predicted to undergo similar CO_2-induced changes to those predicted in the open ocean although coastal and shelf sea pelagic, and benthic activities and riverine input are important factors in contributing to a greater variability [10,40]. Continued uptake of CO_2 by the ocean is predicted to cause some areas in the ocean to be completely outside their natural ranges by the year 2050 [10,40].

4. IMPACTS

4.1 PAST OBSERVATIONS

There have been several prominent ocean acidification events in Earth's deep past, caused by massive input of carbon into the ocean; the best studied occurred 55 Ma ago [16]. The subsequent acidification led to a substantial extinction of microscopic seabed dwelling calcifiers [41]. In parts of the ocean, red clay instead of white carbonate was deposited for more than 100 000 a indicating the magnitude of the dissolution effect on global biogeochemical cycles and the duration of the recovery [42]. It is noteworthy that the timescale of the current anthropogenic input is about 10 times faster than what is thought to have occurred 55 Ma ago [16].

4.2 CURRENT OBSERVATIONS

To date, there are still relatively few observations of trends in biology that can be attributed to be a direct result of ocean acidification. This is in part a result of a lack of long-term coincident chemical and biological data with which to make coherent observations of change in this relatively young field of research; but it may also be an artefact of organisms' ability to cope with short-term variability in pH. Seawater pH in coastal and shelf sea locations can fluctuate by >0.9 depending on time, season, position in water column and freshwater influence [43,44]. That is, there may be only short periods of low pH with the periods of high pH allowing the organisms to recover [45]. Impacts may not become apparent until they are subjected to longer periods of lower pH, or the whole pH range that they experience is reduced. In contrast, seasonal pH variation of open ocean surface waters is around 0.07 (Fig. 3(b)), which may make these regions more sensitive to current and future acidification. Indeed, the detected change in pH (−0.1) since the preindustrial already exceeds the open ocean seasonal variation. Observed changes, for example, in species distributions, which have been attributed to changes in climate, pollution, ecosystem deterioration and so on, may be masking the role of ocean acidification.

One of the most concerning observational impacts of ocean acidification to date is the significantly reduced hatching success and larval development found across the oyster hatcheries on the West Coast of North America, which coincides with the upwelling events bringing corrosive waters to the hatchery inlets. Barton et al. [46] demonstrated larval growth and production was negative correlated with aragonite saturation state, and subsequently management strategies have been implemented to sustain the hatchery economy.

The California Current Ecosystem, which is subjected to the periodic upwelling events bringing corrosive waters to the nearshore and surface [13], has shown to produce regional complexities in organisms' tolerance or sensitivity to exposure to undersaturated conditions [30]. For instance, pteropods showed a strong positive relationship between shell dissolution and undersaturated conditions [47]. Off the west coast of South America, upwelling events similar to those found in North America [48] have also been found to potentially impact the sensitivity of organisms to CO_2 stress, such that organisms from areas where upwelling or large river inputs occur are more tolerant to high CO_2 when exposed in the laboratory [49]. Similar natural observations of pelagic organisms that vertically migrate across large CO_2 gradients also appear to be more tolerant to CO_2 when tested in laboratories [50,51]. Pteropods in the Southern Ocean, as well as the Arctic, have been shown to have increased shell dissolution at present oceanic CO_2 and pH levels in these lower saturation state regions [52,53]. Therefore, pteropods may represent a bellwether for observing ocean acidification impacts, and long-term monitoring of their status will be increasingly important [54].

Coccolithophores, microscopic plants that secrete $CaCO_3$ platelets called liths, occur over a variety of environmental conditions throughout the world's oceans. Areas known to have an extremely large seasonal cycle of calcite saturation states, with wintertime values declining to >1, appear to be areas where coccolithophores are absent [55], implying that the saturation state may have a large influence on their distribution. pH and mixed layer irradiance have also been able to explain most of the variation in coccolithophore distribution and community composition in the subarctic North Atlantic [56].

There have been observed differences in the cold-water coral ecosystems between the North Atlantic and the North Pacific, which may be indicative of biological shifts due to natural changes in ocean chemistry [57]; however on-going acidification impacts on cold-water corals are still unresolvable. Coral on the Great Barrier Reef (GBR), Australia, have shown a 21% decrease in net calcification and 30% decrease in growth over the period 1988–2003 [58], which cannot be solely attributed to increased sea-surface temperature, and is likely to result from a combination of impacts, including acidification. More recent analysis of present and historical data sets from one location on the GBR demonstrates that there has been a reduction in net calcification, without a significant change in coral cover over the past 40 years [59]. The decrease in net calcification is likely a consequence of an almost threefold increase in dissolution rates, but exactly how much ocean acidification contributes to this dissolution, compared to bioerosion and warming, has yet to be entirely resolved [59]. However, reefs in the Red Sea have shown correlated responses in net calcification rate to natural fluctuations in saturation state and temperature [60].

Natural marine volcanic CO_2 vent sites have been useful for assessing diversity shifts along a CO_2 gradient. A number of key ecosystem changes are apparent, for example, calcareous algae were replaced by noncalcareous algae and sea grasses with the latter increasing their primary production, and a loss of calcified invertebrates at low pH [61], although this appears to be a result of competitive interaction as well as direct impact from the acidified conditions [62].

4.3 EXPERIMENTAL OBSERVATIONS

To date, the majority of controlled laboratory experiments have been conducted on single organisms or in larger volume sediment or seawater mesocosms enriched with CO_2 containing mixed populations. There are many important biological processes within the lifecycle of an organism, between organisms and within ecosystems. Therefore, an impact on a process, be it at the cellular level or ecosystem level, may have a negative impact on the ultimate successful functioning of the ecosystem (Fig. 5). Several reviews have described the relative sensitivity of different taxonomic groups to ocean acidification [63,64]; instead here we summarize the response of processes in Table 1.

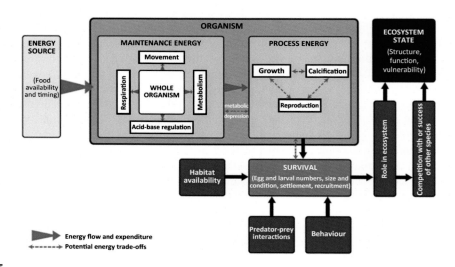

FIGURE 5

Schematic of processes, energy flows and expenditures involved in organisms being able to survive, interacting with external factors that can also influence survival and the consequential impacts on the ecosystem state. Red full arrows indicate energy flows, dashed arrows represent potential energy trade-offs. Ocean acidification or other stressors can cause shifts in the energy distribution within an organism, but could also impact on the external factors, such as food availability, predator–prey interactions, and behaviour. *(Adapted from Turley [141].)*

4.4 COMBINED IMPACTS

Temperature already provides limits to the survival of organisms; it alters many physiological processes by acting on the rates at which these processes occur (e.g. speeding up metabolism, enzyme activity, etc.). However, organisms are acclimatized to a certain temperature range. Acidification may act to narrow these ranges [9,110]. Increasing temperature will also drive many species toward the poles, either as a result of biogeographic range expansion (by temperate and tropical species) or as a result of contraction (by boreal and polar species). However, ocean acidification may act in the opposite direction, as the polar waters will be most affected by increasing CO_2 [6,31,111]. This could lead to a complete disappearance of boreal and polar species and may restrict the ability of temperate and tropical species to migrate. Where experiments have tested the impact of temperature and CO_2, the majority of studies suggest that shifts in temperature have a greater impact than changes in CO_2 or pH. Of the phytoplankton studies, approximately 12 out of about 54 ocean acidification experiments also considered an addition impact, either temperature, light or nutrient availability, and all revealed complicated responses different to those found when just elevating CO_2.

Table 1 Main acidification impacts on specific biological or ecological processes

Process	Main acidification impacts
Photosynthesis	• Phytoplankton C:N:P ratios can shift in response to elevated CO_2 (for a full review, see [65]). • Nearly all coccolithophores and diatoms increase photosynthetic rate and carbon assimilation with increasing CO_2 (with a few exceptions). • There are differences in response between phytoplankton size fractions, and in mixed communities there appears to be a shift toward smaller cells in a more stratified high CO_2 ocean (e.g. [66]).
Calcification	• High variability amongst experiments: Some studies show no change or increased net calcification, although most calcifying species studied to date show reduced net calcification rates (see sensitivity analysis by [64]), with largest effects being found in molluscs (−40%), corals (−32%) and coccolithophores (−23%). • In many instances, the reduction in net calcification could be being dominated either by a direct increase in dissolution [67] and/or as a result of costs associated with changes in an organism's energetic budget [68].
Acid-base regulation and metabolic regulation	• For normal function of an organism, internal pH must be kept within narrow ranges to maintain processes such as enzyme function, protein phosphorylation, and chemical reactions [9]. • Different organisms have different capabilities at regulating their acid−base balance; it appears that those species with high ability to regulate and buffer changes to internal conditions are less sensitive to acute OA stress [69]. • Ion regulation is an energy-consuming process, hence over longer periods, OA will likely alter an organism's energy budget [70]; and increased energy demand could be evidenced through increased metabolic rates [71,72]. • Organisms less able to compensate for acid−base disturbances show metabolic depression [73], which ultimately lowers energy availability for other biological processes. • Knowledge of the physiological type and capabilities of an organism does provide an assessment of its potential sensitivity to OA [51].
Early life stage (fertilization, embryo and larval development)	• Many benthic marine invertebrates produce free-swimming larvae, which develop through several larval stages in the plankton before settling into the adult form. Large numbers of larvae are often produced because of high rates of mortality, for example, coastal estuarine bivalves experience more than 98% mortality during settlement [74]. • Oyster [75], echinoderm [76,77], and fish [78] larvae, shark hatching conditions and survival [79], as well as barnacle [80−82], and copepod eggs and nauplii [51,83] have all been found to either be increasingly malformed, have slower development rates or reduced survival when under OA stress.

Continued

Table 1 Main acidification impacts on specific biological or ecological processes—cont'd

Process	Main acidification impacts
Communication	• Chemical cues are used for marine communication and can have strong influences on habitat selection and predator–prey interactions as well as courtship and mating, species recognition, and symbiotic relationships [84]. • Settlement of oyster larvae can be induced or inhibited by pH (e.g. [85–87]). • Weakly acidic environments impaired the ability of juvenile salmon to detect and respond to alarm cues [88]. • Littorinid snails on rocky shores switched from thickening shells in the presence of predators to avoidance behaviour when exposed to OA [89], while the predatory snail *Nucella* showed decreased chemosensory abilities, taking longer to locate food [90]. • Impacts on chemosensory function when exposed to acidification have also been observed in crabs [91] and fish [92–95].
Interactions	• The responses that occur within one individual can lead to changes in how it interacts with others and its environment. • Burrowing brittle stars, important prey for commercial fish and aid nutrient and oxygen cycling between the sediment and the overlying water [96], had increased muscle wastage in their arms as compensation for increasing their calcified material under OA conditions [97]. Reduced muscle may lead to reduced ability to feed themselves, lower quality of food for predators and reduce nutrient flow. • Food limitation likely exacerbates impacts of stressors including OA, and to date, experiments have shown that when more food is available organisms are able to overcome the negative impacts from OA [98–100]. • Food web interactions may not be so simple because of changing abundance of both the food and the consumer, as well as the interaction. For example, Russell et al. [101] showed that rocky short biofilms increase growth under elevated CO_2 but grazers consumed less. • Interactions between microbes and larger organisms need to be considered. Bleaching of corals can be induced by increased viral activity [102], however, some evidence suggests that viral activity may decrease within OA [103]. The ability of a host to have an immune response to viral attack is critical for its health. Evidence suggests that OA induced poorer health in mussels [104], and they were unable to induce normal immune responses [105,106]. Immune suppression has also been found to be affected by OA in sea urchins [107].
Habitat/Ecosystem — builders	• Mixed results are apparent for coral reefs (both cold- and warm-water corals). Although predominantly the trend shows a decrease in net calcification with increasing OA [108]. • The major impact from OA on coral ecosystems may therefore be the increased dissolution and erosion rates [109].

Available oxygen is also a significant factor in controlling the distribution of organisms in marine environments. Eutrophication events and warming of waters decrease the oxygen content causing hypoxia. As mentioned previously, hypoxia is nearly always accompanied by an elevation of CO_2 (and thus a decrease in pH) and will compound the impacts [112].

Corals are again a good example of the effects of multiple stresses. They are affected by both ocean acidification and by warming of ocean surface waters leading to declining calcification and increase in bleaching [113,114]. Other climate change factors (sea level rise, storm impact, aerosols, ultraviolet irradiation) and nonclimate factors (overfishing, invasion of non-native species, pollution, disease, nutrient and sediment load) add multiple impacts on coral reefs, increasing their vulnerability and reducing their resilience [57].

5. BIOGEOCHEMICAL CYCLING AND FEEDBACK TO CLIMATE
5.1 CHANGES TO THE OCEAN CARBON CYCLE

Over several thousands of years, around 90% of the anthropogenic CO_2 emissions will end up in the ocean [5]. Because of the slow mixing time of the ocean, the current oceanic uptake fraction is only about 30% of this value [1], without which atmospheric CO_2 would be about 65 $\mu mol\ mol^{-1}$ (65 ppm) higher today than what is currently observed (about 400 $\mu mol\ mol^{-1}$).

The Southern Ocean is estimated to account for around 25% of the anthropogenic CO_2 taken up by all the oceans while the North Atlantic is estimated to account for 40% [1]. Unlike the Southern Ocean, which has a strong biological pump, the North Atlantic CO_2 sink is thought to be mainly due to the physical pump, with the 'biological pump' contributing only around 10% [115]. As the surface ocean CO_2 concentrations continue to increase, the ocean's ability to absorb more CO_2 from the atmosphere will slow down. Whilst there were indications that this might be occurring in the analysis of 1990s oceanographic cruises by Sabine et al. [1], more recent analysis of CO_2 in the NE Atlantic [116] and Southern Ocean [11] show a decrease in CO_2 uptake over the last 1–2 decades. Whether this decrease in the efficiency of the ocean sink for anthropogenic CO_2 is decadal variation awaits further long time series study. If the ocean CO_2 sink is becoming less efficient, then more CO_2 will remain in the atmosphere exacerbating global warming.

The 'biological pump' removes carbon from surface waters to the deep ocean via the organic or 'soft' tissue pump (which decreases CO_2 of surface water, increasing its ability to absorb atmospheric CO_2) while the inorganic or 'hard' $CaCO_3$ pump increases CO_2 of the surface water and decreases its ability to absorb atmospheric CO_2. Decreasing calcification and $CaCO_3$ export rates, especially from the decrease in calcification rates from coccolithophores, has been suggested as a possible feedback on the sequestration of carbon to the deep ocean. However, summarizing the responses in coccolithophores and using model evaluation, Ridgwell et al. [117] suggested that the impact of ocean acidification will be relatively minor compared to other global carbon feedbacks. There have been limited attempts to address the contribution of calcification made from other organisms and the feedback from ocean acidification impacts on those species to carbon cycling, e.g. fish carbonate production, which may increase under a high CO_2 future [118], echinoderm contribution to carbonate production [119]; however, the extrapolations from these studies need to be further explored.

5.2 CHANGES TO OCEAN NUTRIENT CYCLES

Shifts in organic ratios in marine phytoplankton have been found to occur as a response to increased CO_2 in several studies over the past few years [57,120,121]. Increasing C:N ratios would lower the nutritional value of organic matter produced by primary producers thereby having further implications for marine ecosystem dynamics. A study linking diatom fatty acid composition to copepod growth under high CO_2 conditions illustrated that shifts in the organic content of food sources (in this case diatoms) can significantly impact growth and reproduction of the consumers [122].

The speciation of nutrients is also dependent on pH conditions, as well as biological processes (bacterial remineralization of nutrients) that can be influenced by pH. For instance, a decrease in pH of 0.3 could reduce the fraction of ammonia (NH_3) by around 50% [123,124]. In addition, the key process of nitrification is sensitive to pH with rates reduced by $\sim$50% at pH 7 [125]. This may result in a reduction of ammonia oxidation rates and the accumulation of ammonia instead of nitrate. Indeed, experiments on ammonia oxidation have shown water column rates to decrease with decreasing pH, while sediment ammonia oxidation rates remain constant, likely due to sediment buffering [126]. A shelf sea ecosystem model was used to assess the implications of these results and projected a 20% decrease in pelagic nitrification by 2100 [10]. *Trichodesmium* cyanobacteria play a key role in sustaining primary production in the large low nutrient areas of the world's oceans through nitrogen fixation and show a >35% increase in rates of nitrogen fixation under elevated CO_2 of 750 μmol mol^{-1} (750 ppm) [127].

5.3 CHANGES TO FLUX OF OTHER CLIMATE REACTIVE GASES FROM THE OCEAN

Phytoplankton are major producers of dimethyl sulphide (DMS), which may have a role in climate regulation via the production of cloud condensation nuclei [128]. DMS has been observed to decrease under predicted year 2100 levels of CO_2, ranging from $\sim$35% [129] to $\sim$60% [130,131] relative to ambient controls. These decreases, when applied globally, may be sufficiently large to influence climate [132]. However, the pattern of decreasing DMS with decreased pH is not consistent amongst all experiments. Kim et al. [133] reported an $\sim$80% increase in DMS under high CO_2, and Hopkins and Archer [134] reported consistent increases up to 225% at 1000 μmol mol^{-1} (1000 ppm) CO_2. These differences are likely explained by the different phytoplankton and bacterial communities captured within the experiments, which can have a large influence on the rate, production and consumption of these gases [134].

6. ADAPTATION, RECOVERY AND MITIGATION

It is difficult to predict if marine organisms and ecosystems will adapt to or recover from the rapid changes to ocean carbonate chemistry. An optimistic view may be that for organisms with short generation times, microevolutionary adaptation could be rapid and that species adversely affected by high CO_2 could be replaced by more CO_2-tolerant strains or species, with minimal impacts up the food chain. However, other assessments are that CO_2-sensitive groups, such as some marine calcifiers, are unlikely to be unable to compete ecologically, resulting in widespread extinctions with profound ramifications to marine food webs and ecosystems [15,135].

6.1 ADAPTATION

While there is some suggestion from past observations that calcifying species may have the ability to adapt to future high CO_2 scenarios by continuing in a noncalcified state, such as some coccolithophore species and some Scleractinian corals [57], there is no evidence yet to suggest whether these adaptive mechanisms will be favoured going into the future. More recent experimental studies and studies from natural CO_2 vent seeps [61] have attempted to address multigenerational responses to OA [136–138], however, few have been able to differentiate the response between genetic adaptation and phenotypic plasticity [139], and studies attempting to address this issue show that both strategies are equally viable to tackle changes in CO_2 [140]. In cases of organisms with shorter life cycles it is an important consideration to take into account, with effects of OA being partially or fully alleviated over several generations [139]. While these experiments provide evidence that adaptation capacity is present within these species, exactly how different selection pressures in the wild will interact to exert pressure on organisms is a long way from being understood. Long-lived organisms are likely to have to rely on species phenotypic plasticity or migration [57,141], rather than adaptation, to cope with environmental shifts.

6.2 RECOVERY

Ocean carbon models and the sediment record both indicate that chemical recovery from projected CO_2 emissions will require thousands of years (chemical equilibration with carbonate minerals) to hundreds of thousands of years (equilibration with the carbonate–silicate cycle) [5]. This means that the chemical effects of CO_2 released from anthropogenic sources are not confined to a century time scale.

Diversity of the seabed dwelling organisms after the acidification event 55 Ma ago took several hundreds of thousands of years to recover. In contrast, there is evidence that planktonic calcifiers tracked their habitat during this event (e.g. tropical species migrated toward the poles), thereby avoiding extinction [142]. The geological record also shows that Scleractinian corals have survived several mass extinction events, likely due to perturbations in the carbon cycle, but they took several millions of years to recover [143–145]. These lessons from the past indicate that should increasing ocean acidification lead to significant loss of biodiversity and even extinction, biological systems may not 'recover' to preindustrial ecosystems but rather may 'transition' to a new state.

6.3 MITIGATION

As concerns over climate change grow, there are increasing numbers of geoengineering solutions proposed. However, they often do not take into account or resolve the issue of ocean acidification (e.g. addition of sulphur dioxide into the stratosphere to deflect some of the sun's energy or ocean pumps of deep water rich in nutrients to increase productivity and draw down CO_2) nor do they look at potential deleterious impacts on the marine environment (adding quicklime to the ocean to soak up CO_2, iron or urea fertilization to increase ocean productivity and draw down CO_2, [146].)

Currently, expert opinion is that the only method of reducing the impacts of ocean acidification on a global scale is through urgent and substantial reductions in anthropogenic CO_2 emissions and/or removal of atmospheric CO_2 [113,146]. A threshold of no more than a 0.2 pH decrease has been recommended to avoid aragonite under saturation in surface waters [147]. In terms of atmospheric CO_2 concentration this would be between the RCP2.6 and RCP4.5 pathways (Fig. 6). However, it should be noted that some upwelling and polar regions would experience aragonite undersaturation even at RCP2.6; biological thresholds may be different than this chemical threshold; and ocean acidification, warming or deoxygenation acting together could further influence the "allowable CO_2 threshold" before ocean ecosystem impacts are substantial [149].

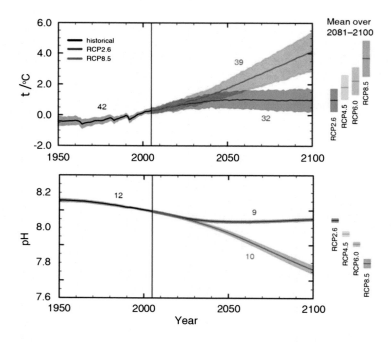

FIGURE 6

CMIP5 multi-model simulated time series from 1950 to 2100 for (top) change in global annual mean surface temperature (t in °C) relative to 1986–2005 and (bottom) global mean ocean surface pH. Time series of projections and a measure of uncertainty (shading) are shown for scenarios RCP2.6 (blue) and RCP8.5 (red). Black (grey shading) is the modelled historical evolution using historical reconstructed forcings. The mean and associated uncertainties averaged over 2081–2100 are given for all RCP scenarios as coloured vertical bars. The number of CMIP5 models used to calculate the multi-model mean is indicated. *(Adapted from IPCC [2].)*

7. CONCLUSIONS

The ocean has been buffering climate change by absorbing about 30% of the CO_2 emitted into the atmosphere from anthropogenic sources. This has resulted in the measurable alteration of surface ocean concentrations of CO_2, HCO_3^-, CO_3^{2-} and pH as well as the reduction of the saturation state and shoaling of the saturation horizons of $CaCO_3$ minerals. Since preindustrial times ocean pH has decreased by a global average of 0.1, and it has been estimated that unmitigated CO_2 emissions will cause ocean pH to decrease by as much as 0.4 by the year 2100 and 0.77 by 2300. These will be the most rapid and greatest changes in ocean carbonate chemistry experienced by marine organisms over the past tens of millions of years. Laboratory experiments, field observations of natural CO_2-rich seawater 'hot spots' and studies of previous ocean acidification events in Earth's history indicate that these changes are a threat to the survival of many marine organisms but particularly organisms that use $CaCO_3$ to produce shells, tests and skeletons (e.g. coccolithophores, pteropods, foraminifera, corals, calcareous macroalgae, mussels, oysters, echinoderms and crustacean). The ASH is already shoaling, bringing increasingly corrosive waters to the productive, shallower shelf seas along the western coast of North America, and models predict that polar and some subpolar waters will be undersaturated this century while saturation states in the tropical surface ocean will be substantially reduced. Recent experiments reveal that other important biological processes (productivity, internal physiology, fertilization, embryo development, larval settlement and communication) are also vulnerable to future changes in ocean chemistry. There could also be changes to ocean carbon and nutrient cycles, but, because of their complexity, it is hard to predict what the implications of the changes to biology will be on marine food webs, ecosystems and the services they provide. However, examination of previous episodes in Earth's history indicates that unmitigated CO_2 emissions are likely to result in widespread extinctions. It will take tens of thousands of years for the changes in ocean chemistry to be buffered through neutralization by calcium carbonate sediments, and the level at which ocean pH will eventually stabilize will be lower than it currently is. The only way of reducing the impacts of ocean acidification on a global scale is through urgent and substantial reductions in anthropogenic CO_2 emissions. Ocean acidification is a key argument for united global societal action in future climate change negotiations.

REFERENCES

[1] C.L. Sabine, R.A. Feely, N. Gruber, R.M. Key, K. Lee, J.L. Bullister, R. Wanninkhof, C.S. Wong, D.W.R. Wallace, B. Tilbrook, F.J. Millero, T.H. Peng, A. Kozyr, T. Ono, A.F. Rios, Science 305 (2004) 367–371.

[2] IPCC, Climate change 2013: the physical science basis, in: T.F. Stocker, D. Qin, G.-K. Plattner, M. Tignor, S.K. Allen, J. Boschung, A. Nauels, Y. Xia, V. Bex, P.M. Midgley (Eds.), Contribution of Working Group I to the Fifth Assessment Report of the Intergovernmental Panel on Climate Change, Cambridge University Press, Cambridge, United Kingdom and New York, NY, USA, 2013, 1535 pp.

[3] L. Bopp, L. Resplandy, J.C. Orr, et al., Multiple stressors of ocean ecosystems in the 21st century: projections with CMIP5 models, Biogeosciences 10 (2013) 6225–6245.

[4] K. Caldeira, M.E. Wickett, Nature 425 (2003) 365.

[5] D.E. Archer, H. Kheshgi, E. Maier-Reimer, Glob. Biogeochem. Cycles 12 (1998) 259–276.

[6] J.C. Orr, V.J. Fabry, O. Aumont, L. Bopp, S.C. Doney, R.A. Feely, A. Gnanadesikan, N. Gruber, A. Ishida, F. Joos, R.M. Key, K. Lindsay, E. Maier-Reimer, R. Matear, P. Monfray, A. Mouchet, R.G. Najjar, G.-K. Plattner, K.B. Rodgers, C.L. Sabine, J.L. Sarmiento, R. Schlitzer, R.D. Slater, I.J. Totterdell, M.-F. Weirig, Y. Yamanaka, A. Yool, Nature 437 (2005) 681–686.

[7] IPCC, Climate change 2014: synthesis report, in: Core Writing Team, R.K. Pachauri, L.A. Meyer (Eds.), Contribution of Working Groups I, II and III to the Fifth Assessment Report of the Intergovernmental Panel on Climate Change, IPCC, Geneva, Switzerland, 2014.

[8] L. Cao, K. Caldeira, A.K. Jain, Geophys. Res. Lett. 34 (2007) L05607, http://dx.doi.org/10.1029/2006GL028605.

[9] H.O. Pörtner, M. Langenbuch, B. Michaelidis, J. Geophys. Res. Oceans 110 (2005) C09S10, http://dx.doi.org/10.1029/2004JC002561.

[10] J.C. Blackford, F.J. Gilbert, J. Mar, Systems 64 (2007) 229–241.

[11] C. Le Quéré, G.P. Peters, R.J. Andres, R.M. Andrew, T. Boden, P. Ciais, P. Friedlingstein, R.A. Houghton, G. Marland, R. Moriarty, S. Sitch, P. Tans, A. Arneth, A. Arvanitis, D.C.E. Bakker, L. Bopp, J.G. Canadell, L.P. Chini, S.C. Doney, A. Harper, I. Harris, J.I. House, A.K. Jain, S.D. Jones, E. Kato, R.F. Keeling, K. Klein Goldewijk, A. Körtzinger, C. Koven, N. Lefèvre, A. Omar, T. Ono, G.-H. Park, B. Pfeil, B. Poulter, M.R. Raupach, P. Regnier, C. Rödenbeck, S. Saito, J. Schwinger, J. Segschneider, B.D. Stocker, B. Tilbrook, S. van Heuven, N. Viovy, R. Wanninkhof, A. Wiltshire, S. Zaehle, C. Yue, Earth Syst. Sci. Data Discuss 6 (2013) 689–760, http://dx.doi.org/10.5194/essdd-6-689-2013.

[12] R. Torres, D.R. Turner, N. Silva, J. Rutllant, Deep Sea Res. 46 (1999) 1161–1179.

[13] R.A. Feely, C.L. Sabine, J.M. Hernandez-Ayon, D. Ianson, B. Hales, Science (2008) 1–4, http://dx.doi.org/10.1126/science.1155676.

[14] P.N. Pearson, M.R. Palmer, Nature 406 (2000) 695–699.

[15] O. Hoegh-Guldberg, R. Cai, E.S. Poloczanska, P.G. Brewer, S. Sundby, K. Hilmi, V.J. Fabry, S. Jung, in: V.R. Barros, C.B. Field, D.J. Dokken, M.D. Mastrandrea, K.J. Mach, T.E. Bilir, M. Chatterjee, K.L. Ebi, Y.O. Estrada, R.C. Genova, B. Girma, E.S. Kissel, A.N. Levy, S. MacCracken, P.R. Mastrandrea, L.L. White (Eds.), Climate Change 2014: Impacts, Adaptation, and Vulnerability. Part B: Regional Aspects. Contribution of Working Group II to the Fifth Assessment Report of the Intergovernmental Panel on Climate Change, Cambridge University Press, Cambridge, United Kingdom and New York, NY, USA, 2014, pp. 1655–1731.

[16] A. Ridgwell, D.N. Schmidt, Nat. Geosci. 3 (2010) 196–200.

[17] B. Hönisch, A. Ridgwell, D.N. Schmidt, et al., Science 335 (2012) 1058–1063, http://dx.doi.org/10.1126/science.1208277.

[18] D. Archer, J. Geophys. Res. 110 (2005) C09S05, http://dx.doi.org/10.1029/2004JC002625.

[19] S. Barker, H. Elderfield, Science 297 (2002) 833–836.

[20] M. Rhein, S.R. Rintoul, S. Aoki, E. Campos, D. Chambers, R.A. Feely, S. Gulev, G.C. Johnson, S.A. Josey, A. Kostianoy, C. Mauritzen, D. Roemmich, L.D. Talley, F. Wang, Observations: ocean, in: T.F. Stocker, D. Qin, G.-K. Plattner, M. Tignor, S.K. Allen, J. Boschung, A. Nauels, Y. Xia, V. Bex, P.M. Midgley (Eds.), Climate Change 2013: The Physical Science Basis. Contribution of Working Group I to the Fifth Assessment Report of the Intergovernmental Panel on Climate Change, Cambridge University Press, Cambridge, United Kingdom and New York, NY, USA, 2013.

[21] N.R. Bates, J. Geophys. Res. Oceans 112 (2007) C09013.

[22] N.R. Bates, Biogeosciences 9 (2012) 2649–2659.

[23] J.M. Santana-Casiano, M. González-Dávila, M.-J. Rueda, O. Llinás, E.-F. González-Dávila, Glob. Biogeochem. Cycles 21 (2007) GB1015, http://dx.doi.org/10.1029/2006GB002788.

[24] J.E. Dore, R. Lukas, D.W. Sadler, M.J. Church, D.M. Karl, Proc. Natl. Acad. Sci. U.S.A. 106 (2009) 12235–12240.

[25] J. Olafsson, S.R. Olafsdottir, A. Benoit-Cattin, M. Danielsen, T.S. Arnarson, T. Takahashi, Biogeosciences 6 (2009) 2661–2668.

[26] M. González-Dávila, J.M. Santana-Casiano, M.J. Rueda, O. Llinas, Biogeosciences 7 (2010) 3067–3081.

[27] T. Midorikawa, M. Ishii, S. Saito, D. Sasano, N. Kosugi, T. Motoi, H. Kamiya, A. Nakadate, K. Nemoto, H.Y. Unoue, Tellus B 62 (2010) 649–659.

[28] R.H. Byrne, S. Mecking, R.A. Feely, X.W. Liu, Geophys. Res. Lett. 37 (2010) L02601.

[29] T. Takahashi, S.C. Sutherland, R. Wanninkhof, C. Sweeney, R.A. Feely, D.W. Chipman, B. Hales, G. Friederich, F. Chavez, C. Sabine, A. Watson, D.C.E. Bakker, U. Schuster, N. Metzl, H. Yoshikawa-Inoue, M. Ishii, T. Midorikawa, Y. Nojiri, A. Kortzinger, T. Steinhoff, M. Hoppema, J. Olafsson, T.S. Arnarson, B. Tilbrook, T. Johannessen, A. Olsen, R. Bellerby, C.S. Wong, B. Delille, N.R. Bates, H.J.W. de Baar, Deep Sea Res. Pt. 56 (2009) 554–577.

[30] G.E. Hofmann, T.G. Evans, M.W. Kelly, J.L. Padilla-Gamino, C.A. Blanchette, L. Washburn, F. Chan, M.A. McManus, B.A. Menge, B. Gaylord, T.M. Hill, E. Sanford, M. LaVigne, J.M. Rose, L. Kapsenberg, J.M. Dutton, Biogeosciences 11 (2014) 1053–1064.

[31] M. Steinacher, F. Joos, T.L. Folicher, G.-K. Plattner, S.C. Doney, Biogeosciences 6 (2009) 515–533.

[32] M. Yamamoto-Kawai, F.A. McLaughlin, E.C. Carmack, S. Nishino, K. Shimada, Science 326 (2009) 1098–1100.

[33] M. Yamamoto-Kawai, F.A. McLaughlin, E.C. Carmack, Geophys. Res. Lett. 38 (2011) L03601, http://dx.doi.org/10.1029/2010GL045501.

[34] J.T. Mathis, R.S. Pickart, R.H. Byrne, C.L. McNeil, G.W.K. Moore, L.W. Juranek, X. Liu, J. Ma, R.A. Easley, M.M. Elliot, J.N. Cross, S.C. Reisdorph, F. Bahr, J. Morison, T. Lichendorf, R.A. Feely, Geophys. Res. Lett. 39 (2012) L07606, http://dx.doi.org/10.1029/2012GL051574.

[35] C.B. Field, V.R. Barros, K.J. Mach, M.D. Mastrandrea, M. van Aalst, W.N. Adger, D.J. Arent, J. Barnett, R. Betts, T.E. Bilir, J. Birkmann, J. Carmin, D.D. Chadee, A.J. Challinor, M. Chatterjee, W. Cramer, D.J. Davidson, Y.O. Estrada, J.-P. Gattuso, Y. Hijioka, O. Hoegh-Guldberg, H.-Q. Huang, G.E. Insarov, R.N. Jones, R.S. Kovats, P. Romero Lankao, J.N. Larsen, I.J. Losada, J.A. Marengo, R.F. McLean, L.O. Mearns, R. Mechler, J.F. Morton, I. Niang, T. Oki, J.M. Olwoch, M. Opondo, E.S. Poloczanska, H.-O. Pörtner, M.H. Redsteer, A. Reisinger, A. Revi, D.N. Schmidt, M.R. Shaw, W. Solecki, D.A. Stone, J.M.R. Stone, K.M. Strzepek, A.G. Suarez, P. Tschakert, R. Valentini, S. Vicuña, A. Villamizar, K.E. Vincent, R. Warren, L.L. White, T.J. Wilbanks, P.P. Wong, G.W. Yohe, Technical summary, in: C. Field, et al. (Eds.), Climate Change 2014: Impacts, Adaptation and Vulnerability. Contribution of Working Group II to the Fifth Assessment Report of the Intergovernmental Panel on Climate Change, Cambridge University Press, Cambridge, 2014.

[36] L. Bopp, L. Resplandy, J.C. Orr, S.C. Doney, J.P. Dunne, M. Gehlen, P. Halloran, C. Heinze, T. Ilyina, R. Seferian, J. Tjiputra, M. Vichi, Biogeosciences 10 (2013) 6225–6245.

[37] F. Joos, T.L. Frölicher, M. Steinacher, G.-K. Plattner, Impact of climate change mitigation on ocean acidification projections, in: J.-P. Gattuso, L. Hansson (Eds.), Ocean Acidification, Oxford University Press, Oxford, 2011, pp. 272–290 (Chapter 14).

[38] R.G.J. Bellerby, A. Olsen, T. Furevik, L.G. Anderson, in: H. Drange (Ed.), The Nordic Seas: An Integrated Perspective Geophysical Monograph Series 158, American Geophysical Union, Washington, DC, 2005, pp. 189–197.

[39] T.F. Stocker, D. Qin, G.-K. Plattner, L.V. Alexander, S.K. Allen, N.L. Bindoff, F.-M. Bréon, J.A. Church, U. Cubasch, S. Emori, P. Forster, P. Friedlingstein, N. Gillett, J.M. Gregory, D.L. Hartmann, E. Jansen, B. Kirtman, R. Knutti, K. Krishna Kumar, P. Lemke, J. Marotzke, V. Masson-Delmotte, G.A. Meehl, I.I. Mokhov, S. Piao, V. Ramaswamy, D. Randall, M. Rhein, M. Rojas, C. Sabine, D. Shindell, L.D. Talley, D.G. Vaughan, S.-P. Xie, Technical summary, in: T.F. Stocker, D. Qin, G.-K. Plattner, M. Tignor, S.K. Allen, J. Boschung, A. Nauels, Y. Xia, V. Bex, P.M. Midgley (Eds.), Climate Change 2013: The Physical Science Basis. Contribution of Working Group I to the Fifth Assessment Report of the Intergovernmental Panel on Climate Change, Cambridge University Press, Cambridge, United Kingdom and New York, NY, USA, 2013.

[40] Y. Artioli, J. Blackford, G. Nondal, R.G.J. Bellerby, S.L. Wakelin, J.T. Holt, M. Butenschon, J.I. Allen, Biogeosciences 11 (2014) 601–612.

[41] E. Thomas, in: S. Monechi, R. Coccioni, M.R. Rampino (Eds.), Large Ecosystem Perturbations: Causes and Consequences 424, Geological Society of America, Special Paper, Boulder, CO, 2007, pp. 1–23.

[42] R.E. Zeebe, A. Ridgwell, Past changes of ocean carbonate chemistry, in: J.-P. Gattuso, L. Hansson (Eds.), Ocean Acidification, Oxford University Press, Oxford, 2011.

[43] K.R. Hinga, Mar. Ecol. Prog. Ser. 238 (2002) 281–300.

[44] J.T. Wootten, C.A. Pfister, J.D. Forester Proc, Nat. Acad. Sci. 105 (2008) 18848–18853.

[45] H.S. Findlay, Y. Artioli, J. Moreno Nava, S.J. Hennige, L.C. Wicks, V.A.I. Huvenne, E.M. Woodward, J.M. Roberts, Glob. Change Biol. 19 (2013) 2708–2719.

[46] A. Barton, B. Hales, G.G. Waldbusser, C. Langdon, R.A. Feely Limnol, Oceanography 57 (2012) 698–710.

[47] N. Bednaršek, R.A. Feely, J.C.P. Reum, B. Peterson, J. Menkel, S.R. Alin, B. Hales, Proc. R. Soc. B 281 (2014) 20140123, http://dx.doi.org/10.1098/rspb.

[48] R. Torres, S. Pantoja, N. Harada, H.E. Gonzalez, G. Daneri, M. Frangopulos, J.A. Rutlant, C.M. Duarte, S. Ruiz-Halpern, E. Mayol, M. Fukasawa, J. Geophys. Res. 116 (2011) C09006, http://dx.doi.org/10.1029/2010JC006344.

[49] C.A. Vargas, V.M. Aguilera, V. San Martin, P.H. Manriquez, J.M. Navarro, C. Duarte, R. Torres, M.A. Lardies, N.A. Lagos, Estuaries Coasts 38 (2014) 1163–1177, http://dx.doi.org/10.1007/s12237-014-9873-7.

[50] A.E. Maas, K.F. Wishner, B.A. Seibel, Biogeosciences 9 (2012) 747–757.

[51] C.N. Lewis, K.A. Brown, L.A. Edwards, G. Cooper, H.S. Findlay, Proc. Nat. Acad. Sci. (2013) E4960–E4967.

[52] N. Bednarsek, G.A. Tarling, D.C.E. Bakker, S. Fielding, E.M. Jones, H.J. Venables, P. Ward, A. Kuzirian, B. Leze, R.A. Feely, E.J. Murphy, Nat. Geosci. 5 (2012) 881–885.

[53] S. Comeau, J.-P. Gattuso, A.-M. Nisumaa, J. Orr, Proc. R. Soc. B 279 (2012), http://dx.doi.org/10.1098/rspb. 2011.0910.

[54] D.L. Mackas, M.D. Galbraith, ICES J. Mar. Sci. 69 (2012) 448–459.

[55] T. Tyrrell, B. Schneider, A. Charalampopoulou, U. Riebesell, Biogeosciences 5 (2008) 485–494.

[56] A. Charalampopoulou, A.J. Poluton, T. Tyrrell, M.I. Lucas, Mar. Ecol. Prog. Ser. 431 (2011) 25–43.

[57] C. Turley, H.S. Findlay, in: T.M. Letcher (Ed.), Climate Change: Observed Impacts on Planet Earth, Elsevier, Oxford, 2009. ISBN: 978-0-444-53301-2.

[58] T.F. Cooper, G. De'Ath, K.E. Fabricius, J.M. Lough, Glob. Change Biol. 14 (2008) 538–539.

[59] J. Silverman, D.I. Kline, L. Johnson, T. Rivlin, K. Schneider, J. Erez, B. Lazar, K. Caldeira, J. Geophys. Res. 117 (2012) G03023, http://dx.doi.org/10.1029/2012JG001974.

[60] J. Silverman, B. Lazar, J. Erez, J. Geophys. Res. 112 (2007) C05004, http://dx.doi.org/10.1029/2006JC003770.

[61] J.M. Hall-Spencer, R. Rodolfo-Mpi, S. Martin, R. Ransome, M. Fine, S.M. Turner, S.J. Rowley, D. Tedesco, M.-C. Buia, Nature 454 (2008) 96–99.

[62] K.J. Kroeker, F. Micheli, M.C. Gambi, Nat. Clim. Change 3 (2012) 156–159, http://dx.doi.org/10.1038/NCLIMATE1680.

[63] P. Williamson, C. Turley, Phil. Trans. R. Soc. A 370 (2012) 4317–4342.

[64] K.J. Kroeker, R.L. Kordas, R. Crim, I.E. Hendriks, L. Ramajo, G.S. Singh, C.M. Duarte, J.-P. Gattuso, Glob. Change Biol. 19 (2013) 1884–1896.

[65] D.A. Hutchins, M.R. Mulholland, F.-X. Fu, Oceanography 22 (2009) 128–145.

[66] Z.V. Finkel, M. Katz, J. Wright, O.M.E. Schofield, P.G. Falkowski, Proc. Natl. Acad. Sci. U.S.A. 102 (2005) 8927–8932.

[67] A.J. Andersson, I.B. Kuffner, F.T. Mackenzie, P.L. Jokiel, K.S. Rodgers, A. Tan, Biogeosciences 6 (2009) 1811–1823.

[68] H.S. Findlay, H.L. Wood, M.A. Kendall, J.I. Spicer, R.J. Twitchett, S. Widdicombe, Mar. Biol. Res. 7 (2011) 565–575.

[69] F. Melzner, M.A. Gutowska, M. Langenbuch, S. Dupont, M. Lucassen, M.C. Thorndyke, M. Bleich, H.-O. Portner, Biogeosciences 6 (2009) 2313–2331.

[70] M. Stumpp, J. Wren, F. Melzner, M.C. Thorndyke, S. Dupont, Comp. Biochem. Physiol. Part A: Mol. Integr. Physiol. 160 (2011) 331–340.

[71] J. Thomsen, F. Melzner, Mar. Biol. 157 (2010) 2667–2676, http://dx.doi.org/10.1007/s00227-010-1527-00.

[72] G. Lannig, S. Eilers, H.-O. Portner, I.A. Sokolova, C. Bock, Mar. Drugs 8 (2010) 2318–2339.

[73] B. Michaelidis, C. Ouzounis, A. Paleras, H.O. Portner, Mar. Ecol. Prog. Ser. 293 (2005) 109–118.

[74] M.A. Green, M.E. Jones, C.L. Boudreau, P.L. Moore, B.A. Westman, Limnol. Oceanogr. 49 (2004) 727–734.

[75] H. Kurihara, S. Kato, A. Ishimatsu, Aquat. Biol. 1 (2007) 91–98.

[76] H. Kurihara, Y. Shirayama, Mar. Ecol. Prog. Ser. 274 (2004) 161–169.

[77] S. Dupont, J. Havenhand, W. Thorndyke, L.S. Peck, M. Thorndyke, Mar. Ecol. Prog. Ser. 373 (2008) 285–294.

[78] T. Kikkawa, J. Kita, A. Ishimatsu, Mar. Pollut. Bull. 48 (2004) 108–110.

[79] R. Rosa, M. Baptista, V.M. Lopes, M.R. Pegado, J.R. Paula, K. Trubenbach, M. Costa Leal, R. Calado, T. Repolho, R. Soc. Proc. B 281 (2014) 20141738, http://dx.doi.org/10.1098/rspb.

[80] H.S. Findlay, M.A. Kendall, J.I. Spicer, C. Turley, S. Widdicombe, Aquat. Biol. 3 (2008) 51–62.

[81] H.S. Findlay, M.A. Kendall, J.I. Spicer, S. Widdicombe, Mar. Ecol. Prog. Ser. 389 (2009) 193–202.

[82] H.S. Findlay, M.A. Kendall, J.I. Spicer, S. Widdicombe, Estuarine Coastal Shelf Sci. 86 (2010) 675–682.

[83] H. Kurihara, S. Shimode, Y. Shirayama, Mar. Pollut. Bull. 49 (2004) 721–727.

[84] A.W. Decho, K.A. Browne, R.K. Zimmer-Faust, Limnol. Oceanogr. 43 (1998) 1410–1417.

[85] S.L. Coon, M. Walch, W.K. Fitt, R.M. Weiner, D.B. Bonar, Biol. Bull. 179 (1990) 297–303.

[86] D.B. Bonar, S.L. Coon, M. Walch, R.M. Weiner, W. Fitt, Bull. Mar. Sci. 46 (1990) 484–498.

[87] M.J. Anderson, Biol. Bull. 190 (1996) 350–358.

[88] A.O.H.C. Leduc, E. Roh, M.C. Harvey, G.E. Brown, Can. J. Fish. Aquat. Sci. 63 (2006) 2356–2363.

[89] R. Bibby, P. Cleal-Harding, S. Rundle, S. Widdicombe, J.I. Spicer, Biol. Lett. 3 (2007) 699–701.

[90] A.M. Queiros, J.A. Fernandes, S. Faulwetter, J. Nunes, S.P.S. Rastrick, N. Mieszkowska, T. Artioli, A. Yool, P. Calosi, C. Arvanditids, H.S. Findlay, M. Barange, W.W.L. Cheung, S. Widdicombe, Glob. Change Biol. 21 (2014) 130–143, http://dx.doi.org/10.1111/gcb.12675.

[91] K.L. de La Haye, J.I. Spicer, S. Widdicombe, M. Briffa, Anim. Behav. 82 (2011) 495–501.

[92] D.L. Dixson, P.L. Munday, G.P. Jones, Ecol. Lett. 12 (2009) 1–8.

[93] I.L. Cripps, P.L. Munday, M.I. McCormick, PLoS One 6 (2011) e22736, http://dx.doi.org/10.1371/journal.pone.0022736.

[94] M.C.O. Ferrari, R.P. Manassa, D.L. Dixson, P.L. Munday, M.I. McCormick, M.G. Meekan, A. Sih, D.P. Chivas, PLoS One 7 (2012) e2589, http://dx.doi.org/10.1371/journal.pone.0031478.

[95] D.L. Dixson, D. Abrego, M.E. Hay, Science 345 (2014) 892–897.

[96] K. Vopel, D. Thistle, R. Rosenberg, Limnol. Oceanogr. 48 (2003) 2034–2045.

[97] H.L. Wood, J.I. Spicer, S. Widdicombe, Proc. R. Soc. B 275 (2008) 1767–1773.

[98] J. Thomsen, I. Casties, C. Pansch, A. Körtzinger, F. Melzner, Glob. Change Biol. 19 (2013) 1017–1027.

[99] F. Melzner, P. Stange, K. Trubenbach, J. Thomsen, I. Casties, U. Panknin, S.N. Gorb, M.A. Gutowska, PLoS One 6 (2011) e24223, http://dx.doi.org/10.1371/journal.pone.0024223.

[100] A. Hettinger, E. Sanford, T.M. Hill, J.D. Hosfelt, A.D. Russell, B. Gaylord, Biogeoscience 10 (2013) 6629–6638.

[101] B.D. Russell, S.D. Connell, H.S. Findlay, K. Tait, S. Widdicombe, N. Mieszkowska, Phil. Trans. R. Soc. B 368 (2013) 20120438, http://dx.doi.org/10.1098/rstb.2012.0438.

[102] R. Danovaro, L. Bongiorni, C. Corinaldesi, D. Giovannelli, E. Damiani, P. Astolfi, L. Greci, A. Pusceddu, Environ. Health Perspect. 116 (2008) 441–447.

[103] J.B. Larsen, A. Larsen, R. Thyrhaug, G. Bratbak, R.-A. Sandaa, Biogeosciences 5 (2008) 523–533.

[104] A. Beesley, D.M. Lowe, C.K. Pascoe, S. Widdicombe, Clim. Res. 37 (2008) 215–225.

[105] R. Bibby, S. Widdicombe, H. Parry, J.I. Spicer, R. Pipe, Aquat. Biol. 2 (2008) 67–74.

[106] C.L. Mackenzie, S.A. Lynch, S.C. Culloty, S.K. Malham, PLoS One 9 (2014) e99712, http://dx.doi.org/10.1371/journal.pone.0099712.

[107] B. Hernroth, S. Baden, M. Thorndyke, S. Dupont, Aquat. Toxicol. 103 (2011) 222–224.

[108] N.C.S. Chan, S.R. Connolly, Glob. Change Biol. 19 (2013) 282–290.
[109] A.J. Andersson, D. Gledhill, Ann. Rev. Mar. Sci. 5 (2013) 321–348.
[110] K. Walther, F.J. Sartoris, C. Bock, H.-O. Portner, Biogeosciences 6 (2009) 2207–2215.
[111] E.H. Shadwick, T.W. Trull, H. Thomas, J.A.E. Gibson, Sci. Rep. 3 (2013), http://dx.doi.org/10.1038/srep02339.
[112] J.S. Gray, R.S.-S. Wu, Y.Y. Or, Mar. Ecol. Prog. Ser. 238 (2002) 249–279.
[113] A. Fischlin, G.F. Midgley, J. Price, R. Leemans, B. Gopal, C. Turley, M. Rounsevell, P. Dube, J. Tarazona, A. Velichko, Climate Change 2007: Climate Change Impacts, Adaptation and vulnerability Cambridge University Press, Cambridge, 2007, 211–272. (Chapter 4: Ecosystems, Their Properties, Goods and Services, Forth Assessment Report of the Intergovernmental Panel on Climate Change.)
[114] O. Hoegh-Guldberg, P.J. Mumby, A.J. Hooten, R.S. Steneck, P. Greenfield, E. Gomez, C.D. Harvell, P.F. Sale, A.J. Edwards, K. Caldeira, N. Knowlton, C.M. Eakin, R. Iglesias-Prieto, N. Muthiga, R.H. Bradbury, A. Dubi, M.E. Hatziolos, Science 318 (2007) 1737–1742.
[115] B. Pasquer, G. Laruelle, S. Becquevort, V. Schoemann, H. Goosse, C. Lancelot, J. Sea Res. 53 (2005) 93–108.
[116] U. Schuster, A.J. Watson, J. Geophys. Res. 112 (2007) C11006, http://dx.doi.org/10.1029/2006JC003941.
[117] A. Ridgwell, I. Zondervan, J.C. Hargreaves, J. Bijma, T.M. Lenton, Biogeosciences 4 (2007) 481–492.
[118] R.W. Wilson, F.J. Millero, J.R. Taylor, P.J. Walsh, V. Christensen, S. Jennings, M. Grosell, Science 323 (2009) 359–362.
[119] M. Lebrato, D. Iglesias-Rodriguez, R.A. Feely, D. Greenley, D.O.B. Jones, N. Suarez-Bosche, R.S. Lampitt, J.E. Cartes, D.R.H. Green, B. Alker, Ecol. Monogr. 80 (2010) 441–467.
[120] U. Riebesell, K.G. Schulz, R.G.J. Bellerby, M. Botros, P. Fritsche, M. Meyerhöfer, C. Neill, G. Nondal, A. Oschlies, J. Wohlers, E. Zöllner, Nature 450 (2007) 545–548.
[121] A. Silyakova, R.G.J. Bellerby, K.G. Schulz, J. Czerny, T. Tanaka, G. Nondal, U. Riebesell, A. Engel, T. De Lange, A. Ludvig, Biogeosciences 10 (2009) 4847–4859.
[122] D. Rossoll, R. Bermudez, H. Hauss, K.G. Schulz, U. Riebesell, U. Sommer, M. Winder, PLoS One 7 (2012) e34737, http://dx.doi.org/10.1371/journal.pone.0034737.
[123] J.A. Raven, in: T. Platt, W.K.W. Li (Eds.), Photosynthetic Picoplankton, vol. 214, 1986, pp. 1–583. Canad. Bull. Fish. Aquat. Sci., Ottawa.
[124] N.J. Wyatt, V. Kitidis, E.M.S. Woodward, A.P. Rees, S. Widdicombe, M. Lohan, J. Plankton Res. 32 (2010) 631–641.
[125] M.H. Huesemann, A.D. Skillman, E.A. Crecelius, Mar. Pollut. Bull. 44 (2008) 142–148.
[126] V. Kitidis, B. Laverock, L.C. McNeill, A. Beesley, D. Cummings, K. Tait, M.A. Osborn, S. Widdicombe, Geophys. Res. Lett. 38 (2011) L21603, http://dx.doi.org/10.1029/2011GL49095.
[127] D.A. Hutchins, F.-X. Fu, Y. Zhang, M.E. Warner, Y. Feng, K. Portune, P.W. Berhardt, Limnol. Oceanogr. 52 (2007) 1293–1304.
[128] R.J. Charlson, J.E. Lovelock, M.O. Andreae, S.G. Warren, Nature 326 (1987) 655–661.
[129] S.D. Archer, S.A. Kimmance, J.A. Stephens, F.E. Hopkins, R.G.J. Bellerby, K.G. Schulz, J. Piontek, A. Engel, Biogeosciences 10 (2013) 1893–1908.
[130] F.E. Hopkins, S.M. Turner, P.D. Nightingale, M. Steinke, P.S. Liss, P. Natl, Acad. Sci. U.S.A. 107 (2010) 760–765.
[131] V. Avgoustidi, P.D. Nightingale, I. Joint, M. Steinke, S.M. Turner, F.E. Hopkins, P.S. Liss, Environ. Chem. 9 (2012) 399–404.
[132] K.D. Six, S. Kloster, T. Ilyina, S.D. Archer, K. Zhang, Nat. Clim. Change 3 (2013) 975–978.
[133] J.-M. Kim, K. Lee, E.J. Yang, K. Shin, J.H. Noh, K.-T. Park, B. Hyun, H.-J. Jeong, J.-H. Kim, K.Y. Kim, M. Kim, H.-C. Kim, P.-G. Jang, M.-C. Jang, Environ. Sci. Technol. 44 (2010) 8140–8143.
[134] F.E. Hopkins, S.D. Archer, Biogeosciences 11 (2014) 4925–4940.

[135] S. Hennige, J.M. Roberts, P. Williamson, T. Aze, J. Barry, R. Bellerby, L. Brander, M. Byrne, J.-P. Gattuso, S. Gibbs, L. Hansson, C. Hattam, C. Hauton, J. Havenhand, J.H. Fosså, C. Kavanagh, H. Kurihara, R. Matear, F. Mark, F. Melzner, P. Munday, B. Niehoff, P. Pearson, K. Rehdanz, S. Tambutté, C. Turley, A. Venn, M. Warnau, J. Young, An updated synthesis of the impacts of ocean acidification on marine biodiversity. Secretariat of the Convention on Biological Diversity, in: S. Hennige, J.M. Roberts, P. Williamson (Eds.), An Updated Synthesis of the Impacts of Ocean Acidification on Marine Biodiversity, Technical Series No. 75, Montreal, 2014, p. 99. http://www.cbd.int/doc/publications/cbd-ts-75-en.pdf.

[136] J.M. Sunday, P. Calosi, S. Dupont, P.L. Munday, J.H. Stillman, T.B.H. Reusch, Trends Ecol. Evol. 29 (2014) 117–125.

[137] J.P. Lim, A. Brunet, Trends Genet. 29 (2013) 176–186.

[138] P.L. Munday, R.R. Warner, K. Monro, J.M. Pandolfi, D.J. Marshall, Ecol. Lett. 16 (2013) 1488–1500.

[139] P. Thor, S. Dupont, Glob. Change Biol. 21 (2014) 2261–2271, http://dx.doi.org/10.1111/gcb.12815.

[140] P. Calosi, S.P.S. Rastrick, C. Lombardi, H.J. de Guzman, L. Davidson, M. Jahnke, A. Giangrande, J.D. Hardege, A. Schulze, J.I. Spicer, M.C. Gambi, Phil. Trans. R. Soc. B 368 (2013) 20120444.

[141] C. Turley, The risk of ocean acidification to ocean ecosystems, in: A. Fischer (Ed.), Open Ocean Technical Assessment Report for the GEF Transboundary Water Assessment. UNEP, IOC-UNESCO, in press.

[142] S.J. Gibbs, P.R. Bown, J.A. Sessa, T.J. Bralower, P.A. Wilson, Science 314 (2006) 1770–1773, http://dx.doi.org/10.1126/science.1133902.

[143] G.D. Stanley, D.G. Fautin, Science 291 (2001) 1913–1914.

[144] G.D. Stanley, Earth Sci. Rev. 60 (2003) 195–225.

[145] J.E.N. Veron, A Reef in Time: The Great Barrier Reef from Beginning to End, Belknap Press, USA, 2008, 1–282.

[146] P. Williamson, C. Turley, Phil. Trans. R. Soc. A Math. Phys. Eng. Sci. 370 (2012) 4317–4342.

[147] German Advisory Council on Global Change, The Future Oceans – Warming up, Rising High, Turning Sour Special Report, WBGU, Berlin, (2006), 1–110, ISBN:3-936191-14-X, http://www.wbgu.de.

[148] C. Turley, J. Blackford, S. Widdicombe, D. Lowe, P.D. Nightingale, A.P. Rees, in: H.J. Schellnhuber, W. Cramer, N. Nakicenovic, T. Wigley, G. Yohe (Eds.), Avoiding Dangerous Climate Change, Cambridge University Press, Cambridge, 2006, pp. 65–70.

[149] J.-P. Gattuso, A. Magnan, R. Billé, W.W.L. Cheung, E.L. Howes, F. Joos, D. Allemand, L. Bopp, S. Cooley, C.M. Eakin, O. Hoegh-Guldberg, R.P. Kelly, H.-O. Pörtner, A.D. Rogers, J.M. Baxter, D. Laffoley, D. Osborn, A. Rankovic, J. Rochette, U.R. Sumaila, S. Treyer, C. Turley, Science, 349 (2015), http://dx.doi.org/10.1126/science.aac4722.

LICHENS

19

André Aptroot[1], Norbert J. Stapper[2], Alica Košuthová[3], Marcela E.S. Cáceres[4]

[1]*ABL Herbarium, Gerrit van der Veenstraat, Soest, The Netherlands;* [2]*Büro für Ökologische Studien, Verresbergerstraße, Monheim am Rhein, Germany;* [3]*Institute of Botany, Department of Cryptogams, Slovak Academy of Science, Bratislava, Slovak Republik; Faculty of Science, Department of Botany and Zoology, Masaryk University, Brno, Czech Republik;* [4]*Departamento de Biociências, Universidade Federal de Sergipe, Av. Vereador Olimpio Grande s/n, Itabaiana, Sergipe, Brazil*

CHAPTER OUTLINE

1. INTRODUCTION

Lichens have been observed to respond rapidly to climate change. So far the changes are as expected with a rather rapid increase of (sub)tropical species in temperate areas and a gradual decrease of some boreo-alpine elements [1]. In recent years, many publications have addressed the issue of lichens in connection with global warming [2], but the relation is not clear in all cases and especially observations in boreal to (Ant-)Arctic regions sometimes show contradictory trends [3–9], suggesting that competition between lichens and other plants is an important factor, i.e. adverse effects on other plant groups may lead to indirect effects on lichens. No lichens have so far been reported to be seriously threatened by climate change. Marked shifts in occurrence and distribution have been predicted based on known habitat preferences and projected climate change [10].

Lichens, like most cryptogams, tend to be widespread, much more so than phanerogams or land animals. Also, many of the species seem to be capable of rather rapid dispersal, as shown by the recent arrival of some (sub)tropical species in a temperate area [1].

Climate Change. http://dx.doi.org/10.1016/B978-0-444-63524-2.00019-1

2. PREDICTED EFFECTS

As a result of the attention paid to the effects of global warming on various groups of organisms and various ecosystems, some lichenologists have addressed the question of what effects global warming might have or have had on lichens. Nash and Olafsen [11] predict that global warming in arctic areas may have a positive effect on lichens with cyanobacteria as photobiont, because the conditions for nitrogen fixation will improve. They reasoned that, under field conditions of optimal water hydration, lichen photosynthesis is primarily light-limited and nitrogen fixation is temperature-limited in both *Peltigera canina* and *Stereocaulon tomentosum* at Anaktuvuk Pass, Alaska. Thus, they continued, 'where duration of optimal hydration conditions remains unchanged from the present-day climate, the anticipated temperature increases in the Arctic may enhance nitrogen fixation in these lichens more than carbon gain. Because nitrogen frequently limits productivity in Arctic ecosystems, the results are potentially important to the many Arctic and subarctic ecosystems in which such lichens are abundant. The expected effect will be a spread of these species at the cost of other lichens and/or plants. So far this has not been unequivocally observed; rather the contrary: lichens have recently decreased in arctic regions, probably due to the increase in phanerogams' [8,12].

Insarov and Schroeter [13] and Insarov and Insarova [14] predict that lichens might, like other groups of organisms, show a response to global warming. As lichens are generally swift colonizers that disperse well, not only negative changes (extinctions) might be observed but also new invasions of more warmth-loving species in areas where they have not occurred before. In order to detect such changes, they installed some baseline monitoring transects across steep climatic gradients, but so far no results have been reported.

Ellis and co-workers [10,15] predict the response, in terms of changed distribution on the British Isles, of groups of lichens with different current distribution patterns and known ecological preferences, based on the current distribution and on several different climate scenarios. Although numerous historic data are also available, no unequivocal correlation between global warming and past changes in the lichen flora of the British Isles has been shown.

Zotz and Baader [16] described the different projected scenarios regarding lichens and bryophytes in the different biomes in the world. They show that a changing climate leaves species with three options: (1) they remain stationary and evolve in situ, (2) they track appropriate niches spatially and migrate, or else (3) they go extinct. For epiphytes, we can expect severe problems with tropical lowland species because their carbon balance is already precarious and could easily become negative at higher temperatures. At higher latitudes, range shifts may bring about new species assemblages.

Several authors [17,18] predict a disastrous effect that global warming might have on biological soil crusts, which are partly dominated by crustose lichens, in arid areas, thus destabilizing these habitats.

Finally, as a result of widespread melting of glaciers, new habitats for (especially) stone-inhabiting lichens are being formed. However, only the pioneer species can be expected to benefit from this.

3. OBSERVED EFFECTS

So far, a number of studies have demonstrated a correlation between global change and change in lichen habitat. The study by van Herk et al. [1] was the first and only one reported in the metanalysis by Parmesan and Yohe [19] in their study of 'globally coherent fingerprint of climate change impacts across natural systems'. The lichen study was based on a long-term (22 years) monitoring involving all the 329 epiphytic and terrestrial lichen species occurring in the Netherlands and were considered in relation to their world distribution. The investigation focussed on the exposed wayside trees in the province Utrecht in the Netherlands. The research was initially begun to document changes resulting from changes in sulphur dioxide air pollution levels. When the levels dropped, the effects on the lichens were clearly visible. However, the pattern was disturbed by a new emergent air pollution problem – ammonia from increasingly intensive cattle farming. As different lichens show different responses to this pollutant, the lichen monitoring was continued for a different purpose, viz. a detailed mapping of the areas with problematic ammonia pollution. Changes between 1995 and 2001, however, could not be explained in terms of air pollution variables alone. Analysis, however, showed a positive correlation with temperature, oceanity and nutrient demand, indicating a recent and significant shift toward species preferring warmer circumstances, independent from, and concurrent with, changes due to nutrient availability. In short, warmth-loving, oceanic lichens are expanding and boreal lichens are diminishing.

This effect, which appears to be restricted to temperate regions in Europe, has recently been reported from neighbouring countries [20–27]. German lichenologists currently are working on developing guidelines to assess regional ecological consequences of global climate change using epiphytic lichens as sensitive monitoring organisms. The Draft Directive (VDI 3957 Part 20 VE) specifies 45 so-called 'climate change indicators', all of which epiphytic lichens meet certain criteria related to Wirth's revised ecological indicator figures [28], or if no indicator figure has been published, lichens with temperate-Mediterranean and sub-Atlantic-Mediterranean distribution in Europe.

In northwestern Germany, the mean number of these indicator species monitored on phorophytes that were selected according to a standardized protocol has increased significantly [29] (see Figs 1 and 2), and even in Bavaria, where the mean temperature is lower due to higher geographical altitude, an increase is reported for sampling sites in the warm Main Valley in the northwestern part of the federal state [27].

The lichens that are expanding most dramatically are those with the green algae *Trentepohlia* as their photobiont. As these lichen species (i.e. the mycobiont) belong to different unrelated taxonomic groups and the effect has been observed in different ecosystems (exposed trees, forests), Aptroot and van Herk [30] argue that it seems likely that the effect of the global warming is in fact directly related to the alga, and all lichens with this alga can profit from the expansion of their photobiont. The process as described here is continuing and probably even accelerating and was also observed in neighbouring countries [20]. A study by van Herk [31] shows that most of the recent changes now can be attributed to global warming (see Figs 3 and 4).

Global warming will most probably aggravate the consequences of urban heat island formation that, e.g. in Düsseldorf, leads, at night-time, to a 9 K temperature difference between the warmed-up city centre as compared to the outskirts close to the Rhine river [32]. As pointed out by Stapper and John [25], epiphytic lichens are an appropriate tool to assess the biological effects of urban heat and of urban planning actions in order to attenuate heat build-up.

Taking into account the changing climate, contrasting responses of cryptogams to recent changes have been reported. Monitoring studies showed that abundance of mosses in high arctic tundra either increased over the last decades [33] or remained relatively stable [34]. Similarly, lichen biomass was reported to increase in a short-term warming experiment in Siberia [35], while the cover of mosses and lichens decreased in standardized warming experiments across the tundra biome [36], although the

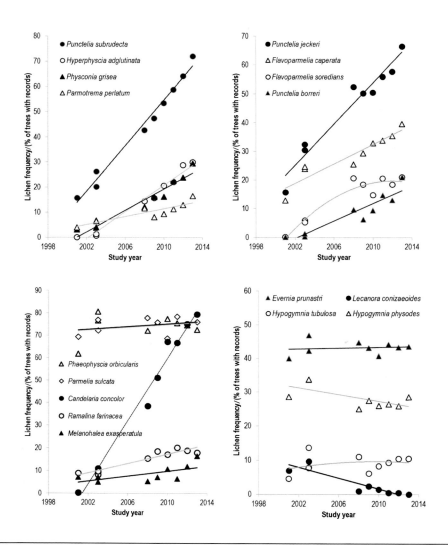

FIGURE 1

Change in the frequency of selected lichen species from 2001 to 2013 in North Rhine-Westphalia, Germany. Frequency is given as the percentage of standard trees on which the lichen species was recorded during the survey. All plotted species of the genera *Flavoparmelia*, *Parmotrema* and *Punctelia* are listed as 'climate change indicator species' in the Draft Directive VDI 3957 Part 20 VE. 2001. *(Data from Ref. [29] obtained on examination of 787 standard trees in the western part of the federal state North Rhine-Westphalia with location at GK-R < 2 590 000. 2003 to 2013: Lichen frequencies on 1015 trees in the state capital Düsseldorf (2003) and up to 211 trees (2003–2013) at four permanent observation sites in Düsseldorf ([25]; figure taken from Ref. [26]).)*

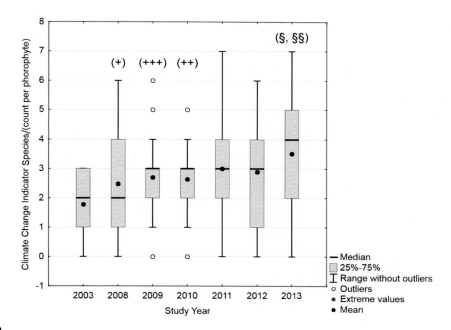

FIGURE 2

Changes in the frequency of climate change indicator species on phorophytes at the sampling station 'south' in Düsseldorf (the capital of North Rhine-Westphalia, Germany) from 2003 to 2013. Frequency is given as the number of indicator species recorded on 27 trees in annually repeated surveys from 2003 to 2013. Significances, as given for some data points: (+), $p < 0.05$ versus 2003; (++), $p < 0.01$ versus 2003; (+++), $p < 0.01$ versus 2003; (§), $p < 0.01$ versus 2012; (§§), $p < 0.002$ versus 2010; Wilcoxon-Test for paired data.

response of the mosses was less clear than that of lichens [37]. On the other hand, southerly vascular and cryptogam species might invade into recently colder ecosystems and outcompete cold-tolerant plants. An indication for this scenario is the recent migration of southerly distributed lichens into the Netherlands where arctic-alpine and boreo-montane species are declining [1]. Indirect effects of climate change may also trigger changes in species composition. As a combined effect of both increased temperature and increased soil nutrient mineralization, bigger and faster-growing vascular plant species might outcompete lichens and at least some bryophyte species [38], leading to drastic changes in plant community composition [4,39,40].

Related or co-occurring species may differ significantly in their response to global warming, as is shown in Figs 5 and 6, based on the work reported by Košuthová et al. [41]. It shows the current occurrence in terms of latitude and longitude in central Europe, as compared to the total distribution range [42,43]. The longitudinal centroids are fully in accordance, but the latitudinal centroids of three out of eight investigated species differ markedly between the world and current central European distributions. These are the two arctic-alpine lichens *Cladonia rangiferina* and *Cladonia stellaris* and boreal *Cladonia arbuscula* ssp, *mitis* [41,44]. These taxa occur in central Europe mostly at higher elevation and in case of *C. stellaris* limited in lowland areas. Arctic-alpine taxa are adapted to tolerate

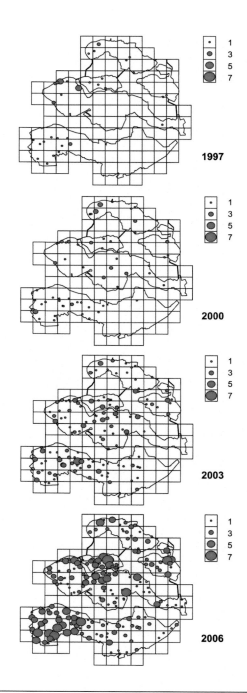

FIGURE 3

The distribution of lichen species with *Trentepohlia* phycobiont in the Netherlands province of Zeeland, in 1997, 2000, 2003 and 2006. The dot size refers to the number of species per site [12].

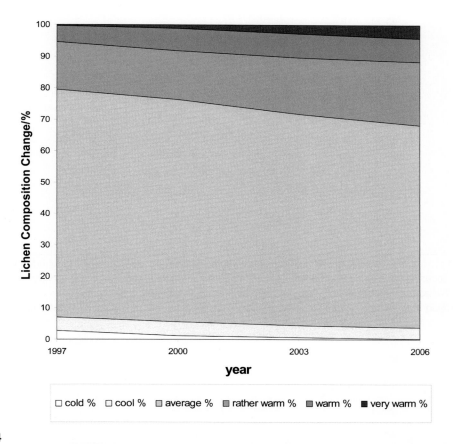

FIGURE 4

Changes of the epiphytic lichen composition in the Netherlands province of Zeeland in relation to temperature preference. The percentages are derived from species' frequencies per year. The total number of species for which a temperature preference is known is given as 100% [12].

extremely cold winter temperatures and warmer weather during the winter can cause damage [45]. *C. stellaris* is known to avoid the most oceanic regions [46] and to occur in continental areas [47]. This example underlines the importance of minimum winter temperature [41]. It would support the model of Ellis et al. [10,15], which predicts an increase in boreal and decrease in oceanic lichens in suboceanic regions in response to a drier climate change.

The axis with recent central European distribution is based on research conducted between 2006 and 2011 in four selected regions differing in climate, distributed from subcontinental Slovakia through inland Poland up to the suboceanic Baltic Sea region. Four regions were selected to cover the continentality gradient in Central-Northern Europe stretching from (1) the Central European subcontinental region (Borská nížina lowland, southwestern Slovakia, at the margin of the Pannonian lowland) via (2) the Central-Northern European subcontinental region (Masovian Landscape Park,

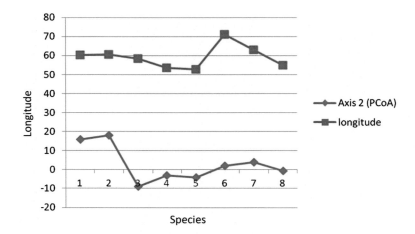

FIGURE 5

Correlation of selected lichen taxa with well-known distributions between average world longitude [42,43] versus the recent data from Central Europe [41]. The species referred to are: 1-*Cetraria islandica* ssp. islandica, 2-*Cladonia arbuscula* ssp. mitis, 3-*Cladonia arbuscula* ssp. squarossa, 4-*Cladonia ciliata* var. tenuis, 5-*Cladonia portentosa*, 6-*Cladonia rangiferina*, 7-*Cladonia stellaris*, 8-*Cladonia uncialis* ssp. uncialis.

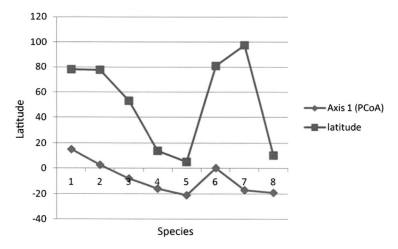

FIGURE 6

Correlation of selected lichen taxa with well-known distributions between average world latitude [42,43] versus the recent data from Central Europe [41].

central Poland, close to Warsaw), (3) the Central-Northern European suboceanic region (Bory Tucholskie National Park, northern Poland) to (4) the Baltic Sea coast suboceanic region (Słowiński National Park, northern Poland, on the coast of the Baltic Sea). The data set contains 400 releves of lichens from lichen-rich dry acidophilus Scots pine forests. The taxa were selected according to their score higher than 0.01. Principal coordinates analysis (PCoA) was used to describe the data structure and to reveal main compositional gradients using total species composition of micro-samples using the Jaccard index of similarity. For this purpose, the R program [48] was used, percentage cover data were logarithmically transformed and rare species were down weighted using the vegan library. The PCoA was performed with total species composition, cryptogams only, as well as with lichens only. The same position of centroids of particular regions along the main axes was obtained in all three analyses (data not shown). Hence, only one ordination is used, which was based on total species composition.

The climate sensitivity of lichens has also been demonstrated by experiments focused on physiological differences between species [49], including both mycobiont and photobiont partners [50,51], which might explain the preference of a species for particular environmental factors [52,53]. On a large scale, the processes between environmental factors and symbionts in lichens have been invoked to explain species biogeographic distributions and to project sensitivity of lichen species to climate changes [54,55].

4. UNCERTAIN EFFECTS

Some observed and reported changes in the lichen flora cannot be unequivocally attributed to global warming. There are several reasons for this but the most common one is that comparable historic or background data are wanting. Also climate change is not the only change taking place, and some of the changes occurring locally may interact or even counteract. Examples are isolated finds of warmth-loving species in more boreal countries, like *Flavoparmelia caperata* in Denmark, reported by Søchting [56] and attributed by him to global warming.

Another source of uncertainty is the intermittent and sometimes devastating effects of El Niño on coastal lichens along the Pacific coast of South America and on the Galapagos Islands. These have been documented, for example, by Follmann [57] and attributed directly to El Niño. The question remains whether the intensity of the El Niño effects is changing due to global warming, or not. In any event, lichens appear to have suffered more during the past few decades than ever before.

In some cases the patterns and processes are confused. An example is the reported work by Cezanne and co-workers [58] claiming that observed differences in lichens along an altitude gradient would prove the aptitude of certain lichens as biomonitors for climate changes. Their conclusions, however, are only based on a single study of poorly defined substrates and data analyzed with inappropriate statistical methods.

5. HABITATS WITH VULNERABLE LICHENS

There are four main habitats where lichens are potentially most vulnerable to climate change in the form of global warming, changes in precipitation and changes in the incidence of fog.

5.1 LOW LEVEL ISLANDS WITH ENDEMIC LICHENS

Examples of such islands include Porto Santo [59–61] and Bermuda [62]. The fact that some of the lichens on these islands are endemic suggests that they are either not capable of dispersal and/or their ecological requirements are not met elsewhere. In the event of a marked temperature rise or a change in the incidence of fog, the climate may become unsuitable for these species, and the chances of reaching a suitable substitute location are remote. The risks are highest at islands without mountains, as no suitable habitat will become available higher up the mountains.

5.2 EXTENDED REGIONS WITH SIMILAR CLIMATE BUT LOCAL ENDEMISM

The main examples are the extensive tropical rain forests. Although the climate and the physiognomy of the vegetation can be very similar over large areas, there can be a considerable amount of local endemism. This endemism is concentrated on the higher tree trunks, and not in the canopy (where wind moves the diaspores), and not at the various habitats at ground level (where bryophytes usually dominate and light conditions are poor). The endemic species usually have large ascospores, on the order of 0.1-mm diameter. The risks to these lichen involved in a climate change are that large expanses of habitat will change simultaneously, and the species with large diaspores have little chance of reaching a relatively remote suitable habitat. Incidentally, this risk may be small compared to the more direct and imminent risk of habitat destruction by logging. Furthermore, it has been pointed out by Zotz and Baader [16] that if tropical coastal regions become warmer, no species may continue to exist that are capable of occupying the habitats that become available as a result of other species shifting to higher elevations.

5.3 THE (ANT-)ARCTIC AND TUNDRA REGIONS

These areas are very rich in lichens, which often dominate the vegetation, both in biomass and in species diversity. Some lichens have been shown to decline, possibly indirectly as a result of global warming, due to increases in vascular plant biomass [12]. This is a potential threat to the rich (Ant-) Arctic lichen flora, but cannot be considered as an immediate one, as most (Ant-)Arctic lichens are relatively abundant and widespread, and a major impact will only occur in the unlikely event of the whole (Ant-)Arctic biome collapsing.

Sancho et al. [63] show that Antarctic lichens grow much faster in areas with milder climate, thus providing a tool for measuring cumulative warming by means of measuring lichen growth. There have been many studies on the effects of climate change on lichens in boreal to (Ant-)Arctic regions, both experimental and statistical [3–9]. Although the results vary, among these studies there were surprisingly many that showed a net positive effect of a warmer climate on lichens in (Ant-)Arctic regions.

5.4 HIGH GROUND IN THE TROPICS

High mountains in tropical areas sustain a rather depauperate lichen flora consisting predominantly of species widespread in boreo-alpine areas elsewhere in the world, but also including local endemics. These species have nowhere to go, other than literally in air, in the case of global warming. The mountains in New Guinea are examples of this group. They are among the most isolated biomes, as they are not connected to temperate regions, as, for example, the Andes are. Mount Wilhelm, reaching about 4500 m, is the highest mountain in Oceania, and from a lichenological point of view is the best investigated high mountain in New Guinea. It is also the richest in lichen species, as several other mountaintops are grass-covered. This is an isolated mountain, of which only less than 100 km^2 lie above the treeline and is at least partly suitable for boreo-alpine terricolous and saxicolous lichen growth. Among these are many cosmopolitan species [64,65]. The species, virtually on the equator, must be considered as 'boreo-alpine' or 'temperate' in a climatic (not geographic) sense. They cannot be considered as 'circumpolar' or 'bipolar' as is often stated [66]. For these New Guinean lichens their next closest localities are in Taiwan, over 4000 km away, and in the Himalayas, more than 5000 km away. How the species actually arrived remains unknown, although the presence of relatively many species that are associated with bird perching suggests that birds may have played an important role as vector of lichen diaspores, next to or even instead of wind and air currents.

The alpine lichen zone on Mount Wilhelm is restricted to a narrow altitudinal belt, above the tree limit at 3900 m to about 4300 m. This belt is known in botanical and tourism descriptions of the vegetation and the climb as the 'dead lichen zone' because the abundant *Thamnolia* is mistaken for dead lichens. The area consists of granite bedrock with large boulders, vertical cliffs and horizontal stretches with some soil compaction supporting heath-like dwarf shrub vegetation. This is a small zone where the recently described endemic *Sticta alpinotropica* [67] occurs on rocks, and the equally endemic *Thamnolia juncea* [68] is found in the (sub-)alpine grassland. The known world populations of both species amount to only a few square metres. Below the tree limit, the availability of various substrates for lichens is much wider, and the lichen diversity in the cloud forest belt is very high. This is the zone where numerous endemic species occur, e.g. of the genera *Anzia* [69] and *Menegazzia* [70].

Ecological niche modelling (ENM) tools are sometimes used to estimate the distribution of the suitable habitats of species. These tools allow identifying the locations of possible refugia of this species during the last glacial maximum (26 KA to 19 KA) as well as to predict the further changes of potential ranges due to the climate change. In lichenology, ENM tools were used so far for *Fuscopannaria confusa* [71] and in a local study on ecological response of *Vulpicida pinastri* to climate change and loss of its primary habitat [72].

ACKNOWLEDGEMENTS

Kok van Herk is warmly thanked for comments on an early version of this chapter.

REFERENCES

[1] C.M. van Herk, A. Aptroot, H.F. van Dobben, Lichenologist 34 (2002) 141–154.
[2] A. Aptroot, Bull. Br. Lichen Soc. 96 (2005) 14–16.
[3] N. Malmer, T. Johansson, M. Olsrud, T.R. Christensen, Global Change Biol. 11 (2005) 1895–1909.
[4] U. Molau, J.M. Alatalo, Ambio 27 (1998) 322–329.
[5] S. Semenov, G. Insarov, Eurasap Newsl. 44 (2002) 2–24.
[6] A.K. Jägerbrand, J.M. Alatalo, D. Chrimes, U. Molau, Oecologia 161 (2009) 601–610.
[7] C.J. Ellis, R. Yahr, in: T.R. Hodkinson (Ed.), Climate Change, Ecology and Systematics, Cambridge University Press, Cambridge, 2011, pp. 457–489.
[8] J.M.G. Hudson, G.H.R. Henry, J. Ecol. 98 (2010) 1035–1041.
[9] M. Hauck, Global Change Biol. 15 (2009) 2653–2661.
[10] C.J. Ellis, B.C. Coppins, T.P. Dawson, Biol. Conserv. 135 (2007) 396–404.
[11] T.H. Nash, A.G. Olafsen, Lichenologist 27 (1995) 559–565.

[12] J.H.C. Cornelissen, T.V. Callaghan, J.M. Alatalo, A. Michelsen, E. Graglia, A.E. Hartley, D.S. Hik, S.E. Hobbie, M.C. Press, C.H. Robinson, G.H.R. Henry, G.R. Shaver, G.K. Phoenix, D.G. Jones, S. Jonasson, F.S. Chapin III, U. Molau, C. Neill, J.A. Lee, J.M. Melillo, B. Sveinbjörnsson, R. Aerts, J. Ecol. 89 (2001) 984–994.

[13] G. Insarov, H. Schroeter, in: P.L. Nimis, C. Scheidegger, P.A. Wolseley (Eds.), Monitoring with Lichens – Monitoring Lichens, Kluwer, Dordrecht, 2002.

[14] G. Insarov, I. Insarova, Bibliotheca Lichenologica 82 (2002) 209–220.

[15] C.J. Ellis, B.C. Coppins, T.P. Dawson, M.R.D. Seaward, Biol. Conserv. 140 (2007) 217–235.

[16] G. Zotz, M.Y. Baader, Prog. Bot. 70 (2008) 147–170.

[17] R.D. Evans, J. Belnap, F. Garcia-Pichel, S.L. Phillips, in: J. Belnap, O. Lange (Eds.), Biological Soil Crusts: Structure, Function, and Management, Ecological Studies, Springer-Verlag, Berlin, Heidelberg, 2001, pp. 417–429.

[18] F.T. Maestre, C. Escolar, M.L. De Guevara, J.L. Quero, R. Lázaro, M. Delgado-Baquerizo, V. Ochoa, M. Berdugo, B. Gozalo, A. Gallardo, Global Change Biol. 19 (2013) 3835–3847.

[19] C. Parmesan, G. Yohe, Nature 421 (2003) 37–42.

[20] D. van den Broeck, Dumortiera 98 (2010) 6–10.

[21] U. Kirschbaum, R. Cezanne, M. Eichler, K. Hanewald, U. Windisch, Environ. Sci. Eur. 24 (2012) 19.

[22] N.J. Stapper, I. Franzen-Reuter, J.P. Frahm, Gefahrstoffe Reinhaltung der Luft 71 (2011) 173–178.

[23] S. Vogel, Archive for Lichenology 2 (2009) 1–15.

[24] N.J. Stapper, Bibliotheca Lichenologica 108 (2012) 221–240.

[25] N.J. Stapper, V. John, Gefahrstoffe Reinhaltung der Luft 73 (2013) 167–168.

[26] N.J. Stapper, V. John, Pollution Atmosphérique, 2015, in press.

[27] U. Windisch, A. Vorbeck, M. Eichler, R. Cezanne, Untersuchung der Wirkung des Klimawandels auf biotische Systeme in Bayern mittels Flechtenkartierung. Abschlussbericht 2011, überarbeitete Fassung vom Januar 2012, Bayerisches Landesamt für Umwelt, Augsburg, 2012.

[28] V. Wirth, Herzogia 23 (2010) 229–248.

[29] I. Franzen, N.J. Stapper, J.-P. Frahm, Ermittlung der lufthygienischen Situation Nordrhein-Westfalens mitepiphytischen Flechten und Moosen als Bioindikatoren. Gutachten im Auftrag des Ministeriums für Umweltund Naturschutz, Landwirtschaft und Verbraucherschutz Nordrhein-Westfalen, MUNLV, Düsseldorf, 2002, pp. 41.

[30] A. Aptroot, C.M. van Herk, Environ. Pollut. 146 (2007) 293–298.

[31] C.M. van Herk, Bibliotheca Lichenologica 99 (2008) 207–225.

[32] S. Düsseldorf, Klimaanalyse für die Landeshauptstadt Düsseldorf, Umweltamt, Landeshauptstadt Düsseldorf, 1995.

[33] J.M.G. Hudson, G.H.R. Henry, Ecology 90 (2009) 2657–2663.

[34] K. Prach, J. Košnar, J. Klimešová, M. Hais, Polar Biol. 33 (2010) 635–639.

[35] C. Biasi, H. Meyer, O. Rusalimova, R. Hämmerle, C. Kaiser, C. Baranyi, H. Daims, N. Lashchinsky, P. Barsukov, A. Richter, Plant Soil 307 (2008) 191–205.

[36] M.D. Walker, C.H. Wahren, R.D. Hollister, G.H.R. Henry, L.E. Ahlquist, J.M. Alatalo, M.S. Bret-Harte, M.P. Calef, T.V. Callaghan, A.B. Carroll, H.E. Epstein, L.S. Jonsdottir, L.A. Klein, B. Magnusson, U. Molau, S.F. Oberbauer, S.P. Rewa, C.H. Robinson, G.R. Shaver, K.N. Suding, C.C. Thompson, A. Tolvanen, O. Totland, P.L. Turner, C.E. Tweedie, P.J. Webber, P.A. Wookey, Proc. Natl. Acad. Sci. U.S.A. 103 (2006) 1342–1346.

[37] M.T. van Wijk, K.E. Clemmensen, G.R. Shaver, M. Williams, T.V. Callaghan, F.S. Chapin, J.H.C. Cornelissen, L. Grough, S.E. Hobbie, S. Johansson, J.A. Lee, A. Michelsen, M.C. Press, S.J. Richardson, H. Rueth, Global Change Biol. 10 (2003) 105–123.

[38] L.E. Rustad, J.L. Campbell, G.M. Marion, R.J. Norby, M.J. Mitchell, A.E. Hartley, J.H.C. Cornelissen, J. Gurevitch, Oecologia 126 (2001) 543–562.

[39] F.S. Chapin III, G.R. Shaver, Ecology 66 (1985) 564–576.

[40] G.R. Shaver, S.M. Bret-Harte, M.H. Jones, J. Johnstone, L. Gough, J. Laundre, F.S. Chapin, Ecology 82 (2001) 3163–3181.

[41] A. Košuthová, I. Svitková, I. Pišút, D. Senko, M. Valachovič, P.T. Zaniewski, M. Hájek, Fungal Ecol. 14 (2015) 8–23.

[42] B. Litterski, T. Ahti, Symbolae botanicae Upsalienses 34 (2004) 205–336.

[43] I. Kärnefelt, Opera Botanica 46 (1979) 55–97.

[44] A.D. Košuthová, J. Šibík, Ecol. Indic. 34 (2013) 246–259.

[45] J.W. Bjerke, Environ. Exp. Bot. 72 (2010) 404–408.

[46] T. Ahti, Annales Societatis Zoologica-Botanica Fennica 32 (1961) 1–160.

[47] M. Chytrý, J. Danihelka, S. Kubešová, P. Lustyk, N. Ermakov, M. Hájek, P. Hájková, M. Kočí, Z. Otýpková, J. Roleček, M. Řezníčková, P. Šmarda, M. Valachovič, D. Popov, I. Pišút, Plant Ecol. 196 (2008) 61–83.

[48] J. Oksanen, F.G. Blanchet, R. Kindt, P. Legendre, P.R. Minchin, R.B. O'Hara, G.L. Simpson, P. Solymos, M.H.H. Stevens, H. Wagner, Vegan: Community Ecology Package, R Package Version 2.0-10, 2013.

[49] J. Vondrák, J. Kubásek, Lichenologist 45 (2012) 1–10.

[50] M. Bačkor, O. Peksa, P. Škaloud, M. Bačkorová, Ecotoxicol. Environ. Saf. 73 (2010) 603–612.

[51] P. Škaloud, O. Peksa, Biologia 63 (2008) 873–880.

[52] L. Muggia, L. Vančurová, P. Škaloud, O. Peksa, M. Wedin, M. Grube, Microb. Ecol. 85 (2013) 313–323.

[53] O. Peksa, P. Škaloud, Mol. Ecol. 20 (2011) 3936–3948.

[54] A.D. Košuthová, J. Steinová, T. Spribille, Acta Biologica Cracoviensia, Series Botanica 55 (2013) 59.

[55] V. Lisewski, C.J. Ellis, Fungal Ecol. 4 (2011) 241–249.

[56] U. Søchting, Graphis Scripta 15 (2004) 53–56.

[57] G. Follmann, Cryptogamic Bot. 5 (1995) 224–231.

[58] R. Cezanne, M. Eichler, U. Kirschbaum, U. Windisch, Sauteria 15 (2008) 159–174.

[59] H. Krog, H. Østhagen, Norw. J. Bot. 27 (1980) 185–188.

[60] H. Krog, Lichenologist 22 (1990) 241–247.

[61] R. Haugan, Mycotaxon 44 (1992) 45–50.

[62] L.W. Riddle, Bull. Torrey Bot. Club 43 (1916) 145–160.

[63] L.G. Sancho, T.G.A. Green, A. Pintado, Flora 202 (2007) 667–673.

[64] A. Aptroot, P. Diederich, E. Sérusiaux, H.J.M. Sipman, Bibliotheca Lichenologica 64 (1997) 1–220.

[65] H. Streimann, Bibliotheca Lichenologica 22 (1986) 1–145.

[66] D.J. Galloway, A. Aptroot, Cryptogamic Bot. 5 (1995) 184–191.

[67] A. Aptroot, Lichenologist 40 (2008) 419–422.

[68] R. Santesson, Symbolae botanicae Upsalienses 34 (1) (2004) 393–397.

[69] I. Yoshimura, H.J.M. Sipman, A. Aptroot, Bibliotheca Lichenologica 58 (1995) 439–469.

[70] P.W. James, A. Aptroot, P. Diederich, H.J.M. Sipman, E. Sérusiaux, Bibliotheca Lichenologica 78 (2001) 91–108.

[71] T. Carlsen, M. Bendiksby, T.H. Hofton, S. Reiso, V. Bakkestuen, R. Haugan, H. Kauserud, E. Timdal, Lichenologist 44 (2012) 565–575.

[72] M.D. Binder, C.J. Ellis, Lichenologist 40 (2008) 63–79.

COASTLINE DEGRADATION AS AN INDICATOR OF GLOBAL CHANGE

20

Robert J. Nicholls[1,2], Colin Woodroffe[3], Virginia Burkett[4]

[1]*Faculty of Engineering and the Environment, University of Southampton, Southampton, UK;* [2]*Tyndall Centre for Climate Change Research, University of Southampton, Southampton, UK;* [3]*School of Earth and Environmental Sciences, University of Wollongong, NSW, Australia;* [4]*U.S. Geological Survey, Sunrise Valley Dr., Reston, VA, USA*

CHAPTER OUTLINE

1. INTRODUCTION

Coastal degradation has been widely reported around the world's coasts over the past century, especially in recent decades as discussed later in this chapter [1,2]. This degradation can be attributed to the intensification of a wide range of drivers of coastal change that are linked directly and indirectly to an expanding global population and economy. The twentieth century was also characterized by recognition of human-induced climate change and sea level rise, which constitutes an additional set of coastal drivers [3,4]. This chapter explores the relative contribution of climate change to observed coastal changes, focusing particularly on the extent to which climate change can be attributed as a significant driver of the change.

An analytical framework is adopted, based on a systems view of coasts as defined in Fig. 1. Comprising the narrow interface between land and sea, coastal systems are influenced by both marine and land surface processes. Coastal systems include intertidal zones and adjacent coastal lowlands and bays, lagoons, estuaries and nearshore waters. The connectivity of coasts with both marine and terrestrial systems is responsible, in part, for the high variability and complexity among coastal system

Climate Change. http://dx.doi.org/10.1016/B978-0-444-63524-2.00020-8

types. In contrast to terrestrial systems that have physical gradients that can stretch over tens or thousands of kilometres, coastal biotic and abiotic gradients are often relatively short, particularly along steep rocky shores. Many coastal areas support large and growing populations and high economic activity [5–7], which are changing coastal environments. River catchments feeding to the coast are increasingly modified, such that coastal systems are also influenced by these external changes [8]. Hence, few of the world's coastlines are now beyond the influence of human pressures [9], with many being dominated by human activities [10]; most coastal systems include elements of human development that interact with environmental changes associated with a warming climate.

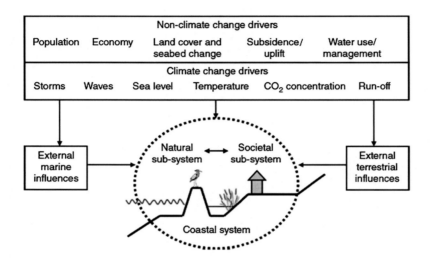

FIGURE 1

The coastal system showing how it is directly and indirectly impacted by climate change and other drivers. *(From Nicholls et al. [11].)*

Global warming through the twentieth century has caused a series of changes with important implications for coastal areas (Fig. 1). These include rising temperatures (both air and sea surface temperatures), rising sea level, increasing CO_2 concentrations with an associated reduction in seawater pH, and more intense precipitation on average (with substantial regional variation). It has also been argued that tropical storms have become more intense [12]. The tragic impacts of Hurricane Katrina on the Gulf of Mexico coast of the United States in 2005, of Cyclone Nargis on Myanmar in 2008, Superstorm Sandy on the East coast of the United States in 2012 and Typhoon Haiyan in the Philippines in 2013 all emphasize the enormous devastation, and loss of life in some cases, that these events can cause. However, it cannot be shown that these individual events were more intense as a result of climate change, and no firm conclusions on intensification of storms can be drawn at present.

Sea level rise is one of the most widely cited outcomes of global warming. Rising global sea level due to thermal expansion and the melting of land-based ice is already being observed with a global-mean rise of 19 cm $\pm$ 2 cm from 1900 to 2010 [12] and a slow accelerating trend [13]. Higher sea level will directly impact coastal areas, including some of the most densely populated and economically active land areas on Earth.

In this chapter we outline historical climate and sea level change and discuss how this impacts coasts, but we also recognize that coastal systems are subject to many other drivers, most especially the impacts of human development. We further discuss the need to discriminate whether coastal degradation can be attributed to the effects of climate or to what degree they are related to non-climate drivers.

2. SEA LEVEL RISE AND COASTAL SYSTEMS

Since the peak of the last glacial maximum about 20 000 a (where 'a' refers to years) ago, sea level has risen approximately 125 m [14,15]. Geologic evidence indicates inundation of coastal lowlands and retreat of shorelines during periods of rapid sea level rise, such as major meltwater pulses (Fig. 2). This pattern of sea level rise was experienced around the world, driven by the melt of the large ice sheets that appears to have ceased 7500–6000 a ago. The level of the sea has risen less than 3 m over the past 6000 a, and regional variations of sea level on timescales of a few hundred years or longer are likely to have been less than 0.3 m – 0.5 m [16]. As sea level stabilized, extensive coastal plains were formed, especially the major deltas, and the first evidence of early civilizations appeared on the plains [17,18].

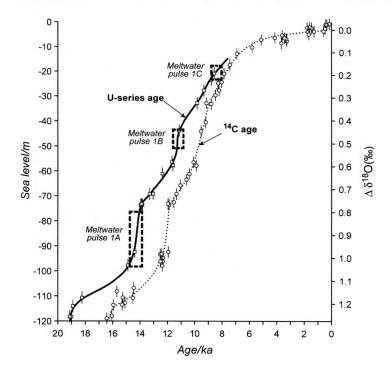

FIGURE 2

Sea level history since the peak of the last glacial maximum with boxes indicating the timing of melt water pulses. Abbreviations: MWP, meltwater pulse. MWP-1A0, c. 19 000 a ago, MWP-1A, (14 600–13 500) a ago, MWP-1B, (11 500–11 000) a ago, MWP-1C, ~8200–7600 a ago. *(Source: Murray–Wallace and Woodroffe [15]. Permission has been granted.)*

Coastline location and stability is intimately linked with changes in mean sea level. However, even under conditions of relatively stable mean sea level, coasts are extremely dynamic systems, involving co-adjustment of form and process at different time and space scales, termed morphodynamics [19,20]. Hence, erosion and deposition of coasts are naturally occurring due to short-term wave and tide conditions, as well as seasonal and longer-term climatic variability. The El Niño phenomenon, for example, has been shown to influence wave processes that shape beaches in the southwest Pacific [21] and cliffs in the eastern Pacific [22].

3. CLIMATE CHANGE AND GLOBAL/RELATIVE SEA LEVEL RISE

The impacts and responses of coasts to sea level rise are a product of *relative* (or local) sea level rise rather than global changes alone [15,23]. Relative sea level rise takes into account global-mean sea level rise, regional trends in the absolute elevation of the ocean surface, and geological uplift or subsidence and related processes that change the position of the land/sea boundary. Relative sea level rise is only partly a response to climate change and can vary significantly among coastal systems (Fig. 3). Abrupt changes may occur, for example the Great East Japan Earthquake caused sudden subsidence of up to 1.2 m along the Pacific coast of northeast Japan [4]. Sea level is presently falling

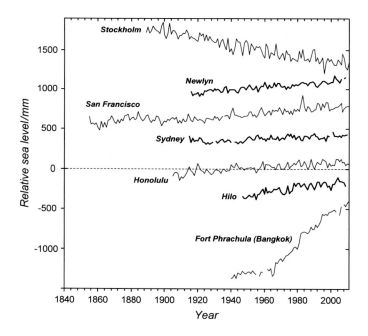

FIGURE 3

Selected relative sea level records for the twentieth century, illustrating different types of trend. The records are from the Permanent Service for Mean Sea Level (PSMSL, http://www.psmsl.org/data/obtaining/), offset for display purposes. (*Source: Murray-Wallace and Woodroffe, 2014 [15]. See also Holgate et al., 2013 [28]. This figure has been reproduced with permission.*)

due to ongoing glacial isostatic adjustment (rebound) in some high-latitude locations that were formerly sites of large (kilometre-thick) glaciers, such as Hudson Bay and the northern Baltic (see Stockholm, Fig. 3). In many places, relative sea level rise is similar to global trends (see Newlyn, San Francisco, Sydney, Hilo, Honolulu, Fig. 3). Sea level is rising more rapidly than global-mean trends on subsiding coasts, such as deltas [24–26]. Most dramatically, human-induced subsidence of susceptible areas due to drainage of organic soils and withdrawal of groundwater can produce dramatic rises in relative sea level, especially in susceptible coastal areas and cities built on recently deposited deltaic landforms [23,27]. Four noteworthy examples over the twentieth century are parts of Tokyo and Osaka, which subsided up to 4 m and 3 m, respectively, most of Shanghai, which subsided up to 3 m, and nearly all of Bangkok, which subsided up to 2 m (mainly since 1970, see Fort Phrachula, Bangkok; Figs 3 and 4). As a management response to human-induced subsidence, stopping shallow sub-surface fluid withdrawals can reduce subsidence.

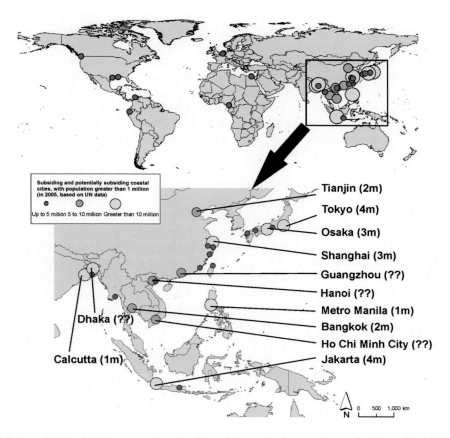

FIGURE 4

Subsiding and potentially subsiding coastal cities, indicating the concentration in Asia. The maximum observed subsidence (in metres) is shown for cities with populations exceeding 5 million people, where known. Maximum subsidence is reported as data on average subsidence is not available. (*From Nicholls [29].*)

4. INCREASING HUMAN UTILIZATION OF THE COASTAL ZONE

Human use of the coast increased dramatically during the twentieth century [4,30,31]. It has been estimated that 37% of the world's population lives within 100 km, and 49% lives within 200 km, of the coast [32]; the greatest number of people live at low elevations and population densities in coastal regions are about three times higher than the global average [5]. Almost two-thirds of urban settlements with populations greater than 5 million occur at low elevations in the coastal zone (less than 10 m above mean sea level), and there are at least 136 urban centres with population exceeding 1 million [33]. A disproportionate number of the countries with a large share of their population in low elevation coastal zones are small island countries. Most of the human population in this zone, however, resides in large countries with densely populated deltas, and migration of people to urban settings in coastal regions is widespread [6]. Tourism is also predominantly a coastal activity and tourist development is causing additional coastal utilization and infrastructure development in many coastal regions, such as around the Mediterranean and in many small islands such as the Maldives [34].

The expansion of human settlements and associated infrastructure (roads, buildings, ports, etc.) has directly altered land cover and land surface processes in large parts of the world's tropical and mid-latitude coastal landscapes. This rapid urbanization has many consequences; for example, land claim (e.g. Singapore, Dubai), enlargement of natural coastal inlets and dredging of waterways for navigation, port and harbour facilities, and pipelines exacerbate saltwater intrusion into surface and ground waters. Increasing shoreline retreat and consequent risk of flooding of coastal cities in many parts of the world have been attributed in part to the degradation of coastal ecosystems by human activities, as well as subsidence as already discussed [3,4]. As a result of this, cities often progressively move to artificial defensive and drainage systems as they develop/expand and their influence on their environs increases.

The natural ecosystems within watersheds have been fragmented and the downstream flow of water, sediment and nutrients to the coast disrupted [35]. Land-use change and hydrological modifications have had downstream impacts, in addition to localized influences, including human development on the coast. Hillslope erosion has increased the sediment load reaching the coast; for example, suspended loads in the Huanghe (Yellow) River have increased 2–10 times over the past 2000 a [36]. In contrast, damming and channelization has greatly reduced the supply of sediments to the coast on other rivers through retention of sediment in dams [8], and this effect dominated through the twentieth century [37,38].

The structure and ecological functions of natural systems are altered as a result of population growth, and ecological services provided by coastal systems are often disrupted directly or indirectly by human activities. For example, tropical and subtropical mangrove forests provide ecosystem services because they accumulate and transform nutrients, support rich ecological communities of fish and crustaceans, attenuate waves and storm surge impacts, and their root systems trap and bind sediments [39,40]. Large-scale conversions of coastal mangrove forests to shrimp aquaculture have occurred during the past three decades along the coastlines of Asia and Central America [41], and the decline or loss of mangrove forests reduces all of these ecosystem services [42]. Similar reductions of temperate salt marshes and wetlands in deltas are often linked to direct land use change [43,44]. Hence, on those developed coasts that have experienced disproportionately rapid expansion of settlements, urban centres and tourist resorts, the direct impacts of human activities on the coastal zone are profound, with more widespread indirect effects of human activities, which are often not fully recognized.

5. CLIMATE CHANGE, SEA LEVEL RISE AND RESULTING IMPACTS

Relative sea level rise has a wide range of effects on the natural system; the main effects are summarized in Table 1. Flooding/submergence, ecosystem change and erosion have received significantly more attention than salinization and rising water tables. Rising sea level alters all coastal processes. The immediate effect is submergence and increased flooding of coastal land, as well as saltwater intrusion into surface waters. Longer-term effects also occur as the coast adjusts to the new environmental conditions, including wetland loss and change in response to higher water tables and increasing salinity, erosion of beaches and soft cliffs, and saltwater intrusion into groundwater. These lagged changes interact with the immediate effects of sea level rise and generally exacerbate them. For instance, coastal erosion will tend to degrade or remove natural protective features (e.g. saltmarshes, mangroves and sand dunes) that in turn increase the likelihood of coastal flooding during storms.

Table 1 The main natural system effects of relative sea level rise, including climate and non-climate interacting factors. Some interacting factors (e.g., sediment supply) appear twice as they can be influenced both by climate and non-climate factors

Natural system effect			Interacting factors	
			Climate	Non-climate
1. Inundation (Including flood and storm damage)	a. Surge (from the sea)		Wave/storm climate, erosion, sediment supply.	Sediment supply, flood management, erosion, land reclamation.
	b. Backwater effect (from rivers)		Run-off.	Catchment management and land use.
2. Morphological change	a. Wetland loss (and change)		CO_2 fertilization of biomass production, sediment supply, migration space.	Sediment supply, migration space, land reclamation (i.e. direct destruction).
	b. Erosion (of beaches and soft cliffs)		Sediment supply, wave/storm climate.	Sediment supply.
3. Hydrological change	a. Saltwater intrusion	i. Surface waters	Run-off.	Catchment management (overextraction), land use.
		ii. Ground-water	Rainfall.	Land use, aquifer use (overpumping).
	b. Rising water tables/impeded drainage		Rainfall, run-off.	Land use, aquifer use, catchment management.

A rise in mean sea level also has a net effect of intensifying flooding during extreme storm events [45]. Changes in storm characteristics could have also influenced extreme water levels. Increases in tropical cyclone intensity in the North Atlantic over the past three decades are consistent with the observed changes in sea surface temperatures [12] and wave data in the North Atlantic support this observation [46]. However, it is difficult to prove if this is a systematic change or a component of cyclic variations in the frequency and intensity of tropical storms. Changes in storm tracks might also result from global climate change. In this context, Cyclone Catarina was the first documented hurricane in the South Atlantic, striking the coast of Brazil in March 2004 as a Category 2 storm on the Saffir-Simpson Hurricane Scale [47,48]. The cyclone killed at least three people and caused an estimated US 350×10^6 in damage in Brazil. It was followed by Tropical Storm Anita in 2010, although it is now recognized that earlier tropical events prior to Catarina have occurred in the south Atlantic [49]. Hence, these represent unusual events, rather than a new trend as a consequence of global warming.

Changes in the natural system due to sea level rise have many important direct socioeconomic impacts on a range of sectors with the effect being overwhelmingly negative. For instance, flooding can damage key coastal infrastructure, the built environment and agricultural areas, and in the worst case lead to significant mortality as occurred in 2008 when Cyclone Nargis devastated southern Myanmar. Erosion can lead to losses of the built environment and related infrastructure and have adverse consequences for sectors such as tourism and recreation. In addition to these direct impacts, there are indirect impacts such as negative effects on human health. For example, mental health problems increase after a flood [50], or the release of toxins from eroded landfills and waste sites that are commonly located in low-lying coastal areas, especially around major cities (e.g., Ref. [51]). Thus, sea level rise has the potential to trigger a cascade of direct and indirect human impacts.

6. RECENT IMPACTS OF SEA LEVEL RISE

Sea level was relatively stable in the sixteenth to eighteenth centuries; it started to rise in the nineteenth century and rose at 1.7 mm a^{-1} ± 0.2 mm a^{-1} from 1900 to 2010 and at 3.2 mm a^{-1} ± 0.4 mm a^{-1} from 1993 to 2010 [12]. Although this change may seem small, it has had many significant effects, most particularly in terms of the return periods of extreme water levels [12,45,52]. Worldwide there are many coasts that have been observed to be eroding [53]. However, attributing particular impacts such as erosion to sea level rise is difficult as erosion can be promoted by processes other than sea level rise (Table 1). As already discussed, many of these non-climate drivers of change operated over the twentieth century. While sea level rise is often inferred as an underlying cause of widespread retreat of sandy shorelines [54,55], negative sediment budgets at all scales also lead to erosion [20]. Human reduction in sediment supply to the coast has contributed to observed changes through activities such as construction of levees, dikes and dams on rivers that drain to the coast [8,56]. In deltas, both erosion and accretion can be apparent reflecting the large sediment supply [57–59]. Lastly, land claim can be confused with natural processes [60].

Changes in flood frequency and flood risk are also difficult to attribute to global sea level rise. Firstly, coastal flood exposure is dynamic [61]. Further, flood defences have often been upgraded substantially through the twentieth century, especially in those (wealthy) places where there are sea level measurements [62]. Most of this defence upgrade reflects expanding populations on the coastal plains and changing attitudes to risk. In many places, relative sea level rise has rarely even been explicitly considered in the design of past coastal infrastructure.

The accelerated rate of sea level rise observed since the late 1800s has been accompanied by coastal erosion and rapid wetland losses in many low-lying coastal regions. On the US east coast, relative sea levels have risen at rates of between 2 mm a^{-1} and 4 mm a^{-1} over the twentieth century due to varying patterns of subsidence caused by glacial isostatic adjustment. Both rates of sea level rise and coastal retreat have been measured, providing the opportunity to explore shoreline response to sea level rise. Away from inlets and engineered shores, the shoreline retreat rate is 50–100 times the rate of sea level rise, as might be anticipated using the concept of the Bruun Rule [55,63]. Near inlets, the indirect effects of sea level rise that cause the associated estuary/lagoon to trap beach sediment can have much larger erosional effects on the neighbouring open coasts than predicted by the Bruun Rule [64]. So whereas a simple heuristic like the Bruun Rule describes the response for some shores, more general relationships are required to fully understand coastal change, taking account of sea level change, sediment supply and coastal morphology [20].

In coastal Maryland and Louisiana, for example, wetland losses and shoreline retreat have led to a rapid restructuring of coastal ecosystems [45,65,66]. In Florida, a decline in coastal cabbage palm forests since the 1970s has been attributed to salt water intrusion associated with sea level rise [67,68]. Due to extensive human development along these coastlines, it is not possible to quantitatively isolate climate change effects versus changes due to other human development activities.

Human responses to sea level rise are even more difficult to document. There are many anecdotal descriptions of environmental refugees due to sea level rise, but these cannot be substantiated. People are certainly being forced to migrate due to coastal change, but as we have already discussed, we cannot with confidence link this to sea level rise or climate change. Further, human migration is a complex phenomenon with multiple interacting drivers [69]. A rare example where explicit links can be made is human abandonment of low-lying islands in the Chesapeake Bay, USA, during the late nineteenth/early twentieth century, which seems to have been triggered by the acceleration of sea level rise and resulting land loss [70].

There have certainly been impacts from relative sea level rise resulting from large rates of subsidence, such as the Mississippi delta where relative sea level rise approaches 1 cm a^{-1} (see Grand Isle, Fig. 3). Between 1978 and 2000, 1565 km^2 of intertidal coastal marshes and adjacent lands were converted to open water, due to sediment starvation and increases in the salinity and water levels of coastal marshes as a result of human development activities coupled with high rates of relative sea level rise [71]. The flooding in New Orleans during Hurricane Katrina was significantly exacerbated by subsidence compared to earlier flood events such as Hurricane Betsy in 1965 [72,73]. Coastal retreat has occurred due to subsidence, such as south of Bangkok where retreat of the shoreline has been more than a kilometre (Fig. 5). However, all the major cities that were impacted by subsidence have been

FIGURE 5

A line of telegraph poles south of Bangkok, in the Gulf of Thailand: built on land that was subsiding, they are now up to 1 km out to sea.

defended, even when the change in relative sea level rise was several metres. Post-Katrina, New Orleans has new upgraded defences costing about US $15 billion, and major new defences are also being implemented for Jakarta, Indonesia.

Hence, while global sea level rise has been a pervasive process, it is difficult to unambiguously link it to impacts, except in some special cases; most coastal change in the twentieth century was a response to multiple drivers of change. However, changes in two contrasting environments, polar coasts and tropical reefs, do appear to be directly exacerbated by warmer temperatures.

7. GLOBAL WARMING AND COASTS AT LATITUDINAL EXTREMES

In coastal areas, sea surface temperature (SST) has significantly warmed along more than 70% of the world's coastlines over the last 30 years [74]. While the changes are highly heterogeneous, both spatially and seasonally, the average rate is 0.18°C ± 0.16°C per decade. These values are larger than in the global ocean where the average of change is 0.11°C ± 0.0°C per decade in the upper 75 m of the ocean during the 1971–2010 period [75]. Importantly, extreme high temperature events have also been reported, such as unprecedented nearshore temperatures along the western Australian coast during 2010 and 2011 [76].

Rising air and sea temperatures pose a particular threat to coasts at the latitudinal extremes, polar coasts and coral reefs. Polar coasts are experiencing permafrost melt and a decrease in the extent of sea ice as a result of warming, which is leading to stronger wave generation causing a significant acceleration in erosion rates. Rapid shoreline erosion has been occurring on parts of the Arctic coast over recent decades, attributed in part to reduced sea ice cover allowing more wave activity to reach the shoreline [77]. Reduction in thickness of near-coastal ice, more rapid ice movement and retreat of the glacier fronts in Greenland appears related to warmer temperatures [78,79], with increased meltwater discharge. Similar trends to Greenland have been reported from the Antarctic Peninsula [80,81].

Parts of the Alaska coastline on the Beaufort Sea have retreated as much as 0.9 km in the past 50 a [82,83]. This coastal region is exposed to a combination of factors relating to climate change – sea level rise, the thawing of permafrost and the reduction in sea ice that protects that coastline from erosion during part of the year – all of which are contributing to rapid shoreline retreat (Fig. 6) [84]. Erosion at the coastline has led to the breaching of thermokarst lakes, causing initial draining followed by an increase in marine flooding that alters plant and animal community structure [82]. Similar retreat is occurring at sites in Arctic Canada [83], and the Mackenzie River delta front is retreating at $(1-10)$ m a^{-1} or more [62]. Evidence documented from traditional ecological knowledge also points to widespread change of coastlines across the North American Arctic from the Northwest Territories, Yukon and Alaska in the west to Nunavut in the east [84]. However, the impacts associated with human settlement along polar coasts are relatively very low due to the low population in these regions.

Within the tropics, widespread coral bleaching was detected on an unprecedented scale around the globe in response to El Niño-related warming in 1998 [86,87]. Further bleaching occurred across much of the Great Barrier Reef off northeast Australia in 2002 [88] and in the Caribbean in 2005 [89]. Bleaching occurs when warmer-than-usual sea surface temperatures lead to expulsion of the symbiotic zooxanthellae from within the coral tissue; the coral surface becomes pale, in many cases leading to mortality. It seems that temperatures ~1°C above the monthly maximum experienced by the coral result in bleaching, and that persistently high temperatures, or temperatures more than 2°C above this threshold, can cause the coral to die. Threshold temperatures above which corals bleach have evidently

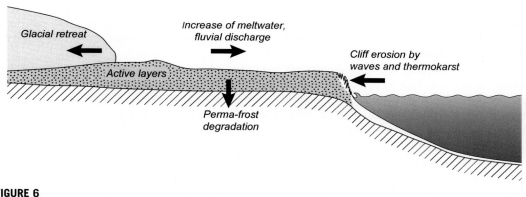

FIGURE 6

Changes in high-latitude coastal environments attributed to a warmer climate. *(Adapted from Kroon [85].)*

been occurring more frequently [90–92], and the prospect of further global warming implies that reefs may bleach with a frequency that exceeds their ability to recover between events.

Coral reefs are also susceptible to many other stresses, and there are many reefs that are severely degraded as a consequence of human activities, particularly overfishing and pollution [93]. As with other considerations of coastal degradation it is difficult to disentangle the effects of human-induced pressures from those that result directly from climate change. The synergistic effects of various pressures, including storms, combine to affect reefs, but the occurrence of bleaching on remote reefs well away from direct human development [94], and its incontrovertible association with increased sea surface temperature, provides a salutary warning of the likely consequences should global warming continue unabated. Human impacts, such as overfishing, appear to be exacerbating the stresses on reef systems and, at least on a local scale, exceeding the thresholds beyond which coral is replaced by other organisms [95]. Nevertheless, as with polar coasts, it is difficult to avoid the conclusion that these remote coastlines are changing for the worst as a consequence of rising sea surface temperatures.

8. THE CHALLENGE TO UNDERSTAND CONTEMPORARY IMPACTS

While significant coastal degradation has occurred over the twentieth century and early twenty-first century, it is difficult to unambiguously attribute the relative role of climate change. Most degradation has occurred on coasts that are influenced by one or more non-climate related drivers such as ongoing tectonic or isostatic adjustments, or, increasingly often, as a result of human activities. Further, the magnitude of climate change to date remains relatively small. In the next few decades, global warming will continue but at rates not dissimilar to the recent past. After this, it is expected to accelerate under higher emission trajectories (most especially RCP8.5), resulting in climate-induced impacts becoming more apparent.

In some coastal regions it is possible to discriminate between those effects that can already be attributed to climate change. Rising air and sea surface temperatures have resulted in detectable impacts on polar and tropical coasts. There is an emerging consensus that the increased frequency of bleaching on coral reefs is related to higher sea surface temperatures. Melting of sea ice and permafrost in high latitudes results from increased temperatures, and this is related to rapid erosion of polar coasts.

However, these coasts were already experiencing extensive erosion, and it is difficult to determine how much additional retreat has occurred because of climate change. Ocean acidification has not been discussed here. Basic chemistry suggests that this will cause significant changes, but observations of effects are not available as the data are too limited [4].

A significant component of global-mean sea level rise also results from global warming, reflecting thermal expansion and ice melt. Discriminating the impacts of the global-mean sea level component at regional and local scales where other contributions to relative sea level change are of variable importance remains problematic. This presents a challenge to further test and refine our understanding about the impacts of climate on coasts so that better predictions can be made and appropriate management plans put in place to respond to the anticipated impacts.

To meet this challenge, it will be necessary to continue and expand monitoring of coastlines, including both the climate and non-climate drivers, and the responses of coastal systems. Climate change is a global phenomenon, and therefore this monitoring and analysis needs to consider changes over broad scales. There will be an increasing role for more sophisticated remote sensing, which will be an important tool [44,82]. Comparative studies offer the opportunity to assess sensitivity, comparing those coasts with intense human pressures with more pristine counterparts in less densely populated regions. However, as indicated above, the indirect effects of human modification of the Earth are leaving a pervasive signal in even these remote places; global sea level change affects those coasts that are uninhabited as well as those that are intensively developed. Studies of analogues of climate change and sea level rise are also relevant, such as relative sea level rise on subsiding coasts. These can provide insights into outcomes expected more widely in response to global warming–induced sea level rise such as erosion due to sea level rise [55].

In parallel with data collection and monitoring, coastal impact modelling is also important to explore and test different ideas about how and why coasts are responding to different drivers, including climate and non-climate effects. In models, these drivers can be separated and their independent and interacting effects analyzed. As confidence in the model grows, so model experiments would allow attribution in a similar way to the link between growing greenhouse gas emissions and rising temperatures [96]. To date, this method has not been applied to coastal impacts.

9. CONCLUDING REMARKS

Finding a climate change signal on coasts is more problematic than often assumed. Coasts undergo natural dynamics at many scales, with erosion and recovery in response to climate variability such as El Niño, or extreme events such as storms and infrequent tsunamis (a non-climate effect). Additionally, humans have had enormous impacts on most coasts, overshadowing most changes that we can presently attribute directly to climate change.

Using the geographic examples cited in this paper, various impacts can be inferred on coasts as a consequence of changes in climate. However, each area of coast is experiencing its own pattern of relative sea level change and climate change, making discrimination of the component of degradation that results from climate change problematic. The best examples of a climate influence are related to temperature rise at low and high latitudes, as seen by the impacts on coral reefs and polar coasts, respectively. Observations through the twentieth century and early twenty-first century demonstrate the importance of understanding the impacts of sea level rise and climate change in the context of multiple drivers of change; this will remain a challenge under a more rapidly changing climate.

Nevertheless, there are emerging signs that climate change provides a global threat – sea ice is retreating – permafrost in coastal areas is widely melting – reefs are bleaching more often – and the sea is rising, amplifying widespread trends of subsidence in many densely populated areas and threatening low-lying areas. From this analysis some important lessons about the response to these challenges become evident. To devise successful adaptation strategies for coastal degradation it will be important to understand coastal changes in the context of integrated assessment and multiple drivers of change, with climate only being part of the problem [30,31,97]. To enhance the sustainability of coastal systems, management strategies will also need to address this challenge, focusing on the drivers that are dominant at each section of coast.

REFERENCES

[1] C.J. Crossland, H.H. Kremer, H.J. Lindeboom, J.I. Marshall Crossland, M.D.A. Le Tissier, Coastal Fluxes in the Anthropocene. The Land-ocean Interactions in the Coastal Zone Project of the International Geosphere-biosphere Programme Series: Global Change – The IGBP Series, 2005.

[2] I. Valiela, Global Coastal Change, Blackwell, Malden, MA, USA, 2006.

[3] R.J. Nicholls, P.P. Wong, V.R. Burkett, J.O. Codignotto, J.E. Hay, R.F. McLean, S. Ragoonaden, C.D. Woodroffe, Coastal systems and low-lying areas, in: M.L. Parry, O.F. Canziani, J.P. Palutikof, P.J. van der Linden, C.E. Hanson (Eds.), Climate Change 2007: Impacts, Adaptation and Vulnerability. Contribution of Working Group II to the Fourth Assessment Report of the Intergovernmental Panel on Climate Change, Cambridge University Press, Cambridge, UK, 2007, pp. 315–356.

[4] P.P. Wong, I.J. Losada, J.-P. Gattuso, J. Hinkel, A. Khattabi, K.L. McInnes, Y. Saito, A. Sallenger, Coastal systems and low-lying areas, in: C.B. Field, V.R. Barros, D.J. Dokken, K.J. Mach, M.D. Mastrandrea, T.E. Bilir, M. Chatterjee, K.L. Ebi, Y.O. Estrada, R.C. Genova, B. Girma, E.S. Kissel, A.N. Levy, S. MacCracken, P.R. Mastrandrea, L.L. White (Eds.), Climate Change 2014: Impacts, Adaptation, and Vulnerability. Part A: Assessment Report of the Intergovernmental Panel on Climate Change, Cambridge University Press, Cambridge, United Kingdom and New York, NY, USA, 2014, pp. 361–409.

[5] C. Small, R.J. Nicholls, J. Coast. Res. 19 (2003) 584–599.

[6] G. McGranahan, D. Balk, B. Anderson, Environ. Urbanization 19 (2007) 17–37.

[7] M. Lichter, A.T. Vafeidis, R.J. Nicholls, G. Kaiser, J. Coast. Res. 27 (2011) 757–768.

[8] J.P.M. Syvitski, C.J. Vörösmarty, A.J. Kettner, P. Green, Science 308 (2005) 376–380.

[9] R.W. Buddemeier, S.V. Smith, D.P. Swaney, C.J. Crossland, The role of the coastal ocean in the disturbed and undisturbed nutrient and carbon cycles, in: LOICZ Reports and Studies Series No. 24, 2002.

[10] K.F. Nordstrom, Beaches and Dunes of Developed Coasts, Cambridge University Press, Cambridge, UK, 2000.

[11] R.J. Nicholls, M.J.F. Stive, R.S.J. Tol, Coping with coastal change, in: G. Masselink, R. Gehrels (Eds.), Coastal Environments and Global Change, John Wiley & Sons, Chichester, 2014, pp. 410–431.

[12] J.A. Church, P.U. Clark, A. Cazenave, J.M. Gregory, S. Jevrejeva, A. Levermann, M.A. Merrifield, G.A. Milne, R.S. Nerem, P.D. Nunn, A.J. Payne, W.T. Pfeffer, D. Stammer, A.S. Unnikrishnan, Sea level change, in: T.F. Stocker, D. Qin, G.-K. Plattner, M. Tignor, S.K. Allen, J. Boschung, A. Nauels, Y. Xia, V. Bex, P.M. Midgley (Eds.), Climate Change 2013: The Physical Science Basis. Contribution of Working Group I to the Fifth Assessment Report of the Intergovernmental Panel on Climate Change, Cambridge University Press, Cambridge, United Kingdom and New York, NY, USA, 2013, pp. 1137–1216.

[13] P.L. Woodworth, N.J. White, S. Jevrejeva, S.J. Holgate, J.A. Church, W.R. Gehrels, Int. J. Clim. 29 (2009) 777–789.

[14] K. Lambeck, J. Chappell, Science 292 (2001) 679–686.

[15] C.V. Murray-Wallace, C.D. Woodroffe, Quaternary Sea-Level Changes: A Global Perspective, Cambridge University Press, Cambridge, 2014, 504 pp.

[16] P.A. Pirazzoli, Sea-Level Changes: The Last 20,000 Years, Wiley, Chichester, 1996.

[17] D.J. Stanley, A.G. Warne, Nature 363 (1993) 435–438.

[18] J.W. Day, M. Moerschbaecher, D. Pimentel, C. Hall, A. Yánez-Arancibia, Sustainability and place: how emerging mega-trends of the 21st century will affect humans and nature at the landscape level, Ecol. Eng. 65 (2014) 33–48.

[19] C.D. Woodroffe, Coasts, Form, Process and Evolution, Cambridge University Press, Cambridge, 2003.

[20] M.J.F. Stive, P.J. Cowell, R.J. Nicholls, Impacts of global environmental change on beaches, cliffs and deltas, in: O. Slaymaker, T. Spencer, C. Embleton-Hamann (Eds.), Geomorphology and Global Environmental Change, International Association of Geomorphologists. Cambridge University Press, Cambridge, 2009, pp. 158–179.

[21] R. Ranasinghe, R. McLoughlin, A.D. Short, G. Symonds, Mar. Geol. 204 (2004) 273–287.

[22] C.D. Storlazzi, G.B. Griggs, Geol. Soc. America Bull. 112 (2000) 236–249.

[23] R.J. Nicholls, S.E. Hanson, J.A. Lowe, R.A. Warrick, X. Lu, A.J. Long, WIREs Clim. Change 5 (2014) 129–150.

[24] J.P. Ericson, C.J. Vorosmarty, S.L. Dingman, L.G. Ward, M. Meybeck, Glob. Planet. Change 50 (2005) 63–82.

[25] C.D. Woodroffe, R.J. Nicholls, Y. Saito, Z. Chen, S.L. Goodbred, Landscape variability and the response of Asian megadeltas to environmental change, in: N. Harvey (Ed.), Global Change and Integrated Coastal Management: The Asia-Pacific Region, Springer, New York, 2006, pp. 277–314.

[26] J.P.M. Syvitski, A.J. Kettner, I. Overeem, E.W.H. Hutton, M.T. Hannon, et al., Sinking deltas due to human activities, Nat. Geosci. 2 (2009) 681–686.

[27] World Bank, Climate Risks and Adaptation in Asian Coastal Megacities: A Synthesis Report, World Bank Group, Washington, DC, 2010, 97 pp.

[28] S.J. Holgate, A. Matthews, P.L. Woodworth, L.J. Rickards, M.E. Tamisiea, E. Bradshaw, P.R. Foden, K.M. Gordon, S. Jevrejeva, J. Pugh, J. Coast. Res. 29 (2013) 493–504.

[29] R.J. Nicholls, Adapting to sea level rise, in: J.T. Ellis, D.J. Sherman (Eds.), Coastal and Marine Hazards, Risks and Disasters, Elsevier, Amsterdam, 2014, pp. 243–270.

[30] R.J. Nicholls, P.P. Wong, V.R. Burkett, C.D. Woodroffe, J.E. Hay, Sustainability Sci. 3 (2008) 89–102.

[31] R.J. Nicholls, C.D. Woodroffe, V. Burkett, J. Hay, P.P. Wong, L. Nurse, Scenarios for coastal vulnerability assessment, in: M. van den Belt, R. Costanza (Eds.), Ecological Economics of Estuaries and Coasts, Elsevier, Amsterdam, 2011, pp. 289–303.

[32] J.E. Cohen, C. Small, A. Mellinger, J. Gallup, J. Sachs, Science 278 (1997) 1211.

[33] S. Hanson, R.J. Nicholls, N. Patmore, S. Hallegatte, J. Corfee-Morlot, C. Herweijer, J. Chateau, Climatic Change 140 (2011) 89–111.

[34] J. Connell, Islands at Risk? Environments, Economies and Contemporary Change, Edward Elgar, Cheltenham, 2013.

[35] C. Nilsson, C.A. Reidy, M. Dynesius, C. Revenga, Science 308 (2005) 405–408.

[36] X. Jiongxin, Environ. Manage. 31 (2003) 328–341.

[37] P.H. Gleick, H. Cooley, D. Katz, E. Lee, G.H. Wolff, M. Palaniappan, J. Morrison, A. Samulon, The World's Water 2006–2007: The Biennial Report on Freshwater Resources, Island Press, Washington, 2006, p. 368.

[38] J.P.M. Syvitski, Y. Saito, Glob. Planet. Change 57 (2007) 261–282.

[39] D.R. Cahoon, P. Hensel, Hurricane Mitch: A Regional Perspective on Mangrove Damage, Recovery and Sustainability, USGS Open File Report, 2002, pp. 03–183.

[40] B.B. Lin, J. Dushoff, Manage. Environ. Qual. 15 (2004) 131–142.

[41] M. Spalding, D. Blasco, C. Field, World Mangrove Atlas, The International Society for Mangrove Ecosystems, The University of the Ryukyus, Okinawa, Japan, 1997.

[42] D.M. Alongi, Est. Coast. Shelf Sci. 76 (2008) 1–13.

[43] V.R. Burkett, D.A. Wilcox, R. Stottlemeyer, W. Barrow, D. Fagre, J. Baron, J. Price, J.L. Neilsen, C.D. Allen, D.L. Peterson, G. Ruggerone, T. Doyle, Ecol. Complexity 2 (2005) 357–394.

[44] J.M. Coleman, O.K. Huh, D. Braud, J. Coast. Res. 24 (2008) 1–14.

[45] M. Menéndez, P.L. Woodworth, J. Geophys. Res. 115 (2010) C10011.

[46] P.D. Komar, J.C. Allan, J. Coast. Res. 24 (2008) 479–488.

[47] A.B. Pezza, I. Simmonds, Geophys. Res. Lett. 32 (2005) L15712.

[48] R. McTaggart-Cowan, L.F. Bosart, C.A. Davis, E.H. Atallah, J.R. Gyakum, K.A. Emanuel, Monthly Weather Rev. 134 (2006) 3029–3053.

[49] M. Gan, J.L. Beven, A. Pezza, G. Holland, A. Pereira, R. McTaggart-Cowan, F. de Assis Diniz, M. Seluchi, H.J. Braga, Topic 2a: The Catarina Phenomenon. The Sixth WMO International Workshop on Tropical Cyclones (IWTC-VI), World Meteorological Organization, San José, Costa Rica, 2006, pp. 329–360.

[50] R. Few, M. Ahern, F. Matthies, S. Kovats, Floods, Health and Climate Change: A Strategic Review. Working Paper 63, Tyndall Centre for Climate Change Research, University of East Anglia, 2004, p. 138.

[51] T.J. Flynn, S.G. Walsh, J.G. Titus, M.C. Barth, Implications of sea level rise for hazardous waste sites in coastal floodplains, in: M.C. Barth, J.G. Titus (Eds.), Greenhouse Effect and Sea Level Rise: A Challenge for This Generation, Van Nostrand Reinhold, New York, USA, 1984.

[52] K. Zhang, B.C. Douglas, S.P. Leatherman, J. Clim. 13 (2000) 1748–1761.

[53] E.C.F. Bird, Coastline Changes, Wiley Interscience, Chichester, 1985.

[54] S.P. Leatherman, Social and economic costs of sea level rise, in: B.C. Douglas, M.S. Kearney, S.P. Leatherman (Eds.), Sea Level Rise, History and Consequences, Academic Press, London, 2001, pp. 181–223.

[55] N. Mimura, Sea level rise caused by climate change and its implications for society, Proc. Jpn. Acad. Ser. B 89 (7) (2013) 281–301.

[56] E.C.F. Bird, Submerging Coasts: The Effects of a Rising Sea Level on Coastal Environments, Wiley, Chichester, 1993.

[57] Centre for Environmental and Geographic Information Systems (CEGIS), in: Final Report on Monitoring Planform Developments in the EDP Area Using Remote Sensing, Dhaka, Bangladesh, 2009, 79 pp.

[58] M. G.M. Sarwar, C.D. Woodroffe, Rates of shoreline change along the coast of Bangladesh, J. Coast. Conserv. 17 (2013) 515–526.

[59] C.A. Wilson, S.L. Goodbred Jr., Construction and maintenance of the ganges-brahmaputra-meghna delta: linking process, morphology and stratigraphy, Annu. Rev. Mar. Sci. 7 (2015) 67–88.

[60] N. Biribo, C.D. Woodroffe, Historical area and shoreline change of reef islands around Tarawa Atoll, Kiribati, Sustain. Sci. 8 (2013) 345–362.

[61] B. Jongman, P.J. Ward, J.C.J.H. Aerts, Global exposure to river and coastal flooding: long term trends and changes, Glob. Environ. Change 22 (2012) 823–835.

[62] A.C. Ruocco, R.J. Nicholls, I.D. Haigh, M.P. Wadey, Nat. Hazards 59 (2011) 1773–1796.

[63] K. Zhang, B.C. Douglas, S.P. Leatherman, Clim. Change 64 (2004) 41–58.

[64] M.J.F. Stive, Clim. Change 64 (2004) 27–39.

[65] K.W. Krauss, J.L. Chambers, J.A. Allen, D.M. Soileau Jr., A.S. DeBosier, J. Coast. Res. 16 (2000) 153–163.

[66] USGS, Synthesis of U.S. Geological Survey Science for the Chesapeake Bay Ecosystem and Implications for Environmental Management, US Geological Survey, Reston, Virginia, 2007. Circular 1316, p. 71.

[67] K. Williams, K.C. Ewel, R.P. Stumpf, F.E. Putz, T.W. Workman, Ecology 80 (1999) 2045–2063.

[68] K. Williams, M. MacDonald, L. da Silveira Lobo Sternberg, J. Coast. Res. 19 (2003) 1116–1121.

[69] R. Black, S.R.G. Bennett, S.M. Thomas, J.R. Beddington, Nature 478 (2011) 447–449.

[70] S.J.A. Gibbons, R.J. Nicholls, Glob. Environ. Change 16 (2006) 40–47.

[71] J. Barras, S. Beville, D. Britsch, S. Hartley, S. Hawes, J. Johnston, P. Kemp, Q. Kinler, A. Martucci, J. Porthouse, D. Reed, K. Roy, S. Sapkota, J. Suhayda, Historical and Projected Coastal Louisiana Land Changes: 1978–2050, Open File Report 03-334, U.S. Geological Survey, 2003, p. 39.

[72] P. Grossi, R. Muir-Wood, Flood Risk in New Orleans: Implications for Future Management and Insurability, Risk Management Solutions (RMS), London, UK, 2006.

[73] L.E. Link, Ocean Eng. 37 (2010) 4–12.

[74] F.P. Lima, D.S. Wethey, Nat. Commun. 3 (2012), http://dx.doi.org/10.1038/ncomms1713. Article 74.

[75] M. Rhein, S.R. Rintoul, S. Aoki, E. Campos, D. Chambers, R.A. Feely, S. Gulev, G.C. Johnson, S.A. Josey, A. Kostianoy, C. Mauritzen, D. Roemmich, L.D. Talley, F. Wang, Observations: ocean, in: T.F. Stocker, D. Qin, G.-K. Plattner, M. Tignor, S.K. Allen, J. Boschung, A. Nauels, Y. Xia, V. Bex, P.M. Midgley (Eds.), Climate Change 2013: The Physical Science Basis. Contribution of Working Group I to the Fifth Assessment Report of the Intergovernmental Panel on Climate Change, Cambridge University Press, Cambridge, United Kingdom and New York, NY, USA, 2013, pp. 255–315.

[76] A.F. Pearce, M. Feng, J. Marine Syst. 111-112 (2013) 139–156.

[77] O.M. Johannessen, L. Bengtsson, M.W. Miles, S.I. Kuzmina, V.A. Semenov, G.V. Aleekseev, A.P. Nagurnyi, V.F. Zakharov, L. Bobylev, L.H. Pettersson, K. Hasselmann, H.P. Cattle, Arctic Climate Change – Observed and Modeled Temperature and Sea Ice Variability, Technical Report 218, Nansen Environmental and Remote Sensing Centre, University of Bergen, Norway, 2002.

[78] W. Krabill, E. Hanna, P. Huybrechts, W. Abdalati, J. Cappelen, B. Csatho, E. Frederick, S. Manizade, C. Martin, J. Sonntag, R. Swift, R. Thomas, J. Yungel, Geophys. Res. Lett. 31 (2004) L24402.

[79] E. Rignot, D. Braaten, P. Gogineni, W. Krabill, J.R. McConnell, Geophys. Res. Lett. 31 (2004) L10401.

[80] E. Rignot, G. Casassa, P. Gogineni, W. Krabill, A. Rivera, R. Thomas, Geophys. Res. Lett. 31 (2004) L18401.

[81] A.J. Cook, A.J. Fox, D.G. Vaughan, J.G. Ferrigno, Science 308 (2005) 541–544.

[82] J.C. Mars, D.W. Houseknecht, Geology 35 (2007) 583–586.

[83] G.K. Manson, S.M. Solomon, D.L. Forbes, D.E. Atkinson, M. Craymer, Geo-Marine Lett. 25 (2005) 138–145.

[84] D.L. Forbes (Ed.), State of the Arctic Coast 2010 – Scientific Review and Outlook. International Arctic Science Committee, Land-Ocean Interactions in the Coastal Zone, Arctic Monitoring and Assessment Programme, International Permafrost Association. Helmholtz-Zentrum, Geesthacht, Germany, 2011, 178 p. http://arcticcoasts.org.

[85] A. Kroon, High-latitude coasts, in: G. Masselink, R. Gehrels (Eds.), Coastal Environments and Global Change, John Wiley & Sons, Chichester, 2014, pp. 338–353.

[86] T. Spencer, K.A. Teleki, C. Bradshaw, M.D. Spalding, Marine Pollut. Bull. 40 (2000) 569–586.

[87] J.M. Lough, Geophys. Res. Lett. 27 (2000) 3901–3904.

[88] R. Berkelmans, G. De'ath, S. Kininmouth, W.J. Skirving, Coral Reefs 23 (2004) 74–83.

[89] J.P. McWilliams, I.M. Cote, J.A. Gill, W.J. Sutherland, A.R. Watkinson, Ecology 86 (2005) 2055.

[90] T.P. Hughes, A.H. Baird, D.R. Bellwood, M. Card, S.R. Connolly, C. Folke, R. Grosberg, O. Hoegh-Guldberg, J.B.C. Jackson, J. Kleypas, J.M. Lough, P. Marshall, M. Nystrom, S.R. Palumbi, J. M.Pandolfi, B. Rosen, J. Roughgarden, Science 301 (2003) 929–933.

[91] O. Hoegh-Guldberg, Symbiosis 37 (2004) 1–31.

[92] O. Hoegh-Guldberg, J. Geophys. Res. 110 (2005) C09S06.

[93] J.M. Pandolfi, R.H. Bradbury, E. Sala, T.P. Hughes, K.A. Bjorndal, R.G. Cooke, D. McArdle, L. McClenachan, M.J.H. Newman, G. Paredes, R.R. Warner, J.B.C. Jackson, Science 301 (2003) 955–958.

[94] G. De'ath, K. Fabricius, H. Sweatman, M. Puotinen, The 27-year decline of coral cover on the Great Barrier Reef and its causes, Proc. Natl. Acad. Sci. 109 (2012) 17995–17999.

[95] R.W. Buddemeier, J.A. Kleypas, R.B. Aronson, Coral Reefs and Global Climate Change: Potential Contributions of Climate Change to Stresses on Coral Reef Ecosystems, Pew Center on Global Climate Change, Arlington, Virginia, USA, 2004.

[96] P.A. Stott, N.P. Gillett, G.C. Hegerl, D.J. Karoly, D.A. Stone, X. Zhang, F. Zwiers, WIREs Clim. Change 1 (2010) 192–211.

[97] S. Brown, R.J. Nicholls, S. Hanson, G. Brundit, J.A. Dearing, M.E. Dickson, S.L. Gallop, S. Gao, I.D. Haigh, J. Hinkel, J.A. Jimenez, R.J.T. Klein, W. Kron, A.N. Lazar, C.F. Neves, A. Newton, C. Pattiaratchi, A. Payo, K. Pye, A. Sanchez-Arcilla, M. Siddall, A. Shareef, E.L. Tompkins, A.T. Vafeidis, B. van Maanen, P.J. Ward, C.D. Woodroffe, Shifting perspectives on coastal impacts and adaptation, Nat. Clim. Change 4 (2014) 752–755.

PLANT PATHOGENS AS INDICATORS OF CLIMATE CHANGE

21

K.A. Garrett[1,2], M. Nita[3], E.D. De Wolf[2], P.D. Esker[4], L. Gomez-Montano[2], A.H. Sparks[5]

[1]*Institute for Sustainable Food Systems and Plant Pathology Department, University of Florida, Gainesville, FL, USA;*
[2]*Department of Plant Pathology, Kansas State University, Manhattan, KS, USA;* [3]*Department of Plant Pathology, Physiology, and Weed Science, AHS Jr. AREC, Virginia Polytechnic Institute and State University, Winchester, VA, USA;*
[4]*Escuela de Agronomía, Universidad de Costa Rica, San Pedro Montes de Oca, San José, Costa Rica;* [5]*International Rice Research Institute (IRRI), Los Baños, Laguna, Philippines*

CHAPTER OUTLINE

1. INTRODUCTION

Plant disease risk is strongly influenced by environmental conditions [1]. While some animal hosts may provide their pathogens with a consistent range of body temperatures, plant pathogens are generally much more exposed to the elements. Plant disease will tend to respond to climate change, though a number of interactions take place among host, pathogen, and potential vectors. In some cases, the actions of land managers may also complicate interpretation of climate change effects. In this chapter we present a brief introduction to plant disease and a synthesis of research in plant pathology related to climate change. We discuss the types of evidence for climate change impacts ('climate change fingerprints') that might be observed in plant disease systems and evaluate potential evidence of climate change fingerprints.

The battle to protect plant health is ongoing, and plant disease management is essential for our continued ability to feed a growing human population. The Great Famine in Ireland is one striking example of the impact of plant disease: in 1846–1851 around 1 million Irish people died during an extremely destructive epidemic of potato late blight [2]. Plant diseases continue to cause serious problems in global food production. Approximately 800 million people do not have adequate food and 10%–16% of global food production is lost to plant disease [3,4]. Not only does plant disease affect

Climate Change. http://dx.doi.org/10.1016/B978-0-444-63524-2.00021-X

human food production but it also impacts natural systems [5]. Introduced diseases such as chestnut blight in the Eastern US, and more recently the increasing occurrence of sudden oak death, have resulted in the rapid decline of dominate tree species and triggered major impacts on forest systems [6]. Furthermore, the genetic diversity of many plant pathogens greatly increases the risk of a large-scale epidemic, as evidenced by the threat posed by the emergence of race Ug99 of the wheat stem rust fungus [7].

Plant pathogen groups include fungi, prokaryotes (bacteria and mycoplasmas), oomycetes, viruses and viroids, nematodes, parasitic plants, and protozoa. The very different life histories of this diverse group of organisms and their different interactions with host plants produce a wide range of responses to environmental and climatic drivers, each with potentially different mechanisms of response under climate change. As such there is great need for more quantitative information related to each type of important disease [8]. For example, viruses may be present in hosts while symptom expression is dependent on temperature [9]; thus, even the difficulty of detection of pathogens varies with climate. Fungal pathogens are often strongly dependent on humidity or dew [10], so changes in these environmental factors are likely to shift disease risk. Managing plant pathogens is also made more complicated due to genetic variation within pathogen populations [3]. Pathogen species may quickly develop resistance to pesticides or adapt to overcome plant disease resistance, and may also adapt to environmental changes, where the rate of adaptation depends on the type of pathogen [11]. Many plant pathogens have a high reproduction potential and pathogen populations may increase rapidly when weather conditions are favourable for disease development [12,13]. The rapid onset of disease makes it difficult to anticipate the best timing of management measures, especially in areas with high levels of interannual variability in climatic conditions. There is also a complex set of interactions among microorganisms, especially those that make up soil biodiversity [14–16], making it challenging to define the temporal and spatial scales required to adequately study disease responses to climate [17].

2. CLIMATIC VARIABLES AND PLANT DISEASE

Understanding the factors that trigger the development of plant disease epidemics is essential if we are to create and implement effective strategies for disease management [18]. This has motivated a large body of research addressing the effects of weather or climate on plant disease [18,19]. Plant disease occurrence is generally driven by three factors: a susceptible host, the presence of a competent pathogen (and vector if needed) and conducive environment [12,13]. All three factors must be in place, at least to some degree, for disease to occur (Fig. 1). A host resistant to local pathogen genotypes, or unfavourable weather for pathogen infection, will each reduce disease intensity. The synchronous interaction among host, pathogen and environment governs disease development. These interactions can be conceptualized as a continuous sequence of cycles of biological events including dormancy, reproduction, dispersal and pathogenesis [1]. In plant pathology this sequence of events is commonly referred to as a disease cycle. Although plant pathologists have long realized the importance of the disease cycle, its component events and the apparent relationships with environment, the quantification of these interactions did not begin in earnest until the 1950s [18]. The following decades of research have established a vast body of literature documenting the impact of temperature, rainfall amounts and frequency, and humidity, on the various components of the disease cycle [18].

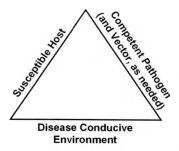

FIGURE 1

Plant disease results from the interaction of host, pathogen and environment. Climatic features such as temperature, humidity and leaf surface wetness are important drivers of disease, and inappropriate levels of these features for a particular disease may be the limiting factor in disease risk.

The quantification of the relationship between the disease cycle of a given plant disease and weather is also the foundation of many predictive models that can be used to advise growers days or weeks before the onset of an increase in disease incidence or severity [1]. Such prediction tools can allow a grower to respond in a timely and efficient manner by adjusting crop management practices. Given enough time to respond, a disease prediction might allow a grower to alter the cultivar they select for planting, the date on which the crop is sown, or the scheduling of cultural practices such as fertilization or irrigation. A prediction of low disease risk may also result in reduced pesticide use with positive economic and environmental outcomes. Larger-scale predictions of disease risk, such as the typical risk for regions or countries based on climatic conditions, can be used to form policy and priorities for research (e.g. Refs [20,21]).

Interestingly, the quantification of these relationships and application of this information as part of disease prediction models has also facilitated the simulation of potential impacts of climate change. For example, Bergot et al. [22] have used models of the impact of weather variables on the risk of infection by the oomycete *Phytophthora cinnamomi* to predict the future distribution of disease caused by this pathogen in Europe under climate change scenarios. As more detailed climate change predictions are more readily available, many plant disease forecasting systems may be applied or adapted for application in climate change scenario analyses (e.g. Refs [23,24]).

Some relationships between climate and disease risk are obvious, such as some pathogens' inability to infect without sufficient surface moisture (i.e. dew or rain droplets) [10] or other pathogens' or vectors' inability to overwinter when temperatures go below a critical level. Other effects of climate may be more subtle. For example, a given pathogen may only be able to infect its host(s) when the plants are in certain developmental stages. This also means that in order to maximize their chance of infection, the life cycle of pathogen populations must be in sync with host development. Here we discuss a few examples where host phenology is key for disease development.

Some pathogens depend on flower tissues as a point of entry to the host. For example, the fungus *Botrytis cinerea*, which causes grey mould of strawberry and other fruits (producing a grey fuzz-balled strawberry, which you may have seen at a grocery store or in your refrigerator), infects the strawberry at the time of flowering [25]. The fungus stays in flower parts until the sugar level of the berry increases, and then causes grey mould disease. Another example is Fusarium head or ear blight, also called scab of wheat and barley, which causes large yield losses, reductions in grain quality and contamination with mycotoxins [26,27]. Mycotoxins are toxic substances produced by the fungi, which can be more important than simple yield reductions, such that climate change effects on mycotoxin production are an important concern in themselves [28]. Several fungal species cause Fusarium head blight, where *Fusarium graminearum* (teleomorph: *Gibberella zeae*) is responsible for the most aggressive form. The anthesis (flowering) period seems to be the critical time for infection [27,29]. An important bacterial disease of apple and pears, called fire blight, also utilizes flowers as a major point of entry [30]. The causal agent (*Erwinia amylovora*) can be disseminated by pollinating insects such as bees and moves into the tree through flowers, causing rapid wilting of branch tips.

Certain hosts become more resistant after a particular developmental stage, some exhibiting a trait referred to as adult plant resistance. There are many examples of genes that follow this pattern in wheat, including leaf rust (caused by the fungus *Puccinia triticina*) resistance genes *Lr13* and *Lr34* [31] and stripe rust (caused by *Puccinia striiformis* f. sp. *tritici*) resistance gene *Yr39* [32]. These genes are activated by a combination of wheat developmental stage and temperature changes. In grape, there are many cases of ontogenic (or age-related) resistance against pathogens. Once grape fruit tissue matures, certain fungal pathogens such as *Erysiphe necator* (causing powdery mildew) [33], or *Guignardia bidwellii* (causing black rot) [34] or the oomycete pathogen *Plasmopara viticola* (causing downy mildew) [35] are less successful at infecting plants.

Host development patterns may be altered with changes in climate. For the examples above, the timing and duration of flowering in wheat are a function of the average daily temperature. Heavy rain and/or strong wind events can shorten flowering duration in strawberry and apple through flower damage. Some pathogen species may be able to maintain their synchrony with target host tissue while others may become out of sync. Thus, there are some efforts to modify disease prediction systems to accommodate potential impacts from climate change. For example, in efforts to predict the risk of apple scab (caused by the fungus *Venturia inaequalis*), the concept of ontogenic resistance was utilized along with inoculum production [36] because tissues become less susceptible as the rate of tissue expansion decreases.

Pathogen dispersal is another aspect of epidemiology that can be influenced by climate change. Pathogens such as rust fungi often overwinter in warmer regions and migrate annually via wind to cooler regions during crop production [37,38]. Pathogens and other microbes can be spread in dust storms [39]. As climate shifts, so may the patterns of wind dispersal of pathogens. For pathogens dispersed by insects, new patterns of insect movement and encounters with potential new vectors as ranges shift may also alter epidemics [40]. Combined with potential changes in cropland area due to climate and other global change factors, the resulting new patterns of invasion, reinvasion and saturation may result in epidemic evolution with new patterns of risk [41,42]. The ability of managers to track geographically the new management requirements will be a key issue in determining disease outcomes because there are important lags in responding to new needs through methods such as crop breeding for resistance, development of new pesticides and simply communicating effectively about management [43–50].

There is no doubt that weather influences plant disease; that relationship is fundamental to the modelling of plant disease epidemiology. Thus it is fairly straightforward to predict that where climate change leads to weather events that are more favourable for disease, there will be increased disease pressure. But the relationship between climate change and associated weather events, and resulting changes in disease development, will generally not be a simple one-to-one relationship (Fig. 2).

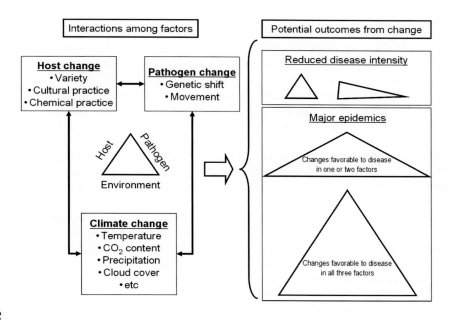

FIGURE 2

Interactions among components of the disease triangle and potential outcomes. Amount of disease (incidence, severity, etc.) or risk is indicated by the area of the triangle. Changes in host, pathogen and climate can increase or decrease the amount of disease as a result of their interactions.

The impacts will tend to be most dramatic when climatic conditions shift above a threshold for pathogen reproduction, are amplified through interactions, or result in positive feedback loops that decrease the utility of disease management strategies [51]. For example, the Karnal bunt pathogen, *Tilletia indica*, which reduces wheat quality, will tend to have lower reproductive rates per capita when populations are low because individuals of different mating types must encounter each other for reproductive success [52]. If climatic conditions change to favour pathogen reproduction, the pathogen will be released from this constraint and show a larger response to the change than would otherwise have been anticipated. The trend toward greater global movement of humans and materials also produces new types of interactions as pathogens are introduced to new areas and may hybridize to produce new pathogens [53,54].

3. EVIDENCE THAT SIMULATED CLIMATE CHANGE AFFECTS PLANT DISEASE IN EXPERIMENTS

Next we consider two types of evidence for effects of *changes* in climate on plant disease. The first is evidence that simulated climate change affects plant disease in experimental settings. The effect of simulated climate change has been studied in experiments with altered heat treatments, altered precipitation treatments and carbon-enrichment treatments. Where there are apparent effects from these treatments, this implies that, to the extent that the simulations do successfully represent future climate scenarios, plant disease will respond. The second type of evidence is for changes in observed patterns of plant disease in agricultural or wild-land systems that can be attributed to climate change with some level of confidence, discussed in Section 4. In this case, the changes in plant disease might be taken as fingerprints of climate change. We also discuss what types of plant disease scenarios might qualify as fingerprints of climate change in this sense.

The range of possibilities for climate change simulations can be characterized in terms of the scale of the effect being considered [55,56]. For many well-studied pathogens and vectors, the temperature ranges that support single infection events or survival are fairly well characterized. The effects of plant water stress and relief from water stress on disease risk have also been studied in controlled experiments for some pathogens and may be quite relevant to scenarios where patterns of drought occurrence are changing. Advances in the development of technologies for studying transcriptomes make it possible to study weather effects on plant gene expression in the field, including genes that may be important for disease resistance. Drawing conclusions about larger-scale processes from plot-level experiments may be challenging, however, since additional forms of interactions are important at larger scales [44].

Field experiments that incorporate simulations of changes in temperature and/or precipitation are becoming increasingly common in both agricultural and natural systems, often associated with long-term study systems such as the US National Science Foundation's Long-term Ecological Research sites. For example, in montane meadows Roy et al. [57] studied the impact of heating treatments on a suite of plant diseases. They found that higher temperatures favoured some diseases but not others. This type of 'winners and losers' scenario is likely to be common as more systems are evaluated; the overall level of disease under climate change may be buffered in some environments as some diseases become less common and others become more common.

The impact of elevated CO_2 on plant disease has been evaluated for a number of plant diseases, but results can be challenging to categorize [58–60]. Compared to studies in experimental chambers, free-air CO_2 enrichment (FACE) experiments allow more realistic evaluations of the effects of elevated CO_2 levels in agricultural fields or natural systems such as forests. Higher CO_2 levels may favour disease through denser, more humid plant canopies and increased pathogen reproduction but may reduce disease risk by enhancing host disease resistance [61], so the outcome for any given host–pathogen interaction is not readily predictable a priori. Elevated ozone levels can also affect plant disease risk (reviewed in Chakraborty et al. [59]).

In addition to the more direct influences of the abiotic environment on plant disease, climate change may also affect plant disease through its impact on other microbes that interact with pathogens. While certain microbes affect plant pathogens strongly enough to be used as biocontrol agents, a number of microbial interactions probably also have more subtle effects. As the effect of climate change on microbial communities is better understood through new experiments and new high-throughput sequencing approaches [14,15,62], this additional form of environmental interaction can be included in models of climate and disease risk.

4. EVIDENCE THAT PLANT DISEASE PATTERNS HAVE CHANGED DUE TO CLIMATE CHANGE

If patterns of plant disease in an area have shifted at the same time that changes in climate are observed, when can this correlation be taken as evidence of climate change impacts on disease?

The number of factors that interact to result in plant disease complicates such an analysis [17]. For example, if a disease becomes important in an area in which it was not important in the past, there are several possible explanations. The pathogen populations may have changed so that they can more readily infect and damage hosts. The pathogen species or particular vectors of the pathogen may be newly introduced to the area. In agricultural systems, host populations may have changed as managers have selected new cultivars based on criteria other than resistance to the disease in question. Management of the abiotic environment may have changed, such as changes in how commonly fields are tilled (tillage often reduces disease pressure) or changes in planting dates (which may result in more or less host exposure to pathogens). To rule out such competing explanations for changes in plant disease patterns, the argument for climate change as an important driver is strongest when (1) the pathogen is known to have been present throughout the area during the period in question, (2) the genetic composition of the pathogen and host populations has apparently not shifted to change resistance dynamics, (3) management of the system has not changed in a way that could explain the changes in disease pattern, (4) the climatic requirements of the pathogen and/or vector are well understood and better match the climate during the period of greater disease pressure, and (5) the change in disease pattern has been observed long enough to establish a convincing trend beyond possible background variation.

Even though the impact of changes in temperature, humidity and precipitation patterns has been quantified, scenario analyses of the potential impact of climate change are limited for many diseases by the lack of needed data and models. Real evidence for the impact of climate change on plant disease could come from verification of the accuracy of scenario analyses. This would require long-term records of disease intensity for the regions projected to be impacted and control regions. Long-term monitoring of pathogens and other plant-associated microbes is necessary in general to understand their ecology and to develop predictions of their impact on plant pathology [63]. The lack of availability of long-term data about disease dynamics in natural systems, and even in agricultural systems, limits opportunities for analysis of climate change effects on plant disease [64,65]. New analyses, using databases indicating where diseases have been observed [66], provide stimulating new assessments of potential poleward movement of diseases and insect pests in recent decades [67]. Use of these types of data and other crowd-sourcing approaches opens many new possibilities for analysis but also requires grappling with the limits to interpretation when there are generally not observations of absence and there are different sampling approaches in different regions [68–71].

Interannual variation in climatic conditions can have important effects on disease risk. For wheat stripe rust (caused by *Puccinia striiformis* Westend. f. sp. *tritici* Eriks.) in the US Pacific Northwest, disease severity was lower in El Niño years than in non-El Niño years [72]. If climate change alters the frequency and/or the intensity of El Niño events [73] or other extreme weather events, it will also alter patterns of disease risk; knowledge of the associations between disease and climate cycles is needed to inform predictions about plant disease epidemics under climate change [72]. The effects of patterns of weather fluctuation and extremes may produce new disease risk scenarios, including new management challenges [74,75].

Some general historical analyses of the relationship between disease and environmental factors have been developed. For example, the first annual appearance of wheat stem rust (caused by *Puccinia graminis* Pers.:Pers. f. sp. *tritici* Eriks. & E. Henn.) was compared for cool (1968–1977) and warm (1993–2002) periods in the US Great Plains, but a significant difference in arrival date was not observed [76]. In the UK, the abundance of two different wheat pathogens shifted in close correlation with patterns of SO_2 pollution during the 1900s [48,77,78]. For potato light blight, Zwankhuizen and Zadoks [79] have analyzed epidemics in the Netherlands from 1950 to 1996 using agronomic and meteorological variables as predictors of disease severity. They found that some factors were associated with enhanced disease, such as greater numbers of days with precipitation, greater numbers of days with temperatures between 10°C and 27°C and a relative humidity >90% during the growing season. Temperatures above 27°C and higher levels of global radiation in the Netherlands appeared to reduce disease risk [79]. Baker et al. [80] evaluated late blight risk in central North America and found that the trends in climatic conditions should result in increased risk. Hannukkala et al. [81] evaluated late blight incidence and first appearance in Finland 1933–2002, concluding that there was higher risk in more recent years. The comparison of years is complicated in this case by changes in the pathogen population and management practices. Increases in fungicide use were consistent with increased disease risk; records of pesticide use or other management change are one potential form of evidence for climate change impacts. Diplodia (or Sphaeropsis) shoot blight of pines emerged in France, with epidemics probably driven in part by more frequent conducive temperature and precipitation in recent years [82].

Pathogens and insect pests of lodgepole pine (*Pinus contorta*) have been well studied and offer an interesting example of a potential climate change fingerprint. Lodgepole pine is the most widely distributed pine species in natural (unmanaged) forests in western North America [83], including forests in British Columbia with more than 14 million ha of lodgepole pine [84]. Due to a lack of natural or human mediated disturbances, lodgepole pine has been increasing in abundance in British Columbia since the 1900s [84,85]. Recently, there have been increased cases of decline of lodgepole pines in these forests and researchers are evaluating the potential effects of climate change on these events.

Mountain pine beetle (*Dendroctonus ponderosae*) is a bark beetle and native to western North American forests [86]. This beetle can infest many pine species, and lodgepole pine is a preferred host [83,85]. The distribution range has not been limited by availability of the host but by the temperature range required for beetle survival through the winter [83,87]. The beetle causes physiological damage on the host trees by creating tunnels (insect galleries) underneath the bark, and in addition, microorganisms, such as the blue-stain fungi complex, can take advantage of these wounds to cause secondary infestation that may further reduce plant health [83,86]. Dead pines are not marketable and also can facilitate the spread of wild fire [88]. Beetle populations can be very low for many decades, but when there is an outbreak, a large area of susceptible hosts may be killed. The beetle has been known to be native to British Columbia [85], but, probably due to low winter temperatures, outbreak events were not common. However, the frequency of outbreaks appeared to be increasing, and 8 million hectares in British Columbia were affected in 2004 [85,88]. Carroll et al. [87] evaluated the shift in infestation range and concluded that the trend toward warmer temperatures more suitable for the beetle was part of the reason for this series of outbreaks. Further, in a study by Mock et al. [85], genetic markers did not reveal any significant differences among beetle genotypes from inside and outside of British Columbia, indicating the beetle population had not changed. Thus, other factors including climate change are likely to be the reason why there have been more outbreaks in northern areas.

Dothistroma needle blight is a fungal disease (causal agent *Dothistroma septosporum*) of a variety of pine species worldwide [89], including lodgepole pines. The disease is associated with mild temperature ranges (18°C is the optimum temperature for sporulation [90]) and rain events with 10 or more hours of wetness [89,91]. It causes extensive defoliation, mortality and a reduced growth rate in pine [89,92]. As with the mountain pine beetle, Dothistroma needle blight has been found in British Columbia in the past, but damage due to this disease was relatively minor. However, the number of cases and intensity of epidemics in this region has increased since the late 1990s [92]. A study by Woods et al. [92] evaluated the relationship between these disease outbreaks and (1) regional climate change and (2) long-term climate records (utilizing the Pacific Decadal Oscillation, PDO, as an indicator variable). Although they did not find a substantial increase in regional temperature nor a significant correlation between PDO and directional increase of precipitation or temperature, increased mean summer precipitation in the study area was observed. Furthermore, a recent study by Welsh et al. [93] indicated that trends in minimum temperatures in August and increasing precipitation in April could be linked to the spread of the disease in British Columbia. On the other hand, in some locations, up to 40% of forest stands became dominated by lodgepole pine due to plantation development [92]. Thus, a combination of increased rain events and the abundance of the favoured host was the likely cause of increased disease occurrence.

More cases of Dothistroma needle blight have been reported in European countries as well [94,95]. However, when pathogen genetic diversity in Estonia, Finland and the Czech Republic was evaluated [96], similarly high levels of genetic diversity were found in all three populations. Therefore, there was no strong evidence of northward introduction of the pathogen. Since Dothistroma needle blight was present in Russia in the 1950s [97], and was considered to be a minor disease in France in the 1960s–1980s, this pathogen could have been present in these counties for a long time prior to recent outbreaks in the 1990s [94], indicating more evidence for a role of climate change.

For both mountain pine beetle and Dothistroma needle blight, it is reasonable to assume that climate has influenced pathogen and pest behaviour; however, at the same time, there has been a substantial increase in the abundance of the host (lodgepole pine) in British Columbia [84,85]. Widely available and genetically similar hosts often increase plant disease risk, and these factors may also explain at least part of the change in risk observed for lodgepole pine. Also, the rapid expansion of a host may encourage the movement of nursery trees without obvious symptoms, even those potentially infected. Movement of planting materials may help to explain the high genetic diversity among isolates from European countries [96].

Another important disease that has exhibited recent changes in its pattern of occurrence is wheat stripe rust (or yellow rust, caused by the fungus *Puccinia striiformis* f. sp. *tritici*). This disease decreased and then increased in importance in the US during the past century. Stripe rust was economically important in the 1930s–1960s, but the development of resistant wheat varieties successfully reduced the number of epidemic events. However, several epidemic events have been observed since 2000 [98,99]. The disease can cause 100% yield loss at a local scale [99], and epidemics in 2003 in the US resulted in losses estimated to total $300 million. Are these changes related to climate change?

Historically, *P. striiformis* f. sp. *tritici* was known to be active at relatively lower temperature ranges. Under favourable conditions (i.e. with dew or free water on plant surfaces), its spores can germinate at 0°C [100], and the temperature range for infection was measured as between 2°C and 15°C with an optimum temperature of 7°C to 8°C [101,102]. Spores could be produced between 0°C and 24.5°C [100]. This pathogen species was not well adapted for higher temperature conditions, and disease development declined at temperatures above 20°C [101–103], while spores produced at 30°C were shown to be nonviable [100].

However, more recent populations of *P. striiformis* f. sp. *tritici* were adapted to warmer temperature ranges [104]. Isolates from the 1970s to 2003 and newer (post-2000) isolates had a significantly higher germination rate and shorter latent period (period between infection and production of spores) than older isolates when they were incubated at 18°C, whereas isolate effects were not different when incubation took place at 12°C. In a follow-up study, Markell and Milus [105] examined isolates from the 1960s to 2004 with genetic markers and morphological comparisons and found that isolates collected pre- and post-2000 could be classified into two different groups. Although within a population group less than nine polymorphic markers were identified, when pre-and post-2000 populations were compared there were 110 polymorphic markers [105]. The large difference between pre- and post-2000 groups led the authors to conclude that post-2000 isolates were introduced from outside of the US rather than resulting from mutations in pre-2000 isolates.

Results from annual race surveys conducted by the United States Agricultural Research Service in Pullman, WA, indicated that pre-2000 isolates were not commonly collected in surveys after 2000 [105]. Thus, it seems that post-2000 isolates took the place of pre-2000 isolate types. The question remains whether the success of post-2000 isolates is due to the change in climatic conditions (i.e. increase in overall temperature) or something else. Since post-2000 isolates were better adapted to a warmer temperature range, climate change might have played a role in selection for the new isolates, but there is another important factor for post-2000 isolates. All post-2000 isolates examined were virulent (able to cause disease) on wheat plants with resistance genes *Yr8* and *Yr9*, while these resistance genes worked very well against pre-2000 isolates [98,105]. There are other wheat varieties that are resistant to post-2000 isolates, but these varieties were less commonly grown since they were not effective against older isolates. Thus, the ability of new isolates to overcome these resistance genes was most likely the major factor behind the drastic change in populations of *P. striiformis* f. sp. *tritici* and recent epidemic events.

Pierce's disease (PD) of grape is caused by a gram-negative, xylem-limited, fastidious bacterium, *Xylella fastidiosa* (Wells) [106,107]. *X. fastidiosa* is vectored by sharpshooter leafhoppers (Hemiptera: Cicadellidae), such as glassy-winged sharpshooter, *Homalodisca vitripennis* (Germar), which has been identified as a major PD vector in California and southeastern regions of the US [108–111]. PD symptoms include interveinal chlorosis, marginal necrosis, uneven lignification of shoots or 'green islands', and leaf abscission that results in characteristic 'matchstick petioles' [112]. Diseased vines suffer decline, yield loss, and even death. With severe disease on a susceptible cultivar, a vine will be killed within a few years [109].

The geographic distribution of PD is concentrated in California and spans from Texas and Florida up to Virginia (VA) in the US [113]. Anas et al. found that presence of PD was limited by daily minimum temperatures below −9.4°C for four or more days or −12.2°C for two days [114] because low temperatures can reduce or kill bacterial populations in the vine [115]. Therefore, areas north of central VA had not been considered a high-risk area for PD when risk was based on the 25-year average between 1972 and 1997 [114,116]. However, when the data from 1997 to 2005 were used, the very high-risk area moved northward considerably to include almost the entire state of VA. The same group conducted a survey in North Carolina and Georgia, and PD was identified in 82% and 75% of surveyed vineyards, respectively. Moreover, a 2006 VA vineyard survey found 70% of surveyed vines PD-positive [117]. However, since the recent reintroduction of wine grapes in the Eastern US resulted in rapid expansion of vineyards since the 1980s, it is not clear to what extent the trend of northward disease detection is due to climate-facilitated disease spread versus simply coinciding with the northward expansion of vineyards.

In summary, there is no doubt that plant disease responds to weather and that changes in weather events due to climate change are likely to shift the frequency and intensity of disease epidemics. Simulated climate change experiments reveal changes in plant disease intensity and the profile of plant diseases. When evidence is sought for climate change based on changes in plant disease patterns, conclusions are less clear. Since the search for fingerprints of climate change is correlative by nature, there may always be alternative predictors for the changes, but this seems particularly true for plant disease. It is a typical biological irony that, while plant disease risk may be particularly sensitive to climatic variables and climatic shifts, plant disease may also be particularly difficult to use as an indicator of climate change because of the many interactions that take place to result in disease. However, as more data sets are collected and synthesized [65,67], and climate patterns exhibit greater changes over a longer period, the impacts of climate change on plant disease are likely to become clearer.

ACKNOWLEDGEMENTS

We appreciate support by the CGIAR Consortium Research Program for Roots, Tubers and Bananas (RTB) and the CGIAR Research Program on Climate Change, Agriculture and Food Security (CCAFS), by USDA NC RIPM Grant 2010-34103-20964 and USDA APHIS Grant 11-8453-1483-CA, by the US National Science Foundation (NSF) through Grant DEB-0516046 and NSF Grant EF-0525712 as part of the joint NSF-National Institutes of Health (NIH) Ecology of Infectious Disease program, by the US Agency for International Development (USAID) to the OIRED at Virginia Tech for the SANREM CRSP under Award No. EPP-A-00-04-00013-00 and for the IPM CRSP under Award No. EPP-A-00-04-00016-00, and by the Kansas Agricultural Experiment Station. The views expressed herein can in no way be taken to reflect the official opinion of these agencies.

REFERENCES

[1] E.D. De Wolf, S.A. Isard, Annu. Rev. Phytopathol. 45 (2007) 203–220.
[2] C. Kinealy, A Death-Dealing Famine: The Great Hunger in Ireland, Pluto Press, 1997.
[3] R.N. Strange, P.R. Scott, Annu. Rev. Phytopathol. 43 (2005) 83–116.
[4] S. Chakraborty, A.C. Newton, Plant Pathol. 60 (2011) 2–14.
[5] J.J. Burdon, P.H. Thrall, L. Ericson, Annu. Rev. Phytopathol. 44 (2006) 19–39.
[6] T. Emiko Condeso, R.K. Meentemeyer, J. Ecol. 95 (2007) 364–375.
[7] R.P. Singh, D.P. Hodson, J. Huerta-Espino, Y. Jin, S. Bhavani, P. Njau, S. Herrera-Foessel, P.K. Singh, S. Singh, V. Govindan, Annu. Rev. Phytopathol. 49 (2011) 465–481.
[8] J. Luck, M. Spackman, A. Freeman, P. Trebicki, W. Griffiths, K. Finlay, S. Chakraborty, Plant Pathol. 60 (2011) 113–121.
[9] J. DeBokx, P. Piron, Potato Res. 20 (1977) 207–213.
[10] L. Huber, T.J. Gllespie, Annu. Rev. Phytopathol. 30 (1992) 553–577.
[11] B.A. McDonald, C. Linde, Ann. Rev. Phytopathol. 40 (2002) 349–379.
[12] G.N. Agrios, Plant Pathology, Academic Press, San Diego, 2005.
[13] J.E. Van Der Plank, Plant Diseases: Epidemics and Control, Academic Press, New York and London, 1963.
[14] K.A. Garrett, A. Jumpponen, C. Toomajian, L. Gomez-Montano, Can. J. Plant Pathol. 34 (2012) 349–361.
[15] S. Chakraborty, I.B. Pangga, M.M. Roper, Global Change Biol. 18 (2012) 2111–2125.
[16] S.G. Pritchard, Plant Pathol. 60 (2011) 82–99.
[17] K.A. Garrett, G.A. Forbes, S. Savary, P. Skelsey, A.H. Sparks, C. Valdivia, A.H.C. van Bruggen, L. Willocquet, A. Djurle, E. Duveiller, H. Eckersten, S. Pande, C.V. Cruz, J. Yuen, Plant Pathol. 60 (2011) 15–30.

[18] C.L. Campbell, L.V. Madden, Introduction to Plant Disease Epidemiology, John Wiley & Sons, New York, 1990.

[19] L.V. Madden, G. Hughes, F. van den Bosch, The Study of Plant Disease Epidemics, APS press, St. Paul, MN, 2007.

[20] R.J. Hijmans, G.A. Forbes, T.S. Walker, Plant Pathol. 49 (2000) 697–705.

[21] A.H. Sparks, G.A. Forbes, R.J. Hijmans, K.A. Garrett, Global Change Biol. 20 (2014) 3621–3631.

[22] M. Bergot, E. Cloppet, V. Pearnaud, M. Deque, B. Marcais, M.-L. Desprez-Loustau, Global Change Biol. 10 (2004) 1539–1552.

[23] A.H. Sparks, G.A. Forbes, R.J. Hijmans, K.A. Garrett, Ecosphere 2 (2011) art90.

[24] R. Ghini, W. Bettiol, E. Hamada, Plant Pathol. 60 (2011) 122–132.

[25] D.R. Cooley, W.F. Wilcox, J. Kovach, S.G. Schloemann, Plant Dis. 80 (1996) 228–237.

[26] M. McMullen, R. Jones, D. Gallenberg, Plant Dis. 81 (1997) 1340–1348.

[27] J.C. Sutton, Can. J. Plant Pathol. 4 (1982) 195–209.

[28] N. Magan, A. Medina, D. Aldred, Plant Pathol. 60 (2011) 150–163.

[29] A.L. Anderson, Phytopathology 38 (1948) 595–611.

[30] S.V. Beer, D.C. Opgenorth, Phytopathology 66 (1976) 317–322.

[31] J.A. Kolmer, Annu. Rev. Phytopathol. 34 (1996) 435–455.

[32] F. Lin, X. Chen, Theor. Appl. Genet. 114 (2007) 1277–1287.

[33] A. Ficke, D.M. Gadoury, R.C. Seem, Phytopathology 92 (2002) 671–675.

[34] L.E. Hoffman, W.F. Wilcox, D.M. Gadoury, R.C. Seem, D.G. Riegel, Phytopathology 94 (2004) 641–650.

[35] M.M. Kennelly, D.M. Gadoury, W.F. Wilcox, P.A. Magarey, R.C. Seem, Phytopathology 95 (2005) 1445–1452.

[36] D.M. Gadoury, R.C. Seem, A. Stensvand, New York Fruit Quarterly 2 (1995) 5–8.

[37] X. Li, P.D. Esker, Z. Pan, A.P. Dias, L. Xue, X.B. Yang, Plant Dis. 94 (2010) 796–806.

[38] J.K.M. Brown, M.S. Hovmoller, Science 297 (2002) 537–541.

[39] C. Gonzalez-Martin, N. Teigell-Perez, B. Valladares, D.W. Griffin, The global dispersion of pathogenic microorganisms by dust storms and its relevance to agriculture, in: S. Donald (Ed.), Advances in Agronomy, Academic Press, 2014, pp. 1–41.

[40] R.A.C. Jones, M.J. Barbetti, CAB Rev. 7 (2012) 1–33.

[41] K.A. Garrett, S. Thomas-Sharma, G.A. Forbes, J. Hernandez Nopsa, Climate change and plant pathogen invasions, in: L. Ziska, J. Dukes (Eds.), Invasive Species and Climate Change, CABI Publishing, 2014, pp. 22–44.

[42] M.R. Sanatkar, C. Scoglio, B. Natarajan, S. Isard, K.A. Garrett, Phytopathology 105 (2015) 947–955.

[43] M. Pautasso, T.F. Doring, M. Garbelotto, L. Pellis, M.J. Jeger, Eur. J. Plant Pathol. 133 (2012) 295–313.

[44] M. Pautasso, K. Dehnen-Schmutz, O. Holdenrieder, S. Pietravalle, N. Salama, M.J. Jeger, E. Lange, S. Hehl-Lange, Biol. Rev. 85 (2010) 729–755.

[45] K.A. Garrett, Eur. J. Plant Pathol. 133 (2012) 75–88.

[46] M.W. Shaw, T.M. Osborne, Plant Pathol. 60 (2011) 31–43.

[47] R.N. Sturrock, S.J. Frankel, A.V. Brown, P.E. Hennon, J.T. Kliejunas, K.J. Lewis, J.J. Worrall, A.J. Woods, Plant Pathol. 60 (2011) 133–149.

[48] B.D.L. Fitt, B.A. Fraaije, P. Chandramohan, M.W. Shaw, Plant Pathol. 60 (2011) 44–53.

[49] S. Savary, A. Nelson, A.H. Sparks, L. Willocquet, E. Duveiller, G. Mahuku, G.A. Forbes, K.A. Garrett, D. Hodson, J. Padgham, S. Pande, M. Sharma, J. Yuen, A. Djurle, Plant Dis. 95 (2011) 1204–1216.

[50] R.R. McAllister, C.J. Robinson, K. Maclean, A.M. Guerrero, K. Collins, B.M. Taylor, P.J. De Barro, Ecol. Soc. 20 (2015) 67.

[51] K.A. Garrett, Climate Change and Plant Disease Risk, Global Climate Change and Extreme Weather Events: Understanding the Contributions to Infectious Disease Emergence, National Academies Press, Washington, DC, 2008, 143–155.

[52] K.A. Garrett, R.L. Bowden, Phytopathology 92 (2002) 1152–1159.

[53] P.K. Anderson, A.A. Cunningham, N.G. Patel, F.J. Morales, P.R. Epstein, P. Daszak, Trends Ecol. Evol. 19 (2004) 535–544.

[54] C.M. Brasier, BioScience 51 (2001) 123–133.

[55] K.A. Garrett, S.P. Dendy, E.E. Frank, M.N. Rouse, S.E. Travers, Annu. Rev. Phytopathol. 44 (2006) 489–509.

[56] S.M. Coakley, H. Scherm, S. Chakraborty, Annu. Rev. Phytopathol. 37 (1999) 399–426.

[57] B.A. Roy, S. Gusewell, J. Harte, Ecology 85 (2004) 2570–2580.

[58] D.M. Eastburn, A.J. McElrone, D.D. Bilgin, Plant Pathol. 60 (2011) 54–69.

[59] S. Chakraborty, J. Luck, G. Hollaway, A. Freeman, R. Norton, K.A. Garrett, K. Percy, A. Hopkins, C. Davis, D.F. Karnosky, CAB Rev. 3 (2008).

[60] I.B. Pangga, J. Hanan, S. Chakraborty, Plant Pathol. 60 (2011) 70–81.

[61] S. Chakraborty, Australas. Plant Pathol. 34 (2005) 443–448.

[62] M.P. Waldrop, M.K. Firestone, Microb. Ecol. 52 (2006) 716–724.

[63] C.D. Harvell, C.E. Mitchell, J.R. Ward, S. Altizer, A.P. Dobson, R.S. Ostfeld, M.D. Samuel, Science 296 (2002) 2158–2162.

[64] H. Scherm, Can. J. Plant Pathol. 26 (2004) 267–273.

[65] M.J. Jeger, M. Pautasso, New Phytol. 177 (2008) 8–11.

[66] N.M. Pasiecznik, I.M. Smith, G.W. Watson, A.A. Brunt, B. Ritchie, L.M.F. Charles, EPPO Bull. 35 (2005) 1–7.

[67] D.P. Bebber, M.A.T. Ramotowski, S.J. Gurr, Nat. Clim. Change 3 (2013) 985–988.

[68] H. Scherm, C.S. Thomas, K.A. Garrett, J.M. Olsen, Annu. Rev. Phytopathol. 52 (2014) 453–476.

[69] D.P. Bebber, T. Holmes, D. Smith, S.J. Gurr, New Phytol. 202 (2014) 901–910.

[70] K.A. Garrett, Nat. Clim. Change 3 (2013) 955–957.

[71] J.R. Rohr, A.P. Dobson, P.T.J. Johnson, A.M. Kilpatrick, S.H. Paull, T.R. Raffel, D. Ruiz-Moreno, M.B. Thomas, Trends Ecol. Evol. 26 (2011) 270–277.

[72] H. Scherm, X.B. Yang, Int. J. Biometeorol. 42 (1995) 28–33.

[73] A. Timmermann, J. Oberhuber, A. Bacher, M. Esch, M. Latif, E. Roeckner, Nature 398 (1999) 694–697.

[74] K.A. Garrett, A.D.M. Dobson, J. Kroschel, B. Natarajan, S. Orlandini, H.E.Z. Tonnang, C. Valdivia, Agric. For. Meteorol. 170 (2013) 216–227.

[75] C. Rosenzweig, A. Iglesias, X.B. Yang, P.R. Epstein, E. Chivian, Global Change Hum. Health 2 (2001) 90–104.

[76] H. Scherm, S.M. Coakley, Australas. Plant Pathol. 32 (2003) 157–165.

[77] S.J. Bearchell, B.A. Fraaije, M.W. Shaw, B.D.L. Fitt, Proc. Natl. Acad. Sci. U.S.A. 102 (2005) 5438–5442.

[78] M.W. Shaw, S.J. Bearchell, B.D.L. Fitt, B.A. Fraaije, New Phytol. 177 (2008) 229–238.

[79] M.J. Zwankhuizen, J.C. Zadoks, Plant Pathol. 51 (2002) 413–423.

[80] K.B. Baker, W.W. Kirk, J.M. Stein, J.A. Anderson, HortTechnology 15 (2005) 510–518.

[81] A.O. Hannukkala, T. Kaukoranta, A. Lehtinen, A. Rahkonen, Plant Pathol. 56 (2007) 167–176.

[82] B. Fabre, D. Piou, M.-L. Desprez-Loustau, B. Marçais, Global Change Biol. 17 (2011) 3218–3227.

[83] G.D. Amman, The role of the mountain pine beetle in lodgepole pine ecosystem: impact on succession, in: W.J. Wattson (Ed.), The Role of Arthropods in Forest Ecosystems, Springer-Verlag, New York, 1978.

[84] S.W. Taylor, A.L. Carroll, Disturbance, forest age, and mountain pine beetle outbreak dynamics in BC: a historicalperspective, in: T.L. Shore, J.E. Brooks, J.E. Stone (Eds.), Mountain Pine Beetle Symposium: Challenges and Solutions. (October 30–31, 2003), Natural Resources Canada, Canadian Forest Service, Pacific Forestry Centre, Information Report BC-X-399, Victoria, BC, Kelowna, British Columbia, 2004, pp. 41–51.

[85] K.E. Mock, B.J. Bentz, E.M. O'Neill, J.P. Chong, J. Orwin, M.E. Pfrender, Mol. Ecol. 16 (2007) 553–568.

[86] W. Cranshaw, D. Leatherman, B. Jacobi, L. Mannex, Insects and Diseases of Woody Plants in the Central Rockies, Colorado State University Bulletin no. 506A, Colorado State University Cooperative Extension, Fort Collins, CO, 2000.

[87] A.L. Carroll, S.W. Taylor, J. Régnière, L. Safranyik, Effects of climate change on range expansion by the mountain pine beetle in British Columbia, in: T.L. Shore, J.E. Brooks, J.E. Stone (Eds.), Mountain Pine Beetle Symposium: Challenges and Solutions. (October 30–31, 2003), Natural Resources Canada, Canadian Forest Service, Pacific Forestry Centre, Information Report BC-X-399, Victoria, BC, Kelowna, British Columbia, 2004, pp. 223–232.

[88] Mountain pine beetle action plan 2006–2011. Ministry of forests and range – province of British Columbia.

[89] I.A.S. Gibson, Annu. Rev. Phytopathol. 10 (1972) 51–72.

[90] M.H. Ivory, Trans. Brit. Mycol. Soc 50 (1867) 563–572.

[91] D. Hocking, D.E. Etheridge, Ann. Appl. Biol. 59 (1967) 133–141.

[92] A. Woods, K.D. Coates, A. Hamann, BioScience 55 (2005) 761–769.

[93] C. Welsh, K.J. Lewis, A.J. Woods, Can. J. For. Res. 44 (2014) 212–219.

[94] D. Villebonne, F. Maugard, Rapid development of Dothistroma needle blight (*Scirrhia pini*) on Corsican pine (*Pinus nigra* subsp. *laricio*) in France, Le Département de la Santé des Forêts, Paris, France, 1999.

[95] J.A. Nowakowska, A. Tereba, T. Oszako, Folia For. Pol. Ser. A For. 56 (3) (2014) 157–159.

[96] R. Drenkhan, J. Hantula, M. Vuorinen, L. Jankovský, M.M. Müller, Eur. J. Plant Pathol. 136 (2013) 71–85.

[97] J.S. Murray, S. Batko, Forestry 35 (1962) 57–65.

[98] X. Chen, M. Moore, E.A. Milus, D.L. Long, R.F. Line, D. Marshall, L. Jackson, Plant Dis. 86 (2002) 39–46.

[99] X.M. Chen, Can. J. Plant Pathol. 27 (2005) 314–337.

[100] H. Tollenaar, B.R. Houston, Phytopathology 56 (1965) 787–790.

[101] E.L. Sharp, Phytopathology 55 (1965) 198–203.

[102] C. de Vallavieille-Pope, L. Huber, M. Leconte, H. Goyeau, Phytopathology 85 (1995) 409–415.

[103] M.V. Wiese (Ed.), Compendium of Wheat Diseases, APS Press, St. Paul, MN, 1987.

[104] E.A. Milus, E. Seyran, R. McNew, Plant Dis. 90 (2006) 847–852.

[105] S.G. Markell, E.A. Milus, Phytopathology 98 (2008) 632–639.

[106] J. Wells, B. Raju, H. Hung, W. Weisburg, L. Mandelco-Paul, D. Brenner, Int. J. Syst. Bacteriol. 37 (1987) 136–143.

[107] S. Chatterjee, R. Almeida, S. Lindow, Annu. Rev. Phytopathol. 46 (2008) 243–271.

[108] R. Hernandez-Martinez, K.A. de la Cerda, H.S. Costa, D.A. Cooksey, F.P. Wong, Phytopathology 97 (2007) 857–864.

[109] D.L. Hopkins, A.H. Purcell, Plant Dis. 86 (2002) 1056–1066.

[110] J.T. Sorensen, R.J. Gill, Pan-Pac. Entomol. 72 (1996) 160–161.

[111] W.C. Adlerz, D.L. Hopkins, J. Econ. Entomol. 72 (1979) 916–919.

[112] R.C. Pearson, A.C. Goheen, Compendium of Grape Diseases, American Phytopathological Society, St. Paul, Minnesota, 1988.

[113] R.P.P. Almeida, F.E. Nascimento, J. Chau, S.S. Prado, C.-W. Tsai, S.A. Lopes, J.R.S. Lopes, Appl. Environ. Microbiol. 74 (2008) 3690–3701.

[114] O. Anas, U.J. Harrison, P.M. Brannen, T.B. Sutton, Plant Health Prog. (2008), http://dx.doi.org/10.1094/PHP-2008-0718-01-RS.

[115] J.H. Lieth, M.M. Meyer, K.H. Yeo, B.C. Kirkpatrick, Phytopathology 101 (2011) 1492–1500.

[116] M.S. Hoddle, Crop Prot. 23 (2004) 691–699.

[117] A.K. Wallingford, S.A. Tolin, A.L. Myers, T.K. Wolf, D.G. Pfeiffer, Plant Health Prog. (2007), http://dx.doi.org/10.1094/PHP-2007-1004-01-BR.

MODELLING CLIMATE CHANGE

STATISTICAL MODELLING OF CLIMATE CHANGE

22

Ben Powell

School of Mathematics, University of Bristol, Bristol, UK

CHAPTER OUTLINE

1. INTRODUCTION

The statistics research community accommodates a healthy plurality of competing personalities and philosophies, all actively producing books, papers and theses on modelling in principle and in practice. Further, the statistical literature borders, and is informed by, those from fields such as the philosophy of science, numerical analysis and computer science. Consequently, manageable texts on the statistical modelling of any phenomena, including climate, will necessarily be incomplete. Having said that, Refs [1] (the Fifth Assessment Report of the Intergovernmental Panel on Climate Change) and [2] by Carbone, stand out as particularly well-referenced reports on the current state of the art of climate modelling and are highly recommended as overviews of the field. In this chapter we concentrate on a subset of aspects of climate modelling that deserve special attention. This will give us an opportunity to discuss some of the limitations of current practices and the problems that prevent them from being addressed. In doing so, we also hope to provide an insight into a range of active areas of statistical research while making a case to the wider scientific community, that is to the readers of this book, for the adoption of a particular way of thinking about statistical modelling.

Although many of the ideas considered here are relevant to scientific modelling at a general level, it is useful to highlight two specific characteristics of climate science that greatly influence statistical modelling of climate, and the rest of this chapter. The first characteristic is the unshakable faith that climate scientists have in a small set of equations that underpin their understanding of climate, namely

Climate Change. http://dx.doi.org/10.1016/B978-0-444-63524-2.00022-1

the Navier–Stokes equations that describe fluid dynamics. So strong is the confidence in these equations that predictions of climate change from a model that neglects them are unlikely to be considered at all credible. Indeed, the past 150 years of scientific inquiry have consistently reaffirmed the predictive power of the Navier–Stokes equations, this success being almost unrivalled across the other sciences, with the notable exception of quantum electrodynamics. The second characteristic is that the type of quantities that tend to matter to us are so far downstream of the Navier–Stokes equations – that is to say, they tend to be nontrivial properties of the solutions of the equations considered over vast time scales and huge domains, with many peripheral systems that couple to them in highly complex ways. Together, these two characteristics mean that numerical simulators that approximate solutions to the Navier–Stokes equations are integral to climate modelling, and that these simulators unavoidably involve some very demanding computations.

Before proceeding further into a discussion of climate modelling, it is helpful to establish a few definitions. Firstly, the idea of *climate* itself – here climate is defined as a probability distribution over meteorological, or weather, events. With this definition, climate becomes a summary or mathematical expression of a range of different weather events resulting from a corresponding range of different physical conditions from amongst which one cannot be committed to with certainty. Such a range of physical conditions could correspond to the atmospheric conditions on individual days over the past 100 years or to the set of hypothesised Earths for which certain physical constants, like the viscosity of water, are slightly different. In this way, the notion of climate is shaped by the nature and scale of an individual's enquiries. The implication here being that climate becomes a subjective construct insofar as meaning different things to different people with different perspectives and interests.

A second concept that requires clarification is that of a *climate model*, which must be differentiated from a *climate simulator*. A climate simulator is simply a mathematical function mapping a set of inputs to a set of outputs, just as multiplication or addition are functions producing one set of numbers from another set of numbers. Climate simulators are, of course, vastly more complex and difficult to develop an intuition for, but are essentially seen in the same way as these more basic mathematical devices. In statistical research literature, statisticians commonly describe their ideas in slightly abstracted contexts in which climate simulators are considered to be 'black boxes', meaning that they assume that it is possible to execute the simulator function but not to take it apart. We will tend to refer mathematically to a simulator using the notation

$$s(\cdot), \tag{1}$$

where the fact that we are referring to a function, rather than a specific value of that function, is emphasized by including brackets that invite an input quantity, whose place is held with a dot in Eqn (1).

A climate simulator only becomes part of a climate model when that function is used to express meaningful statements of belief for weather events. Often, though not always, this development is signified by equating the simulator output s with a real-world quantity x after the addition of an error or discrepancy term ϵ:

$$x = s(\cdot) + \epsilon. \tag{2}$$

It is important to understand that Eqn (2) encodes not just one output or one meteorological event but a whole set of events constituting a climate, because implicitly x can take a range of values corresponding to a range of plausible arguments for the simulator function. These could be different sets of plausible past states at which to initialize the simulator, different future dates to which to run it, or different sets of physical constants describing alternative worlds. Importantly, x can also take different values due to different values of ϵ, which describe different ways or extents for the simulator to be wrong. We will return to all these sets in the following sections, identifying them as sources of uncertainty for climate predictions.

This introductory section ends with the flowchart of Fig. 1 (see also Table 1), which for the time being ought to be understood on an informal basis as a sketch for suggesting relationships between important quantities to be discussed as we move through the following sections.

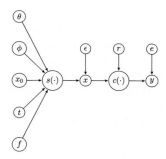

FIGURE 1

A schematic illustrating the structure of a possible model for understanding the climate. The nodes on the far left of the figure represent inputs required by a climate simulator. In this chapter we will consider exactly how the significance of the simulator outputs, s, and real-world meteorological variables, x, are contingent on the formulation of a cost, y, which must also be modelled.

Table 1 Key for Fig. 1	
Symbol	**Object**
θ	Simulator parameters associated with experimentally measurable variables.
ϕ	Simulator parameters understood as parameterizations.
x_0	Initial climate conditions.
t	The time at which the simulated climate system is evaluated.
f	Climate forcings.
$s(\cdot)$	The climate simulator's output.
ϵ	The simulator discrepancy term.
x	Future weather conditions.
$c(\cdot)$	A function measuring weather conditions in terms of the expected costs they entail.
r	Cost function parameters.
e	Cost function discrepancy.
y	The actual cost incurred given certain weather conditions are instantiated.

2. BAYESIAN MODELLING
2.1 BAYESIAN MODELLING IN THEORY

Despite involving a significant degree of abstraction from the climatological context, this section is important for describing the capacity for modern statistical methods to make sense of scientific theory and observational data, in conjunction with issues of doubt and loss, in a way that is particularly relevant to climate modelling. The manner in which climate models were introduced in the preceding section is informed by what is known as the *Bayesian* paradigm for statistical learning. Although variants of the paradigm exist, the notion at the centre of them all is the idea that statistical models are mathematical expressions of reasoning processes rather than physical processes. Their role is to approximate the conclusions one would make if it were possible to think with the high speed and precision of a computer and the rigour of a mathematician's proof.

The mathematical machinery that allows us to perform classical Bayesian inference, that is to derive informed probability statements from prior beliefs and related observations, is constructed from the axioms of probability (of which Bayes' theorem is in fact a consequence). With the adoption of the Bayesian paradigm comes the realization that making predictions with a statistical model essentially involves exploring the logical consequences of beliefs about different quantities, and that the truth of a model consists in its ability to reconcile those beliefs. The Bayesian paradigm is also accompanied by a perpetual reluctance to equate mechanisms within a model, like a climate simulator, to an unchanging and objective reality. The scepticism here provides a statistician with the flexibility to criticize and re-evaluate the components of a model. However, given this reluctance, one may ask why and when that statistician ought to pay any attention to the numbers it spits out. The answer is that statistical models ought to be seen as consisting of a hierarchy of nested models. Within each nesting one imagines a user who accepts the model there as truly representative of his beliefs about the world; as he steps up the hierarchy he acknowledges the inadequacies of the model below and can augment it slightly with, say, an additional error term. From a purely intellectual point of view there is nothing determining how many levels of the hierarchy he ought to ascend. In practice, however, there are clearly times when prompt and definitive action, based on the best information to hand, is required. Thus, when constructing a statistical model, one needs to keep in mind the decisions and actions that the model will inform. In this way, notionally objective scientific knowledge and practical, context-specific decision requirements interact. To deal with this, it is useful to define probability statements, like climate predictions, as statements of intent rather than statements of fact in a way the following example should illustrate.

In the sixteenth century, European bankers began to issue paper money in the form of receipts for gold deposited in their vaults, and today British bank notes are still inscribed with a promise from the Bank of England to 'pay the bearer on demand the sum of v pounds'. Let us imagine now that $v = 1$ and that the promise is appended with the condition '...if event A occurs'. Most importantly, let us consider how the condition alters the value of that note. How much would you be willing to pay, for example, for the note for which the conditioning event is that it rains tomorrow? Or that the global average temperature rises by more than two degrees in the next 100 years? We will choose to use this price, your price, to define your probability for an event occurring. So, as money may be used as a device for operationalizing notions of value, through this gamble, money also becomes a device for operationalizing notions of probability. It is important here to acknowledge that defining value and probability in terms of money can be a very dangerous thing to do when its subjective nature is forgotten. Money is not value, but a device for expressing it, allowing for a model of the values of an individual or community values to be constructed.

The implications of this subjective, or personalistic, interpretation of probability and statistics are numerous. For example, it implies that, as much as describing a quantified reasoning process, a statistical model describes an investment strategy. As such, it is necessarily associated with an investment action and an investment agent. Just as a price of an object is meaningless if there is no one willing to buy it for that amount, the meaning of a probability is contingent on someone's willingness to invest resources in anticipation of a certain event occurring. This brings us to another definition, that of an *expert*. This will be a person, committee or agent whose belief statements, or investment strategies, one is willing to accept as one's own. The authors of the International Panel on Climate Change (IPCC) reports are considered experts by the overwhelming majority of the scientific community, and upon making misleading predictions they stand to lose (at least) the professional standing they have achieved, for example.

The considerations discussed here serve to guide statistical modelling in a holistic way. They imply that, on a foundational level, modelling a phenomenon, like climate, involves the reconciliation of expert knowledge and observational data, with the goal of constructing a coherent investment strategy for avoiding loss. In the next section we will move onto more concrete descriptions of model calibration and validation, understanding these tasks as the elicitation and evaluation of expert knowledge that has been distilled into a climate simulator. We will then return to the discussion of loss and cost in Section 4.

2.2 BAYESIAN STATISTICS IN PRACTICE

Bayes' formula itself may be written as:

$$P(A|B) = \frac{P(B|A) \times P(A)}{P(B)}, \tag{3}$$

where each term denotes a probability of the event within its brackets occurring, and a pair of events separated by a vertical line, like A|B, is to be read as 'event A occurs given event B occurs'. The formula tells us how the a priori probability of the event A ought to be revised in light of the related event B occurring.

The practical methodological implications of Bayes' formula are most easily demonstrated by looking at the following example, and the accompanying sketch in Fig. 2, which describes how 'ensemble-type' climate predictions are made.

To begin with, a statistician will consult relevant experts (people or literature of authority sufficient that he is prepared to adopt their belief specifications as his own) in order to elicit a set of input parameters (the θ, ϕ, x_0 and f of Fig. 1) considered a priori plausible. He will also attempt to elicit prior probabilities or weights that express their relative belief in, or willingness to commit to, the vectors. Then, imagining that he has many computers that he can run in parallel, he plugs each parameter vector into a simulator and runs them forwards in time through a period for which observational or proxy data for weather events are available. The prior weight for each simulation is then re-weighted according to its compatibility with, or closeness to, the observed data. The function for quantifying this compatibility is referred to as a likelihood function and tends to be interpreted as an un-normalized probability of realized measurements being recorded given that the simulation accurately describes the truth.

The reweighted ensemble is then propagated forward in time again, outside the domain for which observations are available, to produce a weighted ensemble of predictions (values for the quantity s in Fig. 1). Now, instead of simulating particular values for the simulator discrepancy quantity ϵ and adding them to the simulators' state vectors so that the ensemble multiplies in size, it is common to specify regions or distributions around each ensemble member describing all the potential simulation/discrepancy combinations. These regions are illustrated with semitransparent collars around the simulator trajectories in Fig. 2, and represent a type of blurring of the simulations. The statistician now has a weighted ensemble of blurred simulations constituting a distribution of future states of the planet, a distribution synonymous with his climate prediction.

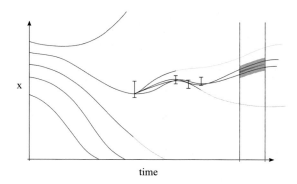

time

FIGURE 2

A sketch for interpreting climate prediction by Monte Carlo simulation. The vertical axis is used to measure an unspecified meteorological variable. The plotted curves correspond to simulated climate systems evolving through time, while the vertical bars correspond to observations of meteorological variables. Light blue curves show simulations that are down-weighted or filtered out of the ensemble, and the semitransparent regions toward the right represent distributions for simulator discrepancies within a region of time of particular interest.

The process just described is known as a Monte Carlo simulation and is used throughout the quantitative sciences. By imagining this ensemble of simulators evolving through time, one can visualize how large prior uncertainties for parameters and initial conditions give rise to diffuse ensembles of simulated worlds and, consequently, to vague climate predictions. One can also see how observational data, by down-weighting certain members of the ensemble almost to zero, serves to reduce the size of the ensemble, and so to sharpen climate predictions.

The example is particularly useful for highlighting the way in which different parts of a climate model interact. A more varied prior parameter ensemble, a less severe data compatibility weighting, or a broader simulator discrepancy distribution will all lead to less precise climate predictions but may do so in different ways. One of the key roles of the statistician is to investigate the dominant causes of imprecision so as to identify the most effective way to sharpen predictions.

Two more definitions are useful here: firstly, model *calibration* is defined as the reweighting of the ensemble of simulations. The actual calculation for applying the weights is trivial. Choosing an appropriate likelihood function to inform the weights, however, can be very difficult and just as subjective as the specification of the set of a priori plausible parameters. The authors of Ref. [3], for example, express doubt that meaningful weighting of climate simulations (in all but extreme cases) is possible given observed data on the sort of scales available. Secondly, model *validation* consists in performing diagnostic checks for which one asks if any of the ensemble members are scoring highly enough on the data compatibility measures, that is, if any of the simulated worlds pass close enough to the data for the data compatibility measures to be reconcilable with anticipated, or tolerated, deficiencies in simulators or measuring instruments within the model. Model validation is thus also clearly highly dependent on the specification of the compatibility measure, but whereas calibration depends on its relative values, validation depends on its absolute values.

Consensus on *the* Bayesian approach to model validation is hard to find. This is because standard Bayesian statistics provides a mechanism only for reallocating probability to events within a known universe of alternatives. In the example just described, this means constantly renormalizing the ensemble weights so that they sum to one. There is no obvious way to allow probability or weight to leak out of the ensemble in reflection of none its members representing plausible descriptions of the world. This represents a fundamental difficulty for Bayesian statisticians and, as such, a topic of serious contention. Readers interested in the statistical community's responses to the validation problem are recommended to consult Ref. [4] and the discussion paper that follows it, if only as a stepping stone to the work of other authors on this topic.

3. MODELLING CLIMATE CHANGE

Next, having established a framework for appreciating models and predictions, and having sketched out a general method for producing climate predictions, we will look at some of the challenges faced when putting the method into action. Further, we will discuss some specific sources of uncertainty that lead to imprecise climate predictions, noting that they tend to do this either by adding members to the ensemble of evolving simulations or by reducing our ability to down-weight ensemble members using observed data. Understanding the nature of these uncertainties will make it possible to partially justify the imprecision of climate predictions and to explain how it might be reduced.

3.1 SPECIFYING THE ENSEMBLE

Let us consider the uncertainties that contribute to the range of ensemble members. Benchmark references for informing the design and treatment of an ensemble of simulations include Refs [5] and [6]. While the first is a textbook of considerable size, the second is a concise and accessible paper in which the authors present a list of sources of uncertainty typically involved when producing predictions using a complex computer simulator. This list informs the titles of the following subsections.

3.1.1 The Ensemble of Simulators

One ought to appreciate that all the members of the ensemble are not necessarily running the same code; the ensemble could consist of a variety of simulators. Some of the best-known climate simulators are also the biggest – in particular, the General Circulation Models (GCMs) that contribute to the IPCC's assessment reports require some of the most complex and demanding computations that have ever been proposed, taking days or months to run on cutting-edge supercomputers. These are simulators developed and maintained by scientific institutions from around the world. More often than not, they are in fact constructed from interacting components, or submodels, written by teams whose members are experts in particular processes. There might be a simulator for the ice caps coupled with an ocean simulator, coupled with a simulator for the biosphere, for example. Often, the GCMs can be run at lower resolutions or with certain components deactivated. So these high-specification simulators give rise to an extensive hierarchy of simulators of varying complexity, and consequently of varying speed. As such, precise classification of climate simulators is neither straightforward nor entirely useful.

It is, however, worth mentioning the simplest, fastest simulators toward the bottom of the hierarchy. It is here that the energy balance models (EBMs) are found. These treat parts of the Earth as static blocks that can absorb, store and emit energy. All the physical properties of the mechanisms at work inside each block are summarized by its ability to do each of these. The numerical values that quantify these abilities are examples of physical parameterizations. They are numbers that are needed to run the climate simulator that do not correspond directly to quantities that can be pinned down by careful laboratory experiments; rather they are the cumulative effect of multiple complex interacting systems, many of which may only be partially understood. Therefore, although one can include in an ensemble many more members from simple and fast simulators, a significant proportion are likely to be naively specified and will be down-weighted almost out of existence by the data compatibility function.

Additionally, the outputs of simple simulators inform necessarily low-resolution summaries of weather. This means that they produce fewer quantities to validate and calibrate against data, and produce less precise predictions. Furthermore, in order to reweight an ensemble of simulators of varying resolution, one must specify extra statistical models for fine-scale variation that can be integrated out in order to define expected likelihood weights for simulators that only produce coarse-scale output, or else to project the output of more sophisticated simulators, as well as real-world data, onto the output space of the simpler models.

3.1.2 The Ensemble of Physical Parameters

For meteorological processes that are considered well understood in principle, many physical parameters, such as the freezing point of water, may be treated as known. Others, however, like the albedo effect of certain particulates in the atmosphere, may be less well constrained. It is here that modellers must elicit a set of plausible parameters from climate scientists. Such elicitation may be considered a science in itself. Reference [7] provides an interesting discussion of a wide variety of situations in which expert belief elicitation is performed, commenting on the statistical and psychological aspects of different techniques.

Typically, ranges or distributions are elicited for each physical parameter individually, with the implication that the uncertainties for the parameters' values constitute independent random variables.

The problem of contending with a large number of independent random variables is known amongst statisticians as the curse of dimensionality. This ominous name refers specifically to the way volumes grow as they are extended into higher dimensions, as opposed to being elongated in any particular direction. The curse is best illustrated in terms of the corners of a parameter space – one may imagine testing the behaviour of a simple climate simulator with one adjustable parameter by running it with parameter settings at the extremes of a notionally plausible range. The analogous strategy for a simulator with two adjustable parameters would involve performing four simulations corresponding to the corners of a square. Scaling the strategy to 20 parameters would mean looking in the 2^{20} corners of a hypercube; this would mean performing over 1 million simulations. It becomes clear that such a thorough investigation of the parameter space would become extremely demanding, if not totally infeasible.

To help us explore large parameter spaces efficiently, it is helpful to turn to the literature on numerical optimization and integration. Of specific relevance are quadrature grids and pseudo-random space-filling designs such as Latin hypercubes, Sobol or Halton sequences. These are names of parameter ensemble designs that are in some way optimized in anticipation of a simulator's response surface being of an approximately known form (like a polynomial). When this anticipation is justified, these designs are orders of magnitude more parsimonious than those that are randomly drawn from a prior distribution. By themselves, however, they cannot overcome the curse of dimensionality, which one should see not so much as a problem to be solved, as an unavoidable property of high-dimensional spaces. There is no trick to overcoming it other than to bring the dimensionality back down. The most satisfactory way to do this appears to be to insist that very high-order parameter interactions are negligibly small. This assumption serves to reduce the number of degrees of freedom in which the simulator function may vary and is the key assumption behind recent developments in sparse quadrature grids, to which Ref. [8] gives a good introduction, that show great promise in providing good approximations to calculations that may once have seemed impossible.

3.1.3 The Ensemble of Parameterizations

It is very difficult to assign weights or probabilities a priori to specific values for parameterizations since they correspond to no individually meaningful quantity. It is similarly very difficult to anticipate a simulator's functional dependency on them. Consequently, the techniques mentioned in the previous subsection to reduce the size of initial ensembles of parameterizations are of little use. Therefore, in general one will have to rely on being able to run simulations with large ensembles of parameterizations and learn about their effects empirically.

3.1.4 The Ensemble of Climate Forcings

Central to the whole endeavour of climate modelling is the need to assess the impact human activity has on the climate system. Consequently, it is necessary to include in the ensemble simulated worlds in which this impact is more or less extreme. The range of possible impacts could be described to an arbitrary level of detail, but, primarily to facilitate comparison between modelling exercises, it is most often represented using a small set of values referred to as 'emissions scenarios', or more recently 'greenhouse gas concentration pathways', endorsed by relevant authorities such as the IPCC as being suitably realistic given their beliefs for economic and technological trends. Significantly, the review of the climate modelling literature given by Ref. [2] reveals a broad consensus that the different pathways are responsible for the greatest part of the variability in climate simulations of the late twenty-first century.

3.1.5 Missing Physics

Although highly developed in many respects, there are limits to the scientific community's understanding of certain meteorological processes. The roles played in climate change by cloud formation, the biosphere, the sulphur and carbon cycles, and medium-scale oceanic eddies are all good examples of such processes. Reluctance to fully endorse current theoretical models of these processes induces a new type of uncertainty, a structural uncertainty that is very difficult to explore. Adding to the ensemble modified versions of the simulators, with these partially understood mechanisms modelled in a selection of different ways, is rarely practical. Instead, this uncertainty is parameterized by increasing the width of the simulator discrepancy distribution.

Similarly, there are also limits to the scientific community's mathematical ability. Perhaps the most important example of such a limit is the inability to solve equations for fluid dynamics. The nonlinear nature of these equations can give rise to chaotic solutions that show extreme sensitivity to initial conditions. This sensitivity is enough to amplify tiny uncertainties for the current state of the climate system into large uncertainties for its future state. Similarly, errors in the approximation of short-term solutions to the equations are also amplified. This is the reason for accurate weather forecasting being virtually impossible over significant periods of time. It is also an indicator of the diminishing returns on investments in extra computing power and observation systems.

3.2 RUNNING SIMULATIONS

At this point one must recognize that with limited computing power and considerable uncertainties for simulator parameters, it is unlikely that all the simulations in an idealized ensemble can be performed. Potential for increasing our ability to run a great number of simulations comes from advances in distributed computing and the parallelization of computational tasks. This is an area in which statisticians and computer scientists are hard at work. The climateprediction.net project is one interesting example of a research group utilizing some of the latent computing power distributed across the Internet, rather than using a supercomputer. At the time of writing, members of the public can donate their home computer's processing power to the project for a short time so that modelling experiments, such as the one presented in Ref. [9], can be performed.

3.3 RECORDING THE SIMULATIONS

The key to efficiently using the simulated data that can be computed is to recognize that many of a simulator's outputs exhibit continuity properties, with the effect that simulations with similar parameter settings tend to produce similar outputs. This idea is formalized by proposing an additional statistical model for a climate simulator's output. Such a model is referred to as an emulator or meta-model. The emulator acts as a device that allows simulator output to inform one's beliefs for simulations that are yet to be performed. In practice, this means empirical trends can be used to approximate simulations, removing the need for their evaluation. Developing the emulation idea further, Goldstein and Rougier [10] suggest that the emulator could also be used to model beliefs for simulators that do not yet exist. The intention here would be to make statements about a simulator that is not practically feasible to produce due to mathematical or computational constraints. They call this simulator the reified simulator and use it to structure a climate model such that its output is sufficient for all theoretical knowledge of the physical mechanisms affecting the weather. As a statistical tool, the reified simulator is undoubtedly useful for its potential to disentangle parameter uncertainties and structural uncertainties, for example.

3.4 ASSESSING THE SIMULATIONS

In this section we return to the ideas of simulator calibration and validation in order to discuss some issues affecting them in practice. A particularly interesting set of issues arises when one considers ensembles of simulations consisting of a range of different climate simulators, and whether it is appropriate to consider those simulators as independent experts. As we saw in Section 3.1.1, climate simulators are best thought of as members of a hierarchy or family tree in which they are grouped together according to the physical laws that inform them and the numerical devices used to find solutions to those laws. Accordingly, related simulators are likely to have similar strengths and weaknesses, and consequently are likely to make the same mistakes. It follows that the discrepancy terms for different simulators are correlated and that one cannot rely on the sort of intuition that would have us simply average out the outputs of multiple simulators because implicit to this plan is the assumption is that their errors will tend to cancel out to produce an answer more accurate than any individual simulator.

A review of published work on multi-model strategies is provided by Ref. [11]. Even more sophisticated strategies can be derived, however, when the family tree of simulators is formalized using a graphical model or Bayesian network in which the reified simulator is represented as a node separating all other simulators from real-world quantities. The literature on Bayesian networks is well developed; see Ref. [12], for example, which describes the mathematical properties that make them particularly amenable to efficient computation. Informally, by looking at the strength of the connections between simulators in the network, one can assess whether their output ought to be down-weighted to avoid effectively double counting them and, by studying the structure of the network, one can formalize how a simulator, or a set of simulators, renders others redundant. Only recently in Ref. [13] for example, have the consequences of these complex simulator interdependences been investigated.

This is also a good place to return to the climate model discrepancy term. Often, modellers will find that almost all of the simulated worlds in their ensemble are more discrepant from real-world observations than the expected variance of measurement errors would suggest is likely. To avoid throwing away a great deal of the simulations that they have invested so much effort in producing, it is tempting for them to re-evaluate the size and nature of the simulator discrepancy term. Discrepancies that are more exotic than additive terms may, for example, come in the form of projections corresponding to smoothing procedures, as discussed in Ref. [14], or the reduction of the outputs to only certain summary statistics. In general, treating the properties of the discrepancy term as inferable quantities is difficult both computationally and conceptually – computationally, because the induced correlation structure between uncertain input parameters and the discrepancy size may be highly complex, and conceptually because, when understood as an expression of tolerance for error, it is not clear that it is the place of the statistician to adjust it. Indeed, when the discrepancy is rephrased as a margin of error a stakeholder can afford rather than one he expects, it may be constrained by the type of cost considerations that we discuss in Section 4. Despite this, discrepancy adjustment or correction procedures are common in practice; in Ref. [15], for example, the authors describe some current correction methods along with their extreme concern for their appropriateness.

Before moving on, it is interesting to look at some partial or approximate Bayesian methods that are gaining traction, especially amongst statisticians working with complex systems. In many fields, genetics and climate science being particularly notable cases, the difficulties involved in applying Bayesian methods have already been identified. Most often this is because problem stakeholders cannot specify a likelihood function or prior ensemble of parameters. Approximate Bayesian Computation and Synthetic Likelihood methods, to which Refs [16] and [17] provide introductions, respectively, are simulation-based inference strategies for dealing with computationally intractable likelihood functions. Linear Bayesian methodology on the other hand, as defined in Ref. [18] and developed into a calibration-type procedure known as *history matching*, addresses the problem of specifying likelihood functions and prior probabilities from a more foundational level, building up a geometrical framework for projecting properties of unknown quantities onto data.

4. MODELLING THE COST OF CLIMATE CHANGE

Until now we have ignored the part of Fig. 1 to the right of the weather variable x, since it is the objects to the left that are traditionally seen as the domain of the climate scientist. As we discussed in Section 2, however, statistical models are fundamentally based on the specification of meaningful costs and prices that someone would actually be willing to pay. Additionally, when calibrating an ensemble of simulations, modellers need to know which variables are most important so that they can inform relevant data compatibility metrics. As such, and as recognized by Mearns [19], to model climate in a purposeful way one needs to be aware of the expected costs incurred given that different meteorological events are realized, and of the decisions that lead to them.

Elicitation of coherent and consensual descriptions of global economic mechanisms and parameter values from expert economists that can inform expected cost models is likely to be considerably harder than the elicitation of physical parameter estimates from climate scientists. Further, elicited beliefs are likely to be subjected to intense scrutiny from stakeholders with vested interests in the policy decisions they may inform, potentially affecting the elicitation process. Even if a sufficiently authoritative and cooperative committee can be convened, modelling large-scale economic mechanisms is a problem of great complexity – arguably just as great as modelling beliefs for the climate itself.

Specifically, one needs to ask how costly the meteorological events brought about by climate change are expected to be, before weighing these costs against the cost of climate change mitigation and adaptation. The Stern review [20], commissioned by the UK government and released in 2006, set out to evaluate such expected costs. Inevitably, the report met with a great deal of criticism, from Refs [21] and [22], for example. In Ref. [21], Nordhaus highlights the sensitivity of the report's conclusions to interest or discount rates. Weitzman [22], too, raises this concern as well as pointing out that important uncertainties are often very difficult to quantify with conventional models. These two points, significant sensitivity to unknown or unforeseeable parameters and the imperative to include external discrepancy terms in models, highlight areas in which expertise gained from working with models for physical systems is applicable to models for economic systems, yet the two sets of models are rarely approached simultaneously.

Similar investigations to those carried out by Stern are also being carried out by corporate and third-sector institutions. The Carbon Tracker Initiative, for example, is an organization actively striving to engage with legislators, business leaders and consumers, with the aim of quantifying costs associated with climate change and disseminating this information. By doing so, they begin to open up the rightmost pathways in Fig. 1, so that quantitative knowledge can be shared between climate scientists and policy makers. Arguably, however, this part of the flowchart is still the bottleneck in the flow of information that would enable coherent climate policy. That is, even if climate uncertainty due to deficiencies in physical theory, observational data and computational power were to be reduced by orders of magnitude, and if the statistical sophistication of climate modellers were to improve to the extent that these uncertainties could be accurately quantified, coherent climate policy would still not be possible – the reason being that policy makers are more often than not unable or unwilling to specify models describing the costs that they will take responsibility for and seek to minimize.

Issues of politics and ethics are integral to the specification of cost models because decision makers must decide which costs and whose costs they are willing to account for. It is recognized in Ref. [23] that the world's poorest people are simultaneously those with least influence on climate change policy and those at greatest risk from it, in the form of food insecurity, for example. Will leaders of the richest, most carbon-hungry nations take into account the costs of poorer foreigners from whom they have no mandate, and for whom they have no statutory responsibility?

One might argue that this is where the reasoning machinery of the statistician must be paused, that perhaps we need to reflect on the suitability of money, or an abstraction of it, as the device for guiding our modelling and, ultimately, our decision-making. Are we prepared to adopt a totally utilitarian philosophy? Would we be deluding ourselves to pretend that we can avoid employing such a philosophy anyway? If this is the case, we leave important calculations implicit, unspoken, unchecked and open to costly errors that could be corrected.

Understanding at least the presence of these concerns, the statistical arena becomes a meeting place for physical and social scientists, politicians and philosophers, and producers and consumers – a place for beliefs and costs to be weighed against each other in order that rational collective action can be taken. For this meeting of experts to be productive it is necessary that participants at all levels demonstrate statistical literacy, and that all of them understand the reasoning process in which they are involved.

5. CONCLUSIONS

The machinery for quantitative or statistical learning is well established; it is constructed from the laws of probability. These laws in conjunction with the Bayesian paradigm allow us to pool knowledge and weigh the expected outcomes of our actions.

Operating this machinery, however, is computationally demanding, and the statistical community continues to develop techniques for doing so. More problematic to our modelling efforts is the reluctance or inability of stakeholders to specify and commit to explicit expressions of their knowledge and their assessments of cost. Though challenging, statistical methodology suited to these situations is also being developed and refined.

We have seen that marshalling our knowledge, and lack of it, is difficult. Summarizing and communicating it is difficult too. It is possible that the scientific and political communities must re-evaluate what they really mean by probability, significance and cost, but what is clear is that their fields are inextricably entwined and that statistical models need to be derived that encompass them both.

ACKNOWLEDGEMENTS

I would like to thank Jonty Rougier for stimulating discussion and valuable advice during the preparation of this chapter.

REFERENCES

[1] T. Stocker, D. Qin, G. Plattner, M. Tignor, S. Allen, J. Boschung, A. Nauels, Y. Xia, V. Bex, P. Midgley (Eds.), Climate Change 2013: The Physical Science Basis. Contribution of Working Group I to the Fifth Assessment Report of the Intergovernmental Panel on Climate Change, Cambridge University Press, Cambridge, 2013.

[2] G.J. Carbone, Phys. Geogr. 35 (2014) 22–49.

[3] A.P. Weigel, R. Knutti, M.A. Liniger, C. Appenzeller, J. Clim. 23 (2010) 4175–4191.

[4] G.E.P. Box, J. R. Stat. Soc. Ser. A 143 (1980) 383–430.

[5] T. Santner, B. Williams, W. Notz, The Design and Analysis of Computer Experiments, in: Springer Series in Statistics, Springer, 2003.

[6] M.C. Kennedy, A. O'Hagan, J. R. Stat. Soc. Ser. B Stat. Methodol. 63 (2001) 425–464.

[7] A. O'Hagan, C.E. Buck, A. Daneshkhah, J.R. Eiser, P.H. Garthwaite, D.J. Jenkinson, J.E. Oakley, T. Rakow, Uncertain Judgements: Eliciting Experts' Probabilities, John Wiley & Sons, Chichester, 2006.

[8] H.-J. Bungartz, M. Griebel, Acta Numer. 13 (2004) 147–269.

[9] D. Ackerley, E.J. Highwood, D.J. Frame, J. Geophys. Res. Atmos. 114 (2009), http://dx.doi.org/10.1029/2008JD010532.

[10] M. Goldstein, J. Rougier, J. Stat. Plan. Inference 139 (2009) 1221–1239.

[11] C. Tebaldi, R. Knutti, Phil. Trans. R. Soc. A Math. Phys. Eng. Sci. 365 (2007) 2053–2075.

[12] O. Pourret, P. Naim, B. Marcot (Eds.), Bayesian Networks: A Practical Guide to Applications, Wiley, Chichester, 2008.

[13] R. Knutti, D. Masson, A. Gettelman, Geophys. Res. Lett. 40 (2013) 1194–1199.

[14] J. Räisänen, J.S. Ylhäisi, J. Clim. 24 (2011) 867–880.

[15] U. Ehret, E. Zehe, V. Wulfmeyer, K. Warrach-Sagi, J. Liebert, Hydrol. Earth Syst. Sci. Discuss. 9 (2012) 5355–5387.

[16] P. Fearnhead, D. Prangle, ArXiv e-prints (2010).

[17] S.N. Wood, Nature 466 (2010) 1102–1104.

[18] M. Goldstein, D. Wooff, Bayes Linear Statistics, Theory and Methods, Wiley, Chichester, 2007.

[19] L.O. Mearns, Philos. Sci. 77 (2010) 998–1011.

[20] N.H. Stern, The Stern Review: The Economic Effects of Climate Change, H.M. Treasury, 2006.

[21] W.D. Nordhaus, J. Econ. Lit. 45((2007) 686–702.

[22] M.L. Weitzman, J. Econ. Lit. 45 (2007) 703–724.

[23] C. Field, V. Barros, D. Dokken, K. Mach, M. Mastrandrea, T. Bilir, M. Chatterjee, K. Ebi, Y. Estrada, R. Genova, B. Girma, E. Kissel, A. Levy, S. MacCracken, P. Mastrandrea, L. White (Eds.), Climate Change 2014: Impacts, Adaptation, and Vulnerability. Part A: Global and Sectoral Aspects. Contribution of Working Group II to the Fifth Assessment Report of the Intergovernmental Panel on Climate Change, Cambridge University Press, Cambridge, 2014.

A MODELLING PERSPECTIVE OF FUTURE CLIMATE CHANGE

23

Babatunde J. Abiodun[1], Akintayo Adedoyin[2]

[1]*Climate System Analysis Group, Department of Environmental and Geographical Sciences, University of Cape Town, Cape Town, South Africa;* [2]*Department of Physics, University of Botswana, Gaborone, Botswana*

CHAPTER OUTLINE

1. INTRODUCTION

This chapter provides an insight into the science of climate modelling and gives an overview of future climate change projections from climate models. Although its focus is on future climate projections, to put the projections into the right perspective, the chapter also addresses some technical issues on climate modelling. Before discussing the projections, the chapter introduces different types of climate models, including their formulations, limitations and the levels of their usefulness for climate change studies. It also shows how well contemporary climate models can simulate past climate changes and describes the procedure for obtaining future climate projections. After presenting the future climate change projections, the chapter addresses the issue of uncertainty in future climate change projections.

2. CLIMATE MODELS

Climate models are grand tools for understanding the complexity of our Earth's climate system. These models use a set of equations (derived from physical, chemical and biological laws) to replicate the climate system. They can simulate past or future states of a climate system over any time range from a few hours to many centuries [1]. These simulations can supplement observed climate data from a sparse global observation network, i.e. reanalysis [2]. Experiments with these models can reveal how climate systems respond to various climate forcing (i.e. changes in solar radiation, land cover, greenhouse effects etc.). However, the capability and reliability of a climate model to perform these tasks depends on the hierarchy of the model.

2.1 HIERARCHY OF CLIMATE MODELS

Climate models are usually classified into a hierarchy based on the complexity of the equations they used to represent the Earth system. For instance, while some models use a simple equation to represent the climate system (which consists of complex processes and multifaceted interactions among the components: land, oceans, ice etc.), other models adopt several complex equations to describe the system as accurately as possible. The complexity and number of the equations used in a model also determine the computational demands of the models. However, the choice of a model for a climate study depends on the purpose of the study. While a complex model may be required for some studies, a simple model may be adequate for other studies. Hence, the climate community uses a variety of climate models, ranging from simple energy balance models that can be implemented using a simple calculator to complex Earth system models that require a high performance computing infrastructure for the calculations. In this section, we provide a brief description of some of these models.

2.1.1 Energy Balance Models

Energy balance models (EBMs), which may be zero or one-dimensional models, are based on a simple energy balance – this being the basic principle that forms the foundation for all climate models. The principle states that in the long run, the amount of solar energy absorbed by the Earth system must be equal to the amount of terrestrial energy emitted by the Earth system [1]. For a zero-dimensional EBM (EBM0), this energy balance law is a simple equation:

$$4\varepsilon\sigma T_g^4 = (1 - \alpha_p)S_o \tag{1}$$

The right-hand side of Eqn (1) refers to the solar energy that the Earth system absorbs: S_o is the solar energy influx that reaches the Earth system, averaged over the entire spherical Earth surface; the factor α_p is the reflectivity of the Earth system (called planetary albedo; it ranges from 0 to 1). Hence, $1 - \alpha_p$ indicates the proportion of the incoming solar energy (S_o) absorbed by the Earth system. The left-hand side of the equation refers to the terrestrial energy that the Earth system emits: T_g is the global mean surface temperature of the Earth system; σ is the Stefan–Boltzmann constant ($\sigma = 5.67 \times 10^{-8}$ W m^{-2} K^{-4}), which relates temperature to radiant emission; and ε represents the infrared transmissivity, which depends on the concentration of atmospheric greenhouse gases (GHG); the higher the concentration of GHG, the lower the ε. If we substitute the present-day values (i.e. $S_o = 1637$ W m^{-2}; $\alpha_p = 0.3$; $\varepsilon = 0.61$) into Eqn (1), the EBM0 will give a global mean surface temperature (+15.3°C), which is close to observed value (+14.0°C) for the present-day climate [3]. This simple model is also realistically sensitive to the temperature changes caused by external climate forcings (i.e. solar radiation, planetary albedo and greenhouse gases). For instance, Fig. 1 shows that a decrease in the infrared transmissivity (due to increase in concentration of GHG) will cause global warming, and an increase in planetary albedo (i.e. due to enhanced reflectivity) will induce global cooling.

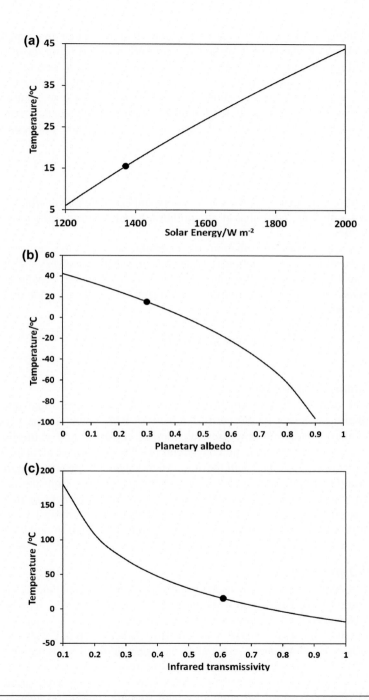

FIGURE 1

The sensitivity of global mean Earth surface temperature to (a) solar energy, (b) planetary albedo and (c) infrared transmissivity of the atmosphere. The black circle in each panel indicates the value for present-day climate.

A main shortcoming of Eqn (1) is that it assumes that a change in climate forcing will produce an instant change in the Earth's surface temperature and will bring the climate system to an immediate equilibrium state; in reality, it may take the Earth several years to attain equilibrium after a change in climate forcing. Hence, the model cannot simulate the time series of the Earth's surface temperature as the climate system moves toward an equilibrium state in response to a change in climate forcing. To overcome this limitation, a tendency term (i.e. the rate at which heat storage changes with time) is introduced into Eqn (1) and the equation for EBM0 becomes:

$$C_p\frac{dT_g}{dt} = \left(1 - \alpha_p\right)S_o - \varepsilon\sigma T_g^4 \qquad (2)$$

where C_p is the specific heat capacity of the Earth system and $\frac{dT_g}{dt}$ is the rate at which the global mean temperature changes with time. Note that at an equilibrium state the tendency term becomes zero, making Eqn (2) identical to Eqn (1).

To account for the geographical distribution of temperature at the Earth's surface, Eqn (2) can be extended to include one dimension (i.e. latitude) by adding a transport term $k_T(T_l - T_g)$ to represent the net effect of heat input and output of horizontal transport. Here, T_l is the zonally averaged surface temperature at a latitude and k_T is the heat transport coefficient. Hence, the equation in one-dimensional EBM (i.e. EBM1) becomes:

$$C_p\frac{dT_l}{dt} = \left(1 - \alpha_p\right)S - 4\varepsilon\sigma T_{lat}^4 + k_T\left(T_g - T_l\right) \qquad (3)$$

2.1.2 Radiative-Convective Model

Radiative convective models (RMs), which can be one- or two-dimensional models, focus on simulating atmospheric processes in the vertical direction. RMs calculate the vertical profile of temperature by simulating radiation processes (absorption, diffusion and scattering) and vertical transports of energy by vertical motion and convection [1]. These models calculate heat absorption and monitor the temperature lapse rate in all the atmospheric layers. If the temperature lapse rate in a layer exceeds a certain threshold, the models will transfer heat energy from the layer to the layer above (i.e. convective adjustment). Hence, RMs use convective adjustment to re-establish a predetermined lapse rate in each layer of the atmosphere. However, in contrast to one-dimensional EBMs, one-dimensional RMs only transfer energy in the vertical direction and not in the horizontal direction.

2.1.3 Statistical-Dynamic Models

Statistical-dynamic models (SDMs), which are usually two-dimensional models (i.e. vertical and latitude), combine the tasks of EBMs and RMs because they transfer heat energy in horizontal and vertical directions. However, the horizontal transfer of energy (i.e. from the equator to the poles) is usually more accurately simulated in SDMs than in EBMs because while EBMs use empirical and theoretical relationships of atmospheric circulation between latitudes to represent the horizontal energy transfer, SDMs use laws of motion to simulate the transfer of energy between equator and the poles [4]. However, SDMs still use statistical relationships to define wind speed and direction. Nevertheless, SDMs are very useful for simulating and studying horizontal energy flows, and for understanding processes that disrupt the energy flows. These models are used in some Earth models of intermediate complexity.

2.1.4 Earth Models of Intermediates Complexity

Earth models of intermediate complexity (EMICS), which can be two- or three-dimensional models, bridge the gap between the simple models and complex models. Unlike simple models, EMICs use complex equations to describe most of Earth system processes and the interaction among the Earth system components, but in a more reduced (i.e. parameterized) form when compared with those in complex models. Some EMICs even include model components that are used in a more complex Earth system model, but at coarser numerical grid resolution. The representation of the geography varies among the EMICS. Some EMICs use a zonally average representation of atmosphere and ocean, but they always make a distinct difference between Atlantic, Pacific and Indian basins because the ocean circulations are different [1]. However, like simple models, EMICs are computationally efficient for long-term (i.e. several millennia) simulations involving many sensitivity experiments.

2.1.5 Global Circulation Models

Global circulation models (GCMs) have the most complex and most detailed representation of the climate system because they represent as many of the Earth system processes as possible. They produce the three-dimensional image and time evolution of all the Earth system components. However, GCMs exist in different forms: ocean general circulation models (OGCMs), atmospheric general circulation models (AGCMs), coupled ocean-atmospheric general circulation models (AOGCMs), and Earth system models (ESMs). AGCM and OGCMs are independent atmosphere and ocean circulations models (respectively) usually used for testing and evaluation purposes, while AOGCMs are coupled atmosphere-ocean models used for understanding the dynamics of the climate system [3]. ESMs are extensions of AOGCMs to include various biogeochemical cycles in the Earth system [3]. Hence, ESMs have components to represent all the Earth system components (atmosphere, land, sea ice etc.) and couple them together to create a system that is close to reality. In ESMs, the atmospheric and ocean components are similar in that they solve similar primitive equations of motion. ESMs and AOGCMs are the standard and credible climate models for making future climate projections based on greenhouse and aerosol forcings, with the ESMs being the state-of-the-art climate models. The simulation results presented in this chapter are from AOGCMs and ESMs.

2.1.6 Regional Climate Models

Regional climate models (RCMs), which are usually three-dimensional models, have a similar structure to GCMs, except that they are designed to simulate a portion of the globe, using global data sets from reanalyses or GCMs as boundary conditions. Hence, RCMs dynamically downscale the global data set (that provides the boundary condition) to a higher resolution over the limited area of interest. Hence, the resolution of RCMs is always higher than that of the forcing global data set. It is important to note that RCMs are structurally different from their counterpart, statistical downscaling models (SDMs), which use empirical relationships to link large-scale atmospheric variables (predictors) with regional or local variables (predictands). Discussions on regional downscaling with either RCM or SDM are beyond the scope of this chapter.

3. EVALUATION OF CLIMATE MODELS

Climate model evaluation is a process of determining how well a climate model represents reality. It is crucial to evaluate a climate model before applying it to project future climate changes. This is because our confidence in the predictive power of a climate model grows with the capability of the model to simulate past climate. For the climate change projections from a climate model to be useful, the model must realistically reproduce many of the observed past changes in the climate system. However, it is difficult to directly relate the performance of a climate model in simulating the past climate to a degree of confidence, to the ability of the model to simulate future climate changes. In addition, the aspects of the observed climate that a climate model must simulate correctly to ensure reliable future predictions are also not clear. Nevertheless, the evaluation metrics can act as a guide on the strengths and weaknesses of individual climate models, as well as the overall state of climate modelling [5].

3.1 TECHNIQUES AND EXPERIMENTS FOR EVALUATING CLIMATE MODELS

There are various techniques and special experiments for evaluating climate models [3]. We described some of them briefly in this section.

3.1.1 Isolating Processes

This technique isolates and evaluates climate processes in climate models. The aim is to ensure that models give a good representation of individual physical processes. The technique can help in identifying and understanding the source and cause of errors in models. There are several ways to implement this technique. For example, simulations from a single column version of GCMs may be compared to simulations from a more detailed process model or to measurements from field campaigns, designed to study climate processes in a particular location. Another way is to average the model results within a distinct physical regime of a climate system (i.e. circulation regimes) and compare the average with corresponding observations. Using this technique, several experiments have been conducted and coordinated at local and international levels, and the data and results are publicly available at the websites related to the experiments. Examples of such experiments include the Cloud-Associated Parameterizations Testbed (CAPT: www-pcmdi.llnl.gov/projects/model_testbed.php) and, the VAMOS Ocean-Cloud-Atmosphere-Land Study (VOCALS: www.eol.ucar.edu/field_projects/vocals).

3.1.2 Evaluating Model Components

This technique evaluates individual components of a GCM (i.e. atmosphere, ocean, land, sea ice) separately. To evaluate a component, the boundary conditions at the interface of the component with the other components should be well defined. For a fair evaluation, the boundary condition and climate forcing used in the model simulation must represent that of the observed condition to which the simulation is to be compared. Examples of internationally coordinated experiments that implemented this technique include the Atmospheric Model Intercomparison Project (AMIP: www-pcmdi.llnl.gov/projects/amip) and the Arctic Ocean Model Intercomparison Project (OMIP: www.whoi.edu/projects/AOMIP). Their websites give detailed information on the experiments, participating models, simulation data and publications from the experiments.

3.1.3 Evaluating Overall Model Results

This is the most direct and most commonly practised technique. It evaluates results of a coupled model (i.e. EMIC, AOGCM or ESM). The technique will expose any error or problem that may be concealed by formulation of boundary condition when components of the models run individually. Nevertheless, evaluation of a model should not be based only on this technique because it is not sufficient that a coupled model provides reasonable results. It is essential to still evaluate each process and each component of the model individually to ensure that the coupled model produces the correct results for the right reason and not because several errors in its various processes or components are cancelling each other out.

Two types of runs are usually performed in evaluating a coupled model using this technique. The first type is a control run, in which the model is forced with a repeating annual cycle of solar radiation at the top of the atmospheric component. Other climate forcings (i.e. concentration of greenhouse gasses, aerosol) that affect the atmosphere may also have a repeating annual cycle or remain constant in time. The values of the forcings may be from preindustrial time (usually 1850) or from present-day conditions. The results of the models are then compared with appropriate observations. The second type is the twentieth-century run, in which the time series of the climate forcings (i.e. solar radiation, concentration of greenhouse gases, natural and anthropogenic aerosol, and historical changes in land use) from 1850 to the present are used in the model simulations. Then, the capability of the models to reproduce the observed changes in climate variables (i.e. temperature, precipitation etc.) over the last 150 years will be assessed. However, most of the validations usually focus on the last 50 years of the simulations. The twentieth-century run is also useful in determining the cause of the Earth's warming. For this, additional simulation is performed, and in the simulation, the levels of anthropogenic greenhouse and aerosol are kept at the preindustrial condition. The model results for these experiments usually indicate that global warming is due to anthropogenic forcing. Examples of experiments that implement this technique include the Coupled Model Intercomparison Projects (i.e. CMIP3 and CMIP5; http://cmip-pcmdi.llnl.gov/). The data from these experiments are publicly available and the results have been published in many papers, some of which are reviewed in Ref. [3] and reproduced here in Figs 3 and 4.

3.1.4 Evaluating Models Ensemble

A climate model ensemble is a collection of climate simulations, each representing different possible realizations of the climate system. Owing to the different approaches to modelling complex climate system processes, there are uncertainties in the model structure, and hence in the simulations from the climate models. Since the climate ensemble is a way of characterizing the uncertainties, evaluation of the ensemble mean will give good insight to the general performance of a climate model. There are two main types of climate ensembles: perturbed parameter ensemble (PPE) and multi-model ensemble (MME). PPE is a group of climate simulations from the same model, and the simulations are generated with slightly different parameters, physics or initialization. Evaluation of PPE gives a better assessment of the performance of a model than using a single simulation. MME is a group of climate simulations from different climate models originating from diverse modelling centres, but the simulations are performed for similar climate conditions. Evaluation of MME provides a good indication of how well we are able to model the climate system. In this chapter, we use the MME approach to present and discuss the model simulations.

3.2 FACTORS INFLUENCING MODEL PERFORMANCE

The results of model performance can be influenced by various factors. The major factor is the uncertainty in how climate processes are represented (i.e. parameterization) in the models. For example, the representation of cloud processes and their interaction with local and large-scale circulations remains a major source of uncertainty and errors in most atmospheric models [3]. The other factor is that the simulation of some climate processes may be sensitive to the horizontal and vertical resolution of climate models. For instance, various studies have shown that increasing model resolution improves the simulation of tropical cyclones, precipitation over complex terrain and extreme precipitation events [3]. Furthermore, uncertainty in observational data is another major factor in model evaluation. For example, insufficient spatial or temporal resolutions of observational data may influence the results of model evaluation of the simulated spatial or temporal variability, extreme events and trends. The quality of the observational data also influences the evaluation. Hence, a change in the resolution or quality observational data set can alter the results of model evaluation [3]. Other factors include uncertainties in how models are forced (i.e. greenhouse gases, land use, aerosols etc.) and how the parameters used for the evaluation (i.e. droughts, extreme events etc.) are defined.

3.3 PAST CLIMATE SIMULATION

Here we review how well contemporary climate models reproduce the twentieth-century climate. The review is largely based on the model evaluation report of the Intergovernmental Panel on Climate change (IPCC, AR5) [3], but our focus here is on important climate variables, processes and extreme events. In addition, we emphasize performance of the CMIP5 multi-model ensemble mean, rather than individual climate model results, to provide a general view of the capability of climate model in simulating the climate system.

3.3.1 Simulation of Climate Variables

In evaluating how well models simulate climate variables, we consider the most important two climate variables: temperature and precipitation. These variables are the most commonly evaluated climate variables. While temperature is one of the easiest variables to simulate well in models, precipitation is the most difficult variable to reproduce because it depends on many processes that are still parameterized in the models. In addition, low-resolution models struggle to simulate precipitation over complex topography.

Figure 2 compares the interannual variability of the simulated and observed global surface temperatures. The figure shows that the models give realistic simulations of the interannual temperature and reproduces all the essential features as determined from observation. For instance, the simulations captured the observed cooling after the Pinatubo eruption in 1991 and the rapid warming over the past few decades. The maximum error from any of the models is less than 0.5°C. The ensemble mean of the model closely follows the observations and generally performs better than any individual model. However, the amplitude of the variability is lower in the ensemble mean than in the models because the averaging of the model results in reducing the simulated variability. The global variation of simulated temperature from the multi-model ensemble mean agrees well with observation (i.e. reanalysis), except for some bias (Fig. 3). The bias, which is generally lower than 2°C in many locations, is more than 3°C over ocean upwelling regions off the west coasts of South America and Southern Africa [3]. The amplitude of the seasonal cycle in the simulated temperature, obtained as the absolute difference between December–February and June–August temperatures, also agrees well with the reanalysis, although there are some biases (Fig. 3). The largest bias occurs over lands at high latitudes and over northern India.

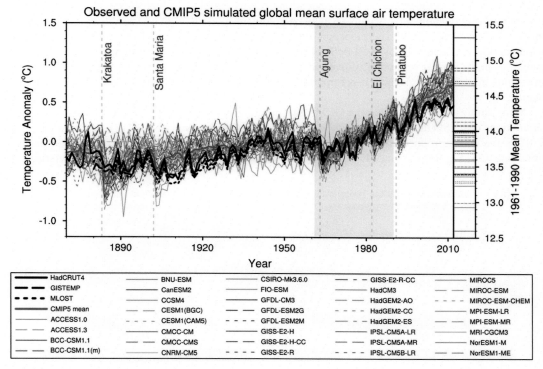

FIGURE 2

Observed and simulated (CMIP5 models) time series of the annual mean surface temperature, t, and anomalies with reference from 1961 to 1990 mean, Δt. The observations are in black tick lines, the CMIP5 ensemble mean is in thick red lines, while single simulation for individual model is in thin lines. *(Adapted from Fig. 9.8(a) of Ref. [3].)*

The multi-model ensemble captures all the essential features in the observed precipitation pattern but with some bias (Fig. 4). For instance, the ensemble mean reproduces the maximum precipitation over the convergence zones near the equator in the central and eastern tropical Pacific, the minimum precipitation in Northern Africa, and the dry areas over the eastern subtropical ocean basins [6]. However, the simulated precipitation pattern shows a dry bias (about 3 mm d^{-1}, where d refers to day) along the equator in the Western Pacific and over South America, and shows wet bias (about 3 mm d^{-1}) along the tropical Atlantic, Indian and Pacific Oceans [7,8]. Reference [9] showed that several models overestimate the frequency of occurrence of light rain events.

3.3.2 Simulation of Climate Phenomena

In evaluating the models' ability to simulate climate phenomena, we consider two phenomena: the El Niño Southern Oscillation Index ENSO (ENSO) and Monsoon systems. The ENSO is the interaction between the ocean and atmosphere over the Pacific Ocean. The interaction, which has a periodic below-normal and above-normal sea surface temperature, induces alternating wet and dry conditions over many regions in the world (i.e. teleconnection). It is essential to evaluate how well climate models simulate ENSO phenomenon because it plays a dominant mode in climate variability on seasonal to interannual timescales [10]. The monsoon, on the other hand, is the annual reversal of the low-level winds and is characterized by well-defined dry and wet seasons [11]. The monsoon plays a dominate roles in the annual variability of tropical precipitation [11,12] and influences weather and climate in various tropical areas. Hence, it is important to investigate how well climate models simulate this feature.

Many studies have shown that the GCMs give credible simulation of the ENSO because the model simulations of ENSO have many characteristics similar to the observed ENSO [13]. For example, most CMIP5 models reproduce the observed maximum variability of ENSO at a two- to seven-year timescale [3]. See Figs 3 and 4. However, the models still exhibit some bias in simulating the

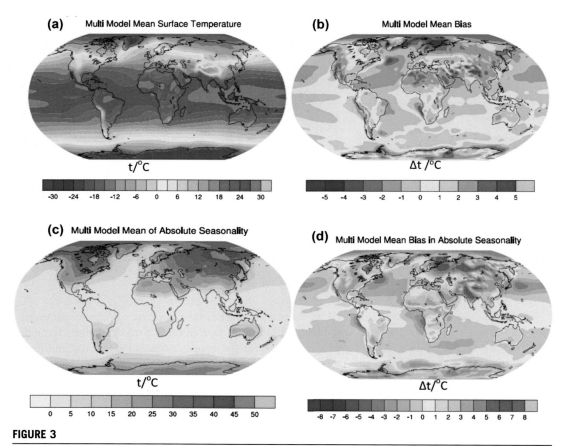

FIGURE 3

Global distribution of the simulated (CMIP5 multi-model ensemble mean) surface temperature: (a) the simulated annual mean temperature, t; (b) the bias in simulated annual mean temperature, Δt; (c) the simulated absolute seasonal temperature, t; and (d) the bias in simulated absolute seasonal temperature, Δt. *(Adapted from Figs. 9.2 (a), 9.2 (b), 9.3 (b), and 9.3 (d) of Ref. [3].)*

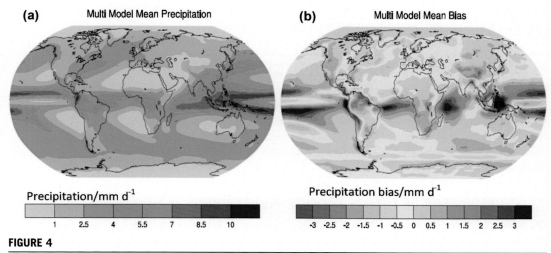

(a) Multi Model Mean Precipitation   **(b)** Multi Model Mean Bias

Precipitation/mm d⁻¹

1   2.5   4   5.5   7   8.5   10

Precipitation bias/mm d⁻¹

-3 -2.5 -2 -1.5 -1 -0.5 0 0.5 1 1.5 2 2.5 3

FIGURE 4

Global distribution of the simulated (CMIP5 multi-model ensemble mean) (a) annual mean precipitation and (b) the bias in simulated annual mean precipitation. *(Adapted from Figs 9.4 (a) and 9.4 (b) of Ref. [3].)*

amplitude, period, irregularity, degree of skew, spatial patterns and teleconnections of ENSO. Many models also capture the observed characteristics of monsoon systems, but they underestimate the horizontal extent and the precipitation intensity of the systems over Asia and North America. Some models capture the weakening of some summer monsoon circulations (especially over South Asia) as observed over the past half century, although the models fail to replicate the magnitude of observed precipitation [14].

3.3.3 Simulation of Extreme Events

It is important to evaluate how well climate models simulate the extremes in climate variability because climate change often induces changes in the intensity, frequency, duration, spatial extent and timing of extreme events like heat waves, extreme precipitation events and droughts [3]. If a model underestimates the statistics of these extreme events in the past climate, it is possible that the model may also underestimate the statistics of the extreme events in the future climate projections. Here, we focus on how well CMIP models simulate temperature extremes, extremely heavy precipitation and droughts. In simulating the extreme temperature events, the multi-model ensemble mean features a bias of a few degrees over most of the Earth [15] but captures well the observed positive trend in temperature extremes (especially in the second half of the twentieth century); and the inter-model spread is within the spread of the observational data sets used for the evaluation [16]. In simulating the extreme precipitation, the results of the multi-model compare well with the observation in the extratropics (midlatitudes beyond the tropics), but the discrepancy among the observation data is too large to assess the performance of the models within the tropics [3]. However, some studies have found that climate models generally underestimate the sensitivity of extreme precipitation intensity to changes in temperature [3]. This suggests that the models may underestimate the future increase in extreme precipitation within the tropics. There is no comprehensive evaluation of how well the models simulate drought. However, Ref. [16] found that climate models simulate well the magnitude and distribution of consecutive dry days. Reference [17] also showed that most CMIP5 models correctly reproduce the observed drought patterns over Southern Africa.

4. FUTURE CLIMATE CHANGE PROJECTIONS

This section presents the future climate projections for the twenty-first century from CMIP5 models. The review is largely based on the IPCC (AR5) report [18]. As done previously, our focus is on the projected changes in some important climate variables, processes and extreme events. In addition, the emphasis is on the future climate change projections from CMIP5 multi-model ensemble mean. However, before discussing the future projection, it is essential to shed light on how the projections are developed.

4.1 DEVELOPMENT OF FUTURE CLIMATE PROJECTIONS

The basis for future climate projection is that climate variations are driven by changes in climate forcing. Therefore, to simulate future climate (i.e. twenty-first century), we need to specify the future climate forcing in the models. Climate forcing can be classified into two types: natural and anthropogenic. Natural climate forcings are due to solar radiation changes and volcanic eruptions, while anthropogenic changes are due to human activities that alter the properties of the Earth system (i.e. greenhouse gasses, aerosols and land cover changes). Various studies have shown that past climate variations (i.e. in the twentieth century) have been largely driven by the anthropogenic climate forcing. Therefore, apart from unprecedented huge natural events in the future, the future climate (i.e. twenty-first century) will likely be driven by anthropogenic forcing. Hence, future climate forcings are usually estimated by developing different future scenarios based on human activities (i.e. emission of greenhouse gases, aerosols, land use etc.) and are used in climate models to simulate the future climate. By convention, these simulations are usually called climate projections (and not climate predictions) because these plausible scenarios are developed based on many uncertain parameters. Two approaches have been adopted in developing future projections. The first approach is based on the Special Report on Emission Scenarios (SRES) and the second approach is based on representative concentration pathways (RCPs).

The SRES approach was used for the fourth assessment reports of IPCC (AR4) [19]. In this approach, future changes in anthropogenic forcing (emission of greenhouse gases, aerosols, land use etc.) were estimated based on future population growth, economic activity, technology changes etc. [20]. The CO_2 emission and atmospheric CO_2 concentration are used to drive climate models in making the future climate projections. The RCP approach was adopted for the fifth assessment of IPCC (AR5) [21]. In this approach, four RCP scenarios (RCP3-PD, RCP4.5, RCP6.0 and RCP8.5) that cover a wide range of future changes in radiation forcing were selected. The emission and concentration of greenhouse gases corresponding to these four RCPs were provided to the climate-modelling community for climate projection [20]. The future climate projections discussed in this chapter are based on the RCP scenarios.

4.2 PROJECTED CHANGES IN CLIMATE VARIABLES

Figure 5 presents the long-term future projections (i.e. twenty-first century) for the most important two climate variables (temperature and precipitation). The projections show that the global warming

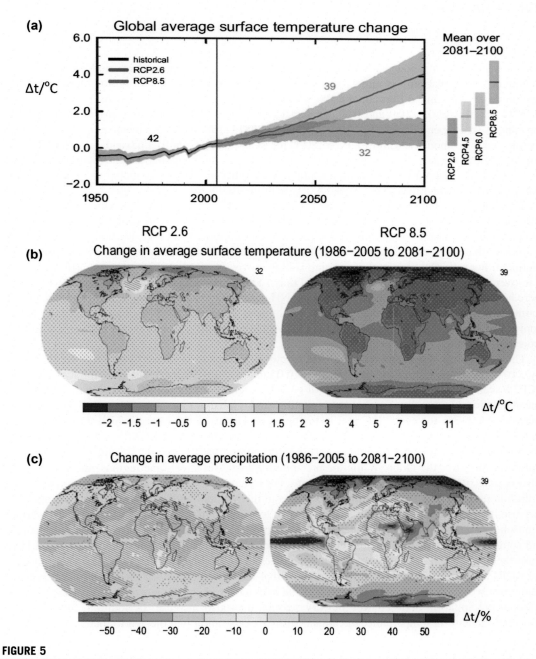

FIGURE 5

Simulations from CMIP5 multi-model ensemble for different climate forcing scenarios. Panel (a) shows the time series of global annual mean surface temperature (Δt) change from 1950–2100 to 1986–2005. Panel (b) shows the global distribution of the annual mean surface temperature in 2081–2100 relative to 1986–2005, while panel (c) shows the corresponding precipitation. *(Adapted from IPCC, Summary for policymakers, in: T.F. Stocker, D. Qin, G.-K. Plattner, M. Tignor, S.K. Allen, J. Boschung, A. Nauels, Y. Xia (Eds.), Climate Change 2013: The Physical Science Basis. Contribution of Working Group I to the Fifth Assessment Report of the Intergovernmental Panel on Climate Change, 2013.)*

continues in the twenty-first century for all the RCP scenarios [18]. During the period 2005–2025, the magnitude of the warming is almost the same for all the scenarios, but in the long-term scale the magnitude of warming differs among the scenarios. While RCP2.6 produces the lowest warming (0.18°C per decade till 2050 and 0°C per decade after), the RCP8.5 scenario produces the highest warming (>0.3°C per decade) (Fig. 5(a)). Hence, by 2081–2100, the global mean temperature change is projected to be about 1.0°C for RCP2.6, 2.0°C for RCP4.5, 2.5°C for RCP6.0 and 4.0°C for RCP6.0. The projections suggest that only with the RCP2.6 scenario can the global mean temperature be kept below 2.0°C throughout the twenty-first century. With other scenarios, the global mean temperature may exceed 2.0°C and may even exceed 4.0°C with the RCP8.5 scenario.

The global distribution of the warming is not uniform (Fig. 5(b)). There is a widespread warming over the globe, except in the North Atlantic Ocean where there is a weak cooling as a result of the reduction in deepwater formations and shifts in ocean currents [18]. The warming is higher over continents than over the oceans because the heat capacity of the ocean is higher than that of the continent. For a similar reason, the warming is higher over the northern hemispheric (NH) land mass than over the southern hemispheric (SH) land mass. Since SH oceans are larger than the NH oceans, more heat is stored in the SH oceans than in the NH oceans. Over the continents, the magnitude of the warming increases inland. Furthermore, the warming is greater in the NH polar region than in the tropics and greater over the Arctic than over the Antarctic regions. However, the highest warming is projected over the NH polar region.

The projections show a gradual increase in global precipitation in the twenty-first century [18]. By 2100, the precipitation increase is about 0.05 mm d^{-1} ($\sim$2% of global precipitation) for the RCP2.6 scenario and 0.15 mm d^{-1} ($\sim$5% of global precipitation) for the RCP8.5 scenario. There is a linear relationship between global precipitation and global temperature such that precipitation increases by 1–3% for every 1°C rise in temperature. However, the rate at which the precipitation increases with the warming is higher for the RCP2.6 scenario than for the RCP4.5 scenario. But, global distribution of the projected precipitation (Fig. 5(c)) shows that the magnitude and signs of the precipitation vary over the globe. While the precipitation increases over some regions, and decreases in other regions, it does not change in other regions. For instance, the annual mean precipitation increases at high latitudes and at the equator, but decreases in many midlatitude arid and subarid regions. The increase in precipitation at the high latitudes is due to increase in the specific humidity of the warmer atmosphere and the transport of water from the tropics. The decrease in the precipitation in the arid and semiarid regions is because the changes in circulations over the regions suppress rainfall [18]. Nevertheless, the magnitudes of these precipitation changes are higher in the RCP8.5 scenario than in the RCP4.5 scenario. The projections also show that the contrast between the annual mean precipitation over wet and dry regions may increase; and over most regions of the world the contrast between seasonal precipitation between wet and dry seasons may also increase. In addition, following the projected increase in temperature, the annual surface evaporation is projected to increase.

4.3 PROJECTED CHANGES IN CLIMATE PHENOMENA

It is projected that the warming alters the characteristics of various climate phenomena; however, for brevity and for consistency with Section 2.1, only projected changes in ENSO and monsoon are discussed here. The projections indicate that the ENSO may continue to be the dominant mode of climate variability in future, and that its influence on regional precipitation variability may be stronger than it is at present. However, the changes in the teleconnection impacts are not global but regional. For instance, the projections show that it shifts the associated temperature and precipitation eastward over the North Pacific and North America, but not over other regions. It is difficult to conclude how the amplitude and pattern of ENSO may change in the future because for all the scenarios, the projected natural variability of the ENSO amplitude and spatial coverage overshadows the projected changes. In addition, the magnitude of the projected changes in the amplitude to El Niño is small compared to spread among the model projections.

In the future projections, the monsoons cover wider areas and produce higher summer precipitation, but the circulation of the monsoons (East Asian monsoon) becomes weaker. Only the East Asian monsoon circulation is projected to be stronger. In addition, the projections suggest that the monsoon onset may be earlier while the cessation may be delayed, indicating longer monsoon periods. The exception to this is the West Africa monsoon (WAM). The models project a delay in the onset of WAM rainy season and intensified late season. Since the skill of the models in simulating the WAM is low, the confidence in the projected WAM is also low. Furthermore, the projections show that interannual variability of summer precipitation from monsoons may be higher, and in all monsoon regions, the extreme precipitation event associated with the system may increase.

4.4 PROJECTED CHANGES IN EXTREME EVENTS

Future projections show that the global warming induces changes in different types of extreme events in the twenty-first century [18,22]. For temperature extremes, the projected changes are toward warmer conditions in the twenty-first century, although the magnitude of the changes depends on future climate scenarios. The projections indicate an increase in warm extreme events but a decrease in cold extreme events. On the global scale, the increase in temperature of the coldest night of the year (CNY) is higher than that of the hottest day of the year (HDY), and the increase in the frequency of tropical nights (i.e. temperature $>20°C$) is greater that than of warmest days. On a regional scale, the largest increase in temperature of CNY is projected over the high latitudes of the Northern Hemisphere (under the RCP8.5 scenario). Furthermore, the largest increase in HDY is simulated over the subtropics and midlatitudes, and the highest increase in the frequency of warm days and warm nights is in the tropics [16]. However, the number of frost days (i.e. temperature $<0°C$) decreases over the regions, with the largest decrease projected over the eastern part of North America and over Scandinavia.

For extreme precipitation, the future projections show an increased frequency and intensity of extreme precipitation over the continents. On a global scale (i.e. average over the continents in wet seasons), the increase in precipitation of extremely wet five-day periods ranges from 5% (for RCP2.6) to 20% (for RCP8.5) by the end of the twenty-first century. On regional scales, the increase in intensity precipitation is most prominent in the higher latitudes of the northern hemisphere and in the wet tropical regions. In general, a shift to more intense individual storms and fewer weak storms is projected for the warmer future. Furthermore, as the climate warms, extreme precipitation rates on daily timescales may increase faster than the time average, and the changes in local extremes on daily and sub-daily timescales are projected to increase by roughly 5%–10% per 1°C. For drought events, the projection indicates a substantial increase in meteorological drought over the Mediterranean, Central America, Brazil, South Africa and Australia but a decrease over the high latitudes in the northern hemisphere. This projection is consistent with changes in the Hadley Circulation and increased temperature. The general picture from the projections is that, as the climate warms up, wet regions get wetter and dry regions get drier.

4.5 UNCERTAINTY IN THE FUTURE CLIMATE PROJECTIONS

The future climate projections are uncertain. Typically, future climate change projections cannot be deterministic like a weather forecast because they are associated with many uncertainties. The uncertainty in the climate projections are mainly from three sources. Firstly, the projections depend on scenarios of future anthropogenic and natural forcings that are not certain. Secondly, the climate models are not accurate because there is no complete understanding of the complex climate system. And thirdly, the existence of internal climate variability (noise) weakens the projected climate change signal. Hence, the term *climate projection* is tacitly used to imply the uncertainty.

5. CONCLUSIONS

This chapter has reviewed the studies of past and future climate simulations from contemporary climate models. The main message from the review is that the climate models give realistic simulations of past climate changes (though with some biases), and they show that the ongoing global changes are largely due to anthropogenic climate forcing. The models project that for all future forcing scenarios, the warming would continue into the twenty-first century for all future climate scenarios. The warming, which may be under 2.0°C throughout the twenty-first century, as described by the RCP4.5 scenario, may exceed 4.0°C at the end of the century if the RCP8.5 scenario is correct. For all scenarios, the warming is projected to increase the intensity and frequency of hot days, warm nights, extreme precipitation events and droughts. It could also alter the characteristics of various atmospheric phenomena and make wet regions wetter and dry regions drier. However, these future climate change projections are not deterministic due to biases in the model simulations, uncertainty in the future forcing scenarios and natural variability in the climate system.

REFERENCES

[1] K. McGuffie, A. Henderson-Seller, A Climate Modelling Primer, third ed., John Wiley & Sons, Ltd, West Sussex, England, 2005, p. 287.
[2] K.I. Hodges, B.J. Hoskins, J. Boyle, C. Thorncroft, Mon. Weather Rev. 9 (2003) 2012–2037.
[3] G. Flato, J. Marotzke, B. Abiodun, P. Braconnot, S.C. Chou, W. Collins, P. Cox, F. Driouech, S. Emori, V. Eyring, C. Forest, P. Gleckler, E. Guilyardi, C. Jakob, V. Kattsov, C. Reason, M. Rummukainen, Evaluation of climate models, in: T.F. Stocker, D. Qin, G.-K. Plattner, M. Tignor, S.K. Allen, J. Boschung, A. Nauels, Y. Xia, V. Bex, P.M. Midgley (Eds.), Climate Change 2013: The Physical Science Basis. Contribution of Working Group I to the Fifth Assessment Report of the Intergovernmental Panel on Climate Change, Cambridge University Press, Cambridge, UK and New York, NY, USA, 2013, pp. 741–866.
[4] B. Saltzman, Adv. Geophys. 20 (1978) 183–304.
[5] P.J. Gleckler, K.E. Taylor, C. Doutriaux, J. Geophys. Res. 113 (2008), online publication D06104, http://dx.doi.org/10.1029/2007JD008972.
[6] A. Dai, G.A. Meehl, W.M. Washington, T.M.L. Wigley, J.M. Arblaster, Bull. Am. Meteorol. Soc. 82 (2011) 2377–2388.

[7] J.-L. Lin, Interdecadal variability of ENSO in 21 IPCC AR4 coupled GCMs, Geophys. Res. Lett. 34 (2007) L12702.

[8] R. Pincus, C.P. Batstone, R.J.P. Hofmann, K.E. Taylor, P.J. Glecker, J. Geophys. Res. 113 (2008) D14209.

[9] G.L. Stephens, T. L'Ecuyer, R. Forbes, A. Gettlemen, J.-C. Golaz, A. Bodas-Salcedo, K. Suzuki, P. Gabriel, J. Haynes, J. Geophys. Res. 115 (2010) D24211.

[10] C.F. Ropelewski, M.S. Halpert, Mon. Weather Rev. 115 (1987) 1606–1626.

[11] B. Wang, Q. Ding, Dyn. Atmos. Oceans 44 (2008) 165–183.

[12] K.E. Trenberth, D.P. Stepaniak, J.M. Caron, J. Clim. 13 (2000) 3969–3993.

[13] E. Guilyardi, A. Wittenberg, A. Fedorov, M. Collins, C. Wang, A. Capotondi, G.J. van Oldenborgh, T. Stockdale, Bull. Am. Meteorol. Soc. 90 (2009) 325–340.

[14] F. Fan, M.E. Mann, S. Lee, J.L. Evans, J. Clim. 23 (2010) 5193–5205.

[15] V.V. Kharin, F.W. Zwiers, X. Zhang, G.C. Hegerl, J. Climate, 20 (2007) 1419–1444, http://dx.doi.org/10.1175/JCLI4066.1.

[16] J. Sillmann, V.V. Kharin, F.W. Zwiers, X. Zhang, D. Bronaugh, J. Geophys. Res. Atmos. 118 (2013) 2473–2493.

[17] E.L. Ujeneza, B.J. Abiodun, Clim. Dynam. 44 (2014) 1595–1609.

[18] M. Collins, R. Knutti, J. Arblaster, J.-L. Dufresne, T. Fichefet, P. Friedlingstein, X. Gao, W.J. Gutowski, T. Johns, G. Krinner, M. Shongwe, C. Tebaldi, A.J. Weaver, M. Wehner, 2013: Long-term climate change: projections, commitments and irreversibility, in: T.F. Stocker, D. Qin, G.-K. Plattner, M. Tignor, S.K. Allen, J. Boschung, A. Nauels, Y. Xia, V. Bex, P.M. Midgley (Eds.), Climate Change 2013: The Physical Science Basis. Contribution of Working Group I to the Fifth Assessment Report of the Intergovernmental Panel on Climate Change, Cambridge University Press, Cambridge, UK and New York, NY, USA, 2013, pp. 1029–1136.

[19] Core Writing Team, R.K. Pachauri, A. Reisinger (Eds.), Climate Change 2007: Synthesis Report. Contribution of Working Groups I, II and III to the Fourth Assessment Report of the Intergovernmental Panel on Climate Change, IPCC, Geneva, Switzerland, 2007, p. 104.

[20] R.H. Moss, J.A. Edmonds, K.A. Hibbard, M.R. Manning, S.K. Rose, D.P. van Vuuren, T.R. Carter, S. Emori, M. Kainuma, T. Kram, G.A. Meehl, J.F.B. Mitchell, N. Nakicenovic, K. Riahi, S.J. Smith, R.J. Stouffer, A.M. Thomson, J.P. Weyant, T.J. Wilbanks, Nature 463 (2010) 747–756.

[21] Climate Change 2013: Working Group I Contribution to the IPCC Fifth Assessment Report Climate Change 2013: The Physical Science Basis, Final Draft Underlying Scientific-Technical Assessment, 2013, p. 103.

[22] M.G. Donat, L.V. Alexander, H. Yang, I. Durre, R. Vose, R.J.H. Dunn, K.M. Willett, E. Aguilar, M. Brunet, J. Caesar, B. Hewitson, C. Jack, A.M.G. Klein Tank, A.C. Kruger, J. Marengo, T.C. Peterson, M. Renom, C. Oria Rojas, M. Rusticucci, J. Salinger, A.S. Elrayah, S.S. Sekele, A.K. Srivastava, B. Trewin, C. Villarroel, L.A. Vincent, P. Zhai, X. Zhang, S. Kitching, J. Geophys. Res. Atmos. 118 (2013) 2098–2118.

POSSIBLE ROLES IN CAUSING CLIMATE CHANGE

THE ROLE OF ATMOSPHERIC GASES

24

Richard P. Tuckett

School of Chemistry, University of Birmingham, Edgbaston, Birmingham, UK

CHAPTER OUTLINE

1. INTRODUCTION

I wrote the opening chapter for the first edition of *Climate Change: Observed Impacts on Planet Earth in 2008* about the role of atmospheric gases in global warming [1]. My opening sentence stated that *If the general public in the developed world is confused about what the greenhouse effect is, what the important greenhouse gases are, and whether greenhouse gases really are the predominant cause of the recent rise in temperature of the earth's atmosphere, it is hardly surprising.* Although the science was relatively mature, it seemed to me then that it was not possible to prove, in an absolute and scientific sense, any correlation between CO_2 global concentrations and average global temperatures, with the net result that mixed messages were being portrayed to the general public. Perhaps I was being too honest because, whilst there are some inconsistencies in the data, my belief was and remains that the rise in global temperature of the planet is genuine and that humankind is the predominant cause through contributing to enhanced concentrations of carbon dioxide by the burning of excess fossil fuels. Furthermore, the public did not then, and still does not understand the concept of error or uncertainty, and any chink in the scientific data was leapt upon by climate sceptics with the voracity of a hungry animal in the wild. However, I also made the point that people should not become obsessed with CO_2 concentrations because there are other gaseous components in the Earth's atmosphere, especially CH_4, which have the capacity to cause as much damage to the environment as CO_2; Shine has made this same point consistently in the literature [2]. I highlighted also the possible future problem of gaseous compounds that have exceptionally long lifetimes in the atmosphere, many hundreds if not thousands of years; perfluorinated compounds, such as CF_4, SF_6,

Climate Change. http://dx.doi.org/10.1016/B978-0-444-63524-2.00024-5

375

CF_3SF_5 and NF_3, fall into this category. Whilst their concentrations currently might be fairly small, they have the capacity to cause serious problems if their concentrations are allowed to increase in an unchecked manner.

In this second edition of *Climate Change*, I update the chapter, but with enhanced experience of leading talks and discussions on this topic to many schools, colleges, universities and voluntary organizations of intelligent lay people over the last 10 years. I pose throughout three slightly different questions to what I wrote about in 2008. First: has the basic science and knowledge base changed in the last six years? Second: have public perceptions about possible global warming and climate change changed, and if so, in what direction? Third: assuming that there is a global problem brewing in the near future, have direct actions or policies been or are being developed to counter the perceived threat? I conclude the chapter with what I perceive as some simple-to-implement, some difficult-to-implement and some incredibly complex issues that must be addressed by us all if this huge threat to civilization, as we know it on this planet, is to be controlled.

2. MYTHS, FACTS, LIES AND OPINIONS ABOUT THE GREENHOUSE EFFECT

I have come to the conclusion that there are several myths that have grown into the conscience of many of the general public on this subject. Like all myths, they need to be corrected as quickly as possible. The first is that the greenhouse effect is all 'bad news'. As I show in Section 3, nothing could be further from the truth, and without the greenhouse effect the average temperature of our planet would be that of winter in Siberia, i.e. down to $-17°C$ (256 K). Secondly, the greenhouse and ozone depletion, if not quite the same effect, have similar scientific explanations and causes. Whilst understandable up to a point because some chemicals, such as gaseous chlorofluorocarbon molecules, contribute to both effects, the basic science of the two effects is very different; furthermore, the former is a property of the troposphere (altitude $h = 0$ km–15 km) whilst the latter is a property of the stratosphere ($h = 15$ km–50 km). Thirdly, and perhaps the most serious issue, is that the large majority of the world's population regard *weather* and *climate* as the same phenomenon. This is not true. The former is a short-term phenomenon on which we base our daily actions; at its simplest and most banal, what is the weather forecast for tomorrow, so what clothes do we wear? The latter is a very long-term phenomenon, taking data from the past to model patterns for the future, the timescales being tens to hundreds of years in both cases. So it is nonsense to say, for example, that just because the winter of 2012–2013 in the UK was very cold, there is no problem ahead. Yet I hear this sort of prejudice from friends and neighbours with disturbing regularity.

What facts are indisputable? In decreasing order of certainty, first, I believe nobody can argue with the observation, made at many observation points around the world (e.g. Fig. 1), that average CO_2 concentrations are increasing slowly, year on year, and the value for 2014 was 400 parts per million by volume (ppmv; or 400 μmol mol^{-1}), the highest ever recorded; the value in preindustrial Britain was *ca.* 280 ppmv. Furthermore, the concentrations of other long-lived greenhouse gases such as CH_4 and N_2O are also increasing year on year, as reported on an annual basis since 2006 with great accuracy by the World Meteorological Organization Greenhouse Gas Bulletins [3]. Secondly, I believe the evidence is strong that average global temperatures are also rising; they have risen somewhere between $0.6°C$ and $1.0°C$ since the preindustrial era, but the certainty level in this statement is lower simply because of the greater uncertainty in the data. Thirdly, the region of the

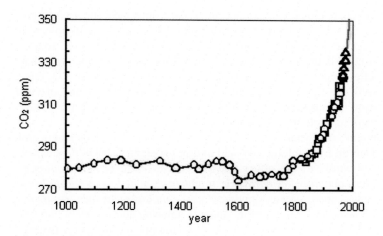

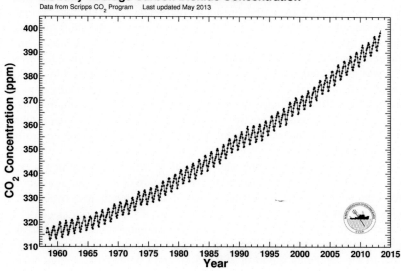

FIGURE 1

The increasing levels of CO_2 in the Earth's atmosphere. The upper picture comes from www.ems.psu.edu (with permission), showing the approximate constant concentration of CO_2 from AD 1000 to *ca*. AD 1750, the start of the Industrial Revolution, of about 280 µmol mol^{-1} (or 280 ppmv); there is a slow but consistent rise thereafter. The lower picture shows data for the last 55 years, recorded at the Mauna Loa Observatory in Hawaii, highlighting the relentless increase in concentration year by year. *(With permission from www.esrl. noaa.gov.)* The small oscillations every six months are caused by seasonal changes due to photosynthetic activity of vegetation, which consumes CO_2; the extent of the oscillation is reduced when the data are recorded in regions, such as Antarctica, with smaller amounts of vegetation. Note: the *y*-axis label should read: CO_2 concentration/µmol mol^{-1}.

Earth's atmosphere where global warming occurs is the troposphere, the first 15 km of altitude above the Earth's surface. Yet this is the region of the atmosphere where both homogeneous and heterogeneous processes take place, including reactions on aerosol surfaces, and I believe that the chemistry of this region is the least well understood. Fourthly, whilst the large majority of scientists in the world do believe that there is a correlation between the increasing CO_2 concentration and the very probable rise in the Earth's temperature, from a mathematical point of view it is difficult to *prove* the correlation because there is simply not the signal-to-noise ratio and/or resolution in the data. I discuss this point further in Section 3.

What are opinions and what are downright untruths in this increasingly political subject? It is rumoured that George Bush infamously said during his eight-year US presidency at the start of the twenty-first century words to the effect that *Global warming is not occurring. Even if it is, it is unrelated to man's activities on earth.* He might have added, *and certainly not us in the USA!* More comically, he was once quoted as saying, *I have opinions of my own, strong opinions, but I don't always agree with them.* Whilst this second quote is amusing and harmless, the first is verging on an untruth because nearly all the scientific evidence from the first decade of the twenty-first century does not support this statement. Yet, such opinions, coming from one of the most powerful and influential people in the world, can be dangerous if they turn out to be untrue because millions of North Americans will take their lead from the president of the day. In this way, prejudices are born. Conversely, on this side of the Atlantic Ocean, Professor David King, the chief scientific advisor to the UK Government, was making statements at the same time to the effect that *Global warming is the most serious phenomenon affecting the world's security and prosperity, more so than terrorism* [4]. King has since claimed he was misquoted and the statement was taken out of context, but the fact remains that, whoever said what or meant what, the US and the UK were taking diametrically opposite positions on this major area of public policy. Such divergence between two countries normally joined at the hip has been very rare in the last 50 years. The two countries are now led by Barack Obama and David Cameron, and they are exploring much more similar areas of policy for controlling greenhouse emissions. This must be welcomed.

What do scientists think? The huge majority working in different disciplines now do believe that the correlation between CO_2 concentrations and the temperature of the planet is as good as proven, and the temperature of the planet will rise from pre-Industrial levels anywhere between 2 K and 5 K by the end of the twenty-first century; the lower end of this range if CO_2 emissions can be stabilized then reduced, the higher end if no controls are put in place. The language of the Fifth Assessment Report of the United Nations Intergovernmental Panel on Climate Change (IPCC) from 2013 is that much stronger than that of the preceding report of 2007 [5,6], and I believe these huge documents have the support of at least 99% of the world's scientists. This was probably not the case for the earlier UN reports in the late twentieth century. A very small minority of scientists believe that, whilst the temperature of the planet may be increasing, this global warming is *not* due to humankind's activities since the Industrial Revolution but to a natural cycle of ice ages with warm periods in between; in other words, we are currently in a warm period between ice ages, and coincidentally this is happening at the same time as the global CO_2 concentration is increasing. This has been refuted by world experts – see Chapter 25. An even smaller number of people, who tend to be loud and articulate nonscientists, deny the existence of global warming and climate change at all. In a democracy everyone is entitled to their opinions, but in time I am convinced such people will be seen as the ultimate 'flat earthers' who will deny whatever evidence is presented to them.

3. ORIGIN OF THE GREENHOUSE EFFECT: 'PRIMARY' AND 'SECONDARY' EFFECTS

Much of this section is unchanged from the first edition because I contend that the basic science of the greenhouse effect has not changed. All that has changed are improvements in data relating to individual gases that contribute to the overall (secondary) greenhouse effect, and some new greenhouse gases have been discovered in the atmosphere since 2008.

The different regions of the Earth's atmosphere are shown in Fig. 2. The total gas pressure in the atmosphere decreases exponentially with altitude, h. We can write that $p_h = p_0 \exp(-h/h_0)$, with the scale height, h_0, having a value of $ca.$ 8500 m or 8.5 km. The temperature drops at a uniform rate through the troposphere ($0 < h < ca.$ 10 km) from $ca.$ 298 K to 210 K, and visible photochemistry (wavelength, $\lambda > ca.$ 390 nm) dominates in this region. A temperature inversion occurs over the next 40 km of altitude through the stratosphere ($ca.$ $10 < h < ca.$ 50 km), but the total gas pressure keeps dropping at an exponential rate. The temperature inversion leads to a very stable gas-phase environment, ozone depletion takes place in this region of the atmosphere, and ultraviolet (UV) photochemistry ($200 < \lambda < ca.$ 390 nm) dominates. The mesosphere and ionosphere lie above the stratosphere, where reactions of charged particles such as cations, anions and free electrons can be important; vacuum-UV photochemistry ($\lambda < ca.$ 200 nm), especially at the Lyman-α wavelength of 121.6 nm, can also be an important process in these two regions. The Earth is a planet in dynamic

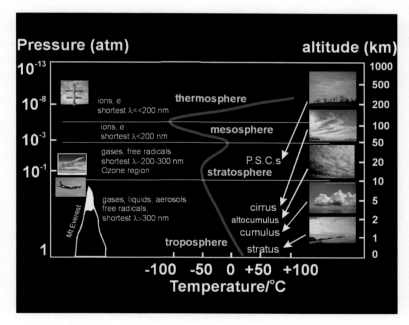

FIGURE 2

Different regions of the Earth's atmosphere. About 90% of the gases exist in the troposphere, of which $ca.$ 99% are either N_2 or O_2. Visible photochemistry dominates in the troposphere ($\lambda > ca.$ 390 nm), UV in the stratosphere ($200 < \lambda < ca.$ 390 nm), and vacuum-UV in the mesosphere and thermosphere/ionosphere ($\lambda < 200$ nm). Note that 1 atm of pressure is equivalent to 101 325 Pa. The left-hand y-axis label should read: Pressure/atm and the right-hand y-axis label should read Altitude/km.

equilibrium since it continually absorbs and emits electromagnetic radiation. As described above, it receives or absorbs vacuum-UV, UV and visible radiation from the sun, and photochemistry of gaseous molecules can occur in different regions of the atmosphere. To maintain energy balance and a constant temperature the Earth must emit electromagnetic radiation, which it does in the form of infrared radiation. By energy balance, 'energy in' must equal 'energy out', and this equality determines what the average temperature of the planet should be.

Both the sun and planet Earth are black body emitters of electromagnetic radiation. That is, they are bodies capable of emitting and absorbing all frequencies (or wavelengths) of electromagnetic radiation uniformly according to the laws of quantum physics. The distribution curve of emitted energy per unit time per unit area versus wavelength for a black body was determined by Planck in the first part of the twentieth century, and is shown pictorially in Fig. 3. Without mathematical detail, two points are relevant. Firstly, the total energy emitted per unit time integrated over all wavelengths is proportional to T^4, where the temperature has units of K. Secondly, the wavelength of the maximum in the emission distribution curve varies inversely with T, i.e. $\lambda_{max} \alpha \ T^{-1}$; these are Stefan's and Wien's laws, respectively. Comparing the black body curves of the sun and the Earth, the sun emits UV/visible radiation with a peak at *ca.* 500 nm characteristic of $T_{sun} = 5780$ K. The temperature of the Earth is a factor of 20 lower, so the Earth's black body emission curve peaks at a wavelength that is 20 times longer or *ca.* 10 μm. Thus the Earth emits infrared radiation with a range of wavelengths spanning *ca.* 4 μm–50 μm, with the majority of the emission being in the range 5 μm–25 μm (or 400 cm^{-1}–2000 cm^{-1}).

The solar flux energy intercepted per second by the Earth's surface from the sun's emission can be written as $F_s(1 - A)\pi R_e^2$, where F_s is the solar flux constant outside the Earth's atmosphere (1368 J s^{-1} m^{-2}), R_e is the radius of the Earth (6.38 × 10^6 m) and A is the Earth's albedo, corresponding to the reduction of incoming solar flux by absorption and scattering of radiation by aerosol particles (average value 0.28). The infrared energy emitted per second from the Earth's surface is $4\pi R_e^2 s T_e^4$, where s is Stefan's constant (5.67 × 10^{-8} J s^{-1} m^{-2} K^{-4}) and $4\pi R_e^2$ is the surface area of the Earth. At equilibrium, the temperature of the Earth, T_e, can be written as:

$$T_e = \left[\frac{F_s(1 - A)}{4s}\right]^{1/4} \tag{1}$$

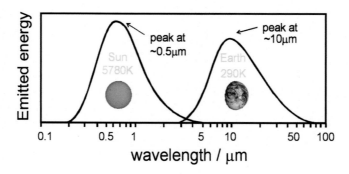

FIGURE 3

Black body emission curves from the sun ($T \sim 5780$ K) and the earth ($T \sim 290$ K), showing the operation of Wien's law that $\lambda_{max} \alpha \ (1/T)$. The two graphs are not to scale. *(With permission, and adapted from A.M. Holloway, R.P. Wayne, Atmospheric Chemistry, RSC Publishing, 2010.)*

Using the data above, the equation yields a value for T_e of *ca.* 256 K. Mercifully, the average temperature of the Earth is not a Siberian $-17°C$ (256 K), otherwise life would be a very unpleasant experience for the majority of humans on this planet. The one-quarter power in Eqn (1) means that any errors in the values of F_s and A are strongly diluted, so there is not much associated error in the value of T_e. The reason why our planet has a hospitable higher average value of *ca.* 290 K is the greenhouse effect. For thousands of years, absorption of some of the emitted infrared radiation by molecules in the Earth's atmosphere (mostly CO_2, O_3 and H_2O) has trapped this radiation from escaping out of the Earth's atmosphere (just as a garden greenhouse operates), some is reradiated back toward the Earth's surface, thereby causing an elevation of the temperature of the surface of the Earth. Thus, it is the greenhouse effect that has maintained our planet at this average temperature, and we should all be grateful. This phenomenon is often called the 'primary' greenhouse effect. It is therefore a complete fabrication of the truth, or a myth, to portray *all* aspects of the greenhouse effect as bad news, and it is the reverse that is true. A relatively simple calculation can show that about 30 K of the 34 K rise in temperature due to primary greenhouse gases is due to H_2O water vapour in the atmosphere, about 3 K is due to CO_2 and about 1 K is due to all the other primary greenhouse gases such as CH_4, N_2O and O_3. Thus the major greenhouse gas by a long way is H_2O vapour and not CO_2. It should also never be forgotten that 99% of the Earth's atmosphere is due to N_2 and O_2, and neither absorbs infrared radiation (see Section 4), so the greenhouse effect is all due to gases that comprise only *ca.* 1% of the Earth's atmosphere. Put another way, our atmosphere is very fragile and sensitive to perturbations in concentration of any of the trace species in the atmosphere that are greenhouse gases.

There is therefore nothing new about the primary greenhouse effect; it has been present for thousands of years, but it is only relatively recently that we have definitive scientific evidence for its presence. Figure 4 shows data from the Nimbus 4 satellite circumnavigating the Earth in 1979 at an altitude outside the Earth's atmosphere. The infrared emission spectrum in the range 6 μm–25 μm

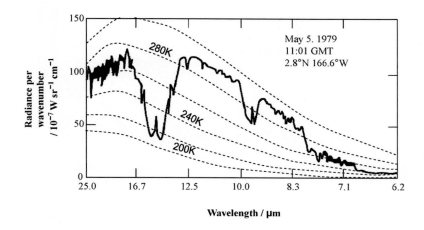

FIGURE 4

Infrared emission spectrum as observed by the Nimbus 4 satellite outside the Earth's atmosphere. Absorptions due to CO_2 between 12 μm–17 μm, O_3 (around 9.6 μm) and H_2O ($\lambda < 8$ μm) are shown. *(With permission from A.M. Holloway, R.P. Wayne, Atmospheric Chemistry, RSC Publishing, 2010; original from R.E. Dickinson, W.C. Clark (Eds.), Carbon Dioxide Review, 1982 (OUP).)*

escaping from Earth represents a black body emitter with a temperature of *ca.* 290 K, with absorptions (i.e. dips) in the radiance per wavenumber data between 12 μm–17 μm, around 9.6 μm, and $\lambda < 8$ μm. These wavelengths correspond to infrared absorption bands of CO_2, O_3 and H_2O, respectively, the three major primary greenhouse gases in the atmosphere.

Of course, the argument that the primary greenhouse gases have maintained our planet at a constant temperature of *ca.* 290 K presupposes that their concentrations have remained approximately constant over very long periods of time. As far as we know, this was the case for the primary greenhouse gases, certainly CO_2, up to the start of the Industrial Revolution, *ca.* 1750. However, concentrations of CO_2, CH_4 and, to a lesser extent, O_3 have increased significantly over the following 264 years, and it is the increases of these and newer greenhouse gases that have caused a 'secondary' greenhouse effect to occur over this time window, leading to the temperature rises that the majority of scientists believe we are experiencing today. (Although the concentration of H_2O vapour is much higher, it is not believed that it has changed significantly over the last 260 years, so H_2O is not classed as a secondary greenhouse gas.) That is the main argument of the proponents of the *greenhouse gases, mostly CO_2, equal global warming* school of thought. There is no doubt that the concentration of CO_2 in our atmosphere has risen from *ca.* 280 ppmv to current levels of *ca.* 393 ppmv over the last 264 years. (For the physical chemist, 1 ppmv is equivalent to a number density of 2.46×10^{13} molecules cm^{-3} for a pressure of 100 kPa and a temperature of 298 K). It is also not in doubt that the average temperature of our planet has risen by *ca.* 0.6 °C–1.0 °C over this same time window (Fig. 5, the famous *hockey stick*

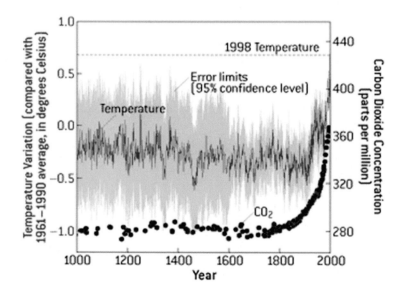

FIGURE 5

The average temperature of the Earth and the concentration level of CO_2 in the Earth's atmosphere during the last 1000 years. *(With permission from www.env.gov.bc.ca/air/climate/indicat/images/appendnhtemp.gif and www.env.gov.bc.ca/air/climate/indicat/images/appendCO2.gif.)* Note: The right-hand-side y-axis label should read: CO_2 concentration/μmol mol^{-1}; and the left-hand-side y-axis label should read: Temperature variation compared to the 1961–1990 average/°C.

graph, about which whole books have been written). In my opinion, what has not yet been proven in a mathematical and scientific sense is that there is a cause-and-effect correlation between these two facts, the main problem being that there is not sufficient structure or resolution with time in either the CO_2 concentration or the temperature data. Even more recent data of the last 100 years (Figure 6), where the correlation *seems* to be better established, will not convince the sceptic. Furthermore, if I was devil's advocate and people did not know what was being plotted on the two axes, I would contend that most, including many scientists, would say there was no clear correlation between the y and the x data in either Figs 5 or 6. That said, as demonstrated with increasing clarity by the recent IPCC reports, the consensus of world scientists, and certainly physical scientists, is that a strong correlation *does* exist even if it is not possible to prove it mathematically.

By contrast, an excellent example in atmospheric science of sufficient resolution being present to confirm a correlation between two sets of data was published in 1989. The concentrations of O_3 and the ClO free radical in the stratosphere were shown to have a strong anti-correlation effect when data were collected by an aircraft as a function of latitude in the Antarctic (Fig. 7) [7]. There was not only

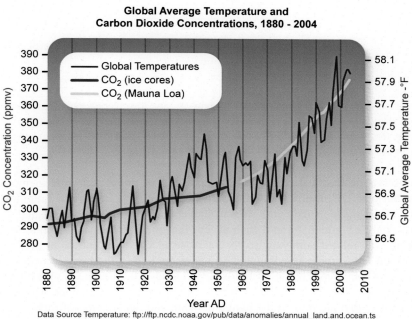

FIGURE 6

The average temperature of the Earth and the concentration level of CO_2 in the Earth's atmosphere during the 'recent' history of the last 100 years. *(With permission from the web sites shown in the figure.)* Note, The left-hand-side y-axis label should read: Concentration/μmol mol^{-1}; the right-hand-side y-axis label should read: Global average temperature/$^\circ$F. Here $t_F/^\circ F = (9/5)(t/^\circ C) + 32$.

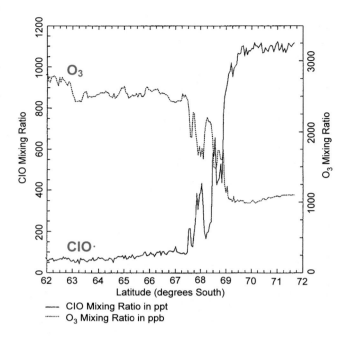

FIGURE 7

Clear anti-correlation between the concentrations of ozone, O_3, and the chlorine monoxide radical, $ClO\cdot$, in the stratosphere above the Antarctic during their spring season of 1987. *(With permission from Anderson et al., J. Geophys. Res. D 94 (1989) 11465.)* 1 ppb is equivalent to one part in 10^9, 1 ppt to one part in 10^{12}. Note the label on the *x*-axis should read: Latitude/degrees South.

the general observation that a decrease of O_3 concentration correlated with an increase in ClO concentration but also the resolution was sufficient to show that at certain latitudes dips in O_3 concentration corresponded exactly with rises in ClO concentration. When presented with the data, even the most doubting scientist could then accept that the decrease in O_3 concentration in the Antarctic spring was related somehow to the increase in ClO concentration, and over the next 20 years this result led to more research and an understanding of the heterogeneous chemistry of chlorine-containing compounds on polar stratospheric clouds. Unfortunately, such high resolution is not present in the data (e.g. Figs 5 and 6) for the 'CO_2 versus T' argument. This has led to the multitude of theories that are now in the public domain, and the creation of lobby groups on both sides of the argument.

I believe it would be very surprising if there was not some relationship between the rapid increases in CO_2 concentration and the temperature of the planet. In making this statement, the basic assumption remains that the firn measurements from ice core samples of CO_2 concentrations extrapolated back in time are accurate. Nevertheless, there are two aspects of Fig. 5 that remain unanswered by proponents of such a simple theory. First, the data suggest that the temperature of the Earth actually decreased between 1750 and *ca.* 1920 whilst the CO_2 concentration increased from 280 to *ca.* 310×10^{-6} over this time window. Second, the drop in temperature around 1480 AD in the 'little ice age' is not mirrored by a similar drop in CO_2 concentration. All that said, the apparent mirroring of increases in both CO_2 levels and T_e over the last 50 years is striking. The most likely explanation surely is that there are a multitude of effects, one of the dominant being the concentrations of greenhouse gases in the atmosphere, contributing to the temperature of the planet. At certain times of history, these effects have been 'in phase' (as now), at other times they may have been in 'anti-phase' and working against each other. That was my position when the first edition of *Climate Change* was published, and it has not changed.

4. THE PHYSICAL CHEMISTRY PROPERTIES OF GREENHOUSE GASES

The basic science of what constitutes an effective greenhouse gas has not changed in the last six years. What follows in this section is therefore only a summary of what I wrote in the first edition. The fundamental physical property of a greenhouse gas is that it must absorb infrared radiation via one or more of its vibrational modes in the infrared range of 5 μm–25 μm. Furthermore, since the primary greenhouse gases of CO_2, O_3 and H_2O absorb in the range 12 μm–17 μm (or 590 cm^{-1}–830 cm^{-1}), 9.6 μm (1040 cm^{-1}) and $\lambda < 8$ μm (>1250 cm^{-1}), an effective secondary greenhouse gas is one that absorbs infrared radiation strongly *outside* these ranges of wavelengths (or wavenumbers), yet *inside* the range of 5 μm–25 μm where infrared radiation is present. A molecular vibrational mode is only active in the infrared if the motion of the atoms generates a dipole moment. That is, $d\mu/dQ \neq 0$, where μ is an instantaneous dipole moment and Q a displacement coordinate representing the vibration of interest. This is the reason why neither N_2 nor O_2 absorbs infrared radiation since their sole vibrational mode is infrared inactive; they therefore play no part in the greenhouse effect. It is only trace gases in the atmosphere (Table 1) such as CO_2 (4×10^{-2})%, CH_4 (2×10^{-4})%, O_3 (3×10^{-6})%, chlorofluorocarbons such as CF_2Cl_2 (5×10^{-8})% and stable fluorinated molecules such as SF_6 (6×10^{-10})% that contribute to the greenhouse effect. I have already commented that the Earth's atmosphere is particularly fragile if only 1% of its molecules can have such a major effect on humans living on the planet. Furthermore, the most important molecular trace gas, CO_2, absorbs via its ν_2 bending vibrational mode at 667 cm^{-1} or 15.0 μm, which coincidentally is very close to the peak of the Earth's black body curve; the spectroscopic properties of CO_2 have not been particularly kind to the environment. Thus, infrared spectroscopy of gas-phase molecules, in particular at what wavelengths and how strongly a molecule absorbs such radiation, will clearly be important properties to determine how effective a trace pollutant will be as a greenhouse gas.

The second property of interest is the lifetime of the greenhouse gas in the Earth's atmosphere – clearly the longer the lifetime, the greater contribution a greenhouse gas will make to global warming. The main removal processes in the troposphere and stratosphere are reactions with OH free radicals and electronically excited oxygen atoms, $O*(^1D)$, and photo-dissociation in the range 200 nm–390 nm in the stratosphere or $\lambda > 390$ nm in the troposphere. Thus, the reaction kinetics of greenhouse gases with OH and $O*(^1D)$ and their photochemical properties in the UV/visible, will yield important parameters to determine their (deleterious) effectiveness. All these data can be incorporated into a dimensionless number, the global warming potential (GWP), sometimes called the greenhouse potential, of a greenhouse gas. All values are calibrated with respect to CO_2 whose GWP value is 1. A molecule with a large GWP is therefore one with strong infrared absorption in the windows where the primary greenhouse gases such as CO_2 etc. do not absorb, long lifetimes, and rising concentrations due to human presence on the planet. GWP values of some of the most

Table 1 Main Constituents of Ground-Level Clean Air in the Earth's Atmosphere

Molecule	x or %	μmol mol^{-1} (ppmv)[a] (2014)	μmol mol^{-1} (ppmv) (1750)
N_2	0.78 or 78%	780 900	780 900
O_2	0.21 or 21%	209 400	209 400
H_2O	0.03 (100% humidity, 298 K)	30 000	31 000
H_2O	0.01 (50% humidity, 298 K)	10 000	16 000
Ar	0.01 or 1%	9300	9300
CO_2	3.8×10^{-4} or 0.038%	393	280
Ne	1.8×10^{-5} or 0.002%	18	18
CH_4	1.77×10^{-6} or 0.0002%	1.80	0.72
N_2O	3.2×10^{-7} or 0.00003%	0.32	0.27
O_3[b]	3.4×10^{-8} or 0.000003%	0.034	0.025
All CFCs[c]	8.7×10^{-10} or 8.7×10^{-8}%	0.0009	0
All HCFCs[d]	1.9×10^{-10} or 1.9×10^{-8}%	0.0002	0
All PFCs[e]	8.3×10^{-11} or 8.3×10^{-9}%	0.00008	0
All HFCs[f]	6.1×10^{-11} or 6.1×10^{-9}%	0.00006	0

[a] *Parts per million by volume 1×10^{-6}; 1 ppmv is equivalent to a number density of 2.46×10^{13} molecules cm^{-3} for a pressure of 100 kPa and a temperature of 298 K.*
[b] *The concentration level of O_3 is very difficult to determine because it is poorly mixed in the troposphere. It shows large variation with both region and altitude.*
[c] *Chlorofluorocarbons (e.g. CF_2Cl_2).*
[d] *Hydrochlorofluorocarbons (e.g. $CHClF_2$).*
[e] *Perfluorinated molecules (e.g. CF_4, C_2F_6, SF_5CF_3, SF_6, NF_3).*
[f] *Hydrofluorocarbons (e.g. CH_3CF_3).*

important secondary greenhouse gases are given in the bottom row of Table 2. Note that CO_2 has the lowest GWP value of the seven greenhouse gases shown.

Information in the previous two paragraphs is described in qualitative terms. The data can be quantified and a mathematical description is now presented. The term that characterizes the infrared absorption properties of a greenhouse gas is the *radiative efficiency*, a_o. It measures the strength of the absorption bands of the greenhouse gas, x, integrated over the infrared black body region of *ca.* 400 cm^{-1}–2000 cm^{-1}. It is a (per molecule) microscopic property and is usually expressed in the strange units of W m^{-2} ppbv^{-1} (where ppbv refers to parts per 10^9 by volume). If this value is multiplied by the change in concentration of pollutant over a defined time window, usually the 264 years from the start of the Industrial Revolution to the current day, the macroscopic radiative forcing in units of W m^{-2} is obtained. Note that a pollutant whose concentration has not changed over this long time window, such as H_2O, will have a macroscopic radiative forcing of zero. The IPCC 2013 Assessment Report quotes the radiative forcing for CO_2 and CH_4, the two most serious secondary greenhouse gases, as 1.82 ± 0.19 and 0.48 ± 0.05 W m^{-2}, respectively, out of a total for long-lived greenhouse gases of 2.83 ± 0.29 W m^{-2} [5]. (The values in IPCC 2007 were 1.66 W m^{-2}, 0.48 W m^{-2} and 2.63 W m^{-2} [6].) These two

Table 2 Examples of Secondary Greenhouse Gases, and Their Contributions to Global Warming [5,11]

Greenhouse Gas	CO_2	O_3	CH_4	N_2O	CF_2Cl_2 [all CFCs]	SF_6	SF_5CF_3	NF_3
Concentration (2014)/μmol mol^{-1} or/ppmv	393	0.036	1.80	0.32	0.0005 [0.0009]	7.3×10^{-6}	ca. 1.6×10^{-7}	9×10^{-7}
ΔConcentration (1748—2014)/μmol mol^{-1} or/ppmv	113	0.011	1.08	0.05	0.0005 [0.0009]	7.3×10^{-6}	ca. 1.6×10^{-7}	9×10^{-7}
Radiative efficiency, a_o/W m^{-2} ppbv^{-1}	1.37×10^{-5}	3.33×10^{-2}	3.63×10^{-4}	3.00×10^{-3}	0.32 [0.20—0.32]	0.57	0.59	0.20
Total radiative forcing[a]/W m^{-2}	1.82	ca. 0.35[b]	0.48	0.17	0.17 [0.26]	4.1×10^{-3}	ca. 9.4×10^{-5}	2.0×10^{-4}
Contribution from long-lived greenhouse gases, excluding ozone, to overall greenhouse effect/%[c]	64 (57)	0 (11)	17 (15)	6 (5)	6 [9] (5 [8])	0.14 (0.13)	0.003 (0.003)	0.007 (0.006)
Lifetime, τ[d]/a	ca. 50—200[e]	ca. days—weeks[f]	12.4	121	100 [45—1020]	3200	800	500
Global warming potential (100 year projection)	1	—[g]	28	265	10200 [4660—13 900]	23 500	17 400	16 100

Here ppmv is identical to μmol mol^{-1} and ppbv is identical to nmol mol^{-1}.

[a]Due to change in concentration of long-lived greenhouse gas from the preindustrial era to the present time.

[b]An estimated positive radiative forcing of 0.40 W m^{-2} in the troposphere is partially cancelled by a negative forcing of 0.05 W m^{-2} in the stratosphere [5].

[c]Assumes the latest value for the total radiative forcing of 2.83 ± 0.29 W m^{-2} [5]. The values in brackets show the percentage contributions when the estimated radiative forcing for ozone is included in the value for the total radiative forcing.

[d]Assumes a single-exponential decay for removal of greenhouse gas from the atmosphere.

[e]CO_2 does not show a single-exponential decay [9].

[f]O_3 is poorly mixed in the troposphere, so a single value for the lifetime is difficult to estimate. It is removed by the reaction $OH + O_3 \rightarrow HO_2 + O_2$. Its concentration shows large variations both with region and altitude.

[g]GWP values are generally not applied to short-lived, i.e. unmixed, pollutants in the atmosphere, due to serious inhomogeneous changes in their concentration.

molecules therefore contribute 81% in total (64% and 17%, individually) to the global warming effect. Effectively, the radiative forcing value of a greenhouse gas gives a current-day estimate of how serious it is to the environment, but one should appreciate that the value uses concentration data *from the past*.

Looking into the future, the overall effect of one molecule of pollutant on the Earth's climate is described by its GWP value. It measures the radiative forcing, A_x, of a pulse emission of the greenhouse gas over a defined time period, t, usually 100 years, relative to the time-integrated radiative forcing of a pulse emission of an equal mass of CO_2:

$$GWP_x(t) = \frac{\int_0^t A_x(t).dt}{\int_0^t A_{CO_2}(t).dt} \tag{2}$$

The GWP value therefore informs how important one molecule of pollutant x is to global warming via the greenhouse effect compared to one molecule of CO_2, which is defined to have a GWP value of unity. It is an attempt to project into the future how serious the presence of a long-lived greenhouse gas will be in the atmosphere. Thus, when the media state that CH_4 is 28 times as serious as CO_2 for global warming, what they are saying is that the GWP_{100} value of CH_4 is 28; one molecule of CH_4 is therefore expected to cause 28 times as much 'damage' as one molecule of CO_2. Within the approximation that the greenhouse gas, x, follows a single-exponential time decay in the atmosphere, it is possible to parameterize Eqn (2) to give an exact analytical expression for the GWP of x over a time period t [8]:

$$\frac{GWP_x(t)}{GWP_{CO_2}(t)} = \frac{MW_{CO_2}}{MW_x} \cdot \frac{a_{0,x}}{a_{0,CO_2}} \cdot \frac{\tau_x}{K_{CO_2}} \cdot \left[1 - \exp\left(\frac{-t}{\tau_x}\right) \right] \tag{3}$$

In this simple form, the GWP only incorporates values for the radiative efficiency of greenhouse gases x and CO_2, $a_{0,x}$ and a_{0,CO_2}; the molecular weights of x and CO_2; the lifetime of x in the atmosphere, τ_x; the time period into the future over which the effect of the pollutant is determined; and the constant K_{CO_2}, a measure of the non-single value of the lifetime of CO_2, which can be calculated for any value of t [9]. It can be seen that the GWP value scales with both the lifetime and the microscopic radiative forcing of the greenhouse gas, but it remains a microscopic property of one molecule of the pollutant. The *recent* rate of increase in concentration of a pollutant (e.g. the rise in concentration per annum over the last decade), one of the factors of most concern to policymakers, does not contribute directly to the GWP value. This and other factors have caused criticism by Shine et al. of the use of GWPs in policy formulation [9]. Note that a similar equation to (3) given by Mason et al. has numerous typographical errors and should be disregarded [10].

Data for eight greenhouse gases are shown in Table 2. (Although H_2O vapour is the most abundant greenhouse gas in the atmosphere, it is neither long-lived nor well mixed: concentrations range from 0%–3% (i.e. 0–30 000 $\times$ 10^{-6} by volume) over different parts of the Earth, and the average lifetime is only a few days. Since its average global concentration has not changed significantly since the middle of the eighteenth century, it has zero radiative forcing and is not included in this table.) CO_2 and O_3 are naturally occurring greenhouse gases whose concentration levels ideally would have remained constant at pre-Industrial Revolution levels. The a_o value of O_3 is over three orders of magnitude greater than that of CO_2, but its tropospheric concentration is four orders of magnitude lower. CH_4 and N_2O constitute naturally occurring greenhouse gases with a_o values intermediate between that of CO_2 and O_3. The CH_4 concentration, although small, has increased by *ca.* 150% since preindustrial times. After CO_2, it is the second most important greenhouse gas, and its current total radiative forcing is *ca.* 26%

that of CO_2. N_2O concentration has increased only by *ca.* 18% over this same time period. It has the fourth highest total radiative forcing of all the naturally occurring greenhouse gases, following CO_2, CH_4 and O_3. Dichlorofluoromethane, CF_2Cl_2, is one of the most common of chlorofluorocarbons. These are man-made, anthropogenic chemicals that have grown in concentration from zero in pre-industrial times to a current total concentration of 0.9 ppbv; 1 ppbv is equivalent to a number density of 2.46×10^{10} molecules cm^{-3} at a pressure of 100 kPa and a temperature of 298 K. Their concentration is now decreasing due to the 1987 Montreal and later international protocols introduced to halt the destruction of stratospheric ozone. It is ironic that these decisions were taken with no regard to their (beneficial) effect on the issue of global warming. SF_6 and SF_5CF_3 are two long-lived halocarbons with currently very low concentration levels, but with high annual percentage increases and exceptionally long lifetimes in the atmosphere. They have very high a_o and GWP values, essentially because of their large number of strong infrared-active vibrational modes and their long lifetimes. NF_3 is a long-lived fluorinated compound that was discovered in the atmosphere since the first edition of this book was written [12]. Its properties are shown in the last column of Table 2. It was not included in the Kyoto Protocols of 1997 listing which greenhouse gases should be included for long-term monitoring [13], but there is now near-universal agreement that it should be included in future protocols from follow-up meetings, e.g. Paris in 2015.

It is noted that CO_2 and CH_4 have the lowest GWP values of all the greenhouse gases listed. Why, then, is there such concern about levels of CO_2 in the atmosphere and, with the possible exception of CH_4, no other greenhouse gas is mentioned in the media? The answer is that the overall contribution of a pollutant to the greenhouse effect, present *and* future, involves a convolution of its concentration with the GWP value. Thus CO_2 and CH_4 currently contribute most to the greenhouse effect (third bottom row of Table 2) simply due to their large change in atmospheric concentration since the Industrial Revolution; note, however, that the a_o and GWP values of both gases are relatively low. By contrast, the SF_5CF_3 molecule has the highest microscopic radiative efficiency of any known greenhouse gas (earning it the title 'super' greenhouse gas [8,14]), even higher than that of SF_6. SF_6 is an anthropogenic chemical used extensively as a dielectric insulator in high-voltage industrial applications, and the variations of concentration levels of SF_6 and SF_5CF_3 with time in the last 50 years have tracked each other closely [15]. The GWP of these two molecules is very high, SF_6 being slightly higher because its atmospheric lifetime, *ca.* 3200 years [16], is about four times greater than that of SF_5CF_3. However, the contribution of these two molecules to the overall greenhouse effect is still very small because their atmospheric concentrations, despite rising rapidly at the rate of *ca.* 6%–7% per annum, are still very low, at the level of parts per 10^{12} (trillion) by volume; 1 pptv is equivalent to a number density of 2.46×10^7 molecules cm^{-3} at 100 kPa and 298 K.

In conclusion, the *macroscopic* properties of greenhouse gases, such as their method of production, their concentration and their annual rate of increase or decrease, are mainly controlled by environmental and sociological factors such as industrial and agricultural methods – ultimately, I believe, population levels on the planet (see Section 5.3). The *microscopic* properties of these compounds, however, are controlled by factors that students worldwide learn about in science degree courses: infrared spectroscopy, reaction kinetics and photochemistry. Data from such lab-based studies determine values for two of the most important parameters for determining the effectiveness of a greenhouse gas: the microscopic radiative efficiency, a_o, and the atmospheric lifetime, τ.

In the first edition of *Climate Change*, a major section of the chapter described how the lifetime of a greenhouse gas was defined, particularly for a long-lived pollutant such as a perfluorocarbon molecule where lifetimes are quoted as several hundreds to thousands of years. In 1994, Ravishankara and Lovejoy made the bold statement that 'all long-lived molecules should be considered guilty [on their potential impact on the earth's atmosphere] until proven otherwise' [17]. Their example to justify this policy was that of anthropogenic chlorofluorocarbon molecules, produced in increasingly large quantities from the 1930s for the next four decades for industrial and domestic purposes when these molecules were thought to be innocuous. The pioneering work of Molina and Rowland in the 1960s and 1970s showed that these long-lived molecules with lifetimes of several hundreds of years were unfortunately having an unforeseen deleterious effect on the ozone layer in the stratosphere [18], and ultimately led to the Montreal Protocol of 1989 and the phasing out of these molecules from production [19]. In many ways, this was a wonderful example of the power of science and scientists to convince politicians that action was needed, and the latter responded accordingly. The latest predictions are that the ozone layer in the stratosphere will recover to its levels of *ca.* 1950 within the next 50 years–100 years, and the problem created by these molecules will have been reversed [20a–20c]. Many scientists now believe that the issue of carbon dioxide concentrations and global temperature is the modern-day scientific equivalent, but are thinking words to the effect *if only the problem was so easy to solve with carbon dioxide and global warming.*

In the final section of this chapter, I move away from the science of greenhouse gases and global warming. I contend that the scientific case is now so strong and accepted by 99%+ of the world's scientists that, as a population of *ca.* 7 billion people living on this planet, we must take ownership of this issue and come up with potential solutions. Of course, there are still inconsistencies in some of the data, and some aspects of the infamous hockey stick graph (Fig. 5) are unexplained. But this should not blinker us to the major environmental issue that needs to be addressed.

5. HAS ANYTHING CHANGED IN THE LAST DECADE?

This is a deliberately provocative question to ask. I answer it in three sections, as it is applied to (1) the science of greenhouse gases and possible global warming, (2) public perceptions of greenhouse gases etc. and (3) action at the private and the political level.

5.1 HAS THE SCIENCE CHANGED?

The simple answer must be no. The data presented in the two tables are taken, in the main, from the IPCC Fifth Assessment Report of 2013. Some of the radiative efficiencies, radiative forcing, lifetimes and GWP values have improved and become more accurate with lower quoted errors. A few new long-lived greenhouse gases, such as NF_3 and perfluorotributylamine, $N(C_4F_9)_3$, have been discovered in the atmosphere [12,21]. But the essential message remains that the net radiative forcing of the atmosphere due to long-lived greenhouse gases is increasing slowly (2.43 ± 0.24 W m^{-2} from the third United Nations IPCC report of 1998, 2.63 ± 0.25 W m^{-2} from the fourth report of 2007, and now 2.83 ± 0.29 W m^{-2} from the fifth report of 2013), with the predominant contributors being CO_2 (*ca.* 60%) and CH_4 (*ca.* 18%). Whilst all other long-lived greenhouse gases contribute a not-insignificant *ca.* 20% and keep atmospheric scientists busy with requests for more money to study their properties, the radiative forcing budget is dominated by CO_2 and CH_4 emissions. Furthermore, the increasing concentration of CO_2 in the atmosphere shows no sign of slowing down, with the current value very close to the emotive level of

400×10^{-6} by volume (400 ppmv). There is nothing special about this number per se, but the general view of climate scientists is that if this value gets close to 500×10^{-6} (500 ppmv) by volume, the Earth's atmosphere will have reached the point of 'no return', and it will be close to impossible to stabilize the temperature of the planet; this is often referred to as the *runaway greenhouse effect*, caused by positive feedback of increasing temperature causing increasing concentrations of water vapour, which cause ever-increasing temperature rises. Put in starker terms that nonscientists may find easier to understand, the concentration of CO_2 in the atmosphere has already increased over halfway from pre-Industrial Revolution times, *ca.* 280 ppmv, to the level, *ca.* 500 ppmv, that will have major consequences on the way the huge majority of us can live on this planet. However, modellers also predict that if the CO_2 concentrations can be stabilized at the current levels of *ca.* 400 ppmv by 2020, then reduced to less than half of 1990 levels by 2050, and continue to be cut thereafter, then the rise in temperature of the Earth from preindustrial times to the end of the twenty-first century may be limited to around 2 K. It is generally believed that most countries should be able to adapt to cope with this increase. Anything higher and the future will become increasingly bleak, which is a sobering thought.

5.2 HAVE PUBLIC PERCEPTIONS OF GREENHOUSE GASES, ETC. CHANGED?

This is a more difficult question to answer, and any response must be subjective. But my general view is that this *is* an issue that is getting into the psychology of the general public, even if opinions can swing with alarming rapidity. The international convention in Copenhagen, Denmark, in 2009 attracted huge publicity worldwide, even if it did not result in much tangible action [22]. However, my overriding impression from television remains that of President Obama jetting into the country in Air Force 1, the US presidential aircraft, and jetting out 24 h later, rather missing the point that air travel is a major component of carbon emissions. The power of the Internet increases exponentially with time and is now one major component of attracting multimillion petitions putting pressure on national governments to act. Unfortunately, bad publicity can halt any positive momentum that has been built up, so reports in 2010 that the University of East Anglia in the UK might have suppressed emails and reports suggesting that the issues around global warming were not as serious as was being made out, were hugely damaging. This was then reflected in opinion polls immediately afterward that global warming was now not a major issue of concern to the individual or the state, thereby reversing the trend that polls had been showing post-Copenhagen.

At a national level, Europe is leading the way and the UK has much to be proud of. For example, the UK government of the day in 2008 legislated to commit the country to a target of reducing greenhouse gas emissions by the year 2050 to less than 20% of the levels they were at in 1990, with an interim target of reducing CO_2 emissions by 2020 to less than 74% of the level in 1990 [23]. Targets are all very well and good, but nobody seems to have said what will happen if this target is not met 36 years from now, or it is clear by, say 2025, that the target will not be met. In October 2014, the European Union Council of Ministers, of which the UK is one of 27 members, committed to reduce CO_2 emission to less than 60% of 1990 levels by the year 2030, and to produce at least 27% of its energy from renewable sources and not from fossil-based fuels [24]. The former target is not dissimilar to that enshrined in UK law in 2008, but it applies to a much larger population and countries with a range of economies so its impact should not be dismissed. The EU hopes that this will lead to the same reduction as legislated by the UK, i.e. a reduction by 2050 to less than 20% of the levels in 1990. The quoting of statistics in such percentage terms can seem rather bland and is at times not very useful. These commitments are therefore expressed slightly differently in units of *metric tonnes of CO_2 emitted per person per year*. This is not exactly the most SI of units, but it is convenient to use because the absolute values are finite, involve no large powers of 10, and should therefore be understandable by the public. In these units the UK average at the moment is about 14, the US average about 23, and the global average about 6. The UK's target for 2050 is about 3 and the EU's target for 2030 is about 4. Modellers predict that to avoid the worst effects of climate change in the next 100 years, i.e. to limit the increase in global temperatures to less than 2 K above that in preindustrial times, the present global average emission must drop from about 6 to below 1.5. *By any standards, these are huge changes.* In November 2014 President Obama finally committed the USA to a major reduction of greenhouse gases, but only time will tell whether his policies can be worked through to positive action given that for the last two years of his presidency the US Senate will be controlled by the opposition Republican party. The two largest populations in the world, China and India, have yet to declare any binding targets. This is sad, because the overriding reaction I hear from Europeans who think about these issues is words to the effect: *what difference will anything I do as an individual make to this global problem, when China's emissions [especially] are so huge compared to those of Europe?* I believe that a binding target set by these countries would help Europeans believe that they were taking the issue seriously and might help individuals in Europe do more themselves. One should not forget that it is the rich and industrialized first-world countries in the West that have, in the main, created the problem.

But perhaps most encouraging for the future, the mantra of the developed world for the last 50+ years has surely been that we must maximize *growth*, however that is defined, in all countries; only then will we prosper. For the first time, in the last decade when the science of global warming has become increasingly robust, I believe that many influential people are publically challenging this premise. Such people are asking *why* is growth the paramount factor if it is leading to a planet that will be a very unpleasant place to live within the next 100 years? This is almost a heretical view to take of criteria we should use (or *not* use) to define our position in the global world, and it turns the world of politics and economics on its head. I contend, however, that potential global warming is the *ultimate* global issue simply because it has the potential to affect every person on this planet. Therefore, it is right to think anew and, if necessary, challenge the criteria on which countries have based their policies and lifestyles in the past.

5.3 WHAT ACTIONS HAVE BEEN TAKEN AT THE PRIVATE AND THE POLITICAL LEVEL?

Here my perception is that the message is mixed. An outpouring of guilt certainly will get us nowhere. To counter such negativity, many examples of excellent practice are emerging at an individual level, certainly in the UK. For example, conservation of energy through double glazing and roof insulation of housing, generation of solar electricity through roof-mounted photovoltaic panels, the trend to driving smaller and more fuel-efficient cars (and perhaps electric cars will be the norm in 50 years' time), and the increase in bicycling and walking as a long-term lifestyle change that healthy people should be making are just four examples. But I suspect that these examples are just scratching at the scale of the problem, and almost inevitably it is the educated 'converted' class who are taking these actions. I do believe that national policies must be imposed, and although it goes against the instincts of politicians of all colours to *tell people how to live their lives*, I fear that this is exactly what they must do. And because this is an issue with the potential to affect the lives of every person on this planet, unlike any other world phenomenon in my lifetime except perhaps the Cuba crisis and the possible outbreak of nuclear war in 1961, global solutions are needed and the normal 'rules' of economics cannot apply. So I fail to see how any attempts to trade in carbon, i.e. the 'transfer' through payment of emissions to other countries, can possibly succeed. It is a short-term solution of dubious morality to a long-term global problem, and is doomed to fail.

Others in this book may write much more knowledgeably than I can about possible ways to (1) change our energy policy to become less reliant on the burning of fossil fuels, (2) ways to trap emissions of greenhouse gases, and possibly (3) engage in geoengineering to reduce incoming radiation from the sun as a means to control our increasing temperature. Under category (1) must fall a renaissance in nuclear energy, and possibly the huge expansion of fracking, which is the release of shale gas reserves from deep within rocks. Whenever the former policy seems to be gaining favour, a serious accident, such as that at Fukushima in northern Japan in 2011, can set the clock back by at least a decade. Soon after this event, Germany changed dramatically to a nuclear-free energy policy, and the UK has not yet committed to a big expansion in this technology that seemed likely in the preceding 10 years before 2011. The risks involved in following the latter policy of fracking are significant, if only because of the huge increases in methane gas in the atmosphere that are likely to happen; nothing can change the basic science that one molecule of CH_4 has the potential to cause about 28 times as much damage to the world's climate as one molecule of CO_2 over the next 100 years. However, the potential benefits are considerable. Categories (2) and (3) can be interpreted as possible solutions to a problem that has been allowed to develop unchecked. I believe it is more sensible to follow the advice of Ravishankara and Lovejoy [17], and reduce the amounts of emissions of damaging greenhouse gases into the atmosphere in the first place. In simple terms, use less energy.

I divide possible solutions into three sections: (1) relatively easy to implement, however painful, (2) much harder to implement, but surely possible if the world is serious about this issue, and (3) incredibly complex world issues that must be addressed, probably by the United Nations.

5.3.1 Easy to Implement and Solve

Six years ago, I wrote that nobody wants to or can turn back the clock on scientific progress [1]. The challenge therefore to reducing our dependence on fossil fuels and save energy is to devise policies that may seem retrospective but do not reduce the standard of living of the population and negate all the benefits that technology has brought us in the last 200 years. An excellent book, *Sustainable Energy – Without the Hot Air* written by MacKay in 2009, available free on the Internet, shows where the UK emissions come from at a personal level [25]. On average, every person in the UK uses 125 kWh of energy per day, after metric tonnes of CO_2 per person per year the most commonly used unit of carbon currency. It is surely stating the obvious that any policies advocated cannot possibly apply to every person in a developed country, and in general the young, the old, the disabled and the infirm will be exempted. That said, MacKay estimates that wearing more clothes and turning down thermostats by a degree or two both at home and work might reduce this figure by about 20 such units; stopping flying might cause a reduction of 35; generally modifying our means of transport within the UK by driving less and biking or walking more might reduce this figure by about 20; avoiding packaging and the buying of clutter, however that is defined, might, to my surprise, cause a huge reduction of 20; and becoming vegetarian might cause a reduction of 10 kWh per person per day. These are all big percentage changes, even though one accepts that there are huge errors in the numbers estimated. It is a reasonable question, however, to ask which of these could be turned into UK national policies, with exceptions built in for vulnerable groups.

In the first edition of *Climate Change*, I advocated a reversal of the 1991 change of legislation in the UK that allowed for Sunday trading for 6 h d^{-1} (where d refers to day) in all shops. I noted that Sunday closing was still the law in Switzerland, but it remains one of the richest European countries by whatever criterion is applied. It is accepted that lifestyles in the UK would have to adapt from what many of us have become used to in the last 23 years, but I believe this is one of the easier national policies to make and implement. It would presumably reduce the energy consumption of UK shops by about one-seventh or 14%. I also advocated a reduction in domestic air travel within a small country such as the UK, with a corresponding increase and investment in rail travel. That is indeed what is happening with the proposed High Speed 2 (and possibly HS3) train routes from London to the north of England and possibly Scotland. Only time will tell if this leads to a big reduction in domestic air travel. I also cannot be the only person to question whether so much international long-distance air travel for business meetings is really necessary, and whether technology can assist. The concept of Skype for 1:1 face-to-face discussions can surely be extended so that 100 people can meet remotely without the need for travel, and indeed early versions of such software (e.g. Visimeet) are now available [26].

Many other policies could be rolled out quickly. Two examples might be free insulation of roofs and double glazing in *all* domestic housing, and a huge investment in cycle routes to make the bicycle a safer and more child-friendly means of local transport. It also seems reasonable to ask whether the minimum working temperature for employees could be reduced from its current level of 16°C (289 K) by a degree or two; often the working temperature in offices is higher. MacKay estimates the savings in energy could be substantial [25], and he is effectively asking whether it is necessary to live in shirtsleeves and the female equivalent for most of our waking hours. Has health and safety legislation become so sacrosanct that external packaging on much food sold in the supermarkets is quite un-necessary, leading to what MacKay calls excess 'clutter'? In the first few years of this century, the UK government announced that civil servants in all the major departments of the state would ensure that future legislation was checked for its impact on the environment. I see little evidence that this has happened. It is also a pity that this policy did not extend back to legislation passed in the last 30 years–50 years. For example, I do not believe that anyone thought of the environmental effects of allowing a free-for-all when the provision of compulsory state education was deregulated by the 1988 Education Act. This decision effectively led to the abolishment of local catchment areas for schools in the cities to which pupils walked, and the 'school run' by car became part of the UK vocabulary. The present UK government of a different colour, when elected in 2010, claimed to become one of the greenest ever elected. Unfortunately, I see little evidence that this claim from either can be substan-tiated. By my criteria, the above changes in the last two paragraphs are relatively easy to make, they may lose a few votes and be painful for some individuals, but ultimately I believe that they have to be made if we wish to control carbon emissions.

5.3.2 Moderately Difficult to Implement

The unit of carbon emission that everyone would understand is the cost to their financial pocket. Perhaps all developed countries should move rapidly to a system of taxation whereby the principle of *the polluter pays* becomes paramount. In our technological world, this could mean that everyone has a carbon credit card, paying, presumably, a premium for excess gas and electricity domestic bills, for petrol and use of the roads, and certainly for air travel. Again, this was an idea proposed by the Labour Government in the UK about 10 years ago, but was dropped when it was decided that it would be too difficult to implement and public reaction was negative, to say the least, from the day it was tentatively suggested. The prime minister of the day infamously said that climate change was not going to be solved by everyone stopping flying, a statement that might have won him support from many newspapers but caused many scientists to raise eyebrows in some exasperation.

A different issue concerns food production, what we eat, and where the food comes from. The more anyone looks at the issue of the supply chain, and especially the huge number of miles that food often travels between source and consumption, the more baffled they become. The organic movement and localism have started to address these issues at a microscopic level, but I sense that their contributions will be negligible unless the large multinational supermarkets embrace these concepts. For the last 50 years, I would suggest that the concept of the individual/customer having the paramount right to eat food at the cheapest price has swept aside any environmental consequences, and I believe that this has to change over the next 50 years. We will almost certainly have to pay more for food in real terms, but that is the price to pay for addressing the issue of excess use of fossil fuels for unnecessary travel. One should then address what we eat. Cattle use a lot of our limited land for grazing, and there is an argument that we should reduce meat consumption, if not become vegetarians of whatever strictness – a policy effectively being advocated by MacKay [25] – thereby also reducing methane emissions. The growth of genetically modified crops must surely be allowed to continue, as I do not believe that the perceived risks have materialized. We may then reduce our dependency on cattle as a source of food.

These are two areas of policy that apply to all developed countries, and different places will implement different means to tackle them. On my scale, however, these qualify as moderately difficult issues to solve, but such painful projects must be addressed if the planet is going to be a pleasant place to live for the majority of its population.

5.3.3 Very Difficult to Solve

There is one overriding issue that dominates all else in this category and that is the population of the planet. The figures are stark. Fifty years ago the population was 3.3 billion (3.3×10^9), today it is 7.3 billion, and it is predicted to rise close to 11 billion by 2100, with the large majority of this expected growth occurring in Africa [27]. Whilst currently 75% of the world's population live in Asia and Africa, that figure is predicted to grow to 82% by the end of the century. Conversely, the population of Europe is predicted to fall from its current level of 11% (or 0.74 billion) to 6% (or 0.64 billion). All of these people will need to be housed and fed. CO_2 and CH_4 currently contribute about 81% of the total radiative forcing of long-lived secondary greenhouse gases (Table 2), but I believe that it is too simplistic to say that control and reduction of CO_2 levels will be the complete solution. It is my personal belief that CO_2 levels in the atmosphere correlate loosely with lifestyle of many of the population, and with serious effort, especially in the developed world, huge reductions are possible; examples are given in Sections 5.3.1 and 5.3.2 above. CH_4 levels, however, in my opinion pose just as serious a threat to our planet as CO_2 simply because they will be much harder to reduce. Whilst it remains surprising and unclear why the radiative forcing of methane, 0.48 W m^{-2}, has been unchanged over the last two decades [5], a major component of methane emissions correlates *strongly* with the number of animal livestock, which itself is dependent on the population of the planet.

If this point is accepted, then population control on a worldwide scale must be openly discussed and the subject cannot be avoided if we are to control methane emissions. This is one area of policy where even the most outspoken of politicians in any country is reticent to go, and apparently simple policies can often have unforeseen consequences decades later. For example, China introduced its one-child-per-family policy over 30 years ago; it has had some of the desired effects on population levels in their large cities, but there are social consequences that are only now beginning to manifest themselves; for example, who will look after the old as their population ages? The message, if there is one, coming from Western countries, is mixed. Europe and the US have always believed in the absolute right of individuals to make this choice independent of the state, but it is not difficult to see how governments in any country could influence peoples' way of thinking by limiting financial access to the (welfare) state once families get above a certain size. That said, family sizes in the West decreased significantly once contraception became freely available in the 1960s, but no government wants population levels to drop too much because of the inevitable loss of revenue from taxation; Japan is currently worried about how few children are being born in their country for just this reason. Conversely, the leaders of the Catholic Church, at least in public, will not discuss the matter, believing in the absolute sanctity of life and refusal to accept any form of contraception, whilst the huge majority of its members, if surveys are to be believed, mercifully leave its rulings on the floor as they close the bedroom door.

Controlling the increase of, let alone reducing, world population levels is a huge area of policy that calls for intergovernmental agreements at all levels. It will call for much patience and understanding of how others lead their lives in different continents, a 'one-size-fits-all' policy simply will not work, and many compromises from positions that are currently viewed as *nonnegotiable* or *the red line over which we will not step* will be needed. For all its faults and decreasing respect with which it is viewed as a global organization, I see no alternative to the United Nations leading on such issues, and surely this should be their major policy directive for the next few decades. It is simply not possible to separate the issue of person-made climate change/global warming from that of world population, and true leadership at the world level is surely needed to help bring about this change in public perception.

ACKNOWLEDGEMENTS

I thank Dr Harriet Martin of the Quaker Living Witness, http://www.livingwitness.org.uk, for many enlightening discussions.

REFERENCES

[1] R.P. Tuckett, Climate Change: Observed Impacts on Planet Earth (Chapter 1), Elsevier, 2008. ISBN:978-0-444-53301-2.
[2] K.P. Shine, W.T. Sturges, Science 315 (2007) 1804–1805.
[3] http://www.wmo.int/pages/prog/arep/gaw/ghg/GHGbulletin.html.
[4] http://news.bbc.co.uk/1/hi/sci/tech/3381425.stm.
[5] Intergovernmental Panel on Climate Change (IPCC), 5th Assessment Report, Working Group 1, Chapters 1, 2 and 8, Cambridge University Press, Cambridge, 2013.
[6] Intergovernmental Panel on Climate Change (IPCC), 4th Assessment Report, Working Group 1, Chapters 1 and 2, Cambridge University Press, Cambridge, 2007.

[7] J.G. Anderson, W.H. Brune, M.H. Proffitt, J. Geophys. Res. D 94 (1989) 11465–11479.

[8] R.P. Tuckett, Adv. Fluor. Sci. 1 (2006) 89–129. Elsevier, ISBN:0-444-52811-3.

[9] K.P. Shine, J.S. Fuglestvedt, K. Hailemariam, N. Stuber, Clim. Change 68 (2005) 281–302.

[10] N.J. Mason, A. Dawes, R. Mukerji, E.A. Drage, E. Vasekova, S.M. Webb, P. Limao-Vieira, J. Phys. B 38 (2005) S893–S911.

[11] T.J. Blasing, Carbon Dioxide Information Analysis Centre, Oak Ridge National Laboratory, 2014. http://cdiac.ornl.gov/pns/current_ghg.html.

[12] T. Arnold, C.M. Harth, J. Muhle, A.J. Manning, P.K. Salameh, J. Kim, D.J. Ivy, L.P. Steele, V.V. Petrenko, J.P. Severinghaus, D. Baggenstos, R.F. Weiss, Proc. Nat. Acad. Sci. U.S.A. 110 (2013) 2029–2034 (and references therein).

[13] http://www.kyotoprotocol.com/resource/kpeng.pdf.

[14] R.P. Tuckett, Educ. Chem. 45 (2008) 17–21. Royal Society of Chemistry UK.

[15] W.T. Sturges, T.J. Wallington, M.D. Hurley, K.P. Shine, K. Sihra, A. Engel, D.E. Oram, S.A. Penkett, R. Mulvaney, C.A.M. Brenninkmeijer, Science 289 (2000) 611–613.

[16] A.R. Ravishankara, S. Solomon, A.A. Turnipseed, R.F. Warren, Science 259 (1993) 194–199.

[17] A.R. Ravishankara, E.R. Lovejoy, J. Chem. Soc. Faraday Trans. 90 (1994) 2159–2169.

[18] F.S. Rowland, M.J. Molina, Rev. Geophys. 13 (1975) 1–35.

[19] http://ozone.unep.org/new_site/en/Treaties/treaties_decisions-hb.php?sec_id=5.

[20] [a] E.C. Weatherhead, S.B. Andersen, Nature 441 (2006) 39–45;
[b] J.R. Ziemke, S. Chandra, Atmos. Chem. Phys. 12 (2012) 5737–5753;
[c] W. Chehade, M. Weber, J.P. Burrows, Atmos. Chem. Phys. 14 (2014) 7059–7074.

[21] A.C. Hong, C.J. Young, M.D. Hurley, T.J. Wallington, S.A. Maburg, Geophys. Res. Lett. 40 (2013) 1–6.

[22] http://www.c2es.org/international/negotiations/cop-15/summary.

[23] http://www.legislation.gov.uk/ukpga/2008/27/pdfs/ukpga_20080027_en.pdf.

[24] http://ec.europa.eu/clima/policies/2030/index_en.htm.

[25] D.J.C. MacKay, Sustainable Energy – Without the Hot Air, UIT Cambridge, UK, 2009. ISBN:978-0-9544529-3-3.

[26] www.iocom.com.

[27] http://esa.un.org/wpp, http://www.worldometers.info/world-population.

THE VARIATION OF THE EARTH'S MOVEMENTS (ORBITAL, TILT AND PRECESSION)

25

Lucas J. Lourens

Department of Earth Sciences, Faculty of Geosciences, Utrecht University, Utrecht, The Netherlands

CHAPTER OUTLINE

1. INTRODUCTION

The climate of the Earth is characterized by trends, aberrations and quasi-periodic oscillations varying over a broad range of timescales [1]. The trends are largely controlled by plate tectonics, and thus tend to change gradually on a million year (Ma) timescale. Aberrations occur when certain thresholds are passed and are manifested in the geological record as unusually rapid (less than a few thousands of years) or extreme changes in climate. The quasi-periodic oscillations are mostly astronomically paced; they are driven by astronomical perturbations that affect the Earth's orbit around the sun and the orientation of the Earth's rotation axis with respect to its orbital plane. These perturbations are described by three main astronomical cycles: eccentricity (shape of the Earth's orbit), precession (date of perihelion) and obliquity (angle between the equator and orbital plane), which together determine the spatial and seasonal pattern of insolation received by the Earth, eventually resulting in climatic oscillations of tens to hundreds of thousands of years. The expression of these astronomical-induced climate oscillations is found in geological archives of widely different ages and environments.

Computation of the orbital solution of the Earth is complex because the Earth's motion is perturbed by the moon and all planets of our solar system. Much of our knowledge of the planetary orbits dates back to the investigations of Kepler (1571–1630) and the universal gravitational theory of Newton (1643–1727). The first approximate solutions were established by Lagrange [2,3] and Pontécoulant [4]. Shortly after, Agassiz [5] formulated a sweeping theory of ice ages that initiated the investigations directed at the causal relationship between large-scale climatic changes and variations of the Earth's astronomical parameters. Adhémar [6] proposed that glaciations originated from the precession of the

Climate Change. http://dx.doi.org/10.1016/B978-0-444-63524-2.00025-7

Earth's rotation axis that alters the lengths of the seasons. He suggested that when the lengths of the winters last longer, ice ages would occur. According to his theory, the northern hemisphere (NH) and the southern hemisphere (SH) would be glaciated during the opposing phases of the precession cycle. He evidenced his idea with the present Antarctic ice sheet and the fact that the NH is essentially not glaciated at this moment. Croll [7] proposed, after the publication of a more precise solution of the Earth by Le Verrier [8], that variations in the Earth's eccentricity were also important for understanding past climates through its modulation of precession. He elaborated Adhémar's idea that winter insolation is critical for glaciations but argued that the large continental areas covered with snow would turn into ice sheets because of a positive ice-albedo feedback. Pilgrim [9] computed for the first time the obliquity variations due to secular changes in the motion of the Earth's orbital plane. This basis was used by Milankovitch [10] to establish his (mathematical) theory of the ice ages. Since then, the understanding of the climate response to orbital forcing has evolved and is the subject of this chapter.

2. ASTRONOMICAL PARAMETERS

The most precise astronomical solution for the past evolution of the orbital motion of the Earth and its rotational axis has been published recently by Laskar and co-workers [11]. This La2010 solution represents the last improvement to the continuous development of orbital solutions that have been computed at the *Observatoire de Paris*, since Bretagnon [12] computed for the first time terms of second order and third degree in the secular (mean) equations. This orbital solution was used by Berger [13,14], who computed the precession and insolation quantities for the Earth following Sharav and Boudnikova [15,16]. Berger's publications have since been extensively used for paleoclimate reconstructions and climate modelling under the acronym Ber78. In 1984, Laskar computed in an extensive way the secular equations giving the mean motion of the whole solar system [17–19]. It was clear from his computations that the traditional perturbation theory could not be used for the integration of the secular equations, due to strong divergences that became apparent in the system of the inner planets [17]. This difficulty was overcome by switching to a numerical integration of the secular equations with steps of 500 years (1 year). These computations provided a much more accurate solution for the orbital motion of the solar system over 10 Ma [19,20]. Extending his integration to 200 Ma, Laskar [21,22] demonstrated that the orbital motion of the planets, and especially that of the terrestrial planets, is chaotic, with an exponential divergence corresponding to an increase of the error by a factor 10 every 10 Ma. The orbital part of this La90 solution was considered to be reliable over 10 Ma–20 Ma [23]. With the publication of the numerical La2004 [24] and La2010 [11] solutions, this precision has been extended down to ~40 Ma and 50 Ma, respectively, providing a solid base for tuning the Neogene Period (0 Ma–23 Ma) [25,26] of the Geologic Timescales 2004 and 2012 [27,28].

A comparison between the La90 solution and the first numerical integration of the solar system by Quinn and co-workers [29] revealed that the main obliquity and precession periods of the two solutions diverge with time over the past 3 Ma [23]. In the QTD91 solution, a term was introduced that describes the change in the speed of rotation of the Earth as a result of the dissipation of energy by the tides. After applying the same present-day value in La90, the discrepancy with QTD91 was almost completely removed. The resulting La90 solution with tidal dissipation set to the present-day value is now generally termed as the La93 solution [23]. The La93 solution also enabled the modification of a second term, referring to the change in the dynamical ellipticity of the Earth, which may strongly depend on the build-up and retreat of large ice caps [30–33] and/or on long-term mantle convection processes [34]. Similar to the tidal dissipation term, a small change in the dynamical ellipticity of the Earth will change the main precession and obliquity frequencies.

The uncertain values of the tidal dissipation and dynamical ellipticity of the Earth are considered as the most limiting factors to obtain accurate solutions for the precession and obliquity time series of the Earth over a time span of millions of years [35]. At present, there exists only one possible way to test the extent of change of both parameters in the (geological) past. This test involves a statistical comparison between the obliquity-precession interference patterns in the insolation time series and those observed in geological records [36]. Lourens and co-workers [37] showed for instance by using a record of climate change from the eastern Mediterranean, that over the past 3 Ma the decline in the speed of rotation was on average smaller than the average value obtained for the present day; this is probably a result of the large ice caps that dominated Earth's climate from the Late Pliocene to present.

2.1 ECCENTRICITY

The Earth's orbit around the sun is an ellipse. The plane in which the Earth moves around the sun is called the ecliptic of date, Ec_t (Fig. 1). The sun is roughly located in one of its two foci. The eccentricity (e) of the Earth's orbit is defined as:

$$e = \frac{c}{a} = \frac{\sqrt{a^2 - b^2}}{a} \tag{1}$$

where a is the ellipse semi-major axis, b the semi-minor axis and c the distance from focus to centre. The current eccentricity is 0.0167 (Figure 2(b)), but in the past hundred millions of years eccentricity has varied from about 0.0669 to almost 0.0001, i.e. a near-circular orbit [24]. For the past 15 Ma, the three most important periods in the series expansion for eccentricity are about 405 ka, 124 ka and 95 ka (Fig. 2(e)), corresponding to the resonances between the secular frequencies of Venus (g2) and Jupiter (g5), Mars (g4) and Venus (g2), and Mars (g4) and Jupiter (g5), respectively [24]. Here ka refers to a thousand years (kyr).

2.2 PRECESSION AND OBLIQUITY

The locations along the Earth's orbit where the sun is perpendicular to the equator at noon are called equinoxes (Fig. 1(a)). Then the night lasts as long as the day at all latitudes. Today this occurs on March 20 (vernal equinox, NH spring) and on September 23 (autumnal equinox, NH autumn). The summer (winter) solstice is defined as the location of the Earth when the sun appears directly overhead at noon at its northernmost (southernmost) latitude, i.e. the tropics of Cancer at 23.44°N (Capricorn at 23.44°S), which occurs on June 21 (December 22).

The Earth's rotational axis (φ) revolves around the normal (n) to the orbital plane like a spinning top (ψ in Fig. 1). This rotation causes a clockwise movement of the equinoxes and solstices along the Earth's orbit, called precession. The quasi-period of precession is 25.672 ka relative to the stars, but because the Earth's orbit rotates in a counterclockwise direction with respect to the reference-fixed ecliptic (Ec_0) at Julian date J2000, the net period of precession is about 21.7 ka. The general precession in longitude ψ is thus defined by $\psi = \Lambda - \Omega$, where Ω is the longitude of the ascending node (N), and Λ the inclination of Ec_t (Fig. 1). The angle between the Earth's equatorial plane (Eq_t, Fig. 1)

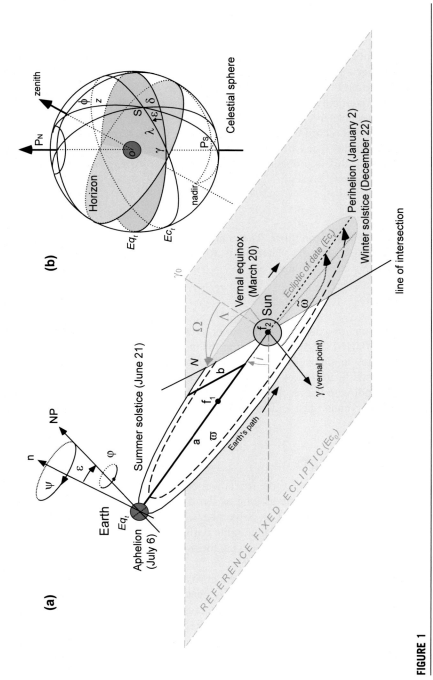

FIGURE 1

Astronomical configuration of the Earth. (a) Elements of the Earth's orbital parameters. *(Modified after Ref. [109].)* (b) Position of a point (S) on the celestial sphere. *(Modified after Ref. [39].)* See text for explanations.

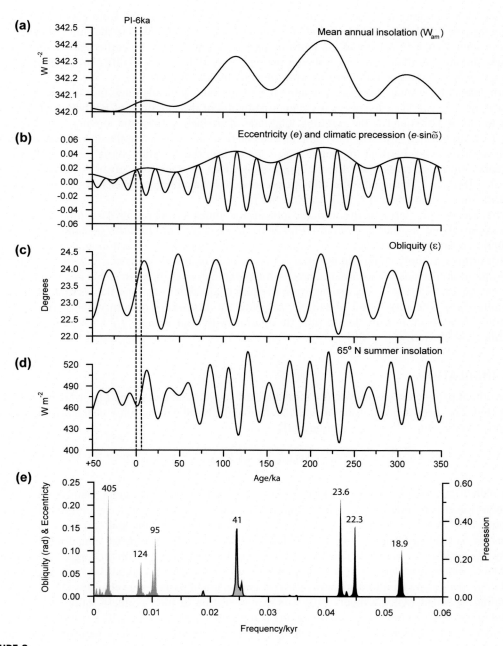

FIGURE 2

Variations of the Earth's orbital parameters between 350 ka and 50 000 years into the future according to the La04 solution [24]. (a) Mean annual insolation including a solar constant of 1368 W m^{-2}. (b) Eccentricity (solid) and climatic precession (dotted line). (c) Obliquity. (d) 65°N summer (May 21–July 20) insolation. (e) Combined power spectrum of eccentricity (grey), obliquity (green) and precession (black) from the past 15 Ma.

and Ec_t is the obliquity (ε). The current value for ε is 23.44° but it varied from about 22° to 24.5° during the past 15 Ma with a main period of about 41 ka (Fig. 2(c)), linked to the sum frequency of ψ and the main secular frequency of the Earth-moon barycentre (s3). In the earlier episodes of Earth's history, obliquity oscillated at a much shorter period (i.e. ~29 ka at 500 Ma BP [38]). This is because the Earth's rate of rotation has declined with time due to tidal friction.

For paleoclimate studies, the usual quantity that relates more to insolation is the climatic precession index $e \sin \overline{\omega}$, where $\overline{\omega} = \varpi + \psi$. ϖ is the approximation of the longitude of perihelion of the Earth from the fixed J2000 and $\overline{\omega}$ the resulting longitude of perihelion from the moving vernal equinox (Fig. 1(a)). In practice, however, the numerical calculations are done using the direction in which the sun is seen from the Earth at the beginning of the spring, the so-called vernal point γ (Fig. 1) as reference. In most cases the climatic precession index is therefore given by $e \sin \tilde{\omega}$, where $\tilde{\omega} = \overline{\omega} - 180°$ [14]. This implies that climatic precession is at a minimum when NH summer solstice is in perihelion, so that when $\tilde{\omega}$ is 270° (or $\overline{\omega}$ is 90°).

The eccentricity term in the climatic precession index is operating as a modulator of the precession-related insolation changes (Fig. 2(b)). In case of a circular orbit (eccentricity is zero), perihelion is undefined and there is no climatic effect associated with precession, while in case the Earth's orbit is strongly elongated the effect of precession on insolation is at a maximum. The three most important periods of the climatic precession parameter over the past 15 Ma are about 23.6 ka, 22.3 ka and 18.9 ka (Fig. 2(e)), arising from the sum frequencies of ψ and the secular frequencies of Jupiter (g5), Venus (g2) and Mars (g4), respectively. Just as for obliquity, the periods of precession shorten back in time due to tidal dissipation with approximately 3 ka–4 ka over the past 500 Ma (500 million years) [38].

2.3 INSOLATION

If the orbital parameters are known, the insolation for any latitude and at any time of the year can be computed. The mean annual insolation at the surface of the Earth depends only on the eccentricity and is represented by the following equation [23]:

$$W_{am} = \frac{S_0}{4\sqrt{(1 - e^2)}} \tag{2}$$

S_0 is called the 'solar constant'. In fact, the intensity of the sun varies along with the number of sunspots. Recent observations have shown that when sunspots are numerous (scarce) the solar constant is about 1368 W m^{-2} (1365 W m^{-2}). In literature, the various astronomical computations include values for S_0 ranging from 1350 W m^{-2} [23] to 1360 W m^{-2} [39]. The mean annual insolation variations are very small, as they depend on the square of the eccentricity, with the largest mean annual insolation values reached during eccentricity maxima (Fig. 2(a)). Orbital-induced mean annual insolation changes are therefore not seen as the primary cause of past climate changes. On the contrary, according to Milankovitch's theory [10], summer insolation at high northern latitudes (Fig. 2(d)) played in particular a crucial role on the waxing and waning of the ice sheets. The theory states that in case insolation in summer was not high enough, ice sheets could expand. It is therefore important to compute the daily (or monthly) insolation at any given point on the Earth. Following Berger and

Loutre [39], insolation W received on a horizontal surface at latitude ϕ and a given time (H) during the course of the year (λ) is described by:

$$W(\phi, \lambda, H) = S_0 \left(\frac{a}{r}\right)^2 \cos z \tag{3}$$

where r is the distance to the sun and z the solar zenith angle (or zenith distance). The horizontal surface of the position of the observer (o) refers to the plane perpendicular to the direction of the local gravity, while the zenith is the point vertically upward (Fig. 1(b)). The zenith distance z of any point S (i.e. the position of the sun seen in the sky from the observer at time H) on the celestial sphere is the angular distance from the zenith measured along the vertical circle through the given point (Fig. 1(b)). It varies from $0°$ to $180°$. The point S can also be calculated from the angle between the meridian (great circle through the celestial poles P_n and P_s, the zenith and the nadir) and the secondary great circle through the point and the poles (Fig. 1(b)). This angle is called the hour angle, H. This gives the following relationship:

$$\cos z = \sin \phi \sin \delta + \cos \phi \cos \delta \cos H \tag{4}$$

The declination δ is the angular distance of point S measured from the equator on the secondary great circle. The latitude ϕ is the angular distance from the equator to the zenith measured on the meridian. The declination δ is related to the true longitude λ of the Earth by:

$$\sin \delta = \sin \lambda \sin \varepsilon \tag{5}$$

Over one year, λ varies from $0°$ to $360°$ while δ varies between $-\varepsilon$ and $+\varepsilon$. The Earth-sun distance r is given by the ellipse equation:

$$r = \frac{a(1 - e^2)}{1 + e \cos \nu} \tag{6}$$

with ν being the true anomaly related to the true longitude λ of the Earth by:

$$\nu = \lambda - \tilde{\omega} \tag{7}$$

Combining Eqns (4–7), Eqn (3) can be rewritten as:

$$W(\phi, \lambda, H) = S_0 \frac{(1 + e \cos(\lambda - \tilde{\omega}))^2}{(1 - e^2)^2} (\sin \phi \sin \lambda \sin \varepsilon + \cos \phi \cos \delta \cos H) \tag{8}$$

Over one year, ε, $\tilde{\omega}$ and e are assumed to be constant. Over a given day, λ and δ are assumed to be constant, while H varies from 0 at solar noon to 24 h ($0°$–$360°$). The long-term behaviour of each factor in Eqn (8) is thus governed by a different orbital parameter. The obliquity ε drives $\cos z$, the precession $\tilde{\omega}$ drives $(1 + e \cos (\lambda - \tilde{\omega}))^2$ and the eccentricity e drives $(1 - e^2)^{-2}$. Note that the eccentricity appears as $(1 - e^2)^{-2}$ while in the mean annual insolation it appears as $(1 - e^2)^{-\frac{1}{2}}$; see Eqn (2).

In climate models [40,41], the true longitude is calculated by setting March 21, the vernal equinox, as the origin of time (t = 0); $\lambda_0 = 0$. Hence, the mean longitude at t = 0, λ_{m0}, which is not zero at the same time as λ_0 due to inclination, equals:

$$\lambda_{m0} = \lambda_0 - 2 \left[\begin{array}{l} \left(\frac{1}{2}e + \frac{1}{8}e^3\right)(1 + \beta)\sin(\lambda_0 - \tilde{\omega}) \\[2ex] -\frac{1}{4}e^2\left(\frac{1}{2} + \beta\right)\sin 2(\lambda_0 - \tilde{\omega}) + \frac{1}{8}e^3\left(\frac{1}{3} + \beta\right)\sin 3(\lambda_0 - \tilde{\omega}) \end{array} \right] \tag{9}$$

and

$$\beta = \sqrt{1 - e^2} \tag{10}$$

Each value of λ_m can be now determined through an increment $\Delta\lambda_m$ because the rate of increase is constant. In case of daily increments:

$$\lambda_m = \lambda_{m0} + l\frac{\pi}{180} \tag{11}$$

where l is the number of days since the vernal equinox. This mean longitude is used to find the true longitude:

$$\lambda = \lambda_m + \left(2e - \frac{1}{4}e^3\right)\sin(\lambda_m - \tilde{\omega}) + \frac{5}{4}e^2 \sin 2(\lambda_m - \tilde{\omega}) + \frac{13}{12}e^3 \sin 3(\lambda_m - \tilde{\omega}) \tag{12}$$

Another point of consideration is that Eqn (8) has to be adapted for latitudes north/south of the polar circles at $90° - \varepsilon$ N/S. In these areas there is no sunset in summer time and no sunrise in wintertime ($W = 0$).

To illustrate the influence of precession and obliquity on insolation, a plot is given of the monthly averaged zonal insolation differences between (1) a climatic precession minimum and maximum (Fig. 3(a)) and (2) an obliquity maximum and minimum (Fig. 3(b)). This comparison shows that high northern (southern) latitudes receive (dispatch) more than 100 W m^{-2} of additional insolation during summer ($\sim 20\%$) when they occur in perihelion (aphelion). In contrast, the accompanying NH (SH) winters receive less (more) insolation. Thus during a precession minimum seasonal contrasts at the NH increase, whereas they decrease at the SH. In contrast, a change in obliquity causes a simultaneous shift in seasonal contrasts at both hemispheres. From an obliquity minimum to maximum situation high latitudes receive more than 50 W m^{-2} of additional insolation during summer while winters gain considerably (~ 15 W m^{-2}) less insolation. In contrast to precession, obliquity influences the mean annual insolation at certain latitude. When obliquity increases, the poles receive more energy in summer but stay in the polar night during winter. The annual mean insolation therefore increases symmetrically at the poles and consequently the annual insolation must decrease around the equator (Fig. 3(c)) because the global annual insolation does not depend on obliquity (Eqn (2)).

3. ORBITAL-INDUCED CLIMATE CHANGE

The stable oxygen isotope ratio between ^{16}O and ^{18}O (denoted by δ^{18}O) of calcareous (micro) fossil shells has been extensively studied since the pioneering work of Emiliani [42] to improve our understanding of paleoceanographic and paleoclimate changes. Emiliani used the δ^{18}O of planktic foraminiferal shells to reconstruct glacial-interglacial variations in seawater temperature over the past 500 ka. His study gave strong support to the Milankovitch hypothesis and revolutionized ideas about the history of the oceans and the role of orbital forcing. Shackleton [43] showed however that the isotopic signal was, besides temperature, partly determined by ice volume changes. When ice caps grow, ^{16}O is preferentially stored on the continents resulting in heavier oxygen isotope values (^{18}O-enriched) of the ambient seawater in which the calcareous organisms thrive.

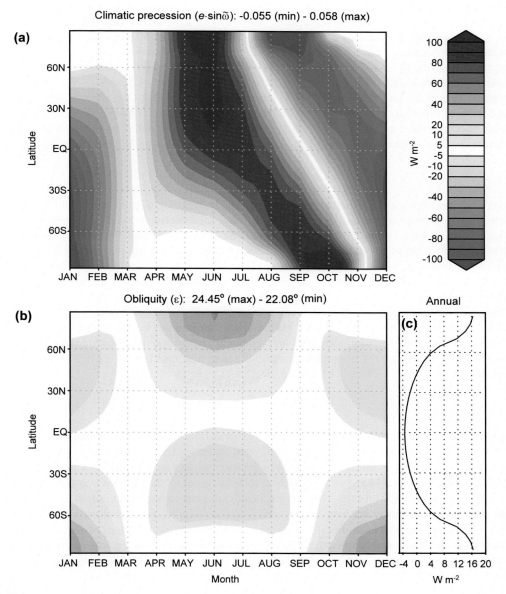

FIGURE 3

Monthly incoming differences in W m^{-2} at the top of the atmosphere. (a) Insolation differences between a minimum (−0.055) and a maximum (+0.058) climatic precession configuration. (b) Insolation differences between a maximum obliquity (Tilt = 24.45°) and a minimum obliquity (Tilt = 22.08°) configuration with zero eccentricity. (c) As in (b), but now plotted the annual incoming insolation differences. Solar constant = 1360 W m^{-2}.

Over the past decades, the inventory of high-resolution oxygen isotope records across the Cenozoic, 0 Ma–65 Ma, has grown, because of the greater availability of high-quality sediment cores. A compilation of these records showed that global climate cooled over the past 50 Ma with maximum temperature conditions occurring between 50 Ma and 55 Ma [1]. The first permanent ice caps start to occur on Antarctica around the Eocene-Oligocene transition, ~34 Ma. Also, recently extensive ice-rafted debris, including macroscopic dropstones, were found in the late Eocene to early Oligocene sediments from the Norwegian–Greenland Sea, indicating already severe glaciations of East Greenland at that time [44]. Orbital-induced variations in $\delta^{18}O$ were also detected superimposed on this long-term trend but revealed different spectral characteristics pending on the background climate state. An overview of the main characteristics with emphasis on icehouse and greenhouse conditions is given in the following sections.

3.1 ICE AGES

Through the development of radioisotopic dating methods, robust power spectra could be obtained from the oxygen isotope records in the time domain. These spectra clearly demonstrate that for the past 700 ka, major climate cycles have followed variations in obliquity and precession, although the dominant power occurs at ~100 ka [45,46]. Understanding the mechanisms that control this long-term variability remains an outstanding question in climate sciences [47]. The most widely adopted explanation is that it originates from a nonlinear response to the precession forcing [46,48]. Other hypotheses relate the 100 ka glacial rhythm directly to eccentricity [49], but the insolation changes that may have caused this are probably too small to be of much climatic relevance (Fig. 2(a)). A third category of theories attributes the 100 ka glaciations to an internal oscillation of the Earth's ice-atmosphere-ocean climate system [50], which is nonlinear phase-locked to the external Milankovitch forcing [51,52]. An historically important example of a nonlinear oscillator is the (ice-sheet) model of Imbrie and Imbrie [48], which may be written as:

$$\frac{dV}{dt} = \frac{1 \pm b}{T_m}(X - V) \tag{13}$$

where X is the model's forcing function (i.e. the 65°N summer insolation), T_m is a mean time constant of the ice sheet response and b a nonlinearity coefficient that switches sign depending on whether ice volume is increasing or decreasing (Fig. 4(a)). For the late Pleistocene, a T_m of 17 ka and b of 0.6 were estimated, which result in a 4 ka–5 ka lag (ice sheet response time) for the precession components and an ~8 ka lag for obliquity [53]. In the latest marine benthic oxygen isotope stack, LR04, of Lisiecki and Raymo [54], the same model has been applied (Fig. 4(b)). Evidently, changes in the model output and marine $\delta^{18}O$ record lag 65°N summer insolation (Fig. 4(a)) with a few thousands of years as a result of the adopted values for T_m and b. Since the $\delta^{18}O$ record preserves not only an ice volume signal but also a deepwater temperature component, care should be taken regarding their individual phase response [55].

One way to separate the $\delta^{18}O$ record into ice volume and temperature contributions is the use of inverse modelling techniques in combination with 3D ice sheet models [56,57]. The latest estimates include ice sheet fluctuations at Antarctica, Greenland, Eurasia and North America. A plot of the total ice volume component, expressed in terms of sea level equivalent, is shown in Fig. 4(b) as an overlay to the LR04 $\delta^{18}O$ record. The calculated eustatic sea level changes share similar orbital-induced features as independent sea level reconstructions from the Red Sea [58,59], although the modelled values are at the upper confidence limits of the reconstructions (Fig. 4(c)). In addition, they clearly lack short-term millennial-scale changes, due to smoothing and the on-average lower resolution. The calculated deep-sea temperature has been translated into annual surface air temperature (ΔT_{NH}) over the continents (40–80°N) with respect to the present day, using a simple linear equation and a 3 ka time lag to correct for the different

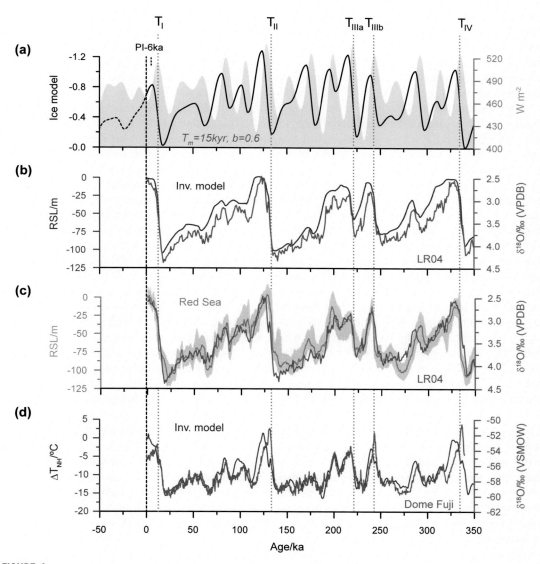

FIGURE 4

Comparison between orbital cycles and glacial-related climatic proxy records for the past 350 ka (late Pleistocene). (a) The outcome (black line) of simple ice sheet model [48] including a mean time constant of the ice sheet response (T_m) of 15 ka and a nonlinearity coefficient (b) of 0.6 and the 65°N summer insolation (grey area) as input [54]. (b) LR04 δ^{18}O benthic stacked record (red line) [54] plotted as overlay to its reconstructed global sea level curve (blue line) through inverse modelling [56,57]. (c) LR04 δ^{18}O benthic stacked record (red line) plotted as overlay to the Red Sea RSL record (grey line, incl. 95% conf. limits) [58,59]. (d) Comparison between the modelled annual (mean over the continents 40–80°N) surface air temperature deviation from 0 ka (blue line) [56,57] and the Dome Fuji δ^{18}O$_{ice}$ record (red line) [60]. Major terminations (I–IV) are defined by maximum Red Sea ΔRSL values (Figure 6(d)).

response times between the deep ocean and the atmosphere (Fig. 4(d)). Clearly, ΔT_{NH} leads ice volume increases up to a few thousands of years because ice sheets will only start to grow (inception) below a certain temperature threshold ($-5\ ^\circ C$) and the slow rate at which mass is gained through snow accumulation [56]. During deglaciations, surface air temperature and sea level increase almost in concurrence, presumably the result of the rapid meltdown of the large ice sheets, with modelled (reconstructed) sea level rises of over 1.5 cm a^{-1} (2 cm a^{-1}) during the major terminations T_{I-IV} (Fig. 5(d)).

The estimated ΔT_{NH} component of the marine $\delta^{18}O$ record resembles the Antarctic $\delta^{18}O_{ice}$ record of Dome Fuji [60], although the latter record tends to lead ΔT_{NH} on the order of a few thousands of years (Fig. 4(d)). There are several explanations that may account for the discrepancy between the insolation-induced response times of ΔT_{NH} and the Antarctic $\delta^{18}O_{ice}$ record. First, there is some uncertainty in the parameterization of the deepwater-atmosphere temperature coupling in the model. Secondly, the $\delta^{18}O_{ice}$ record depends, besides local temperature, on a variety of factors such as seawater $\delta^{18}O$ and the temperature of the water vapour source area [60]. Another part of the discrepancy may arise from uncertainties in the chronologies of either Dome Fuji or the LR04 $\delta^{18}O$ record. Notably, the LR04 $\delta^{18}O$ chronology is not directly constrained by radioisotopic measurements but relies on its correlation to Thorium-230- and Protactinium-231-dated sea level reconstructions from coral terraces [61–64], i.e. thus without separating the temperature and ice volume components of the $\delta^{18}O$ signal. Although the chronology of the last glacial cycles is well constrained, conflicting estimates were found for the age of the penultimate and earlier deglaciations, which argued in favour [65] or against [66–68] an NH summer insolation forcing mechanism. The Dome Fuji chronology on the other hand is based on tuning of the O_2/N_2 ratio of the trapped ice to the local variations in summer insolation (December 21 at $77^\circ S$). The O_2/N_2 ratio lacks a strong 100 ka response, which makes this proxy more suitable for tuning than the $\delta^{18}O_{atm}$ record applied previously by Shackleton [55]. Dating uncertainties in this timescale range from 0.8 ka to 2.9 ka at the tie points [60]. Given these uncertainties, the increases in Antarctic temperature and atmospheric carbon dioxide concentration coincide with the rising phase of NH summer insolation during the last four terminations (Fig. 4(d)), thereby supporting the Milankovitch theory [60].

The role of obliquity is less highlighted in glacial theories despite the fact that from about 1 Ma to 3 Ma ago and also during older geological periods, such as the Middle Miocene (14 Ma–15 Ma), smaller ice sheets varied at an almost metronomic 41 ka rhythm [69–72]. There are several mechanisms proposed to explain the obliquity-dominated climate cycles. The most obvious explanation is that when the Earth's axis reaches minimum tilt, values and high-latitude (annual and summer) insolation are low (Fig. 3(b) and (c)), ice caps may grow, albedo will increase, and hence global mean temperatures will decline [10]. Another possibility is that during obliquity minima the meridional gradient of insolation during the summer half-year of both hemispheres increases, causing an increased moisture transport to the poles and hence the build-up of large ice caps [73]. Evidently, most periods of maximum sea level lowering or ice sheet growth over the past 350 ka occur during obliquity minima. Huybers and Wunsch [74] presented simple stochastic and deterministic models that describe the timing of the late Pleistocene glacial terminations even purely in terms of obliquity forcing, although their findings are not yet confirmed by, for instance, the new results of Dome Fuji [60].

To summarize, the development of glacial-independent chronologies has become one of the major challenges in climate sciences to further unravel the Milankovitch theory of the ice ages. These chronologies could provide new insights to key issues such as the cause-and-effect relationships between climate change and the atmospheric carbon dioxide concentration, the feedback mechanisms associated with the build-up and retreat of large ice sheets, or whether terminations are caused by internal or external processes. Extrapolating the simple ice sheet model of Imbrie and Imbrie [48] into the future (Fig. 4(a)) suggests that a new ice age may develop within 25 ka and that our present interglacial period may come soon to an end as was already discussed during a working conference at Brown University,

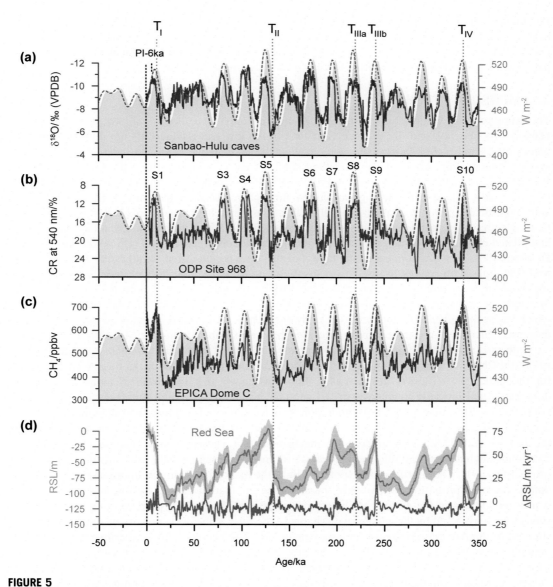

FIGURE 5

Comparison between orbital cycles and monsoon-related climatic proxy records for the past 350 ka (late Pleistocene). (a) The composite speleothem $\delta^{18}O$ record of the Sanbao-Hulu caves in China (blue line) ka [86,87] as overlay to 65°N summer insolation (grey area) and its 3 ka lagged equivalent (red dashed line). (b) Colour reflectance data of ODP Site 968 (blue line) [88] as overlay to 65°N summer insolation (grey area) and its 3 ka lagged equivalent (red dashed line). Sapropels (s) are marked by very low (dark) values. (c) The methane record of EPICA Dome C (blue line) [100,101] as overlay to 65°N summer insolation (grey area) and its 3-ka lagged equivalent (red dashed line). (d) Red Sea RSL (grey line, incl. 95% conf. limits) and ΔRSL (red line) [58,59]. Major terminations (I–IV) are defined by maximum Red Sea ΔRSL values.

Providence, Rhode Island, on January 26–27, 1972 [75]. More recent studies point, however, that we are heading for a long (>50 ka) interglacial period comparable to, for instance, marine isotope stage 11 (~410 ka), which with the addition of a human-induced greenhouse perturbation may not develop as the past few interglacial periods, since the forcings (i.e. CO_2) are not similar [76].

3.2 MONSOON STRENGTH

The expression of orbital-induced climate oscillations is not restricted to glacial-interglacial variability. Data and models revealed that climate variations in the low and midlatitudes are dominated by the precession cycle [77–81]. In particular, they regulate the strength of the African-Indian-Asian monsoon system through time. Many of the modelling studies were carried out within the framework of the Paleoclimate Modelling Intercomparison Project (PMIP) to simulate the Mid-Holocene (6 ka) climate [82]. Recently, a state-of-the-art high-resolution atmosphere-ocean-coupled model, EC-Earth [83,84], with sophisticated parameterizations of small-scale atmospheric, hydrological and surface processes (i.e. thoroughly tested in operational weather forecasts) was used to simulate in detail the monsoon response and related circulation systems such as the Inter Tropical Convergence Zone (ITCZ) during the Mid-Holocene [85]. The astronomical configuration of this time period differs significantly from preindustrial (PI) times with $e \sin \tilde{\omega}$ having negative values and $e \cdot$ and ε close to maximum values (Fig. 2). As shown by previous model simulations, the MH-PI anomaly indicates more than 4.5 mm d^{-1} of increased precipitation during summer (JAS) over Central-North Africa, India and the Himalaya (Fig. 6), which has been explained by the increased landward monsoon winds and moisture advection as well as decreased (increased) moisture convergence over the oceans (land) during the Mid-Holocene [85].

High-resolution and U/Th-dated speleothem oxygen isotope records of the Sanbao and Hulu caves in central China support the dominant precession control on the strength of the East Asian Monsoon for the past 350 ka [86,87]. In particular, Fig. 5(a) shows that (sub)tropical monsoons respond dominantly and instantly to changes in NH summer insolation on orbital timescales, though with a consistent ~3.0 ka time lag [88]. Another indication for enhanced past monsoon strength is the cyclic occurrence of sapropels (organic-rich layers) in the marine sediments of the Mediterranean throughout the last 13 Ma [25,89]. The underlying mechanism that caused their formation gave rise to a contentious debate over the relative importance of anoxia caused by stable stratification [90] versus productivity [91]. The stratification hypothesis links the reduced oxygen conditions of the deepwaters during sapropel formation to a weaker thermohaline circulation caused by lowered surface water density conditions in the eastern Mediterranean. Rossignol-Strick [92] proposed that these circumstances were triggered by the enhanced discharge of the river Nile during precession minima when the strength of the African monsoon is at a maximum [78]. Climate modelling experiments, including a regional ocean model for the Mediterranean Sea, revealed, however, that the precession-induced increase in net precipitation over the Mediterranean Sea itself is of equal or greater importance than the increase in runoff from the bordering continents [93]. Evidence for enhanced primary productivity has been gathered by a variety of geochemical and micropaleontological proxy records [91,94–98]. Possible causes for the enhanced nutrient supply to the mixed layer may include a reversal in the flow directions of the nutrient-poor surface and nutrient-enriched deepwaters, increased runoff, and the development of a deep chlorophyll maximum (DCM).

A plot of the colour reflectance at 540 nm from Ocean Drilling Program Site 968 of the past 350 ka clearly resembles the speleothem oxygen isotope pattern in central China (Fig. 6(b)), suggesting that changes in the strength of the North African and Indian-Asian summer monsoons operated in conjunction [88]. This correlation is confirmed by radiometric dates of the youngest sapropel in the Mediterranean (S1), which indicates a 3 ka time lag between the last precession minimum at 11.5 ka and the midpoint of the S1 dated at 8.5 ka [36]. The lag of the Chinese speleothem $\delta^{18}O$ records was

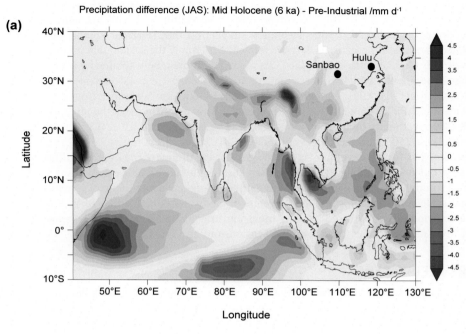

Precipitation difference (JAS): Mid Holocene (6 ka) - Pre-Industrial /mm d⁻¹

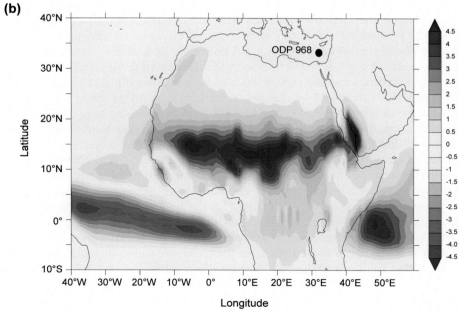

FIGURE 6

EC-Earth simulation showing the average precipitation difference (mm d^{-1}) for the months July, August and September (JAS) between the Mid-Holocene (MH) and Preindustrial (PI). (a) India-East Asia. (b) North Africa. For PI, $e = 0.016{,}725$, $\varepsilon = 23.446°$ and $\tilde{\omega} = 102.04°$, and for MH, $e = 0.018{,}682$, $\varepsilon = 24.105°$ and $\tilde{\omega} = 0.087°$ [85]. Also indicated are the locations of the Sanbao and Hulu caves and ODP Site 968.

interpreted as being in-phase with July instead of June insolation [86], with reference that modern monsoon maximum occur in July. However, climate modelling experiments indicate that summer monsoon strength depends on June insolation, indicating that seasonal signals cannot simply be extrapolated to orbital timescales as they are related to inertia in the system that produces lags of a few weeks [79,99]. On the other hand, the 3 ka time lag cannot be explained by current (transient) modelling experiments, which suggest that tropical monsoons respond instantly to changes in NH summer insolation [78,79,99]. Further support for the ~3 ka time lag comes from the methane record of the Antarctic EPICA Dome C ice core [100,101]. This record reflects the strength of the tropical monsoons with secondary input from boreal sources [100] and reveals a 2 ± 0.85 ka precession-related lag over the past 250 ka on the EDC3 chronology [102] (Fig. 5(c)).

In summary, as for the development of ice ages and the timing of major terminations, uncertainties still exist in our understanding of the processes causing the observed (~3 ka) time lag between astronomical forcing and precession-bound low-latitude climate changes. One explanation could be that millennial-scale cold spells (e.g. Heinrich events) or meltwater pulses in the North Atlantic (Fig. 5(d)) could have temporally suppressed summer monsoon strength and hence led to an apparent systematic delay of low-latitude climate changes to precession forcing [59,88].

3.3 PRE-PLEISTOCENE

During the late Oligocene and early Miocene (18 Ma–27 Ma), when the polar regions were only partially ice covered, benthic isotope records exhibit, besides a dominant obliquity component, a strong response to eccentricity forcing [103]. In the absence of permanent ice caps between 35 Ma and 65 Ma, the imprint of eccentricity seems even more prominent, although the benthic isotope records currently available for the early Cenozoic lack adequate resolution to fully characterize obliquity variance [1]. The pronounced eccentricity imprint can be explained by filtering effects of the precession forcing due to continental geography and differences in land-sea heating, especially in the tropics [81]. A variety of processes have been suggested for exporting the signals to higher latitudes, including changes in ocean and atmospheric circulation, heat transport, precipitation, or the global carbon cycle and pCO_2. Evidence of changes in the carbon cycle are given for instance by the Oligocene-Miocene carbon isotope ($\delta^{13}C$) records, which exhibit pervasive large-amplitudes of 100 ka and 400 ka oscillations that are highly coherent with the benthic oxygen isotope records [103,104].

Also during the late Paleocene and early Eocene (~50 Ma–60 Ma), eccentricity has significantly modulated the carbon isotope records of the Atlantic and Pacific Oceans [105]. Cramer and co-workers [105] identified several short-lived $\delta^{13}C$ depletions, which they linked to maxima in the Earth's orbital eccentricity cycle. They also linked, however, the much larger carbon isotope excursion (CIE) that marks the Paleocene/Eocene boundary to a minimum in the 400 ka component of eccentricity, thereby excluding orbital forcing as triggering mechanism for the Paleocene/Eocene Thermal Maximum (PETM). This is in contrast to the more recent findings of Lourens and co-workers [106], showing on the basis of the more complete successions from the southern Atlantic Walvis Ridge depth transect [107] that the PETM and Eocene Thermal Maximum 2 do correspond to 400 ka and 100 ka eccentricity maxima. They suggested that the critical conjunction of short, long and very long eccentricity cycles and the long-term late Paleocene to early Eocene warming trend may have favoured the build-up of a significant methane hydrate reservoir before its release during both hyperthermal events. Moreover, the covariance between light $\delta^{13}C$ values and severe dissolution horizons in the deep sea during these extreme greenhouse conditions of the late Paleocene and early Eocene indicate that changes in the carbon cycle through orbital forcing have had an important impact on ocean acidification and the position of the lysocline and calcite compensation depth [108].

Although the proposed orbital control as forcing mechanism of Paleogene hyperthermal events should be confirmed, it is evident that eccentricity has left its mark on the global carbon cycle. Moreover, the appearance of this modulation became more visible in the geological archives when the impact of the obliquity-dominated glacial cycles is at a minimum. Evidence that these long-term changes in the carbon cycle determined global climate has not yet been solved. In particular, cross-spectral comparison between the Oligocene-Miocene $\delta^{18}O$ and $\delta^{13}C$ records revealed a time lag of more than 20 000 years in the 405 ka eccentricity band, suggesting a response rather than a forcing to global climate change [103]. In addition, the conspicuous absence of the long eccentricity signal in the Pleistocene glacial cyclicity raised the so-called '400 ka problem' [48], though recent modelling experiments have shown their existence in Antarctic ice sheet fluctuations [57].

4. CONCLUSION

The role of orbital forcing in climate change has been unequivocally shown by their characteristic patterns in sedimentary archives, ice cores and proxy records. Although our knowledge of orbital forcing is concerned with long-term natural climate cycles, it is of fundamental importance to assess and remediate global climate change problems on short-term periods. In particular, the integration of climate modelling experiments with geological observations will provide these insights required for a better understanding of climate change in the past and near future. Considerable challenges will have to be addressed before the full spectrum of orbital-induced climatic variability has been unravelled, including the phase behaviour of different parts of the climate system, feedback mechanisms and the impact on ecosystem dynamics. With the fast-rising CO_2 concentrations in the atmosphere, general orbital theories dealing with the icehouse world will probably not account for future predictions. However, through integrating our knowledge of geological times when greenhouse gas conditions were those as being predicted, we will be able to decipher the role of orbital forcing in future climate change scenarios.

ACKNOWLEDGEMENTS

I am grateful to Erik Tuenter and Joyce Bosmans for allowing the use of (1) their data in Figs 3 and 5, respectively, and (2) the comprehensive overview in their theses of the equations and variables used to compute the insolation variations, which is presented in modified form in Section 2.3.

REFERENCES

[1] J.C. Zachos, M. Pagani, L. Sloan, E. Thomas, K. Billups, Science 292 (2001) 686–693.
[2] J.L. Lagrange, Memoirs of Berlin Academy, Oeuvre complètes t. V, Gauthier-Villars, Paris, 1781, 1870 pp. 125–207.
[3] J.L. Lagrange, Memoirs of Berlin Academy, Oeuvre complètes t. V, Gauthier-Villars, Paris, 1782, 1870 pp. 211–344.
[4] G. de Pontécoulant, Théorie Analytique of Systéme of Monde, t III, Bachelier, Paris, 1834.

[5] L. Agassiz, Etudes sur les glaciers, Paris, 1840.

[6] J. Adhémar, Révolutions de la mer, Bachelier, Paris, 1842.

[7] J. Croll, Philos. Mag. 28 (1864) 121–137.

[8] U. Le Verrier, Ann. Obs. Paris II, Mallet-Bachelet, Paris, 1856.

[9] L. Pilgrim, Jahreshefte fur vaterlandische Naturkunde in Wurttemberg 60 (1904).

[10] M. Milankovitch, Royal Serb. Acad. Spec. Publ. 133 (1941) 1–633.

[11] J. Laskar, A. Fienga, M. Gastineau, H. Manche, Astron. Astrophys. 532 (2011) 15.

[12] P. Bretagnon, Astron. Astrophys. 30 (1974) 141–154.

[13] A. Berger, Astron. Astrophys. 51 (1976) 127–135.

[14] A. Berger, J. Atmos. Sci. 35 (1978) 2362–2367.

[15] S.G. Sharav, N.A. Boudnikova, Bull. I.T.A XI-4 127 (1967) 231–265.

[16] S.G. Sharav, N.A. Boudnikova, Trud I.T.A. XIV (1967) 48–84.

[17] J. Laskar, Observatoire de Paris, 1984.

[18] J. Laskar, Astron. Astrophys. 144 (1985) 133–146.

[19] J. Laskar, Astron. Astrophys. 157 (1986) 59–70.

[20] J. Laskar, Astron. Astrophys. 198 (1988) 341–362.

[21] J. Laskar, Nature 338 (1989) 237–238.

[22] J. Laskar, Icarus 88 (1990) 266–291.

[23] J. Laskar, F. Joutel, F. Boudin, Astron. Astrophys. 270 (1993) 522–533.

[24] J. Laskar, P. Robutel, F. Joutel, M. Gastineau, A.C.M. Correia, B. Levrard, Astron. Astrophys. 428 (2004) 261–285.

[25] L.J. Lourens, F.J. Hilgen, N.J. Shackleton, J. Laskar, D. Wilson, in: F. Gradstein, J. Ogg, A. Smith (Eds.), A Geologic Time Scale 2004, Cambridge University Press, UK, 2004, pp. 409–440.

[26] F.J. Hilgen, L.J. Lourens, J.A. Van Dam, A.G. Beu, A.F. Boyes, R.A. Cooper, W. Krijgsman, J.G. Ogg, W.E. Piller, D.S. Wilson, The Geol. Time Scale 2012 1–2 (2012) 923–978.

[27] F.M. Gradstein, J.G. Ogg, A.G. Sith, W. Bleeker, L.J. Lourens, Episodes 27 (2004) 83–100.

[28] F.M. Gradstein, J.G. Ogg, F.J. Hilgen, Newsl. Stratigr. 45 (2012) 171–188.

[29] T.R. Quinn, S. Tremaine, M. Duncan, Astron. J. 101 (1991) 2287–2305.

[30] K. Lambeck, The Earth's variable rotation, Cambridge University Press, Cambridge, 1980.

[31] W.R. Peltier, X. Jiang, Geophys. Res. Lett. 21 (1994) 2299–2302.

[32] J.X. Mitrovica, A.M. Forte, Geophys. J. Int. 121 (1995) 21–32.

[33] J.X. Mitrovica, A.M. Forte, R. Pan, Geophys. J. Int. 128 (1997) 270–284.

[34] A.M. Forte, J.X. Mitrovica, Nature 390 (1997) 676–680.

[35] J. Laskar, Phil. Trans. R. Soc. Lond. A 357 (1999) 1735–1759.

[36] L.J. Lourens, A. Antonarakou, F.J. Hilgen, A.A.M.v. Hoof, C. Vergnaud-Grazzini, W.J. Zachariasse, Paleoceanography 11 (1996) 391–413.

[37] L. Lourens, J.,R. Wehausen, H.J. Brumsack, Nature 409 (2001) 1029–1032.

[38] A. Berger, M.F. Loutre, J. Laskar, Science 255 (1992) 560–565.

[39] A. Berger, M.F. Loutre, Long-Term Climatic Variations I22 (1994) 107–151.

[40] E. Tuenter, Institute of Earth Sciences, Utrecht University, 2004, 147.

[41] J.H.C. Bosmans, Utrecht Stud. Earth Sci. 55 (2014) 152.

[42] C. Emiliani, J. Geol. 63 (1955) 538–578.

[43] N.J. Shackleton, Nature 215 (1967) 15–17.

[44] J.S. Eldrett, I.C. Harding, P.A. Wilson, E. Butler, A.P. Roberts, Nature 446 (2007) 176–179.

[45] J.D. Hays, J. Imbrie, N.J. Shackleton, Science 194 (1976) 1121–1132.

[46] J. Imbrie, A. Berger, E.A. Boyle, S.C. Clemens, A. Duffy, W.R. Howard, G. Kukla, J. Kutzbach, D.G. Martinson, A. McIntyre, A. Mix, B. Molfino, J.J. Morley, L.C. Peterson, N.G. Pisias, W.L. Prell, M.E. Raymo, N.J. Shackleton, J.R. Toggweiler, Paleoceanography 8 (1993) 699–735.

[47] B. Saltzman, Int. Geophys. Ser. 80 (2001) 320.

[48] J. Imbrie, J.Z. Imbrie, Science 207 (1980) 943–952.

[49] R. Benzi, G. Parisi, A. Sutera, A. Vulpiani, Tellus 34 (1982) 10–16.

[50] H. Gildor, E. Tziperman, Paleoceanography 15 (2000) 605–615.

[51] K.A. Maasch, B. Saltzman, J. Geophys. Res. 95 (1990) 1955–1963.

[52] E. Tziperman, M.E. Raymo, P. Huybers, C. Wunsch, Paleoceanography 21 (2006). PA4206.

[53] J. Imbrie, J.D. Hays, D.G. Martinson, A. McIntyre, A.C. Mix, J.J. Morley, N.G. Pisias, W.L. Prell, N.J. Shackleton, in: Milankovitch and Climate, Understanding the Response to Astronomical Forcing, vol. 26 part 1, 1984, p. 510.

[54] L.E. Lisiecki, M.E. Raymo, Paleoceanography 20 (2005) 1003.

[55] N.J. Shackleton, Science 289 (2000) 1897–1902.

[56] R. Bintanja, R.S.W. van de Wal, J. Oerlemans, Nature 437 (2005) 125–128.

[57] B. De Boer, L.J. Lourens, R.S.W. van de Wal, Nat. Commun. 5 (2014).

[58] K.M. Grant, E.J. Rohling, M. Bar-Matthews, A. Ayalon, M. Medina-Elizalde, C.B. Ramsey, C. Satow, A.P. Roberts, Nature 491 (2012) 744–747.

[59] K.M. Grant, E.J. Rohling, C.B. Ramsey, H. Cheng, R.L. Edwards, F. Florindo, D. Heslop, F. Marra, A.P. Roberts, M.E. Tamisiea, F. Williams, Nat. Commun. 5 (2014).

[60] K. Kawamura, F. Parrenin, L.E. Lisiecki, R. Uemura, F. Vimeux, J.P. Severinghaus, M.A. Hutterli, T. Nakazawa, S. Aoki, J. Jouzel, M.E. Raymo, K. Matsumoto, H. Nakata, H. Motoyama, S. Fujita, K. Goto-Azuma, Y. Fujii, O. Watanabe, Nature 448 (2007) 912–916.

[61] W. Broecker, D.L. Thurber, J. Goddard, T.L. Ku, R.K. Matthews, K.J. Mesolella, Science 159 (1968) 297–300.

[62] R.L. Edwards, J.H. Chen, T.-L. Ku, G.J. Wasserburg, Science 236 (1987) 1547–1553.

[63] R.L. Edwards, H. Cheng, M.T. Murrell, S.J. Goldstein, Science 276 (1997) 782–786.

[64] E. Bard, B. Hamelin, R.G. Fairbanks, Nature 346 (1990) 456–458.

[65] T.D. Herbert, J.D. Schuffert, D. Andreasen, L. Heusser, M. Lyle, A. Mix, A.C. Ravelo, L.D. Stott, J.C. Herguera, Science 293 (2001) 71.

[66] G.M. Henderson, N.C. Slowey, Nature 404 (2000) 61–66.

[67] C.D. Gallup, H. Cheng, F.W. Taylor, R.L. Edward, Science 295 (2002) 310–313.

[68] I.J. Winograd, et al., Quaternary Res. 48 (1997) 141–154.

[69] N.J. Shackleton, J. Backman, H. Zimmerman, D.V. Kent, M.A. Hall, D.G. Roberts, D. Schitker, J. Baldauf, Nature 307 (1984) 620–623.

[70] W.F. Ruddiman, M.E. Raymo, D.G. Martinson, B.M. Clement, J. Backman, Paleoceanography 4 (1989) 353–412.

[71] M.E. Raymo, W.F. Ruddiman, J. Backman, B.M. Clement, D.G. Martinson, Paleoceanography 4 (1989) 413–446.

[72] A. Holbourn, W. Kuhnt, M. Schulz, H. Erlenkeuser, Nature 438 (2005) 483–487.

[73] M.E. Raymo, K. Nisancioglu, Paleoceanography 18 (2003) 1011.

[74] P. Huybers, C. Wunsch, Nature 434 (2005) 491–494.

[75] G.J. Kukla, R.K. Matthews, Science 178 (1972) 190–191.

[76] A.L. Berger, M.-F. Loutre, Science 297 (2002) 1287–1288.

[77] E.M. Pokras, A.C. Mix, Nature 326 (1987) 486–487.

[78] J.E. Kutzbach, Science 214 (1981) 59–61.

[79] J.E. Kutzbach, X. Liu, Z. Liu, G. Chen, Clim. Dyn. 30 (2008) 567–579.

[80] D.A. Short, J.G. Mengel, Nature 323 (1986) 48–50.

[81] D.A. Short, J.G. Mengel, T.J. Crowley, W.T. Hyde, G.R. North, Quat. Res. 35 (1991) 157–173.

[82] P. Braconnot, B. Otto-Bliesner, S. Harrison, S. Joussaume, J.Y. Peterchmitt, A. Abe-Ouchi, M. Crucifix, E. Driesschaert, T. Fichefet, C.D. Hewitt, M. Kageyama, A. Kitoh, M.F. Loutre, O. Marti, U. Merkel, G. Ramstein, P. Valdes, L. Weber, Y. Yu, Y. Zhao, Clim. Past 3 (2007) 279–296.

[83] W. Hazeleger, C. Severijns, T. Semmler, S. Ştefănescu, S. Yang, X. Wang, K. Wyser, E. Dutra, J.M. Baldasano, R. Bintanja, P. Bougeault, R. Caballero, A.M.L. Ekman, J.H. Christensen, B. van den Hurk, P. Jimenez, C. Jones, P. Kållberg, T. Koenigk, R. McGrath, P. Miranda, T. Van Noije, T. Palmer, J.A. Parodi, T. Schmith, F. Selten, T. Storelvmo, A. Sterl, H. Tapamo, M. Vancoppenolle, P. Viterbo, U. Willén, Bull. Am. Meteor. Soc. 91 (2010) 1357–1363.

[84] W. Hazeleger, X. Wang, C. Severijns, S. Ştefănescu, R. Bintanja, A. Sterl, K. Wyser, T. Semmler, S. Yang, B. van den Hurk, T. van Noije, E. van der Linden, K. van der Wiel, Clim. Dyn. 39 (2012) 2611–2629.

[85] J.H.C. Bosmans, S.S. Drijfhout, E. Tuenter, L.J. Lourens, F.J. Hilgen, S.L. Weber, Clim. Past 8 (2012) 723–740.

[86] Y. Wang, H. Cheng, R.L. Edwards, X. Kong, X. Shao, S. Chen, J. Wu, X. Jiang, X. Wang, Z. An, Nature 451 (2008) 1090–1093.

[87] H. Cheng, R.L. Edwards, W.S. Broecker, G.H. Denton, X. Kong, Y. Wang, R. Zhang, X. Wang, Science 326 (2009) 248–252.

[88] M. Ziegler, E. Tuenter, L.J. Lourens, Quat. Sci. Rev. 29 (2010) 1481–1490.

[89] F. Hilgen, Newsl. Stratigr. 17 (1987) 109–127.

[90] E. Olausson, Rep. Swedish Deep Sea Expedition 1947–1948 8 (1961) 353–391.

[91] S.E. Calvert, Oceanol. Acta 6 (1983) 255–267.

[92] M. Rossignol-Strick, Nature 304 (1983) 46–49.

[93] P.T. Meijer, E. Tuenter, J. Marine Syst. 68 (2007) 349–365.

[94] E.J. Rohling, W.W.C. Gieskes, Paleoceanography 5 (1989) 531–545.

[95] D. Castradori, Paleoceanography 8 (1993) 459–471.

[96] J.P. Sachs, D.J. Repeta, Science 286 (1999) 2485–2488.

[97] A.E.S. Kemp, r.B. Pearce, I. Koizumi, J. Pike, S. Jea Rance, Nature 398 (1999) 57–61.

[98] S.J. Schenau, A. Antonarakou, F.J. Hilgen, L.J. Lourens, I.A. Nijenhuis, C.H. van deer Weijden, W.J. Zachariasse, Marine Geol. 153 (1999) 117–135.

[99] E. Tuenter, S.L. Weber, F.J. Hilgen, L.J. Lourens, A. Ganopolski, Clim. Dyn. 24 (2005) 279–295.

[100] L. Loulergue, A. Schilt, R. Spahni, V. Masson-Delmotte, T. Blunier, B. Lemieux, J.-M. Barnola, D. Raynaud, T.F. Stocker, J. Chappellaz, Nature 453 (2008) 383–386.

[101] R. Spahni, J. Chappellaz, T.F. Stocker, L. Loulergue, G. Hausammann, K. Kawamura, J. Fluckiger, J. Schwander, D. Raynaud, V. Masson-Delmotte, J. Jouzel, Science 310 (2005) 1317–1321.

[102] F. Parrenin, J.M. Barnola, J. Beer, T. Blunier, E. Castellano, J. Chappellaz, G. Dreyfus, H. Fischer, S. Fujita, J. Jouzel, K. Kawamura, B. Lemieux-Dudon, L. Loulergue, V. Masson-Delmotte, B. Narcisi, J.R. Petit, G. Raisbeck, D. Raynaud, U. Ruth, J. Schwander, M. Severi, R. Spahni, J.P. Steffensen, A. Svensson, R. Udisti, C. Waelbroeck, E. Wolff, Clim. Past 3 (2007) 485–497.

[103] J.C. Zachos, N.J. Shackleton, J.S. Revenaugh, H. Pälike, B.P. Flower, Science 292 (2001) 274–278.

[104] H.A. Paul, J.C. Zachos, B.P. Flower, A. Tripati, Paleoceanography 15 (2000) 471–485.

[105] B.S. Cramer, J.D. Wright, D.V. Kent, M.-P. Aubry, Paleoceanography 18 (2003) PA000909.

[106] L.J. Lourens, A. Sluijs, D. Kroon, J.C. Zachos, E. Thomas, U. Röhl, J. Bowles, I. Raffi, Nature 435 (2005) 1083–1087.

[107] J.C. Zachos, D. Kroon, P. Blum, et al., Ocean Drilling Program Initial Rep. 208 (2004) 208.

[108] J.C. Zachos, G.R. Dickens, R.E. Zeebe, Nature 451 (2008) 279–283.

[109] L. Hinnov, in: F. Gradstein, J. Ogg, A. Smith (Eds.), A Geologic Time Scale 2004, Cambridge University Press, UK, 2004, pp. 55–62.

THE ROLE OF VOLCANIC ACTIVITY IN CLIMATE AND GLOBAL CHANGE

26

Georgiy Stenchikov

Division of Physical Sciences and Engineering, King Abdullah University of Science and Technology, Thuwal, Saudi Arabia

CHAPTER OUTLINE

1. INTRODUCTION

Volcanic activity is an important natural cause of climate variations because tracer constituents of volcanic origin impact the atmospheric chemical composition and optical properties. We focus here on the recent period of the Earth's history and do not consider the cumulative effect of the ancient volcanic degassing that formed the core of the Earth's atmosphere billions of years ago. At present, a weak volcanic activity results in gas and particle effusions in the troposphere (lower part of atmosphere), which constitute, on average, the larger portion of volcanic mass flux into the atmosphere. However, the products of tropospheric volcanic emissions are short lived and contribute only moderately to the emissions from large anthropogenic and natural tropospheric sources. This study focuses instead on the effects on climate of the Earth's explosive volcanism. Strong volcanic eruptions with a volcanic explosivity index (VEI) [1] equal to or greater than 4 could inject volcanic ash and sulphur-rich gases into the clean lower stratosphere at an altitude about 25 km–30 km, increasing their concentration there by two to three orders of magnitude in comparison with the background level. The flood lava eruptions may also have significant climatic impact as they could maintain activity for extended time. The well-known Laki eruption in Iceland in 1783 erupted for eight months, injecting sulphur-containing gases both in the troposphere and in the lower stratosphere. Although tropospheric emissions produced a vast environmental effect, the long-term climate effects were mostly caused by the long-lived stratospheric aerosols that were formed by about 10 explosive events during the course of the eruption.

Chemical transformations and gas-to-particle conversion of volcanic tracers form a volcanic aerosol layer that remains in the stratosphere for two to three years after an eruption, thereby impacting

Climate Change. http://dx.doi.org/10.1016/B978-0-444-63524-2.00026-9

the Earth's climate because volcanic aerosols cool the surface and the troposphere by reflecting solar radiation, and warm the lower stratosphere, absorbing thermal IR and solar near-IR radiation [2]. Fig. 1 shows stratospheric optical depth for the visible wavelength of 0.55 μm. It roughly characterizes the portion of scattered solar light. Three major explosive eruptions occurred in the second part of the twentieth century, as depicted in Fig. 1: Agung of 1963, El Chichon of 1982 and Pinatubo of 1991.

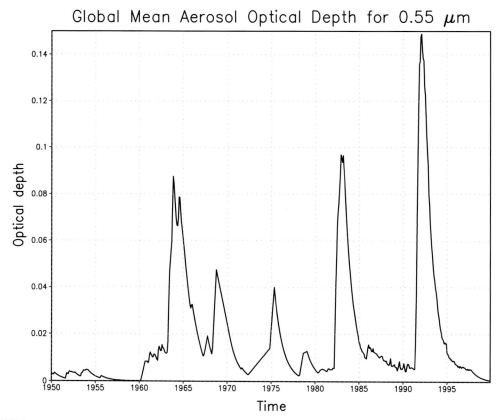

FIGURE 1

The total normal optical depth τ of stratospheric aerosols for the Pinatubo period for the visible wavelength of 0.55 μm as a function of time in years. It causes attenuation of direct solar visible light with a factor of exp(-τ/cos ζ), where cos ζ is a cosine of zenith angle. Note: Optical depth is defined as the negative natural logarithm of the fraction of radiation (e.g. light) that is not scattered or absorbed. It is a dimensionless quantity.

Volcanic eruptions, like the Mt. Pinatubo eruption in 1991, with global visible optical depth maximizing of about 0.16, cause perturbations of the globally averaged radiative balance at the top of the atmosphere reaching -3 W m^{-2} and cause a decrease of global surface air temperature of about 0.5 K. The radiative impact of volcanic aerosols also produces changes in atmospheric circulation, forcing a positive phase of the Arctic Oscillation (AO) and counterintuitive boreal winter warming in middle and high latitudes over Eurasia and North America [3–8]. In addition, stratospheric aerosols affect stratospheric chemistry serving as surfaces for heterogeneous reactions liberating anthropogenic chlorine and causing ozone depletion.

It was traditionally believed that volcanic impacts produced mainly short-term transient climate perturbations. However, the ocean integrates volcanic radiative cooling, and different components of the ocean respond over a wide range of timescales. Volcanically induced tropospheric temperature anomalies vanish in about 7 years–10 years, while volcanically induced sea ice extent and volume changes have a relaxation timescale closer to a decade. Volcanically induced changes in interior ocean temperature, the meridional overturning circulation and steric height have even longer relaxation times, from several decades to a century. Because of their various impacts on climate systems, volcanic eruptions play a role of natural tests, providing an independent means of assessing multiple climate feedback mechanisms and climate sensitivity [7–11].

There are several excellent reviews devoted to volcanic impacts on climate and weather written before the first edition of this book [12–19] and more recently [20]. The present study provides an overview of available observations of volcanic aerosols and discusses their radiative forcing and large-scale effects on climate. It focuses on recently discovered forced stratosphere-troposphere dynamic interaction (recently discussed intensively in literature) and long-term ocean response to volcanic forcing, and aims to add information to that already presented in the previous reviews.

2. AEROSOL LOADING, SPATIAL DISTRIBUTION AND RADIATIVE EFFECT

Volcanic emissions comprised of gases (H_2O, CO_2, N_2, SO_2, H_2S) and solid (mostly silicate) particles, are usually referred to as volcanic ash. Volcanic ash particles are relatively large, exceeding two microns in diameter, and therefore deposit relatively quickly, i.e. within a few weeks. They are responsible for short-term regional-to-continental perturbations of the Earth's radiative balance and meteorological parameters. H_2O, CO_2 and N_2 are abundant in the Earth's atmosphere, so individual volcanic perturbations of their concentrations are negligible. But SO_2 and H_2S, which quickly oxidize to SO_2 if erupted in the stratosphere, could significantly affect stratospheric chemical composition and optical properties. SO_2 gas absorbs UV and IR radiation, producing very strong localized stratospheric heating [21–23]. However, it completely disappears in about half a year and the major long-term impact of volcanic eruptions on climate is due to long-lived sulphate aerosols formed by oxidizing of SO_2 with a characteristic conversion time of about 35 days. Sulphate volcanic aerosols (submicron droplets of highly concentrated sulphuric acid) are transported globally by the Brewer–Dobson stratospheric circulation and eventually fall out in two to three years. A significant amount of volcanic aerosols penetrated to the troposphere through the tropopause folds is washed out in storm tracks. Aerosols deposited in downward branches of the Brewer–Dobson circulation in the polar regions are preserved in the polar ice sheets, recording the history of the Earth's explosive volcanism for thousands of years [24–26]. However, the atmospheric loadings calculated using volcanic time series from high-latitude ice records suffer from uncertainties in observation data and poor understanding of atmospheric transport and deposition processes. The global instrumental observations of volcanic aerosols have been conducted during the last 25 years by a number of remote sensing platforms. Total Ozone Mapping Spectrometer (TOMS) instrumentation onboard the Nimbus-7 provided SO_2 loadings from November 1978 until May 6, 1993 [27]. Prata et al. [28] recently developed a new retrieval technique to obtain SO_2 loadings from TOMS data. The Advanced Very High Resolution Radiometer (AVHRR) provides aerosol optical depth over oceans with 1-km spatial resolution in several visible and near-IR wavebands. However, column observations are not sufficient to reliably separate tropospheric and stratospheric aerosols.

The Stratospheric Aerosol and Gas Experiment (SAGE) and Stratospheric Aerosol Measurement (SAM) projects have provided more than 20 years of vertically resolved stratospheric aerosol spectral extinction, the longest such record. The 3-D observations are most valuable to understand stratospheric aerosols transformations and transport. However, there are significant gaps in the temporal-spatial coverage, e.g. the eruption of El Chichón in 1982 (the second most important in the twentieth century after Mt. Pinatubo) is not covered by SAGE observations because the SAGE I instrument failed in 1981, and SAGE II was only launched in 1984. Fortunately, instruments aboard the Stratosphere Mesosphere Explorer (SME) fill the gap of 1982–1984 in 3-D aerosol observations. The saturation periods when the SAGE instrument could not see the direct sunlight through the dense areas of aerosol cloud also could be partially reconstructed using lidar and mission observations [29,30]. It is important to utilize observations from the multiple platforms to improve data coverage, e.g. combining SAGE II and Polar Ozone and Aerosol Measurement (POAM) data could help to fill in the polar regions. Randall et al. [31,32] have extensively compared the POAM and SAGE data and normalized them, combining them into a consistent data set.

Cryogenic Limb Array Etalon Spectrometer (CLAES), Improved Stratospheric and Mesospheric Sounder (ISAMS), and Halogen Occultation Experiment (HALOE) instruments launched on the Upper Atmosphere Research Satellite (UARS) provide additional information for the post-Pinatubo period. These instruments measure the aerosol volume extinction (HALOE) and volume emission (CLAES, ISAMS) in the near-IR and IR bands. These three infrared instruments provide better horizontal coverage than SAGE but do not penetrate lower than the 100 hPa level. They started operating in September 1991. CLAES and ISAMS stopped working after 20 months. The SAGE III instrument aboard the Russian Meteor III-3M satellite continued the outstanding SAGE aerosol data record [33,34] from 2001 to 2007. The new Moderate Resolution Imaging Spectroradiometer (MODIS) and Multiangle Imaging SpectroRadiometer (MISR) instruments have superior spatial and spectral resolutions but mostly focus on the tropospheric aerosols and surface characteristics, providing column average observations.

Available satellite and ground-based observations were used to construct volcanic aerosol spatiotemporal distribution and optical properties [2,35–40]. Hansen [38] improved a Goddard Institute for Space Studies (GISS) volcanic aerosols data set for 1850–1999, providing zonal mean vertically resolved aerosol optical depth for visible wavelength and column average effective radii. Amman et al. [35] developed a similar data set of total aerosol optical depth based on evaluated atmospheric loadings distributed employing a seasonally varying diffusion-type parameterization that also could be used for paleoclimate applications (if aerosol loadings are available). Amman et al. [35], however, used a fixed effective radius of 0.42 μm for calculating aerosol optical properties and, in general, provided higher values of optical depth than in Hansen [38]. Stenchikov et al. [40] used UARS observations to modified effective radii from Hansen [38], implementing its variations with altitude, especially at the top of the aerosol layer where particles became very small. They conducted Mie calculations for the entire period since 1850 and implemented these aerosol characteristics in the new Geophysical Fluid Dynamics Laboratory (GFDL) climate model. The sensitivity calculations with different effective radii show that total optical depth varies as much as 20% when effective radius changes are in the reasonable range. The study of Bauman et al. [36,37] provides a new approach for calculating aerosol optical characteristics using SAGE and UARS data. Bingen et al. [41,42] have calculated stratospheric aerosols size distribution parameters using SAGE II data. A new partly reconstructed and partly hypothesized climate forcing time series for 500 years, which includes greenhouse gas (GHG) and volcanic effects, was developed by Robertson et al. [43].

Aerosol optical properties include aerosol optical depth (see Fig. 1), single scattering albedo (to characterize aerosol absorptivity) and asymmetry parameter (to define the scattering anisotropy). Using these aerosol radiative characteristics one can evaluate aerosol radiative effect on climate system – aerosol radiative forcing at the top of the atmosphere. Fig. 2 shows the total forcing and its short-wave (SW) and long-wave (LW) components. Increase of reflected SW radiation ranges from 3 W m^{-2} to 5 W m^{-2} but is compensated by aerosol absorption of outgoing LW radiation, so total maximum cooling of the system ranges from 2 W m^{-2} to 3 W m^{-2}.

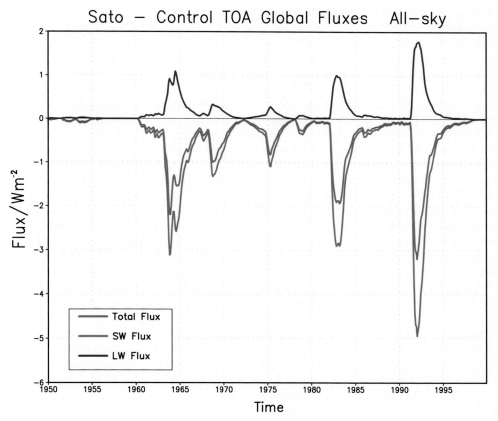

FIGURE 2

Volcanic aerosol total, short-wave (SW) and long-wave (LW) radiative forcing (W m^{-2}) at the top of the atmosphere for all-sky conditions. Positive sign of the forcing corresponds to heating the climate system.

Unfortunately, the above 'empirical' estimates of radiative forcing carry some level of uncertainty due to insufficient observations of aerosol optical properties, their spatial distribution and aerosol abundance. This hampers our ability to assess climate perturbations from volcanic impacts and quantify the associated important climate mechanisms. During the period since 2009, when the first edition of this book was published, a few attempts were made to improve this input information.

Arfeuille et al. [44] used an improved 'gap-filled' SAGE II Version 6 aerosol product [33] to calculate volcanic aerosol characteristics after the 1991 Pinatubo eruption. They compared several approaches of calculating aerosol optical characteristics including that proposed by Stenchikov et al. [2], and built an aerosol data set with high vertical resolution that captures fine structures missing from the data set constructed in Stenchikov et al. [2]. However, they failed to conserve the observed total aerosol optical depth in their calculations and did not provide the aerosol radiative forcing for comparison. Their general circulation model (GCM) calculations, using these data, significantly overestimated the stratospheric temperature responses. This once more emphasizes the difficulties of building precalculated aerosol data sets and the limitations of existing observations.

Another approach, which potentially could shed light on the physical mechanisms governing the development of volcanic plumes and help to resolve some of the problems, is based on GCM simulations interactively accounting for the development of the stratospheric volcanic aerosol layer [45–50].

Timmreck et al. [49], Oman et al. [48] and Aquila et al. [45] used so-called 'bulk' aerosol models when they calculated SO_2 to sulphate conversion and tracked their bulk concentration. The aerosol size distributions were prescribed and optical properties were precalculated. This approach provided a spatiotemporal evolution of an aerosol cloud that is in reasonable agreement with observations; e.g. it showed that on their way to the pole a significant amount of volcanic aerosols are deposited in the midlatitude storm tracks [51], despite the widespread belief that they mostly subside to the pole troposphere in the downward brunches of the Brewer–Dobson circulation. Aquila et al. [45], in the NASA GEOS-5 GCM, specifically studied the effect of aerosol radiative feedback on dispersion of the aerosol cloud after the Pinatubo eruption. Their simulations exhibited quite realistic features although underestimated transport of the aerosol plume into the southern hemisphere. They found that aerosol has been transported from the equatorial reservoir to the north through a lower stratospheric pathway and to the southern hemisphere through a middle stratosphere pathway.

Next in their physical complexity and computational efficiency, modal aerosol models approximate the aerosol number-density distributions with a few log-normal modes with the prescribed width and varying modal radii, accounting for coagulation, condensation growth and size-dependent gravitational settling [47,50].

Niemeier et al. [47] used the Max Planck for Meteorology (MAECHAM5) model to calculate responses to the Pinatubo and hypothetical 10x Pinatubo eruptions accounting both for sulphate and ash aerosols. They were able to obtain fairly realistic spatiotemporal evolution of volcanic cloud and aerosol size distribution. Because of strong absorption, ash increases stratospheric heating by 20 K d^{-1} (where d refers to day) within the ash layer and has a strong impact at the initial stage of the volcanic cloud evolution during the first month after the eruption. But large ash particles rapidly subside, separate from the sulphate layer and do not significantly affect the long-term evolution of a volcanic plume.

The aerosol sectional microphysical models are the most accurate but computationally demanding. English et al. [46] simulated the Pinatubo, hypothetical 10x Pinatubo and Toba (100x Pinatubo) eruptions employing the Whole Atmosphere Community Climate Model (WACCM [52]) coupled with the aerosol microphysical module based on the sectional Community Aerosol and Radiation Model for Atmospheres (CARMA [53]), which is assumed to be the best available tool for microphysical calculation within a general circulation model. This is a remarkable exercise despite that they did not account for the radiative effect of aerosols on the atmospheric circulation. This supposedly produces a moderate effect for the Pinatubo-size eruption but is crucially important for the Toba super-eruption case [54]. We therefore focus here on the simulation of the Pinatubo eruption, especially because this could teach us of the model deficiencies, as the results could be compared with observations. It appears that WACCM/CARMA were able to quite reliably describe the building of the Pinatubo aerosol layer and predicted correct timing when particles reach maximum size in about six months after the eruption, but overestimate the rate of decline of the particle size during the following three years. This could be an effect of neglecting aerosol radiative feedback on circulation, lack of interactive quasi-biannual oscillation (QBO), or caused by a too fast poleward aerosol transport, which is a known deficiency of the relatively coarse resolution GCMs [48]. The total optical depth, which is most important for assessing a volcanic climate impact, shows similar features picking the correct time but declining too rapidly. The model overestimates the descent of the aerosol equatorial reservoir, decreasing the altitudes at which most of the aerosol is transported to poles. The simulated volcanic plume in the southern hemisphere is weaker than observed. The model overestimates the optical depth at 1.024 μm and places too much aerosols in high latitudes where they deposit more rapidly. Overall, despite exiting achievements on the way of interactive simulation of volcanic plumes, they are still far from producing completely reliable results.

3. VOLCANOES AND CLIMATE

The perturbations of the Earth's radiative balance caused by strong volcanic eruptions dominate other forcings for two to three years. Their effect is seen in the atmosphere for about five to seven years, and, as was recently discovered, for much longer in oceans [55–60]. Volcanic perturbations have been used for years as natural experiments to test models and to study climate sensitivity and feedback mechanisms. Many of these studies have focused on simulating the aftermath of the Mt. Pinatubo eruption in the Philippines at 15.1°N, 120.4°E in June 1991 that was both the largest eruption of the twentieth century and the eruption for which the stratospheric aerosol has been best observed [61–67]. During this eruption about 17 Tg (17×10^{12} g) of SO_2 were injected into the lower stratosphere and subsequently converted into sulphate aerosols. There are three main foci of such studies addressed in the present study: analysis of the simulation of atmospheric temperature and precipitation response; simulation of the response of the extratropical circulation in the northern hemisphere (NH) winter season; and, as recently emerged, analysis of volcanic impact on ocean.

The use of volcanic simulations as tests of model climate feedback and sensitivity is somewhat hampered by weather and climate fluctuations because any climate anomalies observed in the aftermath of these eruptions will also reflect other internally generated variability in the atmosphere-ocean system (e.g. El Niño/Southern Oscillation (ENSO), Quasi Biannual Oscillation (QBO), or chaotic weather changes). Due to limited observations one has to use models to better understand the physical processes forced in the climate system by volcanic impacts. With model simulations, one can perform multiple realizations to clearly isolate the volcanic climate signal, but the real-world data are limited to the single realization during the period since quasi-global instrumental records have been available.

Models of different complexity were traditionally used to analyze volcanic climate impacts. Those models might simplify description of atmospheric and/or ocean processes [68] or mimic radiative effect of volcanic aerosol by decreasing of solar constant [69]. In the present study, to illustrate mechanisms of volcanic impacts on climate, a comprehensive coupled climate model, CM2.1, is used. Developed at the National Oceanic and Atmospheric Administration's (NOAA)

Geophysical Fluid Dynamic Laboratory (GFDL), CM2.1 was used in the IPCC AR4 study [56,70]. This model calculates both atmosphere and ocean, and accounts interactively for volcanic aerosol radiative forcing. It is composed of four component models: atmosphere, land, sea ice and ocean. The coupling between the component models occurs at 2-h intervals. The atmospheric model has a grid spacing of 2.5° longitude by 2° latitude and 24 vertical levels. The dynamical core is based on the finite volume scheme of Lin [71]. The model contains a completely updated suite of model physics compared to the previous GFDL climate model, including new cloud prediction and boundary layer schemes, and diurnally varying solar insolation. The radiation code allows for explicit treatment of numerous radiatively important trace gases (including tropospheric and stratospheric ozone, halocarbons etc.), a variety of natural and anthropogenic aerosols (including black carbon, organic carbon, tropospheric sulphate aerosols and volcanic aerosols), and dust particles. Aerosols in the model do not interact with the cloud scheme, so that indirect aerosol effects on climate are not considered. The land model is described in Milly and Shmakin [72]. Surface water is routed instantaneously to ocean destination points on the basis of specified drainage basins. The land cover type in the model uses a classification scheme with 10 different land cover types. The ocean model [73,74] has a nominal grid spacing of 1° in latitude and longitude, with meridional grid spacing decreasing in the tropics to 1/3° near the equator, and uses a tripolar grid to avoid polar filtering over the Arctic. The model has 50 vertical levels, including 22 levels with 10-m thickness each in the top 220 m. A novel aspect is the use of a true freshwater-flux boundary condition. The sea ice model is a dynamical model with three vertical layers and five ice thickness categories. It uses the elastic-viscous-plastic rheology to calculate ice internal stresses, and a modified Semtner three-layer scheme for thermodynamics [75]. The aerosol optical characteristics were calculated following Stenchikov et al. [2] using optical depth from Sato et al. [39] and Hansen [38]. The aerosol size distribution was assumed lognormal with fixed width of 1.8 μm [40].

In this study, various volcanic impacts on climate are illustrated using results from the model experiments and available observations. In each case, twin ensembles of volcano and control runs are conducted, and the response of the climate system calculated to volcanic forcing as the ensemble mean over the volcano runs minus the ensemble mean over the control runs. The variability within ensembles is used to estimate the statistical significance of climate signals.

3.1 TROPOSPHERIC COOLING AND STRATOSPHERIC WARMING

The analysis for the Pinatubo case is easier than for other big eruptions because aerosols were well observed and the climate responses were relatively well documented. However, Pinatubo erupted in an El Niño year and both volcanic and SST effects overlapped at least in the troposphere. ENSO events that occurred near the times of volcanic eruptions could either mask or enhance the volcanic signal. Adams et al. [76] even argued that changing atmospheric circulation caused by volcanic eruptions could cause El Niño. Santer et al. [77] conducted a comprehensive analysis of the ENSO effect on the modelled and observed global temperature trends. Shindell et al. [4] addressed the issue of interfering volcanic and ENSO signals by specific sampling of eruptions so the SST signal will average out in the composite. Yang and Schlesinger [78,79] used singular value decomposition (SVD) analysis to separate spatial patterns of the ENSO and volcanic signals in the model simulations and observations. They showed that ENSO signal is relatively weak over Eurasia but strong over North America, contributing about 50% of the responses after the 1991 Mt. Pinatubo eruption.

The ENSO variability issue is addressed in the present study by comparing simulated and observed responses after extracting the El Niño contribution from the tropospheric temperature. Santer et al. [77] developed an iterative regression procedure to separate a volcanic effect from an El Niño signal using microwave sounding unit (MSU) brightness temperature observations from the lower tropospheric channel 2LT [80]. The globally averaged synthetic 2LT temperature for the Pinatubo ensemble runs is calculated using model output and compared with the response from Santer et al. [77]. The simulated anomaly is calculated with respect to the mean over the corresponding control segments that have the same developing El Niños as in the perturbed runs. It is probably an ideal way to remove the El Niño effect from the simulations because the exact El Niño signal that would have developed in the model if the volcanic eruption did not occur is subtracted. This procedure, however, only works well for the initial El Niño when perturbed runs 'remember' their oceanic initial conditions. Fig. 3 shows a comparison of synthetic ENSO-subtracted anomaly with the observed anomaly from Santer et al. [77] with ENSO removed statistically. Yellow shading shows doubled standard deviation variability for the 10-member ensemble mean. The observed MSU 2LT anomaly itself has much higher variability (not shown) because there is only one natural realization. Thus, the simulated Pinatubo signal in the lower tropospheric temperature reaches -0.7 K; it is statistically significant at 99% confidence level, and the

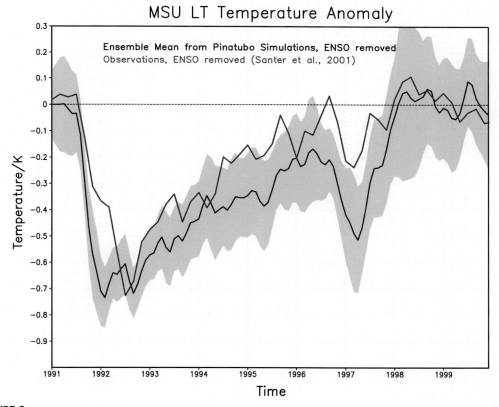

FIGURE 3

The observed Lower Tropospheric MSU 2LT temperature anomaly (K) caused by the Pinatubo eruption from Santer et al. [77] with ENSO effect removed, and the simulated synthetic 2LT ensemble mean temperature anomaly (K) calculated from the Pinatubo ensemble with the El Niño 1991 effect removed; yellow shading shows $\pm 2\sigma$ ensemble mean variability.

difference between simulated and observed responses is below the variability range. The lower tropospheric temperature anomaly reduces below the noise level in about seven years, which corresponds approximately to the thermal response time of the ocean mixed layer [81].

For the lower stratosphere, a similar comparison was conducted as for the lower troposphere, but without removing ENSO because its effect in the lower stratosphere is fairly small. However, the stratospheric response to volcanic forcing might be affected by the phase of a QBO [7,82]. Fig. 4 compares the simulated synthetic MSU channel 4 temperature for the lower stratosphere with the MSU 4 observations. The stratospheric warming is produced by aerosol IR and near-IR absorption. Ramaswamy et al. [83] discussed that the MSU lower stratospheric temperature tends to level in a few years after the Pinatubo eruption; therefore we calculate the anomalies in Fig. 4 with respect to the 1994–1999 mean both in the model and in the observation. The yellow shading shows the $\pm 2\sigma$ ensemble mean variability. The simulated signal compares well with the observation albeit slightly overestimates the stratospheric warming in the second year after the eruption. In the real world, the observed signal could be offset by the easterly phase of QBO in 1992/1993 but not in the model, which lacks QBO. The atmospheric response in the lower stratosphere follows the volcanic forcing and disappears in three years, as expected, when volcanic radiative forcing vanishes.

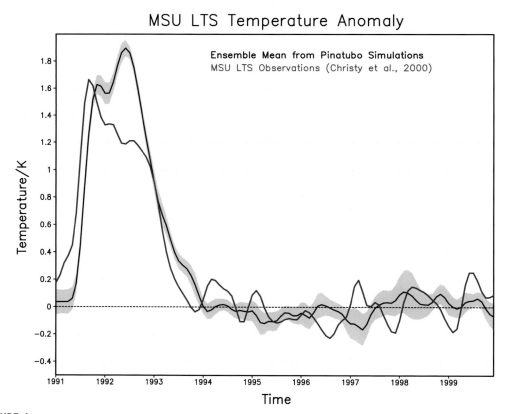

FIGURE 4

The observed MSU 4 Lower Stratospheric temperature anomaly (K) caused by the Pinatubo eruption, and the simulated synthetic channel 4 ensemble mean temperature anomaly (K) calculated from the Pinatubo ensemble; yellow shading shows $\pm 2\sigma$ ensemble mean variability.

3.2 EFFECT ON HYDROLOGICAL CYCLE

Precipitation is more sensitive to variations of solar short-wave radiation than thermal IR radiation because SW radiation directly affects the surface energy budget and links to global precipitation changes through evaporation. Therefore, one could expect that volcanic aerosols might decrease global precipitation for the period of two to three years when volcanic SW radiative forcing remains significant. This effect was detected in observations [84] and in model simulations [85,86]. The Pinatubo case study analysis shows that in the ensemble mean results the global precipitation anomalies (Fig. 5) could be seen for almost five to six years because ocean cools and sea surface temperature (SST) relaxes for about seven years and affects the global hydrologic cycle. The precipitation anomalies over land and over ocean have different dynamics. The land precipitation drops during the first year because of rapid land radiative cooling. The ocean cooling and decrease of precipitation over ocean are delayed and reach maximum values in three to four years after the eruption when the sea surface temperature is coldest. The cold SST tends to shift precipitation over land and the land precipitation goes up, compensating in part the decrease of precipitation over ocean. Geographically, the precipitation anomalies are located in low-latitude monsoon regions and could cause significant disruptions of food production in those regions with very high population density. It must be emphasized that ENSO contributes significantly to the observed precipitation anomalies. When ENSO signal is removed in the model results the amplitude of the precipitation anomalies significantly decreases although temporal behaviour does not change qualitatively.

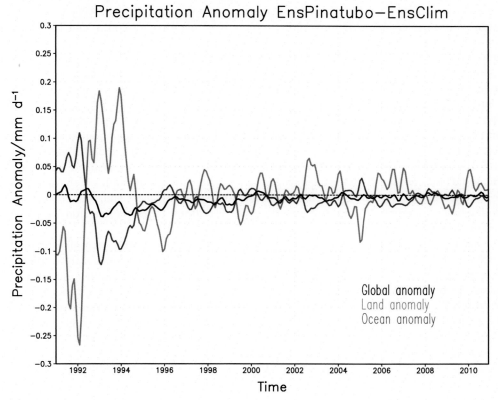

FIGURE 5

Plot of the ensemble mean precipitation anomaly in (mm d^{-1}) against time caused by the Pinatubo eruption averaged over ocean, land and globally, calculated with respect to a climatological mean.

Joseph and Zeng [87] employed a coupled atmosphere-ocean-land-vegetation model of intermediate complexity to study the volcanic aerosol effect on tropical precipitation. They found that the precipitation response over the ocean is slower and weaker than over the land due to the ocean's larger thermal capacity. This land/ocean difference in thermal capacities generates a temperature contrast during the relaxation process that affects precipitation. However, in their model this leads to reduction of rainfall in tropical regions for about two years after an eruption. In contrast with these results, Anchukaitis et al. [88] used well-validated proxy reconstructions of Asian droughts and pluvials to reveal significantly wetter conditions in the year of an eruption over mainland southeast Asia and drier conditions in central Asia. A recent study [89] employed both the model output and observations to examine robust features of the global precipitation response. In agreement with our results they found that the response over oceans remains significant for five years (compare with Fig. 5). The response over land declines in three years. In the tropics, climatologically wet regions become drier, and climatologically dry regions become wetter.

3.3 VOLCANIC EFFECT ON ATMOSPHERIC CIRCULATION

In the two years following major eruptions, the NH winter tropospheric circulation has typically been observed to display features characteristic of an anomalously positive AO index situation. This has a zonal-mean expression with low pressure at high latitudes and a ring of anomalously high pressure in the midlatitudes. This basic zonal-mean pattern is modulated by a very strong regional structure with an intensified high-pressure anomaly over the North Atlantic and Mediterranean sectors called North Atlantic Oscillation (NAO). Consistent with this are poleward shifts in the Atlantic storm track and an increased flow of warm air to Northern Europe and Asia, where anomalously high winter surface temperatures are observed [7,8,40]. It seems that only low-latitude volcanic eruptions could affect the AO/NAO phase and the AO/NAO remain fairly insensitive to the high-latitude NH eruptions [90].

The mechanisms that govern the climate response to volcanic impacts very likely play an important role in global climate change [4,40,91]. The northern polar circulation modes (NAO/AO) experienced significant climate variations during the recent two decades and also are sensitive to volcanic forcing. The southern annular mode (SAM similar to AO) shows recently a very significant climate trend but is not sensitive to volcanic forcing [92]. Supposedly, dynamics of the annular mode in the southern hemisphere are different than in the northern hemisphere and, to a great extent, are controlled by ocean processes and stratospheric ozone variations. In the present study, discussion is limited to NAO/AO.

The most robust effect on atmospheric temperature produced by volcanic aerosols is in the lower stratosphere. It is known that low-latitude explosive eruptions produce anomalously warm tropical lower stratospheric conditions and, in the NH winter, an anomalously cold and intense polar vortex. The tropical temperature anomalies at 50 hPa (Fig. 6) are a direct response to the enhanced absorption of terrestrial IR and solar near-IR radiation by the aerosols. The high-latitude winter perturbations at 50 hPa are a dynamical response to the strengthening of the polar vortex or polar night jet. This is due to stronger thermal wind produced by increasing of the equator-to-pole temperature gradient in the lower stratosphere [6,82,93–97].

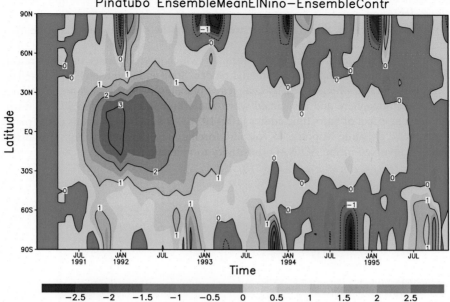

FIGURE 6

Zonal and ensemble mean stratospheric temperature anomaly (K) at 50 hPa (at about 25 km) calculated with respect to control experiment.

The strengthening of the polar jet is amplified by a positive feedback between the polar NH winter vortex and vertical propagation of planetary waves. The stronger vortex reflects planetary waves decreasing deceleration and preserving axial symmetry of the flow. Stenchikov et al. [8] also found that tropospheric cooling caused by volcanic aerosols can affect storminess and generation of planetary waves in the troposphere. This tends to decrease the flux of wave activity and negative angular momentum from the troposphere into the polar stratosphere, reducing wave drag on the vortex. To show this, Stenchikov et al. [8] conducted experiments with only solar, mostly tropospheric and surface cooling (no stratospheric warming). In these experiments, a positive phase of the AO was also produced because aerosol-induced tropospheric cooling in the subtropics decreases the meridional temperature gradient in the winter troposphere between 30°N and 60°N. The corresponding reduction of mean zonal energy and amplitudes of planetary waves in the troposphere decreases wave activity flux into the lower stratosphere. The resulting strengthening of the polar vortex forces a positive phase of the AO.

The high-latitude eruptions cannot warm lower stratosphere, and cannot cool subtropics as much as can low-latitude eruptions. Oman et al. [90] used GISS Model-E to simulate a climate impact of the 1912 Katmai eruption in Alaska. They calculated a 20-member ensemble of simulations and found that the volcanic aerosol cloud spread mostly north of 30°N could not produce a significant winter warming pattern even if it produced a higher hemispheric optical depth than that of the 1991 Pinatubo eruption.

Stenchikov et al. [8] also partitioned the dynamic effect of polar stratospheric ozone loss, caused by heterogeneous chemistry initiated by volcanic aerosols in the post-Pinatubo period. They found that ozone depletion caused a positive phase of the AO in late winter and early spring by cooling the lower stratosphere in high latitudes, strengthening the polar night jet and delaying the final warming.

With respect to the dynamical mechanisms through which perturbations of the stratospheric annular circulation can influence tropospheric annular modes, Song and Robinson [98] pointed out that tropospheric westerlies can be strengthened by changes of planetary wave vertical propagation and/or reflection within the stratosphere and associated wave-zonal flow interaction [93,97,99], downward control or the nonlinear effect of baroclinic eddies [100–103]. All these mechanisms could play a role in shaping tropospheric dynamic response to volcanic forcing. The diagram in Fig. 7 schematically shows the processes involved in the AO/NAO sensitivity to volcanic forcing.

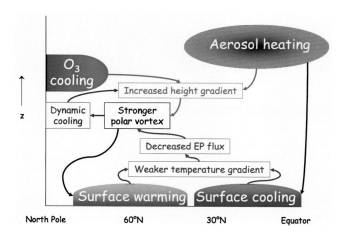

FIGURE 7

Schematic diagram depicting how the stratospheric and tropospheric gradient mechanisms are triggered by volcanic aerosol clouds in the tropical stratosphere. The wave feedback mechanism amplifies the response.

The up-to-date climate models formally include all those processes shown in Fig. 7 but cannot produce the observed amplitude of the AO/NAO variability [40,91]. Shindell et al. [6] reported that the GCM has to well resolve processes in the middle atmosphere in order to reproduce stratospheric influence to the troposphere. Stenchikov et al. [40] composited responses from nine volcanic eruptions using observations and IPCC AR4 model runs. They showed that all models produce a stronger polar vortex in the NH as a response to volcanic forcing but the dynamic signal penetrated to the troposphere is much weaker in the models than in observations. Fig. 8 shows simulations by different models in the course of the IPCC AR4 study, and observed winter warming from Stenchikov et al. [40] that was caused by a poleward shift of tropospheric jet and more intensive transport of heat from ocean to land. The model tends to produce winter warming but significantly underestimates it.

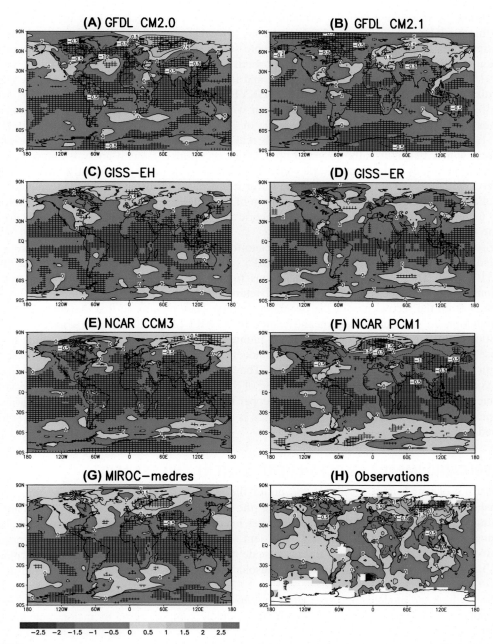

FIGURE 8

Surface winter (December, January, February) air temperature anomalies (K) composited for nine major volcanic eruptions from 1883 until present and averaged for two seasons and all available ensemble members: IPCC AR4 model simulations (a–g); observations from HadCRUT2v data set (h). Hatching shows the areas with at least 90% confidence level calculated using a two-tailed local t-test.

It should be mentioned that the dynamic response to volcanic forcing could interact with the QBO that modulates the strength of polar vortex – it weakens and destabilizes the polar vortex in its easterly phase and makes it stronger and more stable in its westerly phase. The Mt. Pinatubo eruption of 1991 again provides a unique opportunity to test this interaction because in the winter of 1991/92 the QBO was in its easterly phase and in the winter of 1992/93 in its westerly phase. Stenchikov et al. [7] developed a version of the SKYHI troposphere-stratosphere-mesosphere model that effectively assimilates observed zonal mean winds in the tropical stratosphere to simulate a very realistic QBO and performed an ensemble of 24 simulations for the period June 1, 1991, to May 31, 1993. The model produced a reasonably realistic representation of the positive AO response in boreal winter that is usually observed after major eruptions. Detailed analysis shows that the aerosol perturbations to the tropospheric winter circulation are affected significantly by the phase of the QBO, with a westerly QBO phase in the lower stratosphere resulting in an enhancement of the aerosol effect on the AO. Improved quantification of the QBO effect on climate sensitivity helps to better understand mechanisms of the stratospheric contribution to natural and externally forced climate variability.

Driscoll et al. [104] have revisited the Volcano-to-AO connection using model output from the Coupled Model Intercomparison Project 5 (CMIP5 [105,106]) conducted in the scope of the IPCC AR5 [107,108]. They have analyzed simulations from 13 CMIP5 models that represent tropical eruptions from the nineteenth and twentieth centuries and separately sampled the best-observed volcanoes from the second part of the twentieth century. They mentioned that the volcanic aerosol representation did not change significantly since IPCC AR4 [40] and has quite coarse latitudinal resolution in the number of models. The general findings for new CMIP5 simulations are quite similar to those formulated in Boer et al. [9]. The new models are generally unable to produce the observed strength of the volcanic 'winter warming' effect. They underestimate the strength of the post-volcanic polar vortex and tend to overestimate the cooling of the tropical troposphere. The latter suggests that volcanic forcing in the current models is at its upper bound and could not be blamed for insufficient dynamic response in the high latitudes. The reason is probably model deficiencies that have yet to be rectified.

3.4 VOLCANIC IMPACT ON OCEAN HEAT CONTENT AND SEA LEVEL

The Earth's oceans comprise almost the entire thermal capacity of the climate system. Their thermal inertia delays full-scale response of the Earth's surface temperature to greenhouse warming [109]. The rate at which heat accumulates in oceans is an important characteristic of global warming and ultimately defines a climate transient sensitivity (e.g. [110]). It is a complex process that involves slow energy diffusion and large-scale transport in meridional overturning circulation (MOC), as well as faster vertical mixing by seasonal thermohaline convection and by wind-driven gyres.

Observations and model simulations show that the ocean warming effect of the relatively steadily developing anthropogenic forcings is offset by the sporadic cooling caused by major explosive volcanic eruptions [55,57,58]. Delworth et al. [56] conducted a series of historic runs from 1860 to 2000 in the framework of the IPCC AR4 study using GFDL CM2.1, and partitioning contributions of different forcings. Fig. 9 shows the ensemble mean ocean heat content anomalies in the 0 m–3000 m depth range for a subset of the runs from Delworth et al. [56], calculated accounting for all the time varying forcing agents ('ALL') and for volcanic and solar forcings only ('NATURAL'). However, the solar effect for this period is small compared to the volcanic effect. The 'ALL' compares well with the Levitus et al. [111] and Willis et al. [112] observations shown in Fig. 9, and, even better, with the improved analyses from Carton and Santorelli [113] and Dominigues et al. [114] (not shown). Both 'ALL' and 'NATURAL' anomalies are highly statistically significant and far exceed the 'CONTROL' variability shown by yellow shading. The cumulative cooling effect of natural forcings reaches 10^{23} J by year 2000, which is right between the estimates obtained by Church et al. [55] and Gleckler et al. [57] who conducted similar analyses, and offsets about one-third of 'ALL' minus 'NATURAL' ocean warming. The volcanic signal exceeds two standard deviations of unforced variability (yellow shading) throughout the entire run since the Krakatau eruption in 1883. This result suggests that the observed frequency and strength of the Earth's explosive volcanism in the nineteenth and twentieth centuries [1] was sufficient to produce a 'quasi-permanent' signature in the global oceans. Also, ocean warming (cooling) causes expansion (contraction) of water and therefore affects sea level or, so-called, thermosteric height. This effect comprises a significant portion of the observed contemporary sea level rise.

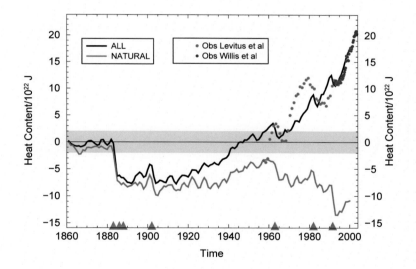

FIGURE 9

The ensemble mean ocean heat content anomalies in the 0 m–3000 m ocean layer. 'ALL' refers to the ensemble mean calculated with all the time varying forcing agents: well-mixed greenhouse gases, anthropogenic aerosols, stratospheric and tropospheric ozone, land use, solar irradiance and volcanic aerosols. 'NATURAL' refers to the ensemble calculated accounting for volcanic and solar forcings only. The red and purple circles depict observational estimates based on, respectively, over 0 m–3000 m layer [111] and over 0 m–750 m layer [112]. Constant offsets have been added to the observed values so that their means are the same as the model results over the period of overlap. The green triangles along the time axes denote the times of major volcanic eruptions. The yellow shading shows plus or minus two standard deviations of ocean heat content estimated from a 2000-year control run of the climate model with forcings fixed at the 1860 level.

To better quantify volcanic impact on ocean, Stenchikov et al. [59,60] calculated a 10-member ensemble of volcano and control 20-year experiments for the Pinatubo period 1991–2010. They found that in contrast to the atmospheric temperature responses, the ocean heat content and the steric height remain well above noise level for decades. Figs 10 and 11 show anomalies of the global ocean heat content and the steric height for the Pinatubo ensembles calculated for the whole-depth ocean and for the upper 300-m layer. Shading depicts the 'two-sigma' ensemble mean variability. The ocean integrates the surface radiative cooling from the volcanic eruption. Since the volcanic aerosols and associated cooling persist for about three years, the anomalies in Figs 10 and 11 reach their maximum value after about this time when the volcanic radiative forcing vanishes. The maximum heat content and sea level decrease in our Pinatubo simulation is 5×10^{22} J and 9 mm, respectively.

The characteristic time, defined as e-folding time for ocean heat content or steric height, is about 40 years–50 years. Assuming that the complete relaxation requires two to three relaxation times, this might take more than a century, and that length of time is sufficiently long for another strong eruption to happen. Therefore, the 'volcanic' cold anomaly in the ocean never disappears at the present frequency of the Earth's explosive volcanism.

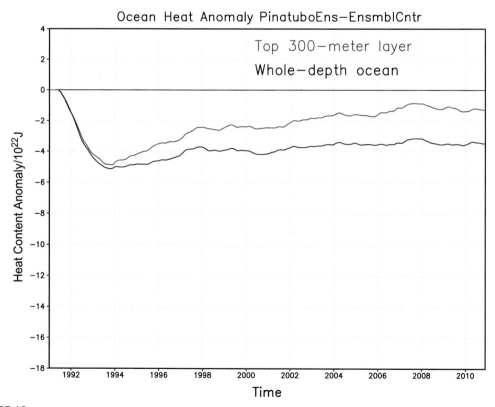

FIGURE 10

The global and ensemble mean ocean heat content (10^{22} J) anomaly for 300 m and whole depth ocean for the Pinatubo ensemble calculated with respect to ensemble control.

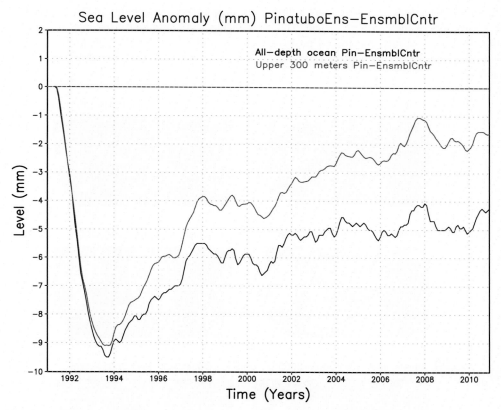

FIGURE 11

The global and ensemble mean thermosteric height anomalies (mm) for 300 m and whole depth ocean for the Pinatubo ensemble calculated with respect to ensemble control. The labels along the *x*-axis and *y*-axis should appear as Level/mm and Time/years, respectively.

3.5 STRENGTHENING THE OVERTURNING CIRCULATION

The short-wave cooling from volcanic aerosol results in a cold surface temperature anomaly that develops during the first three years until volcanic aerosols vanish. Cold surface water is gradually transferred into the deeper ocean layers. A volcanically induced cooling leads to reduced precipitation and river runoff at high latitudes of the northern hemisphere, thereby leading to more saline (and hence denser) upper ocean conditions in the higher latitudes of the NH. Both these factors (colder ocean temperature and enhanced salinity) destabilize the water column, making them more prone to ocean convection. The increased ocean convection tends to enhance the MOC. Further, an enhanced positive phase of the Arctic Oscillation also leads to an MOC increase [115].

As a result, the meridional overturning circulation increases in response to the volcanic forcing (see Fig. 12). The maximum increase is 1.8 Sv or about 10% (here Sv refers to Sverdrup and 1 Sv $= 10^6$ m^3 s^{-1}). The MOC has inherent decadal timescales of adjustment and is thus maximum at some 5 years–15 years after the volcanic eruptions. An increase in MOC also could cause in part the asymmetry of the ocean temperature response in the high northern and southern latitudes.

The simulations show a tendency for cooling of the deepwaters in the Southern Ocean and warming in the deepwaters of the Northern Ocean. This asymmetry also could be caused in part by the redistribution of ocean salinity, the forced positive phase of the Arctic Oscillation during a few years following a volcanic eruption, and by a significant increase of sea ice extent and volume in the northern hemisphere.

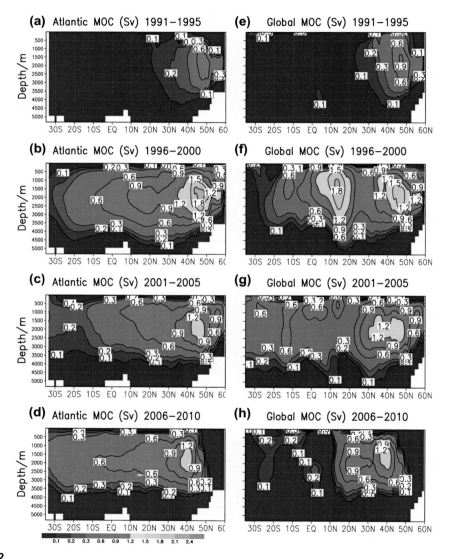

FIGURE 12

The five-year means MOC anomalies (Sverdrup Sv) from the Pinatubo ensemble averaged zonally over Atlantic basin (a–d) and over the globe (e–h).

3.6 **VOLCANIC IMPACT ON SEA ICE**

The effect of volcanic forcing on the sea ice extent in the NH is of great interest because significant loss of perennial sea ice under global warming is occurring in the NH. Therefore, it is very important to better understand what factors could affect them most. Figs 13 and 14 show the anomaly of the northern hemispheric mean annual maximum and minimum ice extent and mass for the Pinatubo runs. The maximum sea ice extent anomalies reach 0.6×10^6 km^2 in the Pinatubo run – it takes at least five years to develop. So, sea ice extent responds more strongly not to the radiative forcing but to ocean temperature and circulation. The sea ice extent relaxes to zero for a decade. It must be mentioned that both observed and simulated ice extent anomalies are not statistically significant, though in simulations they exceed one standard deviation. The minimum ice extent is more sensitive to radiative cooling and ocean temperature, therefore its anomaly is stronger than the anomaly of the maximum ice extent reaching 0.9×10^6 km^2. It builds up in three years when the strongest ocean cooling develops and then declines for about 10 years.

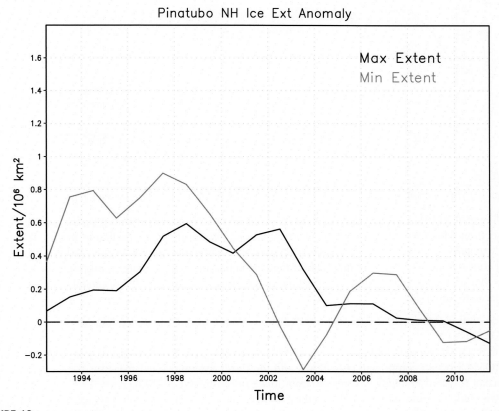

FIGURE 13

The northern hemisphere anomalies of maximum and minimum ice extent (10^6 km^2) for the Pinatubo ensemble.

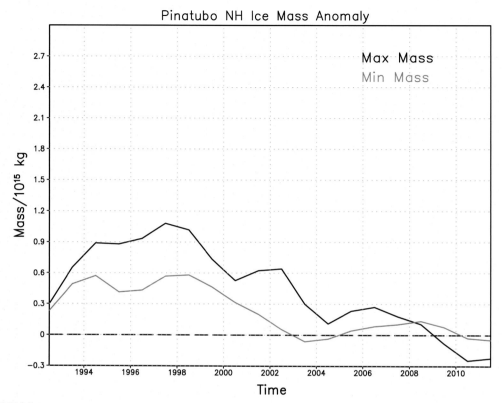

FIGURE 14

The northern hemisphere anomalies of maximum and minimum ice mass (10^{15} kg) for the Pinatubo ensemble.

Ding et al. [116] examined output from eight CMIP5 models to test the findings, discussed above in this section, which were based on our own simulations [59,60] from one particular model. It appears that the results from the CMIP5 project are largely consistent with Stenchikov et al. [59,60]. Ding et al. [116] considered 36 historic simulations for the period that includes the largest eruptions of Krakatau, Santa Maria, Agung, El Chichon and Pinatubo in the last 135 years. The list of models included in the comparison comprises:

The National Centre for Atmospheric Research Community Climate System Model, version 4 (CCSM4)

Three GFDL models including Climate Model version 3 (CM3), the Earth System Model with MOM4.1 ocean component (ESM2M), and the Earth System Model that uses the Generalized Ocean Layered Dynamics isopycnal coordinate ocean model (ESM2G)

Two Hadley Centre models: The Hadley Centre coupled model version 3 (HadCM3) and the Hadley Centre Global Environmental model version 2 – Earth System (HadGEM2-ES)

The Max Planck Institute – Earth System Model – Low Resolution (MPI-ESM-LR) known as ECHAM6

The Japanese Meteorological Research Institute Coupled General Circulation Model version 3 (MRI-CGCM3)

Ding et al. [116] found that all models consistently show an annual average reduction of the surface incoming solar radiation by $1\ W\ m^{-2}$–$5\ W\ m^{-2}$. The global sea surface temperature decreases by 0.1 K–0.3 K. The cool SST signal penetrates in the deep ocean, decreasing temperature in the upper 1000-m layer by roughly 0.03 K. Sea ice extent and mass grow by 5%. Both the deep-ocean and sea ice anomalies persist per decades. Most of the models show strengthening of AMOC with respect to volcanic forcing. The models with the larger AMOC variability (ESM2M, ESM2G, CM3, MPI-ESM-LR, MRI-CGCM3) exhibit the highest AMOC sensitivity. ESM2M and ESM2G are the only models that show warming of the subsurface ocean in the high northern latitudes associated with the strengthening of ocean convection due to increasing salinity.

Iwi et al. [117] also found that change of hydrological cycle and increase of salinity in the high northern latitudes, caused by a volcanic impact, is a driving mechanism of the AMOC intensification. Ottera et al. [118] suggested that volcanic impacts could even synchronize the long-term AMOC variability. Gregory et al. [119] stressed the importance of accounting for preindustrial volcanic forcing to correctly estimate the historical ocean thermal expansion. Balmaseda et al. [120] analyzed ocean heat content (OHC) for the period of 1958–2009 using the improved European Centre for Medium-Range Weather Forecast ocean reanalysis system 4 (ORAS4 [121]). They report the magnitude of volcanic impacts on OHC that is consistent with our model results and emphasized the role of the deeper ocean layers below 700 m in overall ocean heat uptake and its variability.

3.7 EFFECT OF SMALL VOLCANOES, CLIMATE HIATUS AND GEOENGINEERING ANALOGUES

Since the first edition of this book in 2009 two new topics have emerged that have stimulated significant interest and that require more accurate reassessment of climate, microphysical, chemical, and radiative impact of volcanoes of different magnitudes. These topics are of great societal and scientific importance and well deserve a separate full-scale discussion, which is not exactly relevant to the main subject of the current chapter, but we find it useful to briefly mention them in order to put their recent debates in the context of the processes and physical mechanisms discussed here.

The first topic is related to so-called 'climate hiatus' – the observed slowing of global warming since about 2000, despite continued emission of greenhouse gases in the atmosphere [122]. The suggested mechanisms include natural variability associated with increased ocean heat uptake [120,123] and cooling forced by small volcanic eruptions [124–127]. The latter refers to eruptions of Kasatochi in August of 2008, Sarychev in June 2009 and Nabro in June 2011. They were the most significant recent events, however, 15–20 times weaker than Pinatubo. Scaling the observed tropospheric response to Pinatubo (see Fig. 3) we can expect that their maximum global tropospheric cooling effect should not exceed 0.03 K–0.04 K, which is below the natural variability level. More accurate calculations using global models gave estimates ranging from 0.07 K [124,127] to 0.02 K [125]. Therefore, despite that these signals could be detected using a rectified statistical technique [126], it is unlikely that small volcanic eruptions of the twenty-first century could be a primary cause of the climate hiatus.

The second topic is related to solar geoengineering that has been suggested to reduce damage from the anthropogenic climate change [128,129]. Here we solely focus on the technique that is based on deliberate SO_2 injection into the stratosphere. Similarly to volcanic eruptions, SO_2 will be oxidized to produce a permanent sulphate aerosol cloud that reflects solar radiation and therefore diminishes the warming effect of greenhouse gases. This technique is economically and technically quite feasible and has an important advantage above other proposed schemes, as volcanic eruptions provide a natural analogue that helps to better understand and predict possible side effects (e.g. [130–132]). Therefore, understanding of all aspects of stratospheric aerosol microphysics [133], impact on chemical composition [134,135] of the stratosphere, stratospheric circulation [45,136], and tropospheric climate [137–139], which we gain from investigating climate consequences of volcanic eruptions, could be and should be employed for a feasibility assessment of solar geoengineering schemes. The analogy of geoengineering with volcanic eruptions is not complete as it is assumed that the geoengineered stratospheric aerosol layer will provide a quasi-permanent radiative forcing unlike sporadic volcanic forcing. But, similar to volcanic studies, the effect of this forcing on atmospheric circulation, hydrological cycle and ocean heat uptake is of primary interest.

4. SUMMARY

Volcanic eruptions force all elements of the climate system, producing long-term climate signals in ocean. In the atmosphere, however, volcanic signals are masked by meteorological noise in about seven years in the model ensembles and much earlier in the real world. Radiative forcing produced by explosive volcanic events that have occurred in historic periods lasts for about three years. The volcanically induced tropospheric temperature anomalies reduce below noise for approximately seven years. The sea ice responds on the decadal time scale. Deep ocean temperature, sea level, salinity and MOC have relaxation times of several decades to a century. Volcanic eruptions produce long-term impacts on the ocean's subsurface temperature and steric height that accumulate at the current frequency of explosive volcanic events. The vertical distribution of the ocean temperature change signal is asymmetric at high latitudes. A cooling signal penetrates to depth at high southern latitudes, while a warming signal penetrates to depth at high latitudes of the northern hemisphere. This asymmetry is caused in part by an increase in MOC. The decrease of ocean steric height in our simulations, caused by the Pinatubo eruption, reaches 9 mm in comparison with 5 mm estimated by Church et al. [55] from observations. The ocean heat content decreases by 5×10^{22} J. The maximum sea ice extent and ice mass increase are of the order of 0.5×10^{6} km and 21.0×10^{15} kg, respectively. This corresponds to 3% and 5% of the model 'control' maximum extent and mass in the Pinatubo runs. The simulated minimum ice extent is more sensitive to volcanic forcing than the maximum ice extent. The Atlantic MOC strengthens in the Pinatubo runs very significantly by 1.8 Sv or 9% of its maximum value.

Atmospheric temperature anomalies forced by the Pinatubo eruption in the troposphere and lower stratosphere are well reproduced by the models. However, forced AO/NAO responses are underestimated and observed sea level and ocean heat content anomalies are overestimated by all models. Nevertheless, all model results and observations suggest that volcanoes could produce long-lasting impact on ocean heat content and thermosteric level that, in fact, could affect estimates of current climate trends. The quasi-periodic nature of volcanic cooling facilitates ocean vertical mixing and might have an important effect on the thermal structure of the deep ocean. Therefore, it has to be realistically implemented in climate models for calculating 'quasi-equilibrium' initial conditions, climate reconstructions and for future climate projections.

From our evidence the volcanic activity for the historic period has cooled climate and delayed the ocean heat content growth, which is the most robust signature of the Earth's global warming, by about

30%. The relatively weak explosive volcanism of the twenty-first century is unlikely to cause any significant climate effect and would not be considered as a primary cause of the climate hiatus observed in the resent decade that is, most probably, associated with natural climate variability.

REFERENCES

[1] T. Simkin, Annu. Rev. Earth Planet. Sci. 21 (1993) 427–452.

[2] G.L. Stenchikov, I. Kirchner, A. Robock, H.-F. Graf, J.C. Antuña, R.G. Grainger, A. Lambert, L. Thomason, J. Geophys. Res. 103 (1998) 13837–13857.

[3] M. Collins, Volcanism and the earth's atmosphere, in: A. Robock, C. Oppenheimer (Eds.), Predictions of Climate Following Volcanic Eruptions, American Geophysical Union, Washington, DC, 2003, pp. 283–300.

[4] D. Shindell, G. Schmidt, M. Mann, G. Faluvegi, J. Geophys. Res. 109 (2004) D05104.

[5] D. Shindell, G. Schmidt, R. Miller, M. Mann, J. Clim. 16 (2003) 4094–4107.

[6] D.T. Shindell, G.A. Schmidt, R.L. Miller, D. Rind, J. Geophys. Res. 106 (2001) 7193–7210.

[7] G. Stenchikov, K. Hamilton, A. Robock, V. Ramaswamy, M.D. Schwarzkopf, J. Geophys. Res. 109 (2004) D03112.

[8] G. Stenchikov, A. Robock, V. Ramaswamy, M.D. Schwarzkopf, K. Hamilton, S. Ramachandran, J. Geophys. Res. 107 (2002) 4803.

[9] G.J. Boer, M. Stowasser, K. Hamilton, Clim. Dyn. 28 (2007) 481–502.

[10] B.J. Soden, R.T. Wetherald, G.L. Stenchikov, A. Robock, Science 296 (2002) 727–730.

[11] T.M.L. Wigley, C.M. Ammann, B.D. Santer, S.C.B. Raper, J. Geophys. Res. 110 (2005) D09107.

[12] M.L. Asaturov, M.I. Budyko, K.Y. Vinnikov, P.Y. Groisman, A.S. Kabanov, I.L. Karol, M.P. Kolomeev, Z.I. Pivovarova, E.V. Rozanov, S.S. Khmelevtsov, Volcanic Stratospheric Aerosols and Climate (In Russian), Gidrometeoizdat, St. Petersburg, Russia, 1986, pp. 256.

[13] H.W. Elsaesser, Isolating the Climatologic Effects of Volcanoes. Report UCRL-89161, Lawrence Livermore National Laboratory, Livermore, CA, 1983.

[14] K.Y. Kondratyev, Volcanoes and Climate. WCP-54, WMO/TD-166, World Meteorol Org, Geneva, 1988.

[15] K.Y. Kondratyev, I. Galindo, in: A. Deepak (Ed.), Volcanic Activity and Climate, 1997. Hampton, VA, pp. 382.

[16] H.H. Lamb, Philos. Trans. R Soc. London Ser. A 266 (1970) 425–533.

[17] A. Robock, Rev. Geophys. 38 (2000) 191–219.

[18] O.B. Toon, NASA conference publication 2240, in: A. Deepak (Ed.), Volcanoes and climate, in Atmospheric Effects and Potential Climatic Impact of the 1980 Eruptions of Mount St. Helens, 1982, pp. 15–36.

[19] O.B. Toon, J.B. Pollack, Am. Sci. 68 (1980) 268–278.

[20] C. Timmreck, WIREs Clim. Change 3 (2012) 545–564.

[21] M.F. Gerstell, J. Crisp, D. Crisp, J. Clim. 8 (1995) 1060–1070.

[22] D.J. Lary, M. Balluch, S. Bekki, Q. J. R Meteorol. Soc. 120 (1994) 1683–1688.

[23] W. Zhong, J. Haigh, R. Toumi, S. Bekki, Q. J. R Meteorol. Soc. 122 (1996) 1459–1466.

[24] D. Budner, J. Cole-Dai, Volcanism and the earth's Atmosphere, in: A. Robock, C. Oppenheimer (Eds.), The Number and Magnitudes of Large Explosive Volcanic Eruptions between 904 and 1865 A.D.: Quanitative Evidence from a New South Pole Ice Core, American Geophysical Union, Washington, DC, 2003, pp. 165–176.

[25] E. Mosley-Thompson, T.A. Mashiotta, L.G. Thompson, Volcanism and the earth's atmosphere, in: A. Robock, C. Oppenheimer (Eds.), High Resolution Ice Core Records of Late Holocene Volcanism: Current and Future Contributions from the Greenland PARCA Cores, American Geophysical Union, Washington, DC, 2003, pp. 153–164.

[26] K. Yalcin, C. Wake, M. Germani, J. Geophys. Res. 107 (2002) 4012.

[27] A.J. Krueger, S.J. Schaefer, N.A. Krotkov, G. Bluth, S. Barker, Geophys. Monogr. 116 (2000) 25–43.

[28] A. Prata, W. Rose, S. Self, D. O'Brien, Volcanism and the earth's atmosphere, in: A. Robock, C. Oppenheimer (Eds.), Global, Long-term Sulphur Dioxide Measurements from TOVS Data: A New Tool for Studying Explosive Volcanism and Climate, American Geophysical Union, Washington, DC, 2003, pp. 75–92.

[29] J.C. Antuña, A. Robock, G.L. Stenchikov, L.W. Thomason, J.E. Barnes, J. Geophys. Res. Atmos. 107 (2002).

[30] J.C. Antuña, A. Robock, G.L. Stenchikov, J. Zhou, C. David, J. Barnes, L. Thomason, J. Geophys. Res. 108 (2003) 4624.

[31] C.E. Randall, R.M. Bevilacqua, J.D. Lumpe, K.W. Hoppel, J. Geophys. Res. 106 (2001) 27,525–27,536.

[32] C.E. Randall, R.M. Bevilacqua, J.D. Lumpe, K.W. Hoppel, D.W. Rusch, E.P. Shettle, J. Geophys. Res. 105 (2000) 3929–3942.

[33] L. Thomason, T. Peter, SPARC: Assessment of Stratospheric Aerosol Properties (ASAP). WCRP-124, WMO/TD-1295, SPARC Report 4, World Climate Research Program, 2006.

[34] L.W. Thomason, G. Taha, Geophys. Res. Lett. 30 (2003) 1631.

[35] C. Amman, G. Meehl, W. Washington, C. Zender, Geophys. Res. Lett. 30 (2003) 1657.

[36] J. Bauman, P. Russell, M. Geller, P. Hamill, J. Geophys. Res. 108 (2003) 4382.

[37] J. Bauman, P. Russell, M. Geller, P. Hamill, J. Geophys. Res. 108 (2003) 4383.

[38] J. Hansen, J. Geophys. Res. 107 (2002) 4347.

[39] M. Sato, J. Hansen, M.P. McCormick, J. Pollack, J. Geophys. Res. 98 (1993) 22,987–22,994.

[40] G. Stenchikov, K. Hamilton, R.J. Stouffer, A. Robock, V. Ramaswamy, B. Santer, H.F. Graf, J. Geophys. Res. 111 (2006) D07107.

[41] C. Bingen, D. Fussen, F. Vanhellemont, J. Geophys. Res. 109 (2004) D06201.

[42] C. Bingen, D. Fussen, F. Vanhellemont, J. Geophys. Res. 109 (2004) D06202.

[43] A. Robertson, J. Overpeck, D. Rind, E. Mosley-Thomson, G. Zelinski, J. Lean, D. Koch, J. Penner, I. Tegen, R. Healy, J. Geophys. Res. 106 (2001) 14,783–14,803.

[44] F. Arfeuille, B.P. Luo, P. Heckendorn, D. Weisenstein, J.X. Sheng, E. Rozanov, M. Schraner, S. Brönnimann, L.W. Thomason, T. Peter, Atmos. Chem. Phys. 13 (2013) 11221–11234.

[45] V. Aquila, L.D. Oman, R.S. Stolarski, P.R. Colarco, P.A. Newman, J. Geophys. Res. Atmos. 117 (2012) D06216.

[46] J.M. English, O.B. Toon, M.J. Mills, J. Geophys. Res. Atmos. 118 (2013) 1880–1895.

[47] U. Niemeier, C. Timmreck, H.F. Graf, S. Kinne, S. Rast, S. Self, Atmos. Chem. Phys. 9 (2009) 9043–9057.

[48] L. Oman, A. Robock, G. Stenchikov, T. Thordarson, D. Koch, D. Shindell, C. Gao, J. Geophys. Res. 111 (2006) D12209.

[49] C. Timmreck, H.-F. Graf, I. Kirchner, J. Geophys. Res. Atmos. 104 (1999) 9337–9359.

[50] M. Toohey, K. Krüger, U. Niemeier, C. Timmreck, Atmos. Chem. Phys. 11 (2011) 12351–12367.

[51] C.H. Gao, L. Oman, A. Robock, G.L. Stenchikov, J. Geophys. Res. Atmos. 112 (2007) D09109.

[52] R.R. Garcia, D.R. Marsh, D.E. Kinnison, B.A. Boville, F. Sassi, J. Geophys. Res. Atmos. 112 (2007) D09301.

[53] O.B. Toon, R.P. Turco, D. Westphal, R. Malone, M.S. Liu, J. Atmos. Sci. 45 (1988) 2123–2143.

[54] A. Robock, C.M. Ammann, L. Oman, D. Shindell, S. Levis, G. Stenchikov, J. Geophys. Res. Atmos. 114 (2009) D10107.

[55] J. Church, N. White, J. Arblaster, Nature 438 (2005) 74–77.

[56] T.L. Delworth, V. Ramaswamy, G.L. Stenchikov, Geophys. Res. Lett. 32 (2005) L24709.

[57] P.J. Gleckler, T.M.L. Wigley, B.D. Santer, J.M. Gregory, K. AchutaRao, K.E. Taylor, Nature 439 (2006), 675–675.

[58] J. Gregory, J. Lowe, S. Tett, J. Clim. 19 (2006) 4576–4591.

[59] G. Stenchikov, V. Ramaswamy, T. Delworth, in: Impact of Big Tambora Eruption on ENSO, Ocean Heat Uptake, and Sea Level, PP31E-07, Presented at 2007 Fall AGU Meeting, San Francisco, CA, 2007.

[60] G. Stenchikov, T.L. Delworth, V. Ramaswamy, R.J. Stouffer, A. Wittenberg, F. Zeng, J. Geophys. Res. Atmos. 114 (2009) D16104.

[61] A.J. Baran, J.S. Foot, J. Geophys. Res. 99 (1994) 25673–25679.

[62] J.E. Barnes, D.J. Hoffman, Geophys. Res. Lett. 24 (1997) 1923–1926.

[63] G.J.S. Bluth, S.D. Doiron, A.J. Krueger, L.S. Walter, C.C. Schnetzler, Geophys. Res. Lett. 19 (1992) 151–154.

[64] G.J.S. Bluth, W.I. Rose, I.E. Sprod, A.J. Krueger, J. Geol. 105 (1997) 671–683.

[65] A. Lambert, R.G. Grainger, J.J. Remedios, C.D. Rogers, M. Corney, F.W. Taylor, Geophys. Res. Lett. 20 (1993) 1287–1290.

[66] W.G. Read, L. Froidevaux, J.W. Waters, Geophys. Res. Lett. 20 (1993) 1299–1302.

[67] P. Minnis, E.F. Harrison, L.L. Stowe, G.G. Gibson, F.M. Denn, D.R. Doelling, W.L. Smith Jr., Science 259 (1993) 1411–1415.

[68] T. Crowley, Science 289 (2000) 270–277.

[69] A. Broccoli, K. Dixon, T. Delworth, T. Knutson, R. Stouffer, J. Geophys. Res. 108 (2003) 4798.

[70] GFDL, CM2.x References, 2006. Available from: http://nomads.gfdl.noaa.gov/CM2.X/references.

[71] S.-J. Lin, Mon. Weather Rev. 132 (2004) 2293–2307.

[72] P.C.D. Milly, A.B. Shmakin, J. Hydrometeor 3 (2002) 283–299.

[73] A. Gnanadesikan, K.W. Dixon, S.M. Griffies, V. Balaji, M. Barreiro, J.A. Beesley, W.F. Cooke, T.L. Delworth, R. Gerdes, M.J. Harrison, I.M. Held, W.J. Hurlin, H.C. Lee, Z. Liang, G. Nong, R.C. Pacanowski, A. Rosati, J. Russell, B.L. Samuels, Q. Song, M.J. Spelman, R.J. Stouffer, C.O. Sweeney, G. Vecchi, M. Winton, A.T. Wittenberg, F. Zeng, R. Zhang, J.P. Dunne, J. Clim. 19 (2006) 675–697.

[74] S.M. Griffies, A. Gnanadesikan, K.W. Dixon, J.P. Dunne, R. Gerdes, M.J. Harrison, A. Rosati, J.L. Russell, B.L. Samuels, M.J. Spelman, M. Winton, R. Zhang, Ocean Sci. 1 (2005) 45–79.

[75] M. Winton, J. Atmos. Oceanic Technol. 17 (2000) 525–531.

[76] J.B. Adams, M.E. Mann, C.M. Ammann, Nature 426 (2003) 274–278.

[77] B.D. Santer, T.M.L. Wigley, C. Doutriaux, J.S. Boyle, J.E. Hansen, P.D. Jones, G.A. Meehl, E. Roeckner, S. Sengupta, K.E. Taylor, J. Geophys. Res. 106 (2001) 28033–28059.

[78] F. Yang, M. Schlesinger, J. Geophys. Res. 106 (2001) 14,757–14,770.

[79] F. Yang, M. Schlesinger, J. Geophys. Res. 107 (2002) 4073.

[80] J.R. Christy, R.W. Spencer, W.D. Braswell, J. Atmos. Oceanic Technol. 17 (2000) 1153–1170.

[81] S. Manabe, R.J. Stouffer, J. Geophys. Res. 85 (1980) 5529–5554.

[82] I. Kirchner, G. Stenchikov, H. Graf, A. Robock, J. Antuña, J. Geophys. Res. 104 (1999) 19,039–19,055.

[83] V. Ramaswamy, M.D. Schwarzkopf, W. Randel, B. Santer, B.J. Soden, G. Stenchikov, Science 311 (2006) 1138–1141.

[84] K. Trenberth, A. Dai, Geophys. Res. Lett. 34 (2007) L15702.

[85] L. Oman, A. Robock, G.L. Stenchikov, T. Thordarson, Geophys. Res. Lett. 33 (2006) L18711.

[86] A. Robock, Y. Liu, J. Clim. 7 (1994) 44–55.

[87] R. Joseph, N. Zeng, J. Clim. 24 (2011) 2045–2060.

[88] K.J. Anchukaitis, B.M. Buckley, E.R. Cook, B.I. Cook, R.D. D'Arrigo, C.M. Ammann, Geophys. Res. Lett. 37 (2010) L22703.

[89] C.E. Iles, G.C. Hegerl, A.P. Schurer, X. Zhang, J. Geophys. Res. Atmos. 118 (2013) 8770–8786.

[90] L. Oman, A. Robock, G. Stenchikov, G.A. Schmidt, R. Ruedy, J. Geophys. Res. 110 (2005) D13103.

[91] R.L. Miller, G.A. Schmidt, D. Shindell, J. Geophys. Res. 111 (2006) D18101.

[92] A. Robock, T. Adams, M. Moore, L. Oman, G. Stenchikov, Geophys. Res. Lett. 34 (2007) L23710.

[93] K. Kodera, J. Geophys. Res. 99 (1994) 1273–1282.
[94] K. Kodera, Y. Kuroda, Geophys. Res. Lett. 27 (2000) 3349–3352.
[95] K. Kodera, Y. Kuroda, J. Geophys. Res. 105 (2000) 12,361–12,370.
[96] Y. Ohhashi, K. Yamazaki, J. Meteorol. Soc. Jpn. 77 (1999) 495–511.
[97] J. Perlwitz, H.-F. Graf, J. Clim. 8 (1995) 2281–2295.
[98] Y. Song, W.A. Robinson, J. Atmos. Sci. 61 (2004) 1711–1725.
[99] J. Perlwitz, N. Harnik, J. Clim. 16 (2003) 3011–3026.
[100] R. Black, J. Clim. 15 (2002) 268–277.
[101] R.X. Black, B.A. McDaniel, J. Clim. 17 (2004) 3990–4004.
[102] P.H. Haynes, M.E. McIntyre, T.G. Shepherd, C.J. Marks, K.P. Shine, J. Atmos. Sci. 48 (1991).
[103] V. Limpasuvan, D.J.W. Thompson, D.L. Hartmann, J. Clim. 17 (2004) 2584–2596.
[104] S. Driscoll, A. Bozzo, L.J. Gray, A. Robock, G. Stenchikov, J. Geophys. Res. Atmos. 117 (2012) D17105.
[105] K.E. Taylor, R.J. Stouffer, G.A. Meehl, Bull. Am. Met Soc. 93 (2012) 485–498.
[106] CMIP5, CMIP5 Coupled Model Intercomparison Project, 2013. Available from: http://cmip-pcmdi.llnl.gov/cmip5/.
[107] IPCC, Climate Change 2013: The Physical Science Basis. Working Group I Contribution to the Fifth Assessment Report of the Intergovernmental Panel on Climate Change, Cambridge University Press, Cambridge, UK, 2013, pp. 1308 plus annexes.
[108] IPCC, IPCC Fifth Assessment Report (AR5), 2014. Available from: http://www.ipcc.ch/index.htm.
[109] G. Meehl, W. Washington, W. Collins, J. Arblaster, A. Hu, L. Buja, W. Strand, H. Teng, Science 307 (2005) 1769–1772.
[110] T.M. Merlis, I.M. Held, G.L. Stenchikov, F.R. Zeng, L.W. Horowitz, J. Clim. 27 (2014) 7781–7795.
[111] S. Levitus, J. Antonov, T. Boyer, Geophys. Res. Lett. 32 (2005) L02604.
[112] J.K. Willis, D. Roemmich, B. Cornuelle, J. Geophys. Res. 109 (2004) C12036.
[113] J.A. Carton, A. Santorelli, J. Clim. 21 (2008) 6015–6035.
[114] C. Dominigues, J. Church, N. White, P. Gleckler, S. Wijffels, P. Barker, J. Dunn, Nature 453 (2008) 1090–1093.
[115] T.L. Delworth, K.W. Dixon, Geophys. Res. Lett. 33 (2006) L02606.
[116] Y. Ding, J.A. Carton, G.A. Chepurin, G. Stenchikov, A. Robock, L.T. Sentman, J.P. Krasting, J. Geophys. Res. Oceans 119 (2014) 5622–5637.
[117] A.M. Iwi, L. Hermanson, K. Haines, R.T. Sutton, J. Clim. 25 (2012) 3039–3051.
[118] O.H. Ottera, M. Bentsen, H. Drange, L. Suo, Nat. Geosci. 3 (2010) 688–694.
[119] J.M. Gregory, D. Bi, M.A. Collier, M.R. Dix, A.C. Hirst, A. Hu, M. Huber, R. Knutti, S.J. Marsland, M. Meinshausen, H.A. Rashid, L.D. Rotstayn, A. Schurer, J.A. Church, Geophys. Res. Lett. 40 (2013) 1600–1604.
[120] M.A. Balmaseda, K.E. Trenberth, E. Kallen, Geophys. Res. Lett. 40 (2013) 1754–1759.
[121] M.A. Balmaseda, K. Mogensen, A.T. Weaver, Q. J. R. Meteorol. Soc. 139 (2013) 1132–1161.
[122] D.R. Easterling, M.F. Wehner, Geophys. Res. Lett. 36 (2009) L08706.
[123] G.A. Meehl, J.M. Arblaster, J.T. Fasullo, A. Hu, K.E. Trenberth, Nat. Clim. Change 1 (2011) 360–364.
[124] J.C. Fyfe, K. von Salzen, J.N.S. Cole, N.P. Gillett, J.P. Vernier, Geophys. Res. Lett. 40 (2013) 584–588.
[125] J.M. Haywood, A. Jones, G.S. Jones, Atmos. Sci. Lett. 15 (2014) 92–96.
[126] B.D. Santer, C. Bonfils, J.F. Painter, M.D. Zelinka, C. Mears, S. Solomon, G.A. Schmidt, J.C. Fyfe, J.N.S. Cole, L. Nazarenko, K.E. Taylor, F.J. Wentz, Nat. Geosci. 7 (2014) 185–189.
[127] S. Solomon, J.S. Daniel, R.R. Neely, J.P. Vernier, E.G. Dutton, L.W. Thomason, Science 333 (2011) 866–870.
[128] P. Crutzen, Clim. Change 77 (2006) 211–220.
[129] T.M.L. Wigley, Science 314 (2006) 452–454.

[130] B. Govindasamy, K. Caldeira, Geophys. Res. Lett. 27 (2000) 2141–2144.

[131] B. Kravitz, A. Robock, O. Boucher, H. Schmidt, K.E. Taylor, G. Stenchikov, M. Schulz, Atmos. Sci. Lett. 12 (2011) 162–167.

[132] A. Robock, M. Bunzl, B. Kravitz, G.L. Stenchikov, Science 327 (2010) 530–531.

[133] P. Heckendorn, D. Weisenstein, S. Fueglistaler, B.P. Luo, E. Rozanov, M. Schraner, L.W. Thomason, T. Peter, Environ. Res. Lett. 4 (2009) 045108.

[134] G. Pitari, V. Aquila, B. Kravitz, A. Robock, S. Watanabe, I. Cionni, N. De Luca, G. Di Genova, E. Mancini, S. Tilmes, J. Geophys. Res. Atmos. 119 (2014) 2629–2653.

[135] S. Tilmes, R.R. Garcia, D.E. Kinnison, A. Gettelman, P.J. Rasch, J. Geophys. Res. Atmos. 114 (2009) D12305.

[136] V. Aquila, C.I. Garfinkel, P.A. Newman, L.D. Oman, D.W. Waugh, Geophys. Res. Lett. 41 (2014) 1738–1744.

[137] J.M. Haywood, A. Jones, N. Bellouin, D. Stephenson, Nat. Clim. Change 3 (2013) 660–665.

[138] B. Kravitz, K. Caldeira, O. Boucher, A. Robock, P.J. Rasch, K. Alterskjaer, D.B. Karam, J.N.S. Cole, C.L. Curry, J.M. Haywood, P.J. Irvine, D.Y. Ji, A. Jones, J.E. Kristjansson, D.J. Lunt, J.C. Moore, U. Niemeier, H. Schmidt, M. Schulz, B. Singh, S. Tilmes, S. Watanabe, S.T. Yang, J.H. Yoon, J. Geophys. Res. Atmos. 118 (2013) 8320–8332.

[139] S. Tilmes, J. Fasullo, J.F. Lamarque, D.R. Marsh, M. Mills, K. Alterskjaer, H. Muri, J.E. Kristjansson, O. Boucher, M. Schulz, J.N.S. Cole, C.L. Curry, A. Jones, J. Haywood, P.J. Irvine, D.Y. Ji, J.C. Moore, D.B. Karam, B. Kravitz, P.J. Rasch, B. Singh, J.H. Yoon, U. Niemeier, H. Schmidt, A. Robock, S.T. Yang, S. Watanabe, J. Geophys. Res. Atmos. 118 (2013) 11036–11058.

ATMOSPHERIC AEROSOLS AND THEIR ROLE IN CLIMATE CHANGE

27

Jim Haywood

University of Exeter and Met Office Hadley Centre, Exeter, UK

CHAPTER OUTLINE

1. INTRODUCTION

Atmospheric aerosols consist of solid/aqueous particles suspended in the atmosphere and are typically of sizes in the range 0.001 µm–10 µm. Aerosols are generated from a wide range of natural and anthropogenic sources. Natural aerosols include sulphate aerosols that are formed from emissions of sulphur dioxide from volcanic eruptions or from dimethyl sulphide emissions from phytoplankton in the ocean, sea salt particles emitted at the ocean surface, or mineral dust aerosol that is emitted by the effects of wind erosion on arid land. Anthropogenic aerosols, i.e. those that are generated from human activities, include sulphate, nitrate, black carbon and organic carbon aerosols from fossil fuel burning, deforestation fires and burning of agricultural waste.

Aerosols scatter and absorb sunlight and terrestrial radiation and therefore increased concentrations of atmospheric aerosols from anthropogenic activities such as fossil fuel burning and biomass burning lead to modifications of Earth's energy balance and hence modulate climate. Atmospheric aerosols can also act as cloud condensation nuclei and can change the microphysical and radiative properties and hence the lifetime of clouds and further modulate climate. Furthermore, emissions of natural aerosols may be enhanced or reduced in response to changes in climate.

Climate Change. http://dx.doi.org/10.1016/B978-0-444-63524-2.00027-0

For example, should future climate lead to a reduction in soil moisture in a particular area and consequently increase the spatial extent of arid areas, then mineral dust emissions may increase and dust particles will interact with sunlight and terrestrial radiation, further affecting climate. These responses to climate change are known as climate feedback mechanisms, and aerosols are explicitly coupled to the climate system through them.

Here we will describe the life cycle of aerosols in the lowest layer of the atmosphere known as the troposphere and present estimates of their current global distribution (stratospheric aerosols are presented in detail in Chapter 26). We will then use simple approximations to gain a physical understanding of the important parameters associated with atmospheric aerosols in quantifying aerosol-radiation interactions and aerosol-cloud interactions. The role of aerosols in climate feedback mechanisms will be discussed, and recent work on the potential role of aerosols in geoengineering schemes that aim to counterbalance the impacts of global warming will be presented.

2. THE LIFE CYCLE OF TROPOSPHERIC AEROSOLS

Atmospheric aerosols are either *primary* (i.e. emitted directly as particles at source) or *secondary* (i.e. formed from gaseous precursors) and are emitted from a variety of natural and anthropogenic sources. They can form either submicron 'accumulation mode' aerosols or supermicron 'coarse mode' aerosols. Table 1 shows estimates of the global atmospheric emission and global atmospheric burden of aerosols [1–3] showing the aerosol types that are commonly modelled in global general circulation climate models.

Table 1 shows that, in terms of emissions, sea salt, mineral dust and biogenic carbon emissions dominate, but sea salt and mineral dust have the largest column loading as only a small fraction of emissions of biogenic carbon emissions are converted to organic carbon aerosols. The mass loading does *not* give an accurate assessment in terms of the impact of aerosols on climate via the radiative forcing [1,4,5]. This is for two reasons. Firstly, the climate forcing is defined as the perturbation to the Earth's energy balance when compared to preindustrial concentrations. Emissions of industrial pollutants such as sulphur dioxide, NO_x, ammonia, and biomass burning have increased significantly since preindustrial times leading to increased concentrations of sulphate, nitrate and biomass burning aerosols. Secondly, in terms of particle size, sulphate, nitrate and biomass burning aerosols reside in the submicron accumulation mode, which is optically active at solar wavelengths of the electromagnetic spectrum. They are generally efficient cloud condensation nuclei and generate both significant aerosol-radiation interactions (Section 4) and aerosol-cloud interactions (Section 5).

Subsequent to emission, aerosols may undergo complex chemical reactions in the atmosphere and mix with each other either externally (where each particle contains a chemically distinct component) or internally (where each particle contains a combination of different components). Atmospheric aerosols are moved about by the prevailing winds and are removed from the atmosphere either by impacting with the surface of the Earth (dry deposition) or by being washed out by precipitating clouds (wet deposition). Typically, the lifetimes of aerosols in the troposphere range from a few minutes for supermicron particles to several weeks for submicron particles that have been lofted high into the atmosphere.

Table 1 Estimates of the global emission and column loading of aerosols and their precursors [1–3].

	Global emission (Tg a^{-1})	^{2}Burden (mg m^{-2})	Primary or secondary	Accum. or coarse	Anthropogenic sources	Natural sources
Sulphate (as sulphur)	98–127[1]	1.0 ± 0.0	S	A/C	Fossil fuel burning, biomass burning, smelters	Dimethyl sulphide from plankton, SO_2 from volcanoes
Mineral dust	1000–4000[1]	88.2 ± 7.3	P	C	Industrial and agricultural dust from building and ploughing	Windblown dust from arid regions and from deserts
Sea salt	1400–6800[1]	50.3 ± 0.1	P	C	N/A	Breaking waves
Fossil-fuel black carbon	3.6–6^1	0.5 ± 0.0	P	A	Fossil fuel burning, particularly diesel fuels	N/A
Biomass burning	29–85[1]	1.8 ± 0.1	P/S	A	Deforestation, agricultural waste, biofuels	Lightning initiated wildfires
Biogenic carbon	440–720[1]	2.2 ± 0.0	S	A/C	N/A	Monoterpenes, isoprene
Fossil fuel organic carbon	~ 13[2]	0.4 ± 0.0	P/S	A	Fossil fuel burning	N/A
Nitrate (N)	~ 80[3]	0.2 ± 0.0	S	A/C	Mainly in the form NH_4NO_3 from NO_x from fossil fuel burning and NH_x emissions from agriculture	Soils, oceans, lightning

The burdens shown represent a multi-year mean derived from the model and the ± estimates show the variability in the model and do not represent the full variability of the emissions, and inter-model variability. 'P' represents primary, while 'S' represents secondary aerosols. 'A' represents accumulation-mode aerosols and 'C' coarse mode aerosols.

3. THE SPATIAL DISTRIBUTION OF TROPOSPHERIC AEROSOLS

The geographically localized sources and sinks and relatively short atmospheric lifetimes give aerosols an extreme spatial and temporal nonhomogeneity in atmosphere when compared to long-lived greenhouse gases such as carbon dioxide. The geographic distribution of aerosols is frequently shown as either the column integrated mass distribution or, more frequently, as the aerosol optical depth ($\tau_{aer\lambda}$), which is defined by:

$$\tau_{aer\lambda} = \int_0^{TOA} k_{e\lambda}\rho dz \tag{1}$$

Here $k_{e\lambda}$ is the wavelength-dependent specific extinction efficiency, which is a function of the size and composition of a distribution of atmospheric aerosols and has units of $m^2\,g^{-1}$, and ρ is the atmospheric concentration of aerosols in $g\,m^{-3}$. The integral is over the depth, dz, of the atmosphere from the surface to the top of the atmosphere (TOA). Aerosol optical depth or $\tau_{aer\lambda}$ is a dimensionless quantity and is the product of the column loading of the atmospheric aerosol and the optical efficiency of the aerosol. τ_λ may be related to the reduction in the intensity of direct solar radiation from the overhead sun via Beer's law:

$$I(z) = I_0 e^{-\tau_{aer\lambda}} \tag{2}$$

where $I(z)$ is the measured intensity at point z and I_0 is the incident intensity. Figure 1 shows the spatial distribution of aerosols (in terms of their optical depth) derived from models participating in the AEROCOM project [6,7] and from the MACC-II project [8].

Figure 1(a) and (b) shows a very different spatial distribution between the accumulation mode and coarse mode aerosols. Accumulation mode aerosols are preferentially located near sources of industry and areas where agricultural practices and deforestation lead to significant biomass burning activities. Coarse mode aerosols originate mainly from sea salt generation from breaking waves in areas where the near-surface wind speed is high or from mineral dust emissions in arid desert regions. The data shown in Fig. 1(a) and (b) represent annual mean data. However, as demonstrated in Fig. 1(c), there is a high degree of variability at any particular location and at any particular time, owing to, e.g. seasonal cycles in emissions such as biomass burning, or from meteorological variability in, e.g. wind speed that drives emissions of mineral dust and sea salt.

In terms of the vertical profile of aerosols, generally, concentrations of tropospheric aerosols are highest close to the surface of the Earth and decrease with altitude because aerosols are either emitted directly from the surface or from gaseous precursors that are emitted from the surface, but again on an instantaneous basis considerable variability can occur depending on the prevailing meteorological conditions [11].

Both the horizontal and vertical profiles of aerosols are important in determining their climatic effects as we shall see in following sections.

4. AEROSOL-RADIATION INTERACTIONS

Here, instead of describing state-of-the-science radiative transfer models that are used in conjunction with global climate models to determine the impact of aerosols, we present relatively simple conceptual models that provide a physical insight into the impacts of aerosols on climate. It has long been

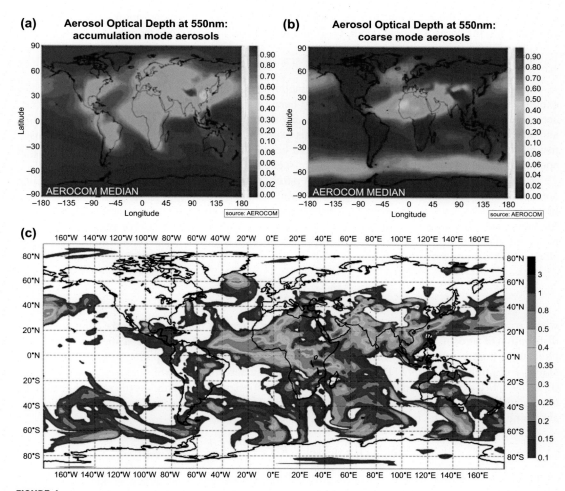

(a) Aerosol Optical Depth at 550nm: accumulation mode aerosols

(b) Aerosol Optical Depth at 550nm: coarse mode aerosols

(c)

FIGURE 1

The global annual mean aerosol optical depth at 550 nm (a dimensionless quantity as defined by Eqn (1)). (a) Shows accumulation mode particles consisting of sulphate, black carbon, organic carbon and biomass burning smoke. (b) Shows coarse mode particles consisting primarily of mineral dust and sea salt aerosols. Values for (a) and (b) are freely available from http://aerocom.met.no/data.html. (c) Shows an instantaneous snapshot of all aerosol components derived from a global model [9] that assimilates aerosol data from the MODIS satellite [10]. Values are shown for 19 January 2015 (freely available from http://www.gmes-atmosphere.eu/). Note that the scales are different between (a and b) and (c).

recognized that anthropogenic aerosols such as sulphate aerosol from sulphur dioxide emissions can exert a climatic impact. A simple expression to determine the annual mean radiative forcing, $dF_{aerosol}$, of purely scattering sulphate aerosols owing to their reflection of sunlight back to outer space may be derived [12]. The expression contains a factor of ½ to account for the fact that any point on the Earth's surface is illuminated approximately 50% of the time and is also a function of the atmospheric transmission (T_{at}), the reflectance of the underlying surface (R_s), the cloud amount (A_c), the solar constant (S_o), the aerosol optical depth, (τ_{aer}) and the aerosol up scattered fraction, β.

$$dF_{aerosol} = -\frac{1}{2}S_o T_{at}^2 (1 - A_c)(1 - R_s)^2 \beta_\lambda \tau_{aer\lambda} \tag{3}$$

Technically, S_o, T_{at} and R_s should also be a function of wavelength, λ, but we ignore this here. This expression can only give a negative forcing and hence purely scattering anthropogenic aerosols (i.e. aerosols that appear white as they do not absorb any sunlight) will lead to a negative radiative forcing and hence a cooling of climate. Additionally, this expression suggests that a purely scattering aerosol residing above a dark scene such as a cloud-free ocean surface (low R_s) will give a much stronger cooling than the same aerosol overlying a bright scene such as a snow surface or a cloud (high R_s).

However, most atmospheric aerosols are not purely scattering owing to the presence of absorbing components such as black carbon (soot) and 'brown' organic carbon from burning of fossil fuels and biomass, or the presence of absorbing mineral components. This absorption makes aerosol appear grey for industrial and biomass burning aerosols or red/orange/yellow in the case of mineral dust. The parameter that determines the degree of absorption/scattering is known as the aerosol single-scattering albedo, $\omega_{o\lambda}$, defined as:

$$\omega_{o\lambda} = \frac{\sigma_{sca\lambda}}{\sigma_{sca\lambda} + \sigma_{abs\lambda}} \tag{4}$$

where σ_{sca} and σ_{abs} are the scattering and absorption efficiencies that are frequently computed for a distribution of aerosols using the scattering theory of Mie [13]. An aerosol single-scattering albedo of 1 indicates a purely scattering (white) aerosol. Aerosols such as industrial pollution typically exhibit $\omega_{o0.55\mu m}$ of around 0.85–0.97 [11] while biomass burning aerosols can exhibit $\omega_{o0.55\mu m}$ in the region 0.75–0.95 [14], depending on the intensity of the combustion, which determines the ratio of absorbing black carbon to less absorbing organic carbon. A simple expression for the forcing from a partially absorbing aerosol can be formulated [15]:

$$dF \approx -\frac{1}{2}S_o T_{at}^2 (1 - A_c)\left[(1 - R_s)^2 - \frac{2R_s}{\beta}\left(\frac{1}{\omega_o} - 1\right)\right]\omega_o \beta \tau_{aer}$$

From this equation (where we have dropped the wavelength dependency of the notation for simplicity), the term in the square bracket determines whether a partially absorbing aerosol exerts a negative forcing (cooling) or positive forcing (warming) of climate. For a cooling of climate, an inequality may be written as:

$$\omega_o > \frac{2R_s}{\beta(1 - R_s)^2 + 2R_s}$$

Thus, an aerosol with a mean β of 0.2 and ω_0 of 0.8 will exert a positive forcing of climate (warming) if it exists above highly reflective cloud that might have an effective R_s of 0.6, while it will exert a negative radiative forcing of climate (cooling) if it exists above a low reflectance ocean surface that typically has a surface reflectance of around 0.05. This effect can readily be observed from satellite retrievals as shown in Fig. 2.

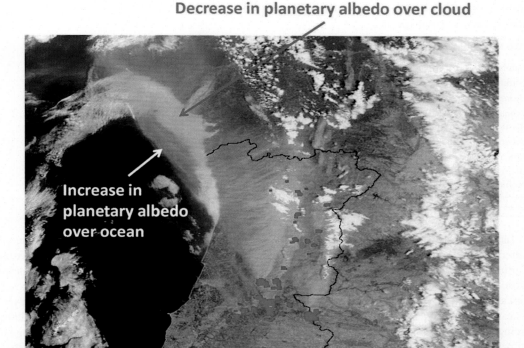

FIGURE 2

A real-colour MODIS satellite image showing the impact of smoke from biomass burning fires off the coast of Portugal on 3 August 2003. Fires are shown by the red spots and the smoke plume is shown in grey.

The impact of aerosols via aerosol-radiation interaction therefore depends to a high degree on the amount of absorption characterized by the single scattering albedo, ω_0. The most recent Intergovernmental Panel on Climate Change report [5] assessed the aerosol-radiation interaction due to increased concentrations of aerosols to give a climate forcing of -0.27 W m^{-2} with a range $+0.23$ W m^{-2} to -0.77 W m^{-2}. The component aerosols tend to exert a negative or neutral radiative forcing of climate with the exception of black carbon, which is the most strongly absorbing species of atmospheric aerosol. The breakdown of species is shown in Table 2.

Table 2 Showing the direct radiative forcing (for the period 1750–2011) from different atmospheric aerosol species.

Aerosol species	Estimate (W m^{-2})	Range (W m^{-2})
Sulphate	−0.4	−0.6 to −0.20
Nitrate	−0.11	−0.30 to −0.03
Biomass burning aerosol	+0	−0.2 to +0.2
Primary organic carbon from fossil fuel and biofuel	−0.09	−0.16 to −0.03
Secondary organic aerosol	−0.03	−0.27 to +0.20
Mineral dust	−0.10	−0.30 to +0.10
Black carbon from fossil fuel and biofuel	+0.40	+0.05 to +0.80

Adapted from Ref. [5].

Table 2 shows that the estimates of the direct radiative forcing tend to be negative for almost purely scattering ($\omega_{o0.55} \sim 1$) aerosols such as sulphate and nitrate. Biomass burning aerosols contain a mixture of primarily scattering inorganic and organic species, mixed with strongly absorbing black carbon aerosols. The $\omega_{o0.55}$ for these aerosols is typically in the range 0.75–0.95 [14], and hence these aerosols exert a direct radiative forcing that is approximately neutral, but significant uncertainty is associated with this estimate (–0.2 W m^{-2} to +0.2 W m^{-2}). Secondary organics also exert a close-to-neutral radiative forcing although again there is significant uncertainty (−0.27 W m^{-2} to +0.20 W m^{-2}) in the range of estimates owing to the presence of significant amounts of 'brown' partially absorbing organic carbon. Black carbon from fossil fuel and biofuel exerts a strong positive direct radiative forcing owing to the low values of the single scattering albedo, which can be as low as $\omega_{o0.55} \sim 0.5$.

The fact that airborne black carbon exerts a positive radiative forcing (i.e. a warming of climate) sets it apart from other aerosol species as anthropogenic emissions of black carbon tend to exacerbate global warming from emissions of greenhouse gases and their precursors. The warming impact of black carbon is further exacerbated by its impact on the reflectance of snow surfaces [16]. When absorbing aerosols such as black carbon are deposited onto bright snow and ice surfaces, they darken them significantly, which reduces the planetary albedo, leading to an additional positive radiative forcing of climate (warming) that needs to be considered when assessing their impact on climate. Because black carbon warms climate, mitigation strategies that aggressively target black carbon emissions have been proposed, offering not just a reduction in the positive radiative forcing that contributes to global warming but improvements to human health. This is because black carbon itself, together with NO_x emissions associated with diesel engines that are a significant source of black carbon, are respiratory irritants that adversely impact air quality and human health [17].

To summarize, the net impact of aerosol-radiation interactions tend to produce a cooling of climate from the associated net negative radiative forcing. However, this simple statement hides a significant complexity with primarily scattering aerosols (i.e. sulphate, nitrate and primary organic carbon) exerting a significant negative radiative forcing, others (i.e. biomass burning, secondary organic aerosols and mineral dust) exerting a close to neutral radiative forcing, and black carbon from fossil fuels and biofuels exerting a significant positive forcing. This complexity leads to a significant uncertainty associated with the aerosol-radiation interaction.

5. AEROSOL-CLOUD INTERACTIONS

Aerosols can also influence climate through their role as cloud condensation nuclei (CCN). Again, we provide a relatively simple conceptual framework here rather than a detailed microphysical evaluation, which necessarily needs to consider complexities of the physical distribution of cloud droplets within clouds, which varies with altitude, updraft velocity, shear, ice water content etc. Our conceptual framework assumes that clouds are made up of a monodisperse distribution of cloud droplets (i.e. droplets of a single size) and is based on the fact that the optical depth of a cloud, τ_{cloud}, can be related to the cloud liquid water path, LWP (units of kg m^{-2}), and the cloud droplet particle size represented by the cloud droplet effective radius, r_{eff} (units of m), by the relationship:

$$\tau_{cloud} \propto \frac{3LWP}{2r_{eff}}$$

Suppose a pristine, nonprecipitating cloud has an effective radius of 16 μm (typical for unpolluted boundary layer clouds). If anthropogenic aerosols are introduced to the system, then if each of the aerosol particles is equally active as a CCN, the cloud liquid water will be distributed equally over a larger number of CCN. Thus, the effective radius, r_{eff}, of the polluted cloud will decrease when compared to the pristine cloud. Clouds made up of smaller cloud droplets have a higher optical depth and hence a higher reflectivity than clouds with identical liquid water contents made up of fewer, larger cloud droplets. This can readily be demonstrated by the following graphic. For the same cloud liquid water content, there are eight times more cloud droplets with effective radius of 8 μm than for those of 16 μm (liquid water content is proportional to the cloud droplet radius cubed), and the cloud made up of 8 μm particles obviously has a higher reflectance as it appears brighter.

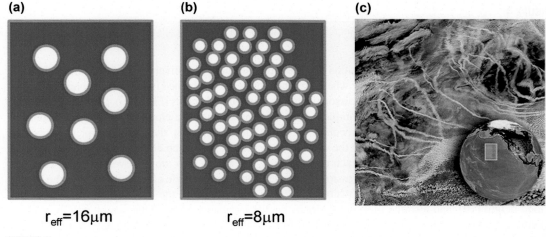

(a)   **(b)**   **(c)**

$r_{eff}=16\mu m$ $r_{eff}=8\mu m$

FIGURE 3

Demonstrating that cloud brightness is inversely proportional to the cloud effective radius, r_{eff}. (a) Shows a schematic of a cloud made up of droplets of 16 μm, while (b) shows a cloud with identical liquid water content made up of clouds of 8 μm. The cloud made up of smaller cloud droplets is obviously 'brighter'. This effect is clearly demonstrated by the 'ship tracks' shown in (c) where clouds are brightened subsequent to the injection of aerosols and their precursors from ships. Note: the label under (a) should read $r_{eff} = 16$ μm and under (b) should read $r_{eff} = 8$ μm.

One further aspect that is important in aerosol-cloud interaction is the concept of *susceptibility*. This arises because there is an asymptotic limit to the impact of injecting aerosols into clouds – as the number of aerosol CCN increases, the total number of cloud droplets in the cloud tends to saturate (Fig. 4). Thus pristine clouds are more susceptible to changes in aerosol concentration, while injecting aerosol into polluted cloud has little effect on the number of cloud droplets and hence the cloud brightness.

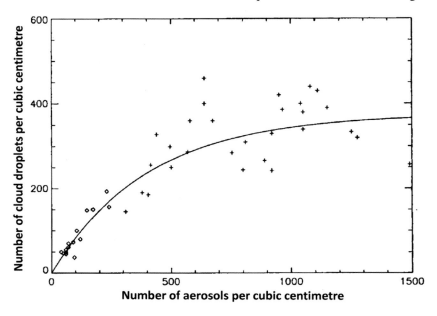

FIGURE 4

Showing that as the number of aerosols increases, the number of cloud droplets tends to asymptote. *(Adapted from Ref. [18].)*

While the impact of aerosol emissions on susceptible clouds is relatively well documented through the regular observation of 'ship tracks' where pristine clouds are brightened in the wake of ships owing to the emission of aerosols and their precursors from the ships funnel stacks (Fig. 3(c)), the overall impacts of aerosols on clouds are far more complex. This is because, once the cloud droplet number concentration is increased, microphysical feedbacks occur. For example, if the cloud droplet number is increased and the radius reduced, then the coalescence efficiency between cloud droplets is reduced and the cloud droplets are less likely to reach the critical size at which they are removed from the cloud in precipitation, thus leading to the suppression of drizzle. Decreasing drizzle effectively reduces the loss of liquid water content from the cloud, which can lead to an increase in the cloud development and the cloud height [19], and also the lifetime of clouds could be enhanced [20]. This complexity in aerosol-cloud interactions makes clear definitive observations of their overall impact extremely difficult as the forcing is coupled to fast feedback processes.

In terms of modelling, high-resolution detailed microphysical models that attempt to represent aerosol-cloud interactions need to be run at resolutions of the order of metres to represent the complexities of cloud updraft, entrainment and aerosol activation. These models are limited to domain sizes that are typically a few tens of kilometres, and they cannot therefore adequately represent the range of different aerosols, clouds and meteorological conditions that are experienced across the globe. Conversely, while global general circulation models can represent the geographic distribution of these features, their spatial resolution of about 100 km (i.e. a single grid box of size about 100 km in the horizontal dimension) means that they cannot adequately represent the complex microphysical processes of aerosol-cloud interaction. While there have been some efforts to capture the sub-grid scale processes within global general circulation models, the problem is far from solved.

The Intergovernmental Panel on Climate Change (IPCC) [1,5] recognizes that earlier assessments had not included the 'fast feedback' terms that are implicit within aerosol-cloud interactions and uses the concept of *effective* radiative forcing to incorporate impacts beyond the change in cloud reflectivity due to increased droplet number concentration. However, they also reiterate the findings of previous assessments, i.e. that there are still considerable uncertainties surrounding the global effective radiative forcing for aerosol-cloud interactions owing to the complexities and difficulties noted above. The effective radiative forcing is estimated as -0.45 W m^{-2} with a range of -0.95 W m^{-2} to $+0.05$ W m^{-2}. Thus, while it is extremely likely that aerosol-cloud interactions contribute a cooling of climate, it is possible (but extremely unlikely given our current knowledge) that the fast feedback impacts could lead aerosols to exert a neutral impact or even a slight warming of climate.

While we have concentrated on the impact of atmospheric aerosols on climate through their influence on the planetary albedo and the top of atmosphere radiative forcing, we should also note that aerosols are also strongly implicated in the observed widespread 'dimming' and 'brightening' phenomena, i.e. changes in the amount of sunlight reaching the surface of the Earth [21].

6. THE NET RADIATIVE FORCING OF AEROSOLS

The net radiative forcing of aerosols can be estimated from the combined aerosol-radiation interaction and the aerosol-cloud interaction. Generally, the approach that has been adopted is a Bayesian statistical approach using a priori assumptions of the probability distribution of the individual forcing components [22]. The total net radiative forcing of greenhouse gases, aerosols and the total anthropogenic radiative forcing from the IPCC [5] is shown in Fig. 5.

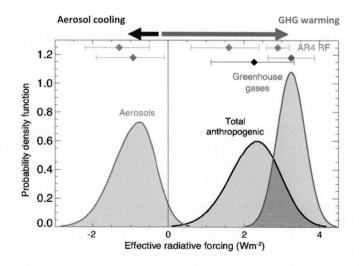

FIGURE 5

The probability density function of the total effective radiative forcing of greenhouse gases (red line), aerosols (blue line), and the net radiative forcing of all anthropogenic forcings (black line). *(Reproduced with permission, courtesy of the IPCC [5].)* Note, the x-axis label should read: effective radiative forcing/W m^{-2}.

From Fig. 4, it is apparent that while increased concentrations of greenhouse gases have certainly led to a positive effective radiative forcing of climate and hence a warming effect, the impact of increased concentrations of anthropogenic aerosols is very likely a smaller counterbalancing cooling of climate. The resulting net effective radiative forcing is a positive net forcing of the climate system and hence a warming, but the uncertainties in the probability density function of atmospheric aerosols feeds through to cause a far greater uncertainty in the net radiative forcing of climate than if only greenhouse gases were considered. This emphasizes the important but poorly quantified role of aerosols in driving the climate of the Earth in the past and suggests that aerosol-radiation and aerosol-cloud interactions are key uncertainties in future climate scenarios.

7. THE ROLE OF AEROSOLS IN CLIMATE FEEDBACK MECHANISMS

In addition to the role of anthropogenic aerosols in forcing the climate system, natural aerosols such as mineral dust, sea salt and biogenic aerosols are intrinsically coupled to the climate system through feedback processes. If, for example, future climate change leads to increased temperatures and decreased precipitation in marginal Sahelian regions of Africa, one might expect a reduction in natural vegetation and soil moisture, leading to increased aridity. The larger fraction of bare soils leads to increased concentrations of atmospheric mineral dust [23]. For another example, if near-surface wind speeds were to increase globally, then over land areas concentrations of mineral dust might be expected to increase, while over oceans sea salt generated from increased wave breaking would also likely increase [24]. Biogenic aerosols generated from, e.g. forests and other vegetation, also play a role in climate feedback mechanisms as changes in vegetation necessarily change emissions of biogenic organic aerosols [25]. For all of these cases, it is important to distinguish between a *forcing*, i.e. a radiative change caused by changes in emissions of anthropogenic aerosol, and a *feedback*, i.e. a change in aerosol concentrations in response to climate change.

8. THE ROLE OF AEROSOLS IN POTENTIAL CLIMATE ENGINEERING SCHEMES

As emphasized in Section 6 and Fig. 5, increased concentrations of atmospheric greenhouse gases cause a warming of climate, and the net impact of aerosols causes a net cooling of climate through aerosol-radiation and aerosol-cloud interactions. While mitigation and adaptation are well-acknowledged strategies for coping with global warming and climate change, recent interest has also turned to potential 'climate engineering' or 'geoengineering' schemes designed to reflect more sunlight out to space and hence exert a cooling to counterbalance the warming from increased concentrations of greenhouse gases. This methodology is frequently referred to as solar radiation management (SRM) as the route for controlling the climate of the Earth is via modifications to the amount of solar radiation (sunlight) that is absorbed by the Earth/atmosphere system.

The simplest model of the Earth's temperature, a 'ball-bearing' Earth, where we neglect the presence of the atmosphere, suggests that, in equilibrium, the total amount of sunlight absorbed by the planet must equal the amount of terrestrial radiation emitted by the planet as determined by the Stefan–Boltzmann law. Thus:

$$\text{Sunlight absorbed} = S_o \pi r^2 \left(1 - \alpha_p\right)$$

$$\text{Terrestrial radiation emitted} = 4\pi r^2 \varepsilon \sigma T^4$$

where S_0 is the solar constant (1370 W m^{-2}), r is the radius of the Earth, α_p is the planetary albedo (i.e. the ratio of the reflected to incident sunlight), ε is the emissivity of the Earth, which we assume to be unity, and σ is the Stefan–Boltzmann constant $= 5.67 \times 10^{-8}$ W m^{-2} K^{-4} This results in an effective temperature of the Earth, T of:

$$T = \sqrt[4]{\frac{S_0(1 - \alpha_p)}{4\varepsilon\sigma}}$$

Solving this equation results in a temperature of the Earth of around 255 K, which is too low compared to observations owing to the lack of the greenhouse effect associated with the atmosphere. However, this equation clearly demonstrates that the effective temperature of the Earth can be reduced by either reducing the solar constant, S_0, or by increasing the planetary albedo, α_p. Increasing concentrations of aerosols leads to an increase in α_p via aerosol-radiation interactions and via aerosol-cloud interactions. Two of the most plausible potential geoengineering proposals in terms of effectiveness, cost and deployability, according to the Royal Society [26], are the deliberate injection of sulphur dioxide or other candidate particles into the stratosphere to mimic explosive volcanic eruptions and the deliberate brightening of low-lying stratocumulus clouds.

Stratospheric aerosol injection simulations from multi-model ensembles under the Geoengineering Model Intercomparison Project [27] suggest that the global mean temperature can indeed be effectively reduced, but there are plenty of potential unwanted side effects [28]. These include the fact that there will be residual impacts on regional temperature and precipitation [29,30], potential impacts on the rainfall associated with the monsoons, particularly if the aerosol were to be injected into a single hemisphere [31], and a rapid cessation of deployment would mean that all of the climate change that had been avoided through geoengineering would be realized in a few years [32]. Simulations of cloud brightening have not yet been studied under a unifying intercomparison project, but the results from stand-alone studies suggest that the regional climate change issues may be more significant than for stratospheric geoengineering owing to the more extreme spatial nonhomogeneity of the achievable forcing [33]. For both forms of solar radiation management geoengineering, the results are clear – no global warming does not mean no regional climate change.

REFERENCES

[1] O. Boucher, D. Randall, P. Artaxo, C. Bretherton, G. Feingold, P. Forster, V.-M. Kerminen, Y. Kondo, H. Liao, U. Lohmann, P. Rasch, S.K. Satheesh, S. Sherwood, B. Stevens, X.Y. Zhang, Clouds and aerosols, in: T.F. Stocker, D. Qin, G.-K. Plattner, M. Tignor, S.K. Allen, J. Boschung, A. Nauels, Y. Xia, V. Bex, P.M. Midgley (Eds.), Climate Change 2013: The Physical Science Basis. Contribution of Working Group I to the Fifth Assessment Report of the Intergovernmental Panel on Climate Change, Cambridge University Press, Cambridge, UK and New York, NY, USA, 2013.
[2] P. Ciais, C. Sabine, G. Bala, L. Bopp, V. Brovkin, J. Canadell, A. Chhabra, R. DeFries, J. Galloway, M. Heimann, C. Jones, C. Le Quéré, R.B. Myneni, S. Piao, P. Thornton, Carbon and other biogeochemical cycles, in: T.F. Stocker, D. Qin, G.-K. Plattner, M. Tignor, S.K. Allen, J. Boschung, A. Nauels, Y. Xia, V. Bex, P.M. Midgley (Eds.), Climate Change 2013: The Physical Science Basis. Contribution of Working Group I to the Fifth Assessment Report of the Intergovernmental Panel on Climate Change, Cambridge University Press, Cambridge, UK and New York, NY, USA, 2013.

[3] N. Bellouin, J. Rae, A. Jones, C. Johnson, J.M. Haywood, O. Boucher, J. Geophys. Res. (2011), http://dx.doi.org/10.1029/2011JD016074.

[4] P. Forster, V. Ramaswamy, P. Artaxo, T. Berntsen, R. Betts, D. Fahey, J.M. Haywood, J. Lean, D. Lowe, G. Myhre, J. Nganga, R. Prinn, G. Raga, M. Schulz, R. Van Dorland, Changes in atmospheric constituents and in radiative forcing, in: S. Solomon, D. Qin, M. Manning, Z. Chen, M. Marquis, K.B. Avery, M. Tignor, H.L. Miller (Eds.), Climate Change 2007: The Physical Science Basis. Contribution of Working Group I to the Fourth Assessment Report of the Intergovernmental Panel on Climate Change, Cambridge University Press, Cambridge, UK and New York, NY, USA, 2007.

[5] G. Myhre, D. Shindell, F.-M. Bréon, W. Collins, J. Fuglestvedt, J. Huang, D. Koch, J.-F. Lamarque, D. Lee, B. Mendoza, T. Nakajima, A. Robock, G. Stephens, T. Takemura, H. Zhang, Anthropogenic and natural radiative forcing, in: T.F. Stocker, D. Qin, G.-K. Plattner, M. Tignor, S.K. Allen, J. Boschung, A. Nauels, Y. Xia, V. Bex, P.M. Midgley (Eds.), Climate Change 2013: The Physical Science Basis. Contribution of Working Group I to the Fifth Assessment Report of the Intergovernmental Panel on Climate Change, Cambridge University Press, Cambridge, UK and New York, NY, USA, 2013.

[6] S. Kinne, M. Schulz, C. Textor, S. Guibert, Y. Balkanski, S.E. Bauer, X. Tie, Atmos. Chem. Phys. Discuss. 5 (5) (2005) 8285–8330.

[7] M. Schulz, C. Textor, S. Kinne, Y. Balkanski, S. Bauer, T. Berntsen, T. Takemura, Atmos. Chem. Phys. 6 (12) (2006) 5225–5246.

[8] N. Bellouin, J. Quaas, J.-J. Morcrette, O. Boucher, Atmos. Chem. Phys. 13 (2013) 2045–2062, http://dx.doi.org/10.5194/acp-13-2045-2013.

[9] J.-J. Morcrette, O. Boucher, L. Jones, D. Salmond, P. Bechtold, A. Beljaars, A. Benedetti, A. Bonet, J.W. Kaiser, M. Razinger, M. Schulz, S. Serrar, A.J. Simmons, M. Sofiev, M. Suttie, A.M. Tompkins, A. Untch, J. Geophys. Res. 114 (2009) D06206, http://dx.doi.org/10.1029/2008JD011235.

[10] A. Benedetti, J.-J. Morcrette, O. Boucher, A. Dethof, R.J. Engelen, M. Fisher, H. Flentje, N. Huneeus, L. Jones, J.W. Kaiser, S. Kinne, A. Mangold, M. Razinger, A.J. Simmons, M. Suttie, J. Geophys. Res. 114 (2009) D13205, http://dx.doi.org/10.1029/2008JD011115.

[11] S.R. Osborne, J.M. Haywood, Atmos. Res. 73 (2005) 173–201.

[12] R.J. Charlson, J. Langner, H. Rodhe, C.B. Leovy, S.G. Warren, Tellus A 43 (4) (1991) 152–163.

[13] G. Mie, Ann. Phys. 330 (1908) 377–445, http://dx.doi.org/10.1002/andp.19083300302.

[14] B.T. Johnson, S.R. Osborne, J.M. Haywood, M.A.J. Harrison, J. Geophys. Res. 113 (2008) D00C06, http://dx.doi.org/10.1029/2007JD009451.

[15] J.M. Haywood, K.P. Shine, Geophys. Res. Lett. 22 (5) (1995) 603–606.

[16] M.G. Flanner, C.S. Zender, J.T. Randerson, P.J. Rasch, J. Geophys. Res. 112 (2007) D11202.

[17] M.Z. Jacobson, W.G. Colella, D.M. Golden, Science 308 (5730) (2005) 1901–1905.

[18] A. Jones, D.L. Roberts, A. Slingo, Nature 370 (1994) 450–453.

[19] R. Pincus, M.B. Baker, Nature 372 (6503) (1994) 250–252.

[20] B.A. Albrecht, Science 245 (4923) (1989) 1227–1230.

[21] M. Wild, J. Geophys. Res. 114 (D10) (2009).

[22] O. Boucher, J.M. Haywood, Clim. Dyn. 18 (2001) 297–302.

[23] D. Rosenfeld, Y. Rudich, R. Lahav, Proc. Natl. Acad. Sci. 98 (11) (2001) 5975–5980.

[24] H. Korhonen, K.S. Carslaw, P.M. Forster, S. Mikkonen, N.D. Gordon, H. Kokkola, Geophys. Res. Lett. 37 (2) (2010).

[25] M. Kulmala, T. Suni, K.E.J. Lehtinen, M. Maso, M. Boy, A. Reissell, P. Hari, Atmos. Chem. Phys. 4 (2) (2004) 557–562.

[26] J.G. Shepherd, Geoengineering the climate: science, governance and uncertainty, Royal Society Policy Document 10/09, R51636, ISBN 978-0-85403-773-5. (2009).

[27] B. Kravitz, A. Robock, O. Boucher, H. Schmidt, K.E. Taylor, G. Stenchikov, M. Schulz, Atmos. Sci. Lett. 12 (2) (2011) 162–167.

[28] A. Robock, Bull. Atomic Scientists. 64 (2) (2008) 14–18, http://dx.doi.org/10.2968/064002006.

[29] B. Kravitz, K. Caldeira, O. Boucher, A. Robock, P.J. Rasch, K. Alterskjær, D. Bou Karam, J.N.S. Cole, C.L. Curry, J.M. Haywood, P.J. Irvine, D. Ji, A. Jones, D.J. Lunt, J.E. Kristjánsson, J. Moore, U. Niemeier, H. Schmidt, M. Schulz, B. Singh, S. Tilmes, S. Watanabe, J.-H. Yoon, J. Geophys. Res. (2013), http://dx.doi.org/10.1002/jgrd.50646.

[30] S. Tilmes, J. Fasullo, J.-F. Lamarque, D.R. Marsh, M. Mills, K. Alterskjær, O. Boucher, J.N.S. Cole, C.L. Curry, J.M. Haywood, P.J. Irvine, D. Ji, A. Jones, D.B. Karam, B. Kravitz, J.E. Kristjansson, J.C. Moore, H.O. Muri, U. Niemeier, P.J. Rasch, A. Robock, H. Schmidt, M. Schulz, Y. Shuting, B. Singh, S. Watanabe, J.-H. Yoon, J. Geophys. Res. (2013), http://dx.doi.org/10.1002/jgrd.508682013.

[31] J.M. Haywood, A. Jones, N. Bellouin, D.B. Stephenson, Nat. Clim. Change 3 (7) (2013) 660–665, http://dx.doi.org/10.1038/NCLIMATE1857.

[32] A. Jones, J.M. Haywood, K. Alterskjær, O. Boucher, J.N.S. Cole, C.L. Curry, P.J. Irvine, D. Ji, B. Kravitz, J.E. Kristjánsson, J. Moore, U. Niemeier, A. Robock, H. Schmidt, B. Singh, S. Tilmes, S. Watanabe, J.-H. Yoon, J. Geophys. Res. 118 (2013), http://dx.doi.org/10.1002/jgrd.50762.

[33] A. Jones, J.M. Haywood, Sea-spray geoengineering in the HadGEM2-ES Earth-system model: radiative impact and climate response, Atmos. Chem. Phys. 12 (2012) 10887–10898, http://dx.doi.org/10.5194/acp-12-10887-2012.

CLIMATE CHANGE AND AGRICULTURE

28

Rattan Lal

Carbon Management and Sequestration Center, The Ohio State University, Columbus, OH, USA

CHAPTER OUTLINE

1. INTRODUCTION

The world population has been growing exponentially since the onset of the Industrial Revolution (Fig. 1) and is projected to reach 9.5 US billion (9.5×10^9) by 2050 [1]. Important global issues of the twenty-first century, driven by the increase in human population, include deforestation of the tropical rainforest, soil degradation and desertification, fossil fuel combustion and the growing energy demand, increase in atmospheric concentration of greenhouse gases (GHGs) and the attendant global warming, food and nutritional insecurity, loss of biodiversity and extinction of species, scarcity and pollution/ eutrophication of natural waters, and urban encroachment etc. Consequently, there is an exponential growth in carbon (C) emissions from fossil fuel combustion, especially since about 1900 (Fig. 2). Expectedly, there is a linear relationship between human population and emission from fossil fuel combustion (Fig. 3(a)). With increasing demand for food production, there has been a large conversion of natural ecosystems (forest, savannah, steppe, wetlands) into agro-ecosystems. Thus, C emissions from land use conversion have increased linearly with time since 1820 (Fig. 2(b)), and also with increase in population (Fig. 3(b)). Whereas the total C emissions (sum of C emission from fossil fuel combustion and land use conversion or deforestation) have increased exponentially over time since 1850 (Fig. 2(c)), there is a linear relationship between total emissions and the world population (Fig. 3(c)). Comparison of Fig. 2(a) and (c) shows higher value of total emissions than those from fossil fuel combustion because of the positive feedback by land use conversion.

Climate Change. http://dx.doi.org/10.1016/B978-0-444-63524-2.00028-2

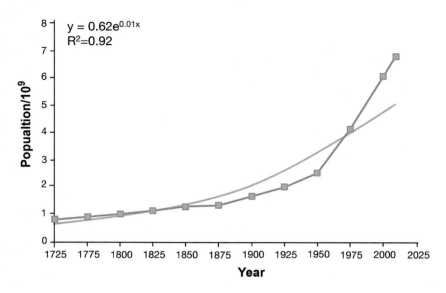

$$y = 0.62e^{0.01x}$$
$$R^2 = 0.92$$

FIGURE 1

Growth of human population. *(Redrawn from Ref. [1].)*

Thus, population-driven energy demand and intensification of agro-ecosystems are strongly impacting the atmospheric chemistry. Concentration of CO_2 in the atmosphere of ~ 280 μmol mol^{-1} (280×10^{-6}) prior to the Industrial Revolution increased to 400 μmol mol^{-1} on 9 May 2013, an increase of $\sim 43\%$ [3]. The radiative forcing of extra CO_2 is 1.88 W m^{-2}, equivalent to 1% of solar energy that reaches the Earth's surface. It traps 23×10^9 MW (23 US billion megawatts) of energy every day, the amount equivalent to that produced in the US during 2013 [3]. The average rate of increase in atmospheric concentration of CO_2 since 2010 is about 2.1 μmol mol^{-1} a^{-1} (2.1×10^{-6} per year). However, the rate of increase was 2.8 μmol mol^{-1} during 2013 [4,5]. For the week of 18–24 January 2015, the atmospheric concentration of CO_2 at Mauna Loa was 399.49 μmol mol^{-1}, compared to that for the same week at 378.35 μmol mol^{-1} in 2005 and 398.11 μmol mol^{-1} in 2014 [6].

Two principal anthropogenic sources of CO_2 emission are fossil fuel combustion (and cement production) and tropical deforestation. Whereas the emission from fossil fuel is increasing, that from deforestation has a decreasing trend. Emissions from land use change, account for $\sim 8\%$ of total CO_2 emissions or 10% of those from fossil fuel combustion and cement production during the decade of 2000–2010 [7]. The gross loss of forests in the humid tropics was 8 Mha a^{-1} (8 million hectare per year) during the 1990s and 7.6 Mha a^{-1} during the 2000s [7]. Estimates of the resultant C losses are 887 Tg(C) a^{-1} (with a range of 646 Tg(C) a^{-1}–1238 Tg(C) a^{-1}) for 1990s and 880 Tg(C) a^{-1} (with a range of 602 Tg(C) a^{-1}–1237 Tg(C) a^{-1}) for 2000s, and the humid regions contributing $\sim 66\%$ of the total [7]. Estimates of the annual C removal are rather small compared with those of emissions and do not include any gains in soil.

Agriculture is an important source of GHGs. Yet, as an industry (in developed countries) and a way of life (in developing countries), agriculture must be integral to any global agenda of climate change adaptation and mitigation. Considering direct and indirect emissions, agriculture accounts for $\sim 30\%$ of the cumulative radiative forcing [8]. Therefore, the objective of this chapter is to discuss the importance of agriculture as a source of GHGs, and the potential of climate-smart agriculture to reducing anthropogenic emissions and developing climate-resilient soils and the agro-ecosystems.

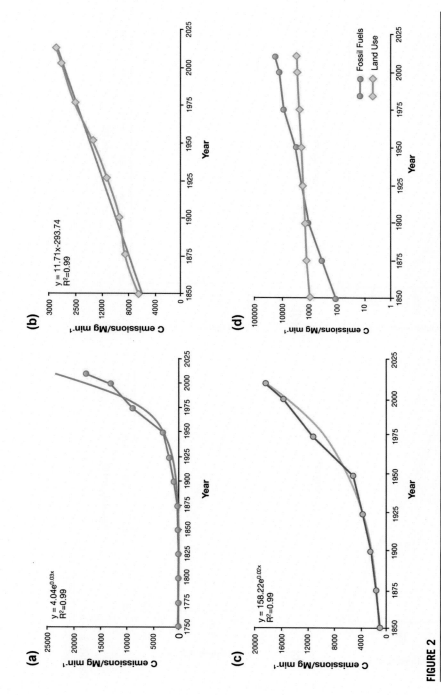

FIGURE 2

Temporal changes in anthropogenic emissions (C emissions/Mg min^{-1}, where min refers to minute) of CO$_2$-C: (a) from fossil fuel combustion since 1750, (b) from land use change since 1850, (c) total emissions from fossil fuel plus land use change, and (d) comparative emissions from fossil fuel and land use conversions. *(These data are computed and drawn from Ref. [1] for population, and Ref. [2] for CO$_2$-C emissions.)*

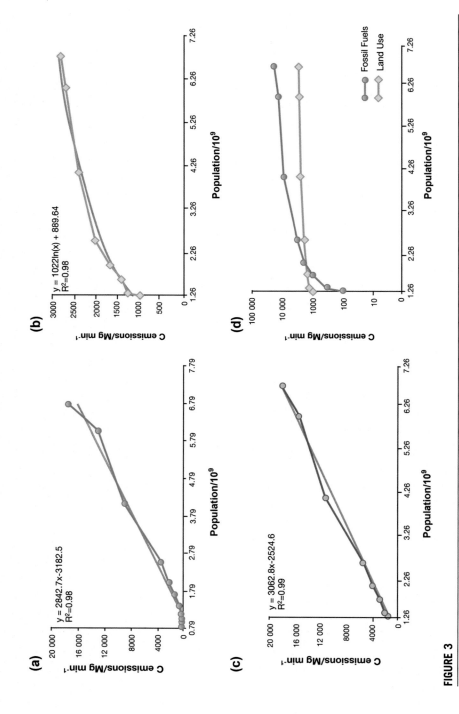

FIGURE 3

Relation between global population and emission of CO_2-C (C emission/Mg min^{-1}, where min refers to minute): (a) from fossil fuel combustion, (b) from land use change, (c) total emissions from fossil fuel combustion plus land use change, and (d) comparative emissions from fossil fuel versus land use change. *(These data are computed and drawn from Ref. [1] for population and from Ref. [2] for CO_2-C emissions.)*

2. AGRICULTURE AND CLIMATE CHANGE

The Anthropocene [9], the era when human impact on the planet Earth is equivalent to that of any geological force, may have begun $\sim 10 \times 10^3$ a (10 millennia) ago with the onset of settled agriculture [10,11]. Land misuse and soil mismanagement not only led to the extinction of numerous once-thriving historic civilizations [12] but were also the principal cause of the Dust Bowl of the 1930s [13]. Despite ever-increasing emissions because of fossil fuel combustion estimated at 10.1 Pg of carbon for 2014, at least 0.9 Pg $\pm$ 0.5 Pg of carbon ($\sim 8.9\%$) was emitted by land use conversion [14].

Estimates of emissions from different agricultural sectors and by continents are shown in Table 1. With total emissions of 1.45 Pg(Ce) a^{-1} in 2011 (where Ce refers to the carbon equivalent), percent emissions by different sectors were 44% for enteric fermentation, 15% for manure left on pastures, 13% for synthetic fertilizers, 10% for rice cultivation, 7% for manure management, and 3% for manure

Table 1 Global agricultural emissions from different sectors in 2011

Source	Emissions	
	Magnitude (Pg(C) a^{-1})	%
I. Sector		
Enteric fermentation	0.58	40
Manure left on pasture	0.22	15
Synthetic fertilizers	0.19	13
Rice cultivation	0.15	10
Manure management	0.10	7
Burning of savannah	0.07	5
Crop residues	0.06	4
Manure applied to soils	0.04	3
Cultivation of organic soils	0.04	3
Burning crop residues	Traces	
II. Continents		
Asia	0.64	44
Americas	0.36	25
Africa	0.22	15
Europe	0.17	12
Oceania	0.06	4
III. Land use		
Forest	0.91	63
Cropland	0.36	25
Grassland	0.16	11
Biomass burning (forest, peat)	0.02	1

Total emissions in 2011 are estimated at 1.45 Pg(Ce).
Recalculated from Ref. [15].

applied to soil. Total contribution by the livestock sector in 2011 was 65% or 0.94 Pg(Ce). In contrast, synthetic fertilizer and rice cultivation contributed 0.19 Pg(Ce) and 0.15 Pg(Ce), respectively. On a regional basis, emissions were 44% (0.64 PgCe) for Asia, 25% (0.36 PgCe) for the Americas and 15% (0.22 Pg(Ce)) for Africa (Table 1). In terms of land use, emissions were in the order of forest > cropland > grassland > biomass burning (Table 1).

While agriculture has been an important source of GHGs since ∼ 10 000 years, it will also be adversely affected by the attendant climate change [16]. The climate change will negatively impact productivity and use efficiency of resources (e.g. water, fertilizer) and also exacerbate the incidence of pests and pathogens. Being a source of GHGs, and also vulnerable to the adverse impact of climate change, agriculture must be integral to any agenda to address the climate change and its impacts on human wellbeing.

3. SOURCES OF EMISSIONS FROM AGRICULTURE

There is a wide range of agricultural and forestry activities that are sources of GHGs (Fig. 4). Principal agricultural activities leading to GHG emissions are ploughing, drainage of wetlands, enteric fermentation, fertilizer use, manure management and fuel consumption [17]. Averaged over the 2001–2010 period, agriculture, forestry and other land use contributed the following [18]: (1) 1.36 Pg(Ce) a^{-1} from crop and livestock production, (2) 1.09 Pg(Ce) a^{-1} from net forest conversion

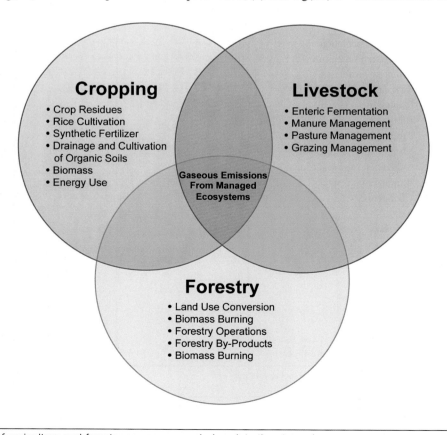

FIGURE 4

Impacts of agriculture and forestry on gaseous emissions into the atmosphere.

to other lands, (3) 0.27 Pg(Ce) a^{-1} from degraded peatlands, and (4) 0.05 Pg(Ce) a^{-1} from biomass fires. In addition to livestock, rice cultivation is another source of CH_4, and the emission may increase with the projected climate change [19]. Persistent growth of CO_2 emission may have strong implication for reaching climate targets [20]. Fires and biomass burning have local [21], and regional effects.

Livestock remains a major source of GHGs [22], especially of CH_4. On average, a cow releases between 70 kg and 120 kg of CH_4 per year [23]. Total livestock-based emissions may be 2.04 Pg(Ce) a^{-1} [23]. Some estimates assess total emissions from livestock at 8.9 Pg(Ce) a^{-1} [24], which seems to be an overestimate. Emission of N_2O from agriculture is another important source of a highly potent GHG. Combined, all agricultural activities may contribute about 25%, 65% and 90% of total anthropogenic emissions of CO_2, CH_4 and N_2O, respectively [25].

Thus far, the Intergovernmental Panel on Climate Change (IPCC) [26] and other organizations have neither adequately considered the implications of current and future land use conversions and of the full range of practices on climate change nor sufficiently considered the potential of terrestrial ecosystems (i.e. cropland, degraded lands) on soil carbon C sequestration to mitigate the anthropogenic emissions. This oversight must be rectified.

3.1 LAND USE CONVERSION

Between one-third and one-half of Earth's land surface has been transformed by human development [27], with a strong impact on the terrestrial C pool and emissions of GHGs into the atmosphere. Since ca. 1500, anthropogenic land use conversion has been a major source of atmospheric CO_2. However, estimates of emissions (Ce) from land use vary widely because of the diffused nature of the source. Similarly, estimates of emissions from agriculture also vary widely because of differences in methodology and the inclusiveness of major sectors concerned. There are also variations due to the use of specific models [28,29], and differences in emission factors used for conversion of inputs into CO_2 equivalent, the area under different types of land use (rice cultivation) and number/type of livestock.

According to one estimate, net terrestrial land use change emissions over the past 300 years may be as much as 250 Pg(C), consisting of 145 Pg from tropical and 105 Pg from regions outside the tropics (extratropics) [30]. Another estimate shows that cumulative anthropogenic emissions from 1870 to 2004 are 545 Pg(C) $\pm$ 55 Pg(C), of which ~75% are by fossil fuel combustion and 25% by land use conversion. Thus, cumulative emissions from land use conversion since 1870 are about 136 Pg(C) [14]. Between 1850 and 1990, changes in land use are estimated to have emitted 124 Pg(C) to the atmosphere, about 50% as much are those from fossil fuel combustion [31]. Of this, 108 Pg were emitted from forests comprising of two-thirds from tropical forest and one-third from temperate zone and boreal forests. Yet another estimate shows that, between 1850 and 2000, net emissions from land use conversion are 210 Pg(C) comprising of 122.5 Pg from the tropical and 87.9 Pg(C) from extra-tropical regions [30]. Above all, emissions (8000 BC to 1850 AD) from land use conversion are estimated at ~320 Pg(C) [10,11]. Thus, total emissions from land use conversion may be as much as 456 Pg(C). The current rate of emissions from land use conversion (for the decade of 2004–2013) is 0.9 Pg(C) a^{-1} $\pm$ 0.5 Pg(C) a^{-1} [14].

Biomass burning, associated with deforestation and crop residue disposal and wildfire, is another factor that is poorly accounted for. Estimates for emissions of GHGs (CO_2, N_2O, NO_X) from fire vary widely (Table 2). Biomass containing 2 Pg(C)–5 Pg(C) is burnt annually producing large amount of GHGs and aerosol particles [32–34]. Yet, there is no standard protocol to account for these missions.

Table 2 Estimates of emission from biomass burning

Region	Period	Average area burnt (10^4 km^2 a^{-1})	Emissions (Tg(C) a^{-1})	References
Africa	2001–2005	195.5–24.0	723 ± 70	[35]
Russia	1998–2010	8.2 ± 0.8	121 ± 28	[36]
Global croplands and forests	–	–	2620 ± 220	[37]
Global fires	2000	350	–	[38]
Global	–	350–450	2000–4000	[39]

Whereas the emissions from deforestation and land use conversion may be decreasing since 1990, those from agricultural inputs have been increasing because of increase in cultivated area and intensification. For the decades of the 1990s and 2000s, total agricultural emissions increased from 1.25 Pg(Ce) a^{-1}–1.36 Pg(Ce) a^{-1} to 1.45 Pg(Ce) a^{-1} in 2011, but those from deforestation decreased from 1.25 Pg(Ce) a^{-1}–1.04 Pg(Ce) a^{-1} to 1.0 Pg(Ce) a^{-1} in 2010 [15]. The FAO-sponsored study also indicated that unless measures of technical efficiency improvements are implemented, future emissions from agriculture may increase by up to 30% by 2050. Indeed, future land use and land cover will remain an important source of emissions.

Over and above the emissions from land use conversion and those from inputs (fertilizer, fuel, irrigation, etc.), land application of waste is another source of GHGs [40], those from forestry operations must also be considered [41] by full cost accounting [42]. Yet emissions caused by ploughing, fertilization, livestock and soil erosion may aggravate the total GHG flux from agro-ecosystems, especially by the contributions from rice paddy and soil cultivation [26,43]. Considering direct and indirect emissions, it is apparent that about one-third of emissions of GHGs come from agriculture [44,45]. In the US, GHG emissions from agriculture have increased by 19% between 1990 and 2012 [46].

3.2 FUEL CONSUMPTION

Fuel consumption by farm operations is an important source of CO_2 emissions (Fig. 5) [47,48]. These operations involve tillage, manufacture and application of fertilizers and pesticides, harvesting, grain drying, heating and cooling, transport, etc. Estimates of C emissions for conventional tillage are 230 kg(C) ha^{-1} for winter wheat, 165 kg(C) ha^{-1} for sugar beet, 116 kg(C) ha^{-1} for adzuki bean and 201 kg(C) ha^{-1} for potato [50]. Grain drying in Ontario, Canada, can consume 33% of the total energy used in corn production [49]. Fuel consumption varies widely among tillage implements and ranges from 4.1 L(diesel) ha^{-1} for wide level disk, 8.7 L(diesel) ha^{-1} for offset disk harrow and 9.3 L(diesel) ha^{-1} for chisel plough [51]. In comparison, fuel consumption is 7.2 L ha^{-1} for row crop planter and 0.02 L ha^{-1} for sprayer [51]. Fuel consumption also varies with depth and speed of ploughing, and has been reported at 16.8 L ha^{-1} of diesel for 20 cm, 22.0 L(diesel) ha^{-1} for 25 cm, and 32.7 L(diesel) ha^{-1} for 30 cm depth, an increase by 31% in fuel use with increase in depth from 20 cm to 30 cm [52]. The average fuel consumption also increases with increase in ploughing speed for a specific tractor, and has been measured at 20.0 L(diesel) ha^{-1}, 24.3 L(diesel) ha^{-1} and 27.2 L(diesel) ha^{-1} for speed of 5.5 km h^{-1}, 6.5 km h^{-1} and 7.5 km h^{-1}, respectively [52]. Therefore, tillage intensity (conventional tillage, minimum tillage, no-till or NT, conservation tillage or CT) has a strong impact

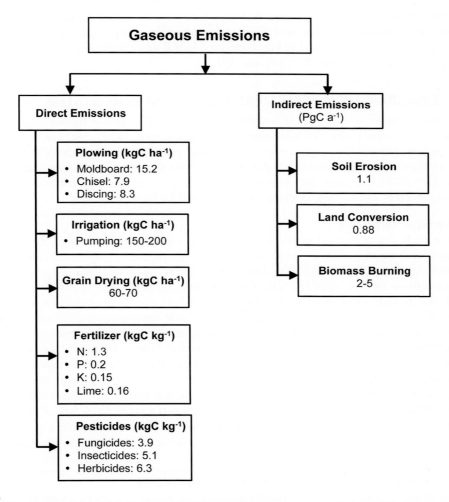

FIGURE 5

Direct and indirect emissions of C from agro-ecosystems. *(Estimates are from Refs [7,32,47–49].)*

on CO_2-C emissions partly because of differences in soil erosion hazard and partly due to different amounts on inputs (herbicides, pesticides, fertilizers, irrigation and seeding rates). A comparative study estimated CO_2-C emissions of 2.0 μmol m^{-2} s^{-1} for NT, 2.9 μmol m^{-2} s^{-1} for reduced tillage and 4.3 μmol m^{-2} s^{-1} for conventional ploughing [53]. In traditional tillage, involving manual hoeing, even a single land hoeing can cause emission of 0.55 kg(C) ha^{-1} (2 kg CO_2 ha^{-1}) because of rapid oxidation of soil organic matter (SOM). Tillage-induced CO_2 emissions are much larger for mechanized systems [54], much higher immediately after ploughing, and lasting for a few days or weeks after the tillage [54,55].

Because different tillage systems result in differences in resource use, choice of reduced/NT or CT can save direct energy input and the amount of machinery time needed. Indeed, total energy input reduction in comparison with conventional tillage may be 26% for reduced tillage and 41% for NT system [56]. Information on C emission of different farm operations for conducting life cycle analyses (LCA) of agro-ecosystems presented in Fig. 4 can be a useful guide in the decision-making process toward the choice of low-C production systems [47].

3.3 N_2O EMISSIONS FROM FERTILIZERS AND MANAGEMENT SYSTEM

Anthropogenic activities have drastically accelerated the Earth's N cycle over the twentieth century by conversion of nonreactive (stable N_2) into reactive forms such as NH_3, NO_3^-, N_2O and NO. During the twentieth century, farmers have replaced legume rotations and traditional sources of N fertility with chemical fertilizers. However, eventual uptake of fertilizers by humans (as amino acids and protein) is hardly 5%–15% of the N applied, and most of the remainder is lost to the environment [57]. The use of nitrogenous fertilizers will increase with increase in demand for food production.

3.3.1 Chemical Fertilizers

While input of N (inorganic and organic) is essential to enhance productivity of agro-ecosystems and advance food security, misuse and mismanagement of reactive N reduces efficiency by plants and increases emissions of N_2O (and NO_X) into the atmosphere. Reactive N is a major source of ever-increasing emissions of N_2O, the GHG with a high radiative forcing (GWP of 310 W m^{-2} compared with 21 W m^{-2} of CH_4 and 1 of CO_2). As a potent GHG, N_2O also depletes the stratospheric ozone. Emissions of N_2O from food systems account for 77% of total anthropogenic emissions and for 87% when those from production of fertilizer N and waste management are also taken into consideration [58]. Emissions of N_2O have increased drastically over the twentieth century [43]. Total emissions are estimated at 4.07 Tg (N_2O-N) for 2010. However, with emissions from fertilizer production and waste, total emissions are 4.4 Tg (N_2O-N) a^{-1} (Table 3). Among land uses, agricultural soils are the principal anthropogenic source because of N fertilizer use [19]. While the rate of N fertilizer input is a useful predictor of N_2O emissions from soils of agro-ecosystems, which account for ~50% of the total global anthropogenic flux, N_2O emission response in relation to the rate of fertilizer input is exponential rather than linear [59]. Furthermore, 50% of total emissions may occur shortly after N application [60].

Table 3 Regional and global estimates of N_2O emissions in 2010

Region	N_2O emissions (Gg (N_2O-N) a^{-1})			
	Direct	**Indirect**	**Total**	**% of total**
Asia	1432	420	1852	45.5
Africa	408	106	514	12.6
Europe	401	127	528	13.0
North America	332	97	429	10.6
Latin America	495	128	623	15.3
Oceania	97	26	123	3.0
Global total	3165	904	4069	100

Total emissions from fertilizer and waste = 4389 Gg.
Adapted and modified from Ref. [58].

Input of N in agro-ecosystems has increased in all continents since the 1960s, except in Sub-Saharan Africa (SSA). On a regional basis, Asia contributes about 45.5% of the total and per-hectare emissions, but has the lowest per capita emission [58]. Rates of fertilizer N application are high in China and increasing in India. However, some satellite observations show a decreasing trend in NO_X emissions from East Asia because of the improvements in technology such as in power plants, transport, etc. [61]. N use in the US increased strongly between 1961 and 2000, with the largest increase in use of inorganic fertilizer [62]. Between 1961 and 1999, N fertilizer use in the US increased nearly four-fold, from 3.1 Tg(N) a^{-1} to 11.2 Tg(N) a^{-1}. Presently, the US uses about 13% of the inorganic-N fertilizer used globally (86 Tg(N) a^{-1}), and the rate per unit area is 2.2 times that of the global average. The consumption of meat protein is the principal driver of the use of inorganic N in agro-ecosystems [62]. In Canada, also, application of synthetic N fertilizers was the largest direct source of soil N_2O. Other direct major sources of N_2O in Canada include crop residues, grazing animals and manure applied to soils. Total emissions from agricultural sources in Canada between 1990 and 2005 were 58.1 Gg(N_2O-N) a^{-1} [63]. A study of the fate of N applied showed that 2.35 Tg(N) entered the Canadian agro-ecosystems in 1996 from fertilizer, biological N fixation (BNF) and atmospheric deposition. However, N leakage into the adjacent environment amounts to almost half of N added, and crop recovery of added N is about 60% [64]. In Australia, fertilizer N use increased from 35 Gg(N) in 1961 to 972 Gg(N) in 2002 (28 times), and most for growing cereals [65]. However, crop uptake (of wheat) is only 41%, and there is a strong need to enhance the N use efficiency. In New Zealand, the single largest source of N_2O emissions is animal excreta deposited during grazing, amounting to 80% of agricultural N_2O emissions [66].

A study of the global and regional surface N balance between 1970 and 1995 indicates that the overall fertilizer nutrient use efficiency (NUE) has slowly increased in industrialized countries but not in developing countries [67]. Between 1955 and 2030, however, a rapid increase in NUE is expected in all countries. Yet, intensification of agro-ecosystems to meet the ever-increasing food demands can increase losses of reactive N into air (NH_3, N_2O, NO) and water (NO_3^-) with severe consequences. Thus, LCA is needed to identify good agricultural practices that enhance NUE and minimize leakage into the environment. For example, an LCA study conducted for wheat production systems in long-term experiments at Rothamsted, UK, showed that agronomically optimal arable farming is not necessarily in conflict with economic and environmental boundary conditions [68]. The fact that N_2O emission increases exponentially with increase in the fertilizer rate is useful in assessment of the fertilizer-derived N_2O emissions, refinement of the global N_2O budget and identification of N_2O mitigation protocols [59]. Such LCA analysis indicates that increasing rate of fertilizer use in low-input Sub-Saharan Africa would cause little increase in N_2O emission, but an equivalent emission in high-input East Asia (China) would have a disproportionately major impact [59].

3.3.2 Tillage Systems

Tillage systems and methods of seedbed preparation also impact N_2O emission. In general, N_2O emissions are greater under NT than conventional tillage systems [69] and may negate any gains of soil organic C (SOC) sequestration. However, several studies also show no significant effect of tillage systems [70], or even a negative effect of NT on N_2O emission [60]. Indeed, tillage effects on N_2O emission are moderated by soil wetness and other factors. Emissions of N_2O can be significantly increased in seasons with high rainfall and in land receiving organic manures, compost and cattle slurry [70].

3.3.3 Rotations

Incorporating legumes in the rotation cycle can also affect N_2O emissions form croplands. Legumes, because of their capacity for Biological nitrogen fixation (BNF), can provide an environmentally friendly and a sustainable cropping system. However, the in-field situation is confounded by several factors, and biologically fixed N may also have equally adverse environmental effects as that of the synthetic N fertilizers. Furthermore, the large human population (7.25×10^9 now and expected to be 9.5×10^9 by 2050) exceeds the carrying capacity of agroecosystems that solely depend on legumes as a source of N [71]. While incorporating leguminous crops in the rotation can reduce input of fertilizer N, the likely production of N_2O during the decomposition of crop residues needs to be considered. Thus, IPCC has revised coefficients for estimating N_2O emissions from legumes, which are 1.0 kg(N) ha^{-1} for annual crops, 1.8 kg(N) ha^{-1} for forage crops and 0.4 kg(N) ha^{-1} for grass legume mixes [72]. Most of the increase in soil N_2O emissions in legume crops may be attributed to the N release from root exudates during the growing season and decomposition of crop residues after harvest rather than from BNF per se [72]. In the US corn belt, rotation corn can reduce N_2O emission by 20% relative to continuous corn [60]. The magnitude of reduction of N_2O emission by legumes in a rotation cycle also depends on the crop species. In Queensland, Australia, for example, replacing wheat with chickpea did not reduce N_2O emission relative to fertilized systems [73]. However, some interventions may be counterproductive due to complex feedback mechanisms.

3.3.4 Cover Crops

Incorporating cover crops in the rotation cycle also affects N_2O emissions. Similar to the effects of NT versus ploughing, there is also not a clear consensus among soil scientists regarding the effects of cover crop on N_2O emissions. Indeed, some cover crops may enhance N_2O emissions, and the magnitude of increase depends on the nature of crop residues produced (grass vs legumes). A long-term study in southern Brazil on a subtropical Acrisol showed that higher soil N_2O emissions occurred within the first 45 days after the cover crop residue management in all rotations. However, the legume-based cover crops had the largest N_2O emissions; this was positively related to the quantity of N applied and negatively to the lignin content of the residue [74]. A metanalysis (a statistical technique for combining findings from independent studies) involving 106 observation of cover crops showed that 40% of cases indicated a decreased N_2O emission and 60% increased it [75].

3.3.5 Soil Properties

Emission of N_2O also differs with other factors such as the amount of synthetic N applied, the percentage of dissolved organic C (DOC) and soil temperature. Relevant managerial factors that reduce N_2O emission by cover cropping include no incorporation of crop residues and use of nonlegume cover crops [75]. Thus, the objective of soil/crop management is the synchronization between release of N through mineralization of residues and the critical period for crop uptake of N. In general, N_2O emissions increase after fertilization of food crops, cultivation of bare soils, ploughing under of grassland and incorporation of crop residues. Relatively high emissions occur soon after two weeks and level off after 30 days–40 days [76]. High concentration of N in the residues (low C:N ratio) increases emissions. Management of cover crop for reducing N_2O emissions remains to be a high researchable priority.

3.3.6 Organic Farming

Organic farming, gaining momentum in Europe and North America, has strong implications to N_2O emission. It is argued that some countries may be strongly dependent on synthetic N fertilizers, but others may have the potential to reduce substantially the use of mineral fertilizers. Organic farming is considered one of the agricultural systems to grow food with environmental constraints of N_2O emissions [77] and of other pollutants. Substantial amounts of organic by-products are produced globally, which can be used as bio-fertilizers, and include plant residues (3.8×10^9 Mg a^{-1}), bio-solids (10×10^7 Mg a^{-1}) and animal manure (7×10^9 Mg a^{-1}) [78]. However, application of organic amendments has important trade-offs that also cause emission of GHGs. By using LCA, some studies have inferred that integrated systems (which use food waste and digestate as fertilizer and utilize some pesticides and NT) have the lowest energy use and global warming potential (GWP) per 1000 kg of wheat produced [79]. Assessment of energy and gaseous emissions of organic and conventional dairy systems in the Netherlands showed that energy use per unit milk in organic dairy is $\sim$25% lower than in conventional dairy and GHG emissions are 5%–10% lower [80].

In contrast, however, energy use and GHG emissions in organic crop production are higher than in conventional crop production because of high-intensity, high-value cash crops, high fertilizer inputs and frequent farm operations [80]. Methods of storage and application of manure affect GHG emissions. Anaerobic digestion may be effective in reducing the overall emission from storage and spreading, and spreading manure during the coolest period of the day may reduce emissions [81]. Method of application of manure, broadcast versus incorporation and solid versus slurry, also impacts GHG emission. In general, abatement is greater from the use of trailing shoe and open slot injection machines than from the trailing hose. With application of manure as solid or slurry, immediate incorporation in the soil is the most effective in reducing emissions [82]. However, ploughing under has numerous trade-offs (e.g. accelerated erosion, increase in decomposition) [47]. Furthermore, incorporation of grass/clover residues in organic rotations may also accentuate N_2O emissions [70]. There exists a large scope for identification of best management practices (BMPs) to reduce emissions from organic farming. Another major but unaccounted source of N_2O is that from biomass burning. The phytodenitrification process of biomass burning causes a sizable loss of fixed N in tropical ecosystems estimated at 10 Tg(N) a^{-1}–20 Tg(N) a^{-1} [32]. Biomass burning is an important source of atmospheric gases including CO_2, H_2, N_2O, CH_3Cl and COS [34], and must be adequately accounted for.

3.4 METHANE (CH$_4$) EMISSION

About 20% of the increase in radiative forcing since 1750 is attributed to CH_4 [83]. It is the second most important well-mixed GHG contributing to anthropogenic climate change. Agriculture, fossil fuel combustion and wetlands are the principal sources of CH_4 emission [84]. Atmospheric CH_4 concentration increased over most of the twentieth century, but the rate of emission decreased during the 1990s and 2000s [85]. Part of the decline in CH_4 emission may have been due to improved methods of rice cultivation in Asia (e.g. reduction in water input and increase in use of chemical fertilizers). However, emissions have been increasing since 2006, due to increase from natural wetland and fossil fuel combustion [86]. Factors affecting the global CH_4 emission and budget are poorly understood [87].

Presently, CH_4 emission from natural sources is estimated at 149 Tg a^{-1}–319 Tg a^{-1} and that from anthropogenic sources at 238 Tg a^{-1}–446 Tg a^{-1}, with total emission of 387 Tg a^{-1}–765 Tg a^{-1}

Table 4 Sources and emissions of CH$_4$

Source	Emissions (Tg (CH$_4$) a^{-1})
I. Natural	149–319
II. Anthropogenic	
• Ruminants	76–92
• Rice cultivation	31–83
• Biomass burning	14–88
• Landfill and waste	35–69
• Fossil fuel	82–114
Sub total	**238–446**
Total	387–765

Adapted from Ref. [90].

(Table 4). However, there are strong interannual variations of emissions, especially from wetlands (about 10 Tg a^{-1} of CH$_4$) and biomass burning (about 7 Tg a^{-1} of CH$_4$) [88]. Atmospheric CH$_4$ concentration in 2011 is 150% more than that in 1750 [89].

3.4.1 Wetlands

Being the principal source, wetlands strongly affect CH$_4$ budget [89]. Tropical wetlands (e.g. flood plains in the Amazon Basin) represent about 50% of the natural wetland CH$_4$ emissions and explain a large fraction of the interannual variability [91]. In addition, hydroelectric reservoirs (covering an area of 34 Mha and comprising about 20% of all reservoirs) are also an important source of GHGs, especially CH$_4$. Hydroelectric reservoirs emit between 48 Tg of C from CO$_2$ and 38 Tg of C from CH$_4$ or about 4% of global C emissions from inland waters [92].

3.4.2 Rice Cultivation

Paddy soils, constituting the largest anthropogenic wetlands, are Anthrosols created by puddling the wet soil. Globally, rice is cultivated on about 155 Mha, and most of it is in Asia. Changes in redox potential through water management influence microbial community structure and function. Emission of GHGs from paddy soils is influenced by high concentration and fluxes of DOC [93]. Estimates of CH$_4$ emissions in China include 8.11 Tg(CH$_4$) a^{-1} from rice paddies, 2.69 Tg(CH$_4$) a^{-1} from natural wetlands and 0.46 Tg(CH$_4$) a^{-1} from lakes including reservoirs and ponds. China accounts for 55% of the global paddy emissions [94]. Emissions from rice paddies can be decreased by improved management including the use of modern varieties, fertilizer management, flooding regime and growing aerobic (upland) rice.

3.4.3 Peatlands

Conversion of wetlands into agro-ecosystems affects CH$_4$ emission and the overall C budget. Indeed, intensive agricultural use of drained/reclaimed peatlands can lead to oxidation and subsidence of the peat soil [95]. Drainage and conversion of peatlands in Southeast Asia for oil palm plantations have strongly affected the ecosystem C budget, and also occasional burning of peatlands, with dire environmental consequences.

3.4.4 Livestock

Livestock management contributes emissions of CH_4 (and N_2O) directly and indirectly (Fig. 6). Livestock contributes about 9% of global anthropogenic emissions (CH_4 and N_2O) [96]. GHG emission ($kgCO_2e\ kg^{-1}$ of edible product) is estimated at 0.37 for milk, 0.40 for poultry, 0.30 for pig, 1.40 for beef and 1.51 for sheep [96]. Livestock contributes about 3% of UK emissions [96] and 8% of total emissions in Canada [97]. Production of CH_4 by ruminants is affected by numerous factors, some of which involve management. Important among these are feed intake, feed composition and digestibility, and prior feed processing [98]. Thus, dietary management is a simple, practical and viable option to mitigate CH_4 emission from livestock. Addition of fats and the use of more starch in animal feed can reduce CH_4 emission [97,99]. Mitigation techniques include improved feed quality, pasture management and manure management [99]. In addition to grazing management and genetic selection, there are some manure treatment options that can reduce CH_4 flux from manure storage, composting, anaerobic digestion, covers and solid–liquid separation [97].

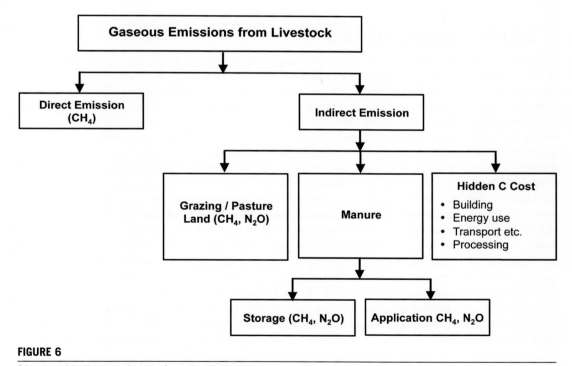

FIGURE 6

Direct and indirect emissions from livestock.

3.4.5 Permafrost

Cryosols or soils under permafrost and glaciers contain a large reserve of C estimated at 1200 Pg [100] and as much as 1500 Pg [101]. Permafrost and glaciers in the Arctic lead to formation of 'Cryosphere Cap' that traps leakage of CH_4 and other gases in these ecosystems. Thus, the Arctic geologic reservoir of CH_4 is large in comparison with the atmospheric pool of merely 5 Pg, and there is a leakage of this pool along boundaries of permafrost thaw and receding glaciers. The risks of CH_4 emissions are more in a warming climate and the attendant disintegration of permafrost, glaciers and the polar ice sheet. There are numerous uncertainties in CH_4 emissions, and its temperature dependence from permafrost and associated wetlands, due to climate change. These uncertainties are attributed to the magnitude of C pool in the permafrost, the rate of thaw and the ratio of $CH_4:CO_2$ emissions upon decomposition [102].

3.4.6 CH₄ Oxidation/Uptake by Soil

Well-aerated soils are CH_4 sink because methanotrophic bacteria can oxidize it [103]. Some soil management practices that enhance oxidation of CH_4 include those that improve aeration and accentuate structure or aggregation. The C:N ratio of crop residues affects oxidation, and those with a wide C:N ratio have no effect and those with narrow C:N ratio reduce oxidation. Perpetual use of ammonium and urea fertilizers can also reduce oxidation probably because of decrease in soil pH. In general, NT farming increases oxidation of CH_4 in soil [104].

4. ACCELERATED SOIL EROSION

Conversion of natural to agricultural ecosystems aggravates soil erosion globally. Land area affected by soil erosion is estimated at 1100 Mha by water and 550 Mha by wind [105]. There are numerous on-site and off-site effects of accelerated erosion. In addition to nonpoint source pollution and sedimentation, emission of GHGs is among a major off-site effect of accelerated erosion [47,48]. On-site soil erosion depletes the SOC pool because of its preferential removal. Being a light fraction, density of $0.6 \, Mg \, m^{-3}$–$0.8 \, Mg \, m^{-3}$ and concentrated in the surface layer, SOC and clay fractions are easily washed away in runoff or blown away by wind. In general, therefore, eroded materials (sediments) are enriched in clay and SOC, and have a high enrichment ratio for these preferentially removed constituents [106]. In Illinois, USA, a comparative assessment for soil C pool in the surface layer of cropland was merely 52% of that in the forestland [107], and the other 48% was either deposited in water bodies/depressional sites or emitted into the atmosphere. Long-term experiments on field runoff plots in western Nigeria indicated severe and rapid depletion of SOC pool on ploughed land even on gentle slope gradient of ~5% [106]. There are numerous uncertainties in determining the exact magnitude of soil C transported by erosion and its fate en route to depositional sites or aquatic ecosystems. There are site-specific factors that influence the pathways of C displaced by erosion and magnitude of erosion-induced emission of CO_2, CH_4 and N_2O [108]. A global modelling study indicated that, considering all four processes of erosion by water (i.e. detachment, breakdown, redistribution and deposition), erosion-induced emission may by equivalent to 1.1 Pg(C) a^{-1} [47,48]. Indeed, accelerated soil erosion leads to gaseous emission by impacting the following processes [108]:

1. Accelerated erosion degrades soil quality on site and adversely impacts agronomic/biomass productivity even with higher use of fertilizers, water and other impacts. The magnitude of reduction in biomass productivity is more severe in tropical soils of low inherent fertility. In addition, erosion also accentuates rates of SOC decomposition on site. Any gain in productivity at depression sites is not large enough to offset losses at the eroded sites.
2. Erosion by water and wind leads to slaking and breakdown of aggregates, and exposing SOC to microbial processes.
3. The high SOC-enrichment ratio of sediments, because of a preferential removal of the light fraction, is partly due to the transport of labile fraction. These labile organic

substances (carbohydrates) are easily mineralized and have a short mean residence time (MRT).

4. Complete removal of surface soil, in case of severe erosion, can expose the carbonaceous materials (carbonates, bicarbonates) to acid deposition and farm input (e.g. acidifying fertilizers). Reactions with acidic materials emit CO_2 into the atmosphere.

5. While deep burial of C-enriched sediments can reduce the rate of mineralization and also reaggregate clay into stable structural units, prevalence of anaerobic conditions can also increase production and emission of CH_4 by methanogenesis and of N_2O by denitrification.

5. MITIGATION POTENTIAL OF AGRICULTURE

For the complex and huge problem that climate change is, all mitigation options (geoengineering, clean coal technology, soils C sequestration and climate-smart agriculture etc.) are on the table for critical appraisal toward reducing emissions by 40–70% by mid-century. There are several technologies of negative emissions including carbon capture and storage (CCS) in geological strata, afforestation, soil C Sequestration etc. [109].

In June 2014, US President Obama proposed a clean power plan of reducing CO_2 emissions from existing US power plants by 30% below 2005 levels by 2030 [110]. Because of the rising cost, US-DOE has again pulled out of FutureGen project [111]. The social cost of C sequestration, the monetary value of reduced climate change damages associated with 1 Mg reduction in CO_2 [112] must be assessed for all options, and this makes the process a difficult choice.

Land use impaction on climate comprises biogeochemical and biophysical processes. The former influences the rate of biogeochemical cycles and changes the chemical composition of the atmosphere. The latter influences the physical parameters that affect the absorption and disposition of energy at the Earth's surface [113]. Thus, choice of an appropriate land use is an important consideration. Despite uncertainties and difficult logistics in implementation at global scale, especially among the resource-poor and small landholders of developing countries, there is a lot of optimism about the capacity of agriculture for climate change adaptation and mitigation [108,114,115]. Furthermore, soil being the essence of all terrestrial life, there is a strong need to place greater emphasis on preserving the fragile and finite soil resource on this ever-changing planet [116].

Biosequestration of atmospheric CO_2, in soils and biota, is an important strategy to offset anthropogenic emissions (Fig. 7) [47,115]. Adoption of BMPs (left side of the graph indicating inputs) has a technical potential of C sequestration of (2.6–5.0) (3.8) Pg(C) a^{-1} in soils, vegetation and wetlands [115]. In addition, it can also lead to industrial outputs (right side of Fig. 6), which upon commercialization (lower side of Fig. 6) provide important services for human wellbeing. Science-based BMPs can (1) reduce the magnitude of emission of GHGs, and (2) sequester C in soil and biomass especially in that of woody perennials. Thus, the strategy is to reduce the sources of emission and enhance the sinks of C in agro-ecosystems. The more stable SOC fraction also depends on C-nutrient stoichiometry. In general, more stable fine fraction of SOC has more N, P, S per unit of C [117]. Thus, inorganic nutrient availability is critical to the humification efficiency and MRT of C in soil. Nutrient requirements for crop growth increase over time because of the CO_2 fertilization effect and increase in temperature [118], and additional food demand of the growing population. In addition to biomass-C, additional nutrients (N, P, S) are required to effectively convert biomass into humus or SOC [119]. In this context, soil temperature may be an important environmental control that can affect root development, nutrient uptake and use efficiency.

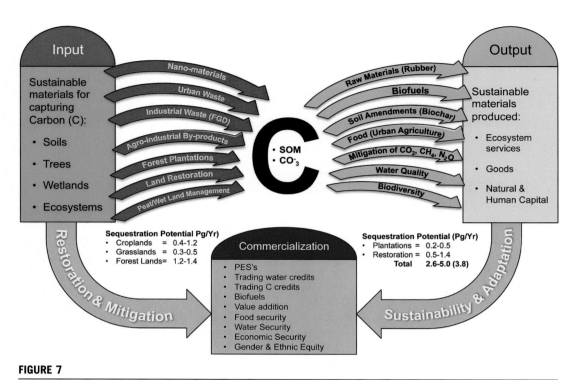

FIGURE 7

Biosequestration of atmospheric CO_2. *(Data from Ref. [115].)*

5.1 SOIL MANAGEMENT

Land use and management systems that can create a positive soil C budget can lead to withdrawal of CO_2 from the atmosphere and long-term sequestration in soil. Potential of C sequestration is relatively high in those soils that have been severely degraded by erosion, nutrient depletion, salinization, acidification and other degradation processes. Conversion from croplands to well-managed pastures, forestlands or any other restorative land use can increase the soil C pool. Conversion of degraded croplands to grasslands and perennial land use can increase SOC pool over decadal timescale [120]. Overall, the soil C sink capacity is finite [121] and its permanence (long MRT) depends on land use and management. Some of the BMPs involve those of sustainable intensification based on reduction or elimination of bare fallow, use of forages and cover crops in the rotation cycle, adopting NT, practising integrated nutrient management for improving soil fertility, and conserving soil and water [104,115,122,123]. Agricultural intensification of cropland can create a positive soil C budget through increase in input of biomass C. Higher production by intensification reportedly avoided emissions by 161 Pg(C) between 1961 and 2005 [124]. The goal of ecological intensification of agriculture is to produce more food per unit of resource use [125]. However, achieving this goal will require a holistic approach to farming. There are specific niches for high-income organic farming as a climate-friendly option [77]. Adoption of conservation agriculture/no-till (CA/NT) farming in the tropics can increase the input of biomass C into the soil by reducing losses through erosion and decomposition [126].

There are also water management options, especially those related to on-site conservation of water in soil (green water), water harvesting and recycling (conversion of blue water into green water), and use of waste water (conversion of grey water into green water). Judicious management of finite water resources, for supplemental drip subirrigation (fertigation) can enhance productivity. However, lifting of groundwater and sprinkling involves energy consumption and the attendant emission of CO_2, which may be as much as 150 kg(C) ha^{-1}–200 kg(C) ha^{-1} [47,48].

5.2 CROP, ANIMAL MANAGEMENT

Emissions of GHGs from crop and animal systems can be avoided through BMPs, especially those of CH_4 and N_2O in rice cultivation and livestock, and through fertilizer management in croplands. Prudent decisions about the rate, timing and mode of application of fertilizers, water and other energy-based inputs are critical to enhancing efficiency, reducing losses into the environment, and avoiding emissions. Integrated systems (e.g. crop-livestock, crop-trees, trees-livestock, cover crops-grain/food crops, crop-tree-livestock, chemical-bio-fertilizers, chemical-bio-pesticides) that mimic nature are environmentally compatible. Management systems involving the holistic agro-ecology-based approach may be suited even for degraded soils managed by resource-poor small landholders of SSA [127].

5.3 LONG-TERM PERSISTENCE AND EFFECTIVENESS OF CARBON SEQUESTRATION IN AGRO-ECOSYSTEMS

The MRT of C in soil and perennial vegetation depends on the land use and management. In general, C sequestered in soil has a longer MRT than that in biomass. Some SOC fractions can persist for millennia, but others only for a few days. It is widely recognized that persistence of SOC does not depend on the molecular structure alone, and there are some significant environmental and biological controls [128]. Thus, MRT of SOC is an ecosystem property. Therefore, choice of ecosystem hotspots for C sequestration is critical. In this context, the projected increase in global temperature would also impact the rate of decomposition and the MRT of SOC.

The SOC pool is temperature sensitive, and microbial respiration increases exponentially with increase in temperature [129]. Indeed, there exists a strong microbial mediation of C-cycle feedback to global warming [130]. There are some critical biochemical mechanisms that are affected by global warming. In general, world soils will accumulate SOC if microbial growth efficiency declines, and will lose C if microbial growth efficiency adapts to warming [131]. There is also a need to understand biological and physicochemical controls of SOC stabilization [132], and the impact of drought stress on C cycling [133]. Vegetation is also affected by climate change, and together (vegetation and soil processes) determine the SOC status even in an alpine intertropical biome [134]. The SOC pool is also determined by textural composition and the water retention characteristics of the soil [135]. Therefore, warming-induced increase in decomposition of SOC pool could be a significant positive feedback to climate change. Indeed, there exists strong evidence of enhanced land-atmosphere feedback with the projected warming [136], especially because of the permafrost thaw. The degree of sensitivity depends on both exogenous and endogenous factors (soil architecture). In this context, stability of micro-aggregates rather than organo-mineral association contributes toward longer MRT of SOC. However, the projected climate change may adversely impact even the so-called 'passive' fraction [137].

5.4 AGRICULTURAL EMISSION AND DIET PREFERENCES

The increasing efficiency of industrial agriculture since the 1960s made meat an affordable everyday food for most of the population in Western nations. The consumption of meat protein has enhanced the input of inorganic N fertilizer. Thus, affordable meat has become a major environmental issue including climate change and water quality [138]. Livestock industry is a major source of CH_4, N_2O and other pollutants. Between 1990 and 2010, an average of 32.8 Tg (CO_2e) a^{-1} (where e refers to equivalent) were embedded in beef, pork and chicken traded internationally, and the traded emissions increased by 19% over this period [139]. The high emission intensity of the beef is attributed to the fact that average CH_4 emission from a cow is 70 kg a^{-1}–120 kg a^{-1} ($\sim$100 kg(CH_4) a^{-1}). However, the importance of human diet as a mitigation strategy cannot be overemphasized. Any increase in preference toward plant-based diet would reduce emissions, especially of CH_4 and N_2O, and decrease the rate of conversion of tropical rainforest [15]. The general public should be made aware of the importance of preference for the plant-based diet to reduce anthropogenic emissions.

6. CONCLUSIONS

The information presented and the literature review support the following conclusions:

1. Agriculture has and continues to be a prominent source of GHGs (CO_2, CH_4, N_2O). Both direct and indirect emissions from agriculture contribute about 30% of anthropogenic emissions. Sources of agricultural emissions are land use conversion, decomposition of soil organic matter (SOM), biomass burning, rice paddy cultivation, enteric fermentation in livestock, manure management, fertilizer use and fuel consumption for farm operations (e.g. ploughing, spraying, harvesting, grain drying).

2. Estimates of direct and indirect emissions from agriculture vary widely, and reliable statistics and a standardized protocol are not available. There is an urgent need to improve and strengthen the database and standardize the methodology of assessing emissions from diverse and diffused sources of GHGs from agro-ecosystems and related industries.

3. Agriculture also has a vast potential to reduce emissions and sequester emissions. Reduction in emissions involves adoption of BMPs of agricultural intensification, which enhance production per unit input of resources (water, nutrients, energy) by decreasing losses (erosion, leaching, volatilization) and enhancing the use efficiency. Conversion from conventional tillage to NT or CT, in conjunction with retention of residue mulch and cover cropping along with integrated nutrient management etc., can reduce use of fossil fuel, minimize risks of soil erosion and improve potential of SOC sequestration. Similarly, drip subfertigation and use of appropriate formulations (slow release, nano-enhanced) can reduce emissions from fertilizer input. Growing aerobic rice with better water management and improved varieties can reduce CH_4 and N_2O emissions from rice paddies. There are numerous options of reducing emissions from livestock industry including improved feed and better systems of manure management. Increasing awareness of the general public about preferences toward plant-based diet can reduce overall resource use in raising livestock.

4. Sequestration of atmospheric CO_2 in soil and woody/perennial biomass can reduce the net anthropogenic emissions. Soil C sequestration is also essential to improving soil quality and use efficiency of input. The critical/threshold level of SOC concentration in the root zone is 1.5%–2.0%,

and most depleted and degraded soils (in South Asia and SSA) have SOC concentration of <1%. Thus, there is a tremendous scope (technical potential) of re-carbonization of soil, forests and other components of the biosphere.

5. The projected climate change may have a strong and positive feedback with enhanced emission from soils, agro-ecosystems and permafrost. Thus, developing climate-smart agriculture and climate-resilient soils through identification and promotion of BMPs is an important strategy.

LIST OF ABBREVIATIONS

BMPs	Best management practice
BNF	Biological nitrogen fixation
CA	Conservation agriculture
CCS	Carbon capture and storage
Ce	Carbon equivalent
CH$_4$	Methane
CO$_2$	Carbon dioxide
Gg	Gigagram (10^9 g)
GHGs	Greenhouse gases
LCA	Life cycle analysis
Mg	Megagram (10^6 g)
Mha	Million hectare
MRT	Mean residence time
N$_2$O	Nitrous dioxide
NT	No-till
NUE	Nutrient use efficiency
Pg	Petagram (10^{15} g)
SOC	Soil organic carbon
SOM	Soil organic matter
SSA	Sub-Saharan Africa
Tg	Teragram (10^{12} g)
US-DOE	US Department of Energy

REFERENCES

[1] UN, World, Population Prospects: The 2012 Revision, 2013. UN New York.

[2] T.A. Boden, R.J. Andres, G. Marland, Global, regional, and National fossil-fuel CO_2 emissions, CDIAC (2013), http://dx.doi.org/10.3334/CDIAC/00001_V2013.

[3] NOAA, 2013. State of the Climate: Carbon Dioxide Tops 400 ppm, 2014.

[4] E.J. Dlugokencky, B.D. Hall, S.A. Montzka, G. Dutton, J. Muhle, J.W. Elkins, Bull. Am. Metrol. Soc. (2013) S33–S34.

[5] U.S.-E.I.A, Monthly Energy Review, U.S. Energy Information Administration, Washington, DC, 2014.

[6] Scripps Institute of Oceanography, CO_2 Now: What the World Needs to Watch, San Diego, CA, 2015.

[7] F. Archard, B. Beuchle, P. Mayaux, H.J. Stibig, C. Bodart, A. Brink, S. Carboni, B. Desclee, F. Donnay, H.D. Eva, A. Lupi, R. Rasi, R. Selinger, D. Simonetti, Global Change Biol. 20 (2014) 2540–2554.

[8] IPCC, Climate Change 2014. Synthesis Report, WMO, Geneva, Switzerland, 2015.

[9] P.J. Crutzen, Nature 415 (2002) 23.

[10] W.F. Ruddiman, Clim. Change 61 (2003) 261–293.
[11] W.F. Ruddiman, Plows, Plagues and Petroleum How Humans Took Control of Climate, Princeton University Press, 2005.
[12] J.D. Diamond, How Societies Choose to Fail of Succeed, Viking Press, New York, 2005, 592 pp.
[13] J. Steinbeck, Grapes of Wrath, Viking Press, New York, 1939.
[14] C. Le Quéré, R. Moriarty, R.M. Andrew, G.P. Peters, P. Ciais, P. Friedlingstein, Earth Syst. Sci. Data Discuss. 7 (2014) 521–610.
[15] F.N. Tubiello, M. Salvatore, R.D. Golec, A. Ferrara, S. Rossi, B. Biancalni, S. Federici, H. Jacobs, A. Flammini, Agriculture, Forestry and Other Land Use Emissions by Sources and Removals by Sinks, FAO, Rome, 2014.
[16] W.R. Cline, Finance & Development, March 2008, pp. 23–27.
[17] K. Paustian, J.M. Antle, J. Sheehan, E.A. Paul, Pew Center on Global Climate Change, 2006. Washington, DC.
[18] FAO, Agriculture's Greenhouse Gas Emissions on the Rise, April 2014. http://www.fao.org/news/story/en/item/216137/icode/.
[19] J.W. Van Groenigen, G.L. Velthof, O. Oenema, K.J. Van Groenigen, C. Van Kessel, Eur. J. Soil Sci. 61 (2010) 903–913.
[20] P. Friedlingstein, R.M. Andrew, J. Rogelj, G.P. Peters, J.G. Canadell, R. Knutti, G. Luderer, M.R. Raupach, M. Schaeffer, D.P. van Vuuren, C. Le Quéré, Nat. Geosci. 7 (2014) 709–715.
[21] M. Krol, W. Peters, P. Hooghiemstra, M. George, C. Clerbaux, D. Hurtmans, D. McInerney, F. Sedano, P. Bergamaschi, M. El Hajj, J.W. Kaiser, D. Fisher, V. Yershov, J.P. Muller, Atmos. Chem. Phys 13 (2013) 4737–4747.
[22] A. Vrbicek, The World's Leading Driver of Climate Change: Animal Agriculture, 2015. http://www.new-harvest.org/2015/01/worlds-leading-driver-climate-change-animal-agriculture/.
[23] FAO, Livestock's Long Shadow: Environmental Issues and Options, FAO, Rome, 2008.
[24] L. Reynolds, World Watch Institute, May 2013. http://www.worldwatch.org/agriculture-and-livestock-remain-major-sources-greenhouse-gas-emissions-0.
[25] J.M. Duxbury, Fert. Res. 38 (1994) 151–163.
[26] IPCC, Summary for policymakers, in: Climate Change 2013: The Physical Science Basis. Contribution of Working Group I to the Fifth Assessment Report of the Intergovernmental Panel on Climate Change, 2013, pp. 1–28.
[27] R.A. Pielke, Science 310 (2005) 1625–1626.
[28] G.J.J. Kreileman, A.F. Bouwman, Water Air Soil Pollut. 76 (1994) 231–258.
[29] J.J. Feddema, K.W. Oleson, G.B. Bonan, L.O. Mearns, L.E. Buja, G.A. Meehl, W.M. Washington, Science 310 (2005) 1674–1678.
[30] S.J. Smith, A.J. Rothwell, BioScience 10 (2013) 6323–6337.
[31] R.A. Houghton, Tellus 51 (1999) 298–313.
[32] P.J. Crutzen, M.O. Andreae, Science 250 (1999) 1669–1678.
[33] W. Seiler, P.J. Crutzen, Clim. Change 2 (1980) 207–247.
[34] P.J. Crutzen, L.E. Haidt, J.P. Krasnec, W.H. Pollock, W. Seiler, Nature 282 (1979) 253–256.
[35] V. Lehsten, A. Spessa, H. Weber, B. Smith, A. Arneth, Biogeochemical Discussions, vol. 5, 2008, 3091–3122.
[36] S.A. Schepaschenko, Contemp. Probl. Ecol. 6 (2013) 683–692.
[37] W. Takeuchi, T. Nakano, S. Ochi, Y. Yasuoka, in: Intl. Geoscience and Remote Sensing Symposia and 24th Canadian Symposia of Remote Sensing, IGRASS, 2002, pp. 2351–2353.
[38] K. Tansey, J.M. Greoire, E. Benaghi, L. Boschetti, et al., Clim. Change 67 (2004) 345–377.
[39] C.A. Powers, Sources of Agricultural Greenhouse Gases, University of Nebraska, 2014, p. 1.
[40] K.C. Cameron, H.J. Di, R.G. McLaren, Aust. J. Soil Res. 35 (1997) 995–1035.
[41] E. Sonne, J. Environ. Qual. 35 (2006) 1439–1450.

[42] G.P. Robertson, P.R. Grace, Environ. Dev. Sustainability 6 (2004) 51–63.
[43] WMO, Greenhouse Gas Bulletin, 9 September 2014.
[44] N. Gilbert, Nature (October 2012), http://dx.doi.org/10.1038/nature.2012.11708.
[45] IPCC, Technical Summary, 2014. Geneva, Switzerland.
[46] USEPA, Sources of Greenhouse Gas Emissions, 2014. http://www.epa.gov/climatechange/ghgemissions/sources.html.
[47] R. Lal, Land Degrad. Dev. 14 (2003) 309–322.
[48] R. Lal, Environ. Int. 29 (2003) 437–450.
[49] M. Hanna, J.E. Sawyer, D. Peterson, Energy Consumption for Row Crop Production, Iowa State University Extension and Outreach, 2012.
[50] N. Koga, H. Tsuruta, H. Tsuji, H. Nakano, Agric. Ecosyst. Environ. 99 (2003) 213–219.
[51] L.A. Yousif, M.H. Dahab, H.R. El Ramlawi, J. Agric. Biotechnol. Sustainable Dev. 5 (2013) 84–90.
[52] A.O. Adewoyin, E.A. Ajav, CIGR J. 15 (2013) 67–74.
[53] K. Kristof, T. Sima, L. Nozdrovicky, P. Findura, Agron. Res. 12 (2014) 115–120.
[54] J.M. Johnson, D.C. Reicosky, R.R. Allmaras, T.J. Sauer, et al., Soil Tillage Res. 83 (2005) 73–94.
[55] N. La Scala, D. Bolonhezi, G.T. Pereira, Soil Tillage Res. 91 (2006) 244–248.
[56] C.G. Sørensen, N. Halberg, F.W. Oudshoom, B.M. Petersen, R. Dalgaard, Biosyst. Eng. 120 (2014) 2–14.
[57] J.W. Erisman, A. Bleeker, J. Galloway, M.S. Sutton, Environ. Pollut. 150 (2007) 140–149.
[58] O. Oenema, J.X. de Klein, et al., Curr. Opin. Environ. Sustainability 9 (2014) 55–64.
[59] I. Shcherbak, N. Millar, G.P. Robertson, Proc. Natl. Acad. Sci. U.S.A. 111 (2014) 9199–9204.
[60] R.A. Omonode, D.R. Smith, A. Gal, T.J. Vyn, Soil Sci. Soc. Am. 75 (2011) 152–163.
[61] M. Reuter, M. Buchwitz, A. Hilboll, A. Ritcher, O. Schneising, M. Hiker, J. Heymann, H. Bovensmann, J.P. Burrows, Nat. Geosci. 7 (2014) 792–795.
[62] R.W. Howarth, E.W. Boyer, W.J. Pabich, J.N. Galloway, AMBIO 31 (2002) 88–96.
[63] P. Rochette, D.E. Worth, E.C. Huffman, J.A. Brierley, B.G. McConkey, J.Y. Yang, J.J. Hutchinson, R.L. Desjardins, R. Lemke, S. Gameda, Can. J. Soil Sci. 88 (2008) 655–669.
[64] H.H. Janzen, K.A. Beauchemin, Y. Bruinsma, C.A. Campbell, R.L. Desjardins, B.H. Ellert, E.G. Smith, Nutr. Cycling Agroecosyst. 67 (2003) 85–102.
[65] D. Chen, H. Suter, A. Islam, R. Edis, J.R. Freney, C.N. Walker, Aust. J. Soil Res. 46 (2008) 289–301.
[66] C.A.M. de Klein, S.F. Ledgard, Nutr. Cycling Agroecosyst. 72 (2005) 77–85.
[67] A.F. Bouwan, G. van Drecht, K.W. van der Hoek, Pedosphere 15 (2005) 137–155.
[68] F. Brentrup, J. Kusters, H. Kuhlmann, J. Lammel, Eur. J. Agron. 20 (2004) 247–264.
[69] H.S. Steinbach, R. Alvarez, J. Environ. Qual. 35 (2006) 3–13.
[70] B.C. Ball, B.S. Griffiths, C.F.E. Topp, R. Wheatley, R.L. Walker, R.M. Rees, C.A. Watson, H. Gordon, P.D. Hallett, B.M. McKenzie, I.M. Nevison, Agric. Ecosyst. Environ. 189 (2014) 171–180.
[71] T.E. Crews, M.B. Peoples, Agric. Ecosyst. Environ. 102 (2004) 279–297.
[72] P. Rochette, H.H. Janzen, Nutr. Cycling Agroecosyst. 73 (2005) 171–179.
[73] B.W. Hütch, Eur. J. Agron. 14 (2001) 237–260.
[74] J. Gomes, C. Bayer, F.D. Costa, M.D. Piccolo, J.A. Zanatta, F.C.B. Vieria, J. Six, Soil Tillage Res. 106 (2009) 36–44.
[75] A.D. Basche, F.E. Miguez, T.C. Kaspar, M.J. Castellano, J. Soil Water Conserv. 69 (2014) 471–482.
[76] E.M. Baggs, R.M. Rees, K.A. Smith, A.J.A. Vinten, Soil Use Manage. 16 (2000) 82–87.
[77] N.E.H. Scialabba, M. Muller-Lindenlauf, Renewable Agric. Food Syst. 25 (2010) 158–169.
[78] R. Thangarajan, N.S. Bolan, G.L. Tian, R. Naidu, A. Kunhikrishnan, Sci. Total Environ. 465 (2013) 72–96.
[79] H.L. Tuomisto, I.D. Hodge, P. Riordan, D.W. Macdonald, J. Environ. Manage. 112 (2012) 309–320.
[80] J.F.F.P. Bos, J. de Haan, W. Sukkel, R.L.M. Schils, NJAS-Wageningen J. Life Sci. 68 (2014) 61–70.
[81] S.M. Novak, J.L. Fiorelli, Agron. Sustainable Dev. 30 (2010) 215–236.
[82] J. Webb, B. Pain, S. Bittman, J. Morgan, Agric. Ecosyst. Environ. 137 (2010) 39–46.

[83] E.G. Nisbet, E.J. Dlugokencky, P. Bousquet, Science 343 (2014) 493–494.

[84] M. Aydin, K.R. Verhulst, E.S. Saltzman, M.O. Battle, S.A. Montzka, D.R. Blake, Q. Tang, M.J. Prather, Nature 476 (2011) 198–201.

[85] F.M. Kai, S.C. Tyler, J.T. Randerson, D.R. Blake, Nature 476 (2011) 194–197.

[86] S. Kirschke, P. Bousquet, P. Ciais, M. Saunois, J.G. Canadell, E.J. Dlugokencky, P. Bergamaschi, D. Bergmann, D.R. Blake, L. Bruhwiler, et al., Nat. Geosci. 6 (2013) 813–823.

[87] P.J. Aselman, J. Crutzen, J. Atmos. Chem. 8 (1989) 307–358.

[88] P. Bergamaschi, S. Houweling, A. Segers, M. Krol, C. Frankenberg, R.A. Scheepmaker, E. Dlugokencky, S.C. Wofsy, E.A. Kort, C. Sweeney, T. Schuck, C. Brenninkmeijer, H. Chen, V. Beck, C. Gerbig, J. Geophys. Res. 118 (2013) 7350–7369.

[89] Q. Zhu, J. Liu, C. Peng, H. Chen, X. Fang, H. Jiang, D. Zhu, W. Wang, X. Zhou, Geosci. Model Dev. 7 (2014) 981–999.

[90] A. van Amstel, J. Integr. Environ. Sci. 9 (2015) 5–30.

[91] B. Ringeval, S. Houweling, P.M. van Bodegom, R. Spahni, R. van Beek, F. Joos, T. Röckmann, Biogeosciences 11 (2014) 1519–1558.

[92] N. Barros, J.J. Cole, L.J. Tranvik, Y.T. Prairie, D. Bastviken, V.L.M. Huszar, P. del Giorgio, F. Roland, Nat. Geosci. 4 (2011) 593–596.

[93] I. Kögel-Knabner, W. Amelung, Z.H. Cao, S. Fiedler, P. Frenzel, R. Jahn, K. Kalbitz, A. Kolbl, M. Schloter, Geoderma 157 (2010) 1–14.

[94] H. Chen, Q. Zhu, C. Peng, N. Wu, Y. Wang, X. Fang, H. Jiang, W. Xiang, J. Chang, X. Deng, G. Yu, Global Change Biol. 19 (2013) 19–32.

[95] J.T.A. Verhoeven, T.L. Setter, Ann. Bot. 105 (2010) 155–163.

[96] M. Gill, P. Smith, J.M. Wilkinson, Animal 4 (2010) 323–333.

[97] E. Kebreab, K. Clark, C. Wagner-Riddle, J. France, Can. J. Animal Sci. 86 (2006) 135–158.

[98] J.A.B. Cardenas, C.L. Flores, Rev. Mex. Cienc. Pecu. 3 (2012) 215–246.

[99] X.P.C. Vergé, C. De Kimpe, R.L. Desjardins, Agric. For. Meteorol. 142 (2007) 255–269.

[100] K.M.W. Anthony, P. Anthony, G. Grosse, J. Chanton, Nat. Geosci. 5 (2012) 419–426.

[101] H.F. Jungkunst, J.P. Krüger, F. Heitkamp, S. Erasmi, S. Glatzel, S. Fiedler, R. Lal, Recarbonization of the Biosphere, Springer, Dordrecht, 2012, pp. 127–157.

[102] F.M. O'Connor, O. Boucher, N. Gednet, C.D. Jones, G.A. Folberth, R. Copell, P. Friedlingstein, W.J. Collins, J. Chappellaz, J. Ridley, C.E. Johnson, Rev. Geophys. 48 (2010) RG4005.

[103] B.W. Hütsch, Eur. J. Agron. 14 (2001) 237–260.

[104] R. Lal, Science 304 (2004) 1567.

[105] L.R. Oldeman, Soil Resilience and Sustainable Land Use, CAB International, Wallingord, 1994, pp. 99–118.

[106] R. Lal, Geoderma 16 (1976) 419–431.

[107] K.R. Olson, A.N. Gennadiyev, A.P. Zhidkin, M.V. Markelov, Soil Sci. 176 (2011) 449–458.

[108] R. Lal, Environ. Int. 30 (2004) 981–990.

[109] S.M. Benson, Science 344 (2014) 1431.

[110] M. Fowlie, L. Goulder, M. Kotchen, S. Borenstein, J. Bushnell, L. Davis, M. Greenstone, C. Kolstad, C. Knittel, R. Stavins, M. Wara, F. Wolak, C. Wolfram, Science 346 (2014) 815.

[111] Anonymous, Science 347 (2015) 697.

[112] W. Pizer, M. Adler, J. Aldy, D. Anthoff, M. Cropper, K. Gillingham, M. Greenstone, B. Murray, R. Newell, R. Richels, A. Rowell, S. Waldhoff, J. Wiener, Science 346 (2014) 1189.

[113] J.J. Feddema, K.W. Olseon, G.B. Bonan, L.O. Mearns, L.E. Buja, G.A. Meehl, W.M. Washington, Science 310 (2005) 1674–1678.

[114] R. Kates, Clim. Change 45 (2000) 5–17.

[115] R. Lal, BioScience 60 (2010) 708–721.

[116] H.H. Janzen, P.E. Fixen, A.J. Franzluebbers, J. Harrey, R.C. Izaurralde, Q.M. Ketterings, D.A. Lobb, W.H. Schlesinger, Soil Sci. Soc. Am. J. 75 (2011) 1–8.

[117] C.A. Kirkby, A.E. Richardson, L.J. Wade, G.D. Batten, C. Blanchard, J.A. Kirkegaard, Soil Biol. Biochem. 60 (2013) 77–86.

[118] M.E. Gavito, P.S. Cutis, T.N. Mikkelsen, I. Jakobsen, J. Exp. Bot. 362 (2001) 1913–1923.

[119] R. Lal, J. Soil Water Conserv. 69 (2014) 186A–192A.

[120] K.K. McLauchlan, S.A. Hobbie, W.M. Post, Ecol. Appl. 16 (2006) 143–153.

[121] D.R. Saurbeck, Nutr. Cycling Agroecosyst. 60 (2001) 1–3.

[122] K. Paustian, O. Andrén, H.H. Janzen, R. Lal, P. Smith, G. Tian, H. Tiessen, M. Van Noordwijk, P.L. Woomer, Soil Use Manage. 13 (1997) 230–244.

[123] K. Paustian, C.V. Cole, D. Sauerbeck, N. Sampson, Clim. Change 40 (1998) 135–162.

[124] J.A. Burney, S.J. Davis, D.B. Lobell, PNAS 107 (2010) 12052–12057.

[125] Z. Hochman, P.S. Carberry, M.J. Robertson, D.S. Gaydon, L.W. Bell, P.C. McIntosh, Eur. J. Agron. 44 (2013) 109–123.

[126] C.E.P. Cerri, G. Sparovek, M. Bernoux, W.E. Easterling, J.M. Melillo, C.C. Cerri, Sci. Agric. 64 (2007) 83–99.

[127] P. Tittonell, E. Scopel, N. Andieu, H. Posthumus, P. Mapfumo, M. Corbeels, G.E. van Halsemn, R. Lahmar, et al., Field Crops Res. 132 (2012) 168–174.

[128] M.W.I. Schmidt, M.S. Torn, S. Abiven, T. Dittmar, G. Guggenberger, I.A. Janssens, M. Kleber, I. Kögel-Knabner, et al., Nature 478 (2011) 49–56.

[129] K. Karhu, M.D. Auffret, J.A.J. Dungait, D.W. Hopkins, J.I. Prosser, B. Singh, et al., Nature 513 (2014) 81–84.

[130] J. Zhou, K. Xue, J. Xie, Y. Deng, L. Wu, X. Cheng, S. Fei, S. Deng, Z. He, J.D. Van Nostrand, Y. Lou, Nat. Clim. Change 2 (2012) 106–110.

[131] W.R. Wieder, G.B. Bonaan, S.D. Allison, Nat. Clim. Change 3 (2013) 909–912.

[132] M. von Lützow, I. Kögel-Knaber, Biol. Fertil. Soils 46 (2009) 1–15.

[133] M.K. van der Molen, A.J. Dolman, P. Ciais, T. Eglin, N. Gobron, B.E. Lawe, P. Meir, et al., Agric. For. Meteorol. 151 (2011) 765–773.

[134] P. Podwojewski, J.L. Janeau, S. Grellier, C. Valentin, S. Lorentz, V. Chaplot, Earth Surf. Processes and Landforms 36 (2011) 911–922.

[135] G. Saiz, M.I. Bird, T. Domingues, F. Schrodt, M. Schwarz, T.R. Feldpausch, E. Veenendaal, G. Djagbletey, F. Hien, H. Compaore, A. Diallo, J. Lloyds, Global Change Biol. 18 (2012) 1670–1683.

[136] P.A. Dirmeyer, B.A. Cash, J.L. Knitter, C. Stan, T. Jung, L. Marx, et al., J. Hydrometeor. 13 (2012) 981–995.

[137] J. Leifeld, J. Fuhrer, Biogeochemistry 75 (2005) 433–453.

[138] H. Steinfeld, P. Gerber, T. Vassenaar, V. Castel, M. Rosales, C. de Haan, Livestocks's Long Shadows: Environmental Issues and Opinions, FAO, Rome, Italy, 2006.

[139] D. Caro, A. LoPresti, S.J. Davis, S. Bastianoni, K. Caldeira, Environ. Res. Lett. 9 (2014) 114005.

WIDESPREAD SURFACE SOLAR RADIATION CHANGES AND THEIR EFFECTS: DIMMING AND BRIGHTENING

29

Shabtai Cohen, Gerald Stanhill

Department of Environmental Physics and Irrigation, Institute of Soil, Water and Environmental Sciences, Agricultural Research Organization, The Volcani Centre, Bet Dagan, Israel

CHAPTER OUTLINE

1. INTRODUCTION – SOLAR RADIATION BASICS

The flux density and wavelength of electromagnetic radiation emitted from a body depend on its temperature. At the Earth's surface the wavebands that contain the most energy, and are therefore of prime interest in the context of climate influences, are those emitted by the Sun and the Earth. The calculation of spectral distributions from Planck's law using their approximate temperatures of 5800 K and 300 K, for Sun and Earth, as shown in Fig. 1, shows that 97% of the energy of solar and >99% of that of terrestrial radiation fall within the wavebands of 0.29 μm–3 μm and 3 μm–100 μm, respectively. Those wavebands are referred to as short-wave (or solar) and long-wave (or terrestrial) radiation [1]. The current ubiquitous, steady increase in atmospheric carbon dioxide concentration influences climate indirectly by its absorption of radiation in the long-wave band, which decreases long-wave radiative losses from the Earth. Since its absorption in the solar spectrum is small, CO_2 has a negligible direct influence on the Earth's solar radiation balance.

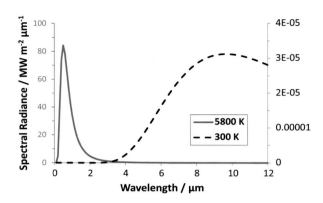

FIGURE 1

Spectral radiance for black bodies at absolute temperatures of 5800 K and 300 K, which represent approximate temperatures of the Sun and Earth, respectively. The curves were computed using Planck's law. Because of the differences in radiance, the left scale is for 5800°, and the right for 300°.

Several of the quantities encountered when studying the Earth's short-wave radiation balance are easily computed. These relationships provide the quantitatively minded reader with a greater understanding of solar radiation and its trends.

Black body radiation is described by the Stefan–Boltzmann equation, i.e.

$$B = \sigma T^4$$

where B is radiant flux density (W m^{-2}) emitted from a black body of absolute temperature, T, and σ is the Stefan–Boltzmann constant, 5.67×10^{-8} W m^{-2} K^{-4}. Taking the Sun as a black body with average surface temperature 5800 K, we can compute the solar output to be 6.42×10^7 W m^{-2}. The total output can be computed from the surface area of the Sun (taking its radius as 6.96×10^8 m), and that radiation spreads out in all directions. At the Earth, a distance of 1 astronomical unit (1.5×10^{11} m) from the Sun, the radiant flux density reaching a surface normal to the Sun's rays, before it is attenuated by the atmosphere, can be computed by multiplying the solar output by the ratio of the area of the two spheres whose radii are that of the Sun and the Sun-Earth distance (above). The resulting flux density, 1380 W m^{-2}, is an estimate of the extraterrestrial 'solar constant', which is very close to the currently accepted value of 1366 W m^{-2}, also known as total solar irradiance (TSI; see below). The top of the atmosphere (TOA) radiation varies during the year by about 3.3% due to the eccentricity of the Earth's orbit. As long as the solar surface temperature and composition do not change, the yearly average, i.e. TSI, will be constant. The ratio of the area of a sphere to that of a circle of the same radius is 4, so the mean TSI for the globe is 342 W m^{-2}.

TOA (or extraterrestrial) solar radiation on a plane parallel to the Earth's surface varies with the solar zenith angle, i.e. the angle between the vertical and the solar vector. Calculation of solar angles and TOA solar radiation is straightforward and given elsewhere [1–4]. TOA values are used to compare with bottom of the atmosphere (BOA) measurements in order to determine atmospheric absorption of radiation, e.g. atmospheric transmission and turbidity and aerosol optical depth.

2. SOLAR RADIATION ABOVE THE ATMOSPHERE
2.1 TOTAL SOLAR IRRADIANCE VARIATIONS

A review of the very extensive literature on TSI [5] shows that its mean value, measured directly from satellites since 1975, is 1366 W m^{-2} with a range of 1 W m^{-2} between its maximum and minimum values, which coincide with those of the 11-year Sun spot cycle. Based on the close relationship between TSI and the number of sunspots, observations of which go back some 300 years, it has been calculated that TSI has increased by 0.69 W m^{-2} since 1750. It is important to note that in absolute rather than proportional terms the changes in TSI are not insignificant when compared with the size of the other radiative forcing factors believed responsible for contemporary climate change [6].

2.2 EARTH'S ALBEDO AND NET SOLAR RADIATION ABOVE THE ATMOSPHERE

The Earth's planetary albedo depends mostly on cloud amount and properties, but also on land use. There is no reason to believe that the planetary albedo has been and will remain constant, and a change of 1% in its value would have a large impact on the Earth's climate system [7]. Although a method of determining the Earth's albedo based on lunar earthshine intensity was suggested by Leonardo de Vinci 500 years ago, accurate measurements of the albedo with his method only began at the end of the last century [8]. Satellite observations made continuously during the past 20 years indicate that albedo is relatively constant at 29% $\pm$ 2% [9,10]. These measurements are close to previous estimates of 30% [11] and 31% [12]. Analyses of earthshine measurements from 1999 and later together with older cloud data from the International Satellite Cloud Climatology Project (ISCCP) shows that from 1984 results are similar [8,13–15]. Taking the current Earth albedo to be 29%, the net solar input into the planet is about 243 W m^{-2} [16].

3. SOLAR RADIATION BELOW THE ATMOSPHERE

From this brief discussion of TOA solar radiation balance we discuss the situation at the surface below the atmosphere where the solar radiation balance is confounded by atmospheric transmissivity and surface albedo. The former depends mostly on cloudiness and cloud properties, but also on dust and other aerosols. The latter, which has a small influence on downward radiation, depends on surface properties, which are influenced by land use and climate.

As solar radiation traverses the atmosphere it is absorbed and reflected by gases and nongaseous particles [17]. Ozone is responsible for absorption of most of the ultraviolet (UV) radiation, i.e. the solar radiation at wavelengths below 0.29 μm; at larger wavelengths oxygen and ozone absorption is negligible. Water vapour is a significant absorber in the infrared portion of the solar spectrum above 0.7 μm. Carbon dioxide absorption of solar radiation is negligible. Aerosols and clouds can scatter and reflect radiation back to space. Radiation reflected from the Earth's surface can be re-reflected back, and so surface albedo can influence the downward flux. Thus BOA solar radiation is always less than that at TOA and is commonly divided into two fluxes: direct radiation coming from a 2.5° to 5° angle centred on the Sun, and diffuse radiation from the rest of the sky hemisphere. The total of these two, i.e. global radiation ($E_g\downarrow$), is the total solar energy available at the surface.

3.1 MEASUREMENT OF SURFACE RADIATION

Total short wave 'solar' radiant flux density on a horizontal surface on the Earth's surface (BOA), i.e. global radiation, $E_g\downarrow$, is measured with a pyranometer. First-class pyranometers measure the temperature difference between an exposed optically black surface and either a white surface (in the older instruments) or the lower shielded surface using a thermopile. In order to exclude thermal radiation and advection losses from the black surface, this is covered with two quartz glass domes that transmit radiation between approximately 0.3 μm and 3 μm wavelength, and a temperature correction circuit is incorporated in the instrument [1] (see http://en.wikipedia.org/wiki/Pyranometer). Another type of pyranometer in common use, due to its lower cost, is based on a semiconductor sensor (e.g. a photodiode), which upon illumination causes an electrical current to flow. The sensor is covered with appropriate filters to measure solar radiation, but the maximum wavelength measured is 1.1 μm, so total solar radiation is determined indirectly by assuming that the ratio of the full spectrum to that below 1.1 μm is constant. In most outdoor conditions the assumption is good enough for many applications, e.g. calculation of crop water requirements, but semiconductor pyranometers are not acceptable for first-class meteorological measurement.

Daily cleaning of the outer pyranometer dome and yearly calibration of sensors is necessary in order to ensure the reliability of measurements. These and other constraints have restricted the size of measurement networks producing reliable data for solar radiation as compared to those measuring air temperature. Most networks began to operate during the International Geophysical Year, 1957–58 and much of their data is available (see below).

A second widely used surface measure of some current interest is sunshine duration, or the amount of time that direct solar radiation exceeds a threshold of 120 W m^{-2}, corresponding approximately to direct irradiance at 3° solar elevation under clear sky conditions [1]. This measure has been shown to be highly correlated with global radiation, both on a single day basis as well as for yearly totals [18,19]. Instruments measuring sunshine duration came into use in the last quarter of the nineteenth century, and some of their history has been recently reviewed [20]. Many measurement series dating back to the nineteenth century are available in various forms, and analysis of these has enabled estimates of variations in solar radiation to be made for more than a century, e.g. [21,22].

In addition to surface measurements, satellite-based sensors have been monitoring Earth radiance in different wavebands for more than two decades. Algorithms have been developed to use these measurements to calculate solar radiation at the surface. These measurements have the advantage of spatial averaging over an area several orders of magnitude larger than the few square centimetres measured by the surface-based sensors, and the ongoing efforts to improve the reliability and accuracy of the satellite measurements have led to their increased acceptance. High-resolution surface measurement stations in use today are often justified by the operators as a means to calibrate and corroborate the satellite measurements [23].

3.2 COMPARING GLOBAL RADIATION ($E_g\!\downarrow$) FROM DIFFERENT SITES

When comparing sites it is convenient to consider annual totals of $E_g\!\downarrow$, since seasonal variations can be large and vary greatly. However, $E_g\!\downarrow$ varies with altitude and latitude. One way to normalize data from different sites is to determine the transmission of a unit atmosphere, which is similar to turbidity [3,24]. Yearly means of $E_g\!\downarrow$ are converted to atmospheric transmittance, τ_m, by dividing by integrated yearly extraterrestrial solar irradiance on a horizontal surface (S_o) computed for the latitude of the measurements, i.e.

$$\tau_m = \frac{\int E_g\!\downarrow}{\int S_o\!\downarrow} \tag{1}$$

Transmittance is also an exponential function of the optical thickness of the atmosphere, k, and the vertical nondimensional air mass, m, such that

$$\tau_m = \exp(-km)$$

or

$$k = -\ln(\tau_m)/m \qquad (2)$$

For a unit air mass ($m = 1$) Eqn (2) yields:

$$\tau_1 = \exp(-k) = \exp(\ln(\tau_m)/m) \qquad (3)$$

Values of τ_1, which represents the yearly average transmittance of a unit atmosphere at the site, are computed for each yearly mean of $E_g\downarrow$, where m is computed from site altitude using a simple altimetric relationship such as:

$$m = \exp\frac{-A}{8200} \qquad (4)$$

where A is site altitude (m) (after reference [3]). A second method to normalize data from different sites is multiple regression of $E_g\downarrow$ on time and site parameters, where the influence of altitude is taken as linear, but site latitude (Φ) is taken as $\cos^3(\Phi)$ [24].

3.3 ARCHIVES OF SURFACE SOLAR RADIATION MEASUREMENTS

Most of the studies of changes in solar radiation around the globe have been based on analysis of data sets of global radiation collected at weather stations. Since most of the scientists doing the analyses did not measure the data themselves, it is important to understand where the data comes from. Solar radiation data measured by the different national weather services and conforming to World Meteorological Organization (WMO) standards are collected in various national archives and are available from national weather services. Much of this data has also been collected in two archives: the World Radiation Data Centre (WRDC) archive in St Petersburg, Russia, which was established by the WMO in 1964 (see http://wrdc-mgo.nrel.gov/) and is in the public domain; and the Global Energy Balance Archive (GEBA) in Zurich, Switzerland [25], which has incorporated much data after quality control filtering [26].

Data from the US is managed by the National Renewable Energy Laboratory's (NREL) Renewable Resource Data Centre (RReDC, www.nrel.gov/rredc). Although solar radiation has been measured in the US for over a century, first-class long-term data are not available for stations in their network.

The World Radiation Monitoring Centre (WRMC, http://www.bsrn.awi.de/) archives data from the Baseline Surface Radiation Network (BSRN, [27]), which is a small number of stations (currently about 60) in contrasting climatic zones, covering a latitude range from 80°N to 90°S, where solar and atmospheric radiation are measured with instruments of high accuracy and with high time resolution (1–3 min). The BSRN program began in the late 1990s and is based on voluntary participation of organizations measuring radiation in different countries.

Maps of the global distribution of stations contributing to the WRDC, BSRN and NREL are given in Fig. 2, which can guide scientists interested in this topic to relatively easily obtained historic data sets. The uneven distribution of stations underscores the problems in using these data to determine global changes in solar radiation and the intrinsic superiority of satellite measurements once those become completely reliable.

(a)

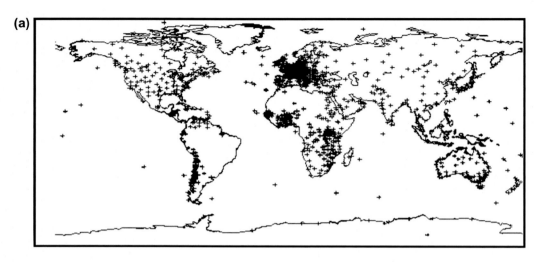

(b)

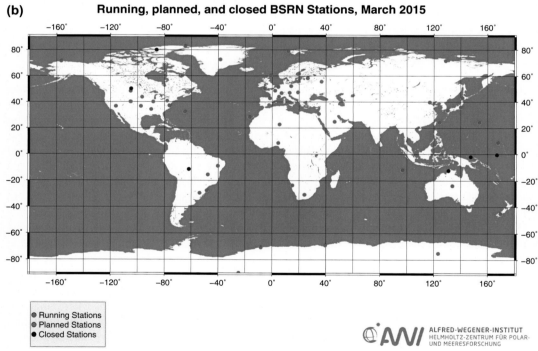

FIGURE 2

(a) World radiometric network (1964–1993) archived in the WRDC. *(From http://wrdc-mgo.nrel.gov/html/ mapap.html.)* Stations are marked with a cross. (b) Active BSRN stations (red, black) and candidate stations (green) as of mid-2013 [23]. (c) Map of weather stations operated in the USA whose solar radiation data is available from the National Renewable Energy Laboratory data centre (http://www.nrel.gov/).

(c)

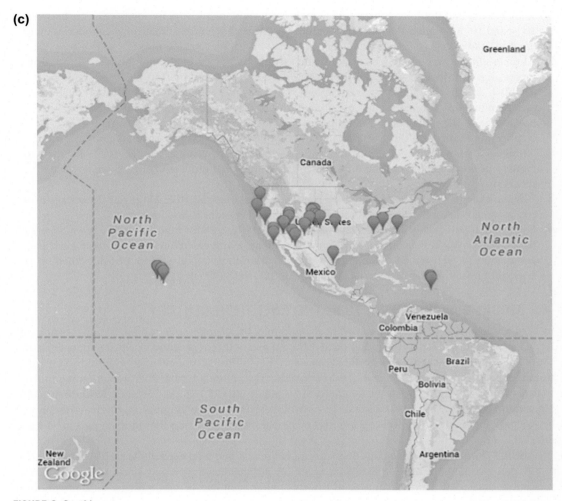

FIGURE 2 Cont'd

4. TRENDS IN SURFACE SOLAR RADIATION, OR GLOBAL DIMMING AND BRIGHTENING

Significant multiyear trends in $E_g\downarrow$ during the first decades that measurements were made were reported by a few scientists during the twentieth century. Many of these decreasing trends, called 'global dimming' [24], were in excess of 1% per decade. They were viewed with considerable scepticism by the scientific community. The reasons for this scepticism are important because they reflect the way climate change science is currently carried out. Here are some possibilities:

1. Previous texts, which were accepted as foundations of climate science, assumed that Earth's solar radiation budget was constant on the timescale of hundreds of years [28], although changes in solar activity and the solar constant were included as possible drivers for long-term (i.e. between 10^3 a and 10^7 a) climate changes (see reference [29] for a review of climate change theories up to the mid-1960s).

2. Climate change science has been dominated by the influence of the ubiquitous and steadily increasing atmospheric greenhouse gases, especially CO_2. Much effort has been made to establish that this change is large enough to warrant worldwide political action to reduce fossil fuel combustion. The magnitude of 'global dimming' was clearly of the same order of magnitude as the greenhouse gas influence [IPCC-AR5]. If large changes were occurring unnoticed to the scientific community, how good was our understanding of climate and climate change? That question may have been viewed as a threat to the attempts to harness political action and the unprecedented funding that climate change science was receiving [30].

3. Climate change science has focused on TOA influences (e.g. radiative forcing at the tropopause) and assumed that the distribution of energy within the system is less important.

4. Solar radiation is highly variable spatially and temporally, and this high variability has hampered integration of worldwide trends. This is in sharp contrast with greenhouse gases, which have been assumed to be well mixed in the atmosphere and whose rate of increase can be discerned on a seasonal basis from measurements at a few carefully selected sites.

4.1 GLOBAL DIMMING REPORTS IN THE TWENTIETH CENTURY

In 1974 Suraqui et al. [31] reported 'severe changes over the years in solar radiation' and issued a call for 'a careful study of incoming radiation at different places throughout the world ... to determine the exact kind, order of magnitude and their causes ...' [31]. The 'severe changes' referred to emerged from the measurements at the site of the Smithsonian Institution's former solar radiation monitoring station on Mt. St. Katherine in the southern Sinai peninsula (28° 31′ N, 33°56′E, 2643 m altitude). Measurements using modern radiometers as well as some of the original instruments employed between 1933 and 1937 showed a 12% loss in global radiation during the intervening four-decade interval.

Atsumu Ohmura, whose background was in glaciology, and who headed the GEBA archive [25], reported at a conference in 1988 that solar radiation was decreasing at many sites in Europe where it was being measured. His colleagues, who were highly sceptical of his findings, discouraged him from pursuing this, and the report was published (or temporarily buried) in a little-known conference proceedings [32]. Russak [33] reported in 1990 decreasing trends of 0.2 W m^{-2} a^{-2}–0.6 W m^{-2} a^{-2} for a few stations in northern Europe. Gerald Stanhill, who used solar radiation measurements for determining evaporation and crop water use in arid environments, was intrigued by the decreasing trends in solar radiation that he found in radiation records. Stanhill and Moreshet [34] in 1992 analyzed data from 45 stations for the years 1958, 1965, 1975 and 1985, and found a statistically significant average worldwide decrease of $E_{g\downarrow}$ totalling 5.3% (or 0.34 W m^{-2} a^{-2}) from 1958 to 1985. Decreasing trends of the same order of magnitude were found for sites in Australia [35], Japan [36], the Arctic [37], Antarctica [38], Israel [39] and Ireland [40]. The largest decrease, found in Hong Kong, was 1.8 W m^{-2} a^{-2}, i.e. a decrease that is in excess of 1% per year [41]. Other groups reported dimming for China [42], the former Soviet Union [43] and Germany [44,45]. Reductions in solar radiation were larger for urban industrial sites, but even at sites remote from pollution, $E_g\downarrow$ was found to be decreasing at a rapid rate [36].

Gilgen et al. [26] reviewed trends found in the GEBA archive. Their 1998 paper, entitled 'Means and trends of short wave irradiance at the surface estimated from Global Energy Balance Archive Data', included analyses of accuracy and biases, and trends in $E_g\downarrow$ for different regions of the world. The final sentence of the abstract noted that 'on most continents, shortwave irradiance decreases significantly in large regions, and significant positive trends are observed only in four small regions'.

Stanhill and Cohen [24] tabulated the negative trends for different sites around the world and coined the term *global dimming* to refer to the widespread decreasing trends in global radiation. Of the 30 stations where detailed analyses of trends had been published, at 28 stations the values of $E_g\downarrow$ had decreased and only at two – Dublin, Ireland, and Griffith, Australia – had $E_g\downarrow$ increased (by 0.56 W m^{-2} a^{-2} and 0.76 W m^{-2} a^{-2}, respectively). They also analyzed solar radiation records from the International Geophysical Year, 1958, and the years 1965, 1975, 1985 and 1992. These records were from between 145 (1958) and 303 (1992) stations whose measurements conformed to WMO standards. Average transmittance of a unit atmosphere for the northern hemisphere was 0.52 in 1957 and declined steadily to 0.44 in 1992 while that for the southern hemisphere averaged 0.57 until 1985 and declined between 1985 and 1992 to 0.52. A spline fit to the latitudinal distribution of $E_g\downarrow$ showed that the decrease during the 34-year period had been especially large in the industrialized region of the northern hemisphere with a centre at $\sim 35°$N and a width of approximately 20°. This feature and an analysis of the various possible reasons for the dimming phenomenon, led to the conclusion that particulate aerosols, and especially those from anthropogenic sources, were the cause of the changes. Similar conclusions were drawn at about the same time by Liepert for data from the US [46].

Many subsequent studies have highlighted similar trends based on data collected from the mid-twentieth century and onward. Trends for individual sites are highly variable, and for some places and some parts of the world no change or increases in solar radiation have been found.

4.2 FROM DIMMING TO BRIGHTENING

Subsequent studies [47,48] found evidence for a reversal in the negative trends in solar radiation, which, for many sites changed to positive trends in the late 1980s and early 1990s. The data sets analyzed were from the GEBA archive [47] and, for the first time, long-term trends in satellite data from 1983 to 2001 [48]. However, there is an inconsistency between the two studies, since the satellite data show brightening over the oceans and no trend over the land surfaces while the surface GEBA and BSRN measurements are land based and show clear brightening during this period. The reversal in the trend is thought to be related to the decreases in air pollution in Europe and other parts of the industrial regions following legislation that limited air pollution. The positive trend has not led to a full recovery in $E_g\downarrow$, and current levels of solar radiation in most places where dimming took place are still below the values measured during the 1950s. A selection of widespread trends reported for $E_g\downarrow$ is given in Table 1a, and a regional breakdown in Table 1b. Since 2004 several international meetings have been held to discuss these topics (Table 2).

Table 1a Selected estimates of widespread trends in surface solar radiation from surface measurements and satellite-based estimates. *(Based on [100]. A more extensive list can be found in [51].)*

Time period	Energy trend per decade (W m^{-2})	Comments	References
From mid-1950s to 1992	−3	Trend analysis of about 30 sites of various lengths, and data from five years from 1957 to 1992 for >145 stations	[24]
1960−1990	−2	Trend analysis of GEBA and US NREL data sets from 1960 to 1990	[101]
From mid-1950s to 1990	−3	Statistics of the GEBA data set based on about 300 sites of various length	[26]
From mid-1950s to 1990	−1.6 −4.3	Analysis of GEBA data to constrain the 'urbanization' effect. Separation of sparsely populated sites (<0.1 million inhabitants) and populated sites (>0.1 million inhabitants)	[54]
1977−1990	−2	Trend analysis of 5 records of the GMD data set from remote sites from South Pole to Barrow, Alaska	[102]
1993−2004	4.7	Trend analysis of 18 BSRN records	[47]
1985−2005	2.2	Decadal change between (1985−1995) and (1995−2005) based on 320 GEBA sites	[47]
1983−2001	1.6 2.4 −0.5	Global. University of Maryland algorithm with ISCCP clouds − Global average Ocean surfaces Land surfaces	[48]
1984−2000	2.4	Global. ISCCP clouds with own RT model	[103]
1984−2000	0.4 1 −1	Global (ISCCP FD) Ocean (ISCCP FD) Land (ISCCP FD)	[60]

Notes: GMD − Global monitoring division of NOAA; ISCCP FD − International Satellite Cloud Climatology Project result data sets.

4.3 VALUES OF $E_g\downarrow$ PRIOR TO THE 1950s

Little is known about $E_g\downarrow$ prior to the 1950s, and since temperature changes then are well documented, such information could be valuable for understanding the influences of $E_g\downarrow$ on climate. Sunshine duration (SSD) is defined by the WMO as the number of hours that solar radiation exceeds 120 W m^{-2}, and routine SSD measurements began in the nineteenth century in some places. Stanhill and Cohen [18,19] used SSD data as proxies for $E_g\downarrow$ based on recent simultaneous measurements of both measures, in order to deduce trends of $E_g\downarrow$ from 1891 to 1987 for the US and from 1890 to 2004 for Japan. SSD was found to be well correlated with $E_g\downarrow$ and therefore can serve as a proxy. The data from the US and Japan were from 106 and 65 stations with at least 70 years and 35 years of data each, respectively. In the US, mean SSD increased from 1891 to the 1930s and then decreased until the

Table 1b An overview of regional trends in surface solar radiation ($E_g\downarrow$) during three periods. Predominant declines ('dimming') occurred in the 1950–1980s, partial recoveries ('Brightening') at many locations, except India, in the 1980s–2000, and data after 2000 show mixed tendencies. Numbers are typical literature estimates for the specified region and period. (Based on various Sources as Referenced in Wild [51]. Adapted from Wild [52].)

Regional trends in surface solar radiation:			
Region	~1957 to ~1980	~1980 to 2000	After 2000
USA	− −	+ +	+ +
Europe	−	+	+
China and Mongolia	− −	+	−
Japan	− −	+ +	0
India	−	− −	− −

− − between −10 W m^{-2} and −5 W m^{-2} per decade; − between −4 W m^{-2} and −1 W m^{-2} per decade; + between 1 W m^{-2} and 4 W m^{-2} per decade; + + between 5 W m^{-2} and 8 W m^{-2} per decade.

Table 2 International meetings held on changing surface solar radiation ($E_g\downarrow$) and related changes in evaporation.

Organizing organization and event	Date	Session title	Location	References
AGU/CGU joint assembly	May 17–18, 2004	Magnitude and causes of decreasing surface solar radiation	Montreal, Canada	[97]
Australian Academy of Science international workshop	November 22–23, 2004	Pan evaporation: An example of the detection and attribution of trends in climate variables	Canberra, Australia	[98]
EGU general assembly	April 15–20, 2007 and yearly since then	Surface radiation budget, radiative forcings and climate change	Vienna, Austria	
AGU fall meeting	December 10–14, 2007	Pan evaporation trends: observations, interpretations, and the ecohydrological implications	San Francisco, CA, USA	
Israel Science Foundation international workshop	February 10–14, 2008	Global dimming and brightening	Ein Gedi, Israel	[99,100]

mid-1940s. In Japan, a similar increase was observed from 1900 to the mid-1940s. This was followed by a decline until the late 1950s. Palle and Butler [22] found a decrease in SSD for four stations in Ireland for the period from 1890 to the 1940s. Sanchez-Lorenzo et al. [49] analyzed SSD for the Iberian Peninsula for 1931–2004 and found a dimming trend from the 1950s to the early 1980s followed by brightening, but the early data (1931–1950) showed no clear trend. Thus, it is possible to obtain estimates of $E_g\downarrow$ for the first half of the twentieth century using the many SSD data sets that exist, but more work is needed to understand this period.

4.4 GLOBAL AND REGIONAL CHANGES

As noted, the name 'global dimming' was originally intended to refer to trends in global radiation. However, at the outset there was some ambiguity in the name because of the similarity to and comparisons with global warming, viewed as a global phenomenon whose energetics are quantitatively similar to those of global dimming although it also exhibits regional variation [6]. All the early work on dimming noted the large variability from site to site and that in some regions no dimming and even brightening occurred. Considering that only part of the globe is covered by the surface radiation measurement network, and the more than 70% covered by oceans is seriously underrepresented, it is difficult to estimate a truly 'global' trend for surface global radiation, except, perhaps using satellite data [48]. We can imagine these trends to occur in patches of the globe. Martin Wild, who has led much of the recent research on this topic, has summarized trends for the dimming and brightening periods in different parts of the globe ([51,52] and Table 1b).

Significant rates of dimming and brightening have been observed at many sites remote from major sources of air pollution, e.g. the polar regions [37,38], but the largest trends have been observed in heavily polluted regions (e.g. Hong Kong [41], India [50] and China [53]), suggesting a significant relationship between pollution rates and global radiation trends. Alpert et al. [54] found that dimming from the 1950s to the 1980s averaged 0.41 W m^{-2} a^{-2} for highly populated sites while for sparsely populated sites, i.e. populations $<0.1 \times 10^6$, dimming was only 0.16 W m^{-2} a^{-2}. In equatorial locations with low population density, there were slightly increasing trends. Since most of the globe is sparsely populated, this implies that the spatially averaged changes in $E_g\downarrow$ are significant but smaller than those obtained by averaging the data, which may be biased toward population centres. However, to date no model has been developed to integrate population density and its influence on $E_g\downarrow$ with the worldwide grid of $E_g\downarrow$ in order to update the estimates of dimming and brightening, and current estimates revolve around those given in Table 1. In addition, investigation of trends for a small country like Israel, where a number of stations have been measuring, shows that these are not correlated with population density, thus highlighting the problems with this approach [55]. Trends observed from satellites are for wide regions ([48]; Table 1a) and it is encouraging that those trends are similar to those computed by averaging data from surface stations.

5. THE CAUSES OF DIMMING AND BRIGHTENING

Stanhill and Cohen [24] reviewed the possible causes for dimming in the context of a simplified expression:

$$E_{g\downarrow} = E_o \exp[-(\tau_r + \tau_g + \tau_w + \tau_a + \tau_c)] \tag{5}$$

where $E_{g\downarrow}$ is estimated from the extraterrestrial irradiance at the top of the atmosphere, E_o, modified by a chain of five transmissivities (τ) that quantify the solar scattering and absorbing properties of the different components of the atmosphere. These include τ_r, representing Rayleigh scattering, and τ_g, permanent gas absorption, τ_w, absorption by water vapour, and τ_a and τ_c, the absorption and scattering by the aerosols and cloud components, respectively. One factor whose known changes and influence on global radiation is large enough to cause changes of the magnitude observed is aerosol loading. Aerosol influences on radiation include direct effects, i.e. absorption, reflection and scattering of radiation by aerosols, and indirect effects, that is changes in cloud characteristics (e.g. albedo changes, the Twomey effect), rain suppression (the Albrecht effect), and cloud lifetime. Natural aerosols can cause sharp declines in $E_{g\downarrow}$ as observed in the year or two following the eruptions of El Chichón in 1983 and Pinatubo in 1991 [56]. Large longer-term changes in $E_{g\downarrow}$ can be pinned to some extent on anthropogenic pollution, as suggested by the large dimming in urban mega-cities and the industrialized zone of the northern hemisphere. The connection between dimming and aerosols has been clearly demonstrated (e.g. [57]), and known changes in aerosol loading of the atmosphere are correlated with the transition from dimming to brightening in the 1980s [58].

Prior to the twenty-first century scientists studying aerosols had suspected that aerosol influences on climate were far larger than was being acknowledged, and Satheesh and Ramanathan [59] demonstrated the magnitude of radiative forcing that can be caused by aerosols. As the evidence for worldwide dimming of a magnitude of several percent mounted, scientists studying aerosol influences began to include the full extent of aerosol influences in models of the Earth's climate (e.g. [60,61]).

5.1 CLOUD TRENDS AND THEIR INFLUENCE ON $E_{g\downarrow}$

Changes in cloudiness during parts of the dimming and brightening periods were studied by Joel Norris [62]. The data were from both surface data sets and satellite observations. The surface set, which was divided into 10×10 cells, was from the Extended Edited Cloud Report Archive (EECRA) and included ground-based cloud observations from land stations (1971–1996) and ship reports (1952–1997). These showed that zonal mean upper-level cloud cover at low and middle latitudes decreased by 1.5% sky cover between 1971 and 1996 over land and by about 1% sky cover between 1951 and 1997 over ocean. The upper level data were closely related to satellite (ISCCP) estimates for an overlapping period. Estimates of the cloud cover influence on solar radiation showed that between 1952 and 1997 over midlatitude oceans cloud changes decreased $E_{g\downarrow}$ by about 1 W m^2, and over northern midlatitude land areas cloud changes increased $E_{g\downarrow}$ slightly. For low-latitude land and ocean regions cloud changes increased $E_{g\downarrow}$ from the 1980s to the mid-1990s. These changes in cloudiness are relatively small, and although they probably played a significant part in global dimming and brightening, they could not be considered to be major players. Similar conclusions, i.e. that cloud trend influences on short-wave radiative forcing could not account for most of the global dimming and brightening, were made by Norris and Wild [63], who subtracted the estimated cloud cover influence on solar radiation from surface $E_{g\downarrow}$ data in the GEBA archive and found that dimming and brightening trends in the residual $E_{g\downarrow}$ were unchanged.

Sunshine duration (SSD) is another climatic measurement well correlated with cloudiness. Stanhill et al. [64] investigated long-term $E_{g\downarrow}$ and SSD data from five sites from a wide range of climates and aerosol emission rates. Basic differences between the two measures were used to separate between cloud and aerosol influences. As opposed to the previous studies, they concluded that, at least for the sites investigated, changes in cloudiness were the major causes of dimming and brightening. Thus, although dimming and brightening are clearly related to aerosols, their causes and mechanisms are still topics of debate.

6. INFLUENCE OF SOLAR RADIATION CHANGES (GLOBAL DIMMING AND BRIGHTENING) ON CLIMATE

6.1 THE EVAPORATION CONUNDRUM – EVAPORATION TRENDS AND THEIR RELATION TO DIMMING AND BRIGHTENING

Potential evaporation rates in many places in the world decreased during the second half of the twentieth century. As with solar radiation measurements, a major client for these measurements is the agricultural community, where evaporation rates are used to determine irrigation scheduling and application rates. Measurement of evaporation is usually done with an evaporimeter of the evaporation pan type, e.g. the US Class-A pan and Russian GGI-3000 pan [65]. Specifications of pan size, deployment and exposure are given in the previous reference. Networks of pans have been established in many parts of the world.

Evaporation of water requires large quantities of energy. Therefore, one model of evaporation is the energy budget of the evaporating surface, i.e.

$$R_n = \lambda E + C + G \quad \text{and} \quad \lambda E = R_n - C - G \tag{6}$$

where R_n is net radiation at the surface, λ is the latent heat of vaporization, E is the evaporation flux, C is convective heat transfer with the environment and G is surface heat flux and/or energy storage. For annual totals, heat flux and energy storage can usually be ignored and evaporation depends only on net radiation and convection.

Widespread reductions in pan evaporation during the second half of the twentieth century were first reported for the former Soviet Union (FSU) and much of the northern hemisphere [66,67]. These reports were considered evidence of global warming, which was thought to be increasing regional evaporation but decreasing pan evaporation due to a feedback influence of increasing regional humidity on local (or pan) potential evaporation [68] (see below). However, Stanhill and Cohen [24] considered decreasing evaporation to be evidence for decreasing solar radiation and Cohen et al. [69] showed that in Israel's arid conditions the overwhelming influence on evaporation is solar radiation. A full analysis of environmental factors showed that decreasing solar radiation was decreasing potential evaporation rates. Qian et al. [53] found a striking correspondence between decreasing $E_g\downarrow$ and pan evaporation in China.

Two Australian biologists, Roderick and Farquhar [70], analyzed worldwide changes in temperature and humidity and their relationship to evaporation rates. If regional evaporation were increasing and causing local pan evaporation to decrease, then the vapour pressure deficit (VPD), which is a driver of local evaporation and depends on air temperature and humidity, should be decreasing. However, there was no evidence that this was occurring worldwide, and Roderick and Farquhar showed that observed climate changes should have stabilized VPD during that period, as observed in climate data from the US, and therefore decreasing evaporation would have to be caused by decreasing net radiation, which is dominated by solar radiation. For a first-order analysis the evaporation equivalent of radiative energy is expressed by λ, whose value is approximately 2.4 MJ kg^{-1}, and 1 kg of water will cover a surface area of 1 m^2 to a depth of 1 mm. For the region of the FSU where both radiation and evaporation trends were available, solar radiation, which was in the range of 3000 MJ m^{-2} a^{-1}–4000 MJ m^{-2} a^{-1}, had declined by approximately 9% or 315 MJ m^{-2} in three decades, which is equivalent to 131 mm of water. This is similar to the average reported evaporation reduction during that period, $\sim$111 mm of water. Thus, the reported reductions in evaporation rates matched those for solar radiation, and the pan evaporation data set corroborated the reported dimming trends in $E_g\downarrow$. Roderick and Farquhar's analysis [70] convinced many scientists that dimming was real and was having a significant impact on the Earth's climate.

Evaporation at most sites in Australia has decreased significantly during the period on record, with no signs of recovery during the 'brightening' era [71]. The climate parameters that could be causing this were investigated by Roderick et al. [72] using a physical model. They found that the primary cause for the reduction in evaporation in Australia was decreasing wind speed, or 'stilling' [73], with some regional contributions from decreasing solar radiation. Stilling of the wind has subsequently been found to have occurred in much of the globe in recent decades with no recovery (see below), and this is also contributing to changes in evaporation.

The question as to whether changes in pan evaporation are similar or opposite to changes in regional evaporation involves the 'complementary' hypothesis [74], which states that when regional evaporation changes, air humidity will change in the same direction causing a feedback to occur that has an opposite effect on local evaporation. The hypothesis [75] considers the sum of regional and local (e.g. pan) evaporation to be equal to a constant value, making them 'complementary'. For example, in the Tibetan plateau, $E_g\downarrow$ and pan evaporation decreased from 1966 to 2003 [76], yet regional evaporation increased [77].

Since global radiation influences both local and regional evaporation similarly, when global radiation changes, the constant of the complementary equation may also change. Nevertheless, when significant changes in air temperature occur, especially if accompanied by changes in wind speed, which have also been noted for many sites, changes in pan evaporation cannot be taken as unambiguous evidence for dimming, brightening, stilling or warming [78].

6.2 SOIL MOISTURE TRENDS

Further evidence for changes in regional evaporation rates has come from the study of soil moisture data from an extensive network of stations in the Ukraine where plant available soil moisture for the top 1 m of soil is determined gravimetrically every 10 days from April–October at 141 stations from fields with crops of either winter or spring cereals. The data, from 1958 to 2002 [79], show that soil moisture increased until approximately 1980 and then levelled off. No trends in rainfall were observed for this region while air temperature increased slightly. As noted above, one of the first reports of dimming was from this region during the period in question [43]. The observed changes in soil moisture were opposite to the predictions that global warming would lead to soil desiccation [80,81]. Thus, Robock and Li [79] concluded that the changes in soil moisture were evidence of dimming and its reduction of regional evaporation rates. Subsequent modelling with a sophisticated land surface model, which included a decreasing trend of solar radiation along with increasing CO_2 and global warming, demonstrated similar increases in soil moisture [82].

6.3 THE HYDROLOGICAL CYCLE

Regional evaporation rates are balanced by rainfall and are a central part of the hydrological cycle, so the question as to whether decreases in pan evaporation indicate decreasing or increasing regional evaporation is of great importance. An accelerating hydrological cycle with increased regional evaporation would lead to increased rainfall rates. It would also increase cloudiness whose feedback influence would cause a decrease in $E_g\downarrow$. As noted above, cloud changes have been relatively small.

Prior to the twenty-first century, it was assumed that global warming would enhance evaporation and lead to an enhancement (or spinning up) of the hydrological cycle. Ramanathan et al. [50] evaluated the influences of anthropogenic aerosols on solar and thermal radiation balances, atmospheric temperature profiles and climate. They found that 'aerosols enhance scattering and absorption of solar radiation and produce brighter clouds that are less efficient at releasing precipitation. These in turn lead to large reductions in the amount of solar irradiance reaching Earth's surface, a corresponding increase in solar heating of the atmosphere, changes in the atmospheric temperature structure, suppression of rainfall, and less efficient removal of pollutants. Thus, these aerosol effects can lead to a weaker hydrological cycle'. A case in point is the Indian subcontinent where anthropogenic aerosol 'brown clouds' can reduce $E_g\downarrow$ by more than 10% and change the regional hydrological cycle. In particular, dark aerosols absorb solar radiation causing enhanced atmospheric warming and decreased $E_g\downarrow$, reducing surface temperatures and evaporation rates. Together, these enhance atmospheric stability and decelerate – 'spin down' – the hydrological cycle [83].

Liepert et al. [84] and Wild et al. [85] also considered that a reduction of $E_g\downarrow$ and related reductions in evaporation rates could be 'spinning down' the hydrological cycle. They argued that reductions in surface solar radiation were only partly offset by enhanced down-welling long-wave radiation from the warmer and moister atmosphere and that the radiative imbalance at the surface leads to weaker latent and sensible heat fluxes and hence to reductions in evaporation and precipitation despite global warming. This is in line with experimental evidence of the influence of aerosols on climate [86].

6.4 DAILY TEMPERATURE RANGE

$E_g\downarrow$ is directly related to maximum midday temperatures since it heats the surface. The same factors that reduce $E_g\downarrow$, i.e. clouds, haze and aerosols, increase down-welling long-wave radiation at night leading to higher night-time, or minimum daily, temperatures. Therefore, it is no surprise that $E_g\downarrow$ is significantly correlated with daily temperature range (DTR, [87]). Various episodes of temperature changes that correspond to sudden changes in atmospheric aerosol loading have been reported. One dramatic demonstration of the influence of aerosol on DTR was shown by Travis et al. [88], who studied climate data for the period of the World Trade Center tragedy in September 2001. During the three days that air traffic in the US was grounded, there were no atmospheric contrails caused, leading to an increase of approximately 1°C in DTR. Stanhill and Moreshet [36] found an average 18% increase in $E_g\downarrow$ during Yom Kippur (the Day of Atonement) in Israel, which is a one-day Jewish holiday in the fall when industries close and car use is minimal. The corresponding increase in DTR from 1963 to 2003 on Yom Kippur was 0.31°C (Stanhill and Cohen, unpublished data). Robock and Mass [89] and Mass and Robock [90] showed that tropospheric aerosol loading from the 1980 Mt. St. Helens volcanic eruption strongly reduced the diurnal temperature range for several days in the region with the volcanic dust, and surface temperature effects under smoke from forest fires were correlated with a reduction in daytime temperatures [91,92].

Global surface temperatures have been increasing since the beginning of the industrial era. Minimum temperatures have been increasing faster than maximum temperatures, and thus DTR has been decreasing. This may also be related to decreasing surface radiation. Wild et al. [93] used daily temperature range to analyze the influence of changes in $E_g\downarrow$ on global temperatures. They contend that global dimming masked global warming until the 1980s and that during the global brightening era until the early 2000s the accelerating temperature increases demonstrate the full (unmasked) global warming that is caused by greenhouse gases.

Wang and Dickenson [94] used an extensive data set of $E_g\downarrow$ and DTR to evaluate the influence of changes in $E_g\downarrow$ on decadal temperature variability. Their analysis is for the land masses where the measurements were made and does not include the oceans, but they show that the 'cooling effect of dimming accounts for the near constant temperature from the 1930s to the 1970s … but neither the rapid increase in temperature from the 1970s to the 1990s nor the slowdown of warming in the early twenty-first century are related to changes in $E_g\downarrow$'.

6.5 WIND SPEED AND THE MONSOON SYSTEM

Another mechanism by which changes in $E_g\downarrow$ can influence climate is sea warming and its influence on wind speed and the monsoon rain system [95]. Xu et al. [96] showed that wind speeds over China have decreased because of dimming. This is related to the increased atmospheric stability caused by aerosol-mediated warming of the atmosphere as surface radiation decreases. Thus, aerosols over China changed the land-ocean temperature contrast, affecting monsoon winds.

As noted above, at most places where wind data has been monitored for a long time and subsequently analyzed, wind speed has decreased in recent decades without any recovery. Reasons for the decrease may include those mentioned above, but also land use changes and changes in global circulation patterns [73].

7. CONCLUSIONS

Global radiation $E_g\downarrow$ decreased significantly (i.e. dimming) from the beginning of widespread measurements in the 1950s to the late 1980s over large parts of the globe and then partly recovered (i.e. brightening) in many places. The real extent of these changes is not certain because of the large spatial variability and the lack of extensive data over the oceans, but the mean trends are evident in satellite estimates of global radiation. The trends have been linked to anthropogenic aerosols, which reduce surface short-wave radiation directly and indirectly through their influence on cloud properties. Changes in $E_g\downarrow$ have played a part in regional and global changes in daily temperature range (positively correlated) as well as soil moisture (negatively correlated) and potential evaporation rates (positively correlated), but in some cases potential evaporation has changed due to other factors, for example wind stilling. Dimming may have offset global warming from the early twentieth century until the 1980s while the more recent partial brightening may have accelerated global warming up until the twenty-first century, but the cessation of global warming since 2000 does not seem to be related to $E_g\downarrow$. It is important to note that little is known about trends in the radiation balance over the three-quarters of the Earth covered by sea. These latter points and the scientific debates sparked by this topic, which are far from resolved, demonstrate the complexity of the climate system and the long road and many surprises that await us as scientists continue to study this topic.

REFERENCES

[1] W.M.O., Measurement of radiation. Chapter 7. Measurement of sunshine duration. Chapter 8, in: Guide to meteorological instruments and methods of observation. WMO-No. 8, Draft seventh ed., World Meteorological Organization, Geneva, 2006.

[2] R.J. List (Ed.), Smithsonian Meteorological Tables, sixth ed., Smithsonian Institute, Washington D.C., 1966.

[3] G.S. Campbell, J.M. Norman, An Introduction to Environmental Biophysics, second ed., Springer-Verlag, N.Y, 1998.

[4] G.W. Paltridge, C.M.R. Platt, Radiative Processes in Meteorology and Climatology, Elsevier, NY, 1976.

[5] L.I. Grey, I. Beer, M. Geller, J.D. Haigh, M. Lockwood, K. Matthess, U. Cubash, D. Fleitmann, G. Harrison, L. Hood, I. Luterbacher, G.A. Meehl, D. Shindell, B. van Geel, W. White, Rev. Geophys. 48 (2010) 1–53.

[6] IPCC, Climate change 2013, in: The Physical Science Basis. Contribution of Working Group I to the Fifth Assessment Report of the Intergovernmental Panel on Climate Change, WMO and UNEP, 2013.

[7] A. Raval, V. Ramanathan, Nature 342 (1989) 758–761.

[8] E. Pallé, P.R. Goode, P. Montanes-Rodriguez, S.E. Koonin, Science 304 (2004) 1299–1301.

[9] B.A. Wielicki, T. Wong, N. Loeb, K. Minnis, K. Priestley, R. Kandel, Science 308 (2005) 825.

[10] V. Ramanathan, iLEAPS Newsl. 5 (2008) 18–20.

[11] B.R. Barkstrom (Ed.), Earth Radiation Budget Measurements: Pre-ERBE, ERBE, and CERES. Long-Term Monitoring of the Earth's Radiation Budget, Proc. SPIE, vol. 1299, 1990, pp. 52–60.

[12] W.B. Rossow, Y.-C. Zhang, J. Geophys. Res. 100 (1995) 1167–1197, http://dx.doi.org/10.1029/94JD02746.

[13] E. Pallé, P. Montañés-Rodriguez, P.R. Goode, S.E. Koonin, M. Wild, S. Casadio, Geophys. Res. Lett. 32 (2005) L21702, http://dx.doi.org/10.1029/2005GL023847.

[14] P.R. Goode, E. Pallé, J. Atmos. Sol. Terr. Phys. 69 (2007) 1556–1568.

[15] E. Palle, P.R. Goode, P. Montanes-Rodriguez, J. Geophys. Res. 114 (2009) D00D03, http://dx.doi.org/10.1029/2008JD010734.

[16] A. Ohmura, New radiation and energy balance of the world and its variability, in: H. Fischer, B. Sohn (Eds.), IRS 2004: Current Problems in Atmospheric Radiation, 2006, pp. 327–330.

[17] J.T. Kiehl, K.E. Trenberth, Bull. Amer. Meteor. Soc. 78 (1997) 197–208.

[18] G. Stanhill, S. Cohen, J. Clim. 18 (2005) 1503–1512.

[19] G. Stanhill, S. Cohen, J. Met. Soc. Jpn. 86 (2008) 57–67.

[20] G. Stanhill, Weather 58 (2003) 3–11.

[21] D.J. Hatch, J. Meteorol. U.K 6 (1981) 101–113.

[22] E. Pallé, C.J. Butler, Int. J. Climatol. 21 (2001) 709–729.

[23] G. König-Langlo, R. Sieger, H. Schmithüsen, A. Bücker, F. Richter, E.G. Dutton, The baseline surface radiation network and its world radiation monitoring Centre at the Alfred Wegener Institute, in: WCRP Report. Global Climate Observing System. Baseline Surface Radiation Network (BSRN), WMO, 2013.

[24] G. Stanhill, S. Cohen, Agric. For. Meteorol. 107 (2001) 255–278.

[25] A. Ohmura, H. Gilgen, M. Wild, Global Energy Balance Archive GEBA, World Climate Program—Water Project A7, Report 1: Introduction. Zuercher Geografische Schriften Nr. 34, Verlag der Fachvereine, Zuerich, 1989.

[26] H. Gilgen, M. Wild, A. Ohmura, J. Clim. 11 (1998) 2042–2061.

[27] A. Ohmura, H. Gilgen, H. Hegner, G. Muller, M. Wild, E.G. Dutton, B. Forgan, C. Frohlich, R. Philipona, A. Heimo, G. Konig-Langlo, B. McArthur, R. Pinker, C.H. Whitlock, K. Dehne, Bull. Amer. Meteor. Soc. 79 (1998) 2115–2136.

[28] M. Budyko, The Heat Balance of the Earth's Surface. (Translated by Nina A. Stepnova from: Teplovoi Balans Zemnoi Poverkhnosti; Gidrometeorologicheskoe Iz Datel'stovo. Leningrad), U S Dept. Commerce, Washington, D.C, 1956.

[29] W.D. Sellers, Physical Climatology, The University of Chicago Press, Chicago, 1965.

[30] G. Stanhill, EOS 80 (1999) 396–397.

[31] S. Suraqui, H. Tabor, W.H. Klein, B. Goldberg, Sol. Energy 16 (1974) 155–158.

[32] A. Ohmura, H. Lang, Secular variation of global radiation over Europe, in: J. Lenoble, J.F. Geleyn (Eds.), Current Problems in Atmospheric Radiation, Deepak, Hampton, VA, 1989, pp. 98–301.

[33] V. Russak, Tellus 42 B (1990) 206–210.

[34] G. Stanhill, S. Moreshet, Clim. Change 21 (1992) 57–75.

[35] G. Stanhill, J.D. Kalma, Aust. Met. Mag. 43 (1994) 81–86.

[36] G. Stanhill, S. Moreshet, Clim. Change 26 (1994) 89–103.

[37] G. Stanhill, Philos. Trans. R. Soc. A 352 (1995) 247–258.

[38] G. Stanhill, S. Cohen, J. Clim. 10 (1997) 2078–2086.

[39] G. Stanhill, A. Ianitz, Tellus 49 B (1997) 112–122.

[40] G. Stanhill, Int. J. Climatol. 18 (1998) 1015–1030.

[41] G. Stanhill, J.D. Kalma, Int. J. Climatol. 15 (1995) 933–941.

[42] X. Li, X. Zhou, W. Li, Acta Met. Sin. 9 (1995) 57–68.

[43] G.M. Abakumova, E.M. Feigelson, V. Bussak, V.V. Stadnik, J. Clim. 9 (1996) 1319–1327.

[44] B.G. Liepert, Int. J. Climatol. 17 (1997) 1581–1593.

[45] B.G. Liepert, G.J. Kukla, J. Clim. 10 (1997) 2391–2400.

[46] B.G. Liepert, GRL 29 (2002) 12.

[47] M. Wild, H. Gilgen, A. Roesch, A. Ohmura, C.N. Long, E.G. Dutton, B. Forgan, A. Kallis, V. Russak, A. Tsvetkov, Science 308 (2005) 847–850.

[48] R.T. Pinker, B. Zhang, E.G. Dutton, Science 308 (2005) 850–854.

[49] A. Sanchez-Lorenzo, M. Brunetti, J. Calbó, J. Martin-Vide, J. Geophys. Res. 112 (2007) D20115, http://dx.doi.org/10.1029/2007JD008677.

[50] V. Ramanathan, P.J. Crutzen, J.T. Kiehl, D. Rosenfeld, Science 294 (2001) 2119.

[51] M. Wild, J. Geophys. Res. 114 (2009), http://dx.doi.org/10.1029/2008JD011470.

[52] M. Wild, Bull. Amer. Meteor. Soc. 93 (2012) 27–37.

[53] Y. Qian, D.P. Kaiser, L.R. Leung, M. Xu, Geophys. Res. Lett. 33 (2006) L01812, http://dx.doi.org/10.1029/2005GL024586.

[54] P. Alpert, P. Kishcha, Y.J. Kaufman, R. Schwarzbard, Geophys. Res. Lett. 32 (2005) L17802, http://dx.doi.org/10.1029/2005GL023320.

[55] G. Stanhill, S. Cohen, J. Geophys. Res. Atmos. 114 (2009) D00D17, http://dx.doi.org/10.1029/2009JD011976.

[56] M. Wild, H. Gilgen, A. Roesch, A. Ohmura, C.N. Long, E.G. Dutton, B. Forgan, A. Kallis, V. Russak, A. Tsvetkov, Science 308 (2005). Supporting online material, http://www.sciencemag.org/cgi/data/308/5723/847/DC1/1.

[57] B.G. Liepert, I. Tegen, J. Geophys. Res. Atm. 107 (D12) (2002) 4153–4168.

[58] D.G. Streets, Y. Wu, M. Chin, L15806, Geophys. Res. Lett. 33 (2006), http://dx.doi.org/10.1029/2006GL026471.

[59] S.K. Satheesh, V. Ramanathan, Nature 405 (2000) 60–63.

[60] A. Romanou, B. Liepert, G.A. Schmidt, W.B. Rossow, R.A. Ruedy, Y.-C. Zhang, Geophys. Res. Lett. 34 (2007) L05713, http://dx.doi.org/10.1029/2006GL028356.

[61] M.M. Kvalevag, G. Myhre, J. Clim. 20 (2007) 4874–4883.

[62] J.R. Norris, J. Geophys, Res. 110 (2005) D08206, http://dx.doi.org/10.1029/2004JD005600.

[63] J.R. Norris, M. Wild, J. Geophys. Res. 112 (2007) D08214, http://dx.doi.org/10.1029/2006JD007794.

[64] G. Stanhill, O. Achiman, R. Rosa, S. Cohen, J. Geophys. Res. Atmos. 119 (2014), http://dx.doi.org/10.1002/2013JD021308.
[65] WMO, Measurement of evaporation. Chapter 10, in: Guide to Meteorological Instruments and Methods of Observation. WMO-no. 8, Draft seventh ed., World Meteorological Organization, Geneva, 2006.
[66] T.C. Peterson, V.S. Golubev, P.Y. Groisman, Nature 377 (1995) 687–688.
[67] V.S. Golubev, J.H. Lawrimore, P.Ya Groisman, N.A. Speranskaya, S.A. Zhuravin, M.J. Menne, T.C. Peterson, R.W. Malone, Geophys. Res. Lett. 28 (2001) 2665–2668.
[68] W. Brutsaert, M.B. Parlange, Nature 396 (1998) 30.
[69] S. Cohen, A. Ianitz, G. Stanhill, Agric. For. Meteorol. 111 (2002) 83–91.
[70] M.L. Roderick, G.D. Farquhar, Science 298 (2002) 1410–1411.
[71] M.L. Roderick, G.D. Farquhar, Int. J. Climatol. 24 (2004) 1077–1090.
[72] M.L. Roderick, L.D. Rotstayn, G.D. Farquhar, M.T. Hobbins, Geophys. Res. Lett. 34 (2007) L17403, http://dx.doi.org/10.1029/2007GL031166.
[73] T.R. McVicar, et al., J. Hydrol 416–417 (2012) 182–205, http://dx.doi.org/10.1016/j.jhydrol.2011.10.024.
[74] R.J. Bouchet, Evapotranspiration Réelle Evapotranspiration Potentielle, Signification Climatique, Symp. Publ. 62, Int. Assoc. Sci. Hydrol., Berkeley, Calif. (1963) 134–142.
[75] W. Brutsaert, Evaporation into the Atmosphere: Theory, History and Applications, D. Reidel Publishing Co., Dordrecht, Holland, 1982.
[76] S.B. Chen, Y.F. Liu, A. Thomas, Clim. Change 76 (2006) 291–319.
[77] Y. Zhang, C. Liu, Y. Tang, Y. Yang, J. Geophys. Res. 112 (2007) D12110, http://dx.doi.org/10.1029/2006JD008161.
[78] W. Brutsaert, Geophys. Res. Lett. 33 (2006) L20403, http://dx.doi.org/10.1029/2006GL027532.
[79] A. Robock, M. Mu, K. Vinnikov, I.V. Trofimova, T.I. Adamenko, Geophys. Res. Lett. 32 (2005) L03401, http://dx.doi.org/10.1029/2004GL021914.
[80] S. Manabe, R.T. Wetherald, J. Atmos. Sci. 44 (1987) 1211–1235.
[81] J.M. Gregory, J.F.B. Mitchell, A.J. Brady, J. Clim. 10 (1997) 662–686.
[82] A. Robock, H. Li, Geophys. Res. Lett. 33 (2006) L20708, http://dx.doi.org/10.1029/2006GL027585.
[83] V. Ramanathan, C. Chung, D. Kim, T. Bettge, L. Buja, J.T. Kiehl, W.M. Washington, Q. Fu, D.R. Sikka, M. Wild, Proc. Natl. Acad. Sci. U.S.A 102 (2005) 5326–5333.
[84] B.G. Liepert, J. Feichter, U. Lohmann, E. Roeckner, Geophys. Res. Lett. 31 (2004) L06207.
[85] M. Wild, A. Ohmura, H. Gilgen, Geophys. Res. Lett. 31 (2004).
[86] D. Rosenfeld, Science 287 (2000) 1796–2793.
[87] K.L. Bristow, G.S. Campbell, Agric. For. Meteorol. 31 (1984) 159.
[88] D.J. Travis, A.M. Carleton, R.G. Lauritsen, Nature 418 (2002).
[89] A. Robock, C. Mass, Science 216 (1982) 628–630.
[90] C. Mass, A. Robock, Mon. Wea. Rev. 110 (1982) 614–622.
[91] A. Robock, Science 242 (1988) 911–913.
[92] A. Robock, J. Geophys. Res. 96 (1991), 869–20, 878.
[93] M. Wild, A. Ohmura, K. Makowski, Geophys. Res. Lett. 34 (2007) L04702, http://dx.doi.org/10.1029/2006GL028031.
[94] K. Wang, R.E. Dickinson, PNAS 110 (37) (2013) 14877–14882.
[95] L.D. Rotstayn, U. Lohmann, J. Clim. 15 (2002) 2103–2116.
[96] M. Xu, C.-P. Chang, C. Fu, Y. Qi, A. Robock, D. Robinson, H. Zhang, J. Geophys. Res. 111 (2006) D24111, http://dx.doi.org/10.1029/2006JD007337.
[97] S. Cohen, B. Liepert, G. Stanhill, EOS 85 (2004) 362.
[98] Pan evaporation: an example of the detection and attribution of trends in climate variables, in: R. Gifford (Ed.), National Committee for Earth System Science. Proceedings of a Workshop, Canberra, 22–23 November 2004, Australian Academy of Science, 2005.

[99] G. Ohring, S. Cohen, J. Norris, A. Robock, Y. Rudich, M. Wild and W. Wiscombe, EOS, 89 (2008) 212 and supplemental material at http://www.agu.org/eos_elec/2008/ohring_89_23.html.

[100] M. Wild, N. Loeb, G. Stanhill, B. Liepert, P. Alpert, J. Calbo, C. Long, G. Ohring, E. Palle, P. Kishcha, Measurements of GDB. Workgroup 1 report at the ISF International workshop on GDB, Ein Gedi, Isr. (2008).

[101] B.G. Liepert, Geophys. Res. Lett. 29/12 (2002), http://dx.doi.org/10.1029/2002GL014910.

[102] E.G. Dutton, D.W. Nelson, R.S. Stone, D. Longenecker, G. Carbaugh, J.M. Harris, J. Wendell, J. Geophys. Res. 111 (2006) D19101, http://dx.doi.org/10.1029/2005JD006901.

[103] N. Hatzianastassiou, C. Matsoukas, A. Fotiadi, K.G. Pavlakis, E. Drakakis, D. Hatzidimitriou, I. Vardavas, Atmos. Chem. Phys. 5 (2005) 2847–2867.

SPACE WEATHER AND COSMIC RAY EFFECTS

30

Lev I. Dorman[1,2]

[1]*Head of Cosmic Ray and Space Weather Centre with Emilio Segrè Observatory on Mt Hermon, affiliated to Tel Aviv University, Golan Research Institute, and Israel Space Agency, Qazrin, Israel;* [2]*Chief Scientist of Cosmic Ray Department of IZMIRAN Russian Academy of Science, Moscow, Troitsk, Russia*

CHAPTER OUTLINE

1. INTRODUCTION

There are a number of space phenomena that influence the Earth's climate and influence its long-term and short-term changes. These include:

- variability in the sun's irradiation flux energy;
- variations in the Earth's orbital characteristics;
- periodicity solar activity (main period about 11 years), variable general solar magnetic field (main period 22 years) together with the related phenomena of variable solar wind, coronal mass ejections and interplanetary shocks in the heliosphere and modulated galactic cosmic rays (CR) – see Section 2;
- solar CR generated during great solar flares – see Section 2.9;
- precipitation of energetic electrons and protons from the Earth's magnetosphere during magnetic disturbances – see Section 2.10;

- variability in the Earth's magnetic field, which influenced on CR cutoff rigidity and changed galactic and solar CR intensity in the Earth's atmosphere – see Section 2.12;
- the moving of the solar system around the galactic centre and crossing the galaxy arms – see Section 3;
- impacts of the solar system with galactic molecular dust clouds – see Section 4;
- impacts of the interplanetary zodiac dust cloud – see Section 5;
- asteroid impacts – see Section 7;
- effects of nearby supernova explosions – see Section 8.

The first phenomenon is the subject of Chapter 29 (see also Sections 2.5 and 2.6 in our chapter), and the second is discussed in Chapter 25. In our chapter the other phenomena are discussed and compared to anthropogenic-induced changes. Details of CR behaviour in the Earth's atmosphere, magnetosphere and in space are the subject of recent monographs by the author [1–4] and by the author with colleagues [5,6]. A comparison and the possible role of the above-mentioned factors in our changing climate are discussed at the end of chapter.

2. SOLAR ACTIVITY, COSMIC RAYS AND CLIMATE CHANGE
2.1 LONG-TERM COSMIC RAY INTENSITY VARIATIONS AND CLIMATE CHANGE

About 200 years ago the famous astronomer William Herschel [7] suggested that the price of wheat in England was directly related to the number of sunspots. He noticed that less rain fell when the number of sunspots was small (Joseph in the Bible, recognized a similar periodicity in food production in Egypt, about 4000 a ago; here 'a' refers to annum). The solar activity level is known from direct observations over the past 450 a, and from data of cosmogenic nuclides (through CR intensity variations) for more than 10 000 a [1,8]. Over this period there is a striking correlation between cold and warm climate periods and high and low levels of galactic CR intensity (low and high solar activity). As an example, Fig. 1 shows the change in the concentration of radiocarbon during the last millennium (a higher concentration of ^{14}C corresponds to a higher intensity of galactic CR and to lower solar activity).

FIGURE 1

The change of CR intensity reflected in radiocarbon concentration during the last millennium. The Maunder minimum refers to the period 1645–1715, when sunspots were rare. *(From Ref. [9].)*

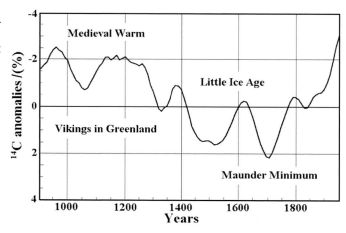

It can be seen from Fig. 1 that during the period 1000–1300 CE, the CR intensity was low and solar activity high, which coincided with a warm medieval period (during this period Vikings settled in Greenland). After 1300, solar activity decreased and CR intensity increased, and a long cold period followed (the so-called Little Ice Age, which included the Maunder minimum, 1645–1715, and lasted until the middle of the nineteenth century).

2.2 THE POSSIBLE ROLE OF SOLAR ACTIVITY AND SOLAR IRRADIANCE IN CLIMATE CHANGE

Friis-Christiansen and Lassen [10,11] found, from 400 years of data, that the filtered solar activity cycle length was closely connected to variations of the average surface temperature in the northern hemisphere. Labitzke and Van Loon [12] showed, from solar cycle data, that the air temperature increased with increasing levels of solar activity. Swensmark [9] also discussed the problem of the possible influence of solar activity on the Earth's climate through changes in solar irradiance. But direct satellite measurements of solar irradiance during the last two solar cycles showed that the variations during a solar cycle were only about 0.1%, corresponding to about 0.3 W m^{-2}. This value is too small to explain the observed changes in the global temperature (Lean et al. [13]). Much bigger changes during a solar cycle occur due to UV radiation (about 10%, which is important in the formation of the ozone layer). High [14] and Shindell et al. [15] suggested that the heating of the stratosphere by ultraviolet (UV) radiation can be dynamically transported into the troposphere. This effect might be responsible for small contributions toward (11 and 22) a cycle modulation of climate but not to the 100 a of climate change that we are presently experiencing.

2.3 COSMIC RAYS AS AN IMPORTANT LINK BETWEEN SOLAR ACTIVITY AND CLIMATE CHANGE

Many authors have considered the influence of galactic and solar CR on the Earth's climate. Cosmic radiation is the main source of air ionization below 40–35 km (only near the ground level, lower than 1 km, are radioactive gases from the soil also important in air ionization) [1]. The first to suggest a possible influence of air ionization by CR on the climate was Ney [16]. Swensmark [9] noted that the variation in air ionization caused by CR could potentially influence the optical transparency of the atmosphere, by either a change in aerosol formation or by influencing the transition between the different phases of water. Many authors considered these possibilities: Ney [16], Dickinson [17], Pudovkin and Raspopov [18], Pudovkin and Veretenenko [19,20], Tinsley [21], Swensmark and Friis-Christiansen [22], Swensmark [23] and Marsh and Swensmark [24,25]. The possible statistical connections between the solar activity cycle and the corresponding long-term CR intensity variations with characteristics of climate change were considered by Dorman et al. [26–28]. Dorman et al. [29] reconstructed CR intensity variations over the last 400 years on the basis of solar activity data and compared the results with radiocarbon and climate change data.

Cosmic radiation plays a key role in the formation of thunderstorms and lightning [1]. Many authors [30–35] have considered atmospheric electric field phenomena as a possible link between solar activity and the Earth's climate.

Also important in the relationship between CR and climate is the influence of long-term changes in the geomagnetic field on CR intensity through the changes of cutoff rigidity [3,36]. One can consider that the general hierarchical relationship as: (solar activity cycles + long-term changes in the geomagnetic field) → (CR long-term modulation in the heliosphere + long-term variation of cutoff rigidity) → (long-term variation of clouds covering + atmospheric electric field effects) → climate change.

2.4 THE CONNECTION BETWEEN GALACTIC COSMIC RAY SOLAR CYCLES AND THE EARTH'S CLOUD COVERAGE

Recent research has shown that the Earth's cloud coverage (observed by satellites) is strongly influenced by CR intensity [9,21,23–25]. Clouds influence the irradiative properties of the atmosphere by both cooling through reflection of incoming short-wave solar radiation and heating through trapping of outgoing long-wave radiation (the greenhouse effect). The overall result depends largely on the height of the clouds. According to Hartmann [36], high optically thin clouds tend to heat while low optically thick clouds tend to cool (see Table 1).

Table 1 Global annual mean forcing due to various types of clouds, from the Earth Radiation Budget Experiment (ERBE). *(According to Hartmann [36].)*

Parameter	High clouds		Middle clouds		Low clouds	
	Thin	Thick	Thin	Thick	All	Total
Global fraction (%)	10.1	8.6	10.7	7.3	26.6	63.3
Forcing (relative to clear sky)	–	–	–	–	–	–
Albedo (SW radiation) (W m^{-2})	−4.1	−15.6	−3.7	−9.9	−20.2	−53.5
Outgoing LW radiation (W m^{-2})	6.5	8.6	4.8	2.4	3.5	25.8
Net forcing (W m^{-2})	2.4	−7.0	1.1	−7.5	−16.7	−27.7

Note: Global fraction of cloudiness is relative to all sky of the Earth (on average 36.7% of the Earth's surface is free from clouds); The positive forcing increases the net radiation budget of the Earth and leads to a warming; negative forcing decreases the net radiation and causes a cooling.

From Table 1 it can be seen that low clouds result in a cooling effect of about 17 W m^{-2}, which means that they play an important role in the Earth's radiation budget [37–39]. The important issue is that even small changes in the lower cloud coverage can result in important changes in the radiation budget and hence have a considerably influence on the Earth's climate (let us remember that the solar irradiance changes during solar cycles is only about 0.3 W m^{-2}).

Figure 2 shows a comparison of the Earth's total cloud coverage (from satellite observations) with CR intensities (from the Climax neutron monitor) and solar activity data over 20 years.

From Fig. 2 it can be seen that the correlation of global cloud coverage with CR intensity is much better than with solar activity. Marsh and Swensmark [24] came to the conclusion that CR intensity relates well with low global cloud coverage but not with high and middle clouds (see Fig. 3).

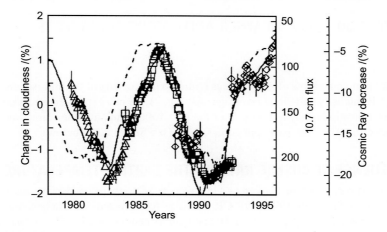

FIGURE 2

Changes in the earth's cloud coverage: triangles – from satellite nimbus 7, CMATRIX project [40];
squares – from the international satellite cloud climatology project, ISCCP [41]; diamonds – from the defence
meteorological satellite program, DMSP [42,43]. Solid curve – CR intensity variation according to climax NM,
normalized to may 1965. Broken curve – solar radio flux at 10.7 cm. All data are smoothed using a 12-month
running mean. *(From Ref. [9].)*

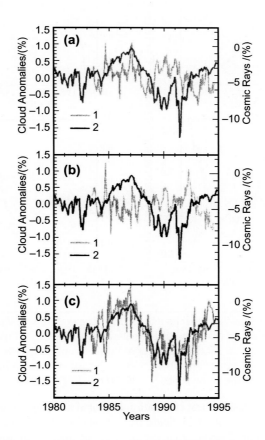

FIGURE 3

CR intensity obtained at the Huancayo/Haleakala neutron monitor, (normalized to October 1965, curve 2) in
comparison with global average monthly cloud coverage anomalies (curve 1) at heights, H, for: (a) high clouds,
H > 6.5 km, (b) middle clouds, 6.5 km > H > 3.2 km, and (c) low clouds, H < 3.2 km. *(From Ref. [24].)*

It is important to note that low clouds lead, as a rule, to the cooling of the atmosphere. It means that with increasing CR intensity and cloud coverage (see Fig. 2), we can expect the surface temperature to decrease. The relationship is shown in Fig. 1 for the past 1000 a, and includes direct measurements of the surface temperature over the past four solar cycles (see Section 2.5, below).

2.5 THE INFLUENCE OF COSMIC RAYS ON THE EARTH'S TEMPERATURE

Figure 4 shows a comparison of an 11-year moving average of northern hemisphere marine and land air temperature anomalies for 1935–1995 with CR intensity (constructed for Cheltenham/Fredericksburg for 1937–1975 and Yakutsk for 1953–1994, [44]) and Climax neutron monitor (NM) data, as well as with other parameters (unfiltered solar cycle length, sunspot numbers, and reconstructed solar irradiance).

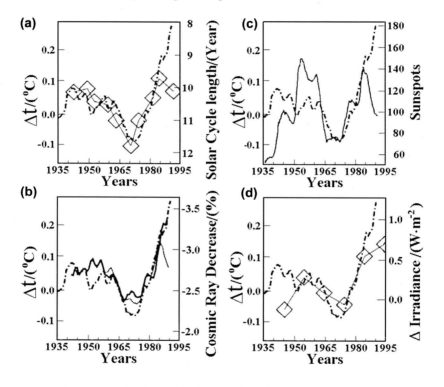

FIGURE 4

Eleven-year average northern hemisphere marine and land air temperature anomalies (δt), (broken curve) compared with: (a) unfiltered solar cycle length; (b) 11-year average CR intensity (thick solid curve – from ion chambers 1937–1994, normalized to 1965, and thin solid curve – from climax NM, normalized to ion chambers); (c) 11-year average of sunspot numbers; and (d) decade variation in reconstructed solar irradiance from Lean et al. [9] (zero level corresponds to 1367 W m⁻²). *(From Swensmark [9].)*

From Fig. 4 it can be seen that the best correlation of global air temperature is with CR intensity, in accordance with the results described in Sections 2–5 above. According to Swensmark [5], the comparison of Fig. 4 with Fig. 2 shows that the increase of air temperature by 0.3°C corresponds to a decrease of CR intensity of 3.5% and a decrease of global cloudiness of 3%; this is equivalent to an increase of solar irradiance on the Earth's surface of about 1.5 W m^{-2} [45] and is about five times bigger than the solar cycle change of solar irradiance, which as we have seen, is only 0.3 W m^{-2}.

2.6 COSMIC RAY INFLUENCE ON WEATHER DURING MAUNDER MINIMUM

Figure 5 shows the situation in the Maunder minimum [8] (a time when sunspots were rare) for: solar irradiance [13,46]; concentration of the cosmogenic isotope ^{10}Be [47] – a measure of CR intensity [1]; and a reconstructed air surface temperature for the northern hemisphere [48].

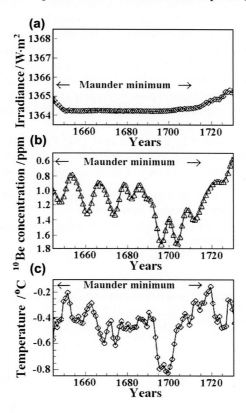

FIGURE 5

Situation in the maunder minimum: (a) reconstructed solar irradiance [12]; (b) cosmogenic ^{10}Be concentration [47]; (c) reconstructed air surface temperature for the northern hemisphere [48]. *(From Swensmark [9].)*

The solar irradiance was almost constant during the Maunder minimum and about 0.24% (or about 0.82 W m^{-2}) lower than the present value (see Panel (a) in Fig. 5), but CR intensity and air surface temperature varied in a similar manner – see above sections; with increasing CR intensity there was a decrease in air surface temperature (see Panels (b) and (c) in Fig. 5). The highest level of CR intensity was in 1690–1700, which corresponds to the minimum of air surface temperature [49] and also to the coldest decade (1690–1700).

2.7 THE INFLUENCE OF LONG-TERM VARIATIONS OF COSMIC RAY INTENSITY ON WHEAT PRICES (RELATED TO CLIMATE CHANGE) IN MEDIEVAL ENGLAND AND MODERN USA

Herschel's observations [7] mentioned in Section 2.1, were based on the published wheat prices [50] and showed that five prolonged periods of sunspot numbers correlated with costly wheat. This idea was taken up by the English economist and logician William Stanley Jevons [51]. He directed his attention to price of wheat from 1259 to 1400 and showed that the time intervals between high prices were close to 10 a–11 a. This work was later published by Rogers [52]. The coincidence of these intervals with the period of the recently discovered 11-year cycle of solar activity led him to suggest that the solar activity cycle was a 'synchronization' factor in the fluctuations of wheat prices (Jevons [53]). As a next step, he extrapolated his theory to stock markets of the nineteenth century in England and was impressed by a close coincidence of five stock exchange panics with five minima in solar spot numbers that preceded these panics. He suggested that both solar and economic activities are subjected to a harmonic process with the same constant period of 11 years. However, the subsequent discovery of the nonharmonic behaviour of solar cycles, with periods varying from 8 a–15 a, and the later observation of lack of coincidence between panics predicted by Jevons [51,53] and the actual ones, destroyed his argument.

The Rogers [52] database was used by Pustil'nik et al. [54], Pustil'nik and Yom Din [55] to search for possible influences of solar activity and CR intensity on wheat prices (through climate changes). The graph of wheat prices as a function of time (Fig. 6) contains two specific features:

1. A transition from 'low price' state to 'high price' state during 1530–1630, possibly as a result of access to cheap silver, recently discovered New World.
2. The existence of two populations in the price sample: noise-like variations with low amplitude bursts and several bursts of large amplitude.

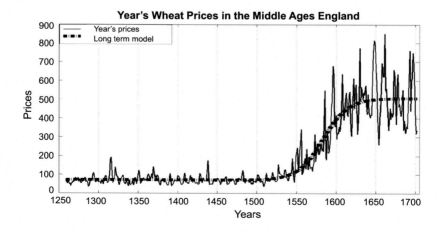

FIGURE 6

Wheat prices in England during 1259–1702 with a price transition at 1530–1630. *(From Refs [54,55].)*

Pustil'nik et al. [54], Pustil'nik and Yom Din [55], analyzed the data and compared the distribution of intervals of price bursts with the distribution of the intervals between minimums of solar cycles (see Fig. 7)

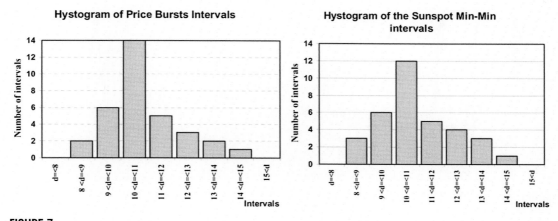

FIGURE 7

Histograms of the interval distribution for price bursts for the period over 450 as (1249–1702), and of minimum−minimum intervals of sunspots during the last 300 a (1700–2000). *(From Refs. [54,55].)*

In their analysis they found that for the sunspot minimum−minimum interval distribution the estimated parameters are: median 10.7 a; mean 11.02 a; standard deviation 1.53 a and for the price burst interval distribution, the estimated parameters are: median 11.0 a; mean 11.14 a; and standard deviation 1.44 a.

The main problem with a comparison between price levels and solar activity is the absence of the time interval, common to sunspot observation data (the years 1700–2001) and wheat price data (the years 1259–1702). However, the discovery of a strong correlation between the concentration of ^{10}Be isotopes in Greenland ice and CR intensity (according to measurements of CR intensity over the last 60 a [1]) sheds a new light on the problem. In Fig. 8, the wheat prices for 1600–1702 are shown and compared to ^{10}Be data [56]. White marks show prices averaged over three-year intervals centred on moments of minimum CR intensity. Black marks correspond to average prices over three-year intervals for maximum CR intensities.

As can be seen from Fig. 8, all prices in the neighbourhood of the seven maxima of CR intensity (corresponding approximately to minima of solar activity) are systematically higher than those in the neighbourhood of the seven minima of CR intensity (maxima of solar activity) in the long-term variation of CR intensity according to ^{10}Be data [56]. A similar result was obtained by Pustil'nik and Yom Din [57] for wheat prices in USA during the twentieth century.

2.8 THE CONNECTION BETWEEN ION GENERATION IN THE ATMOSPHERE BY COSMIC RAYS AND TOTAL SURFACE OF CLOUDS

The time variation of the integral rate of ion generation, q, (approximately proportional to CR intensity) in the middle latitude atmosphere at an altitude between 2 km–5 km was found by Stozhkov et al. [58] for the period January 1984–August 1990 using regular CR balloon measurements. The relative

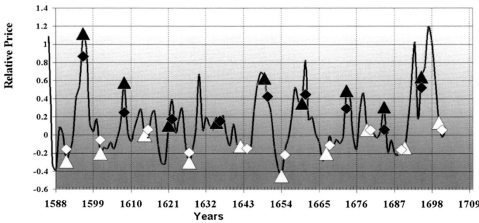

FIGURE 8

Systematic differences in wheat prices at moments of minimum and maximum CR intensity related to [10]Be data [56]. White diamonds show prices averaged for three-year intervals centred on moments of minimum CR intensity (maximum solar activity); black rectangles show prices averaged over three-year intervals centred on moments of maximum CR intensity (minimum solar activity). White and black triangles show prices at moments of minimum and maximum CR intensity. *(From Ref. [55].)*

changes in q, ($\Delta q/q$), have been compared with the relative changes of the total surface of clouds over the Atlantic Ocean $\Delta S/S$, and are shown in Fig. 9 – the correlation coefficient is 0.91 ± 0.04. This result is in good agreement with results described above (see panel (b) in Fig. 4 and panel (c) in Fig. 5) and shows that there is a direct correlation between cloud cover and CR-generated ions.

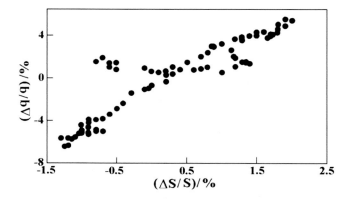

FIGURE 9

The connection of relative changes of total clouds covering surface over Atlantic Ocean $\Delta S/S$ in the period January 1984–August 1990 (according to [22]) with the relative changes of integral rate of ion generation $\Delta q/q$ in the middle latitude atmosphere in the altitude interval 2 km–5 km. *(From Ref. [58].)*

2.9 THE INFLUENCE OF BIG MAGNETIC STORMS (FORBUSH DECREASES) AND SOLAR COSMIC RAY EVENTS ON RAINFALL

A decrease of atmospheric ionization leads to a decrease in the concentration of charge condensation centres. In these periods, a decrease of total cloudiness and atmosphere turbulence together with an increase in isobaric levels was observed [59]. As a result, a decrease of rainfall is also expected. Stozhkov et al. [60–62] and Stozhkov [63] analyzed 70 events of Forbush decreases (defined as a rapid decrease in observed galactic CR intensity, and caused by big geomagnetic storms) observed in 1956–1993, and compared these events with rainfall data over the former USSR. It was found that during the main phase of the Forbush decrease, the daily rainfall levels decreased by about 17%. Similarly, Todd and Kniveton [64,65], investigating 32 Forbush decrease events over the period 1983–2000, found reduced cloud cover of 18% [64] and 12% [65].

During big solar CR events when CR intensity and ionization in the atmosphere significantly increases, an inverse situation is expected and an increase in cloudiness should lead to an increase in rainfall. A study [60–63], involving 53 events of solar CR enhancements between 1942 and 1993, showed a positive increase of about 13% in the total rainfall over the former USSR.

2.10 THE INFLUENCE OF GEOMAGNETIC DISTURBANCES AND SOLAR ACTIVITY ON THE CLIMATE THROUGH ENERGETIC PARTICLE PRECIPITATION FROM INNER RADIATION BELT

The relationship between solar and geomagnetic activity and climate parameters (cloudiness, temperature, rainfall, etc.) considered above is the subject of much ongoing research. The clearly pronounced relationship observed at high and middle latitudes is explained by a decrease of galactic CR (energies in the range of MeV and GeV) with increasing solar and geomagnetic activity, and by the appearance of solar CR fluxes ionizing the atmosphere [66]. This mechanism works efficiently at high latitudes because CR particles with energy up to 1 GeV penetrate this region more easily due to its very low cutoff rigidity. Near the equator, in the Brazilian magnetic anomaly (BMA) region, the main part of galactic and solar CR is shielded by a geomagnetic field. This field is at an altitude of 200 km–300 km and contains large fluxes of energetic protons and electrons trapped in the inner radiation belt. Significant magnetic disturbances can produce precipitation of these particles and subsequent ionization of the atmosphere. The influence of solar-terrestrial connections on climate in the BMA region was studied by Pugacheva et al. [67]. Two types of correlations were observed: (1) a significant short and long timescale correlation between the index of geomagnetic activity Kp (where the Kp-index is a global geomagnetic storm index with a scale of 0–9) and rainfall in Sao Paulo State; (2) the correlation and anti-correlation of rainfalls with the 11- and 22-year cycles of solar activity for 1860–1990 in Fortaleza.

Figure 10 shows the time relationship between Kp-index and rain in Campinas (23°S, 47°W) and in Ubajara (3°S, 41°W), during 1986.

From Fig. 10, it can be seen that, with a delay of 5 days–11 days, almost every significant (>3.0) increase of the Kp-index is accompanied by an increase in rainfall. The effect is most noticeable at the time of the great geomagnetic storm of February 8, 1986, when the electron fluxes of inner radiation belt reached the atmosphere between February 18 and 21 [68] and the greatest rainfall of 1986 was recorded on February 19. Again, after a series of solar flares, great magnetic disturbances were

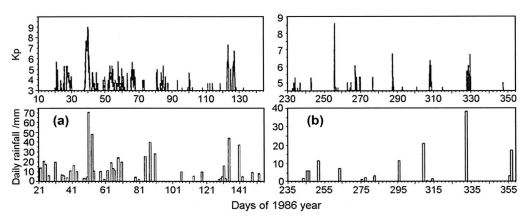

FIGURE 10

The Kp-index of geomagnetic activity (top panels) and rainfall level (bottom panels) in Campinas (left panels (a)) and in Ubajara (right panels (b)) in 1986. According to Pugacheva et al. [67].

registered between March 19 and 22, 1991. On March 22, a São Paolo station showed the greatest rainfall of the year.

The relationship between long-term variations of annual rainfall at Campinas, the Kp-index and sunspot numbers are shown in Figs. 11 and 12.

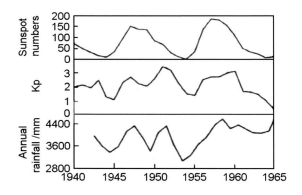

FIGURE 11

Long-term variations of rainfalls (Campinas, the bottom panel) in comparison with variations of solar and geomagnetic activity (the top and middle panels, respectively) for 1940–1965. *(From Ref. [67].)*

From Figs. 11 and 12 the relationship between the double peak structure of rainfall and the Kp-index can be seen. Only during the 1964–1975 solar cycle, the weakest of the six cycles shown in the figures, is there an anti-correlation between rainfall and sunspot numbers. The Kp–rainfall correlation is more pronounced in the regions connected with magnetic lines occupied by trapped particles.

In Fortaleza (4°S, 39°W), which is located in an empty magnetic tube ($L = 1.054$), another kind of correlation is observed (see Fig. 13).

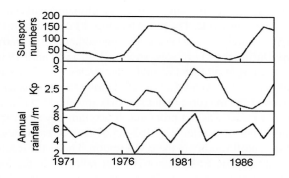

FIGURE 12

The same as in Fig. 11, but for 1971–1990. *(From Ref. [67]).*

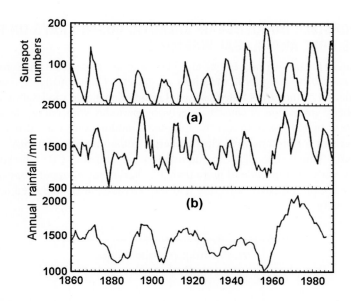

FIGURE 13

The comparison of yearly sunspot numbers long-term variation (the top panel) with 3 and 11 a running averaged rainfalls (Panels (a) and (b), respectively) in Fortaleza (4°S, 39°W) during 1860–1990. *(From Ref. [67].)*

From Fig. 13, it can be seen that a correlation exists between sunspot numbers and rainfall during 1860–1900 (11, 12 and 13 solar cycles) and during the period 1933–1954 (17 and 18 cycles). The anti-correlation was observed during 1900–1933 (cycles 14, 15 and 16) and 1954–1990 years (cycles 19, 20 and 21). As the sunspot numbers mainly anti-correlate with the galactic CR flux, an anti-correlation of sunspot numbers with rainfall could be interpreted as a correlation of rainfall with the CR. The positive and negative phases of the correlation that interchanged several times during the interval 1860–1990 were observed earlier in North America by King [69]. Some climate events have a 22-year periodicity similar to the 22-year solar magnetic cycle. The panel b in Fig. 13 demonstrates the 22-year periodicity of 11-year running averaged rainfalls in Fortaleza. The phenomenon is clearly observed during five periods from 1860 to 1990. During the 11–16 solar cycles (from 1860 until 1930), the maxima rainfall corresponds to a maxima of sunspot numbers of odd solar cycles 11, 13, 15 and minima of rainfalls corresponds to maxima of even solar cycles 12, 14, 16. During the 17th solar cycle the phase of the 22-year periodicity was inverted and the sunspot number maxima of odd cycles 19 and 21 corresponded to the minima of rainfall. The effect was not pronounced (excluding years 1957–1977) in São Paolo.

The difference in results obtained in references [63–65,67] can be easily understood if we take into account the large value of the cutoff rigidity in the BMA region. This is the reason why the variations in galactic and solar CR intensity in the BMA region are not reflected in the ionization of the air and hence do not influenced the climate. However, in the BMA region other mechanisms of solar and magnetic activity can influence climatic parameters such as energetic particle precipitation coming from the inner radiation belt.

2.11 ON THE POSSIBLE INFLUENCE OF GALACTIC COSMIC RAYS ON FORMATION OF CIRRUS HOLE AND GLOBAL WARMING

According to Ely and Huang [70] and Ely et al. [71], there are expected variations of upper tropospheric ionization caused by long-term variations of galactic CR intensity. These variations have resulted in the formation of the cirrus hole (a strong latitude-dependent modulation of cirrus clouds). The upper tropospheric ionization is caused, largely, by particles with energy less than 1 GeV but greater than 500 MeV. In Fig. 14 is shown the long-term modulation of the difference between Mt Washington and Durham for protons with kinetic energy in the range of 650 MeV–850 MeV.

Figure 14 clearly shows the 22-year modulation of galactic CR intensity in the range 650 MeV–850 MeV with an amplitude of more than 3%. Variations of upper tropospheric ionization do have some influence on the cirrus covering and the 'cirrus hole' is expected to correspond to a decrease in CR intensity.

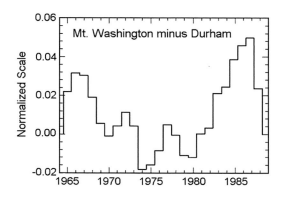

FIGURE 14

The observed 22 a modulation of galactic CR between 1.24 GV and 1.41 GV rigidity (i.e. protons with kinetic energy between 650 MeV and 850 MeV, ionizing heavily in the layer 200 g cm^{-2}–300 g cm^{-2}). *(From Ely et al. [71].)*

According to Ely et al. [71], the 'cirrus hole' was observed in different latitude zones over the whole world between 1962 and 1971, centred at 1966 (see Fig. 15).

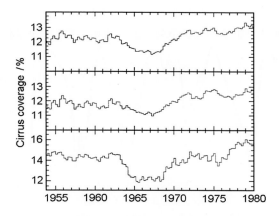

FIGURE 15

The 'cirrus hole' of the 1960s for: the whole world (the top panel); the equatorial zone (30°S–30°N; middle panel); the northern zone (bottom panel). *(From Ely et al. [71].)*

Figure 15 gives the cirrus cloud cover data over a 25-year period, for the whole world, the equatorial zone (30°S−30°N) and the northern zone (30°N−90°N), showing fractional decreases in cirrus coverage of 7%, 4% and 17%, respectively.

The decrease of cirrus covering leads to an increase in heat loss to outer space (note, that only a 4% change in total cloud cover is equivalent to twice the present greenhouse effect due to anthropogenic carbon dioxide). The influence of cirrus hole in the northern latitude zone (30°N−90°N), where the cirrus covering was reduced by 17%, is expected to be significant (this effect of the cirrus hole is reduced in summer by the increase of lower clouds resulting in enhanced insulation). The low temperatures produced from mid- to high latitude significantly increase the pressure of the polar air mass and cause frequent 'polar break troughs' at various longitudes in which, for example, cold air from Canada may go all the way to Florida and freeze the grapefruit [71]. However, when the cirrus hole is not present, the heat loss from mid- to high latitudes is much less, and the switching of the circulation patterns (Rossby waves) is much less frequent.

2.12 DESCRIPTION OF LONG-TERM GALACTIC COSMIC RAY VARIATION BY BOTH CONVECTION-DIFFUSION AND DRIFT MECHANISMS WITH POSSIBILITY OF FORECASTING OF SOME PART OF CLIMATE CHANGE IN NEAR FUTURE CAUSED BY COSMIC RAYS

It was shown in previous sections that CR may be considered as an important factor in understanding climate change. From this point of view it is important to understand the mechanisms of galactic CR long-term variations and using this to forecast expected CR intensity in the future. Dorman [72–74] made a study of the monthly sunspot numbers taking into account the time lag between processes on the Sun and the situation in the interplanetary space as well as the sign of the magnetic field (see Fig. 16). Related work was done by Belov et al. [75] who focused on monthly solar magnetic field data (see Fig. 17).

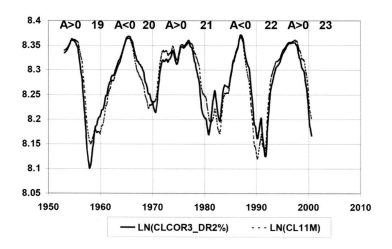

FIGURE 16

A comparison of the observed Climax neutron monitor CR intensity averaged over 11-month periods LN(CL11M-OBS) with predictions of monthly sunspot numbers from a model of convection-diffusion modulation, corrected for drift effects LN(CL11MPRED). The correlation coefficient between the curves is 0.97. *(From Dorman [74].)*

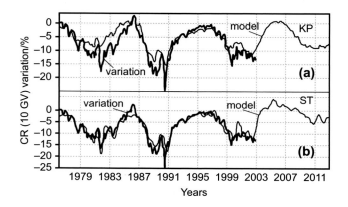

FIGURE 17

The forecast of galactic CR behaviour based on the predicted values of the global characteristics of the solar magnetic field, thick line – data of CR intensity observations (Moscow neutron monitor), thin line – the predicted CR variation up to 2013 based on data of Kitt Peak Observatory (upper panel) and Stanford Observatory (bottom panel). *(From Belov et al. [75].)*

From Fig. 16 it follows that the convection-diffusion and drift models can be used to determine with very good accuracy, galactic CR intensity in the past (where monthly sunspot numbers are known) as well as the CR intensity in future so long as the monthly sunspot numbers can be well predicted [72–74]. According to reference [75], the same can be made with good accuracy on the basis of monthly solar general magnetic field data (see Fig. 17). These results show that it is possible to use CR activity to forecast some part of climate change (see references [72–75]).

2.13 INFLUENCE OF LONG-TERM VARIATION OF MAIN GEOMAGNETIC FIELD ON GLOBAL CLIMATE CHANGE THROUGH COSMIC RAY CUTOFF RIGIDITY VARIATION

A significant change in the geomagnetic field leads to a change of planetary distribution of cutoff rigidities, R_c, and to corresponding change of the i-th component of CR intensity $N_i(R_c, h_o)$ at some level, h_o, in the Earth's atmosphere $\Delta N_i(R_c, h_o)/N_{io} = -\Delta R_c W_i(R_c, h_o)$, where $W_i(R_c, h_o)$ is the coupling function (see details in Chapter 3 in reference [1]). Variations of CR intensity caused by change of R_c are described in detail in reference [3], and here we discuss the results of Shea and Smart [76] on R_c changing over the past 300 and 400 years (see Fig. 18 and Table 2, correspondingly).

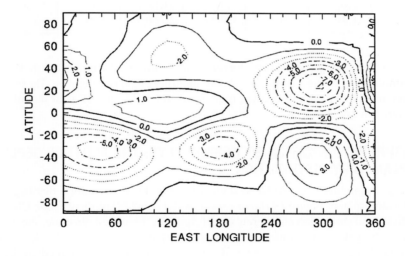

FIGURE 18

Contours of the change in vertical cutoff rigidity values (in GV) between 1600 and 1900. Full lines reflect positive trend (increasing of cutoff rigidity from 1600 to 1900); dotted lines reflect negative trend. According to Shea and Smart [76].

Table 2 shows that the change of geomagnetic cutoffs is not homogeneous – from 14 selected regions, in 5 regions cutoffs increased with a corresponding decrease in CR intensity, and in 9 regions cutoffs decreased with a corresponding increase of CR intensity from 1600 to 1900. From Table 2 it can be seen that at the present time (from 1900 to 2000) there are significant changes in

Table 2 Vertical cutoff rigidities (in GV) for various epochs 1600, 1700, 1800, 1900 and 2000, as well as galactic primary CR (GCR) intensity variation out of the earth's atmosphere from 1600 to 1900 owed to changes of geomagnetic field. *(According to Shea and Smart [76].)*

Lat.	Long. (E)	Epoch 2000	Epoch 1900	Epoch 1800	Epoch 1700	Epoch 1600	Change in GCR flux (1600–1900)	
55	30	2.30	2.84	2.31	1.49	1.31	−48%	Europe
50	0	3.36	2.94	2.01	1.33	1.81	−37%	Europe
50	15	3.52	3.83	2.85	1.69	1.76	−55%	Europe
40	15	7.22	7.62	5.86	3.98	3.97	−58%	Europe
45	285	1.45	1.20	1.52	2.36	4.14	+214%	N. Amer.
40	255	2.55	3.18	4.08	4.88	5.89	+118%	N. Amer.
20	255	8.67	12.02	14.11	15.05	16.85	+68%	N. Amer.
20	300	10.01	7.36	9.24	12.31	15.41	+195%	N. Amer.
50	105	4.25	4.65	5.08	5.79	8.60	+132%	Asia
40	120	9.25	9.48	10.24	11.28	13.88	+76%	Asia
35	135	11.79	11.68	12.40	13.13	14.39	+37%	Japan
−25	150	8.56	9.75	10.41	11.54	11.35	+25%	Australia
−35	15	4.40	5.93	8.41	11.29	12.19	+178%	S. Africa
−35	300	8.94	12.07	13.09	10.84	8.10	−63%	S. Amer.

cutoff rigidities – decreasing (with corresponding increasing of CR intensity) in 10 regions, and increasing (with corresponding decreasing of CR intensity) in three regions. These trends in CR intensity change must be taken into account together with CR 11- and 22-year modulation by solar activity considered in Section 2.12.

2.14 ATMOSPHERIC IONIZATION BY COSMIC RAYS: THE ALTITUDE DEPENDENCE AND PLANETARY DISTRIBUTION

The main process in the link between CR and cloudiness is the air ionization that triggers chemical processes in the atmosphere. Figure 19 shows experimental data [77] of the galactic CR generation of secondary particles and absorption at different cutoff rigidities. Figure 20 illustrates the total ionization of the atmosphere by galactic CR (primary and secondary) as a function of altitude.

The planetary distribution of ionization at an altitude of 3 km, calculated by Usoskin et al. [75], is shown in Fig. 21 for the year 2000; its time variation during 1950–2000 is presented in Fig. 22.

2.15 PROJECT 'CLOUD' AS AN IMPORTANT STEP IN UNDERSTANDING THE LINK BETWEEN COSMIC RAYS AND CLOUD FORMATION

The many unanswered questions in understanding the relationship between CR and cloud formation is being investigated by a special collaboration, within the framework of the European Organization for Nuclear Research, involving 17 institutes and universities [79]. The experiment, named

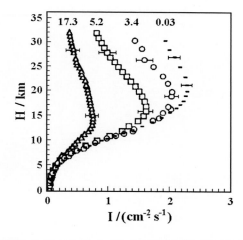

FIGURE 19

The absorption, I, curves of CR in the atmosphere at different cutoff rigidities (numbers at the top in units of GV) as a function of altitude, H. The horizontal bars indicate the standard deviations. *(From Ermakov et al. [77].)*

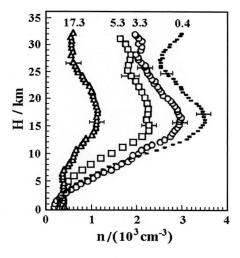

FIGURE 20

The ion concentration, n, profiles as a function of altitude, H, for different geomagnetic cutoff rigidities (numbers at the top are in units of GV). The horizontal bars indicate the standard deviations. *(From Ermakov et al. [77] with permission.)*

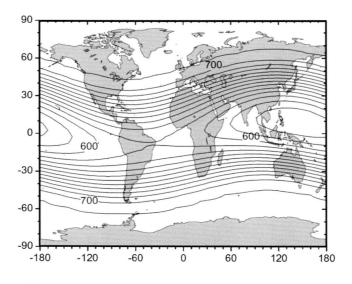

FIGURE 21

Planetary distribution of calculated equilibrium galactic CR-induced ionization at the altitude of 3 km (h = 725 g cm^{-2}) for the year 2000. Contour lines are given as the number of ion pairs per cubic centimetre in steps of 10 cm^{-3}. *(From Usoskin et al. [78].)*

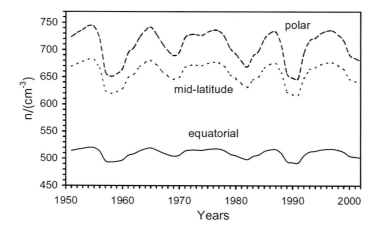

FIGURE 22

Calculated time profiles of the annual ionization, n, at altitude of 3 km (h = 725 g cm^{-2}), induced by galactic CR, for three regions: polar (cutoff rigidity $R_c < 1$ GV), midlatitudes (Rc ~ 6 GV) and equatorial (Rc ~ 15 GV) regions. *(From reference [78].)*

'CLOUD', is based on a Wilson cloud chamber (which is designed to duplicate the conditions prevailing in the atmosphere) and 'cosmic rays' from the CERN Proton Synchrotron. The project will consider the possible links between CR, variable sun intensities and the Earth's climate change (see Fig. 23)

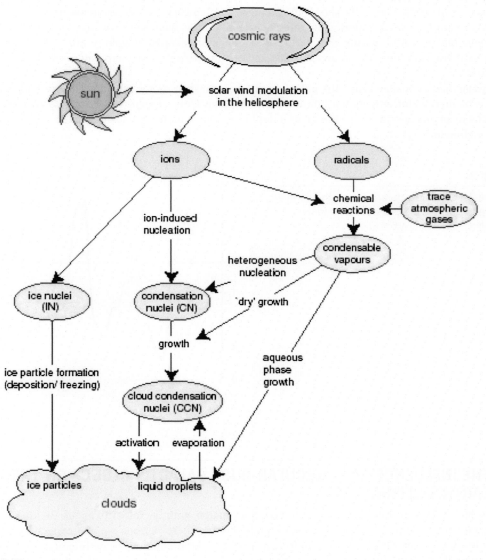

FIGURE 23

Possible paths of solar modulated CR influence on different processes in the atmosphere leading to the formation of clouds and their influence on climate. *(From Ref. [79].)*

3. THE INFLUENCE ON THE EARTH'S CLIMATE OF THE SOLAR SYSTEM MOVING AROUND THE GALACTIC CENTRE AND CROSSING GALAXY ARMS

The influence of space dust on the Earth's climate was considered in details and reviewed by Ermakov et al. [80]. Figure 24 shows the changes in planetary surface temperature over the past 520 Ma according work reported in reference [77]. These data were obtained from the paleo-environmental records. During this period the solar system crossed galaxy arms four times. In doing so, there were four alternating warming and cooling periods with temperature changes of more than 5°C.

The amount of matter inside the galactic arms is more than on the outside. The gravitational influence of this matter attracts the inflow of comets from Oort's cloud to the solar system [82,83]. It results in an increase in concentration of interplanetary dust in the zodiac cloud and a cooling of the Earth's climate [84].

FIGURE 24

Changes of air temperature, Δt, near the Earth's surface for the last 520 Ma according to the paleo-environmental records [81]. *(From Ref. [80].)*

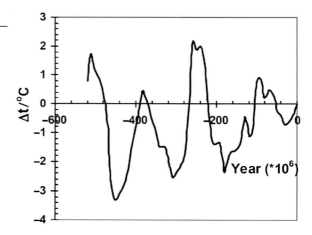

4. THE INFLUENCE OF MOLECULAR-DUST GALACTIC CLOUDS ON THE EARTH'S CLIMATE

The solar system moves relative to interstellar matter with a velocity about $30\,km\,s^{-1}$ and sometimes passes through molecular-dust clouds. During these periods we can expect a decrease in sea level air temperature. According to Dorman [85], the prediction of the interaction of a dust-molecular cloud with the solar system can be performed by measurements of changes in the galactic CR distribution function. From the past we know that the dust between the sun and the Earth has led to decreases of solar irradiation flux resulting in reduced global planetary temperatures (by 5°C–7°C in comparison with the 0.8°C increase from the present greenhouse effect). The plasma in a moving molecular dust cloud contains a frozen-in magnetic field; this moving field can modify the stationary galactic CR distribution outside the heliosphere. The change in the distribution function can be significant, and it should be possible to identify these changes when the distance between the cloud and the sun becomes comparable with the dimension of the cloud. The continuous observation of the time variation of CR distribution function for many years should provide the possibility of determining the direction and the speed of the dust-molecular cloud relative to the sun, as well as its geometry. Therefore, it should be possible to forecast climate changes caused by this molecular-dust cloud.

Figure 25 shows the temperature changes at the Antarctic station Vostok (bottom curve), which took place over the last 420 000 a according to Petit et al. [86]. These data were obtained from isotopic analysis of O and H extracted from the ice cores at a depth 3300 m.

It is seen from Fig. 25 that during this time the warming and cooling periods changed many times and that the temperature changes amounted up to 9°C. Also, there are the data obtained from isotope analysis of ice cores of Greenland holes, which cover the last 100 000 a (Fuhrer et al. [82]). These data confirm the existence of large changes in climate.

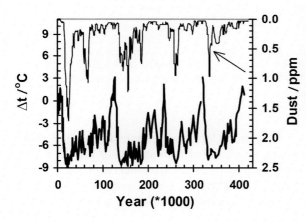

FIGURE 25

Changes of temperature, Δt, relative to the modern epoch (bottom thick curve) and dust concentration (upper thin curve) over the last 420 000 a [86]. *(From reference [80] with permission. The right-hand y-axis should read: dust size/μmol·mol⁻¹, and the x-axis should read: time/1000 a.)*

5. THE INFLUENCE OF INTERPLANETARY DUST SOURCES ON THE EARTH'S CLIMATE

According to Ermakov et al. [80], the dust of zodiac cloud is a major contributory factor to climate changes in the past and at the present time. The proposed mechanism of cosmic dust influence is as follows: dust from interplanetary space enters the Earth's atmosphere during the yearly rotation of the Earth around the sun. The space dust participates in the processes of cloud formation. The clouds reflect some part of solar irradiance back to space. In this way the dust influences climate. The main sources of interplanetary dust are comets, asteroids and meteor fluxes. The rate of dust production is continually changing. The effect of volcanic dust on the Earth's air temperature is illustrated in Fig. 26 [84]. (Note air temperature can be found at ftp://ftp.ncdc.noaa.gov/pub/data/anomalies/global_ meanT_C.all.)

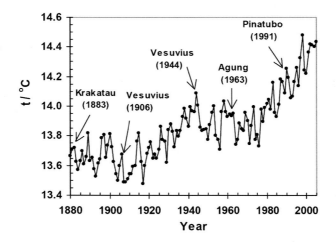

FIGURE 26

Yearly average values of the global air temperature, t, near the Earth's surface for the period 1880 to 2005 [84]. Arrows show the dates of the volcano eruptions with the dust emission to the stratosphere and short times cooling after eruptions. *(From Ref. [80] with permission.)*

According to Ermakov et al. [80], the spectral analysis of global temperature during 1880–2005 shows the presence of several spectral lines that can be identified with periods of meteor fluxes, comets and asteroids. The results of the analysis have been used [80,87] to predict the climate change to the nearest half-century: the interplanetary dust factor of cooling will be more important than warming from greenhouse effect, and as result, cooling is expected to be observed in the coming tens of years.

6. SPACE FACTORS AND GLOBAL WARMING

It has commonly been thought that the current trend of global warming is causally related to the accelerating consumption of fossil fuels by the industrial nations. However, it has also been suggested that this warming is a result of a gradual increase of solar and magnetic activity over the last 100 a. According to Pulkkinen et al. [88], as shown in Figs. 27 and 28, the solar and magnetic activity has been increasing since the year 1900 with decreases in 1970 and post 1980. Figs. 27 and 28, show that the aa index of geomagnetic activity (a measure of the variability of the interplanetary magnetic field, IMF), varies, almost in parallel, with the sunspot activity and with the global temperature anomaly.

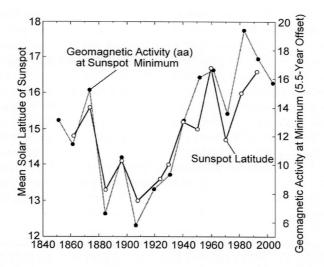

FIGURE 27

The geomagnetic activity (index aa) at the minimum of solar activity and the mean sunspots latitude, from 1840 to 2000. *(From Ref. [88].)*

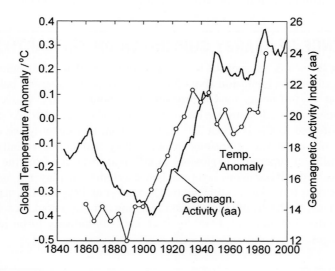

FIGURE 28

The geomagnetic activity (index aa) at the minimum of solar activity variation of the global temperature anomaly, Dt, from 1840 to 2000. *(From Ref. [88].)*

It has been well established that the brightness of the sun varies in proportion to solar activity. The variability of this brightness is very small and cannot explain all of the present global warming. However, the gradual increase of solar activity over the past 100 years has been accompanied by a gradual decrease of CR intensity in interplanetary space [89]. The direct measurements of CR intensity on the ground by the global network of neutron monitors (NM) as well as regular CR intensity measurements from balloons in the troposphere and stratosphere over a period of more than 40 a, show that there is a small negative trend of galactic CR intensity [90] of about 0.08% per year. Extrapolating this trend to 100 years gives a CR intensity decrease of 8%. From Fig. 2 it can be seen that the decreasing of CR intensity by 8% will lead to a decrease of cloud coverage of about 2%. According to Dickinson [17], decreasing cloud coverage by 2% corresponds to increasing the solar radiation falling on the Earth by about 0.5%. Using this information, Stozhkov et al. [58] conclude that the observed global climate change (increasing of average planetary ground temperature by 0.4°C–0.8°C over the last 100 years) may be caused by this negative trend of CR intensity. Sakurai [91] came to the same conclusion on the basis of analyzing data of solar activity and CR intensity.

7. THE INFLUENCE OF ASTEROIDS ON THE EARTH'S CLIMATE

It is well known that asteroids have in the past struck the Earth with sufficient force to make major climatic changes (the famous dinosaur-killing mass extinction at the end of the Cretaceous, which began the Tertiary era, has been convincingly identified with such an asteroid impact [92,93]). However, it is unlikely that our present climate change is due in any way to such events. Fortunately today, with modern methods of astronomy, the trajectory of dangerous asteroids can be determined exactly and, together with modern rocket power, could possibly be deflected.

8. THE INFLUENCE OF NEARBY SUPERNOVA ON THE EARTH'S CLIMATE

It is well known that the sun is a star of the second generation, in that it was born together with the solar system from a supernova explosion about 5 Ga ago. From the energetic balance of CR in the galaxy it follows that the full power for CR production is about 3×10^{33} W. Now it is commonly accepted that the supernova explosions are the main source of galactic CR. At each explosion the average energy transferred to CR is about 10^{43}–10^{44} J. From this quantity we can determine the expected frequency of supernova explosions in our galaxy and in the vicinity of the sun, and estimate: the probability of supernova explosions at different distances from the sun; the expected UV radiation flux (destroyer of our ozone layer and hence a significant player in our Earth's climate); and the expected CR flux. It has been estimated by Dorman et al. [94] and Dorman [85] that if such an event does take place, the levels of CR radiation reaching our Earth could reach levels extremely dangerous to our civilization and biosphere. Such an event is unlikely to be responsible for our present climate changes.

9. DISCUSSION AND CONCLUSIONS

Many factors from space and from anthropogenic activities can influence the Earth's climate. The initial response is that space factors are unlikely to be responsible for most of our present climate change. However, it is important that all possible space factors be considered, and from an analysis of past climate changes, we can identify our present phase and can predict future climates. During the last several hundred million years the sun has moved through the galactic arms several times with resultant climate changes. For example, considering the effects due to galactic molecular-dust concentrated in the galactic arms, as given in Fig. 24, we can see that during the past 520×10^6 a, there were four periods with surface temperatures lower than what we are presently experiencing and four periods with higher temperatures.

On the other hand, during the past 420 000 a (Fig 25) there were four decreases of temperature (the last one was about 20 000–40 000 a ago, the so-called big ice period), and five increases of temperature, the last of which happened a few thousand years ago. At present the Earth is in a slight cooling phase (of the order of 1°C over several thousand years).

When considering CR variations as one of the possible causes of long-term climate change (see Section 2) we need to take into account not only CR modulation by solar activity but also the change of geomagnetic cutoff rigidities (see Table 2). It is especially important when we consider climate change on a scale of between 10^3 and 10^6 a – paleomagnetic investigations show that during the last 3.6×10^6 a the magnetic field of the Earth changed sign nine times, and the Earth's magnetic moment changed, sometimes having a value of only one-fifth of its present value [95], corresponding to increases of CR intensity and decreases of the Earth's surface temperature.

The effects of space factors on our climate can be divided into two types:

- the 'gradual' type, related to changes on timescales ranging from 10^8 a to 11–22 a, producing effects that could be greater than that produced from anthropogenic factors, and
- the 'sudden' type, coming from supernova explosions and asteroid impacts, for example, which may indeed be catastrophic to our civilization. Volcanic and anthropogenic factors are also in a sense, 'sudden' factors in their effect on climate change.

It is necessary to investigate all of the possible 'sudden' factors and to develop methods of forecasting for protecting the biosphere and the Earth's civilization from big changes in climate and environment. We cannot completely exclude the possibility that a supernova explosion, for example, took place 20 a ago at a distance of say 30 light years away. In this case, its influence on our climate and environment will be felt in 10 a time. According to Ellis and Schramm [96], in this case, UV radiation would destroy the Earth's ozone layer over a period of about 300 a. The recent observations of Geminga, PSR J0437–4715 and SN 1987A strengthen the case for one or more supernova extinctions having taken place during the Phanerozoic era. In this case, a nearby supernova explosion would have depleted the ozone layer, exposing both marine and terrestrial organisms to potentially lethal solar UV radiation. In particular, photosynthesizing organisms including phytoplankton and reef communities would most likely have been badly affected.

As Quante [97] noted, clouds play a key role in our climate system. They strongly modulate the energy budget of the Earth and are a vital factor in the global water cycle. Furthermore, clouds significantly affect the vertical transport in the atmosphere and determine, in a major way, the redistribution of trace gases and aerosols through precipitation. In our present-day climate, on average, clouds cool our planet; the net cloud radiative forcing at the top of the atmosphere is about -20 W m^{-2}. Any change in the amount of cloud or a shift in the vertical distribution of clouds can lead to considerable changes in the global energy budget and thus affect climate [97].

Many of the 'gradual' types of space factors are linked to cloud formation. Quante [97] noted that galactic CR [22–25] was an important link between solar activity and low cloud cover (see Figs 2 and 3). However, new data after 1995 show that the problem is more complicated and the correlation no longer holds [98]. Kristjánsson et al. [99] pointed out that still many details are missing for a complete analysis, and a cosmic ray modulation of the low cloud cover seems less likely to be the major factor in our present climate change, but its role in future climate changes must not be ruled out. I, according to our paper [100], agree with this opinion and with results in references [101–103] that there is no simple causal connection between CR and low cloud coverage, that this connection can explain only some part of observed climate change. But the supposition in references [101,102] that the observed long-term correlation between cosmic ray intensity and cloudiness may be caused by parallel separate correlations between CR, cloudiness and solar activity contradicts the existence of the hysteresis effect in cosmic rays caused by the big dimensions of the heliosphere [104,105]. This effect, which formed a time lag of cosmic rays relative to solar activity of more than one year gives the possibility of distinguishing phenomena caused by cosmic rays from phenomena caused directly by solar activity (without time lag; see Fig. 2). The importance of cosmic ray influence on climate in comparison with the influence of solar irradiation can be seen clearly during the Maunder minimum (see Fig. 5). Cosmic ray influence on climate over a very long timescale of many hundreds of years can be seen from Fig. 1 (through variation of ^{14}C).

In this chapter much emphasis has been given to the formation of clouds and the influence CR plays (through ionization and influence on chemical processes in atmosphere) in their formation. This does not imply that CR is the only factor in their formation; dust, aerosols, precipitation of energetic particles from radiation belts and greenhouse gases, all play their part. However, the influence of CR is important and has been demonstrated here through:

- a direct correlation during one solar cycle (Figs. 2 and 3) and also for much longer periods,
- the correlation of CR intensity with the planetary surface temperature (Figs. 1, 4 and 5),
- by the direct relationship between CR intensity and wheat prices in medieval England (Fig. 8),
- by the direct relationship between cloud formation and CR air ionization (Fig. 9),
- by the relationship between geomagnetic activity and rainfall through precipitation of energetic particles from radiation belts (Figs. 10–13), and
- by linking CR intensities with the cirrus holes (Figs. 14 and 15).

The importance of CR cannot be stressed highly enough and it is important to develop methods for determining, with high accuracy, galactic CR intensity variations for the past, the present and the near future.

In this chapter several attempts have been made to explain the present climate change (the relatively rapid warming of the Earth discussed in Section 6 for 1937–1994), using space factors:

- by 11 a (11 years) average CR intensities discussed in Section 2.5,
- by the increasing geomagnetic activity,
- by the decreasing CR intensity (of 8% over the past 100 a), and
- by relating the spectral analysis of Ermakov et al. [80] to global temperature during 1880–2005.

Their results show the presence of several spectral lines that can be identified with the periods of meteor fluxes, comets and asteroids. On the basis of this work, Ermakov et al. [80,87] have predicted a cooling of the Earth's climate over the next half-century, which they believe will be more important than warming from greenhouse effect.

Finally, it appears that our present climate change (including a rapid warming of about 0.8°C over the past 100 a, see Fig. 26) is caused by a collective action of several space factors, volcano activities (with the dust emission rising to the stratosphere, resulting in short-term cooling after eruptions), as well as by anthropogenic factors with their own cooling and warming contributions. The relation between these contributions will determine the final outcome. At present the warming effect is stronger than the cooling effect. It is also very possible that the present dominant influence is anthropogenic in origin.

From Fig. 25 it can be seen that as we are near the maximum global temperature reached over the past 400 000 a, an additional rapid increase of even a few degrees Celsius could lead to an unprecedented and catastrophic situation. It is necessary that urgent and collective action be taken now by the main industrial countries and by the United Nations to minimize the anthropogenic influence on our climate before it is too late. On the other hand, in the future, if the natural change of climate results in a cooling of the planet (see Figs. 24 and 25), then special man-made factors, resulting in warming, may have to be used to compensate the natural cooling.

Let me underline that there is also one additional mechanism by which cosmic rays influence lower cloud formation, raining, and climate change: the nucleation by cosmic energetic particles of aerosol and dust, and through aerosol and dust, increasing of cloudiness. It was shown in references [106,107] in the CLOUD experiment (Section 2.15) at CERN that the irradiation by energetic particles (about 580 MeV) of the air at normal conditions in the closed chamber led to aerosol nucleation (induced by high energy particles), and simultaneously to an increase in ionization.

ACKNOWLEDGEMENTS

My great gratitude to Uri Dai, Isaac Ben Israel, Colin Price and Abraham Sternlieb for constant attention and support of the work of the Israel Cosmic Ray and Space Weather Center and Emilio Ségre Observatory; to L. A. Pustil'nik, Yu. I. Stozhkov, A. W. Wolfendale and I. G. Zukerman for interesting discussions, and to Trevor Letcher for useful comments.

REFERENCES

[1] L.I. Dorman, Cosmic Rays in the Earth's Atmosphere and Underground, Kluwer Academic Publishers, Dordrecht, 2004.

[2] L.I. Dorman, Cosmic Ray Interactions, Propagation, and Acceleration in Space, Springer, Netherlands, 2006.

[3] L.I. Dorman, Cosmic Rays in Magnetospheres of the Earth and Other Planets, Springer, Netherlands, 2008.

[4] L.I. Dorman, Solar Neutrons and Related Phenomena, Springer, Dordrecht/Heidelberg, 2010.

[5] J. Perez Peraza, L. Dorman, I. Libin, Space Sources of Earth's Climate: Natural Science and Economic Aspects of Global Warming, Euromedia, Moscow, 2011 (In English).

[6] L.I. Dorman, I.V. Dorman, Cosmic Ray History, Nova Publishers, New York, 2014.

[7] H. Herschel, Philos. Trans. R. Soc. London 91 (1801) 265–318.

[8] J.A. Eddy, Science 192 (1976) 1189–1202.

[9] H. Swensmark, Space Sci. Rev. 93 (2000) 175–185.

[10] E. Friis-Christiansen, K. Lassen, Science 254 (1991) 698–700.

[11] K. Lassen, E. Friis-Christiansen, J. Atmos. Sol. Terr. Phys. 57 (1995) 835–845.

[12] K. Labitzke, H. van Loon, Ann. Geophys. 11 (1993) 1084–1094.

[13] J. Lean, J. Beer, R. Breadley, Geophys. Res. Lett. 22 (1995) 3195–3198.

[14] J.D. Haigh, Science 272 (1996) 981–984.

[15] D. Shindell, D. Rind, N. Balabhandran, J. Lean, P. Lonengran, Science 284 (1999) 305–308.

[16] E.R. Ney, Nature 183 (1959) 451–452.

[17] R.E. Dickinson, Bull. Am. Met. Soc. 56 (1975) 1240–1248.

[18] M.I. Pudovkin, O.M. Raspopov, Geomagn. Aeronomy 32 (1992) 593–608.

[19] M. Pudovkin, S. Veretenenko, J. Atmos. Sol. Terr. Phys. 57 (1995) 1349–1355.

[20] M. Pudovkin, S. Veretenenko, Adv. Space Res. 17 (11) (1996) 161–164.

[21] B.A. Tinsley, J. Geomagn. Geoelectr. 48 (1996) 165–175.

[22] H. Swensmark, E. Friis-Christiansen, J. Atmos. Sol. Terr. Phys. 59 (1997) 1225–1232.

[23] H. Swensmark, Phys. Rev. Lett. 81 (1998) 5027–5030.

[24] N.D. Marsh, H. Swensmark, Phys. Rev. Lett. 85 (2000) 5004–5007.

[25] N. Marsh, H. Swensmark, Space Sci. Rev. 94 (2000) 215–230.

[26] L.I. Dorman, I.Y. Libin, M.M. Mikalayunas, K.F. Yudakhin, Geomagn. Aeronomy 27 (1987) 303–305.

[27] L.I. Dorman, I.Y. Libin, M.M. Mikalajunas, Regional Hydrometeorol. (Vilnius) 12 (1988) 119–134.

[28] L.I. Dorman, I.Y. Libin, M.M. Mikalajunas, Regional Hydrometeorol. (Vilnius) 12 (1988) 135–143.

[29] L.I. Dorman, G. Villoresi, I.V. Dorman, N. Iucci, M. Parisi, in: Proc. 25th Intern. Cosmic Ray Conf., Durban, vol. 7, 1997, pp. 345–348.

[30] R. Markson, Nature 273 (1978) 103–109.

[31] C. Price, Nature 406 (2000) 290–293.

[32] B.A. Tinsley, Space Sci. Rev. 94 (2000) 231–258.

[33] K. Schlegel, G. Diendorfer, S. Them, M. Schmidt, J. Atmos. Sol. Terr. Phys. 63 (2001) 1705–1713.

[34] L.I. Dorman, I.V. Dorman, N. Iucci, M. Parisi, Y. Ne'eman, L.A. Pustil'nik, F. Signoretti, A. Sternlieb, G. Villoresi, I.G. Zukerman, J. Geophys. Res. 108 (A5) (2003) 1181. SSH 2_1–8.

[35] L.I. Dorman, I.V. Dorman, Adv. Space Res. 35 (2005) 476–483.

[36] D.L. Hartmann, in: P.V. Hobbs (Ed.), Aerosol–Cloud–Climate Interactions, Academic Press, 1993, 151.

[37] G. Ohring, P.F. Clapp, J. Atmos. Sci. 37 (1980) 447–454.

[38] V. Ramanathan, R.D. Cess, E.F. Harrison, P. Minnis, B.R. Barkstrom, E. Ahmad, D. Hartmann, Science 243 (4887) (1989) 57–63.

[39] P.E. Ardanuy, L.L. Stowe, A. Gruber, M. Weiss, J. Geophys. Res. 96 (1991) 18537–18549.

[40] L.L. Stowe, C.G. Wellemayer, T.F. Eck, H.Y.M. Yeh, The Nimbus-7 Team, J. Clim. 1 (1988) 445–470.

[41] W.B. Rossow, R. Shiffer, Bull. Am. Met. Soc. 72 (1991) 2–20.

[42] F. Weng, N.C. Grody, J. Geophys. Res. 99 (1994) 25535–25551.

[43] R.R. Ferraro, F. Weng, N.C. Grody, A. Basist, Bull. Am. Met. Soc. 77 (1996) 891–905.

[44] H.S. Ahluwalia, in: Proc. 25th Int. Cosmic Ray Conf., Durban, vol. 2, 1997, pp. 109–112.

[45] W.B. Rossow, B. Cairns, J. Clim. 31 (1995) 305–347.

[46] J. Lean, A. Skumanich, O. White, Geophys. Res. Lett. 19 (1992) 1591–1594.

[47] J. Beer, G.M. Raisbeck, F. Yiou, in: C.P. Sonett, M.S. Giampapa, M.S. Matthews (Eds.), The Sun in Time, University of Arizona Press, 1991, pp. 343–359.

[48] P.D. Jones, K.R. Briffa, T.P. Barnett, S.F.B. Tett, The Holocene, vol. 8, 1998, pp. 455–471.

[49] M.E. Mann, R.S. Bradley, M.K. Hughes, Nature 392 (1998) 779–787.

[50] A. Smith, An Inquiry into the Nature and Causes of the Wealth of Nations, W. Strahan & T. Cadell, London, 1776.

[51] W.S. Jevons, Nature 19 (1875) 33–37.

[52] J.E.T. Rogers, Agriculture and Prices in England, vols 1–8, Clarendon Press, Oxford, 1887.

[53] W.S. Jevons, Nature 26 (1882) 226–228.

[54] L. Pustil'nik, G. Yom Din, L. Dorman, in: Proc. 28th Intern. Cosmic Ray Conf. Tsukuba, vol. 7, 2003, pp. 4131–4134.

[55] L. Pustil'nik, G. Yom Din, Solar Phys. 223 (2004) 335–356.

[56] J. Beer, S. Tobias, N. Weiss, Solar Phys. 181 (1998) 237–249.

[57] L. Pustil'nik, G. Yom Din, Solar Phys. 224 (2004) 473–481.

[58] YuI. Stozhkov, V.I. Ermakov, P.E. Pokrevsky, Izvestia Russian Academy, Sci. Ser. Phys. 65 (3) (2001) 406–410.

[59] S.V. Veretenenko, M.I. Pudovkin, Geomagn. Aeronomy 34 (1994) 38–44.

[60] YuI. Stozhkov, P.E. Pokrevsky, I.M. Martin, et al., in: Proc. 24th Int. Cosmic Ray Conf. Rome, vol. 4, 1995, pp. 1122–1125.

[61] Yu.I. Stozhkov, J. Zullo, I.M. Martin, et al., Nuovo Cimento C18 (1995) 335–341.

[62] Yu.I. Stozhkov, P.E. Pokrevsky, J. Zullo Jr., et al., Geomagn. Aeronomy 36 (1996) 211–216.

[63] Yu.I. Stozhkov, J. Phys. G 28 (2002) 1–11.

[64] M.C. Todd, D.R. Kniveton, J. Geophys. Res. 106 (D23) (2001) 32031–32042.

[65] M.C. Todd, D.R. Kniveton, J. Atmos. Sol. Terr. Phys. 66 (2004) 1205–1211.

[66] B.A. Tinsley, G.W. Deen, J. Geophys. Res. 96 (D12) (1991) 22283–22296.

[67] G.I. Pugacheva, A.A. Gusev, I.M. Martin, et al., in: Proc. 24th Intern. Cosmic Ray Conf., Rome, vol. 4, 1995, pp. 1110–1113.

[68] I.M. Martin, A.A. Gusev, G.I. Pugacheva, A. Turtelli, Y.V. Mineev, J. Atmos. Terr. Phys. 57 (2) (1995) 201–204.

[69] J.W. King, Astronaut. Aeronaut. 13 (4) (1975) 10–19.

[70] J.T.A. Ely, T.C. Huang, Geophys. Res. Lett. 14 (1) (1987) 72–75.

[71] J.T.A. Ely, J.J. Lord, F.D. Lind, in: Proc. 24th Intern. Cosmic Ray Conf., Rome, vol. 4, 1995, pp. 1137–1140.

[72] L.I. Dorman, Adv. Space Res. 35 (2005) 496–503.

[73] L.I. Dorman, Ann. Geophysicae 23 (9) (2005) 3003–3007.

[74] L.I. Dorman, Adv. Space Res. 37 (2006) 1621–1628.

[75] A.V. Belov, L.I. Dorman, R.T. Gushchina, V.N. Obridko, B.D. Shelting, V.G. Yanke, Adv. Space Res. 35 (3) (2005) 491–495.

[76] M.A. Shea, D.F. Smart, in: Proc. 28th Intern. Cosmic Ray Conf., Tsukuba, vol. 7, 2003, pp. 4205–4208.

[77] V.I. Ermakov, G.A. Bazilevskaya, P.E. Pokrevsky, YuI. Stozhkov, in: Proc. 25th Intern. Cosmic Ray Conf., Durbin, vol. 7, 1997, pp. 317–320.

[78] I.G. Usoskin, O.G. Gladysheva, G.A. Kovaltsov, J. Atmos. Sol. Terr. Phys. 66 (18) (2004) 1791–1796.

[79] B. Fastrup, E. Pedersen, E. Lillestol, et al., (Collaboration CLOUD), Proposal CLOUD, CERN/SPSC, 2000, pp. 2000–2021.

[80] V. Ermakov, V. Okhlopkov, Yu.I. Stozhkov, in: Proc. European Cosmic Ray Symposium, Lesboa, 2006. Paper 1–72.

[81] J. Veizer, Y. Godderis, I.M. Francois, Nature 408 (2000) 698–701.

[82] K. Fuhrer, E.W. Wolff, S.J. Johnsen, J. Geophys. Res. 104 (D24) (1999) 31043–31052.

[83] O.A. Maseeva, Astronomichesky vestnik 38 (4) (2004) 372–382.

[84] J. Hansen, R. Ruedy, J. Glascoe, M. Sato, J. Geophys. Res. 104 (D24) (1999) 30997–31022.

[85] L.I. Dorman, Adv. Geosci. 14 (2008) 281–286.

[86] J.R. Petit, J. Jouzel, D. Raunaud, et al., Nature 399 (1999) 429–436.

[87] V.I. Ermakov, V.P. Okhlopkov, Yu.I. Stozhkov, in: Bull. Lebedev Phys. Inst. Russian Acad. Sci. Moscow, 3, 2006, pp. 41–51.

[88] T.I. Pulkkinen, H. Nevanlinna, P.J. Pulkkinen, M. Lockwood, Space Sci. Rev. 95 (2001) 625–637.

[89] M. Lockwood, R. Stamper, M.N. Wild, Nature 399 (6735) (1999) 437–439.

[90] Yu.I. Stozhkov, P.E. Pokrevsky, V.P. Okhlopkov, J. Geophys. Res. 105 (A1) (2000) 9–17.

[91] K. Sakurai, in: Proc. 28th Intern. Cosmic Ray Conf., Tsukuba, 7, 2003, pp. 4209–4212.

[92] L. Alvarez, W. Alvarez, F. Asaro, H. Michel, Science 208 (1980) 1095–1108.

[93] V. Sharpton, K. Burke, A. Camargo-Zaroguera, et al., Science 261 (1993) 1564–1567.

[94] L.I. Dorman, N. Iucci, G. Villoresi, Astrophys. Space Sci. 208 (1993) 55–68.

[95] A. Cox, G.B. Dalrymple, R.R. Doedl, Sci. Am. 216 (2) (1967) 44–54.

[96] J. Ellis, D.N. Schramm, in: Proc. Nat. Acad. Sci. USA Astron, 92, 1995, pp. 235–238.

[97] M. Quante, J. Phys. IV France 121 (2004) 61–86.

[98] J.E. Kristjánsson, A. Staple, J. Kristiansen, Geophys. Res. Lett. 23 (2002) 2107–2110.

[99] J.E. Kristjánsson, J. Kristiansen, E. Kaas, Adv. Space Res. 34 (2) (2004) 407–415.

[100] L.I. Dorman, Ann. Geophys. 30 (2012) 9–19.

[101] A.D. Erlykin, G. Gyalai, K. Kudela, T. Sloan, A.W. Wolfendale, J. Atmos. Sol. Terr. Phys. 71 (2009) 823–829.

[102] A.D. Erlykin, G. Gyalai, K. Kudela, T. Sloan, A.W. Wolfendale, J. Atmos. Sol. Terr. Phys. 71 (2009) 1794–1806.

[103] A.D. Erlykin, A.W. Wolfendale, J. Atmos. Sol. Terr. Phys. 73 (2011) 1681–1686.

[104] I.V. Dorman, L.I. Dorman, J. Geophys. Res. 72 (1967) 1513–1520.

[105] I.V. Dorman, L.I. Dorman, J. Atmos. Terr. Phys. 29 (1967) 429–449.

[106] M.B. Enghoff, J.O.P. Pedersen, U.I. Uggerhøj, S.M. Paling, H. Svensmark, Geophys. Res. Lett. 38 (2011) L09805, http://dx.doi.org/10.1029/2011GL047036.

[107] J. Kirkby, J. Curtius, J. Almeida, et al., Letter 476 (2011) 429–433, http://dx.doi.org/10.1038/nature10343.

ENGINEERING, SOCIETAL AND FORESTRY ASPECTS OF CLIMATE CHANGE

ENGINEERING ASPECTS OF CLIMATE CHANGE

31

Daniel A. Vallero

Pratt School of Engineering, Duke University, Durham, NC, USA

CHAPTER OUTLINE

1. INTRODUCTION

Engineering, at its core, is the application of science to meet specific needs. Note that this avoids the 'problem solving' moniker so often applied to engineering. Indeed, engineers solve problems. They also prevent problems. They also enhance benefits. Certainly, one way to solve the 'heat problems' of potential warming scenarios is to devise cooled, HAZMAT suits for every human being on earth. Obviously, this approach is unacceptable from many perspectives, including practicality and livability. So then, engineering as a response to large-scale problems, such as those brought on by climate change, requires a comprehensive, systematic approach.

2. THE ROLE OF THE ENGINEER

Much of the engineering-related research and practice associated with societal responses to climate change falls under civil engineering and its subdiscipline, environmental engineering, which seeks innovative ways to control and treat the various types of pollutants, including greenhouse gases, by applying the physical sciences. However, all engineering fields are involved in finding solutions to problems brought about by climate change. Although the principal interest in this section is addressing the pollutants that lead to climate change, engineers must also solve the indirect problems caused by climate change, including providing goods and services to migrating populations, preventing diseases

Climate Change. http://dx.doi.org/10.1016/B978-0-444-63524-2.00031-2

caused by entry of vector populations (e.g. mosquitos, other insects and vermin) to previously uninfected regions, addressing biome shifts and their effects on agriculture (including planting, harvest, transportation, storage), and stresses on infrastructure (e.g. flood control, water supply, wastewater treatment, roads, railway, ports, etc.).

Environmental engineering, for the most part, has supplanted 'sanitary engineering' in the United States and much of the world. The origin of formal sanitary engineering can be traced to as early as the nineteenth century [1]. To the sanitary engineer first and the environment engineer later, the public places its trust in the engineer to ensure a livable and healthful environment [2]. In addition to water quality and quantity, sanitary and environmental engineers, along with chemical engineers, design air pollution control equipment, soil remediation systems, landfills and hazardous waste operations. These systems operate at a vast range of spatial and temporal scales, from single sites to regional, e.g. carbon dioxide (CO_2) sequestration systems in rock formations and deep ocean layers.

Air pollution controls remove and treat emissions, especially products of incomplete combustion. Ironically, a measure of success has been how close the operations achieve complete combustion; that is, the oxidation of organic compounds to produce carbon dioxide and water, both global greenhouse gases (GHGs). As mentioned, in addition to treating air and water pollution, engineers design systems to remove and store CO_2 and other GHGs (sequestration), but engineers also adapt existing processes to prevent and decrease emissions. This is the domain of 'green engineering', i.e. the design, commercialization and use of processes and products that are feasible and economical, while lowering the amount of pollution generated by source and minimizing the health risks and reducing environmental stress [3]. Certainly, structure and mechanics have always been key components of sanitary engineering, but modern environmental engineering adds a systems aspect to environmental solutions.

From a systems perspective engineers must consider the entire life cycle of processes that result in climate change. In addition, engineers must appropriately scale solutions to problems. Since changes in climate are global in scale, the problems associated with these changes must be addressed at very large scales. However, the solutions always include every scale. Indeed, many solutions to climate change problems are aggregations of smaller-scale decisions and actions, such as an individual's decision to purchase a product with fewer or less potent GHG emissions over the product's life cycle than a competing product. Again, engineers must consider the entire system. For example, drinking cup 'A' may release 20% less CO_2 than cup 'B' after disposal in a landfill; however, in making cup 'A' twice as much CO_2 was emitted than in making cup 'B'. Indeed, the systems view requires an even broader perspective. For instance, there are numerous other GHGs that could come into play during different stages of a product's life cycle, e.g. if one of the cups results in emission of large amounts of nitrous oxide (N_2O) during manufacturing, this could tip the scales against buying that product, even if its landfill emissions would be lower. Thus, individual decisions, when aggregated, result in global impacts, harkening the need to 'think globally, but act locally'.

Climate change recommendations tend largely to be devoted to finding ways to decrease emissions of greenhouse gases. However, adaptation and remedies may also be employed. Global-scale interventions, known as geoengineering, have yet to be taken to address climate change but are being considered. Currently, greenhouse gas removal and solar radiation management are the two prominent geoengineering solutions being debated. These include:

- spraying SO_2 into the stratosphere to enhance cloud albedo
- spraying engineering nanoparticles into the stratosphere to enhance cloud albedo
- placing reflective mirrors, discs or particles in earth orbit

- painting roofs and other structures with reflective material
- placing solar reflectors in the desert [4].

Note that some of these can be implemented at smaller scales, e.g. roof painting, and aggregated to achieve results. Others, e.g. reflective mirrors, must begin at substantially large scales, with concomitant risks.

Ideally, engineers begin with small laboratory studies, controlling for each variable, then step up to increasingly larger prototypes, before launching large-scale projects such as these. Often, in environmental science, natural experiments can provide some of the salient information, such as the 1991 Mount Pinatubo eruption that emitted 10×10^6 t (10 million metric tonnes) of SO_2 into the stratosphere, which was converted to sulphate particles. Sulphate particles in the atmosphere increased cloud albedo, i.e. the reflectivity of solar radiation into space. Noting this, Nobel Prize-winning geochemist Paul Crutzen hypothesized that spraying SO_2 into the stratosphere would increase cloud albedo and could result in mean cooling of the Earth by 0.5°C. According to Crutzen, sulphate particles will last one or two years longer if SO_2 were sprayed in the stratosphere rather than the troposphere, and have a greater impact on cloud albedo. To generate this amount of cooling, about 2×10^6 t of SO_2 would need to be sprayed annually. This represents 3.6% of the 55×10^6 t of SO_2 emitted into the atmosphere each year from the burning of fossil fuels. Although there is substantial risk and uncertainty of this geoengineering approach, it can be based on at least a modicum of real-world data.

The atmosphere is complex. Changing one variable, i.e. SO_2 concentrations, will affect other variables. For example, spraying SO_2 into the stratosphere may adversely impact human health and the environment if it interferes with the ozone layer or other chemical balances. In the troposphere, SO_2 is converted into sulphuric acid (H_2SO_4) in the atmosphere and produces acidic precipitation that damages ecosystems, including threatening sensitive plant and animal species.

Sulphur dioxide is certainly not innocuous. Inhalation of SO_2 contributes to respiratory problems, such as airway constriction and asthma exacerbation; as such, most nations regulate the emissions of SO_2. Short-term exposure is associated with increased visits to emergency departments and hospitalization for respiratory problems, especially among young children and asthmatics. So, it may be difficult to justify increasing the global mean concentration of SO_2 while decreasing local concentrations.

There is also the risk of simplifying complex systems. Solving only the cooling problem does not address the increasing concentrations of atmospheric CO_2, which can be associated with other problems. For example, when CO_2 dissolves in seawater, it changes the ionic strength and other chemical properties of the water, which increases the net acidity of oceans and other surface waters. The oceans have increased in acidity by 30% since 1750, due to increases in anthropogenic carbon dioxide in the atmosphere. This trend may decrease the availability of calcium carbonate, threatening species that form shells from this compound, such as molluscs, corals and some types of plankton. A reduction in these species could have wide-ranging impacts on other marine species and ecosystems, since many organisms feed on molluscs or plankton or depend on coral reefs for shelter.

Increasing cloud albedo may also lead to environmental problems. Assuming the global temperatures can be anthropogenically stabilized, the reflectivity increase could affect precipitation patterns, tropical storm activity, temperature distribution and wind. Also, much of the radiation that is reflected by sulphate particles strikes the Earth as diffuse light. This increases the whiteness of the daytime sky and may reduce the efficiency of plant photosynthesis and solar power.

Further, the complexity of the atmosphere means that the expected results will be highly uncertain, whatever actions are taken. Thus, spraying too much SO_2 into the stratosphere or improperly locating mirrors could lead to excessive cooling, droughts, floods or other meteorological events.

3. GLOBAL GREENHOUSE GASES

Carbon dioxide is the most ubiquitous of the GHGs. It is the product of decomposition of organic material, whether biologically or through combustion. The effectiveness of CO_2 as a global warming gas has been known for over 100 years, but the first useful measurements of atmospheric CO_2 were not taken until 1957. The results from Mauna Loa in Hawaii are exceptionally useful since they show that even in the 1950s the CO_2 concentration had increased from the baseline 280 µmol mol^{-1} (280×10^{-6}) to 315 µmol mol^{-1}, and this has continued to climb over the last 50 years at a constant rate of about 1.6 µmol mol^{-1} per year.

Methane (CH_4) is the product of anaerobic decomposition and human food production. One of the highest producers of methane in the world is New Zealand, which boasts 80 million sheep. Methane also is emitted during the combustion of fossil fuels and cutting and clearing of forests. The concentration of CH_4 in the atmosphere has been steady at about 0.75 µmol mol^{-1} for over 1000 years, and then increased to 0.85 µmol mol^{-1} in 1900. Since then, in the space of only 100 years, has skyrocketed to 1.7 µmol mol^{-1}. Methane is commonly removed from the atmosphere by reaction with the hydroxyl radical (OH) as:

$$CH_4 + OH + 9O_2 \rightarrow CO_2 + 0.5H_2 + 2H_2O + 5O_3 \tag{1}$$

As is often the case in environmental and green engineering, one equation or reaction does not completely characterize the problem. In this instance, it creates carbon dioxide, water vapour and ozone, all of which are greenhouse gases. That is, methane is problematic not only as a GHG but as a precursor for other GHGs.

Halocarbons are the same classes of compounds involved in the destruction of atmospheric ozone that are also at work in promoting global warming. The most effective GHGs are CFC-11 and CFC-12, both of which are no longer manufactured, and the banning of these substances has shown a levelling off in the stratosphere.

Nitrous oxide is emitted to the atmosphere predominantly as a result of human activities, especially the cutting and clearing of tropical forests (which, as mentioned, also has deleterious albedo consequences). The greatest problem with nitrous oxide is that there appear to be no natural removal processes for this gas and so its residence time in the stratosphere is quite long. This is another example of the interrelationships and interdependencies between the carbon (C) and nitrogen (N) cycles. The net effect of these global pollutants is still being debated. Various atmospheric models used to predict temperature change over the next 100 years vary widely. They nevertheless agree that some positive change will occur. By the year 2100, even if we do not increase our production of greenhouse gases and international agreements are reached and subsequently followed, the global temperature is likely to be between 0.5°C and 1.5°C warmer than at present. The effect of this on natural systems and dynamics in the oceans and atmosphere could be devastating.

4. ENGINEERING ASPECTS OF THE 'SPHERES'

The National Academy of Engineering has identified the most important challenges to the future of engineering. Both the nitrogen and carbon biogeochemical cycles are explicitly identified among the most pressing engineering needs. The biogeochemical cycle that extracts nitrogen from the air for its incorporation into plants, hence food, has become altered by human activity. With widespread use of fertilizers and high-temperature industrial combustion, humans have doubled the rate at which nitrogen is removed from the air relative to preindustrial times, contributing to smog and acid rain, polluting drinking water and even worsening global warming. Engineers must design countermeasures for nitrogen cycle problems, while maintaining the ability of agriculture to produce adequate food supplies [5]. The C cycle and nutrient cycles are inextricably linked. Like C, as discussed in the next chapter, chemical species of N and the other nutrient elements are essential and toxic, depending on its dose and form. The Academy articulates this challenge:

> The biogeochemical cycle that extracts nitrogen from the air for its incorporation into plants and hence food has become altered by human activity. With widespread use of fertilizers and high-temperature industrial combustion, humans have doubled the rate at which nitrogen is removed from the air relative to pre-industrial times, contributing to smog and acid rain, polluting drinking water, and even worsening global warming. Engineers must design countermeasures for nitrogen cycle problems, while maintaining the ability of agriculture to produce adequate food supplies [5].

Engineers can expect to be increasingly asked to recommend improvements to the biogeochemical cycling of C, such as enhancements to the food life cycles (e.g. animal feeding operations, farmlands, rangelands and groceries). How can engineering innovation improve the efficiency of various human activities related to nitrogen, from making fertilizer to recycling food wastes? Currently, less than half of the fixed nitrogen generated by farming practices actually ends up in harvested crops. And, less than half of the nitrogen in those crops actually ends up in the foods that humans consume. In other words, fixed nitrogen leaks out of the system at various stages in the process from the farm field to the feedlot to the sewage treatment plant. Engineers not only need to identify the leakage points and devise systems to plug them, i.e. the structural and mechanical solutions, but must engage biological solutions, such as understanding the processes that lead to increased C and nutrient emissions, and applying this understanding to modify the processes accordingly [6].

Beyond physical principles, chemical and biological systems must be appreciated. For example, a climate change model would be remiss to exclude biological feedbacks with changing meteorological conditions. For example, warmer, more humid conditions would affect photosynthesis and respiration by organisms, which would change balances, such as those for carbon and nitrogen. That is, the hydrosphere, atmosphere and biosphere are actually all part of a larger system whose components and feedback mechanisms are intricately linked. Everything seems to matter in climate change. The challenge for the engineer is to choose wisely the variables to be optimized and to weight them correctly.

The hydro-atmos-bio system is complex. The behaviour of GHGs and other substances in this system can be to some extent predicted from the inherent properties of the compound of concern. For example, vapour pressure is the pressure exerted by a vapour in a confined space. Similarly, vaporization is the change of a liquid or solid to the vapour phase. This means that if a substance vaporizes it can enter the atmosphere. Thus, volatilization is a function of the concentration of a contaminant in solution and the contaminant's partial pressure.

The proportionality between solubility and vapour pressure can be established for any chemical. Henry's law states that the concentration of a dissolved gas is directly proportional to the partial pressure of that gas above the solution:

$$p_a = K_H[c] \qquad (2)$$

where,

K_H = Henry's law constant
p_a = Partial pressure of the gas
$[c]$ = Molar concentration of the gas

or,

$$p_a = K_H C_W \tag{3}$$

where, C_W is the concentration of gas in water.

Thus, integrating information about the concentration of a dissolved contaminant and its partial pressure in atmosphere at equilibrium is a means of estimating the likelihood that a chemical will move into the atmosphere. Conversely, it is also a means of estimating how long the substance will remain in the atmosphere. A dimensionless version of K_H partitioning is similar to that of sorption, except that instead of the partitioning between solid and water phases, it is between the air and water phases (K_{AW}):

$$K_{AW} = C_A / C_W \tag{4}$$

where C_A is the concentration of gas A in the air.

The relationship between the air/water partition coefficient and Henry's law constant for a substance is:

$$K_{AW} = K_H / RT \tag{5}$$

where, R is the gas constant (8.21×10^{-2} L atm mol^{-1} K^{-1}) and T is the temperature (K).

Henry's law relationships work well for most environmental conditions, representing a limiting factor for systems where a substance's partial pressure is approaching zero. At very high partial pressures (e.g. 30 Pa) or at very high contaminant concentrations (e.g. 1000 μmol mol^{-1}($>1000 \times 10^{-6}$)), Henry's law assumptions cannot be met.[1] Such extreme vapour pressures and concentrations are seldom seen in the troposphere. Thus, in modelling and estimating the tendency for a substance's release in vapour form, Henry's law can be a reasonable metric in compartmental transport models to indicate the fugacity from the water to the atmosphere.

Any sorbed or otherwise bound fraction of the contaminant will not exert a partial pressure, so this fraction should not be included in calculations of partitioning from water to air. For example, it is important to differentiate between the mass of the contaminant in solution (available for the K_{AW} calculation) and that in the suspended solids (unavailable for K_{AW} calculation). This is crucial for many hydrophobic organic contaminants, where they are most likely not to be dissolved in the water column (except as co-solutes), with the largest mass fraction in the water column being sorbed to particles.

An important biospheric partitioning coefficient is the octanol–water coefficient (K_{ow}), which is the ratio of a substance's concentration in octanol ($C_7H_{13}CH_2OH$) to the substance's concentration in water at equilibrium (i.e. the reactions have all reached their final expected chemical composition in a control volume of the fluid). Octanol is a surrogate for lipophilic solvents in general because it has degrees of affinity for both water and organic compounds, that is, octanol is amphiphilic. Since K_{ow} is the ratio [$C_7H_{13}CH_2OH$]: [H_2O], then the larger the K_{ow} value, the more lipophilic the substance. The relationship between K_H and K_{ow} is often used as indicator of environmental persistence, as reflected by the chemical half-life ($t_{1/2}$) of a contaminant. For example, even if the GHG is a specific compound, its

---

[1]Henry's law constants are highly dependent upon temperature, since both vapour pressure and solubility are also temperature dependent. So, when using published K_H values, they must be compared isothermally. Also, when combining different partitioning coefficients in a model or study, it is important either to use only values derived at the same temperature (e.g. sorption, solubility and volatilization all at 20°C) or to adjust them accordingly. A general adjustment is an increase of a factor of 2 in K_H for each 8°C temperature increase [7].

precursor species may exist for many years in reservoirs in various physical states. A compound with a relatively large K_{ow} may be degraded aerobically, resulting in the release of water and carbon dioxide, or degraded anaerobically, resulting in the release of water and methane. Indeed, many other variables determine the actual persistence of a compound after its release. For example, benzene and chloroform have nearly identical values of K_H and K_{ow}, yet benzene is far less persistent in the environment, largely due to the affinity of microbial degradation of benzene compare to the halogenated compound.

Thus, the relative affinity for a substance or its degradation products to reside in air and water can be used to estimate the potential for the substance to partition not only between water and air but more generally between the atmosphere and biosphere, especially when considering the long-range transport of contaminants (e.g. across continents and oceans). Such long-range transport estimates make use of both atmospheric $t_{1/2}$ and K_H. Also, the relationship between octanol–water and air–water coefficients can be an important part of predicting a contaminants transport. For example, Figure 1 provides some general classifications according various substances' K_{AW} and K_{ow} relationships. In general, chemicals in the upper left-hand group have a great affinity for the atmosphere, so unless there are contravening factors, this is where to look for them. Conversely, substances with relatively low K_{AW} and K_{ow} values are less likely to be transported long distance in the air. Since K_{AW} is proportional to K_H, these groupings also apply to Henry's law constants [7].

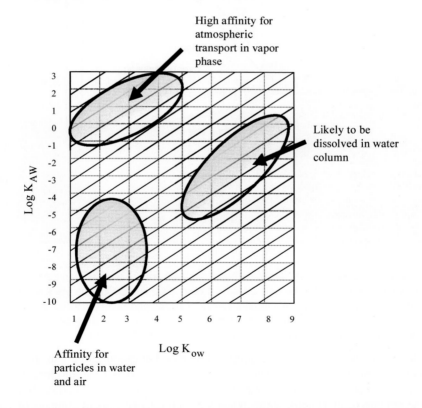

FIGURE 1

Relationship between air–water partitioning and octanol–water partitioning and affinity of classes of contaminants for certain environmental compartments under standard environmental conditions. *(Source: Ref. [8].)*

Arguably, two chemical elements account for much of the anthropogenic GHGs, i.e. carbon and nitrogen. The change in chemical form and cycling of C and N must be understood to address climate change adequately. With sufficient moisture, numerous simultaneous chemical reactions occur, making for a balance among various chemical forms of C and N, as well as those of other macro-nutrients, sulphur and potassium, and numerous micronutrients.

5. ENGINEERING AND THE CARBON CYCLE

Carbon in its many inorganic and organic species is cycled continuously through the environment (See Figure 2). From a mass balance perspective, only a relatively small amount of C-based compounds reside in the atmosphere (See Table 1), but this accounts for much of the planet's greenhouse effect.

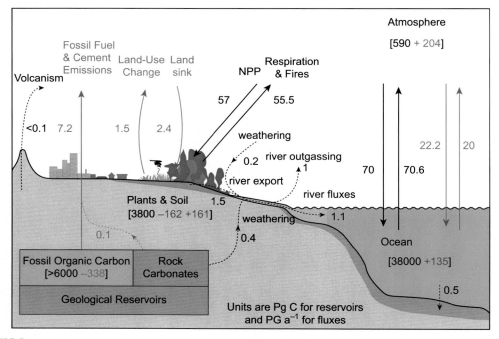

FIGURE 2

Global carbon cycle from 2000 to 2005. Mean carbon storage (in square brackets) and exchanges of CO_2 between different pools of carbon for the time period between 2000 and 2005. The black arrows indicate natural CO_2 exchanges. The red (dark grey in print versions) arrows and numbers indicate additional exchanges and storage of carbon resulting from human activity. The exchanges are in petagrams (i.e. 10^{15} g) of carbon per year (Pg a^{-1}), where a refers to annum. *(Adapted from Ref. [10].)*

Figure 2 demonstrates the diversity of C sinks and sources. For example, if C can remain sequestered in the soil, roots, sediment and other compartments, it is not released to the atmosphere. For example, C that is sequestered is not released. Thus, sequestered C is a means of ameliorating climate change by decreasing the amount of greenhouse gases released. Even relatively small amounts of methane and carbon dioxide can profoundly increase the atmosphere's greenhouse potential. There are many drivers and constraints involved in climate (See Figure 3). The increased amounts of CO_2 will likely affect global temperature, which affects biomes and the kinetics within individual ecosystems. Indeed, most

Table 1 Approximate mass of carbon in the earth's major sinks in 1999.

Carbon sink	Carbon mass/10^9 t (US billions of metric tonnes)
Atmosphere	766
Soil organic matter	1500 to 1600
Ocean	38 000 to 40 000
Marine sediments and sedimentary rocks	66 000 000 to 10 000 000
Terrestrial plants	540 to 610
Fossil fuel deposits	4000

Source: Ref. [9].

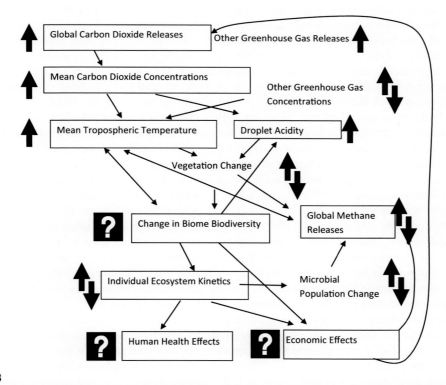

FIGURE 3

Systematic view of changes in tropospheric carbon dioxide. Thick arrows indicate whether this factor will increase (up arrow), decrease (down arrow) or will vary depending on the specifics (e.g. some greenhouse gas releases have decreased, e.g. the chlorofluorocarbons and some gases can cool the atmosphere, e.g. sulphate aerosols). Question mark indicates that the type and/or direction of change are unknown or mixed. Thin arrows connect the factors as drivers toward downstream effects. *(Source: Ref. [11].)*

climate change models indicate that the build-up of CO_2 is the principal driver for warming. Removal of CO_2 includes methods for extracting the gas from the atmosphere and storing or sequestering it.

Biological feedbacks are highly complex and models of living things are always more complicated and uncertain than the abiotic components of ecosystems. Short-term and localized change can occur if CO_2 is sequestered in various ecosystems. Indeed, countless biological feedback and interfaces are possible that will lead to different both functional and structural responses. Although any GHG can affect an ecosystem, carbon molecules are particularly important to an ecosystem's structure, e.g. if CO_2 is sequestered in wetlands with peat soils and plant life, the wetland structure is altered. If the wetlands are part of a larger ecosystem, it can lead to even greater structural change, such as changes in tree associations, which include potentially new species of tree canopies and forest floors. On the global scale, if CO_2 injected into ecosystems leads to net decreases in soil and water oxygen levels, which in turn leads to global increase in anaerobic microbial decomposition, this would translate into greater global releases of CH_4, which would mean increasing global temperatures. However, if greater biological activity and increased photosynthesis is triggered by the increase in CO_2, and wetland depth is decreased, CH_4 global concentrations would fall, leading to less global temperature rise. Conversely, if this increased biological activity and photosynthesis leads to a decrease in forest floor detritus mass, then less anaerobic activity may lead to lower releases of CH_4. Every one of these scenarios assumes, of course, that all other factors are held constant, which is never the case. Even limiting the scope to CO_2 and CH_4, the variability of increases and decreases in concentrations with time and at various scales, predictions of the net effects on a complex, planetary system are highly uncertain.

Acidic deposition is often considered as a regional air pollution problem, but distinct from most climate change discussions. Indeed, much of the concern for acid rain has been rightly concerned about compounds other than GHGs, notably oxides of sulphur and nitrogen. These compounds can dramatically decrease the pH of rain. However, GHGs do factor into the pH of precipitation, since rainfall's natural acidity owes in large part to the carbon cycle, including CO_2 and carbonic acid. The oxides of N and S exacerbate this natural acidity, leading to forest and other ecological stress and harm to fish and wildlife, as well as damage to structures.

Combustion is the principal source of acidic deposition, so decreasing in the amount of combustion would also decrease emissions of GHGs. Air pollution controls are more complex in terms of benefits. For example, thermodynamic efficiency is indirectly related to the release of hydrocarbons, i.e. approaching complete combustion will result in decreasing hydrocarbon emissions. Oxides of nitrogen (NO_x) emissions are different. Most of these result not from the fuel but from the molecular nitrogen (N_2) in the air entering combustion reactor. Thus, the richness of the fuel-air mix will affect NO_x emissions. It is possible for increasing fuel efficiency to result in higher rates of NO_x. Thus, as shown in Figure 4, the highest NO_x emissions are found at air: fuel mixtures between 14 and 16. Combustion at very high temperatures is needed to oxidize the relatively nonreactive N_2.

Another consideration includes the possible effects of transferring GHGs from one environmental component to another. The difference in gas concentrations and the exchange coefficients between the atmosphere and surface waters determines how quickly a molecule of gas can move across the ocean-atmosphere boundary. At the ocean surface, CO_2 equilibrates with atmospheric CO_2, which commonly leads to large atmosphere-ocean differences in CO_2 concentrations. Biota and ocean circulation account for the majority of the difference. The oceans contain vast carbon reservoirs, with which the atmosphere exchanges since CO_2 reacts with water to form carbonic acid and its dissociation products. With the increased atmospheric CO_2 concentrations, the interaction with the ocean surface alters the chemistry of the seawater resulting in ocean acidification [10]. Ocean uptake of anthropogenic CO_2 is primarily a physical response to increasing atmospheric CO_2 concentrations. Increasing the partial pressure of a gas in the atmosphere directly above the body of water causes the gas to diffuse into that

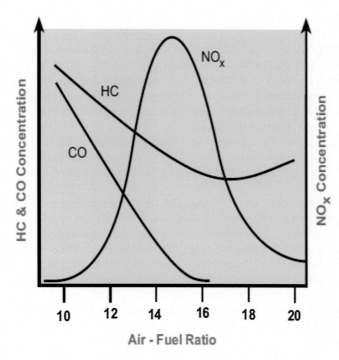

FIGURE 4

Exhaust emissions from an automobile engine related to air-fuel mix ratios. *(Source: Ref. [12].)*

water until the partial pressures across the air–water interface are equilibrated. The effects are complex, e.g. increasing CO_2 also modifies the climate, which in turn may change ocean circulation, which changes the rate of ocean CO_2 uptake. Marine ecosystem changes also alter the uptake [10].

The cycling of C and other elements indicate the complexity in addressing air pollution science and engineering. For example, pollution control efficiency and success have often been based on CO_2 production rates, i.e. the greater the amount of CO_2 leaving a stack, the better air pollution control equipment is operating. This is because complete combustion results in the production of CO_2 and H_2O. Likewise, the production of CH_4 has been an indication of complete anaerobic digestion of organic compounds. Both of these indicators of pollution control success are greenhouse gases.

The effectiveness of a particular gas to promote global warming (or cooling, as is the case with aerosols) is known as forcing. The gases of most importance in forcing are listed in Table 2. Note that

Table 2 Relative forcing of increased global temperature (excluding water vapour).

Greenhouse gas	Percent of relative radiative forcing
Carbon dioxide, CO_2	64
Methane, CH_4	19
Halocarbons (mostly CFCs)	11
Nitrous oxide, N_2O	6

Source: Ref. [13].

the only non-C compound listed is nitrous oxide and that the C-based compounds account for 94% of the radiative forcing.

Engineering decisions involve trade-offs. For example, cleanup may require the excavation and incineration of contaminated soil. The complete incineration will result in generating and emitting greenhouse gases. Determining whether this approach is successful and appropriate depends on the extent and quality of options from which these emissions occur. For example, if dioxin-laden soil is incinerated, this may have been the only viable approach to detoxify a very toxic and persistent compound. Releasing CO_2 in this case is truly a measure of success.

Climate change engineering must address chaos. For example, scientific principles can be used to select actions at various points in an event cascade. Steps in the causal chain occur before larger impacts in terms of space and time. Therefore, preventing the impact calls for actions prior to the larger-scale problem's occurrence.

One of the most frustrating aspects of the global climate change debate is the seeming paucity of proposed solutions. Recently, some engineers and scientists have suggested that it would be prudent to consider engaging in planetary-scale interventions, so-called 'geoengineering', to mitigate global climate change and global warming in particular. Geoengineering differs from other methods for mitigating global warming because it involves a deliberate effort to affect the climate at a global scale. Although geoengineering is not a new idea, it has taken on added significance as a result of difficulties with implementing other proposals to mitigate climate change. While proponents of geoengineering admit that these measures can be associated with significant risks to the environment and public health, they maintain that they are well worth pursuing, given the failure of means of mitigating global warming. Conversely, some environmental groups have voiced strong opposition to all forms of geoengineering. In this chapter, the arguments for and against geoengineering are reviewed and some policy options discussed. It is argued that geoengineering research should continue, but that specific proposals should not be implemented until we have a better understanding of the risks, costs and practical and political problems associated with geoengineering.

Carbon sequestration is one type of geoengineering that has received considerable attention. Actually, sequestration is an ongoing, geophysical process. Carbon compounds, especially CO_2 and CH_4, find their way to the ocean, forests and other carbon sinks. Like many geobiochemical processes, sequestration is one that can be influenced by human activity. Thus, there is a conservation aspect to protecting these mechanisms that are working to our benefit.

The second approach is one that is most familiar to the engineer, that is, we can apply scientific principles to enhance sequestration. These sequestration technologies include either new ways to sequester carbon or to enhance or expedite processes that already exist.

Conservation is an example of a more 'passive' approach. There are currently enormous releases of carbon that, if eliminated, would greatly reduce the loading to the troposphere. For example, anything we can do to protect the loss of forest, woodlands, wetlands and other ecosystems is a way of preventing future problems. In fact, much of the terrestrial fluxes and sinks of carbon involves the soil. Keeping the soil in place must be part of the overall global strategy to reduce greenhouse gases.

Anticipating the continued use of fossil fuels, engineers have explored technological methods of capturing the carbon dioxide produced from fuel burning and sequestering it underground [14]. Sequestration is a biosystematics solution since it is an ongoing process on planet Earth, with myriad interactions between biotic and abiotic factors. C-compounds, especially CO_2 and CH_4, find their way to the ocean, forests and other carbon sinks. Like many biogeochemical processes, sequestration can be influenced by human action.

'Active' approaches include the application of technologies to send carbon to the sinks, including deep rock formations and the oceans. Such technology can be applied directly to sources. For example, fires from China's coal mines presently release about 1×10^9 t (1 US billion tonnes) of CO_2 to the atmosphere every year. Estimates put India's coal mine fire releases to be about 50×10^6 t. This accounts for as much as 1% of all carbon greenhouse releases. This is about the same as the CO_2 released by all of the gasoline-fuel automobiles in the United States. Engineering solutions that reduce these emissions would actively improve the net greenhouse gas global flux.

Some coalmine fires have burned for more than a century. Combustion depends on the presence of a fuel, a heat source and oxygen. Unfortunately, since the fire is in an underground vein open to the atmosphere, fuel is plentiful. Actually, the solid-phase coal is less of a factor than the available CH_4, which is ubiquitous in coalmines. This leaves depriving the fire of O_2 as the only reasonable engineering solution to most coal fires. Flooding the mines has been ineffective, since the fire simply finds alternative pathways in the leaky underground strata. Excavation has to be almost 100% to be effective. Flushing with slurries has the same problems. In fact, miner safety and post-ignition fire suppression can be seen as competing factors in mining. To ensure sufficient oxygen levels and low toxic gas concentrations, the mine's ventilation systems require methane-drainage holes to control methane at the face. In many abandoned mines, cross-measure holes were the most common types. These systems are one reason that oxygen remains available to the fire [15]. Recent studies have shown that certain foams can deprive fires of O_2 over extensive areas. For example, a study sanctioned by the US National Institute of Occupational Safety and Health (NIOSH) showed preliminary success in sealing a coalmine from oxygen inflow, and suppression of the fire with liquid nitrogen and gas-enhanced foam [16]. The technology needs to be advanced to address the very large fires.

Another active engineering approach is an enhancement of existing processes. For example, in addition to conserving present levels of carbon sequestration, technologies can be adapted to increase the rates of sequestration. The scale of such technology can range from an individual source, such as a fossil fuel–burning electricity-generation station that returns its stack gases to an underground rock stratum to an extensive system of collection and injection systems to that which includes a whole network of facilities. The combination of disincentives, like carbon taxes, and application of emerging technologies can decrease the carbon flux to the atmosphere. Thus, green engineering is part of the overall, comprehensive geopolitical strategy.

Even the ocean has its limits in greenhouse gas sequestration. Most of the CO_2 generated by human activities (i.e. anthropogenic) resides in the upper layers of the ocean. Carbon compounds move into and out of oceans predominantly as a function of the solubility of the compound and water temperature. For CO_2, this means that more of the compound will remain in the ocean water with decreasing temperature. Ocean mixing is very slow. Thus, the anthropogenic CO_2 from the atmosphere is predominantly confined to the very top layers. Virtually half of the anthropogenic CO_2 taken up by the ocean for the previous two centuries has stayed in the upper 10% of the ocean. The ocean has removed 48% of the CO_2 released to the troposphere from burning fossil fuels and cement manufacturing [17].

Sequestration not only requires moving carbon to a sink but that the carbon should be held for long time periods. Thus, to keep CO_2 sequestered, it must move to the cooler, deeper parts of the ocean. Otherwise, when CO_2 resides near the warmer surface, it is more likely to be released to the atmosphere. The actual mass of C can be increased by management. For example, certain species of plankton are often limited in growth by metals, especially iron. Thus, increasing the iron concentrations in certain ocean layers could dramatically increase the ability of these organisms to take up and store C. In addition, even if the technological means of moving C to the ocean reservoirs is workable, the sequestration may lead to large-scale and virtually irreversible consequences. In particular, there is already troubling evidence of ocean acidification due to increases in atmospheric CO_2 concentrations, notwithstanding the effects of intentional sequestration.

Obviously, any large-scale endeavour like this must be approached with appropriate caution, accounting for all possible outcomes. Such an approach would likely include tests in laboratories, stepped up to prototypes on as many possible scenarios and species possible, before actual implementation.

With these caveats, C sequestration is promising. The Intergovernmental Panel on Climate Change has identified four basic systems for capturing CO_2 from use of fossil fuels and/or biomass processes [18]:

1. Capture from industrial process streams
2. Postcombustion capture
3. Oxy-fuel combustion capture
4. Precombustion capture.

The likely critical paths of these technologies are shown in Figure 5, and general configurations of the technologies are depicted in Figure 6. Thus, at numerous points in the C biogeochemical cycle, there are numerous ways of conserving and adding to natural sequestration processes that could significantly decrease the net greenhouse gas concentrations in the atmosphere.

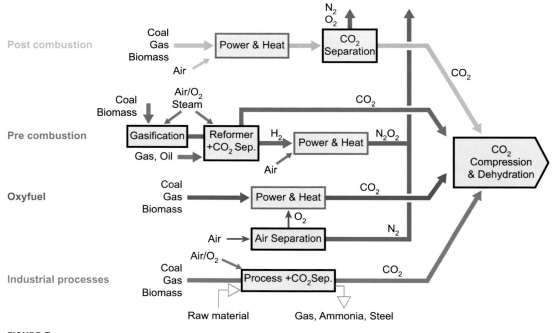

FIGURE 5

Schematic diagram of potential means of capturing carbon dioxide. *(Source: Ref. [18].)*

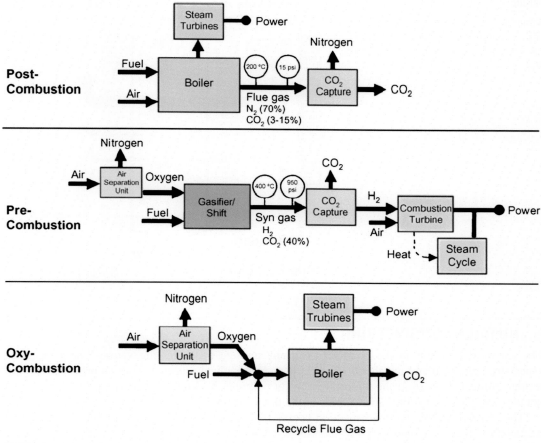

FIGURE 6

Block diagrams illustrating postcombustion, precombustion, and oxy-combustion systems. *(Source: Ref. [19].)*

The sequestration can take place at numerous locations in the C cycle. Organisms have very effective means of sequestering CO_2 during respiration. Technologies can be developed to take advantage of metabolic reactions involving CO_2 in living organisms, particularly enzymatic processes. In one enzyme-based system [19], CO_2 is captured and released by emulating the mammalian respiratory system (see Figure 7). The process employs carbonic anhydrase (CA) contained in a hollow fibre liquid membrane. The CA developing membrane is used to separate CO_2 from the flue gas. Modifying the membrane allows for enhanced diffusion of CO_2 due to the strong interactions between the permeating CO_2 molecules and the amine functional membrane pores should enhance selective diffusion of CO_2. Movement of other gases, such as O_2, N_2 and SO_2, is blocked, which increases selective sequestration of CO_2.

Carbon cycling must be considered with the biogeochemical cycling of other substances. The cycling of C is affected by and affects the types and rates of cycling of many other elements, especially N, S and other nutrients.

FIGURE 7

Schematic of carbozyme permeation process for carbon dioxide sequestration. Here, CA refers to carbonic anhydrase. *(Source: Ref. [19].)*

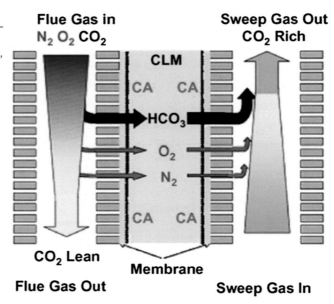

6. NUTRIENT ENGINEERING

The chemical reactions in the N cycle, as in any nutrient cycle, require various energy sources, especially light, heat and metabolic sources. Some biochemical processes with an organism fix molecular nitrogen (N_2) from the atmosphere to form simple N-compounds (e.g. diazotrophs in root nodules), which in turn form amino acids in the tissues of plants and animals.

The mineralization and denitrification occurs by numerous processes, in addition to microbial degradation, including photolysis, hydrolysis and reduction or oxidation. The result is a wide array of conversions of nitrogen-containing organic compounds (e.g. proteins and amino acids) to inorganic (mineral) forms, such as ammonia, ammonium hydroxide, nitrite and nitrate. Note that the gases at the top Figure 8 include those that are important in air pollution. For example, NO is one of the compounds involved in the photochemistry that leads to the formation of the pollutant ozone (O_3) in the troposphere. Note also that macrophytes are central in the figure. Much of the chemistry occurs on the floor in the detritus where microbes degrade complex molecules. Nutrients in the soil are transported by the roots' capillary action to plant cells. Gases are transpired through leaves back to the atmosphere.

The atmospheric speciation at the top of Figure 8 is an oversimplification, with many competing processes. For example, soil NO_2^- compounds can release nitrous acid (HONO) directly to the atmosphere. When soil contains elevated amounts of nitrates due to fertilization, HONO is released to the atmosphere. The more acidic the soil the greater will be the release of HONO. In the troposphere, HONO leads to the formation of hydroxyl radicals, which both degrade and increase the deposition of air pollutants. Large volumes of acids are released from soil continuously to the atmosphere. Soils with high N concentrations form the acids from NO_2^- ions. These anions are first

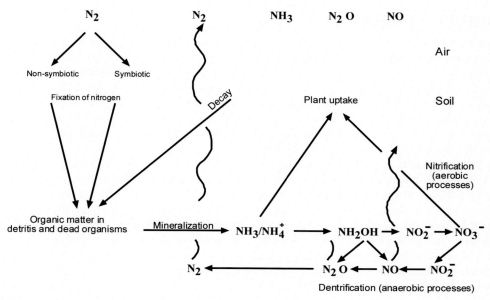

FIGURE 8

Nitrogen cycling in the troposphere. *(Source: D.A. Vallero, Environmental Contaminants: Assessment and Control, Elsevier Academic Press, Burlington, MA, 2004.)*

released into the soil by microbes that have transformed ammonium and nitrate ions into nitrite ions. Increasing soil acidity produces high nitrite concentrations, leading to greater concentrations of HONO emitted to the troposphere [20]. About 30% of the primary OH radical production is attributed to the photolysis of HONO.

An apparent soil-water interaction (fertilizer added to increase crop yield) can actually be a soil-water-biota (microbial) process, followed by an air-water interaction (chemical transformation of air pollutants, followed by precipitation). In this instance, the air pollutants are decreased by increasing concentrations of water pollutants (nitrates and ammonium).

When organic compounds are degraded by microbes, especially nitrifying bacteria, NO_x is released to the atmosphere. Thus, the flux of nitric oxide (NO) and nitrogen dioxide (NO_2) from the soil to the lower troposphere is inversely related to the rate of degradation of organic compounds in the soil.

Ion exchange is actually an example of sorption, that is, movement of a chemical species from the liquid or gas phase to the solid phase.[2] Plants grow as a function of available nutrients and other cycles within the forest ecosystem [21]. With this growth, compounds of N and other nutrients find their way to the atmosphere. The N compounds enter the troposphere by several mechanisms. In addition to the ordinary concentrations of molecular N, various nitrogen compounds are formed from reactions ranging from very fast (especially combustion) to quite slow, multistage (microbial) reaction rates.

---

[2]Movement of a chemical species from solid to liquid phase is dissolution. Movement from the solid phase to gas phase is volatilization.

7. ALBEDO ENGINEERING

In addition to GHG removal and treatment, engineers can take advantage of varying albedo for different substances. Albedo is the fraction of solar energy (i.e. short-wave radiation) reflected from the Earth back into space. It is a measure of the reflectivity of the Earth's surface. The energy not reflected is absorbed and reradiated as heat. Carbon cycling directly affects albedo by affecting the vegetative cover. In effect, the Earth acts as a wave converter, receiving the high-energy, high-frequency radiation from the sun and converting most of it into low-energy, low-frequency heat to be radiated back into space. In this manner, the earth maintains a balance of temperature. Green plants ameliorate the effects of these light-heat exchanges.

To understand albedo-related engineering, let us revisit the processes that determine GHGs. The light energy and the heat energy have to be defined in terms of their radiation patterns. The incoming radiation (light) wavelength has a maximum at around 0.5 nm, and almost all of it is less than 3 nm. The heat energy spectrum, or that energy reflected back into space, has the maximum at about 10 nm and almost all of it is at a wavelength higher than 3 nm. As both the light and heat energy pass through the Earth's atmosphere, they encounter the aerosols and gases surrounding the Earth. These can either allow the energy to pass through, or they can interrupt it by scattering or absorption. If the atoms in the gas molecules vibrate at the same frequency as the light energy, they will absorb the energy and not allow it to pass through. Aerosols will scatter the light and shade the Earth. The incoming radiation is impeded by water vapour and oxygen and ozone, as discussed in the preceding section. Most of the light energy is unimpeded, but the heat energy meets impediments, e.g. water vapour, CO_2. The CH_4, O_3 and N_2O have absorptive wavelengths in the middle of the heat spectrum. Increasing the concentration of any of these will limit heat transfer to space, enhancing the greenhouse effect.

Photophoretic forces occur when there is a temperature differential between an aerosol and the surrounding gas (see Figure 9). One recently proposed approach applies these forces by spraying into the stratosphere disc-shaped, engineered nanoparticles, composed of layers of aluminium oxide, metallic aluminium, and barium titanate. The aluminium layer provides high solar-band reflectivity with high transparency to outgoing thermal infrared radiation, resulting in mass-specific cooling. The Al_2O_3 layer lowers the rate of oxidation of the aluminium surface. The $BaTiO_3$ layer's thickness is determined by the electrostatic torque from the atmospheric electric field so as to orient the disc to optimize levitation, overcoming gravity. The disc is analogous to a colloidal suspension within a mixture of stratospheric gases. Nanoparticles' loss mass and large relative diameters are expected to invoke photophoretic and electromagnetic forces to levitate the discs above the stratosphere.

The relative low reactivity and resistance to oxidation substantially increase the time that such nanoparticles may remain suspended in the atmosphere, compared to the sulphate particles generated from spraying SO_2. Their specific reflective properties could also be controlled through engineering and design, so that they would produce a correct amount of diffuse light. Also, since the nanoparticles would be above the stratosphere, they are less likely to interfere with ozone chemistry and, unlike sulphate, would not produce acid rain.

The efficacy of nanoparticles in reducing atmospheric temperatures is much less certain than that of SO_2. The eruption of Mt Pinatubo indicates the cooling potential of SO_2, but no analogous process has been observed in the troposphere.

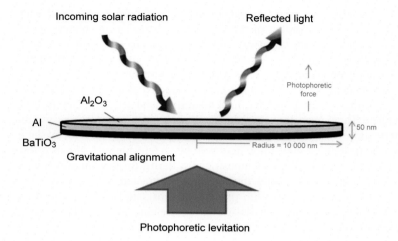

FIGURE 9

Photophoretic levitation using a nanoscale composite disc with an aluminium oxide upper layer barium titanate bottom layer. *(Sources: Ref. [22]; Adapted by Ref. [23].)*

Stabilizing global temperatures at higher CO_2 levels would not address the problem of ocean acidification, and it might impact precipitation patterns, temperature distribution, tropical storms and winds. The risks may be even more uncertain than those for SO_2.

Little is known about the direct and indirect risks from nanoparticles. Indeed, there could be significant environmental and public health risks of spraying nanoparticles into the stratosphere, which are not well understood at this point.

8. ENGINEERING-BASED DECISION-MAKING

To paraphrase Clint Eastwood's character, Harry Callahan, engineers have to know their limitations. Ideally, any decision to implement a large-scale action should be based on a thorough understanding of the benefits and risks. Proposals should only be initiated when there is sufficient evidence that the benefits outweigh the risks, and serious harms can be prevented or avoided. However, there is a substantial lack of understanding of the benefits and risks of most geoengineering approaches, due to their scale and complexity. Such large-scale engineering efforts have no direct precedents and may require a high degree of international cooperation. Large uncertainties about their potential success and potential impacts to human health and the environment may keep such large projects from occurring, because people may not want to take risks that currently are difficult to predict or manage.

Because the benefits and risks of most geoengineering proposals are uncertain at this point, a precautionary approach is warranted. That is, the onus should be that for any action with the potential for severe and irreversible consequences, the engineer must provide adequate evidence of its safety. This evidence may begin with smaller-scale, lower-risk projects preceding increasingly large ones.

Arguably, there is less controversy associated with greenhouse gas removal than with enhancing albedo, due to a better understanding of the benefits and risks of greenhouse gas removal. Indeed, carbon sequestration projects are underway, although these are also controversial as their scale increases, due to questions about the efficiency of storing CO_2 in geological formations and potential escape of gases from deep ocean storage, as well as concerns about costs.

Traditional cost-benefit analyses have not been conducted at the global scale. Even for relatively small-scale projects, these analyses often cannot quantify social scientific factors, such as political and economic variables. Thus, even though reducing greenhouse gas emissions is preferable to either removing them or addressing the climatic changes brought about by their increased concentrations in the troposphere, the benefits of prevention may well be underestimated. Notably, three of the world's largest greenhouse gas emitters have demonstrated a desire to give large weight to economics in climate change decisions, which could translate into protracted time periods during which GHG emissions will continue to increase. With or without substantial reduction in these emissions, the engineering community must continue to design and devise innovative and larger-scale solutions to the direct and indirect problems brought on by climate change.

REFERENCES

[1] J.B. Denton, Sanitary Engineering, E & FN Spon, London, 1877.

[2] National Archives, Records of the Public Health Service [PHS], 1912–1968, 90.8 Records of the Bureau of State Services, 2013, pp. 1948–1963.

[3] U.S. Environmental Protection Agency, Green engineering, USEPA, 2015 [cited 2015]. Available from: http://www.epa.gov/oppt/greenengineering/.

[4] D.B. Resnik, D.A. Vallero, J. Earth Sci. Clim. Change (Suppl. 1:001) (2011).

[5] National Academy of Engineering, Grand Challenges for Engineering: Manage the Nitrogen Cycle, 2009 [cited 2015 January 6, 2015]. Available from: http://www.engineeringchallenges.org/cms/8996/9132.aspx.

[6] R.H. Socolow, Nitrogen management and the future of food: lessons from the management of energy and carbon, Proc. Natl. Acad. Sci. 96 (1999) 6001–6008.

[7] D.A. Vallero, Fundamentals of Air Pollution, fifth ed., Elsevier Academic Press, Waltham, MA, Burlington, Mass, USA, 2014, 999.

[8] D. van de Meent, T. McKone, T. Parkerton, M. Matthies, M. Scheringer, F. Wania, et al., Persistence and Transport Potential of Chemicals in a Multimedia Environment. SETAC Pellston Workshop on Criteria for Persistence and Long-range Transport of Chemicals in the Environment; July 14–19, 1998, Society of Environmental Toxicology and Chemistry, Fairmont Hot Springs, British Columbia, Canada, 1999.

[9] J. Schreiber, The Importance of Carbon for Climate Regulation, 2007 [cited 2014 November 5, 2013]. Available from: http://www.indiana.edu/~geol105b/1425chap8.htm.

[10] National Oceanic and Atmospheric Administration PMEL, Carbon Education Tools, 2013 [cited 2014 November 5, 2013]. Available from: http://www.pmel.noaa.gov/co2/file/Carbon+Cycle+Graphics.

[11] D. Vallero, Environmental Biotechnology: A Biosystems Approach, Academic Press, Burlington, Mass USA, 2010.

[12] W.F. Franek, L. DeRose, in: Agency USEP (Ed.), Principles and Practices of Air Pollution Control, APTI Course 452, USEPA, Research Triangle Park, NC, 2003.

[13] P.A. Vesilind, T.D. DiStefano, Controlling Environmental Pollution: An Introduction to the Technologies, History and Ethics, DEStech Publications, Lancaster, PA USA, 2006.

[14] National Academy of Engineering, Grand Challenges for Engineering: Develop Carbon Sequestration Methods, 2009 [cited 2015 January 6, 2015]. Available from: http://www.engineeringchallenges.org/cms/8996/9077.aspx.

[15] A.C. Smith, Bleederless Ventilation Systems as a Spontaneous Combustion Control Measure in U.S. Coal Mines, United States Department of the Interior, Bureau of Mines, 1994.

[16] M. Trevits, A. Smith, A. Ozment, J. Walsh, M. Thibou, Application of Gas-Enhanced Foam at the Excel No. 3 Mine Fire, 2005. Proc National Coal Show June 7–9, 2005.

[17] National Oceanic and Atmospheric Administration, Study details distribution, impacts of carbon dioxide in the world oceans, NOAA Magazine (2004).

[18] O. Davidson, B. Metz, Special Report on Carbon Dioxide Capture and Storage. International Panel on Climate Change, Switzerland, Geneva, 2005, pp. 1–100.

[19] J.D. Figueroa, T. Fout, S. Plasynski, H. McIlvried, R.D. Srivastava, Int. J. Greenhouse Gas Control 2 (2008) 9–20.

[20] H. Su, Y. Cheng, R. Oswald, T. Behrendt, I. Trebs, F.X. Meixner, M.O. Andreae, P. Cheng, Y. Zhang, U. Pöschl, Science 333 (2011) 1616–1618.

[21] T. Green, G. Brown, L. Bingham, D. Mays, K. Sistani, J. Joslin, et al., Environmental impacts of conversion of cropland to biomass production, in: Bionergy '96, Partnerships to Develop and Apply Biomass Technologies, Nashville, Tennessee, September 15–20, 1996, pp. 918–924.

[22] D.W. Keith, Proc. Natl. Acad. Sci. U.S.A. 107 (2010) 16428–16431.

[23] D.A. Vallero, D.B. Resnik, Geo-engineering: Enhancing Cloud Albedo, McGraw-Hill Yearbook of Science & Technology, New York, 2013.

FURTHER READING

[1] D. Bodansky, May we engineer the climate? Clim. Change 33 (1996) 309–321.

[2] Congressional Budget Office, The Economic Effects of Legislation to Reduce Greenhouse-Gas Emissions, 2009. Available at: http://www.cbo.gov/ftpdocs/105xx/doc10573/09-17-Greenhouse-Gas.pdf (accessed 15.08.11).

[3] P. Crutzen, Albedo enhancement by stratospheric sulphur injections: a contribution to resolve a policy dilemma? Clim. Change 77 (2006) 211–220.

[4] A. Dessler, E. Parson, The Science and Politics of Global Climate Change, Cambridge University Press, Cambridge, 2006.

[5] K. Elliott, Geo-engineering and the precautionary principle, International Journal of Applied Philosophy 24 (2010) 237–253.

[6] EPA, Opportunities to Reduce Greenhouse Gas Emissions through Materials and Land Management Practices, 2009. EPA 530-R-09-017, http://www.epa.gov/oswer/.

[7] EPA, Sulphur Dioxide, 2011. Available at: http://www.epa.gov/oaqps001/sulphurdioxide/ (accessed 17.08.11).

[8] V.J. Fabry, B.A. Seibel, R.A. Feely, J.C. ORR, Impacts of ocean acidification on marine fauna and ecosystem processes, ICES J. Mar. Sci. 65 (2008) 414–432.

[9] Interagency Working Group on Climate Change and Health, A Human Health Perspective on Climate Change, 2010. Available at: http://ehp03.niehs.nih.gov/static/climatechange.action (accessed 12.08.11).

[10] Intergovernmental Panel on Climate Change (IPCC), Climate Change 2007: Synthesis Report, 2007. Available at: http://www.ipcc.ch/publications_and_data/ar4/syr/en/spm.html (accessed 12.08.11).

[11] Intergovernmental Panel on Climate Change (IPCC), Climate Change 2007: Mitigation of Climate Change, 2007. Available at: http://www.ipcc.ch/publications_and_data/ar4/wg3/en/contents.html (accessed 15.08.11).

[12] Intergovernmental Panel on Climate Change (IPCC), Climate Change 2007: Impacts, Adaptation, and Vulnerability, 2007. Available at: http://www.ipcc.ch/publications_and_data/ar4/wg2/en/contents.html (accessed 15.08.11).

[13] D. Jamieson, Intentional climate change, Clim. Change 33 (1996) 323–336.

[14] D. Jamieson, Can space reflectors save us? Why we shouldn't buy into geo-engineering fantasies, Slate (2010). September 23, 2010. Available at: http://www.slate.com/id/2268034/ (accessed 25.07.11).

[15] J. Keihl, Geo-engineering climate change: treating the symptom over the cause? Clim. Change 77 (2006) 227–278.

[16] D.W. Keith, Geo-engineering the climate: history and prospect, Ann. Rev. Energy and Environ. 25 (2000) 245–284.

[17] D.W. Keith, Photophoretic levitation of engineered aerosols for geo-engineering, Proc. Natl. Acad. Sci. 107 (2010) 16428–16431.

[18] D.W. Keith, E. Parson, E.G. Morgan, Research on global sun block needed now, Nature 463 (2010) 426–427.

[19] E. Kintisch, Asilomar 2 takes small steps toward rules for geo-engineering, Science 328 (2010) 22–23.

[20] O. Morton, Why people disagree about geo-engineering, Oliver Morton's Bookblog (2009). October 26, 2009. Available at: http://heliophage.wordpress.com/2009/10/26/why-people-disagree-about-geo-engineering/ (accessed 19.08.11).

[21] National Aeronautics and Space Administration, Global Climate Change, 2010. Available at: http://climate.nasa.gov/ (accessed 12.08.11).

[22] S.H. Schneider, Geo-engineering: could-or-should-we do it, Clim. Change 33 (1996) 291–302.

[23] The Royal Society, Geo-engineering the Climate: Science, Governance, and Uncertainty, The Royal Society, London, 2009.

[24] J. Tolleson, Geo-engineering faces ban, Nature 468 (2010) 13–14.

[25] T.M. Wigley, A combined mitigation/geo-engineering approach to climate stabilization, Science 314 (2006) 452–454.

[26] L.F. Wiley, L.O. Gostin, The international response to climate change: an agenda for global health, J. Am. Med. Assoc. 302 (2009) 1218–1220.

SOCIETAL ADAPTATION TO CLIMATE CHANGE

32

Daniel A. Vallero

Pratt School of Engineering, Duke University, Durham, NC, USA

CHAPTER OUTLINE

1. INTRODUCTION

Homo sapiens is a highly adaptive species. Indeed, one of the features that distinguishes humanity is that humans adapt and control many facets of their environments. The social sciences try to explain how and why humans go about adapting and controlling their environments. The focus ranges from the individual (psychology) to large populations (economics and sociology). Geographers and other spatial scientists differentiate the extent to which humans are influenced by the environment (i.e. environmental determinism) and how the environment is influenced by humans (i.e. anthropogenic factors). Land use planners attempt to structure and control responses to social and physical changes by zoning ordinances, subdivision regulations, open space rules and other land use controls. Engineering is devoted to finding the best means of such adaptation and control at the human scale (see Chapter 31). The biomedical community, especially epidemiologists and other public health experts, must consider necessary steps needed for future disease outbreaks and other shifting and increased stresses (e.g. heat exposure).

Like many other aspects of climate change, discussion of adaptation is considered within the milieu of the possibilities of what may occur under various scenarios. It is folly to project a future without human influence, i.e. a baseline scenario for 2050 without human influence would be ridiculous (not to mention, quite depressing). Thus, the scenarios must consider reasonable possibilities of varying extents of human influence. For example, a straight-line projection of fuel extraction and combustion by the year

Climate Change. http://dx.doi.org/10.1016/B978-0-444-63524-2.00032-4

2050 may well overpredict global greenhouse gas (GHG) emissions, given increased replacement of fossil fuels with alternatives and conservation efforts. Conversely, expecting a 90% reduction in combustion also would likely be unreasonable, given real-world geopolitical and economic scenarios. Thus, most of the discussions to date have addressed plausible ranges of potential emissions.

That society should seek adaptive strategies is not necessarily admitting failure. Indeed, human population migration and settlement have always had an element of adaptation. Nomadic tribes for millennia have remained in regions during feast but moved when famine became intolerable. The transition from hunting and gathering to sustained agriculture-based societies was a type of adaptation, just as modern technological advances have assisted adaptation, as well as precipitated the need for adaptation. An example of the former is the interconnectivity of robotic manufacturing, transportation corridors, express shipping and other advances that have allowed manufacturing in previously inaccessible environments. An example of the latter is the need for high-voltage transmission lines, cell towers and infrastructures that have caused the migration of populations to regions, but also to some extent have driven people away due to concerns about electromagnetic radiation, urban sprawl and crowding, as well as localized air and water pollution. Some of this push-and-pull migration can be attributed to economics, including housing costs that result from the induced value placed by society.

Certainly, the types of adverse outcomes drive the necessary adaptation. Droughts call for agricultural, potable water and food-related actions. Sea level rise calls for plans to address migrations. Shifting biomes call for changes in forestry. Vector-related and other contagious disease movement requires prevention and treatment contingency planning. Changes in micro- and macro-meteorological conditions, e.g. increased or shifting centres for hurricanes, typhoons, flooding, winter storms and tornadoes, call for both better land use planning and enhanced emergency response actions. This is a complicated process. Actions to adapt to one effect could either help or exacerbate other effects. For example, greater use of air conditioning indoors to reduce heat-related stress could mean increases in fossil fuel combustion, migration of populations from coastal areas could increase the need for potable water in areas prone to drought, and applying pesticides to reduce insect and vermin infestations could increase the incidence of chemical exposures and associated health effects.

Most adaptation discussion has focused on how humans will or should respond, but the topic has to a lesser extent considered how natural systems can and will adapt. This ecological adaptation includes whether and how various species will migrate or modify behaviour to survive under different climate change scenarios [1]. The less-resilient species and those whose habitats are more sensitive to changes in temperature, moisture, seasonal fluctuation and other climatic factors will face losses and even extinction. Others may simply follow biome and habitat shifts, whereas some will thrive under the changed conditions. Biological diversity, predator–prey relationships and primary productivity are some of the metrics that will change. Unfortunately, these ecosystem characteristics are influenced by myriad factors and variables, so is very difficult to predict.

2. RISK AND VULNERABILITY

When humans adapt to change, it is often driven by risks. Avoiding and ameliorating risks from disease contagion, pollution, poor nutrition, violence, warfare and other stressors are strong motivations for adaptive behaviour. Complicating this, humans have had to adapt to changes in other species, which are also affected by these stressors. For example, the vulnerability of plant and animal species (e.g. crops, prey, livestock) depends on many factors. For example, the farmers, anglers and hunters may decide whether to intensify, reduce impact on or abandon an area, depending on their perspectives of possible future scenarios [2]. Thus, if climate change induces conditions that are no longer favourable to certain crops, livestock or prey, the land use will change because the farmer, angler and hunter's perception of risk and vulnerability to stressors will have changed.

The Intergovernmental Panel on Climate Change (IPCC) has defined 'vulnerability' as:

> 'the propensity or predisposition to be adversely affected. Vulnerability encompasses a variety of concepts including sensitivity or susceptibility to harm and lack of capacity to cope and adapt' [3].

Vulnerability applies to any system, whether natural or societal. The IPCC has applied the concept to climate change specifically:

> …the extent to which a natural or social system is susceptible to sustaining damage from climate change. Vulnerability is a function of the sensitivity of a system to changes in climate…and the ability to adapt the system to changes in climate…. Under this framework, a highly vulnerable system would be one that is highly sensitive to modest changes in climate, where the sensitivity includes the potential for substantial harmful effects, and one for which the ability to adapt is severely constrained [4].

Thus, vulnerability is inversely related to a system's resilience, discussed below. An aspect of climate change that supports proactive planning and precaution is that at least some of the damage to the environment would be irreversible. Given the time between the release and gradual build-up of GHGs and documentable evidence of damage, the precautionary approach for both would call for an intervention, e.g. lowering GHG emission. However, it would also require thoughtful adaptation strategies matched to contingencies, e.g. shifts in storm tracks, warming and sea level rise.

3. DISEASE OCCURRENCE AND TRANSMISSION

Chief among risks associated with climate change are those associated with the direct and indirect threats to public health. Human diseases can be classified and categorized in a number of ways. Diseases are often distinguished according to the type of causal agent, which could be microbial or otherwise biological (e.g. allergens, such as pollen, cysts, dander etc.), chemical (e.g. pollutants, tobacco smoke etc.), and physical (e.g. electromagnetic radiation). They may also be classified as to the type of exposure (e.g. long-term and chronic versus short-term and acute), or by disease duration (e.g. chronic diseases, such as cancer and cardiovascular effects, versus subchronic or acute diseases, such as influenza or pesticide poisoning).

Large-scale changes in meteorological conditions will affect human health (see Fig. 1). Climate change–induced health effects can be affected directly or indirectly, through multiple causal pathways (see Fig. 2). The frequency and severity of extreme events, such as drought, flooding and storms, will also increase psychological stress and concomitant physiological response, e.g. hospital admissions for cardiovascular and neurological ailments can be indicators of responses to these stresses [5,6]. In addition, climate change–induced loss of wetlands and other natural systems would also lead to diminished water quality, since they filter and otherwise purify water systems. Flooding often leads to increased exposures to water-borne pathogens, in addition to the immediate threats to health and safety, e.g. drowning.

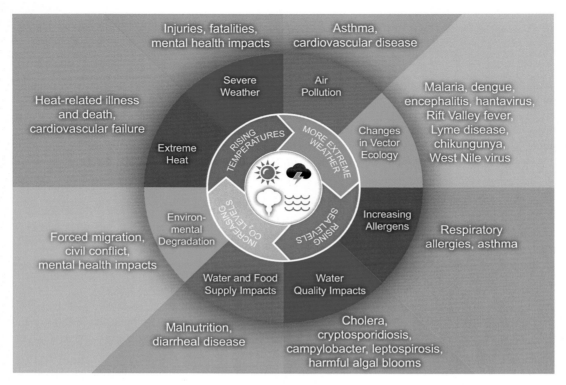

FIGURE 1

Direct and indirect health effects potentially caused by changes in climate. *Source: Ref. [6].*

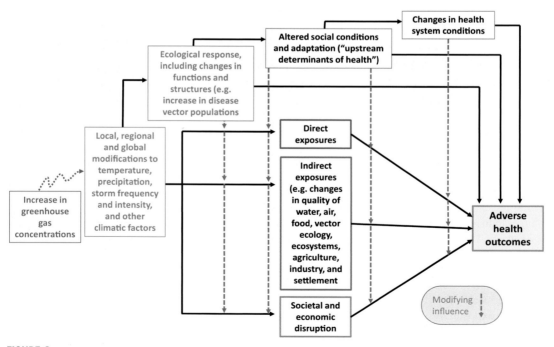

FIGURE 2

Human health causal pathways that may result from global climate change. *Adapted from: Ref. [7].*

Warmer ambient atmospheric temperatures and changes to the hydrological cycle can also lead to changes in ecosystems that adversely affect the health of human populations indirectly. Climatic change can alter seasonality of some allergenic species, and disease vectors would lead to increased incidents of acute and chronic diseases, as would changes that invite insect and other vector migrations, for example, changes in spatial and temporal distributions of malaria, dengue, tick-borne diseases, cholera and some other diarrhoeal diseases [7]. In addition to direct symptoms of these opportunistic diseases, indirect cardiovascular and other effects would likely increase in sensitive populations (e.g. asthmatics, children and elderly).

The build-up of GHGs may create extreme heat events, i.e. those characterized by several days of warm, stable air masses, with daytime temperatures >32°C, and consecutive night-time temperatures above normal. If climate change does indeed increase the number of and severity of heatwaves in the temperate zones, this will be met by greater cardiovascular effects (see Fig. 3).

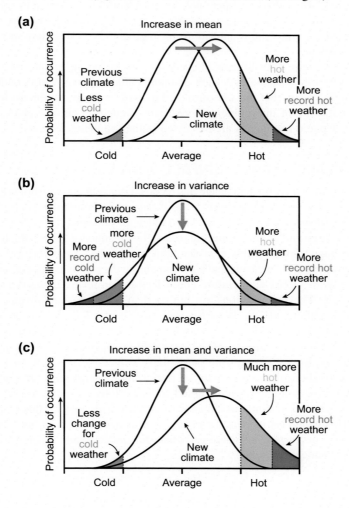

FIGURE 3

Potential shifts in the distribution of cold and hot weather as result of climate change, showing the probabilities of greater hot weather with increases in mean temperature and variance around the mean. *Source: Ref. [8].*

Heatwaves presently are the most deadly weather-related events in the US, leading to a greater number of deaths annually than those attributed to hurricanes, tornadoes, floods and earthquakes combined. Like direct air pollutant exposures, heat-sensitive subpopulations include children and the elderly. Urban settings are generally more susceptible to heat island effects due to locations, e.g. valleys, and albedo of anthropogenic materials (see Chapter 31). Thus, it is likely that climate change will bring more heatwaves to the US. Given the continued migration and demographic trends in the number of people living in cities, as well as population ageing, will further increase heat-related health risks. Studies suggest that, if current emissions hold steady, excess heat-related deaths in the US may increase from the present average of about 700 each year to as many as 5000 per year by 2050 [8].

The indirect health effects demonstrate the complexity of causal links between emissions of air pollutants and health and ecosystem wellbeing. This applies to acute and chronic adverse health, as well as to the condition of ecosystems. For example, ground-level ozone concentrations are likely to increase with increasing atmospheric temperature, so cities would be expected to experience a greater number of unhealthful high-ozone events (i.e. $>80 \times 10^{-9}$) [9]. Other pollutants would also likely increase under changes in temperature and moisture regimes, e.g. particulate matter; however, the predictions are less certain.

As in any complex system, the wellbeing of every species, including humans, is inextricably linked to the wellbeing of the whole system. As such, human health and ecosystem condition are actually components of a single system that must be sustained to avoid widespread threats to a livable planet. Keeping this in mind, some of the problems potentially elicited by climate change (inner circle of Fig. 1) and related adaptation strategies are discussed in the following sections.

4. OCEAN AND LARGE-SCALE SURFACE WATER CHANGES

Rising sea level associated with climate change has received much attention from scientists, journalists and the public at large. The sea level has been rising at rates of 3–6 mm m per year throughout the last century [10]. The rate of sea level rise has increased every decade since 1950. Scientific studies have linked this increase in large part to anthropogenic forcing by greenhouse gases, which is responsible for rising global temperatures. Some of this evidence comes from the strong correlation between atmospheric CO_2 concentrations and sea levels that occurred during the past three glacial–interglacial cycles.

Predicted increases in global temperature could lead to sea levels rising 0.5 m–1.4 m over the next 100 years [11,12]. These predictions are higher than previous ones given a more rapid rate of melting of both Greenland and the West Antarctic ice sheets. This rise in sea level would have serious consequences for people living in low-lying parts of the world. For example, Fig. 4 shows the extent of inundation around Washington, DC, that would result from a 1-m rise in sea level [13].

Sensitive coastal and littoral systems are particularly vulnerable to sea level changes. Problems associated with climate change are multifactorial. For example, coral reefs are not only particularly vulnerable to sea level rise resulting from increased ambient global temperature and consequent polar melting but also to ocean acidification associated with increasing global CO_2 flux from the troposphere to surface waters. This generates increased carbonic acid concentrations in ocean water, which dissociates to bicarbonate, decreasing available carbonate for the coral and other species. This leads to increases in normal chemical reactions:

$$CO_2 + H_2O \rightarrow H_2CO_3^* \quad \text{Carbonic Acid} \tag{1}$$

$$H_2CO_3 \leftrightarrow H^+ + HCO_3^- \quad \text{Bicarbonate ion} \tag{2}$$

$$HCO_3^- \leftrightarrow H^+ + CO_3^{2-} \quad \text{Carbonate ion} \tag{3}$$

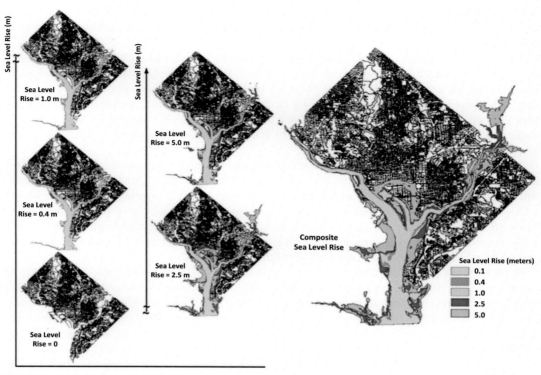

FIGURE 4

Prediction and impact of sea level rise on properties and infrastructure of Washington, DC. *Adapted from: Ref. [13].*

These are part of the normal carbon geochemical cycles, but increasing the amount of CO_2 drives this process toward more acidity. Ninety percent of all inorganic carbon in the ocean is in the form of bicarbonate (HCO_3^-), with 9% as the carbonate ion (CO_3^{2-}) and the last 1% as dissolved CO_2. With increasing atmospheric CO_2, more CO_2 is absorbed into the ocean, driving the reaction toward CO_3^{2-}. The carbonate ion reacts again with large concentration of hydrogen ions, reverting to HCO_3^-. The increase in CO_2 also over time increases the number of H^+ ions, i.e. decreases the pH. The ocean absorbs about one-third of the anthropogenic CO_2, meaning that the amount of ions has increased, so that the pH of the ocean has decreased by about 0.1. Since pH is the negative log of the H^+, this represents a 25% increase in protons. Some estimates expect ocean pH to continue to decrease to 0.3–0.5 by the year 2100 [5].

The rise in sea level is perplexing from both scientific and policy-setting perspectives. The science is complex because sea level rises due to changes in the hydrological cycle. The amount of water on Earth does not vary, but it changes in form and location. Water is stored in the large and small reservoirs, most being in liquid phase in the oceans and seas; other large reservoirs include freshwater systems (lakes, rivers and other surface waters), underground (aquifers), in ice (ice caps and glaciers) and the atmosphere. Changes to the hydrological cycle involve many variables. Predicting the effects of a change, such as melting of ice caps, is not a simple calculation. There are numerous contingencies beyond direct steps from increasing atmospheric temperature to increased melting of polar ice to a rise in the global sea level. Among the variables to consider are regional differences.

The global atmosphere is not a homogeneous, completely mixed thermal system. Indeed, heat is transferred constantly and varies in time and space. For example, even if mean global temperatures rises, it does not necessarily mean that polar temperatures will rise accordingly. They may rise more or less than the global mean, depending on movements of air masses, changes in albedo and other climatological factors. Similarly, the past may not be a good indicator of future climatic conditions, so straight-line projections will not be accurate if the conditions have changed in recent decades.

The policy-setting standpoint may be more complex and frustrating than the science. Policymakers must often make decisions with incomplete information. The first challenge is to decide whether to believe the models on what will occur several decades hence. Does the seemingly inevitable increase in CO_2 atmospheric concentrations necessarily mean a concomitant increase in global temperature? Will other factors also come into play, such as increased vegetative cover, which could attenuate some of the temperature rise? And, even if the sea level were to increase substantially, are there measures that can be taken to adapt to these changes over time [14]?

North Carolina, USA, has almost 6000 km^2 of land below 1-m elevation, more than 500 km of beaches and a shoreline of more than 7000 km, making it particularly vulnerable to sea level rise. Thus, North Carolina has been at the epicentre of sea level–rise debates. Some argue that the scientific consensus about global warming is a logical extension to increase sea level, while others argue that such warming is not occurring. Between these two extreme viewpoints, others argue that warming may occur but that this would not lead to substantial sea level rise beyond past trends. Still others argue that even if we were to see a 1-m rise, humans have a knack for adapting to such changes, so being overly cautious would devastate local economies unnecessarily by placing restrictions on growth and requiring burdensome zoning and land use planning.

Debates on the amount of sea level rise, acidification rates and other ocean impact from climate change are ongoing within the scientific community, as well as between coastal planning authorities, port and transportation authorities, other governmental bodies, environmental groups, commercial enterprises and others. Unanimity will likely not be reached, nor even a dominant consensus on the appropriate adaptation approaches. The costs to public safety, economics and other social values could be quite large. Even well-documented problems are presently ignored or tolerated, such as continuing to rebuild in hurricane-prone areas. Less obvious and incremental problems like sea level rise are even more difficult to plan for, but if not addressed could lead to human suffering, legal liabilities and monetary costs well beyond any immediate benefits, e.g. continued development.

Even if actions are taken, such as strong land use restriction, other environmental costs from sea level rise may be inevitable, such as the increase in salination levels of groundwater from saltwater intrusion. The increase in salinity threatens drinking water quality of many local coastal communities throughout the world, including the northeastern seaboard of the US. This would be disastrous for many coastal communities, especially in rural and unincorporated areas, where well water is the exclusive source of drinking water [14].

Adaptation for sea level change and other climate-induced problems, therefore, must be based on a reliable knowledge base. The data and information must support surveillance systems and predictive models, as well as serve as the basis for appropriate adaptations. The knowledge base must also support interventions to address problems. For example, one approach to addressing sea level rise would be to properly plan and build infrastructures to support migrations. However, the plans should also include measures to prevent and address problems beyond simple abandonment and migration, e.g. changes to shipping, adaptive port facilities, and large-scale engineering and land use planning[1], such as building of water-based settlements (e.g. floating structures). Likewise for droughts, migration would likely occur, but efforts should also be made to address the droughts themselves, including innovative water storage and conservation systems, as well as use of xerophilic crops, where appropriate. In this sense, the adaptation for sea level rise would also ameliorate some of the problems resulting from shifting biomes, which are discussed next.

5. SHIFTING BIOMES AND RESILIENCE

In addition to human population problems, potential climate change scenarios would be expected be accompanied by extensive and severe ecosystem adverse effects (the upper right-hand portion of Fig. 5), such as loss of diversity and abundance of species, as well as changes in primary productivity. Habitats vary with respect to vulnerability and irreversibility [15]. Thus, systems that lack resilience are particularly vulnerable to irreversible damage, e.g. habitats in the lower left-hand portion of Fig. 5 would more likely endure irreversible harm from exposure to climate change that damages habitat (e.g. sea level rise and acid rain).

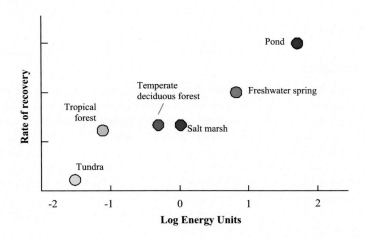

FIGURE 5

System resilience index calculated from bioenergetics for six community types. Rate of recovery units are arbitrary; energy units = energy input per unit standing vegetation. *Sources: Ref. [16]; data from: Refs [15,17].*

---

[1]However, in this instance, the structures would not actually be on land but on water.

Loss of resilience can occur quite gradually. It can take decades for disturbances that gradually contribute to a 'flip' from one equilibrium state to another [18]. Thus, a system that is strongly buffered against such a shift is more resilient than one that is sensitive.

This can be likened to Garrett Hardin's 'Tragedy of the Commons' [19] wherein Hardin hypothesizes a village common where every farmer in the village lets one cow graze. This stress on the system is tolerable and sustainable. However, the villagers perceive a greater value and return on investment by each incrementally adding a cow. The individual farmer gains all of the benefit but shares the cost with the entire village. Thus, the farmer has little incentive to consider the entire system and prefers to focus on immediate and personal gain. Others cannot help but notice the farmer's success and correspondingly add cows. They all benefit, until carrying capacity is exceeded and the system crashes. With the crash, every village, including the farmers, loses the resource, not simply the farmers participating in the operation. Thus, even systems with large buffering and carrying capacities can be overtaxed to the point of catastrophic failure. Thus, an adaptive strategy should include a means to monitor and detect changes that are nearly imperceptible as the climate changes.

Other ecological damage that could increase as a result of climate change includes forest fires and wildfires, pest infestation and plant diseases. Adaptation to these changes can include management of forest and other ecosystems to instil greater resistance to the threats. This may include early detection and treatment of invading species at targeted migration points, tactically locating fuel breaks around vulnerable areas, removal of invading agents, introducing barriers (including introduced protective areas of resistant plant species), pheromone applications and integrated pest management strategies. These efforts are an anthropogenic attempt to instil greater resilience to vulnerable ecosystems [20].

6. EXTREME EVENTS

Most extreme so-called 'natural' events are not completely natural, at least from the perspective of the damage they cause. Human decisions and activities worsen extreme events in at least two ways. First, human activities contribute to increases in their occurrence and/or severity, such as climate change. Second, human decisions lead to greater numbers of people who will experience the harm caused by the event. Recent examples include the completely natural occurrence of the earthquake and tsunami in the Sea of Japan. The quake resulted from Japan's tectonic situation, a completely nonhuman cause. The main island, Honshu Island, is located where the Eurasian, Pacific and Philippine Sea tectonic plates meet and push against each other. The tsunami that occurred was also to be expected, since the first two laws of motion dictate that the release of this seismic energy had to be displaced. The wave is merely the result of energy transport via the waves.

The decision to cite a nuclear power plant within a high hazard zone, the lack of preparedness for cooling of fuel rods, and the weaknesses in evacuation planning and other planning and engineering failures led to the human and ecological disasters that will continue for decades, at least in terms of unacceptable levels of radioisotopes and other contamination. Certainly the natural events triggered and exacerbated the problems, but the problem was worsened by human decisions.

Of all the climate change problems, perhaps the increase and/or relocation of sites of extreme events may be the most amenable to existing adaptive strategies. That is, such events are known to be unpredictable in terms of when and where they will occur, and the extent and severity of damage they may cause. Thus, scientists and the lay public alike have a healthy appreciation for the importance of preparing for such events.

Extreme events include several of the problems already discussed, including heatwaves and wildfires. They also include storms, especially hurricanes, and drought. Although it is well known that the exact location and time of most extreme events are impossible to predict precisely, projecting even the expected shape and size of the distribution of hurricanes, droughts, fires and other extreme events is fraught with uncertainty.

The hydrological cycle is at the root of most climate-induced extreme events. The extremes occur at each end of the water cycle. Excess water in the wrong place at the wrong time can be highly destructive, but not enough water in the wrong place at the wrong time is also problematic. Actually, one of the worst disasters experienced in the US was the Dust Bowl in the 1930s, which affected large portions of the country including western Kansas, eastern Colorado to New Mexico, Texas and Oklahoma. Again, the Dust Bowl resulted from a combination of natural and human-caused factors. Certainly the lack of rain led to drought conditions for a decade, which in turn contributed to erosion and loss of top soil. However, agricultural practises were an important contributor as well. Farmers had supplanted drought-resistant native prairie grasses with drought-sensitive crops like corn and beans. Thus, when the crops died the top soil was easily eroded and lost. Thus, the Dust Bowl's dust that caused respiratory problems and 'black blizzards' should provide a valuable lesson to those deciding on adaptive strategies.

Microbial populations are particularly vulnerable to extreme events. Flooding, scouring and other physical changes can change these populations and, in turn, upset entire habitats. For example, bacteria and fungi require conditions within tight ranges of oxidation. Scouring of sediments may expose anaerobic bacteria to molecular oxygen, which eliminates them and completely alters the microbial ecology. Conversely, aerobic bacteria may be submerged and exposed to anoxic conditions. Either condition can impair ecological processes, e.g. decomposition, which can lead to loss of diversity and sustainability of the ecosystem [21].

Extreme weather event prediction becomes more precise and accurate as the event unfolds. A hurricane forming in the Caribbean Sea may or may not reach landfall on continental North America, but as the storm moves to the north, predictive models become more reliable. Local meteorology is much more amenable to predictive modelling than global climate. Thus, extreme event adaptation begins with statistical inference and parameter selection. That is, an event's rate of occurrence may become more common or less common, depending on numerous variables. A change to the distribution of an extreme event could lead to a shift in central tendencies and/or the one or two of the tails of the distribution. Selecting and weighting a predictive model's parameters determines the shape and size of the distribution.

From these inferences, adaptation strategies would likely include buffer zones and other means of separating at-risk populations from likely landfall locations, flood-prone areas, fire zones and other areas where human habitation is deemed vulnerable. Unfortunately, programmes often do just the opposite. For example, people living in such areas may receive government-subsidized flood insurance for homes that have a high likelihood of being flooded, at least compared to similar inland developments [22]. In this case, one societal value (keeping one's home) militates against another (public safety). The aftermath of Hurricane Katrina in southern Louisiana, for example, should have been an incentive to keep large numbers of population from returning to the high-risk area. However, government programmes stepped up efforts to rebuild in the same areas, albeit with improved infrastructure, levees and flood controls. These may have reduced the risk, but certainly have not eliminated it.

Improper land use planning and zoning will exacerbate climate-induced increases in extreme events in vulnerable areas. Beyond inducements to leave these areas, populations in near-coastal areas have been increasing. Since this trend is likely to continue, any increase in the occurrence of extreme events could be met with accelerated loss of life and property [22].

7. FOOD AND WATER SUPPLY

The food supply is affected by small changes in weather for each growing season and has a very small range of optimal conditions. Thus, any long-term trends affecting the hydrological cycle could be devastating for food crops. The supply is not only dependent on caloric quantities but on nutritional completeness. That is, total volume of food supply could remain constant but would still be unacceptable if the climate change results in a new mix of available crops that would miss key protein, vitamin and other nutritional needs.

The supplied food must be acceptable to society. Some adaptation recommendations have included the creation of hardier and more robust crops [1]. However, for this to be attained at a large scale, it would likely involve genetic enhancements. Genetically modified organisms (GMOs) have been perceived by many to be unacceptable from various perspectives, including health, nutrition and environment.

Besides GMOs in the food, agricultural biotechnology can introduce other insults, such as the use of genetically modified microbes and their release into the environment. Researchers using genetically modified bacteria and other microbes must ensure proper measures for containment are in place. However, similar genetically modified microbes are used for other purposes, such as those for environmental cleanup (e.g. oil spills). Figure 6 shows that an event leading to an environmental disaster is actually quite similar to one resulting in successful environmental cleanup. Thus, it is not the act of release (escape, if it is unintended) that leads to a deleterious or beneficial outcome but all of the steps and contingencies within the system. Indeed, even this scenario can affect the quality of the food supply, e.g. when the genetically modified microbes or their genetic material drift into standing agricultural crops. They may also affect health via the water supply, e.g. if the microbial release is a virulent form of a bacterium, which not only causes a health effect for those who consume contaminated drinking water but which may lead to cross-resistant pathogens, so that existing antibiotics become less effective in treating numerous other diseases. Further, if the next generations of GMOs are not completely sterile, reproduce and become part of the formerly exclusively natural microbial species, the traits of the population may be altered in unknown ways [16].

Thus, if GMOs are to be part of the food supply adaptation, precautions must be followed. Unintended consequences are likely:

> Inserting genes is similar to ecological practises that we thought we understood well, but which held unexpected consequences, such as introducing industrial chemicals to the environment (consider DDT, PCBs), or such as introducing alien species (consider Purple Loosestrife, Kudzu, European Starling).... Regardless of our fundamental ignorance of the genetic mechanisms mentioned above, no one has properly studied the ecological and health ramifications of releasing so many GMOs into farms and grocery stores [23].

(a)

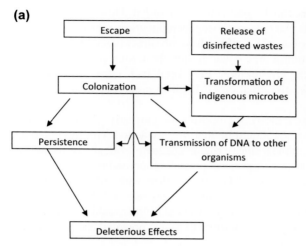

(b)

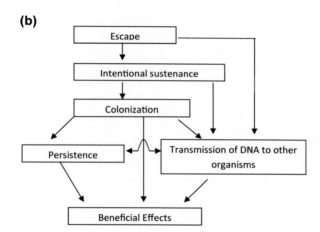

FIGURE 6

Microbial disaster scenario (a) versus a bioremediation success scenario (b). In both scenarios, a microbial population is released into the environment. The principal difference is that the effects in (a) are unanticipated and deleterious, whereas the effects in (b) are desired and beneficial. *Source: Ref. [16].*

Climate change can also introduce indirect risks to food supply, such as changes in microbial growth and release of mycotoxins [24,25].

8. CONCLUSIONS

To paraphrase Clint Eastwood's character, Harry Callahan, engineers have to know their limitations. Ideally, any decision to implement a large-scale action should be based on a thorough understanding of the benefits and risks. Proposals should only be initiated when there is sufficient evidence that the benefits outweigh the risks and serious harms can be prevented or avoided. However, there is a substantial lack of understanding of the benefits and risks of most geoengineering approaches, due to their scale and complexity. Such large-scale engineering efforts have no direct precedents and may require a high degree of international cooperation. Large uncertainties about their potential success and potential impacts to human health and the environment may keep such large projects from occurring because people may not want to take risks that are difficult to predict or manage at this point.

Because the benefits and risks of most geoengineering proposals are currently uncertain, a precautionary approach is warranted. That is, the onus should be that for any action with the potential for severe and irreversible consequences, the engineer must provide adequate evidence of its safety. This evidence may begin with smaller-scale, lower-risk projects preceding increasingly large ones.

Arguably, there is less controversy associated with GHG removal than with enhancing albedo, due to a better understanding of the benefits and risks of greenhouse gas removal. Indeed, carbon sequestration projects are underway, although these are also controversial as their scale increases, due to questions about the efficiency of storing CO_2 in geological formations and potential escape of gases from deep ocean storage, as well as concerns about costs.

Traditional cost-benefit analyses have not been conducted at the global scale. Even for relatively small-scale projects, these analyses often cannot quantify social scientific factors, such as political and economic variables. Thus, even though reducing greenhouse gas emissions is preferable to either removing them or addressing the climatic changes brought about by their increased concentrations in the troposphere, the benefits of prevention may well be underestimated. Notably, three of the world's largest greenhouse gas emitters have demonstrated a desire to give large weight to economics in climate change decisions, which could translate into protracted time periods during which GHG emissions will continue to increase. With or without substantial reduction in these emissions, the engineering community must continue to design and devise innovative and larger-scale solutions to the direct and indirect problems brought on by climate change.

REFERENCES

[1] U.S. Environmental Protection Agency, Adaptation Overview, EPA, 2015. Available from: http://www.epa.gov/climatechange/impacts-adaptation/adapt-overview.html (cited 09.01.15.).

[2] P. Berry, M. Rounsevell, P. Harrison, E. Audsley, Environ. Sci. Policy 9 (2006) 189–204.

[3] Intergovernmental Panel on Climate Change, Definition of Terms Used within the DDC Pages, 2015. Available from: http://www.ipcc-data.org/guidelines/pages/glossary/glossary_uv.html (cited 12.01.15.).

[4] Intergovernmental Panel on Climate Change, Climate change 2014: impacts, adaptation, and vulnerability. Part A: global and sectoral aspects. Contribution of working group II to the fifth assessment report of the intergovernmental panel on climate change, in: C.B. Field, V.R. Barros, D.J. Dokken, K.J. Mach, M.D. Mastrandrea, T.E. Bilir, et al. (Eds.), Fifth Assessment Report of the Intergovernmental Panel on Climate Change, Cambridge University Press, Cambridge, United Kingdom and New York, NY, USA, 2014, pp. 1132.

[5] D.A. Vallero, Fundamentals of Air Pollution, fifth ed., Elsevier Academic Press, Waltham, MA, 2014, p. 999.

[6] Centers for Disease Control and Prevention, Climate Effects on Health, CDC, 2014. Available from: http://www.cdc.gov/climateandhealth/effects/ (cited 13.01.15.).

[7] Intergovernmental Panel on Climate Change. Climate Change, Working group II: impacts, adaptation and vulnerability; current sensitivity and vulnerability, in: IPCC Fourth Assessment Report: Climate Change 2007, 2007.

[8] Centers for Disease Control and Prevention, Heat Waves, 2013. Available from: http://www.cdc.gov/climateandhealth/effects/heat.htm (cited 16.10.15.).

[9] D.J. Jacob, D.A. Winner, Atmos. Environ. 43 (2009) 51–63.

[10] M. Ablain, A. Cazenave, G. Valladeau, S. Guinehut, Ocean Sci. 5 (2009) 193–201.

[11] A. Grinsted, J.C. Moore, S. Jevrejeva, Clim. Dyn. 34 (2010) 461–472.

[12] S. Rahmstorf, Science 315 (2007) 368–370.

[13] B.M. Ayyub, H.G. Braileanu, N. Qureshi, Risk Anal. 32 (2012) 1901–1918.

[14] D.A. Vallero, T.M. Letcher, Unraveling Environmental Disasters, Elsevier Boston, 2013, p. 500.

[15] R. O'Neill, Ecology 57 (1976) 1244–1253.

[16] D. Vallero, Environmental Biotechnology: A Biosystems Approach, Academic Press, Burlington MA, 2010, p. 750.

[17] M. Begon, C.R. Townsend, J.L. Harper, Ecology: Individuals, Populations and Communities, Blackwell Publishing, Oxford, 1996, p. 720.

[18] F. Berkes, C. Folke, Linking Social and Ecological Systems: Management Practices and Social Mechanisms for Building Resilience, Cambridge University Press, New York, 2000.

[19] G. Hardin, Science 162 (1968) 1243–1248.

[20] C.I. Millar, N.L. Stephenson, S.L. Stephens, Ecol. Appl. 17 (2007) 2145–2151.

[21] J. Iliopoulou-Georgudaki, C. Theodoropoulos, D. Venieri, M. Lagkadinou, Int. J. Biol. Life Sci. 5 (2009) 1–8.

[22] C.M. McMillan, Hous. L. Rev. 44 (2007) 471.

[23] C. Picone, D. Anderson, G. Thomas, D. Griffith, Say No to GMOs! (Genetically Modified Organisms), Agenda, , University of Michigan Chapter to the New World Agriculture and Ecology Group, 1999.

[24] M. Miraglia, H. Marvin, G. Kleter, P. Battilani, C. Brera, E. Coni, et al., Food Chem. Toxicol. 47 (2009) 1009–1021.

[25] N. Magan, A. Medina, D. Aldred, Plant Pathol. 60 (2011) 150–163.

CLIMATE IMPACTS AND ADAPTATIONS IN FOREST MANAGEMENT

33

Anna Lawrence, Bruce Nicoll

Forest Research, Midlothian, UK

CHAPTER OUTLINE

1. INTRODUCTION

Forest ecosystems and the communities and industries that depend on them are vulnerable to changes in the climate worldwide. Forests matter to humans in many ways. Climate change may affect biodiversity, recreation and cultural significance, in addition to the production of timber and other materials. But the expected impacts of climate change on forests and the ecosystem services that they provide are influenced by many sources of uncertainty, both those inherent in projected future climates and in the relatively long timescales involved in forest growth cycles. Furthermore, the responses of forests to climate change are mediated by human interventions, which are in turn influenced by perceptions and experiences of the forest and climate.

Forests are a crucial part of the Earth's carbon stocks and flow, but their consideration in global climate policy is relatively recent. They were included for the first time in the fourth Assessment Report of the Intergovernmental Panel on Climate Change, published in 2007. This provided authoritative evidence of how planting and managing woodlands, avoiding deforestation and replacing fossil fuels and carbon-intensive products with wood can make a major contribution to mitigating the effects of climate change. It also examined the impacts of climate change on forests, and the importance of adaptation to make forest ecosystems more resilient [1]. However, forest managers and researchers had given attention to these issues for many years before that.

Climate Change. http://dx.doi.org/10.1016/B978-0-444-63524-2.00033-6

Dual responses to climate change focus on approaches that mitigate climate change by enhancing forest carbon stores (often through motivating actors to invest in additional tree planting) and support adaptation by maintaining compositionally and structurally complex forests [2]. An international expert panel established in 2007, under the International Union of Forest Research Organisations, undertook an assessment of the impacts of climate change on forests and forest-dependent people as well as on management and policy options for effective adaptation to climate change. They highlighted 'severe limitations in current knowledge', especially on the forest-related social and economic impacts of climate change [3]. Politically and practically, mitigation has continued to receive more attention than adaptation; the latter is challenging because it is characterized by complexity and uncertainty [4], and by difficulties in quantifying the processes and outcomes that make activities difficult to incentivize politically [5].

In this chapter we outline this complexity and the uncertainty in forest adaptation, discuss the implications for management, and compare that with what is happening in practice. Our main focus is on Europe, and within that we focus on recent empirical research in the UK to illustrate the challenges associated with adaptation, action to maintain flexibility of management, and options for future change.

2. HOW CLIMATE IS CHANGING AND HOW THAT AFFECTS FORESTS

In most parts of the world, projected increases in air temperature and changes in precipitation type (rain, snow etc.), timing and amount, are expected to have increasing impacts on how forests will survive and grow this century, and on the composition of tree species and other biodiversity growing in those forests [6]. At the same time, extreme events (including droughts, floods, windstorms and fire) are expected to be more severe and more frequent than in the past century [7]. Impacts of the mean changes in climate may include reduced growth, reproduction and survival, although there is debate about this, and some models predict a 'carbon fertilization' effect, i.e. that higher levels of CO_2 will contribute to faster tree growth [8]. It is the more frequent extreme events that are expected to cause widespread disturbance and forest decline [9]. Although it is impossible to attribute any one extreme event to climate change, the observed trends of increasing frequency and severity of disturbance can be separated between climate change origins and other long-term changes such as forest composition and age structure [10]. Forest disturbance through such agents is a natural part of forest ecology, with forest composition continually being in a state of change. But the severity and increasing occurrence of extreme events put new pressures on forest ecosystems that can lead to their decline or collapse.

Each tree species can be mapped onto a bioclimatic zone that it is potentially capable of inhabiting. As the climate changes, this suitable zone will commonly move and often shrink [8]. For example, as the climate warms, the suitable zones for many tree species will move to more northerly latitudes or higher altitudes, and these may be colonized only if species can disperse fast enough or if they are moved by management intervention. However, physical boundaries such as coastlines and mountaintops constrain the edge of these new zones and potentially limit the ability of species to survive by moving location [11]. Adaptation through natural selection, if allowed or encouraged to progress, may allow species to maintain populations in locations despite a changing climate [12]. However, this relies on natural regeneration or managed local regeneration that may not always be possible. Although such regeneration may be effective in providing ecosystem stability in the face of some climatic change or developing threats, it is unlikely to provide resilience to the more extreme events and threats. Where adaptation through natural selection is constrained and species or ecosystem movement is limited either physically or by human inability or reluctance to be adaptive, the ecosystem services provided by the forest will decline over time [11].

3. MODELLING AND UNCERTAINTY

Modelling is an important part of understanding forests and adaption, both of future climate, and of species response to climate. Forest trees must usually be grown and managed for several to many decades, and are therefore expected to live through considerable climate changes from germination or planting through to harvesting. Over the same time, they may need to endure a range of extreme events that might include increasingly severe and frequent droughts, fires, windstorms, and pest and disease outbreaks. So, to manage and adapt their forests, foresters need to have a picture of how the climate will change over the next century or more, and what extremes the forest may be exposed to during this time.

Projections of future climates are based on models using a range of different trajectories of future greenhouse gas (GHG) emissions, but there is considerable uncertainty in how GHG emissions will change over the next century as societies and technology develop [7]. There is further uncertainty in model outputs due to differences in how the various models deal with the physical responses of the climate system to the projected increases in greenhouse gases [13]. Therefore, providing foresters with a single climate trajectory will not give them a realistic picture of future climate in their location and the associated uncertainty that they would need in making appropriate decisions. Multiple trajectories, or at least a trajectory with associated probability shown around it, are needed, regardless of the difficulties that these present in terms of communicating climate change and its impacts [15]. For these reasons, climate projections can be developed based on different emissions scenarios and multiple runs of a range of models, and presented with associated probabilities [14].

Models of tree suitability in relation to climate and site conditions are used to map species distributions and how they will be expected to change with changing climate [8]. These models provide useful indications of how rapidly species would need to move to track their climate space, and what species and provenances would be most suitable at a location with future climates [16]. Decision support tools are now being made available to help forest planners choose appropriate species for future forests.

4. MITIGATION AND ADAPTATION

Foresters and land managers have a range of adaptation measures available to them to help improve the resilience of forests in the face of future climates and the associated increasing frequency of extreme events [6]. These measures are sometimes applied with a concurrent objective of increasing the carbon sequestered in forests, to partially mitigate the effects of climate change. Some studies question the compatibility of the two approaches. For example, in an experimental study researchers [2,17] found trade-offs between the achievement of mitigation and adaptation objectives. Higher stocking density (so, increased carbon stores) is associated with lower stand-level structural and compositional complexity (and therefore lower adaptation potential). However, as new forests are planted either for increased sequestration or more conventional timber production or conservation purposes, and existing forests are harvested and replaced, there are opportunities to change both species and silviculture systems, to manipulate diversity of habitat, stand structure and species.

5. RESILIENT FOREST ECOSYSTEMS AND MANAGEMENT

The word *resilience* has become central to discussions of adaptation, albeit with a wide range of interpretations. The definition from the Stockholm Resilience Centre is widely referred to: 'The capacity of a system to deal with change and continue to develop; withstanding shocks and disturbances (such as climate change or financial crises) and using such events to catalyse renewal and innovation' [18]. In fact climate change is a relatively new context for discussion of resilience: much of the resilience debate in forestry has come out of concerns about tree health [19,20].

Several themes repeatedly characterize interpretations of resilience including stand-level spatial patterns, species and structural diversity [21,22]. Forests with these characteristics appear to be more sustainable in the face of climate change through wider diversity, which buffers them against climatic extremes and spreads the risks, so that even if individual components are lost the forest as a whole may adapt and recover [23].

Many discussions of resilience focus only on the natural ecosystem, but humans are part of the forest ecosystem wherever they are engaged in managing it, and it is important to take into account the social dimensions of resilience. Reviews from within the social sciences emphasize the interdependency of social and ecological systems and the value of social learning, social capital (particularly networks and partnerships) and adaptive management [24,25].

6. THE SILVICULTURE OF RESILIENCE

Silvicultural recommendations for more resilient and adapted forests focus on three dimensions: species composition, stand structure (which is often linked to silvicultural system) and adaptive management. The first and second of these require knowledge of what will work in the expected climatic conditions. The third of these allows managers to continue adapting as they move into the future. It is helpful to think about these options in the context of intensity of forest management. One typology that has proven useful in recent years elucidates a scale of increasing intensity from natural or seminatural mixed native woodlands, through forests that are managed for continuous cover, to multipurpose commercial forests, single-aged monocultures and intensive short-rotation forestry [26]. The more intensive options are preferred where provision to timber or biomass-related industries is the priority, while the extensive approaches become more appropriate where society expects a wider range of ecosystem services.

In a review of evidence for the value of close-to-nature silviculture for managing forests to cope with future climate change, six principles are identified for enhancing the adaptive capacity of European temperate forests in a changing climate [21]:

- increase tree species richness;
- increase structural diversity;
- maintain and increase genetic variation within tree species;
- increase resistance of individual trees to biotic and abiotic stress;

- replace high-risk stands; and
- keep average growing stocks low.

The authors conclude that such approaches have many attributes that enhance forest resilience but highlight the need to also use increased species diversity by encouraging light-demanding tree species, non-native species and nonlocal provenances. Others agree that adaptation may need changes in the composition and structure of forest stands, selection of adapted provenances or assisted migration and substitution of native with non-native species, planting of mixtures, and changes to thinning regimes [20,27,28].

Species choice enables forest planners to ensure tree species and provenances are appropriate for the conditions of the site throughout the expected life of forest stands [29]. This may mean introducing trees from lower latitudes or choosing currently used species that show a wide enough suitability range to allow them to perform well in both the current and possible future climates. Selecting appropriate tree species and provenances for forest planting requires an understanding of not only their suitability for the local conditions and projected climate but also the ecosystem services that a forest or woodland will be expected to deliver [16]. Many tree species are unlikely to disperse fast enough to keep up with their moving climate space through natural dispersion, and tree dispersal is affected more by extreme events than by average climate. Additional strategies such as assisted migration and increasing species diversity may reduce the risks to forest ecosystems, in what have been described as 'neonative' forests, including the use and intermixing of native and non-native tree species as well as nonlocal tree provenances that may adapt better to future climate conditions [30]. These questions are not without controversy and can touch deeply held values and differences between foresters and ecologists, among others.

It is likely that most forest types will require some transformation to maintain their resilience [6]. In managed near-natural systems, disturbances such as wind or fire can be used as opportunities to adapt [31]. At the other end of the spectrum, intensive systems such as short-rotation forestry [32] may be adapted by changing species as climate impacts manifest. Most forest management falls in between these two extremes, however, and scientists advocate planned rather than reactive adaptation [33]. Management interventions to change species, change one forest type to another, or adapt existing forests to be more resilient all have practical limits, and evidently a forest cannot be transformed overnight. For example, species change would usually be made at the point of replanting after harvesting the previous crop. Transformation of single-aged stands (designed for clear-fell harvest and replant) to continuous cover, and potentially to a greater mixture of species, may take several decades to achieve. Therefore, although the forest may not usually be adapted at a rate of more than 1–2% of area per year, clear adaptation objectives and plans are needed early on. These plans need to be flexible depending on the success or otherwise of proactive adaptation measures, and should ideally include contingency plans so that adaptation may be accelerated by being reactive and taking advantage of opportunities to replant and revise management following natural disturbances [34].

This challenge can be summarized as both adapting and being adaptive. Where changes are known to help, they can be made. But in many cases either the evidence is still being gathered or uncertainties about the climate or sociopolitical factors make it imprudent to limit options. This approach requires maintaining options and capacity to change in the future. The situation can be compared to that advocated for conservation, which combines the process-based 'resilience thinking' and the more rationalistic and outcome-oriented approach of 'optimization for conservation' [24]. Partnerships that integrate researchers from multiple disciplines with forest managers and local actors can build a shared understanding of future challenges and facilitate improved decision-making in the face of climate change [6]. This need to combine both knowledge and process requires a strategic combination of research at different scales; for example, one study advocates combining at the international scale, with priority mapping of adaptation strategies at the national to regional scale, and implementation at the local scale [30].

In fact, the local scale may have more to contribute than just implementation. Adaptive management is a systematic approach to learning from the outcomes of previous management actions. There are intentional connections between the planning, monitoring and modification steps [35,36]. Many take the view that, correctly understood, adaptive management is not simply 'trial and error' or 'learning by doing', but a highly structured approach to planning, implementing, monitoring, reviewing and modifying in the light of new evidence in collaboration with relevant stakeholders [37]. It is widely seen as part of an appropriate response to climate change and other environmental change, although most cases where AFM has been implemented has not been explicitly as a response to climate change.

7. ADAPTATION IN PRACTICE

The shift of focus from mitigation to adaptation has taken place largely in the worlds of science and modelling. Research is still heavily focused on assessing impacts and vulnerability, but this knowledge is not seen to be leading to better management decisions [6]. There are difficulties in translating the multiple levels of knowledge into practice. Numerous studies highlight the challenge for highly structured and practical organizations, of working with uncertainty, taking risks, using innovative and flexible approaches, engaging with stakeholders and learning from experience [38–40].

The challenge is not only in practising adaptation but also in measuring it. Little research focuses on understanding the practicalities of adaptation and the degree to which forest owners, managers and planners are changing attitudes and behaviour. The study of practice requires more qualitative modes of social inquiry and is not easily translated from one context to another. Compared with the abstractions of mitigation, adaptation is a more local and contextualized activity; for example, trees planted in the south of England will not prevent floods in Poland. It is a more qualitative, process-based activity, relying on knowledge and judgement to make changes that may make forests more resilient under expected future conditions [5]. There are also challenges in measuring resilience and adaptive capacity within a management context [18].

However, in this novel area a few indications are emerging that outline attitudes in the public and private sectors to adopting new species recommendations, shifting from clear-fell (cutting down trees) to continuous cover systems, and taking a more flexible and devolved approach to forest management decisions.

Uncertainty is a constraint to adaptation in forest management. Much effort has been put into elucidating different types and sources of uncertainty and the difficulties of communicating them [8,41], but much less into understanding how forest managers respond. One study in the UK concluded that forest planners 'actively manage uncertainty' [15] while another concluded that they are more likely to take a passive approach [42]. A survey of forestry professionals in southwest Germany found a majority of respondents (72%) who said they were under informed, but most (83%) view climate change as a reality, human caused and a significant risk [43]. These mixed findings highlight the novelty of the subject and a need for further contextualization.

Social research has a better understanding of forest owners' attitudes to risk because some of this work predates a concern with climate change, and engages particularly with wind risk. This research has become relevant to climate adaptation, and to managing for tree health. One survey found that owners were on the whole optimistic about impacts of climate change on their forest and indicated differences of risk attitudes between owners based on characteristics of their personality (e.g. risk tolerance), forest and economic dependence on the forest [44]. A large survey across three European countries concluded that personal strength of belief and perception of local effects of climate change are strongly correlated with responses to climate change [45].

A range of actors influence forest management decisions, and where the majority of forest is privately owned the role of advisers (known in some countries as consultants or extension officers) is essential. One study in Wales focused on these advisers, many of whom were not generally convinced of a need to adapt [42]. The views of commercial forestry consultants contrasted with those of local government and small-scale owners; while the former were enthusiastic about introducing non-native species, the latter preferred to rely on native genetic diversity.

All the stakeholders in this Welsh study were reluctant to use decision support tools to aid their species choices or other management decisions, and this finding is common to other studies. A survey of British forestry stakeholders identified underlying problems with stakeholder engagement during the development and use of such tools, and found that forest managers regularly prefer to rely on their own professional judgement [46]. There is a much wider literature to fall back on here – if we focus out from the challenge of climate change adaptation, the attitudes and decisions of forest owners have been a favourite topic for decades and can inform an understanding of priorities and objectives, cultural meanings of forest ownership and preferred modes of communication. It is always clear, however, that forest owners are diverse, and that attitudes and modes of action vary according to cultural and ecological context [47–52].

Most knowledge about the adoption of adaptive management focuses on large public programmes in North America and Australia. Many such studies highlight the gap between policy ambition and implementation [53,54]. A key stumbling point is often monitoring; projects have faltered because of the high costs of data collection, poor data management, the long timeframe over which monitoring must occur, and the challenges of designing indicators of complex concepts such as resilience [53,55]. However, a more informal approach to adaptive management makes use of local experiments. Recent work in the UK highlights how old research is being used in new ways, and forest managers are contributing to new research. The boundaries between science and practice become more open in new knowledge creation processes [56]. Published information about alternative conifers and silvicultural systems is an important starting point [57–60]. Much of the existing guidance itself comes from earlier trials, whether formal or not. Private estates and collections provide examples of mature growth for a range of pines, cypresses and spruces, along with trials established in the 1960s and 1970s, some of which have since been suspended, and are now being rediscovered and valued anew [61,62].

A handful of studies examine attitudes to alternative management approaches such as continuous cover or close-to-nature; while the results are mixed, ranging from negative attitudes to sometimes inaccurate knowledge among professions [63,64], the only conclusion that we can draw at this stage is that further research is needed.

8. CHALLENGES FOR THE FUTURE

In less than a decade, much progress has been made with understanding the implications of projected climate change for forest function and distribution, and for clarifying the scope and limits of uncertainty. These efforts have led to active research and debate around the implications for forest management, in particular the management of genetic and stand diversity and the potential for intervention. There are no fixed answers because adaptation is local and context specific, and because many of the issues involve values as well as science, particularly with regard to the movement of species and intensity of intervention in forest management. These debates must continue, but not impede the practice of adaptation, which requires interaction between local and practical experience, advice based on climate modelling and species suitability, and international experimentation. Research and practice need to give attention to the social as well as the ecological aspects of the challenge, and to enrich our understanding of how owners, forest managers and advisers are working with forest ecology to develop a resilient forest human ecosystem.

REFERENCES

[1] IPCC, Climate Change 2007: Synthesis Report. 2007. Contribution of Working Groups to the Fourth Assessment Report of the Intergovernmental Panel on Climate Change, IPCC, Geneva, Switzerland, 2007.

[2] A.W. D'Amato, J.B. Bradford, S. Fraver, B.J. Palik, For. Ecol. Manage. 262 (2011) 803–816.

[3] R. Seppälä, Scand. J. For. Res. 24 (2009) 469–472.

[4] R. Seppälä, A. Buck, P. Katila (Eds.), Adaptation of forests and people to climate change – a global assessment report. IUFRO World Series, vol. 22, IUFRO, Helsinki, 2009, p. 224.

[5] M. Buizer, A. Lawrence, Environ. Sci. Policy 35 (2013) 57–66.

[6] R.J. Keenan, Ann. For. Sci. 72 (2015) 45–167.

[7] IPCC, Climate Change 2014. Synthesis Report. Summary for Policymakers, 2014. Geneva Switzerland, https://www.ipcc.ch/pdf/assessment-report/ar5/syr/AR5_SYR_FINAL_SPM.pdf.

[8] M. Lindner, J.B. Fitzgerald, N.E. Zimmermann, C. Reyer, S. Delzon, E. van der Maaten, M.-J. Schelhaas, P. Lasch, J. Eggers, M. van der Maaten-Theunissen, F. Suckow, A. Psomas, B. Poulter, M. Hanewinkel, J. Environ. Manage. 146 (2014) 69–83.

[9] C.D. Allen, A.K. Macalady, H. Chenchouni, D. Bachelet, N. McDowell, M. Vennetier, N. Cobb, For. Ecol. Manage. 259 (2010) 660–684.

[10] R. Seidl, M.J. Lexer, J. Environ. Manage. 114 (2013) 461–469.

[11] S.N. Aitken, S. Yeaman, J.A. Holliday, T. Wang, S. Curtis-McLane, Evol. Appl. 1 (2008) 95–111.

[12] S. Cavers, J.E. Cottrell, Forestry 88 (2015) 13–26.

[13] J.E. Olesen, T.R. Carter, C.H. Diaz-Ambrona, S. Fronzek, T. Heidmann, T. Hickler, J. Holt, M.I. Miguez, P. Morales, J.P. Palutikof, M. Quemada, M. Ruiz-Ramos, G.H. Rubaek, F. Sau, B. Smith, M.T. Sykes, Clim. Change 81 (2007) 123–143.

[14] J.M. Murphy, D.M. Sexton, D.N. Barnett, G.S. Jones, M.J. Webb, M. Collins, D.A. Stainforth, Nature 430 (2004) 768–772.

[15] M. Petr, L. Boerboom, D. Ray, A. van der Veen, For. Policy Econ. 41 (2014) 1–11.

[16] D. Ray, S. Bathgate, D. Moseley, P. Taylor, B. Nicoll, S. Pizzirani, B. I Gardiner, Reg. Environ. Change 26 (2014) 1–13.

[17] J.B. Bradford, A.W. D'Amato, Front. Ecol. Environ. 10 (2012) 210–216.

[18] L. Rist, J. Moen, For. Ecol. Manage. 310 (2013) 416–427.

[19] R.A. Ennos, Forestry 88 (2015) 41–52.

[20] A.D. Cameron, Forests 6 (2015) 398–415.

[21] P. Brang, P. Spathelf, J.B. Larsen, J. Bauhus, A. Bončína, C. Chauvin, L. Drössler, C. García-Güemes, C. Heiri, G. Kerr, M.J. Lexer, B. Mason, F. Mohren, U. Mühlethaler, S. Nocentini, M. Svoboda, Forestry 87 (2014) 492–503, http://dx.doi.org/10.1093/forestry/cpu018.

[22] D.J. Churchill, A.J. Larson, M.C. Dahlgreen, J.F. Franklin, P.F. Hessburg, J.A. Lutz, For. Ecol. Manage. 291 (2013) 442–457.

[23] H. Jactel, B.C. Nicoll, M. Branco, J.R. Gonzalez-Olabarria, W. Grodzki, B. Långström, F. Moreira, S. Netherer, C. Orazio, D. Piou, H. Santos, M.-J. Schelhaas, K. Tojic, F. Vodde, Ann. For. Sci. 66 (2009) 1–18.

[24] J. Fischer, G.D. Peterson, T.A. Gardner, L.J. Gordon, I. Fazey, T. Elmqvist, A. Felton, C. Folke, S. Dovers, Trends Ecol. Evol. 24 (2009) 549–554.

[25] E.L. Tompkins, W.N. Adger, Ecol. Soc. 9 (2004) 10.

[26] P.S. Duncker, S.M. Barreiro, G.M. Hengeveld, T. Lind, W.L. Mason, S. Ambrozy, H. Spiecker, Ecol. Soc. 17 (2012) 51.

[27] F. Bussotti, M. Pollastrini, V. Holland, W. Brüggemann, Environ. Exp. Bot. 111 (2015) 91–113, http://dx.doi.org/10.1016/j.envexpbot.2014.11.006.

[28] W.L. Mason, M. Petr, S. Bathgate, J. For. Sci. 58 (2012) 265–277.

[29] M.S.J. Broadmeadow, D. Ray, C.J.A. Samuel, Forestry 78 (2005) 145–161.

[30] A. Bolte, C. Ammer, M. Löf, P. Madsen, G.J. Nabuurs, P. Schall, P. Spathelf, J. Rock, Scand. J. For. Res. 24 (2009) 473–482.

[31] B. Buma, C.A. Wessman, For. Ecol. Manage. 306 (2013) 216–225.

[32] M. Weih, Can. J. For. Res. 34 (2004) 1369–1378.

[33] D.J. Read, P.H. Freer-Smith, J.I.L. Morison, N. Hanley, C.C. West, P. Snowdon, Combating climate change: a role for UK forests. An assessment of the potential of the UK's trees and woodlands to mitigate and adapt to climate change, The Stationary Office, Edinburgh, 2009, p. 242.

[34] G.M. Blate, L.A. Joyce, J.S. Littell, S.G. McNulty, C.I. Millar, S.C. Moser, R.P. Neilson, K.O. Halloran, D.L. Peterson, Unasylva 231 (2009) 57–62.

[35] J. Pahl-Wostl, P. Sendzimir, J. Jeffrey, G. Aerts, K. Berkamp, Cross Ecol. Soc. 12 (2007) 30, http://www.ecologyandsociety.org/vol12/iss2/art30/.

[36] B. Nyberg, An Introductory Guide to Adaptive Management for Project Leaders and Participants, British Columbia Forest Service, Vancouver, Canada, 1999.

[37] A. Lawrence, S. Gillett, Human Dimensions of Adaptive Forest Management and Climate Change: A Review of International Experience. Forestry Commission Research Report, Forestry Commission, Edinburgh, 2011, p. 52.

[38] G.B. MacDonald, J.A. Rice, For. Chron. 80 (2004) 391–400.

[39] K.F. Butler, T.M. Koontz, Environ. Manage. 35 (2005) 138–150.

[40] B.T. Bormann, R.W. Haynes, J.R. Martin, BioScience 57 (2007) 186–191.

[41] M. Pasalodos-Tato, A. Mäkinen, J. Garcia-Gonzalo, J.G. Borges, T. Lämås, L.O. Eriksson, For. Syst. 22 (2013) 282–303.

[42] A. Lawrence, M. Marzano, Ann. For. Sci. (2013), http://dx.doi.org/10.1007/s13595-013-0326-4.

[43] R. Yousefpour, M. Hanewinkel, Clim. Change (2015), http://dx.doi.org/10.1007/s10584-015-1330-5.

[44] L. Eriksson, Small Scale For. 13 (2014) 483–500.

[45] K. Blennow, J. Persson, M. Tomé, M. Hanewinkel, PLoS One 7 (2012), http://dx.doi.org/10.1371/journal.pone.0050182.

[46] A. Stewart, D. Edwards, A. Lawrence, Scand. J. For. Res. 29 (2014) 144–153.

[47] T. Hujala, J. Pykäläinen, J. Tikkanen, Scand. J. For. Res. 22 (2007) 454–463.

[48] P. Põllumäe, H. Korjus, T. Paluots, For. Policy Econ. 42 (2014) 8–14.

[49] A. Lawrence, N. Dandy, Land Use Policy 36 (2014) 351–360.

[50] Z. Ma, B.J. Butler, D.B. Kittredge, P. Catanzaro, Land Use Policy 29 (2012) 53–61.

[51] M. Andersson, J. For. Econ. 18 (2012) 3–13.

[52] M.F. Marey-Pérez, V. Rodríguez-Vicente, For. Policy Econ. 13 (2011) 318–327.

[53] C. Allan, A. Curtis, Environ. Manage. 36 (2005) 414–425.

[54] C. Jacobson, et al., Toward more reflexive use of adaptive management, Soc. Nat. Res. 22 (5) (2009) 484–495.

[55] B.J. McAfee, C. Malouin, N. Fletcher, For. Chron. 82 (2006) 321–334.

[56] A. Lawrence, Chartered Forester, April 2015, pp. 26–28.

[57] Forestry Commission Wales, Guidance for the Use of Silvicultural Systems to Increase Woodland Diversity, Forestry Commission Wales, 2012, p. 23.

[58] Forestry Commission Wales, A Guide for Increasing Tree Species Diversity in Wales. Grants and Regulations, Forestry Commission Wales, 2010.

[59] S.M. Wilson, Using Alternative Conifer Species for Productive Forestry in Scotland, Forestry Commission Scotland, Edinburgh, 2011.

[60] Forest Research, Tree Species and Provenance, 2015. Available from: http://www.forestry.gov.uk/fr/tree species.

[61] D.F. Meason, W.L. Mason, Ann. For. Sci. (2013), http://dx.doi.org/10.1007/s13595-013-0300-1.

[62] W.L. Mason, T. Connolly, Forestry 87 (2014) 209–217.

[63] L. Vítková, Á.N. Dhubháin, V. Upton, Scott. For. 68 (2014) 17.

[64] R. Axelsson, P.K. Angelstam, Forestry (2011), http://dx.doi.org/10.1093/forestry/cpr034.

Index

Note: Page numbers followed by "f" and "t" indicate figures and tables, respectively.

595

Printed in the United States
By Bookmasters